Gauge Theories in Particle Physics: A Practical Introduction Volume 2

The fifth edition of this well-established, highly regarded two-volume set continues to provide a fundamental introduction to advanced particle physics while incorporating substantial new experimental results, especially in the areas of Higgs and top sector physics, as well as CP violation and neutrino oscillations. It offers an accessible and practical introduction to the three-gauge theories comprising the Standard Model (SM) of particle physics: quantum electrodynamics (QED), quantum chromodynamics (QCD), and the Glashow-Salam-Weinberg (GSW) electroweak theory.

Volume 2 of this updated edition covers the two non-Abelian gauge theories of QCD and the GSW theory. A distinctive feature is the extended treatment of two crucial theoretical tools: spontaneous symmetry breaking and the renormalization group. The underlying physics of these is elucidated by parallel discussions of examples from condensed matter systems: superfluidity and superconductivity, and critical phenomena. This new edition includes updates to jet algorithms, lattice field theory, CP violation and the CKM matrix, and neutrino physics.

New to the fifth edition:

- Tests of the SM in the Higgs and top quark sectors
- The naturalness problem and responses to it going beyond the SM
- The SM as an effective field theory

Each volume should serve as a valuable handbook for students and researchers in advanced particle physics looking for an accessible introduction to the SM of particle physics.

Ian J.R. Aitchison is Emeritus Professor of Physics at the University of Oxford. He has previously held research positions at Brookhaven National Laboratory, Saclay, and the University of Cambridge. He was a visiting professor at the University of Rochester and the University of Washington, and a scientific associate at CERN and SLAC. Dr. Aitchison has published over 90 scientific papers mainly on hadronic physics and quantum field theory. He is the author of two books and joint editor of further two.

Anthony J.G. Hey is now Honorary Senior Data Scientist at the UK's National Laboratory at Harwell. He began his career with a doctorate in particle physics from the University of Oxford. After a career in particle physics that included a professorship at the University of Southampton and research positions at Caltech, MIT, and CERN, he moved to Computer Science and founded a parallel computing research group. The group was one of the pioneers of distributed memory message-passing computers and helped establish the 'MPI' message passing standard. After leaving Southampton in 2001 he was director of the UK's 'eScience' initiative before becoming a Vice-President in Microsoft Research. He returned to the UK in 2015 as Chief Data Scientist at the U.K.'s Rutherford Appleton Laboratory. He then founded a new 'Scientific Machine Learning' group to apply AI technologies to the 'Big Scientific Data' generated by the Diamond Synchrotron, the ISIS neutron source, and the Central Laser Facility that are located on the Harwell campus. He is the author of over 100 scientific papers on physics and computing and editor of 'The Feynman Lectures on Computation'.

Gauge Theories in Particle Physics: A Practical Introduction Volume 2

Non-Abelian Gauge Theories: QCD and The Electroweak Theory

Fifth Edition

Ian J.R. Aitchison and Anthony J.G. Hey

CRC Press
Taylor & Francis Group
Boca Raton London New York

CRC Press is an imprint of the
Taylor & Francis Group, an **informa** business

Designed cover image: CERN

Fifth edition published 2024
by CRC Press
2385 NW Executive Center Drive, Suite 320, Boca Raton FL 33431

and by CRC Press
4 Park Square, Milton Park, Abingdon, Oxon, OX14 4RN

CRC Press is an imprint of Taylor & Francis Group, LLC

Fourth edition published by CRC Press 2013

Library of Congress Cataloging-in-Publication Data

Names: Aitchison, Ian Johnston Rhind, 1936- author. | Hey, Anthony J. G., author.
Title: Gauge Theories in Particle Physics, 40th Anniversary Edition : A Practical Introduction, Volume 1 : From Relativistic Quantum Mechanics to QED, Fifth Edition / Ian J.R. Aitchison and Anthony J.G. Hey.
Description: Fifth edition. | Boca Raton, FL : CRC Press, 2024. | Includes bibliographical references and index. | Contents: v. 1. From relativistic quantum mechanics to QED -- v. 2. Non-Abelian gauge theories : QCD and The electroweak theory. | Summary: "The fifth edition of this well-established, highly regarded two-volume set continues to provide a fundamental introduction to advanced particle physics while incorporating substantial new experimental results, especially in the areas of Higgs and top sector physics, as well as CP violation and neutrino oscillations. It offers an accessible and practical introduction to the three gauge theories comprising the Standard Model of particle physics: quantum electrodynamics (QED), quantum chromodynamics (QCD), and the Glashow-Salam-Weinberg (GSW) electroweak theory. Volume 1 of this updated edition provides a broad introduction to the first of these theories, QED. The book begins with self-contained presentations of relativistic quantum mechanics and electromagnetism as a gauge theory. Lorentz transformations, discrete symmetries, and Majorana fermions are covered. A unique feature is the elementary introduction to quantum field theory, leading in easy stages to covariant perturbation theory and Feynman graphs, thereby establishing a firm foundation for the formal and conceptual framework upon which the subsequent development of the three quantum gauge field theories of the Standard Model is based. Detailed tree-level calculations of physical processes in QED are presented, followed by an elementary treatment of one-loop renormalization of a model scalar field theory, and then by the realistic case of QED. The text includes updates on nucleon structure functions and the status of QED, in particular the precision tests provided by the anomalous magnetic moments of the electron and muon. The authors discuss the main conceptual points of the theory, detail many practical calculations of physical quantities from first principles, and compare these quantitative predictions with experimental results, helping readers improve both their calculation skills and physical insight. Each volume should serve as a valuable handbook for students and researchers in advanced particle physics looking for an introduction to the Standard Model of particle physics"-- Provided by publisher.
Identifiers: LCCN 2023043693 | ISBN 9781032531717 (hbk ; volume 1) | ISBN 9781032531748 (pbk ; volume 1) | ISBN 9781003410720 (ebk ; volume 1) | ISBN 9781032531700 (hbk ; volume 2) | ISBN 9781032533612 (hbk ; volume 2) | ISBN 9781003411666 (ebk ; volume 2)
Subjects: LCSH: Gauge fields (Physics) | Particles (Nuclear physics) | Weak interactions (Nuclear physics) | Quantum electrodynamics. | Feynman diagrams.
Classification: LCC QC793.3.F5 A34 2024 | DDC 539.7/2--dc23/eng/20231026
LC record available at https://lccn.loc.gov/2023043693

ISBN: 978-1-032-53170-0 (hbk)
ISBN: 978-1-032-53361-2 (pbk)
ISBN: 978-1-003-41166-6 (ebk)

DOI: 10.1201/9781003411666

Typeset in CMR10
by KnowledgeWorks Global Ltd.

Publisher's note: This book has been prepared from camera-ready copy provided by the authors.

To Jessie
and
in memory of Jean

Contents

Preface to Volume 2 of the Fifth Edition

In the decade or more that has passed since the appearance of the fourth edition of this book, the Standard Model (SM) has been subjected to increasingly stringent experimental tests, often involving elaborate theoretical calculations and remarkable experimental precision. At this time of writing, no significant discrepancies have been established between the predictions of the SM and experiments. Our focus has always been on the SM, and we have aimed to combine, where possible, theoretical developments with discussion of relevant experimental results. The success of the SM means that no major changes seemed to be required in the theoretical framework provided in the fourth edition. The motivation for this new edition arises, rather, from the need to take account of the progress made, in the last decade, in confronting the SM with experiment.

The major updates in this new edition concern the rapidly evolving areas of the Higgs sector and top quark physics; relatively minor updates have been necessary in the slower-moving fields of neutrino physics and CP violation. We have (we hope) improved the presentation in various places, for example, in lattice quantum field theory, and we have extended the discussion of jet algorithms. We have to admit that the book has (once again) grown in length — but we did not want to discard much of the earlier experimental discovery material, believing it to have historical and pedagogical value.

We have, however, added one entirely new chapter, which begins with a discussion of recent precision data in the Higgs and top quark sectors and proceeds to describe why, somewhat paradoxically, the very success of the SM has created what is perceived by many theorists to be a problem: namely the 'naturalness' problem. We outline some theoretical approaches to this problem (if such it is), which inevitably take us into Beyond the SM (BSM) territory, and various suggestions as to what new physics may lie at an as yet unexplored higher energy scale. But, faced with no clear experimental support for any one of these ideas, a more modest strategy is being currently pursued: a general parametrization of possible extensions of the SM, known as the Standard Model as an Effective Field Theory (SMEFT). This is introduced in the last section of the book.

Acknowledgements

We repeat here our expression of thanks, adapted from the Preface to the fourth edition, to the many people who helped us with major additions to the third and fourth editions; their inputs remain an essential part of this one.

We are very grateful to Paolo Strolin for providing a list of misprints and a very thorough catalogue of excellent comments for volume 2 of the third edition, which resulted in a large number of improvements. The CP-violation sections were much improved following detailed comments by Abi Soffer, and the neutrino sections similarly benefited greatly from careful readings by Francesco Tramontano and Tim Cohen; we thank all three for their generous help. The eps file for figure 16.11 was kindly supplied by Stephan Dürr. IJRA thanks Michael Peskin and Stan Brodsky for welcoming him as a visitor to the SLAC National Accelerator

Laboratory Particle Theory group (supported by the Department of Energy under contract DE-AC02-76SF00515), and Bill Dunwoodie and BaBar colleagues for very kindly arranging for him to be a BaBar Associate; these connections were invaluable.

We are grateful to the reviewers of the present edition for helpful comments concerning what needed to be updated (and also what was in no need of alteration). We particularly thank Ben Gripaios for detailed suggestions about possible new material, which formed the basis for the development of the new last chapter; he also very kindly read through a draft of the chapter. Naturally, errors and omissions are our responsibility.

Ian J R Aitchison
Anthony J G Hey
August 2023

Part V

Non-Abelian Symmetries

12

Global Non-Abelian Symmetries

12.1 The Standard Model

In the preceding volume, a very successful dynamical theory — QED — has been introduced based on the remarkably simple *gauge principle*, namely, that the theory should be invariant under local phase transformations on the wavefunctions (chapter 2) or field operators (chapter 7) of charged particles. Such transformations were characterized as *Abelian* in section 2.6 since the phase factors commuted. The second volume of this book will be largely concerned with the formulation and elementary application of the remaining two dynamical theories within the Standard Model—that is, QCD and the electroweak theory. They are built on a generalization of the gauge principle, in which the transformations involve more than one state, or field, at a time. In that case, the 'phase factors' become matrices, which generally do not commute with each other, and the associated symmetry is called a *non-Abelian* one. When the phase factors are independent of the space-time coordinate x, the symmetry is a 'global non-Abelian' one; when they are allowed to depend on x, one is led to a non-Abelian gauge theory. Both QCD and the electroweak theory are of the latter type, providing generalizations of the Abelian U(1) gauge theory which is QED. It is a striking fact that all three dynamical theories in the Standard Model are based on a gauge principle of local phase invariance.

In this chapter, we are mainly concerned with two global non-Abelian symmetries, which lead to useful conservation laws but not to any specific dynamical theory. We begin in section 12.1 with the first non-Abelian symmetry to be used in particle physics, the *hadronic isospin* 'SU(2) symmetry' proposed by Heisenberg (1932) in the context of nuclear physics, and now understood as following from QCD and the smallness of the u and d quark masses as compared with the QCD scale parameter $\Lambda_{\overline{\text{MS}}}$ (see section 18.3.3). In section 12.2, we extend this to SU(3)$_f$ flavour symmetry as was first done by Gell-Mann (1961) and Ne'eman (1961)—an extension seen, in its turn, as reflecting the smallness of the u, d, and s quark masses as compared with $\Lambda_{\overline{\text{MS}}}$. The 'wavefunction' approach of sections 12.1 and 12.2 is then reformulated in field-theoretic language in section 12.3.

In the last section of this chapter, we shall introduce the idea of a global *chiral* symmetry, which is a symmetry of theories with massless fermions. This may be expected to be a good approximate symmetry for the u and d quarks. But the anticipated observable consequences of this symmetry (for example, nucleon parity doublets) appear to be absent. This puzzle will be resolved in Part 7 via the profoundly important concept of 'spontaneous symmetry breaking'.

The formalism introduced in this chapter for SU(2) and SU(3) will be required again in the following one, when we consider the local versions of these non-Abelian symmetries and the associated dynamical gauge theories. The whole modern development of non-Abelian gauge theories began with the attempt by Yang and Mills (1954) (see also Shaw 1955) to make hadronic isospin into a local symmetry. However, the beautiful formalism developed by these authors turned out *not* to describe interactions between hadrons. Instead, it describes the interactions between the *constituents* of the hadrons, namely quarks—and this in two respects. First, a local SU(3) symmetry (called SU(3)$_c$) governs the strong interactions of quarks, binding them into hadrons (see Part 6). Secondly, a local SU(2) symmetry (called *weak isospin*) governs the weak interactions of quarks (and leptons); together with QED, this constitutes the electroweak theory (see Part 8). It is important to realise that, despite the fact that each of these two local symmetries is based on the same group as one of the earlier global (flavour) symmetries, the physics involved is completely different. In the case of the strong quark interactions, the SU(3)$_c$ group refers to a new degree of freedom ('colour') which is quite distinct from flavour u, d, s (see chapter 14). In the weak interaction case, since the group is an SU(2), it is natural to use 'isospin language' when talking about it, particularly since flavour degrees of freedom are involved. But we must always remember that it is *weak* isospin, which (as we shall see in chapter 20) is an attribute of leptons as well as of quarks, and hence physically quite distinct from hadronic spin. Furthermore, it is a parity-violating chiral gauge theory.

Despite the attractive conceptual unity associated with the gauge principle, the way in which each of QCD and the electroweak theory 'works' is actually quite different from QED, and from each other. Indeed it is worth emphasizing very strongly that it is, *a priori*, far from obvious why either the strong interactions between quarks, or the weak interactions, should have anything to do with gauge theories at all. Just as in the U(1) (electromagnetic) case, gauge invariance forbids a mass term in the Lagrangian for non-Abelian gauge fields, as we shall see in chapter 13. Thus it would seem that gauge field quanta are necessarily massless. But this, in turn, would imply that the associated forces must have a long-range (Coulombic) part due to the exchange of these massless quanta—and of course in neither the strong nor the weak interaction case is that what is observed.[1] As regards the former, the gluon quanta are indeed massless but the contradiction is resolved by *non-perturbative* effects which lead to *confinement* as we indicated in chapter 1. We shall discuss this further in chapter 16. In weak interactions, a third realization appears: the gauge quanta acquire mass via (it is believed) a second instance of *spontaneous symmetry breaking*, as will be explained in Part 7. In fact, a further application of this idea is required in the electroweak theory, because of the chiral nature of the gauge symmetry in this case: the quark and lepton masses also must be 'spontaneously generated'.

12.2 The flavour symmetry SU(2)$_f$

12.2.1 The nucleon isospin doublet and the group SU(2)

The transformations initially considered in connection with the gauge principle in section 2.5 were just global phase transformations on a single wavefunction

$$\psi' = e^{i\alpha}\psi. \tag{12.1}$$

The generalization to non-Abelian invariances comes when we take the simple step—but one with many ramifications—of considering more than one wavefunction, or state, at a time. Quite generally in quantum mechanics, we know that whenever we have a set of states which are *degenerate* in energy (or mass) there is no unique way of specifying the states:

---

[1] Pauli had independently developed the theory of non-Abelian gauge fields during 1953 but did not publish any of this work because of the seeming physical irrelevancy associated with the masslessness problem (Enz 2002, pages 474-82; Pais 2000, pages 242-5).

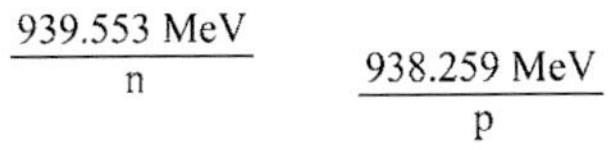

$$\underset{\text{n}}{\underline{939.553 \text{ MeV}}} \qquad \underset{\text{p}}{\underline{938.259 \text{ MeV}}}$$

FIGURE 12.1

Early evidence for isospin symmetry.

any linear combination of some initially chosen set of states will do just as well, provided the normalization conditions on the states are still satisfied. Consider, for example, the simplest case of just two such states—to be specific, the neutron and proton (figure 12.1). This single near coincidence of the masses was enough to suggest to Heisenberg (1932) that as far as the strong nuclear forces were concerned (electromagnetism being negligible by comparison), the two states could be regarded as truly degenerate, so that any arbitrary linear combination of neutron and proton wavefunctions would be entirely equivalent, as far as this force was concerned, for a single 'neutron' or single 'proton' wavefunction. This hypothesis became known as 'charge independence of nuclear forces'. Thus redefinitions of neutron and proton wavefunctions could be allowed, of the form

$$\psi_p \rightarrow \psi_p' = \alpha \psi_p + \beta \psi_n \tag{12.2}$$

$$\psi_n \rightarrow \psi_n' = \gamma \psi_p + \delta \psi_n \tag{12.3}$$

for complex coefficients α, β, γ, and δ. In particular, since ψ_p and ψ_n are degenerate, we have

$$H\psi_p = E\psi_p, \qquad H\psi_n = E\psi_n \tag{12.4}$$

from which it follows that

$$\begin{aligned} H\psi_p' &= H(\alpha\psi_p + \beta\psi_n) = \alpha H\psi_p + \beta H\psi_n \tag{12.5} \\ &= E(\alpha\psi_p + \beta\psi_n) = E\psi_p' \tag{12.6} \end{aligned}$$

and similarly

$$H\psi_n' = E\psi_n' \tag{12.7}$$

showing that the redefined wavefunctions still describe two states with the same energy degeneracy.

The two-fold degeneracy seen in figure 12.1 is suggestive of that found in spin-$\frac{1}{2}$ systems in the absence of any magnetic field; the $s_z = \pm\frac{1}{2}$ components are degenerate. The analogy can be brought out by introducing the *2-component nucleon isospinor*

$$\psi^{(1/2)} \equiv \begin{pmatrix} \psi_p \\ \psi_n \end{pmatrix} \equiv \psi_p \chi_p + \psi_n \chi_n \tag{12.8}$$

where

$$\chi_p = \begin{pmatrix} 1 \\ 0 \end{pmatrix}, \qquad \chi_n = \begin{pmatrix} 0 \\ 1 \end{pmatrix}. \tag{12.9}$$

In $\psi^{(1/2)}$, ψ_p is the amplitude for the nucleon to have 'isospin up', and ψ_n is that for it to have 'isospin down'.

As far as the states are concerned, this terminology arises, of course, from the formal identity between the 'isospinors' of (12.9) and the 2-component eigenvectors (3.60) corresponding to eigenvalues $\pm\frac{1}{2}\hbar$ of (true) spin: compare also (3.61) and (12.8). It is important to be clear, however, that the degrees of freedom involved in the two cases are quite distinct;

in particular, even though both the proton and the neutron have (true) spin$-\frac{1}{2}$, the transformations (12.2) and (12.3) leave the (true) spin part of their wavefunctions completely untouched. Indeed, we are suppressing the spinor part of both wavefunctions altogether (they are of course 4-component Dirac spinors). As we proceed, the precise mathematical nature of this 'spin-1/2' analogy will become clear.

Equations (12.2) and (12.3) can be compactly written in terms of $\psi^{(1/2)}$ as

$$\psi^{(1/2)} \to \psi^{(1/2)'} = \mathbf{V}\psi^{(1/2)}, \quad \mathbf{V} = \begin{pmatrix} \alpha & \beta \\ \gamma & \delta \end{pmatrix} \tag{12.10}$$

where $\mathbf{V}$ is the indicated complex 2×2 matrix. Heisenberg's proposal, then, was that the physics of strong interactions between nucleons remained the same under the transformation (12.10). In other words, a symmetry was involved. We must emphasize that such a symmetry can *only* be exact in the *absence* of electromagnetic interactions. It is therefore an intrinsically approximate symmetry, though presumably quite a useful one in view of the relative weakness of electromagnetic interactions as compared to hadronic ones.

We now consider the general form of the matrix $\mathbf{V}$, as constrained by various relevant restrictions. Quite remarkably, we shall discover that (after extracting an overall phase) $\mathbf{V}$ has essentially the same mathematical form as the matrix $\mathbf{U}$ of (4.33) which we encountered in the discussion of the transformation of (real) spin wavefunctions under rotations of the (real) space axes. It will be instructive to see how the present discussion leads to the same form (4.33).

We first note that $\mathbf{V}$ of (12.10) depends on four arbitrary complex numbers, or alternatively on eight real parameters. By contrast, the matrix $\mathbf{U}$ of (4.33) depends on only three real parameters, which we may think of in terms of two to describe the direction of the axis of rotation, and a third for the angle of rotation. However, $\mathbf{V}$ is subject to certain restrictions, and these reduce the number of free parameters in $\mathbf{V}$ to three, as we now discuss. First, in order to preserve the normalization of $\psi^{(1/2)}$ we require

$$\psi^{(1/2)'\dagger}\psi^{(1/2)'} = \psi^{(1/2)\dagger}\mathbf{V}^\dagger\mathbf{V}\psi^{(1/2)} = \psi^{(1/2)\dagger}\psi^{(1/2)} \tag{12.11}$$

which implies that $\mathbf{V}$ has to be *unitary*:

$$\mathbf{V}^\dagger\mathbf{V} = \mathbf{1}_2, \tag{12.12}$$

where $\mathbf{1}_2$ is the unit 2×2 matrix. Clearly this unitarity property is in no way restricted to the case of two states: the transformation coefficients for n degenerate states will form the entries of an $n \times n$ unitary matrix. A trivialization is the case $n = 1$, for which, as we noted in section 2.6, $\mathbf{V}$ reduces to a single-phase factor as in (12.1), indicating how all the previous work is going to be contained as a special case of these more general transformations. Indeed, from elementary properties of determinants, we have

$$\det\mathbf{V}^\dagger\mathbf{V} = \det\mathbf{V}^\dagger \cdot \det\mathbf{V} = \det\mathbf{V}^* \cdot \det\mathbf{V} = \mid \det\mathbf{V} \mid^2 = 1 \tag{12.13}$$

so that

$$\det\mathbf{V} = \exp(i\theta) \tag{12.14}$$

where θ is a real number. We can separate off such an overall phase factor from the transformations mixing 'p' and 'n', because it corresponds to a rotation of the phase of both p and n wavefunctions by the *same* amount:

$$\psi'_p = e^{i\alpha}\psi_p, \quad \psi'_n = e^{i\alpha}\psi_n. \tag{12.15}$$

The $\mathbf{V}$ corresponding to (12.15) is $\mathbf{V} = e^{i\alpha}\mathbf{1}_2$, which has determinant $\exp(2i\alpha)$ and is therefore of the form (12.1) with $\theta = 2\alpha$. In the field-theoretic formalism of section 7.2, such a symmetry can be shown to lead to the conservation of baryon number $N_u + N_d - N_{\bar{u}} - N_{\bar{d}}$, where bar denotes the antiparticle.

The new physics will lie in the remaining transformations which satisfy

$$\det \mathbf{V} = +1. \tag{12.16}$$

Such a matrix is said to be a *special* unitary matrix, which simply means it has unit determinant. Thus, finally, the $\mathbf{V}$'s we are dealing with are *special, unitary, 2×2 matrices*. The set of all such matrices form a *group*. The general defining properties of a group are given in Appendix M. In the present case, the elements of the group are all such 2×2 matrices, and the 'law of combination' is just ordinary matrix multiplication. It is straightforward to verify (problem 12.1) that all the defining properties are satisfied here; the group is called 'SU(2)', the 'S' standing for 'special', the 'U' for 'unitary', and the '2' for '2×2'.

SU(2) is actually an example of a *Lie group* (see Appendix M). Such groups have the important property that their physical consequences may be found by considering 'infinitesimal' transformations, that is—in this case—matrices $\mathbf{V}$ which differ only slightly from the 'no-change' situation corresponding to $\mathbf{V} = \mathbf{1}_2$. For such an infinitesimal SU(2) matrix $\mathbf{V}_{\text{infl}}$, we may therefore write

$$\mathbf{V}_{\text{infl}} = \mathbf{1}_2 + i\boldsymbol{\xi} \tag{12.17}$$

where $\boldsymbol{\xi}$ is a 2×2 matrix whose entries are all first-order small quantities. The condition $\det \mathbf{V}_{\text{infl}} = 1$ now reduces, on neglect of second-order terms $0(\boldsymbol{\xi}^2)$, to the condition (see problem 12.2)

$$\text{Tr}\boldsymbol{\xi} = 0. \tag{12.18}$$

The condition that $\mathbf{V}_{\text{infl}}$ be unitary, i.e.,

$$(\mathbf{1}_2 + i\boldsymbol{\xi})(\mathbf{1}_2 - i\boldsymbol{\xi}^\dagger) = \mathbf{1}_2 \tag{12.19}$$

similarly reduces (in first order) to the condition

$$\boldsymbol{\xi} = \boldsymbol{\xi}^\dagger. \tag{12.20}$$

Thus $\boldsymbol{\xi}$ is a 2×2 traceless Hermitian matrix, which means it must have the form

$$\boldsymbol{\xi} = \begin{pmatrix} a & b - ic \\ b + ic & -a \end{pmatrix}, \tag{12.21}$$

where a, b, and c are infinitesimal real parameters. Writing

$$a = \epsilon_3/2, \quad b = \epsilon_1/2, \quad c = \epsilon_2/2, \tag{12.22}$$

(12.21) can be put in the more suggestive form

$$\boldsymbol{\xi} = \boldsymbol{\epsilon} \cdot \boldsymbol{\tau}/2 \tag{12.23}$$

where $\boldsymbol{\epsilon}$ stands for the three real quantities

$$\boldsymbol{\epsilon} = (\epsilon_1, \epsilon_2, \epsilon_3) \tag{12.24}$$

which are all first-order small. The three matrices $\boldsymbol{\tau}$ are just the familiar Hermitian Pauli matrices

$$\tau_1 = \begin{pmatrix} 0 & 1 \\ 1 & 0 \end{pmatrix}, \tau_2 = \begin{pmatrix} 0 & -i \\ i & 0 \end{pmatrix}, \tau_3 = \begin{pmatrix} 1 & 0 \\ 0 & -1 \end{pmatrix}, \tag{12.25}$$

here called 'tau' precisely in order to distinguish them from the mathematically identical 'sigma' matrices which are associated with the real spin degree of freedom. Hence a general infinitesimal SU(2) matrix takes the form

$$\mathbf{V}_{\text{infl}} = (\mathbf{1}_2 + i\boldsymbol{\epsilon} \cdot \boldsymbol{\tau}/2), \tag{12.26}$$

and an infinitesimal SU(2) transformation of the p-n doublet is specified by

$$\begin{pmatrix} \psi'_{\text{p}} \\ \psi'_{\text{n}} \end{pmatrix} = (\mathbf{1}_2 + i\boldsymbol{\epsilon} \cdot \boldsymbol{\tau}/2) \begin{pmatrix} \psi_{\text{p}} \\ \psi_{\text{n}} \end{pmatrix}. \tag{12.27}$$

The $\boldsymbol{\tau}$-matrices clearly play an important role since they determine the forms of the three independent infinitesimal SU(2) transformations. They are called the *generators* of infinitesimal SU(2) transformations; more precisely, the matrices $\boldsymbol{\tau}/2$ provide a particular *matrix representation* of the generators, namely the two-dimensional, or 'fundamental' one (see Appendix M). We note that they do not commute amongst themselves: rather, introducing $\mathbf{T}^{(\frac{1}{2})} \equiv \boldsymbol{\tau}/2$, we find (see problem 12.3)

$$[T_i^{(\frac{1}{2})}, T_j^{(\frac{1}{2})}] = i\epsilon_{ijk} T_k^{(\frac{1}{2})}, \tag{12.28}$$

where i, j, and k run from 1 to 3, and a sum on the repeated index k is understood as usual. The reader will recognize the commutation relations (12.28) as being precisely the same as those of angular momentum operators in quantum mechanics:

$$[J_i, J_j] = i\epsilon_{ijk} J_k. \tag{12.29}$$

In that case, the choice $J_i = \sigma_i/2 \equiv J_i^{(\frac{1}{2})}$ would correspond to a (real) spin-1/2 system. Here the identity between the tau's and the sigma's gives us a good reason to regard our 'p-n' system as formally analogous to a 'spin-1/2' one. Of course, the 'analogy' was made into a mathematical identity by the judicious way in which $\boldsymbol{\xi}$ was parametrized in (12.23).

The form for a *finite* SU(2) transformation $\mathbf{V}$ may then be obtained from the infinitesimal form using the result

$$\mathrm{e}^A = \lim_{n \to \infty} (1 + A/n)^n \tag{12.30}$$

generalized to matrices. Let $\boldsymbol{\epsilon} = \boldsymbol{\alpha}/n$, where $\boldsymbol{\alpha} = (\alpha_1, \alpha_2, \alpha_3)$ are three real finite (not infinitesimal) parameters, apply the infinitesimal transformation n times, and let n tend to infinity. We obtain

$$\mathbf{V} = \exp(i\boldsymbol{\alpha} \cdot \boldsymbol{\tau}/2) \tag{12.31}$$

so that

$$\psi^{(1/2)'} \equiv \begin{pmatrix} \psi'_{\text{p}} \\ \psi'_{\text{n}} \end{pmatrix} = \exp(i\boldsymbol{\alpha} \cdot \boldsymbol{\tau}/2) \begin{pmatrix} \psi_{\text{p}} \\ \psi_{\text{n}} \end{pmatrix} = \exp(i\boldsymbol{\alpha} \cdot \boldsymbol{\tau}/2)\psi^{(1/2)}. \tag{12.32}$$

Note that in the finite transformation, the generators appear in the exponent. Indeed, (12.31) has the form

$$\mathbf{V} = \exp(iG) \tag{12.33}$$

where $G = \boldsymbol{\alpha} \cdot \boldsymbol{\tau}/2$, from which the unitary property of $\mathbf{V}$ easily follows:

$$\mathbf{V}^\dagger = \exp(-iG^\dagger) = \exp(-iG) = \mathbf{V}^{-1} \tag{12.34}$$

where we used the Hermiticity of the tau's. Equation (12.33) has the general form

$$\text{unitary matrix} = \exp(i \text{ Hermitian matrix}) \tag{12.35}$$

where the 'Hermitian matrix' is composed of the generators and the transformation parameters. We shall meet generalizations of this structure in the following subsection for SU(2), again in section 12.2 for SU(3), and a field theoretic version of it in section 12.3.

As promised, (12.32) has essentially the same mathematical form as (4.33). In each case, three real parameters appear. In (4.33) they describe the axis and angle of a physical rotation in real three-dimensional space: we can always write $\boldsymbol{\alpha} = |\boldsymbol{\alpha}|\hat{\boldsymbol{\alpha}}$ and identify $|\boldsymbol{\alpha}|$ with the angle θ and $\hat{\boldsymbol{\alpha}}$ with the axis $\hat{\boldsymbol{n}}$ of the rotation. In (12.32) there are just the three parameters in $\boldsymbol{\alpha}$.[2]

In the form (12.32), it is clear that our 2×2 isospin transformation is a generalization of the global phase transformation of (12.1) except that

(a)   there are now *three* 'phase angles' $\boldsymbol{\alpha}$;

(b)   there are non-commuting matrix operators (the $\boldsymbol{\tau}$'s) appearing in the exponent.

The last fact is the reason for the description 'non-Abelian' phase invariance. As the commutation relations for the $\boldsymbol{\tau}$ matrices show, SU(2) is a non-Abelian group in that two SU(2) transformations do not in general commute. By contrast, in the case of electric charge or particle number, successive transformations clearly commute: this corresponds to an Abelian phase invariance and, as noted in section 2.6, to an Abelian U(1) group.

We may now put our initial 'spin-1/2' analogy on a more precise mathematical footing. In quantum mechanics, states within a degenerate multiplet may conveniently be characterized by the eigenvalues of a complete set of Hermitian operators which commute with the Hamiltonian and with each other. In the case of the p-n doublet, it is easy to see what these operators are. We may write (12.4), (12.6), and (12.7) as

$$H_2\psi^{(1/2)} = E\psi^{(1/2)} \tag{12.36}$$

and

$$H_2\psi^{(1/2)'} = E\psi^{(1/2)'}, \tag{12.37}$$

where H_2 is the 2×2 matrix

$$H_2 = \begin{pmatrix} H & 0 \\ 0 & H \end{pmatrix}. \tag{12.38}$$

Hence H_2 is proportional to the unit matrix in this two-dimensional space, and it therefore commutes with the tau's:

$$[H_2, \boldsymbol{\tau}] = 0. \tag{12.39}$$

It then also follows that H_2 commutes with $\mathbf{V}$, or equivalently

$$\mathbf{V}H_2\mathbf{V}^{-1} = H_2 \tag{12.40}$$

which is the statement that H_2 is invariant under the transformation (12.32). Now the tau's are Hermitian, and hence correspond to possible observables. Equation (12.39) implies that their eigenvalues are constants of the motion (i.e., conserved quantities), associated with the invariance (12.40). But the tau's do not commute amongst themselves and so according to the general principles of quantum mechanics, we cannot give definite values to more than one of them at a time. The problem of finding a classification of the states which makes the maximum use of (12.39), given the commutation relations (12.28), is easily solved by making

---

[2]It is not obvious that the general SU(2) matrix can be parametrized by an angle θ with $0 \leq \theta \leq 2\pi$, and $\hat{\boldsymbol{n}}$: for further discussion of the relation between SU(2) and the three-dimensional rotation group, see Appendix M, section M.7.

use of the formal identity between the operators $\tau_i/2$ and angular momentum operators J_i (cf (12.29)). The answer is[3] that the total squared 'spin'

$$(\mathbf{T}^{(1/2)})^2 = \left(\frac{1}{2}\boldsymbol{\tau}\right)^2 = \frac{1}{4}(\tau_1^2 + \tau_2^2 + \tau_3^2) = \frac{3}{4}\mathbf{1}_2 \tag{12.41}$$

and one component of spin, say $T_3^{(1/2)} = \frac{1}{2}\tau_3$, can be given definite values simultaneously. The corresponding eigenfunctions are just the χ_{p}'s and χ_{n}'s of (12.9), which satisfy

$$\frac{1}{4}\boldsymbol{\tau}^2\chi_{\mathrm{p}} = \frac{3}{4}\chi_{\mathrm{p}}, \qquad \frac{1}{2}\boldsymbol{\tau}_3\chi_{\mathrm{p}} = \frac{1}{2}\chi_{\mathrm{p}} \tag{12.42}$$

$$\frac{1}{4}\boldsymbol{\tau}^2\chi_{\mathrm{n}} = \frac{3}{4}\chi_{\mathrm{n}}, \qquad \frac{1}{2}\boldsymbol{\tau}_3\chi_{\mathrm{n}} = -\frac{1}{2}\chi_{\mathrm{n}}. \tag{12.43}$$

The reason for the 'spin' part of the name 'isospin' should by now be clear; the term is actually a shortened version of the historical one 'isotopic spin'.

In concluding this section we remark that, in this two-dimensional n-p space, the electromagnetic charge operator is represented by the matrix

$$\mathbf{Q}_{\mathrm{em}} = \begin{pmatrix} 1 & 0 \\ 0 & 0 \end{pmatrix} = \frac{1}{2}(\mathbf{1}_2 + \tau_3). \tag{12.44}$$

It is clear that although $\mathbf{Q}_{\mathrm{em}}$ commutes with τ_3, it does not commute with either τ_1 or τ_2. Thus, as we would expect, electromagnetic corrections to the strong interaction Hamiltonian will violate SU(2) symmetry.

12.2.2 Larger (higher dimensional) multiplets of SU(2) in nuclear physics

For the single nucleon states considered so far, the foregoing is really nothing more than the general quantum mechanics of a two-state system, phrased in 'spin-1/2' language. The real power of the isospin (SU(2)) symmetry concept becomes more apparent when we consider states of *several* nucleons. For A nucleons in the nucleus, we introduce three 'total isospin operators' $\mathbf{T} = (T_1, T_2, T_3)$ via

$$\mathbf{T} = \frac{1}{2}\boldsymbol{\tau}_{(1)} + \frac{1}{2}\boldsymbol{\tau}_{(2)} + \ldots + \frac{1}{2}\boldsymbol{\tau}_{(A)}, \tag{12.45}$$

which are Hermitian. Here $\boldsymbol{\tau}_{(n)}$ is the $\boldsymbol{\tau}$-matrix for the nth nucleon. The Hamiltonian H describing the strong interactions of this system is presumed to be invariant under the transformation (12.40) for all the nucleons independently. It then follows that

$$[H, \mathbf{T}] = 0. \tag{12.46}$$

Thus the eigenvalues of the $\mathbf{T}$ operators are constants of the motion. Further, since the isospin operators for different nucleons commute with each other (they are quite independent), the commutation relations (12.28) for each of the individual $\boldsymbol{\tau}$'s imply (see problem 12.4) that the components of $\mathbf{T}$ defined by (12.45) satisfy the commutation relations

$$[T_i, T_j] = i\epsilon_{ijk}T_k \tag{12.47}$$

for $i, j, k = 1, 2, 3$, which are simply the standard angular momentum commutation relations, once more. Thus the energy levels of nuclei ought to be characterized—after allowance for

---

[3]See for example Mandl (1992).

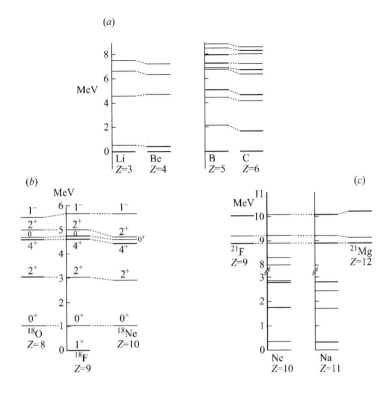

FIGURE 12.2
Energy levels (adjusted for Coulomb energy and neutron-proton mass differences) of nuclei of the same mass number but different charge, showing (a) 'mirror' doublets, (b) triplets, and (c) doublets and quartets.

electromagnetic effects, and correcting for the slight neutron-proton mass difference—by the eigenvalues of $\mathbf{T}^2$ and T_3, say, which can be simultaneously diagonalized along with H. These eigenvalues should then be, to a good approximation, 'good quantum numbers' for nuclei, if the assumed isospin invariance is true.

What are the possible eigenvalues? We know that the $\mathbf{T}$'s are Hermitian and satisfy exactly the same commutation relations (12.47) as the angular momentum operators. These conditions are all that are needed to show that the eigenvalues of $\mathbf{T}^2$ are of the form $T(T+1)$, where $T = 0, \frac{1}{2}, 1, \ldots$, and that for a given T the eigenvalues of T_3 are $-T, -T+1, \ldots, T-1, T$; that is, there are $2T + 1$ *degenerate states* for a given T. These states all have the same A value, and since T_3 counts $+\frac{1}{2}$ for every proton and $-\frac{1}{2}$ for every neutron, it is clear that successive values of T_3 correspond physically to changing one neutron into a proton or vice versa. Thus we expect to see 'charge multiplets' of levels in neighbouring nuclear isobars. These are indeed observed; figure 12.2 shows some examples. These level schemes (which have been adjusted for Coulomb energy differences, and for the neutron-proton mass difference) provide clear evidence of $T = \frac{1}{2}$ (doublet), $T = 1$ (triplet), and $T = \frac{3}{2}$ (quartet) multiplets. It is important to note that states in the same T-multiplet must have the same J^P quantum numbers (these are indicated on the levels for ^{18}F); obviously the nuclear forces will depend on the space and spin degrees of freedom of the nucleons, and will only be the same between different nucleons if the space-spin part of the wavefunction remains the same.

Thus the assumed invariance of the nucleon-nucleon force produces a richer nuclear multiplet structure, going beyond the original n-p doublet. These higher-dimensional multiplets ($T = 1, \frac{3}{2}, \ldots$) are called 'irreducible representations' of SU(2). The commutation relations (12.47) are called the *Lie algebra* of SU(2)[4] (see Appendix M), and the general group theoretical problem of understanding all *possible* multiplets for SU(2) is equivalent to the problem of finding matrices which satisfy these commutation relations. These are, in fact, precisely the angular momentum matrices of dimension $(2T + 1) \times (2T + 1)$ which are generalizations of the $\boldsymbol{\tau}/2$'s, which themselves correspond to $T = \frac{1}{2}$, as indicated in the notation $\mathbf{T}^{(\frac{1}{2})}$. For example, the $T = 1$ matrices are 3×3 and can be compactly summarized by (problem 12.5)

$$(T_i^{(1)})_{jk} = -i\epsilon_{ijk} \tag{12.48}$$

where the numbers $-i\epsilon_{ijk}$ are deliberately chosen to be the *same* numbers (with a minus sign) that specify the algebra in (12.47); the latter are called the *structure constants* of the SU(2) group (see Appendix M, sections M.3 – M.5). In general there will be matrices $\mathbf{T}^{(T)}$ of dimensionality $(2T + 1) \times (2T + 1)$ which satisfy (12.47), and correspondingly $(2T + 1)$ dimensional wavefunctions $\psi^{(T)}$ analogous to the two-dimensional ($T = \frac{1}{2}$) case of (12.8). The generalization of (12.32) to these higher-dimensional multiplets is then

$$\psi^{(T)\prime} = \exp(i\boldsymbol{\alpha} \cdot \mathbf{T}^{(T)})\psi^{(T)}, \tag{12.49}$$

which has the general form of (12.35). In this case, the matrices $\mathbf{T}^{(T)}$ provide a $(2T + 1)$-dimensional matrix representation of the generators of SU(2). We shall meet field-theoretic representations of the generators in section 12.3.

We now proceed to consider isospin in our primary area of interest, which is particle physics.

12.2.3   Isospin in particle physics: flavour SU(2)$_f$

The neutron and proton states themselves are actually only the ground states of a whole series of corresponding $B = 1$ levels with isospin $\frac{1}{2}$ (i.e., doublets). Another series of baryonic levels comes in *four* charge states corresponding to $T = \frac{3}{2}$; and in the meson sector, the π's appear as the lowest states of a sequence of mesonic triplets ($T = 1$). Many other examples also exist, but with one remarkable difference as compared to the nuclear physics case: no baryon states are known with $T > \frac{3}{2}$, nor any meson states with $T > 1$.

The most natural interpretation of these facts is that the observed states are composites of more basic entities which carry different charges but are nearly degenerate in mass, while the forces between these entities are charge-independent, just as in the nuclear (p,n) case. These entities are, of course, the quarks: the n contains (udd), the p is (uud), and the Δ-quartet is (uuu, uud, udd, ddd). The u-d isospin doublet plays the role of the p-n doublet in the nuclear case, and this degree of freedom is what we now call SU(2) isospin flavour symmetry at the quark level, denoted by SU(2)$_f$. We shall denote the u-d quark doublet wavefunction by

$$q = \begin{pmatrix} u \\ d \end{pmatrix} \tag{12.50}$$

omitting now the explicit representation label '($\frac{1}{2}$)', and shortening 'ψ_u' to just 'u', and similarly for 'd'. Then, under an SU(2)$_f$ transformation,

$$q \to q' = \mathbf{V}q = \exp(i\boldsymbol{\alpha} \cdot \boldsymbol{\tau}/2)\, q. \tag{12.51}$$

---

[4]Likewise, the angular momentum commutation relations (12.29) are the Lie algebra of the rotation group SO(3). The Lie algebras of the two groups are therefore the same. For an indication of how, nevertheless, the groups do differ, see Appendix M, section M.7.

The limitation $T \leq \frac{3}{2}$ for baryonic states can be understood in terms of their being composed of three $T = \frac{1}{2}$ constituents (two of them pair to $T = 1$ or $T = 0$, and the third adds to $T = 1$ to make $T = \frac{3}{2}$ or $T = \frac{1}{2}$, and to $T = 0$ to make $T = \frac{1}{2}$, by the usual angular momentum addition rules). It is, however, a challenge for QCD to explain why, for example, states with four or five quarks should not exist (nor states of one or two quarks!), and why a state of six quarks, for example, appears as the deuteron, which is a loosely bound state of n and p, rather than as a compact $B = 2$ analogue of the n and p themselves.

Meson states such as the pion are formed from a quark and an antiquark, and it is therefore appropriate at this point to explain how *antiparticles* are described in isospin terms. An antiparticle is characterized by having the signs of all its additively conserved quantum numbers reversed, relative to those of the corresponding particle. Thus, if a u-quark has $B = \frac{1}{3}, T = \frac{1}{2}$, and $T_3 = \frac{1}{2}$, a ū-quark has $B = -\frac{1}{3}, T = \frac{1}{2}$, and $T_3 = -\frac{1}{2}$. Similarly, the d̄ has $B = -\frac{1}{3}, T = \frac{1}{2}$, and $T_3 = \frac{1}{2}$. Note that, while T_3 is an additively conserved quantum number, the magnitude of the isospin is not additively conserved: rather, it is 'vectorially' conserved according to the rules of combining angular-momentum-like quantum numbers, as we have seen. Thus the antiquarks d̄ and ū form the $T_3 = +\frac{1}{2}$ and $T_3 = -\frac{1}{2}$ members of an SU(2)$_f$ doublet, just as u and d themselves do, and the question arises: given that the (u, d) doublet transforms as in (12.51), how does the $(\bar{u}, \bar{d})$ doublet transform?

The answer is that antiparticles are assigned to the *complex conjugate* of the representation to which the corresponding particles belong. Thus identifying $\bar{u} \equiv u^*$ and $\bar{d} \equiv d^*$ we have[5]

$$q^{*\prime} = \mathbf{V}^* q^*, \quad \text{or} \quad \begin{pmatrix} \bar{u} \\ \bar{d} \end{pmatrix}' = \exp(-i\boldsymbol{\alpha} \cdot \boldsymbol{\tau}^*/2) \begin{pmatrix} \bar{u} \\ \bar{d} \end{pmatrix} \qquad (12.52)$$

for the SU(2)$_f$ transformation law of the antiquark doublet. In mathematical terms, this means (compare (12.32)) that the three matrices $-\frac{1}{2}\boldsymbol{\tau}^*$ must represent the generators of SU(2)$_f$ in the $\mathbf{2}^*$ representation (i.e., the complex conjugate of the original two-dimensional representation, which we will now call $\mathbf{2}$). Referring to (12.25), we see that $\tau_1^* = \tau_1, \tau_2^* = -\tau_2$, and $\tau_3^* = \tau_3$. It is then easy to check that the three matrices $-\tau_1/2, +\tau_2/2$, and $-\tau_3/2$ do indeed satisfy the required commutation relations (12.28), and thus provide a valid matrix representation of the SU(2) generators. Also, since the third component of isospin is here represented by $-\tau_3^*/2 = -\tau_3/2$, the desired reversal in sign of the additively conserved eigenvalue does occur.

Although the quark doublet (u, d) and antiquark doublet $(\bar{u}, \bar{d})$ do transform differently under SU(2)$_f$ transformations, there is nevertheless a sense in which the $\mathbf{2}^*$ and $\mathbf{2}$ representations are somehow the 'same'. After all, the quantum numbers $T = \frac{1}{2}, T_3 = \pm\frac{1}{2}$ describe them both. In fact, the two representations are 'unitarily equivalent', in that we can find a unitary matrix U_C such that

$$U_C \exp(-i\boldsymbol{\alpha} \cdot \boldsymbol{\tau}^*/2) U_C^{-1} = \exp(i\boldsymbol{\alpha} \cdot \boldsymbol{\tau}/2). \qquad (12.53)$$

This requirement is easier to disentangle if we consider infinitesimal transformations, for which (12.53) becomes

$$U_C(-\boldsymbol{\tau}^*)U_C^{-1} = \boldsymbol{\tau}, \qquad (12.54)$$

or

$$U_C \tau_1 U_C^{-1} = -\tau_1, \quad U_C \tau_2 U_C^{-1} = \tau_2, \quad U_C \tau_3 U_C^{-1} = -\tau_3. \qquad (12.55)$$

---

[5]The overbar (ū etc.) here stands only for 'antiparticle', and has nothing to do with the Dirac conjugate $\bar{\psi}$ introduced in section 4.4.

Bearing the commutation relations (12.28) in mind, and the fact that $\tau_i^{-1} = \tau_i$, it is clear that we can choose U_C proportional to τ_2, and set

$$U_C = i\tau_2 = \begin{pmatrix} 0 & 1 \\ -1 & 0 \end{pmatrix} \tag{12.56}$$

to obtain a convenient unitary form. From (12.52) and (12.53) we obtain $(U_C q^{*\prime}) = \mathbf{V}(U_C q^*)$, which implies that the doublet

$$U_C \begin{pmatrix} \bar{u} \\ \bar{d} \end{pmatrix} = \begin{pmatrix} \bar{d} \\ -\bar{u} \end{pmatrix} \tag{12.57}$$

transforms in exactly the same way as (u, d). This result is useful, because it means that we can use the familiar tables of (Clebsch-Gordan) angular momentum coupling coefficients for combining quark and antiquark states together, *provided* we include the relative minus sign between the $\bar{d}$ and $\bar{u}$ components which has appeared in (12.57). Note that, as expected, the $\bar{d}$ is in the $T_3 = +\frac{1}{2}$ position, and the $\bar{u}$ is in the $T_3 = -\frac{1}{2}$ position.

As an application of these results, let us compare the $T = 0$ combination of the p and n states to form the (isoscalar) deuteron, and the combination of (u, d) and $(\bar{u}, \bar{d})$ states to form the isoscalar ω-meson. In the first, the isospin part of the wavefunction is $\frac{1}{\sqrt{2}}(\psi_p \psi_n - \psi_n \psi_p)$, corresponding to the $S = 0$ combination of two spin-$\frac{1}{2}$ particles in quantum mechanics given by $\frac{1}{\sqrt{2}}(|\uparrow\rangle\,|\downarrow\rangle - |\downarrow\rangle\,|\uparrow\rangle)$. But in the second case, the corresponding wavefunction is $\frac{1}{\sqrt{2}}(\bar{d}d - (-\bar{u})u) = \frac{1}{\sqrt{2}}(\bar{d}d + \bar{u}u)$. Similarly, the $T = 1$ $T_3 = 0$ state describing the π^0 is $\frac{1}{\sqrt{2}}(\bar{d}d + (-\bar{u})u) = \frac{1}{\sqrt{2}}(\bar{d}d - \bar{u}u)$.

There is a very convenient alternative way of obtaining these wavefunctions, which we include here because it generalizes straightforwardly to SU(3); its advantage is that it avoids the use of the explicit C-G coupling coefficients and of their (more complicated) analogues in SU(3).

Bearing in mind the identifications $\bar{u} \equiv u^*, \bar{d} \equiv d^*$, we see that the $T = 0$ $\bar{q}q$ combination $\bar{u}u + \bar{d}d$ can be written as $u^*u + d^*d$ which is just $q^\dagger q$ (recall that † means transpose and complex conjugate). Under an $SU(2)_f$ transformation, $q \to q' = \mathbf{V}q$, so $q^\dagger \to q'^\dagger = q^\dagger \mathbf{V}^\dagger$ and

$$q^\dagger q \to q'^\dagger q' = q^\dagger \mathbf{V}^\dagger \mathbf{V} q = q^\dagger q \tag{12.58}$$

using $\mathbf{V}^\dagger \mathbf{V} = \mathbf{1}_2$; thus $q^\dagger q$ is indeed an $SU(2)_f$ invariant, which means it has $T = 0$ (no multiplet partners).

We may also construct the $T = 1$ $q - \bar{q}$ states in a similar way. Consider the three quantities v_i defined by

$$v_i = q^\dagger \tau_i q \quad i = 1, 2, 3. \tag{12.59}$$

Under an infinitesimal $SU(2)_f$ transformation

$$q' = (\mathbf{1}_2 + i\boldsymbol{\epsilon} \cdot \boldsymbol{\tau}/2)q, \tag{12.60}$$

the three quantities v_i transform to

$$v_i' = q^\dagger (\mathbf{1}_2 - i\boldsymbol{\epsilon} \cdot \boldsymbol{\tau}/2)\tau_i(\mathbf{1}_2 + i\boldsymbol{\epsilon} \cdot \boldsymbol{\tau}/2)q, \tag{12.61}$$

where we have used $q'^\dagger = q^\dagger (\mathbf{1}_2 + i\boldsymbol{\epsilon} \cdot \boldsymbol{\tau}/2)^\dagger$ and then $\boldsymbol{\tau}^\dagger = \boldsymbol{\tau}$. Retaining only the first-order terms in $\boldsymbol{\epsilon}$ gives (problem 12.6)

$$v_i' = v_i + i\frac{\epsilon_j}{2}q^\dagger(\tau_i\tau_j - \tau_j\tau_i)q \tag{12.62}$$

where the sum on $j = 1, 2, 3$ is understood. But from (12.28) we know the commutator of two τ's, so that (12.62) becomes

$$
\begin{aligned}
v_i' &= v_i + \mathrm{i}\frac{\epsilon_j}{2} q^\dagger . 2\mathrm{i}\epsilon_{ijk}\tau_k q \quad \text{(sum on } k = 1, 2, 3) \\
&= v_i - \epsilon_{ijk}\epsilon_j q^\dagger \tau_k q \\
&= v_i - \epsilon_{ijk}\epsilon_j v_k,
\end{aligned}
\tag{12.63}
$$

which may also be written in 'vector' notation as

$$
\boldsymbol{v}' = \boldsymbol{v} - \boldsymbol{\epsilon} \times \boldsymbol{v}.
\tag{12.64}
$$

Equation (12.63) states that, under an (infinitesimal) SU(2)$_f$ transformation, the three quantities v_i $(i = 1, 2, 3)$ transform into *specific linear combinations* of themselves, as determined by the coefficients ϵ_{ijk} (the ϵ's are just the parameters of the infinitesimal transformation). This is precisely what is needed for a set of quantities to *form the basis for a representation*. In this case, it is the $T = 1$ representation as we can guess from the multiplicity of three, but we can also directly verify it, as follows. Equation (12.49) with $T = 1$, together with (12.48), tell us how a $T = 1$ triplet should transform: namely, under an infinitesimal transformation (with $\mathbf{1}_3$ the unit 3×3 matrix),

$$
\begin{aligned}
\psi_i^{(1)\prime} &= (\mathbf{1}_3 + \mathrm{i}\boldsymbol{\epsilon} \cdot \mathbf{T}^{(1)})_{ik}\psi_k^{(1)} \quad \text{(sum on } k = 1, 2, 3) \\
&= (\mathbf{1}_3 + \mathrm{i}\epsilon_j T_j^{(1)})_{ik}\psi_k^{(1)} \quad \text{(sum on } j = 1, 2, 3) \\
&= (\delta_{ik} + \mathrm{i}\epsilon_j (T_j^{(1)})_{ik})\psi_k^{(1)} \\
&= (\delta_{ik} + \mathrm{i}\epsilon_j . - \mathrm{i}\epsilon_{jik})\psi_k^{(1)} \quad \text{using (12.48)} \\
&= \psi_i^{(1)} - \epsilon_{ijk}\epsilon_j \psi_k^{(1)} \quad \text{using the antisymmetry of } \epsilon_{ijk}
\end{aligned}
\tag{12.65}
$$

which is exactly the same as (12.63).

The reader who has worked through problem 4.2(a) will recognize the exact analogy between the $T = 1$ transformation law (12.64) for the isospin bilinear $q^\dagger \boldsymbol{\tau} q$, and the 3-vector transformation law (c.f.(4.9)) for the Pauli spinor bilinear $\phi^\dagger \boldsymbol{\sigma} \phi$.

Returning to the physics of v_i, inserting (12.50) into (12.59) we find explicitly

$$
v_1 = \bar{u}d + \bar{d}u, \quad v_2 = -\mathrm{i}\,\bar{u}d + \mathrm{i}\,\bar{d}u, \quad v_3 = \bar{u}u - \bar{d}d.
\tag{12.66}
$$

Apart from the normalization factor of $\frac{1}{\sqrt{2}}$, v_3 may therefore be identified with the $T_3 = 0$ member of the $T = 1$ triplet, having the quantum numbers of the π^0. Neither v_1 nor v_2 has a definite value of T_3, however. Rather, we need to consider the linear combinations

$$
\frac{1}{2}(v_1 + \mathrm{i}v_2) = \bar{u}d \quad T_3 = -1
\tag{12.67}
$$

and

$$
\frac{1}{2}(v_1 - \mathrm{i}v_2) = \bar{d}u \quad T_3 = +1
\tag{12.68}
$$

which have the quantum numbers of the π^- and π^+. The use of $v_1 \pm \mathrm{i}v_2$ here is precisely analogous to the use of the 'spherical basis' wavefunctions $x \pm \mathrm{i}y = r\sin\theta e^{\pm \mathrm{i}\phi}$ for $\ell = 1$ states in quantum mechanics, rather than the 'Cartesian' ones x and y.

We are now ready to proceed to SU(3).

12.3 Flavour $SU(3)_f$

Larger hadronic multiplets also exist, in which strange particles are grouped with non-strange ones. Gell-Mann (1961) and Ne'eman (1961) (see also Gell-Mann and Ne'eman (1964)) were the first to propose $SU(3)_f$ as the correct generalization of isospin $SU(2)_f$ to include strangeness. Like $SU(2)$, $SU(3)$ is a group whose elements are matrices—in this case, unitary 3×3 ones, of unit determinant. The general group-theoretic analysis of $SU(3)$ is quite complicated, but is fortunately not necessary for the physical applications we require. We can, in fact, develop all the results needed by mimicking the steps followed for $SU(2)$.

We start by finding the general form of an $SU(3)$ matrix. Such matrices obviously act on 3-component column vectors, the generalization of the 2-component isospinors of $SU(2)$. In more physical terms, we regard the three quark wavefunctions u, d, and s as being approximately degenerate, and we consider unitary 3×3 transformations among them via

$$q' = \mathbf{W}q \qquad (12.69)$$

where q now stands for the 3-component column vector

$$q = \begin{pmatrix} u \\ d \\ s \end{pmatrix} \qquad (12.70)$$

and $\mathbf{W}$ is a 3×3 unitary matrix of determinant 1 (again, an overall phase has been extracted). The representation provided by this triplet of states is called the 'fundamental' representation of $SU(3)_f$ (just as the isospinor representation is the fundamental one of $SU(2)_f$).

To determine the general form of an $SU(3)$ matrix $\mathbf{W}$, we follow exactly the same steps as in the $SU(2)$ case. An infinitesimal $SU(3)$ matrix has the form

$$\mathbf{W}_{\mathrm{infl}} = \mathbf{1}_3 + \mathrm{i}\boldsymbol{\chi} \qquad (12.71)$$

where $\boldsymbol{\chi}$ is a 3×3 traceless Hermitian matrix. Such a matrix involves *eight* independent parameters (problem (12.7)) and can be written as

$$\boldsymbol{\chi} = \boldsymbol{\eta} \cdot \boldsymbol{\lambda}/2 \qquad (12.72)$$

where $\boldsymbol{\eta} = (\eta_1, \ldots, \eta_8)$ and the $\boldsymbol{\lambda}$'s are eight matrices generalising the $\boldsymbol{\tau}$ matrices of (12.25). They are the generators of $SU(3)$ in the three-dimensional fundamental representation, and their commutation relations define the *algebra of SU(3)* (compare (12.28) for $SU(2)$):

$$[\lambda_a/2, \lambda_b/2] = \mathrm{i}f_{abc}\lambda_c/2, \qquad (12.73)$$

where a, b, and c run from 1 to 8.

The λ - matrices (often called the *Gell-Mann matrices*), are given in Appendix M, along with the *SU(3) structure constants* $\mathrm{i}f_{abc}$; the constants f_{abc} are all real.

A finite $SU(3)$ transformation on the quark triplet is then (cf (12.32))

$$q' = \exp(\mathrm{i}\boldsymbol{\alpha} \cdot \boldsymbol{\lambda}/2)q, \qquad (12.74)$$

which also has the 'generalized phase transformation' character of (12.35), now with *eight* 'phase angles'. Thus $\mathbf{W}$ is parametrized as $\mathbf{W} = \exp(\mathrm{i}\boldsymbol{\alpha} \cdot \boldsymbol{\lambda}/2)$.

As in the case of SU(2)$_f$, exact symmetry under SU(3)$_f$ would imply that the three states u, d, and s were degenerate in mass. Actually, of course, this is not the case: in particular, while the u and d quark masses are of order 2–5 MeV, the s quark mass is greater, of order 100 MeV. Nevertheless it is still possible to regard this as relatively small on a typical hadronic mass scale, so we may proceed to explore the physical consequences of this (approximate) SU(3)$_f$ flavour symmetry.

Such a symmetry implies that the eigenvalues of the $\boldsymbol{\lambda}$'s are constants of the motion, but because of the commutation relations (12.73) only a subset of these operators have simultaneous eigenstates. This happened for SU(2) too, but there the very close analogy with SO(3) told us how the states were to be correctly classified, by the eigenvalues of the relevant complete set of mutually commuting operators. Here it is more involved— for a start, there are 8 matrices λ_a. A glance at Appendix M, section M.4.5, shows that *two* of the $\boldsymbol{\lambda}$'s are diagonal (in the chosen representation), namely λ_3 and λ_8. This means physically that for SU(3) there are *two* additively conserved quantum numbers, which in this case are of course the third component of hadronic isospin (since λ_3 is simply τ_3 bordered by zeros), and a quantity related to strangeness. Defining the hadronic hypercharge Y by $Y = B + S$, where B is the baryon number ($\frac{1}{3}$ for each quark) and the strangeness values are $S(\mathrm{u}) = S(\mathrm{d}) = 0$, $S(\mathrm{s}) = -1$, we find that the physically required eigenvalues imply that the matrix representing the hypercharge operator is $Y^{(\boldsymbol{3})} = \frac{1}{\sqrt{3}}\lambda_8$, in this fundamental (three-dimensional) representation, denoted by the symbol $\boldsymbol{3}$. Identifying $T_3^{(\boldsymbol{3})} = \frac{1}{2}\lambda_3$ then gives the Gell-Mann-Nishijima relation $Q = T_3 + Y/2$ for the quark charges in units of $\mid e \mid$.

So λ_3 and λ_8 are analogous to τ_3; what about the analogue of $\boldsymbol{\tau}^2$, which is diagonalizable simultaneously with τ_3 in the case of SU(2)? Indeed, (cf (12.41)) $\boldsymbol{\tau}^2$ is a multiple of the 2×2 unit matrix. In just the same way one finds that $\boldsymbol{\lambda}^2$ is also proportional to the unit matrix:

$$(\boldsymbol{\lambda}/2)^2 = \sum_{a=1}^{8}(\lambda_a/2)^2 = \frac{4}{3}\mathbf{1}_3, \tag{12.75}$$

as can be verified from the explicit forms of the λ-matrices given in Appendix M, section M.4.5. Thus we may characterize the 'fundamental triplet' (12.70) by the eigenvalues of $(\boldsymbol{\lambda}/2)^2$, λ_3, and λ_8. The conventional way of representing this pictorially is to plot the states in a $Y - T_3$ diagram, as shown in figure 12.3.

We may now consider other representations of SU(3)$_f$. The first important one is that to which the *antiquarks* belong. If we denote the fundamental three-dimensional representation accommodating the quarks by $\boldsymbol{3}$, then the antiquarks have quantum numbers appropriate to the 'complex conjugate' of this representation, denoted by $\boldsymbol{3}^*$ just as in the SU(2) case. The $\bar{q}$ wavefunctions identified as $\bar{u} \equiv u^*, \bar{d} \equiv d^*$, and $\bar{s} \equiv s^*$, then transform by

$$\bar{q}' = \begin{pmatrix} \bar{u} \\ \bar{d} \\ \bar{s} \end{pmatrix}' = \mathbf{W}^*\bar{q} = \exp(-\mathrm{i}\boldsymbol{\alpha} \cdot \boldsymbol{\lambda}^*/2)\bar{q} \tag{12.76}$$

instead of by (12.74). As for the $\boldsymbol{2}^*$ representation of SU(2), (12.76) means that the eight quantities $-\boldsymbol{\lambda}^*/2$ represent the SU(3) generators in this $\boldsymbol{3}^*$ representation. Referring to Appendix M, section M.4.5, one quickly sees that λ_3 and λ_8 are real, so that the eigenvalues of the physical observables $T_3^{(\boldsymbol{3}^*)} = -\lambda_3/2$ and $Y^{(\boldsymbol{3}^*)} = -\frac{1}{\sqrt{3}}\lambda_8/2$ (in this representation) are reversed relative to those in the $\boldsymbol{3}$, as expected for antiparticles. The $\bar{u}, \bar{d}$ and $\bar{s}$ states may also be plotted on the $Y - T_3$ diagram, figure 12.3, as shown.

Here is already one important difference between SU(3) and SU(2): the fundamental SU(3) representation $\boldsymbol{3}$ and its complex conjugate $\boldsymbol{3}^*$ are *not* equivalent. This follows

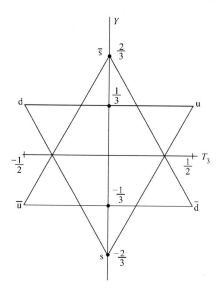

FIGURE 12.3
The $Y - T_3$ quantum numbers of the fundamental triplet **3** of quarks, and of the antitriplet **3**[*] of antiquarks.

immediately from figure 12.3, where it is clear that the extra quantum number Y distinguishes the two representations.

Larger $SU(3)_f$ representations can be created by combining quarks and antiquarks, as in $SU(2)_f$. For our present purposes, an important one is the eight-dimensional ('octet') representation which appears when one combines the **3**[*] and **3** representations, in a way which is very analogous to the three-dimensional ('triplet') representation obtained by combining the **2**[*] and **2** representations of $SU(2)$.

Consider first the quantity $\bar{u}u + \bar{d}d + \bar{s}s$. As in the $SU(2)$ case, this can be written equivalently as $q^\dagger q$, which is invariant under $q \to q' = \mathbf{W}q$ since $\mathbf{W}^\dagger \mathbf{W} = \mathbf{1}_3$. So this combination is an $SU(3)$ *singlet*. The *octet* coupling is formed by a straightforward generalization of the $SU(2)$ triplet coupling $q^\dagger \boldsymbol{\tau} q$ of (12.59),

$$w_a = q^\dagger \lambda_a q \quad a = 1, 2, \ldots 8. \tag{12.77}$$

Under an infinitesimal $SU(3)_f$ transformation (compare (12.61) and (12.62)),

$$\begin{aligned} w_a \to w'_a &= q^\dagger (\mathbf{1}_3 - \mathrm{i}\boldsymbol{\eta} \cdot \boldsymbol{\lambda}/2)\lambda_a(\mathbf{1}_3 + \mathrm{i}\boldsymbol{\eta} \cdot \boldsymbol{\lambda}/2)q \\ &\approx q^\dagger \lambda_a q + \mathrm{i}\frac{\eta_b}{2} q^\dagger (\lambda_a \lambda_b - \lambda_b \lambda_a)q \end{aligned} \tag{12.78}$$

where the sum on $b = 1$ to 8 is understood. Using (12.73) for the commutator of two λ's we find

$$w'_a = w_a + \mathrm{i}\frac{\eta_b}{2} q^\dagger.2\mathrm{i}f_{abc}\lambda_c q \tag{12.79}$$

or

$$w'_a = w_a - f_{abc}\eta_b w_c \tag{12.80}$$

which may usefully be compared with (12.63). Just as in the $SU(2)_f$ triplet case, equation (12.80) shows that, under an $SU(3)_f$ transformation, the eight quantities $w_a(a = 1, 2, \ldots 8)$ transform with specific linear combinations of themselves, as determined by the coefficients f_{abc} (the η's are just the parameters of the infinitesimal transformation).

This is, again, precisely what is needed for a set of quantities to form the basis for a representation—in this case, an eight-dimensional representation of SU(3)$_f$. For a finite SU(3)$_f$ transformation, we can 'exponentiate' (12.80) to obtain

$$\boldsymbol{w}' = \exp(\mathrm{i}\boldsymbol{\alpha} \cdot \mathbf{G}^{(8)})\boldsymbol{w} \tag{12.81}$$

where $\boldsymbol{w}$ is an 8-component column vector

$$\boldsymbol{w} = \begin{pmatrix} w_1 \\ w_2 \\ \vdots \\ w_8 \end{pmatrix} \tag{12.82}$$

such that $w_a = q^\dagger \lambda_a q$, and where (cf (12.49) for SU(2))$_f$) the quantities $\mathbf{G}^{(8)} = (G_1^{(8)}, G_2^{(8)}, \dots G_8^{(8)})$ are 8×8 matrices, acting on the 8-component vector $\boldsymbol{w}$, and forming an 8-dimensional representation of the algebra of SU(3): that is to say, the $\mathbf{G}^{(8)}$'s satisfy (cf (12.73))

$$\left[G_a^{(8)}, G_b^{(8)} \right] = \mathrm{i} f_{abc} G_c^{(8)}. \tag{12.83}$$

The actual form of the $G_a^{(8)}$ matrices is given by comparing the infinitesimal version of (12.81) with (12.80)

$$\left(G_a^{(8)} \right)_{bc} = -\mathrm{i} f_{abc}, \tag{12.84}$$

as may be checked in problem 12.8, where it is also verified that the matrices specified by (12.84) do obey the commutation relations (12.83).

As in the SU(2)$_f$ case, the 8 states generated by the combinations $q^\dagger \lambda_a q$ are not necessarily the ones with the physically desired quantum numbers. To get the $\pi^\pm$, for example, we again need to form $(w_1 \pm \mathrm{i} w_2)/2$. Similarly, w_4 produces $\bar{u}s + \bar{s}u$ and w_5 the combination $-\mathrm{i}\,\bar{u}s + \mathrm{i}\,\bar{s}u$, so the $\mathrm{K}^\pm$ states are $w_4 \mp \mathrm{i} w_5$. Similarly the K^0, $\bar{\mathrm{K}}^0$ states are $w_6 - \mathrm{i} w_7$, and $w_6 + \mathrm{i} w_7$, while the η (in this simple model) would be $w_8 \sim (\bar{u}u + \bar{d}d - 2\bar{s}s)$, which is orthogonal to both the π^0 state and the SU(3)$_f$ singlet. In this way all the pseudoscalar octet of π-partners has been identified, as shown on the $Y - T$ diagram of figure 12.4. We say 'octet of π-partners', but a reader knowing the masses of these particles might well query why we should feel justified in regarding them as (even approximately) degenerate. By contrast, a similar octet of vector ($J^P 1^-$) mesons (the $\omega, \rho, \mathrm{K}^*$, and $\bar{\mathrm{K}}^*$) are all much closer in mass, averaging around 800 MeV; in these states the $\bar{q}q$ spins add to $S = 1$, while the orbital angular momentum is still zero. The pion, and to a much lesser extent the kaons, seem to be 'anomalously light' for some reason. We shall learn the likely explanation for this in chapter 15.

There is a deep similarity between (12.84) and (12.48). In both cases, a representation has been found in which the matrix element of a generator is minus the corresponding structure constant. Such a representation is always possible for a Lie group, and is called the *adjoint*, or *regular*, representation (see Appendix M, section M.5). These representations are of particular importance in gauge theories, as we will see, since gauge quanta always belong to the adjoint representation of the gauged group (for example, the 8 gluons in SU(3)$_c$).

Further flavours c, b, and t of course exist, but the mass differences are now so large that it is generally not useful to think about higher flavour groups such as SU(4)$_f$ etc. Instead, we now move on to consider the field-theoretic formulation of global SU(2)$_f$ and SU(3)$_f$.

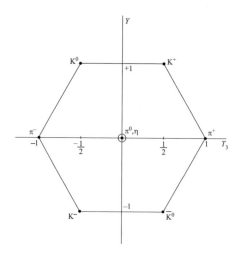

FIGURE 12.4
The $Y - T_3$ quantum numbers of the pseudoscalar meson octet.

12.4 Non-Abelian global symmetries in Lagrangian quantum field theory

12.4.1 SU(2)$_f$ and SU(3)$_f$

As may already have begun to be apparent in chapter 7, Lagrangian quantum field theory is a formalism which is especially well adapted for the description of symmetries. Without going into any elaborate general theory, we shall now give a few examples showing how global flavour symmetry is very easily built into a Lagrangian, generalizing in a simple way the global U(1) symmetries considered in sections 7.1 and 7.2. This will also prepare the way for the (local) gauge case, to be considered in the following chapter.

Consider, for example, the Lagrangian

$$\hat{\mathcal{L}} = \bar{\hat{u}}(\mathrm{i}\,\slashed{\partial} - m)\hat{u} + \bar{\hat{d}}(\mathrm{i}\,\slashed{\partial} - m)\hat{d} \qquad (12.85)$$

describing two free fermions 'u' and 'd' of equal mass m, with the overbar now meaning the Dirac conjugate for the 4-component spinor fields. Note carefully that we are suppressing the spacetime arguments of the quantum fields $\hat{u}(x), \hat{d}(x)$. As in (12.50), we are using the convenient shorthand $\hat{\psi}_\mathrm{u} = \hat{u}$ and $\hat{\psi}_\mathrm{d} = \hat{d}$. Let us introduce

$$\hat{q} = \begin{pmatrix} \hat{u} \\ \hat{d} \end{pmatrix} \qquad (12.86)$$

so that $\hat{\mathcal{L}}$ can be compactly written as

$$\hat{\mathcal{L}} = \bar{\hat{q}}(\mathrm{i}\,\slashed{\partial} - m)\hat{q}. \qquad (12.87)$$

In this form it is obvious that $\hat{\mathcal{L}}$ - and hence the associated Hamiltonian $\hat{\mathcal{H}}$ — is invariant under the global U(1) transformation

$$\hat{q}' = \mathrm{e}^{\mathrm{i}\alpha}\hat{q} \qquad (12.88)$$

(cf (12.1)) which is associated with baryon number conservation. It is also invariant under global SU(2)$_f$ transformations acting in the flavour u-d space (cf (12.32)):-

$$\hat{q}' = \exp(-i\boldsymbol{\alpha} \cdot \boldsymbol{\tau}/2)\hat{q} \tag{12.89}$$

(for the change in sign with respect to (12.31), compare section 7.1 and section 7.2 in the U(1) case). In (12.89), the three parameters $\boldsymbol{\alpha}$ are independent of x.

What are the conserved quantities associated with the invariance of $\hat{\mathcal{L}}$ under (12.89)? Let us recall the discussion of the simpler U(1) cases studied in sections 7.1 and 7.2. Considering the complex scalar field of section 7.1, the analogue of (12.89) was just $\hat{\phi} \to \hat{\phi}' = e^{-i\alpha}\hat{\phi}$, and the conserved quantity was the Hermitian operator $\hat{N}_\phi$ which appeared in the exponent of the unitary operator $\hat{U}$ that effected the transformation $\hat{\phi} \to \hat{\phi}'$ via

$$\hat{\phi}' = \hat{U}\hat{\phi}\hat{U}^\dagger, \tag{12.90}$$

with

$$\hat{U} = \exp(i\alpha\hat{N}_\phi). \tag{12.91}$$

For an infinitesimal α, we have

$$\hat{\phi}' \approx (1 - i\epsilon)\hat{\phi}, \quad \hat{U} \approx 1 + i\epsilon\hat{N}_\phi, \tag{12.92}$$

so that (12.90) becomes

$$(1 - i\epsilon)\hat{\phi} = (1 + i\epsilon\hat{N}_\phi)\hat{\phi}(1 - i\epsilon\hat{N}_\phi) \approx \hat{\phi} + i\epsilon[\hat{N}_\phi, \hat{\phi}]; \tag{12.93}$$

hence we require

$$[\hat{N}_\phi, \hat{\phi}] = -\hat{\phi} \tag{12.94}$$

for consistency. Insofar as $\hat{N}_\phi$ determines the form of an infinitesimal version of the unitary transformation operator $\hat{U}$, it seems reasonable to call it the *generator* of these global U(1) transformations (compare the discussion after (12.27) and (12.35), but note that here $\hat{N}_\phi$ is a quantum field operator, not a matrix).

Consider now the SU(2)$_f$ transformation (12.89), in the infinitesimal case:

$$\hat{q}' = (1 - i\boldsymbol{\epsilon} \cdot \boldsymbol{\tau}/2)\hat{q}. \tag{12.95}$$

Since the single U(1) parameter ϵ is now replaced by the three parameters $\boldsymbol{\epsilon} = (\epsilon_1, \epsilon_2, \epsilon_3)$, we shall need three analogues of $\hat{N}_\phi$, which we call

$$\hat{\boldsymbol{T}}^{(\frac{1}{2})} = (\hat{T}_1^{(\frac{1}{2})}, \hat{T}_2^{(\frac{1}{2})}, \hat{T}_3^{(\frac{1}{2})}), \tag{12.96}$$

corresponding to the three independent infinitesimal SU(2) transformations. The generalizations of (12.90) and (12.91) are then

$$\hat{q}' = \hat{U}^{(\frac{1}{2})}\hat{q}\hat{U}^{(\frac{1}{2})\dagger} \tag{12.97}$$

and

$$\hat{U}^{(\frac{1}{2})} = \exp(i\boldsymbol{\alpha} \cdot \hat{\boldsymbol{T}}^{(\frac{1}{2})}) \tag{12.98}$$

where the $\hat{\boldsymbol{T}}^{(\frac{1}{2})}$'s are Hermitian, so that $\hat{U}^{(\frac{1}{2})}$ is unitary (cf (12.35)). It would seem reasonable in this case too to regard the $\hat{\boldsymbol{T}}^{(\frac{1}{2})}$'s as providing a *field theoretic representation* of the generators of $SU(2)_f$, an interpretation we shall shortly confirm. In the infinitesimal case, (12.97) and (12.98) become

$$(1 - i\boldsymbol{\epsilon} \cdot \boldsymbol{\tau}/2)\hat{q} = (1 + i\boldsymbol{\epsilon} \cdot \hat{\boldsymbol{T}}^{(\frac{1}{2})})\hat{q}(1 - i\boldsymbol{\epsilon} \cdot \hat{\boldsymbol{T}}^{(\frac{1}{2})}), \qquad (12.99)$$

using the Hermiticity of the $\hat{\boldsymbol{T}}^{(\frac{1}{2})}$'s. Expanding the right-hand side of (12.99) to first order in $\boldsymbol{\epsilon}$, and equating coefficients of $\boldsymbol{\epsilon}$ on both sides, (12.99) reduces to (problem 12.9)

$$[\hat{\boldsymbol{T}}^{(\frac{1}{2})}, \hat{q}] = -(\boldsymbol{\tau}/2)\hat{q}, \qquad (12.100)$$

which is the analogue of (12.94). Equation (12.100) expresses a very specific *commutation* property of the operators $\hat{\boldsymbol{T}}^{(\frac{1}{2})}$, which turns out to be satisfied by the expression

$$\hat{\boldsymbol{T}}^{(\frac{1}{2})} = \int \hat{q}^\dagger(\boldsymbol{\tau}/2)\hat{q}\,\mathrm{d}^3\boldsymbol{x} \qquad (12.101)$$

as can be checked (problem 12.10) from the anticommutation relations of the fermionic fields in $\hat{q}$. We shall derive (12.101) from Noether's theorem (Noether 1918) in a little while. Note that if '$\boldsymbol{\tau}/2$' is replaced by 1, (12.101) reduces to the sum of the u and d number operators, as required for the one-parameter U(1) case. The '$\hat{q}^\dagger \boldsymbol{\tau}\hat{q}$' combination is precisely the field-theoretic version of the $q^\dagger \boldsymbol{\tau} q$ coupling we discussed in section 12.1.3. It means that the three operators $\hat{\boldsymbol{T}}^{(\frac{1}{2})}$ themselves belong to a $T = 1$ triplet of $SU(2)_f$.

It is possible to verify that these $\hat{\boldsymbol{T}}^{(\frac{1}{2})}$'s do indeed commute with the Hamiltonian $\hat{H}$:

$$\mathrm{d}\hat{\boldsymbol{T}}^{(\frac{1}{2})}/\mathrm{d}t = -i[\hat{\boldsymbol{T}}^{(\frac{1}{2})}, \hat{H}] = 0 \qquad (12.102)$$

so that their eigenvalues are conserved. That the $\hat{\boldsymbol{T}}^{(\frac{1}{2})}$ are, as already suggested, a field theoretic representation of the generators of $SU(2)$, appropriate to the case $T = \frac{1}{2}$, follows from the fact that they obey the $SU(2)$ algebra (problem 12.11):

$$[\hat{T}_i^{(\frac{1}{2})}, \hat{T}_j^{(\frac{1}{2})}] = i\epsilon_{ijk}\hat{T}_k^{(\frac{1}{2})}. \qquad (12.103)$$

For many purposes it is more useful to consider the raising and lowering operators

$$\hat{T}_\pm^{(\frac{1}{2})} = (\hat{T}_1^{(\frac{1}{2})} \pm i\hat{T}_2^{(\frac{1}{2})}). \qquad (12.104)$$

For example, we easily find

$$\hat{T}_+^{(\frac{1}{2})} = \int \hat{u}^\dagger \hat{d} \, \mathrm{d}^3\boldsymbol{x}, \qquad (12.105)$$

which destroys a d quark and creates a u, or destroys a $\bar{u}$ and creates a $\bar{d}$, in either case raising the $\hat{T}_3^{(\frac{1}{2})}$ eigenvalue by $+1$, since

$$\hat{T}_3^{(\frac{1}{2})} = \frac{1}{2}\int (\hat{u}^\dagger \hat{u} - \hat{d}^\dagger \hat{d})\mathrm{d}^3\boldsymbol{x} \qquad (12.106)$$

which counts $+\frac{1}{2}$ for each u (or $\bar{d}$) and $-\frac{1}{2}$ for each d (or $\bar{u}$). Thus these operators certainly 'do the job' expected of field theoretic isospin operators, in this isospin-1/2 case.

In the U(1) case, considering now the fermionic example of section 7.2 for variety, we could go further and associate the conserved operator $\hat{N}_\psi$ with a *conserved current* $\hat{N}_\psi^\mu$:

$$\hat{N}_\psi = \int \hat{N}_\psi^0\mathrm{d}^3\boldsymbol{x}, \qquad \hat{N}_\psi^\mu = \bar{\hat{\psi}}\gamma^\mu\hat{\psi} \qquad (12.107)$$

where

$$\partial_\mu \hat{N}_\psi^\mu = 0. \tag{12.108}$$

The obvious generalization appropriate to (12.101) is

$$\hat{\boldsymbol{T}}^{(\frac{1}{2})} = \int \hat{\boldsymbol{T}}^{(\frac{1}{2})0} \mathrm{d}^3\boldsymbol{x}, \quad \hat{\boldsymbol{T}}^{(\frac{1}{2})\mu} = \bar{\hat{q}}\gamma^\mu \frac{\boldsymbol{\tau}}{2}\hat{q}. \tag{12.109}$$

Note that both $\hat{N}_\psi^\mu$ and $\hat{\boldsymbol{T}}^{(\frac{1}{2})\mu}$ are of course functions of the space-time coordinate x, via the (suppressed) dependence of the $\hat{q}$-fields on x. Indeed one can verify from the equations of motion that

$$\partial_\mu \hat{\boldsymbol{T}}^{(\frac{1}{2})\mu} = 0. \tag{12.110}$$

Thus $\hat{\boldsymbol{T}}^{(\frac{1}{2})\mu}$ is a *conserved isospin current operator* appropriate to the $T = \frac{1}{2}$ (u, d) system; it transforms as a 4-vector under Lorentz transformations, and as a $T = 1$ triplet under $SU(2)_f$ transformations.

Clearly there should be some general formalism for dealing with all this more efficiently, and it is provided by a generalization of the steps followed, in the U(1) case, in equations (7.6) – (7.8). Suppose the Lagrangian involves a set of fields $\hat{\psi}_r$ (they could be bosons or fermions) and suppose that it is *invariant* under the infinitesimal transformation

$$\delta\hat{\psi}_r = -\mathrm{i}\epsilon T_{rs}\hat{\psi}_s \tag{12.111}$$

for some set of numerical coefficients T_{rs}. Equation (12.111) generalizes (7.5). Then since $\hat{\mathcal{L}}$ is invariant under this change,

$$0 = \delta\hat{\mathcal{L}} = \frac{\partial\hat{\mathcal{L}}}{\partial\hat{\psi}_r}\delta\hat{\psi}_r + \frac{\partial\hat{\mathcal{L}}}{\partial(\partial^\mu\hat{\psi}_r)}\partial^\mu(\delta\hat{\psi}_r). \tag{12.112}$$

But

$$\frac{\partial\hat{\mathcal{L}}}{\partial\hat{\psi}_r} = \partial^\mu\left(\frac{\partial\hat{\mathcal{L}}}{\partial(\partial^\mu\hat{\psi}_r)}\right) \tag{12.113}$$

from the equations of motion. Hence

$$\partial^\mu\left(\frac{\partial\hat{\mathcal{L}}}{\partial(\partial^\mu\hat{\psi}_r)}\delta\hat{\psi}_r\right) = 0 \tag{12.114}$$

which is precisely a current conservation law of the form

$$\partial^\mu\hat{j}_\mu = 0. \tag{12.115}$$

Indeed, disregarding the irrelevant constant small parameter ϵ, the conserved current is

$$\hat{j}_\mu = -\mathrm{i}\frac{\partial\hat{\mathcal{L}}}{\partial(\partial^\mu\hat{\psi}_r)}T_{rs}\hat{\psi}_s. \tag{12.116}$$

Let us try this out on (12.87) with

$$\delta\hat{q} = (-\mathrm{i}\boldsymbol{\epsilon}\cdot\boldsymbol{\tau}/2)\hat{q}. \tag{12.117}$$

As we know already, there are now three ϵ's, and so three T_{rs}'s, namely $\frac{1}{2}(\tau_1)_{rs}, \frac{1}{2}(\tau_2)_{rs}, \frac{1}{2}(\tau_3)_{rs}$. For each one we have a current, for example

$$\hat{T}_{1\mu}^{(\frac{1}{2})} = -\mathrm{i}\frac{\partial\hat{\mathcal{L}}}{\partial(\partial^\mu\hat{q})}\frac{\tau_1}{2}\,\hat{q} = \bar{\hat{q}}\gamma_\mu\frac{\tau_1}{2}\hat{q} \tag{12.118}$$

and similarly for the other τ's, and so we recover (12.109). From the invariance of the Lagrangian under the transformation (12.117) there follows the conservation of an associated symmetry current. This is the quantum field theory version of Noether's theorem.

This theorem is of fundamental significance as it tells us how to relate symmetries (under transformations of the general form (12.111)) to 'current' conservation laws (of the form (12.115), and it constructs the actual currents for us. In gauge theories, the *dynamics* is generated from a symmetry, in the sense that (as we have seen in the local U(1) of electromagnetism) the symmetry currents are the dynamical currents that drive the equations for the force field. Thus the symmetries of the Lagrangian are basic to gauge field theories.

Let us look at another example, this time involving spin-0 fields. Suppose we have three spin-0 fields all with the same mass, and take

$$\hat{\mathcal{L}} = \frac{1}{2}\partial_\mu \hat{\phi}_1 \partial^\mu \hat{\phi}_1 + \frac{1}{2}\partial_\mu \hat{\phi}_2 \partial^\mu \hat{\phi}_2 + \frac{1}{2}\partial_\mu \hat{\phi}_3 \partial^\mu \hat{\phi}_3 - \frac{1}{2}m^2(\hat{\phi}_1^2 + \hat{\phi}_2^2 + \hat{\phi}_3^2). \tag{12.119}$$

It is obvious that $\hat{\mathcal{L}}$ is invariant under an arbitrary rotation of the three $\hat{\phi}$'s among themselves, generalizing the 'rotation about the 3-axis' considered for the $\hat{\phi}_1 - \hat{\phi}_2$ system of section 7.1. An infinitesimal such rotation is (cf (12.64), and noting the sign change in the field theory case)

$$\hat{\boldsymbol{\phi}}' = \hat{\boldsymbol{\phi}} + \boldsymbol{\epsilon} \times \hat{\boldsymbol{\phi}} \tag{12.120}$$

which implies

$$\delta\hat{\phi}_r = -\mathrm{i}\epsilon_a T^{(1)}_{ars}\hat{\phi}_s, \tag{12.121}$$

with

$$T^{(1)}_{ars} = -\mathrm{i}\epsilon_{ars} \tag{12.122}$$

as in (12.48). There are of course three conserved $\hat{\boldsymbol{T}}$ operators again, and three $\hat{\boldsymbol{T}}^{\mu}$'s, which we call $\hat{\boldsymbol{T}}^{(1)}$ and $\hat{\boldsymbol{T}}^{(1)\mu}$ respectively, since we are now dealing with a $T = 1$ isospin case. The $a = 1$ component of the conserved current in this case is, from (12.116),

$$\hat{T}^{(1)\mu}_1 = \hat{\phi}_2 \partial^\mu \hat{\phi}_3 - \hat{\phi}_3 \partial^\mu \hat{\phi}_2. \tag{12.123}$$

Cyclic permutations give us the other components which can be summarized as

$$\hat{\boldsymbol{T}}^{(1)\mu} = \mathrm{i}(\hat{\phi}^{(1)\mathrm{tr}}\,\mathbf{T}^{(1)}\partial^\mu\hat{\phi}^{(1)} - (\partial^\mu\hat{\phi}^{(1)})^{\mathrm{tr}}\,\mathbf{T}^{(1)}\hat{\phi}^{(1)}) \tag{12.124}$$

where we have written

$$\hat{\phi}^{(1)} = \begin{pmatrix} \hat{\phi}_1 \\ \hat{\phi}_2 \\ \hat{\phi}_3 \end{pmatrix} \tag{12.125}$$

and ^{tr} denotes transpose. Equation (12.124) has the form expected of a bosonic spin-0 current, but with the matrices $\mathbf{T}^{(1)}$ appearing, appropriate to the $T = 1$ (triplet) representation of SU(2)$_{\mathrm{f}}$.

The general form of such SU(2) currents should now be clear. For an isospin T-multiplet of bosons we shall have the form

$$\mathrm{i}(\hat{\phi}^{(T)\dagger}\mathbf{T}^{(T)}\partial^\mu\hat{\phi}^{(T)}) - (\partial^\mu\hat{\phi}^{(T)})^\dagger\mathbf{T}^{(T)}\hat{\phi}^{(T)}) \tag{12.126}$$

where we have put the $\dagger$ to allow for possibly complex fields; and for an isospin T-multiplet of fermions we shall have

$$\bar{\hat{\psi}}^{(T)}\gamma^\mu\mathbf{T}^{(T)}\hat{\psi}^{(T)} \tag{12.127}$$

where in each case the $(2T + 1)$ components of $\hat{\phi}$ or $\hat{\psi}$ transforms as a T-multiplet under SU(2), i.e.

$$\hat{\psi}^{(T)\prime} = \exp(-\mathrm{i}\boldsymbol{\alpha} \cdot \mathbf{T}^{(T)})\hat{\psi}^{(T)} \tag{12.128}$$

and similarly for $\hat{\phi}^{(T)}$, where $\mathbf{T}^{(T)}$ are the $2T + 1 \times 2T + 1$ matrices representing the generators of SU(2)$_\mathrm{f}$ in this representation. In all cases, the integral over all space of the $\mu = 0$ component of these currents results in a triplet of isospin operators obeying the SU(2) algebra (12.47), as in (12.103).

The cases considered so far have all been *free* field theories, but SU(2)-invariant interactions can be easily formed. For example, the interaction $g_1\bar{\hat{\psi}}\boldsymbol{\tau}\hat{\psi} \cdot \hat{\boldsymbol{\phi}}$ describes SU(2)-invariant interactions between a $T = \frac{1}{2}$ isospinor (spin-$\frac{1}{2}$) field $\hat{\psi}$, and a $T = 1$ isotriplet (Lorentz scalar) $\hat{\boldsymbol{\phi}}$. An effective interaction between pions and nucleons could take the form $g_\pi\bar{\hat{\psi}}\boldsymbol{\tau}\gamma_5\hat{\psi} \cdot \hat{\boldsymbol{\phi}}$, allowing for the pseudoscalar nature of the pions (we shall see in the following section that $\bar{\hat{\psi}}\gamma_5\hat{\psi}$ is a pseudoscalar, so the product is a true scalar as is required for a parity-conserving strong interaction). In these examples the 'vector' analogy for the $T = 1$ states allows us to see that the 'dot product' will be invariant. A similar dot product occurs in the interaction between the isospinor $\hat{\psi}^{(\frac{1}{2})}$ and the weak SU(2) gauge field $\hat{\boldsymbol{W}}_\mu$, which has the form

$$g\bar{\hat{q}}\gamma^\mu\frac{\boldsymbol{\tau}}{2}\hat{q} \cdot \hat{\boldsymbol{W}}_\mu \tag{12.129}$$

as will be discussed in the following chapter. This is just the SU(2) dot product of the symmetry current (12.109) and the gauge field triplet, both of which are in the adjoint $(T = 1)$ representation of SU(2).

All of the foregoing can be generalized straightforwardly to SU(3)$_\mathrm{f}$. For example, the Lagrangian

$$\hat{\mathcal{L}} = \bar{\hat{q}}(\mathrm{i}\,\slashed{\partial} - m)\hat{q} \tag{12.130}$$

with $\hat{q}$ now extended to

$$\hat{q} = \begin{pmatrix} \hat{u} \\ \hat{d} \\ \hat{s} \end{pmatrix} \tag{12.131}$$

describes free u, d, and s quarks of equal mass m. $\hat{\mathcal{L}}$ is clearly invariant under global SU(3)$_\mathrm{f}$ transformations

$$\hat{q}' = \exp(-\mathrm{i}\boldsymbol{\alpha} \cdot \boldsymbol{\lambda}/2)\hat{q}, \tag{12.132}$$

as well as the usual global U(1) transformation associated with quark number conservation. The associated Noether currents are (in somewhat informal notation)

$$\hat{G}_a^{(\mathrm{q})\mu} = \bar{\hat{q}}\gamma^\mu\frac{\lambda_a}{2}\hat{q} \qquad a = 1, 2, \ldots 8 \tag{12.133}$$

(note that there are eight of them), and the associated conserved 'charge operators' are

$$\hat{G}_a^{(\mathrm{q})} = \int \hat{G}_a^{(\mathrm{q})0}\mathrm{d}^3\boldsymbol{x} = \int \hat{q}^\dagger\frac{\lambda_a}{2}\hat{q} \quad a = 1, 2, \ldots 8, \tag{12.134}$$

which obey the SU(3) commutation relations

$$[\hat{G}_a^{(\mathrm{q})}, \hat{G}_b^{(\mathrm{q})}] = \mathrm{i}f_{abc}\hat{G}_c^\mathrm{q}. \tag{12.135}$$

SU(3)-invariant interactions can also be formed. A particularly important one is the 'SU(3) dot-product' of two octets (the analogues of the SU(2) triplets), which arises in the quark-gluon vertex of QCD (see chapters 13 and 14):

$$-\mathrm{i}g_s \sum_f \bar{\hat{q}}_f \gamma^\mu \frac{\lambda_a}{2} \hat{q}_f \hat{A}^a_\mu. \tag{12.136}$$

In (12.136), $\hat{q}_f$ stands for the $SU(3)_c$ *colour* triplet

$$\hat{q}_f = \begin{pmatrix} \hat{f}_r \\ \hat{f}_b \\ \hat{f}_g \end{pmatrix} \tag{12.137}$$

where '$\hat{f}$' is any of the six quark flavour fields $\hat{u}, \hat{d}, \hat{c}, \hat{s}, \hat{t}, \hat{b}$, and $\hat{A}^a_\mu$ are the 8 ($a = 1, 2, \ldots 8$) gluon fields. Once again, (12.136) has the form 'symmetry current . gauge field' characteristic of all gauge interactions.

12.4.2   Chiral symmetry

As our final example of a global non-Abelian symmetry, we shall introduce the idea of *chiral symmetry*, which is an exact symmetry for fermions in the limit in which their masses may be neglected. We have seen that the u and d quarks have indeed very small masses (≤ 5 MeV) on hadronic scales, and even the s quark mass (~ 100 MeV) is relatively small. Thus we may certainly expect some physical signs of the symmetry associated with $m_u \approx m_d \approx 0$, and possibly also of the larger symmetry holding when $m_u \approx m_d \approx m_s \approx 0$. As we shall see, however, this expectation leads to a puzzle, the resolution of which will have to be postponed until the concept of 'spontaneous symmetry breaking' has been developed in Part 7.

We begin with the simplest case of just one fermion. Since we are interested in the 'small mass' regime, it is sensible to use the representation (3.40) of the Dirac matrices, in which the momentum part of the Dirac Hamiltonian is 'diagonal' and the mass appears as an 'off-diagonal' coupling:

$$\boldsymbol{\alpha} = \begin{pmatrix} \boldsymbol{\sigma} & 0 \\ 0 & \boldsymbol{\sigma} \end{pmatrix}, \quad \beta = \begin{pmatrix} 0 & 1 \\ 1 & 0 \end{pmatrix}. \tag{12.138}$$

Writing the general Dirac spinor ω as

$$\omega = \begin{pmatrix} \phi \\ \chi \end{pmatrix}, \tag{12.139}$$

we have (as in (4.14), (4.15))

$$E\phi = \boldsymbol{\sigma} \cdot \boldsymbol{p}\phi + m\chi \tag{12.140}$$

$$E\chi = -\boldsymbol{\sigma} \cdot \boldsymbol{p}\chi + m\phi. \tag{12.141}$$

We now recall the matrix γ_5 introduced in section 4.2.1

$$\gamma_5 = \mathrm{i}\gamma^0\gamma^1\gamma^2\gamma^3, \tag{12.142}$$

which takes the form

$$\gamma_5 = \begin{pmatrix} 1 & 0 \\ 0 & -1 \end{pmatrix}. \tag{12.143}$$

in this representation. The matrix γ_5 plays a prominent role in chiral symmetry, as we shall see. Its defining property is that it anti-commutes with the γ^μ matrices:

$$\{\gamma_5, \gamma^\mu\} = 0. \tag{12.144}$$

'Chirality' means 'handedness', from the Greek word for hand, $\chi\epsilon\iota\rho$. Its use here stems from the fact that, in the limit $m \to 0$ the 2-component spinors ϕ, χ become helicity eigenstates (cf problem 9.4), having definite 'handedness'. As $m \to 0$ we have $E \to |\boldsymbol{p}|$, and (12.140) and (12.141) reduce to

$$(\boldsymbol{\sigma} \cdot \boldsymbol{p}/|\boldsymbol{p}|)\tilde{\phi} = \tilde{\phi} \tag{12.145}$$
$$(\boldsymbol{\sigma} \cdot \boldsymbol{p}/|\boldsymbol{p}|)\tilde{\chi} = -\tilde{\chi}, \tag{12.146}$$

so that the limiting spinor $\tilde{\phi}$ has positive helicity, and $\tilde{\chi}$ negative helicity (cf (3.68) and (3.69)). In this $m \to 0$ limit, the two helicity spinors are *decoupled*, reflecting the fact that no Lorentz transformation can reverse the helicity of a massless particle. Also in this limit, the Dirac energy operator is

$$\boldsymbol{\alpha} \cdot \boldsymbol{p} = \begin{pmatrix} \boldsymbol{\sigma} \cdot \boldsymbol{p} & 0 \\ 0 & -\boldsymbol{\sigma} \cdot \boldsymbol{p} \end{pmatrix} \tag{12.147}$$

which is easily seen to commute with γ_5. Thus the massless states may equivalently be classified by the eigenvalues of γ_5, which are clearly ± 1 since $\gamma_5^2 = I$.

Consider then a massless fermion with positive helicity. It is described by the 'u'-spinor $\begin{pmatrix} \tilde{\phi} \\ 0 \end{pmatrix}$ which is an eigenstate of γ_5 with eigenvalue $+1$. Similarly, a fermion with negative helicity is described by $\begin{pmatrix} 0 \\ \tilde{\chi} \end{pmatrix}$ which has $\gamma_5 = -1$. Thus for these states chirality equals helicity. We have to be more careful for antifermions, however. A physical antifermion of energy E and momentum $\boldsymbol{p}$ is described by a 'v'-spinor corresponding to $-E$ and $-\boldsymbol{p}$; but with $m = 0$ in (12.140) and (12.141) the equations for ϕ and χ remain the same for $-E, -\boldsymbol{p}$ as for $E, \boldsymbol{p}$. Consider the spin, however. If the physical antiparticle has positive helicity, with $\boldsymbol{p}$ along the z-axis say, then $s_z = +\frac{1}{2}$. The corresponding v-spinor must then have $s_z = -\frac{1}{2}$ (see section 3.4.3) and must therefore be of $\tilde{\chi}$ type (12.146). So the v-spinor for this antifermion of positive helicity is $\begin{pmatrix} 0 \\ \tilde{\chi} \end{pmatrix}$ which has $\gamma_5 = -1$. In summary, for fermions the γ_5 eigenvalue is equal to the helicity, and for antifermions it is equal to minus the helicity. It is the γ_5 eigenvalue that is called the 'chirality'.

In the massless limit, the chirality of $\tilde{\phi}$ and $\tilde{\chi}$ is a good quantum number (γ_5 commuting with the energy operator), and we may say that 'chirality is conserved' in this massless limit. On the other hand, the massive spinor ω is clearly *not* an eigenstate of chirality:

$$\gamma_5 \omega = \begin{pmatrix} \phi \\ -\chi \end{pmatrix} \neq \lambda \begin{pmatrix} \phi \\ \chi \end{pmatrix}. \tag{12.148}$$

Referring to (12.140) and (12.141), we may therefore regard the mass terms as 'coupling the states of different chirality'.

It is usual to introduce operators $P_{\mathrm{R,L}} = \left(\frac{1 \pm \gamma_5}{2}\right)$ which 'project' out states of definite chirality from ω:

$$\omega = \left(\frac{1 + \gamma_5}{2}\right)\omega + \left(\frac{1 - \gamma_5}{2}\right)\omega \equiv P_{\mathrm{R}}\omega + P_{\mathrm{L}}\omega \equiv \omega_{\mathrm{R}} + \omega_{\mathrm{L}}, \tag{12.149}$$

so that

$$\omega_{\mathrm{R}} = \begin{pmatrix} 1 & 0 \\ 0 & 0 \end{pmatrix} \begin{pmatrix} \phi \\ \chi \end{pmatrix} = \begin{pmatrix} \phi \\ 0 \end{pmatrix}, \quad \omega_{\mathrm{L}} = \begin{pmatrix} 0 \\ \chi \end{pmatrix}. \tag{12.150}$$

Then clearly $\gamma_5\omega_{\mathrm{R}} = \omega_{\mathrm{R}}$ and $\gamma_5\omega_{\mathrm{L}} = -\omega_{\mathrm{L}}$; slightly confusingly, the notation 'R', 'L' is used for the *chirality* eigenvalue.

We now reformulate the above in field-theoretic terms. The Dirac Lagrangian for a single massless fermion is

$$\hat{\mathcal{L}}_0 = \bar{\hat{\psi}}\mathrm{i}\,\slashed{\partial}\hat{\psi}. \tag{12.151}$$

This is invariant not only under the now familiar global U(1) transformation $\hat{\psi} \to \hat{\psi}' = \mathrm{e}^{-\mathrm{i}\alpha}\hat{\psi}$, but also under the 'global *chiral* U(1)' transformation

$$\hat{\psi} \to \hat{\psi}' = \mathrm{e}^{-\mathrm{i}\theta\gamma_5}\hat{\psi} \tag{12.152}$$

where θ is an arbitrary (x-independent) real parameter. The invariance is easily verified: using $\{\gamma^0, \gamma_5\} = 0$ we have

$$\bar{\hat{\psi}}' = \hat{\psi}'^{\dagger}\gamma^0 = \hat{\psi}^{\dagger}\mathrm{e}^{\mathrm{i}\theta\gamma_5}\gamma^0 = \hat{\psi}^{\dagger}\gamma^0\mathrm{e}^{-\mathrm{i}\theta\gamma_5} = \bar{\hat{\psi}}\mathrm{e}^{-\mathrm{i}\theta\gamma_5}, \tag{12.153}$$

and then using $\{\gamma^{\mu}, \gamma_5\} = 0$,

$$\begin{aligned} \bar{\hat{\psi}}'\gamma^{\mu}\partial_{\mu}\hat{\psi}' &= \bar{\hat{\psi}}\mathrm{e}^{-\mathrm{i}\theta\gamma_5}\gamma^{\mu}\partial_{\mu}\mathrm{e}^{-\mathrm{i}\theta\gamma_5}\hat{\psi} \\ &= \bar{\hat{\psi}}\gamma^{\mu}\mathrm{e}^{\mathrm{i}\theta\gamma_5}\partial_{\mu}\mathrm{e}^{-\mathrm{i}\theta\gamma_5}\hat{\psi} \\ &= \bar{\hat{\psi}}\gamma^{\mu}\partial_{\mu}\hat{\psi} \end{aligned} \tag{12.154}$$

as required. The corresponding Noether current is

$$\hat{j}_5^{\mu} = \bar{\hat{\psi}}\gamma^{\mu}\gamma_5\hat{\psi}, \tag{12.155}$$

and the spatial integral of its $\mu = 0$ component is the (conserved) chirality operator

$$\hat{Q}_5 = \int \hat{\psi}^{\dagger}\gamma_5\hat{\psi}\mathrm{d}^3\boldsymbol{x} = \int \left(\hat{\phi}^{\dagger}\hat{\phi} - \hat{\chi}^{\dagger}\hat{\chi}\right)\mathrm{d}^3\boldsymbol{x}. \tag{12.156}$$

We denote this chiral U(1) by U(1)$_5$.

It is interesting to compare the form of $\hat{Q}_5$ with that of the corresponding operator $\int \hat{\psi}^{\dagger}\hat{\psi}\mathrm{d}^3\boldsymbol{x}$ in the non-chiral case (cf (7.51)). The difference has to do with their behaviour under a transformation already discussed in section 4.2.1, namely *parity*. Under the parity transformation $\boldsymbol{p} \to -\boldsymbol{p}$ and thus, for (12.140) and (12.141) to be covariant under parity, we require $\phi \to \chi$, $\chi \to \phi$; this will ensure (as we saw in section 4.2.1) that the Dirac equation in the parity-transformed frame will be consistent with the one in the original frame. In the representation (12.138), this is equivalent to saying that the spinor $\omega_{\mathbf{P}}$ in the parity-transformed frame is given by

$$\omega_{\mathbf{P}} = \gamma^0\omega. \tag{12.157}$$

which implies $\phi_{\mathbf{P}} = \chi$, $\chi_{\mathbf{P}} = \phi$.

All this carries over to the field theory case, with $\hat{\psi}_{\mathbf{P}}(\boldsymbol{x}, t) = \gamma^0\hat{\psi}(-\boldsymbol{x}, t)$, as we saw in section 7.5.1. Consider then the operator $\hat{Q}_5$ in the parity-transformed frame:

$$\begin{aligned} (\hat{Q}_5)_{\mathbf{P}} &= \int \hat{\psi}_{\mathbf{P}}^{\dagger}(\boldsymbol{x}, t)\gamma_5\psi_{\mathbf{P}}(\boldsymbol{x}, t)\mathrm{d}^3\boldsymbol{x} = \int \hat{\psi}^{\dagger}(-\boldsymbol{x}, t)\gamma^0\gamma_5\gamma^0\hat{\psi}(-\boldsymbol{x}, t)\mathrm{d}^3\boldsymbol{x} \\ &= -\int \hat{\psi}^{\dagger}(\boldsymbol{y}, t)\gamma_5\hat{\psi}(\boldsymbol{y}, t)\mathrm{d}^3\boldsymbol{y} = -\hat{Q}_5 \end{aligned} \tag{12.158}$$

where we used $\{\gamma^0, \gamma_5\} = 0$ and $\left(\gamma^0\right)^2 = 1$, and changed the integration variable to $\boldsymbol{y} = -\boldsymbol{x}$. Hence $\hat{Q}_5$ is a 'pseudoscalar' operator, meaning that it changes sign in the parity-transformed frame. We can also see this directly from (12.156), making the interchange $\hat{\phi} \leftrightarrow \hat{\chi}$. In contrast, the non-chiral operator $\int \hat{\psi}^\dagger \hat{\psi} \mathrm{d}^3 \boldsymbol{x}$ is a (true) scalar, remaining the same in the parity-transformed frame.

In a similar way, the appearance of the γ_5 in the current operator $\hat{j}_5^\mu = \hat{\bar{\psi}} \gamma^\mu \gamma_5 \hat{\psi}$ affects its parity properties. For example, the $\mu = 0$ component $\hat{\psi}^\dagger \gamma_5 \hat{\psi}$ is a pseudoscalar, as we have seen. Problem 4.4(b) showed that the spatial parts $\hat{\bar{\psi}} \boldsymbol{\gamma} \gamma_5 \hat{\psi}$ behave as an *axial vector* rather than a normal (*polar*) vector under parity, that is, they behave like $\boldsymbol{r} \times \boldsymbol{p}$ for example, rather than like $\boldsymbol{r}$, in that they do *not* reverse sign under parity. Such a current is referred to generally as an 'axial vector current', as opposed to the ordinary vector currents with no γ_5.

As a consequence of (12.158), the operator $\hat{Q}_5$ changes the parity of any state on which it acts. We can see this formally by introducing the (unitary) parity operator $\hat{\mathbf{P}}$ in field theory, such that states of definite parity $|+\rangle, |-\rangle$ satisfy

$$\hat{\mathbf{P}}|+\rangle = |+\rangle, \quad \hat{\mathbf{P}}|-\rangle = -|-\rangle. \tag{12.159}$$

Equation (12.158) then implies that $\hat{\mathbf{P}}\hat{Q}_5\hat{\mathbf{P}}^{-1} = -\hat{Q}_5$, following the normal rule for operator transformations in quantum mechanics. Consider now the state $\hat{Q}_5|+\rangle$. We have

$$\begin{aligned} \hat{\mathbf{P}}\hat{Q}_5|+\rangle &= \left(\hat{\mathbf{P}}\hat{Q}_5\hat{\mathbf{P}}^{-1}\right)\hat{\mathbf{P}}|+\rangle \\ &= -\hat{Q}_5|+\rangle \end{aligned} \tag{12.160}$$

showing that $\hat{Q}_5|+\rangle$ is an eigenstate of $\hat{\mathbf{P}}$ with the opposite eigenvalue, -1.

A very important physical consequence now follows from the fact that (in this simple $m = 0$ model) $\hat{Q}_5$ is a symmetry operator commuting with the Hamiltonian $\hat{H}$. We have

$$\hat{H}\hat{Q}_5|\psi\rangle = \hat{Q}_5\hat{H}|\psi\rangle = E\hat{Q}_5|\psi\rangle. \tag{12.161}$$

Hence for every state $|\psi\rangle$ with energy eigenvalue E, there should exist a state $\hat{Q}_5|\psi\rangle$ with the same eigenvalue E and the opposite parity, that is, chiral symmetry apparently implies the existence of 'parity doublets'.

Of course, it may reasonably be objected that all of the above refers not only to the massless, but also the *non-interacting* case. However, this is just where the analysis begins to get interesting. Suppose we allow the fermion field $\hat{\psi}$ to interact with a U(1)-gauge field $\hat{A}^\mu$ via the standard electromagnetic coupling

$$\hat{\mathcal{L}}_{\text{int}} = q\hat{\bar{\psi}}\gamma^\mu\hat{\psi}\hat{A}_\mu. \tag{12.162}$$

Remarkably enough, $\hat{\mathcal{L}}_{\text{int}}$ is *also* invariant under the chiral transformation (12.152), for the simple reason that the 'Dirac' structure of (12.162) is exactly the same as that of the free kinetic term $\hat{\bar{\psi}}\, \partial\!\!\!/\hat{\psi}$: the 'covariant derivative' prescription $\partial^\mu \to D^\mu = \partial^\mu + \mathrm{i}q\hat{A}^\mu$ automatically means that any 'Dirac' (e.g. γ_5) symmetry of the kinetic part will be preserved when the gauge interaction is included. Thus chirality remains a 'good symmetry' in the presence of a U(1) gauge interaction.

The generalization of this to the more physical $m_\mathrm{u} \approx m_\mathrm{d} \approx 0$ case is quite straightforward. The Lagrangian (12.87) becomes

$$\hat{\mathcal{L}} = \hat{\bar{q}}\,\mathrm{i}\,\partial\!\!\!/\hat{q} \tag{12.163}$$

as $m \to 0$, which is invariant under the γ_5-version of (12.89),[6] namely

$$\hat{q}' = \exp(-\mathrm{i}\boldsymbol{\beta} \cdot \boldsymbol{\tau}/2\gamma_5)\hat{q}. \tag{12.164}$$

There are three associated Noether currents (compare (12.109))

$$\hat{\boldsymbol{T}}_5^{(\frac{1}{2})\,\mu} = \bar{\hat{q}}\gamma^\mu\gamma_5\frac{\boldsymbol{\tau}}{2}\hat{q} \tag{12.165}$$

which are axial vectors, and three associated 'charge' operators

$$\hat{\boldsymbol{T}}_5^{(\frac{1}{2})} = \int \hat{q}^\dagger\gamma_5\frac{\boldsymbol{\tau}}{2}\hat{q}\mathrm{d}^3\boldsymbol{x} \tag{12.166}$$

which are pseudoscalars, belonging to the T $=1$ representation of SU(2). We have a new non-Abelian global symmetry, called chiral SU(2)$_\mathrm{f}$, which we shall denote by SU(2)$_{\mathrm{f}\,5}$. As far as their action in the isospinor u-d space is concerned, these chiral changes have exactly the same effect as the ordinary flavour isospin operators of (12.109). But they are pseudoscalars rather than scalars, and hence they flip the parity of a state on which they act. Thus, whereas the isospin raising operator $\hat{T}_+^{(\frac{1}{2})}$ is such that

$$\hat{T}_+^{(\frac{1}{2})}|d\rangle = |u\rangle, \tag{12.167}$$

$\hat{T}_{+\,5}^{(\frac{1}{2})}$ will also produce a u-type state from a d-type one via

$$\hat{T}_{+\,5}^{(\frac{1}{2})}|d\rangle = |\tilde{u}\rangle, \tag{12.168}$$

but the $|\tilde{u}\rangle$ state will have opposite parity from $|u\rangle$. Further, since $[\hat{T}_{+\,5}^{(\frac{1}{2})}, \hat{H}] = 0$, this state $|\tilde{u}\rangle$ will be degenerate with $|d\rangle$. Similarly, the state $|\tilde{d}\rangle$ produced via $\hat{T}_{-\,5}^{(\frac{1}{2})}|u\rangle$ will have opposite parity from $|d\rangle$, and will be degenerate with $|u\rangle$. The upshot is that we have two massless states $|u\rangle$, $|d\rangle$ of (say) positive parity, and a further two massless states $|\tilde{u}\rangle$, $|\tilde{d}\rangle$ of negative parity, in this simple model.

Suppose we now let the quarks interact, for example by an interaction of the QCD type, already indicated in (12.136). In that case, the interaction terms have the form

$$\bar{\hat{u}}\gamma^\mu\frac{\lambda_a}{2}\hat{u}\hat{A}_\mu^a + \bar{\hat{d}}\gamma^\mu\frac{\lambda_a}{2}\hat{d}\hat{A}_\mu^a \tag{12.169}$$

where

$$\hat{u} = \begin{pmatrix} \hat{u}_\mathrm{r} \\ \hat{u}_\mathrm{b} \\ \hat{u}_\mathrm{g} \end{pmatrix}, \hat{d} = \begin{pmatrix} \hat{d}_\mathrm{r} \\ \hat{d}_\mathrm{b} \\ \hat{d}_\mathrm{g} \end{pmatrix} \tag{12.170}$$

and the 3×3 λ's act in the r-b-g space. Just as in the previous U(1) case, the interaction (12.169) is invariant under the global SU(2)$_{\mathrm{f}\,5}$ chiral symmetry (12.164), acting in the u-d space. Note that, somewhat confusingly, (12.169) is *not* a simple 'gauging' of (12.163): a covariant derivative is being introduced, but in the space of a new (colour) degree of freedom, not in flavour space. In fact, the flavour degrees of freedom are 'inert' in (12.169), so that it is invariant under SU(2)$_\mathrm{f}$ transformations, while the Dirac structure implies that it is also invariant under chiral SU(2)$_{\mathrm{f}\,5}$ transformations (12.164). All the foregoing can be

---

[6]$\hat{\mathcal{L}}_0$ is also invariant under $\hat{q}' = \mathrm{e}^{-\mathrm{i}\theta\gamma_5}\hat{q}$ which is an 'axial' version of the global U(1) associated with quark number conservation. We shall discuss this additional U(1)-symmetry in section 18.1.1.

extended unchanged to chiral SU(3)$_{f5}$, given that QCD is 'flavour blind', and supposing that $m_s \approx 0$.

The effect of the QCD interactions must be to bind the quark into nucleons, such as the proton (uud) and neutron (udd). But what about the equally possible states ($\tilde{u}\tilde{u}d$) and ($\tilde{u}\tilde{d}d$), for example? These would have to be degenerate in mass with (uud) and (udd), and of opposite parity. Yet such 'parity doublet' partners of the physical p and n are not observed, and so we have a puzzle.

One might feel that this whole discussion is unrealistic, based as it is on massless quarks. Are the baryons then supposed to be massless too? If so, perhaps the discussion is idle, as they are evidently by no means massless. But it is not necessary to suppose that the mass of a relativistic bound state has any very simple relation to the masses of its constituents. Its mass may derive, in part at least, from the interaction energy in the fields. Alternatively, one might suppose that somehow the finite mass of the u and d quarks, which of course breaks the chiral symmetry, splits the degeneracy of the nucleon parity doublets, promoting the negative parity 'nucleon' state to an acceptably high mass. But this seems very implausible, in view of the actual magnitudes of m_u and m_d, compared to the nucleon masses.

In short, we have here a situation in which a *symmetry of the Lagrangian* (to an apparently good approximation) does *not* seem to result in the expected *multiplet structure of the states*. The resolution of this puzzle will have to await our discussion of 'spontaneous symmetry breaking', in Part 7.

In conclusion, we note an important feature of the flavour symmetry currents $\hat{T}^{(\frac{1}{2})\mu}$ and $\hat{T}_5^{(\frac{1}{2})\mu}$ discussed in this and the preceding section. Although these currents have been introduced entirely within the context of *strong* interaction symmetries, it is a remarkable fact that exactly these currents also appear in strangeness-conserving semileptonic *weak* interactions such as β-decay, as we shall see in chapter 20. (The fact that *both* appear is precisely a manifestation of *parity violation* in weak interactions, as we noted in section 4.2.1). Thus some of the physical consequences of 'spontaneously broken chiral symmetry' will involve weak interaction quantities.

Problems

12.1 Verify that the set of all unitary 2×2 matrices with determinant equal to $+1$ form a group, the law of combination being matrix multiplication.

12.2 Derive (12.18).

12.3 Check the commutation relations (12.28).

12.4 Show that the T_i's defined by (12.45) satisfy (12.47).

12.5 Write out each of the 3×3 matrices $T_i^{(1)}$ ($i = 1, 2, 3$) whose matrix elements are given by (12.48), and verify that they satisfy the SU(2) commutation relations (12.47).

12.6 Verify (12.62).

12.7 Show that a general Hermitian traceless 3×3 matrix is parametrized by eight real numbers.

12.8 Check that (12.84) is consistent with (12.80) and the infinitesimal form of (12.81), and verify that the matrices $G_a^{(8)}$ defined by (12.84) satisfy the commutation relations (12.83).

12.9 Verify, by comparing the coefficients of ϵ_1, ϵ_2, and ϵ_3 on both sides of (12.99), that (12.100) follows from (12.99).

12.10 Verify that the operators $\hat{T}^{(\frac{1}{2})}$ defined by (12.101) satisfy (12.100). (Note: use the anticommutation relations of the fermionic operators.)

12.11 Verify that the operators $\hat{T}^{(\frac{1}{2})}$ given by (12.101) satisfy the commutation relations (12.103).

13

Local Non-Abelian (Gauge) Symmetries

... The difference between a neutron and a proton is then a purely arbitrary process. As usually conceived, however, this arbitrariness is subject to the following limitations: once one chooses what to call a proton, what a neutron, at one space time point, one is then not free to make any choices at other space time points.

It seems that this is not consistent with the localized field concept that underlies the usual physical theories. In the present paper we wish to explore the possibility of requiring all interactions to be invariant under *independent* rotations of the isotopic spin at all space time points ....

Yang and Mills (1954)

Consider the global SU(2) isospinor transformation (12.32), written here again,

$$\psi^{(\frac{1}{2})\prime}(x) = \exp(\mathrm{i}\boldsymbol{\alpha} \cdot \boldsymbol{\tau}/2)\psi^{(\frac{1}{2})}(x) \tag{13.1}$$

for an isospin doublet wavefunction $\psi^{(\frac{1}{2})}(x)$. The dependence of $\psi^{(\frac{1}{2})}(x)$ on the spacetime coordinate x has now been included explicitly, but the parameters $\boldsymbol{\alpha}$ are independent of x, which is why the transformation is called a 'global' one. As we have seen in the previous chapter, invariance under this transformation amounts to the assertion that the choice of *which* two base states—$(n, p), (u, d), \ldots$—to use is a matter of convention; any such non-Abelian phase transformation on a chosen pair produces another equally good pair. However, the choice cannot be made independently at all space time points, only *globally*. To Yang and Mills (1954) (cf the quotation above) this seemed somehow an unaesthetic limitation of symmetry: 'Once one chooses what to call a proton, what a neutron, at one space-time point, one is then not free to make any choices at other space-time points'. They even suggested that this could be viewed as 'inconsistent with the localised field concept', and they therefore 'explored the possibility' of replacing this global (space-time independent) phase transformation by the local (space-time dependent) one

$$\psi^{(\frac{1}{2})\prime}(x) = \exp[\mathrm{i}g\boldsymbol{\tau} \cdot \boldsymbol{\alpha}(x)/2]\psi^{(\frac{1}{2})}(x) \tag{13.2}$$

in which the phase parameters $\boldsymbol{\alpha}(x)$ are also now functions of $x = (t, \boldsymbol{x})$ as indicated. Notice that we have inserted a parameter g in the exponent to make the analogy with the electromagnetic U(1) case

$$\psi\prime(x) = \exp[\mathrm{i}q\chi(x)]\psi(x) \tag{13.3}$$

even stronger: g will be a coupling strength, analogous to the electromagnetic charge q. The consideration of theories based on (13.2) was the fundamental step taken by Yang and Mills (1954); see also Shaw (1955).

Global symmetries and their associated (possibly approximate) conservation laws are certainly important, but they do not have the *dynamical* significance of local symmetries. We saw in section 7.4 how the 'requirement' of local U(1) phase invariance led almost

DOI: 10.1201/9781003411666-13

automatically to the local gauge theory of QED, in which the conserved current $\bar{\hat{\psi}}\gamma^\mu\hat{\psi}$ of the of the global U(1) symmetry is 'promoted' to the role of dynamical current which, when dotted into the gauge field $\hat{A}^\mu$, gave the interaction term in $\hat{\mathcal{L}}_{\text{QED}}$. A similar link between symmetry and dynamics appears if, following Yang and Mills, we generalize the non-Abelian global symmetries of the preceding chapter to local non-Abelian symmetries, which are the subject of the present one.

However, as mentioned in the introduction to chapter 12, the original Yang-Mills attempt to get a theory of hadronic interactions by 'localizing' the flavour symmetry group SU(2) turned out not to be phenomenologically viable (although a remarkable attempt was made to push the idea further by Sakurai (1960)). In the event, the successful application of a local SU(2) symmetry was to the *weak* interactions. But this is complicated by the fact that the symmetry is 'spontaneously broken', and consequently we shall delay the discussion of this application until after QCD—which *is* the theory of strong interactions, but at the quark, rather than the composite (hadronic) level. QCD is based on the local form of an SU(3) symmetry; once again, however, it is *not* the flavour SU(3) of section 12.2, but a symmetry with respect to a totally new degree of freedom, colour. This will be introduced in the following chapter.

Although the application of local SU(2) symmetry to the weak interactions will follow that of local SU(3) to the strong, we shall begin our discussion of local non-Abelian symmetries with the local SU(2) case, since the group theory is more familiar. We shall also start with the 'wavefunction' formalism, deferring the field theory treatment until section 13.3.

13.1 Local SU(2) symmetry

13.1.1 The covariant derivative and interactions with matter

In this section we shall introduce the main ideas of the non-Abelian SU(2) gauge theory which results from the demand of invariance, or covariance, under transformations such as (13.2). We shall generally use the language of isospin when referring to the physical states and operators, bearing in mind that this will eventually mean *weak* isospin.

We shall mimic as literally as possible the discussion of electromagnetic gauge covariance in sections 2.4 and 2.5 of volume 1. As in that case, no free particle wave equation can be covariant under the transformation (13.2) (taking the isospinor example for definiteness), since the gradient terms in the equation will act on the phase factor $\boldsymbol{\alpha}(x)$. However, wave equations with a suitably defined *covariant derivative* can be covariant under (13.2); physically this means that, just as for electromagnetism, covariance under local non-Abelian phase transformations requires the introduction of a definite force field.

In the electromagnetic case the covariant derivative is

$$D^\mu = \partial^\mu + \mathrm{i}qA^\mu(x). \tag{13.4}$$

For convenience we recall here the crucial property of D^μ. Under a local U(1) phase transformation, a wavefunction transforms as (cf (13.3))

$$\psi(x) \to \psi'(x) = \exp(\mathrm{i}q\chi(x))\psi(x), \tag{13.5}$$

from which it easily follows that the derivative (gradient) of ψ transforms as

$$\partial^\mu\psi(x) \to \partial^\mu\psi'(x) = \exp(\mathrm{i}q\chi(x))\partial^\mu\psi(x) + \mathrm{i}q\partial^\mu\chi(x)\exp(\mathrm{i}q\chi(x))\psi(x). \tag{13.6}$$

Comparing (13.6) with (13.5), we see that, in addition to the expected first term on the right-hand side of (13.6), which has the same form as the right-hand side of (13.5), there is an *extra* term in (13.6). By contrast, the covariant derivative of ψ transforms as (see section 2.4 of volume 1)

$$D^\mu \psi(x) \to D'^\mu \psi'(x) = \exp(iq\chi(x))D^\mu \psi(x) \tag{13.7}$$

exactly as in (13.5), with no additional term on the right-hand side. Note that D^μ has to carry a prime also, since it contains A^μ which transforms to $A'^\mu = A^\mu - \partial^\mu \chi(x)$ when ψ transforms by (13.5). The property (13.7) ensured the gauge covariance of wave equations in the U(1) case; the similar property in the quantum field case meant that a globally U(1)-invariant Lagrangian could be converted immediately to a locally U(1)-invariant one by replacing ∂^μ by $\hat{D}^\mu$ (section 7.4).

In Appendix D of volume 1 we introduced the idea of 'covariance' in the context of coordinate transformations of 3- and 4-vectors. The essential notion was of something 'maintaining the same form', or 'transforming the same way'. The transformations being considered here are gauge transformations rather than coordinate ones; nevertheless it is true that, under them, $D^\mu \psi$ transforms in the same way as ψ, while $\partial^\mu \psi$ does not. Thus the term covariant derivative seems appropriate. In fact, there is a much closer analogy between the 'coordinate' and the 'gauge' cases, which we did not present in volume 1, but give now in appendix N, for the interested reader.

We need the local SU(2) generalization of (13.4), appropriate to the local SU(2) transformation (13.2). Just as in the U(1) case (13.6), the ordinary gradient acting on $\psi^{(\frac{1}{2})}(x)$ does not transform in the same way as $\psi^{(\frac{1}{2})}(x)$. Taking ∂^μ of (13.2) leads to

$$
\begin{aligned}
\partial^\mu \psi^{(\frac{1}{2})'}(x) &= \exp[ig\boldsymbol{\tau} \cdot \boldsymbol{\alpha}(x)/2]\partial^\mu \psi^{(\frac{1}{2})}(x) \\
&\quad + ig\boldsymbol{\tau} \cdot \partial^\mu \boldsymbol{\alpha}(x)/2 \exp[ig\boldsymbol{\tau} \cdot \boldsymbol{\alpha}(x)/2]\psi^{(\frac{1}{2})}(x)
\end{aligned} \tag{13.8}
$$

as can be checked by writing the matrix exponential $\exp[\mathrm{A}]$ as the series

$$\exp[\mathrm{A}] = \sum_{n=0}^{\infty} \mathrm{A}^n/n!$$

and differentiating term by term. By analogy with (13.7), the key property we demand for our *SU(2) covariant derivative* $D^\mu \psi^{(\frac{1}{2})}$ is that this quantity should transform like $\psi^{(\frac{1}{2})}$—i.e. without the second term in (13.8). So we require

$$(D'^\mu \psi^{(\frac{1}{2})'}(x)) = \exp[ig\boldsymbol{\tau} \cdot \boldsymbol{\alpha}(x)/2](D^\mu \psi^{(\frac{1}{2})}(x)). \tag{13.9}$$

The definition of D^μ which generalizes (13.4) so as to fulfil this requirement is

$$D^\mu(\text{acting on an isospinor}) = \partial^\mu + ig\boldsymbol{\tau} \cdot \boldsymbol{W}^\mu(x)/2. \tag{13.10}$$

The definition (13.10), as indicated on the left-hand side, is only appropriate for isospinors $\psi^{(\frac{1}{2})}$; it has to be suitably generalized for other $\psi^{(t)}$'s (see (13.44)).

We now discuss (13.9) and (13.10) in detail. The ∂^μ is multiplied implicitly by the unit 2 matrix, and the $\boldsymbol{\tau}$'s act on the 2-component space of $\psi^{(\frac{1}{2})}$. The $\boldsymbol{W}^\mu(x)$ are *three* independent gauge fields

$$\boldsymbol{W}^\mu = (W_1^\mu, W_2^\mu, W_3^\mu), \tag{13.11}$$

generalizing the single electromagnetic gauge field A^μ. They are called SU(2) gauge fields, or more generally *Yang-Mills fields*. The term $\boldsymbol{\tau} \cdot \boldsymbol{W}^\mu$ is then the 2×2 matrix

$$\boldsymbol{\tau} \cdot \boldsymbol{W}^\mu = \begin{pmatrix} W_3^\mu & W_1^\mu - iW_2^\mu \\ W_1^\mu + iW_2^\mu & -W_3^\mu \end{pmatrix} \tag{13.12}$$

using the τ's of (12.25); the x-dependence of the W^μ's is understood. Let us 'decode' the desired property (13.9), for the algebraically simpler case of an infinitesimal local SU(2) transformation with parameters $\boldsymbol{\epsilon}(x)$, which are of course functions of x since the transformation is local. In this case, $\psi^{(\frac{1}{2})}$ transforms by

$$\psi^{(\frac{1}{2})\prime} = (1 + ig\boldsymbol{\tau} \cdot \boldsymbol{\epsilon}(x)/2)\psi^{(\frac{1}{2})} \tag{13.13}$$

and the 'uncovariant' derivative $\partial^\mu \psi^{(\frac{1}{2})}$ transforms by

$$\partial^\mu \psi^{(\frac{1}{2})\prime} = (1 + ig\boldsymbol{\tau} \cdot \boldsymbol{\epsilon}(x)/2)\partial^\mu \psi^{(\frac{1}{2})} + ig\boldsymbol{\tau} \cdot \partial^\mu \boldsymbol{\epsilon}(x)/2 \, \psi^{(\frac{1}{2})}, \tag{13.14}$$

where we have retained only the terms linear in $\boldsymbol{\epsilon}$ from an expansion of (13.8) with $\boldsymbol{\alpha} \to \boldsymbol{\epsilon}$. We have now dropped the x-dependence of the $\psi^{(\frac{1}{2})}$'s, but kept that of $\boldsymbol{\epsilon}(x)$, and we have used the simple '1' for the unit matrix in the two-dimensional isospace. Equation (13.14) exhibits again an 'extra piece' on the right-hand side, as compared to (13.13). On the other hand, inserting (13.10) and (13.13) into our covariant derivative requirement (13.9) yields, for the left-hand side in the infinitesimal case,

$$D^{\prime\mu}\psi^{(\frac{1}{2})\prime} = (\partial^\mu + ig\boldsymbol{\tau} \cdot \boldsymbol{W}^{\prime\mu}/2)[1 + ig\boldsymbol{\tau} \cdot \boldsymbol{\epsilon}(x)/2]\psi^{(\frac{1}{2})} \tag{13.15}$$

while the right-hand side is

$$[1 + ig\boldsymbol{\tau} \cdot \boldsymbol{\epsilon}(x)/2](\partial^\mu + ig\boldsymbol{\tau} \cdot \boldsymbol{W}^\mu/2)\psi^{(\frac{1}{2})}. \tag{13.16}$$

In order to verify that these are the same, however, we would need to know $\boldsymbol{W}^{\prime\mu}$—that is, the transformation law for the three $\boldsymbol{W}^\mu$ fields. Instead, we shall proceed 'in reverse', and use the *imposed* equality between (13.15) and (13.16) to determine the transformation law of $\boldsymbol{W}^\mu$.

Suppose that, under this infinitesimal transformation,

$$\boldsymbol{W}^\mu \to \boldsymbol{W}^{\prime\mu} = \boldsymbol{W}^\mu + \delta\boldsymbol{W}^\mu. \tag{13.17}$$

Then the condition of equality is

$$[\partial^\mu + ig\boldsymbol{\tau}/2 \cdot (\boldsymbol{W}^\mu + \delta\boldsymbol{W}^\mu)][1 + ig\boldsymbol{\tau} \cdot \boldsymbol{\epsilon}(x)/2]\psi^{(\frac{1}{2})}$$
$$= [1 + ig\boldsymbol{\tau} \cdot \boldsymbol{\epsilon}(x)/2](\partial^\mu + ig\boldsymbol{\tau} \cdot \boldsymbol{W}^\mu/2)\psi^{(\frac{1}{2})}. \tag{13.18}$$

Multiplying out the terms, neglecting the term of second order involving the product of $\delta\boldsymbol{W}^\mu$ and $\boldsymbol{\epsilon}$ and noting that

$$\partial^\mu(\boldsymbol{\epsilon}\psi) = (\partial^\mu\boldsymbol{\epsilon})\psi + \boldsymbol{\epsilon}(\partial^\mu\psi) \tag{13.19}$$

we see that many terms cancel and we are left with

$$ig\frac{\boldsymbol{\tau} \cdot \delta\boldsymbol{W}^\mu}{2} = -ig\frac{\boldsymbol{\tau} \cdot \partial^\mu\boldsymbol{\epsilon}(x)}{2}$$
$$+ (ig)^2\left[\left(\frac{\boldsymbol{\tau} \cdot \boldsymbol{\epsilon}(x)}{2}\right)\left(\frac{\boldsymbol{\tau} \cdot \boldsymbol{W}^\mu}{2}\right) - \left(\frac{\boldsymbol{\tau} \cdot \boldsymbol{W}^\mu}{2}\right)\left(\frac{\boldsymbol{\tau} \cdot \boldsymbol{\epsilon}(x)}{2}\right)\right]. \tag{13.20}$$

Using the identity for Pauli matrices (see problem 3.4(b))

$$\boldsymbol{\sigma} \cdot \boldsymbol{a}\,\boldsymbol{\sigma} \cdot \boldsymbol{b} = \boldsymbol{a} \cdot \boldsymbol{b} + i\boldsymbol{\sigma} \cdot \boldsymbol{a} \times \boldsymbol{b} \tag{13.21}$$

this yields

$$\boldsymbol{\tau} \cdot \delta \boldsymbol{W}^{\mu} = -\boldsymbol{\tau} \cdot \partial^{\mu} \boldsymbol{\epsilon}(x) - g\boldsymbol{\tau} \cdot (\boldsymbol{\epsilon}(x) \times \boldsymbol{W}^{\mu}). \tag{13.22}$$

Equating components of $\boldsymbol{\tau}$ on both sides, we deduce

$$\boxed{\delta \boldsymbol{W}^{\mu} = -\partial^{\mu} \boldsymbol{\epsilon}(x) - g[\boldsymbol{\epsilon}(x) \times \boldsymbol{W}^{\mu}].} \tag{13.23}$$

The reader may note the close similarity between these manipulations and those encountered in section 12.2.3.

Equation (13.23) defines the way in which the SU(2) gauge fields $\boldsymbol{W}^{\mu}$ transform under an infinitesimal SU(2) gauge transformation. If it were not for the presence of the first term $\partial^{\mu} \boldsymbol{\epsilon}(x)$ on the right hand side, (13.23) would be simply the (infinitesimal) transformation law for the $T = 1$ triplet representation of SU(2) — see (12.64) and (12.65) in section 12.1.3. As mentioned at the end of section 12.2, the $T = 1$ representation is the 'adjoint', or 'regular', representation of SU(2), and this is the one to which gauge fields belong, in general. But there is the extra term $-\partial^{\mu} \boldsymbol{\epsilon}(x)$. Clearly this is directly analogous to the $-\partial^{\mu} \chi(x)$ term in the transformation of the U(1) gauge field A^{μ}; here, an independent infinitesimal function $\epsilon_i(x)$ is required for each component $W_i^{\mu}(x)$. If the ϵ's were independent of x, then $\partial^{\mu} \boldsymbol{\epsilon}(x)$ would of course vanish and the transformation law (13.23) would indeed be just that of an SU(2) triplet. Thus we can say that under global SU(2) transformations, the $\boldsymbol{W}^{\mu}$ behave as a normal triplet. But under *local* SU(2) transformations they acquire the additional $-\partial^{\mu} \boldsymbol{\epsilon}(x)$ piece, and thus no longer transform 'properly', as an SU(2) triplet. In exactly the same way, $\partial^{\mu} \psi^{(\frac{1}{2})}$ did not transform 'properly' as an SU(2) doublet, under a local SU(2) transformation, because of the second term in (13.14), which also involves $\partial^{\mu} \boldsymbol{\epsilon}(x)$. The remarkable result behind the fact that $D^{\mu} \psi^{(\frac{1}{2})}$ *does* transform 'properly' under local SU(2) transformations, is that the extra term in (13.23) precisely cancels that in (13.14)!

To summarize progress so far: we have shown that, for infinitesimal transformations, the relation

$$(D'^{\mu} \psi^{(\frac{1}{2})\prime}) = [1 + \mathrm{i} g\boldsymbol{\tau} \cdot \boldsymbol{\epsilon}(x)/2](D^{\mu} \psi^{(\frac{1}{2})}) \tag{13.24}$$

(where D^{μ} is given by (13.10)) holds true if in addition to the infinitesimal local SU(2) phase transformation on $\psi^{(\frac{1}{2})}$

$$\psi^{(\frac{1}{2})\prime} = [1 + \mathrm{i} g\boldsymbol{\tau} \cdot \boldsymbol{\epsilon}(x)/2]\psi^{(\frac{1}{2})} \tag{13.25}$$

the gauge fields transform according to

$$\boldsymbol{W}'^{\mu} = \boldsymbol{W}^{\mu} - \partial^{\mu} \boldsymbol{\epsilon}(x) - g[\boldsymbol{\epsilon}(x) \times \boldsymbol{W}^{\mu}]. \tag{13.26}$$

In obtaining these results, the form (13.10) for the covariant derivative has been assumed, and only the infinitesimal version of (13.2) has been treated explicitly. It turns out that (13.10) is still appropriate for the finite (non-infinitesimal) transformation (13.2), but the associated transformation law for the gauge fields is then slightly more complicated than (13.26). Let us write

$$\mathbf{U}(\boldsymbol{\alpha}(x)) \equiv \exp[\mathrm{i} g\boldsymbol{\tau} \cdot \boldsymbol{\alpha}(x)/2] \tag{13.27}$$

so that $\psi^{(\frac{1}{2})}$ transforms by

$$\psi^{(\frac{1}{2})\prime} = \mathbf{U}(\boldsymbol{\alpha}(x))\psi^{(\frac{1}{2})}. \tag{13.28}$$

Then we require

$$D'^{\mu} \psi^{(\frac{1}{2})\prime} = \mathbf{U}(\boldsymbol{\alpha}(x))D^{\mu} \psi^{(\frac{1}{2})}. \tag{13.29}$$

The left-hand side is

$$(\partial^{\mu} + \mathrm{i} g\boldsymbol{\tau} \cdot \boldsymbol{W}'^{\mu}/2)\mathbf{U}(\boldsymbol{\alpha}(x))\psi^{(\frac{1}{2})}$$
$$= (\partial^{\mu} \mathbf{U})\psi^{(\frac{1}{2})} + \mathbf{U}\partial^{\mu} \psi^{(\frac{1}{2})} + \mathrm{i} g\boldsymbol{\tau} \cdot \boldsymbol{W}'^{\mu}/2\,\mathbf{U}\psi^{(\frac{1}{2})}, \tag{13.30}$$

while the right-hand side is

$$\mathbf{U}(\partial^\mu + ig\boldsymbol{\tau} \cdot \boldsymbol{W}^\mu/2)\psi^{(\frac{1}{2})}. \tag{13.31}$$

The $\mathbf{U}\partial^\mu\psi^{(\frac{1}{2})}$ terms cancel leaving

$$(\partial^\mu\mathbf{U})\psi^{(\frac{1}{2})} + ig\boldsymbol{\tau} \cdot \boldsymbol{W}'^\mu/2\,\mathbf{U}\psi^{(\frac{1}{2})} = \mathbf{U}ig\boldsymbol{\tau} \cdot \boldsymbol{W}^\mu/2\,\psi^{(\frac{1}{2})}. \tag{13.32}$$

Since this has to be true for all (2-component) $\psi^{(\frac{1}{2})}$'s, we can treat it as an operator equation acting in the space of $\psi^{(\frac{1}{2})}$'s to give

$$\partial^\mu\mathbf{U} + ig\boldsymbol{\tau} \cdot \boldsymbol{W}'^\mu/2\,\mathbf{U} = \mathbf{U}ig\boldsymbol{\tau} \cdot \boldsymbol{W}^\mu/2, \tag{13.33}$$

or equivalently

$$\frac{1}{2}\boldsymbol{\tau} \cdot \boldsymbol{W}'^\mu = \frac{i}{g}(\partial^\mu\mathbf{U})\mathbf{U}^{-1} + \mathbf{U}\frac{1}{2}\boldsymbol{\tau} \cdot \boldsymbol{W}^\mu\mathbf{U}^{-1}, \tag{13.34}$$

which defines the (finite) transformation law for SU(2) gauge fields. Problem 13.1 verifies that (13.34) reduces to (13.26) in the infinitesimal case $\boldsymbol{\alpha}(x) \to \boldsymbol{\epsilon}(x)$.

Suppose now that we consider a Dirac equation for $\psi^{(\frac{1}{2})}$:

$$(i\gamma_\mu\partial^\mu - m)\psi^{(\frac{1}{2})} = 0 \tag{13.35}$$

where both the 'isospinor' components of $\psi^{(\frac{1}{2})}$ are 4-component Dirac spinors. We assert that we can ensure *local SU(2) gauge covariance by replacing* ∂^μ *in this equation by the covariant derivative of* (13.10). Indeed, we have

$$\begin{aligned}
\mathbf{U}(\boldsymbol{\alpha}(x))[i\gamma_\mu D^\mu - m]\psi^{(\frac{1}{2})} &= i\gamma_\mu\mathbf{U}(\boldsymbol{\alpha}(x))[D^\mu\psi^{(\frac{1}{2})}] - m\mathbf{U}(\boldsymbol{\alpha}(x)]\psi^{(\frac{1}{2})} \\
&= i\gamma_\mu D'^\mu\psi^{(\frac{1}{2})\prime} - m\psi^{(\frac{1}{2})\prime}
\end{aligned} \tag{13.36}$$

using equations (13.9) and (13.28). Thus if

$$(i\gamma_\mu D^\mu - m)\psi^{(\frac{1}{2})} = 0 \tag{13.37}$$

then

$$(i\gamma_\mu D'^\mu - m)\psi^{(\frac{1}{2})\prime} = 0, \tag{13.38}$$

proving the asserted covariance. In the same way, any free particle wave equation satisfied by an 'isospinor' $\psi^{(\frac{1}{2})}$—the relevant equation is determined by the Lorentz spin of the particles involved—can be made locally covariant by the use of the covariant derivative D^μ, just as in the U(1) case.

The essential point here, of course, is that the locally covariant form includes *interactions* between the $\psi^{(\frac{1}{2})}$'s and the gauge fields $\boldsymbol{W}^\mu$, which are determined by the local phase invariance requirement (the 'gauge principle'). Indeed, we can already begin to find some of the Feynman rules appropriate to tree graphs for SU(2) gauge theories. Consider again the case of an SU(2) isospinor fermion, $\psi^{(\frac{1}{2})}$, obeying equation (13.38). This can be written as

$$(i\,\not\partial - m)\psi^{(\frac{1}{2})} = g(\boldsymbol{\tau}/2)\cdot\boldsymbol{W}\psi^{(\frac{1}{2})}. \tag{13.39}$$

In lowest-order perturbation theory the one-W emission/absorption process is given by the amplitude (cf (8.39)) for the electromagnetic case)

$$-ig\int \bar{\psi}_{\mathrm{f}}^{(\frac{1}{2})}(\boldsymbol{\tau}/2)\gamma_\mu\psi_{\mathrm{i}}^{(\frac{1}{2})} \cdot \boldsymbol{W}^\mu\mathrm{d}^4x \tag{13.40}$$

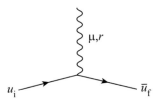

FIGURE 13.1
Vertex for isospinor-W interaction.

exactly as advertised (for the field-theoretic vertex) in (12.129). The matrix degree of freedom in the τ's is sandwiched between the 2-component isospinors $\psi^{(\frac{1}{2})}$; the γ matrix acts on the 4-component (Dirac) parts of $\psi^{(\frac{1}{2})}$. The external $\boldsymbol{W}^\mu$ field is now specified by a spin-1 polarization vector ϵ^μ, like a photon, and by an 'SU(2) polarization vector' $a^r (r = 1, 2, 3)$ which tells us which of the three SU(2) W-states is participating. The Feynman rule for figure 13.1 is therefore

$$-\mathrm{i}g(\tau^r/2)\gamma_\mu \tag{13.41}$$

which is to be sandwiched between spinors/isospinors $u_\mathrm{i}, \bar{u}_\mathrm{f}$ and dotted into ϵ^μ and a^r. (13.41) is a very economical generalization of rule (ii) in Comment (3) of section 8.3.1.

The foregoing is easily generalized to other SU(2) multiplets than doublets. We shall change the notation slightly to use t instead of T for the 'isospin' quantum number, so as to emphasize that it is *not* the hadronic isospin, for which we retain T; t will be the symbol used for the *weak isospin* to be introduced in chapter 20. The general local SU(2) transformation for a t-multiplet is then

$$\psi^{(t)} \to \psi^{(t)\prime} = \exp[\mathrm{i}g\boldsymbol{\alpha}(x) \cdot \mathbf{T}^{(t)}]\psi^{(t)} \tag{13.42}$$

where the $(2t+1) \times (2t+1)$ matrices $T_i^{(t)}$ $(i = 1, 2, 3)$ satisfy (cf (12.47))

$$[T_i^{(t)}, T_j^{(t)}] = \mathrm{i}\epsilon_{ijk}T_k^{(t)}. \tag{13.43}$$

The appropriate covariant derivative is

$$D^\mu = \partial^\mu + \mathrm{i}g\mathbf{T}^{(t)} \cdot \boldsymbol{W}^\mu \tag{13.44}$$

which is a $(2t+1) \times (2t+1)$ matrix acting on the $(2t+1)$ components of $\psi^{(t)}$. The gauge fields interact with such 'isomultiplets' in a *universal* way—only one g, the same for all the particles—which is prescribed by the local covariance requirement to be simply that interaction which is generated by the covariant derivatives. The fermion vertex corresponding to (13.44) is obtained by replacing $\boldsymbol{\tau}/2$ in (13.40) by $\boldsymbol{T}^{(t)}$.

We end this section with some comments:

(a) It is a remarkable fact that only one constant g is needed. This is *not* the same as in electromagnetism. There, each charged field interacts with the gauge field A^μ via a coupling whose strength is its charge $(e, -e, 2e, -5e \ldots)$. The crucial point is the appearance of the quadratic g^2 multiplying the *commutator* of the $\boldsymbol{\tau}$'s, $[\boldsymbol{\tau} \cdot \boldsymbol{\epsilon}, \boldsymbol{\tau} \cdot \boldsymbol{W}]$, in the $\boldsymbol{W}^\mu$ transformation (equation (13.20)). In the electromagnetic case, there is no such commutator — the associated U(1) phase group is Abelian. As signalled by the presence of g^2, a commutator is a non-linear quantity, and the scale of quantities appearing in such commutation relations is not arbitrary. It is an instructive exercise to check that, once $\delta \boldsymbol{W}^\mu$ is given by equation (13.23) — in the SU(2) case — then the g's appearing in $\psi^{(\frac{1}{2})\prime}$ (equation (13.13)) and $\psi^{(t)\prime}$ (via the infinitesimal version of equation (13.42)) must be the *same* as the one appearing in $\delta \boldsymbol{W}^\mu$.

(b) According to the foregoing argument, it is actually a mystery why electric charge should be quantized. Since it is the coupling constant of an Abelian group, each charged field could have an arbitrary charge from this point of view: there are no commutators to fix the scale. This is one of the motivations of attempts to 'embed' the electromagnetic gauge transformations inside a larger non-Abelian group structure. Such is the case, for example, in 'grand unified theories' of strong, weak, and electromagnetic interactions.

(c) Finally we draw attention to the extremely important physical significance of the second term $\delta \boldsymbol{W}^\mu$ (equation (13.23)). The gauge fields themselves are not 'inert' as far as the gauge group is concerned: in the SU(2) case they have 'isospin' 1, while for a general group they belong to the regular representation of the group. This is profoundly different from the electromagnetic case, where the gauge field A^μ for the photon is of course uncharged: quite simply, $e = 0$ for a photon, and the second term in (13.23) is absent for A^μ. The fact that non-Abelian (Yang-Mills) gauge fields carry non-Abelian 'charge' degrees of freedom means that, since they are also the quanta of the force field, *they will necessarily interact with themselves.* Thus a non-Abelian gauge theory of gauge fields alone, with no 'matter' fields, has non-trivial interactions and is not a free theory.

We shall examine the form of these 'self-interactions' in section 13.3.2. First, we need to find the equivalent, for the Yang-Mills field, of the Maxwell field strength tensor $F^{\mu\nu}$, which gave us the gauge-invariant formulation of Maxwell's equations, and in terms of which the Maxwell Lagrangian can be immediately written down.

13.1.2   The non-Abelian field strength tensor

A simple way of arriving at the desired quantity is to consider the commutator of two covariant derivatives, as we can see by calculating it for the U(1) case. We find

$$[D^\mu, D^\nu]\psi \equiv (D^\mu D^\nu - D^\nu D^\mu)\psi = \mathrm{i}eF^{\mu\nu}\psi \tag{13.45}$$

as is verified in problem 13.2. Equation (13.45) suggests that we will find the SU(2) analogue of $F^{\mu\nu}$ by evaluating

$$[D^\mu, D^\nu]\psi^{(\frac{1}{2})} \tag{13.46}$$

where as usual

$$D^\mu(\text{on } \psi^{(\frac{1}{2})}) = \partial^\mu + \mathrm{i}g\boldsymbol{\tau} \cdot \boldsymbol{W}^\mu/2. \tag{13.47}$$

Problem 13.3 confirms that the result is

$$[D^\mu, D^\nu]\psi^{(\frac{1}{2})} = \mathrm{i}g\boldsymbol{\tau}/2 \cdot (\partial^\mu \boldsymbol{W}^\nu - \partial^\nu \boldsymbol{W}^\mu - g\boldsymbol{W}^\mu \times \boldsymbol{W}^\nu)\,\psi^{(\frac{1}{2})}; \tag{13.48}$$

the manipulations are very similar to those in (13.20) – (13.23). Noting the analogy between the right-hand side of (13.48) and (13.45), we accordingly expect the SU(2) 'curvature' or field strength tensor, to be given by

$$\boldsymbol{F}^{\mu\nu} = \partial^\mu \boldsymbol{W}^\nu - \partial^\nu \boldsymbol{W}^\mu - g\boldsymbol{W}^\mu \times \boldsymbol{W}^\nu \tag{13.49}$$

or, in component notation,

$$F_i^{\mu\nu} = \partial^\mu W_i^\nu - \partial^\nu W_i^\mu - g\epsilon_{ijk}W_j^\mu W_k^\nu. \tag{13.50}$$

This tensor is of fundamental importance in a (non-Abelian) gauge theory. Since it arises from the commutator of two gauge-covariant derivatives, we are guaranteed that it itself is gauge covariant—that is to say, 'it transforms under local SU(2) transformations in the way its SU(2) structure would indicate'. Now $\boldsymbol{F}^{\mu\nu}$ has clearly three SU(2) components and must be an SU(2) triplet. Indeed, it is true that under an infinitesimal local SU(2) transformation

$$\boldsymbol{F}'^{\mu\nu} = \boldsymbol{F}^{\mu\nu} - g\boldsymbol{\epsilon}(x) \times \boldsymbol{F}^{\mu\nu} \tag{13.51}$$

which is the expected law (cf (12.64)) for an SU(2) triplet. Problem 13.4 verifies that (13.51) follows from (13.49) and the transformation law (13.23) for the $\boldsymbol{W}^{\mu}$ fields. Note particularly that $\boldsymbol{F}^{\mu\nu}$ transforms 'properly', as an SU(2) triplet should, *without* the ∂^{μ} part which appears in $\delta\boldsymbol{W}^{\mu}$.

This non-Abelian $\boldsymbol{F}^{\mu\nu}$ is a much more interesting object than the Abelian $F^{\mu\nu}$ (which is actually U(1)-gauge *invariant*, of course: $F'^{\mu\nu} = F^{\mu\nu}$). $\boldsymbol{F}^{\mu\nu}$ contains the gauge coupling constant g, confirming (cf comment(c) in section 13.1.1) that the gauge fields themselves carry SU(2) 'charge', and act as sources for the field strength. Appendix N shows how these field strength tensors may be regarded as analogous to geometrical curvatures.

It is now straightforward to move to the quantum field case and construct the SU(2) Yang-Mills analogue of the Maxwell Lagrangian $-\frac{1}{4}\hat{F}_{\mu\nu}\hat{F}^{\mu\nu}$. It is simply $-\frac{1}{4}\hat{\boldsymbol{F}}_{\mu\nu} \cdot \hat{\boldsymbol{F}}^{\mu\nu}$, the SU(2) 'dot product' ensuring SU(2) invariance (see problem 13.5), even under *local* transformation, in view of the transformation law (13.51). But before proceeding in this way we first need to introduce local SU(3) symmetry.

13.2 Local SU(3) Symmetry

Using what has been done for global SU(3) symmetry in section 12.2, and the preceding discussion of how to make a global SU(2) into a local one, it is straightforward to develop the corresponding theory of local SU(3). This is the gauge group of QCD, the three degrees of freedom of the fundamental quark triplet now referring to 'colour', as will be further discussed in chapter 14. We denote the basic triplet by ψ, which transforms under a local SU(3) transformation according to

$$\psi' = \exp[\mathrm{i}g_{\mathrm{s}}\boldsymbol{\lambda} \cdot \boldsymbol{\alpha}(x)/2]\psi, \tag{13.52}$$

which is the same as the global transformation (12.74) but with the 8 constant parameters $\boldsymbol{\alpha}$ replaced by x-dependent ones, and with a coupling strength g_{s} inserted. The SU(3)-covariant derivative, when acting on an SU(3) triplet ψ, is given by the indicated generalization of (13.10), namely

$$D^{\mu}(\text{acting on SU(3) triplet}) = \partial^{\mu} + \mathrm{i}g_{\mathrm{s}}\boldsymbol{\lambda}/2 \cdot \boldsymbol{A}^{\mu} \tag{13.53}$$

where $A_1^{\mu}, A_2^{\mu}, \ldots A_8^{\mu}$ are eight gauge fields which are called *gluons*. The coupling is denoted by 'g_{s}' in anticipation of the application to strong interactions via QCD.

The infinitesimal version of (13.52) is (cf (13.13))

$$\psi' = (1 + \mathrm{i}g_{\mathrm{s}}\boldsymbol{\lambda} \cdot \boldsymbol{\eta}(x)/2)\psi \tag{13.54}$$

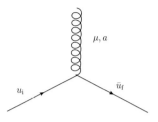

FIGURE 13.2
Quark-gluon vertex.

where '1' stands for the unit matrix in the three-dimensional space of components of the triplet ψ. As in (13.14), it is clear that $\partial^\mu \psi'$ will involve an 'unwanted' term $\partial^\mu \boldsymbol{\eta}(x)$. By contrast, the desired covariant derivative $D^\mu \psi$ should transform according to

$$D'^\mu \psi' = (1 + ig_s \boldsymbol{\lambda} \cdot \boldsymbol{\eta}(x)/2) D^\mu \psi \qquad (13.55)$$

without the $\partial^\mu \boldsymbol{\eta}(x)$ term. Problem 13.6 verifies that this is fulfilled by having the gauge fields transform by

$$A_a'^\mu = A_a^\mu - \partial^\mu \eta_a(x) - g_s f_{abc} \eta_b(x) A_c^\mu. \qquad (13.56)$$

Comparing (13.56) with (12.80) we can identify the term in f_{abc} as telling us that the 8 fields A_a^μ transform as an SU(3) octet, the η's now depending on x, of course. This is the adjoint, or regular representation of SU(3), as we have now come to expect for gauge fields. However, the $\partial^\mu \eta_a(x)$ piece spoils this simple transformation property under local transformations. But it *is* just what is needed to cancel the corresponding $\partial^\mu \boldsymbol{\eta}(x)$ term in $\partial^\mu \psi'$, leaving $D^\mu \psi$ transforming as a proper triplet via (13.55). The finite version of (13.56) can be derived as in section 13.1 for SU(2), but we shall not need the result here.

As in the SU(2) case, the free Dirac equation for an SU(3)-triplet ψ,

$$(i\gamma_\mu \partial^\mu - m)\psi = 0, \qquad (13.57)$$

can be 'promoted' into one which is covariant under local SU(3) transformations by replacing ∂^μ by D^μ of (13.53), leading to

$$(i\,\slashed{\partial} - m)\psi = g_s \boldsymbol{\lambda}/2 \cdot \slashed{A}\psi \qquad (13.58)$$

(compare (13.39)). This leads immediately to the one gluon emission amplitude (see figure 13.2)

$$-ig_s \int \bar{\psi}_f \boldsymbol{\lambda}/2 \gamma^\mu \psi_i \cdot A_\mu d^4 x \qquad (13.59)$$

as already suggested in section 12.3.1: the SU(3) current of (12.133)—but this time in *colour* space—is 'dotted' with the gauge field. The Feynman rule for figure 13.2 is therefore

$$-ig_s \lambda_a/2\, \gamma^\mu. \qquad (13.60)$$

The SU(3) field strength tensor can be calculated by evaluating the commutator of two D's of the form (13.53); the result (problem 13.7) is

$$F_a^{\mu\nu} = \partial^\mu A_a^\nu - \partial^\nu A_a^\mu - g_s f_{abc} A_b^\mu A_c^\nu \qquad (13.61)$$

which is closely analogous to the SU(2) case (13.50) (the structure constants of SU(2) are given by iϵ_{ijk}, and of SU(3) by if_{abc}). Once again, the crucial property of $F_a^{\mu\nu}$ is that, under *local* SU(3) transformations it develops no '$\partial^\mu \eta_a$' part, but transforms as a 'proper' octet:

$$F_a^{\mu\nu} = F_a^{\mu\nu} - g_s f_{abc} \eta_b(x) F_c^{\mu\nu}. \tag{13.62}$$

This allows us to write down a locally SU(3)-invariant analogue of the Maxwell Lagrangian

$$-\frac{1}{4} F_a^{\mu\nu} F_{a\mu\nu} \tag{13.63}$$

by dotting the two octets together.

It is now time to consider locally SU(2)- and SU(3)-invariant quantum field Lagrangians and, in particular, the resulting self-interactions among the gauge quanta.

13.3 Local non-Abelian symmetries in Lagrangian quantum field theory

13.3.1 Local SU(2) and SU(3) Lagrangians

We consider here only the particular examples relevant to the strong and electroweak interactions of quarks: namely, a (weak) SU(2) doublet of fermions interacting with SU(2) gauge fields W_i^μ, and a (strong) SU(3) triplet of fermions interacting with the gauge fields A_a^μ. We follow the same steps as in the U(1) case of chapter 7, noting again that for quantum fields the sign of the exponents in (13.2) and (13.52) is reversed, by convention; thus (12.89) is replaced by its local version

$$\hat{q}' = \exp(-\mathrm{i}g\hat{\boldsymbol{\alpha}}(x) \cdot \boldsymbol{\tau}/2)\hat{q} \tag{13.64}$$

and (12.132) by

$$\hat{q}' = \exp(-\mathrm{i}g_s\hat{\boldsymbol{\alpha}}(x) \cdot \boldsymbol{\lambda}/2)\hat{q}. \tag{13.65}$$

Correspondingly, the ϵ in (13.23) and the η's in (13.56) become field operators, with a reversal of sign.

The globally SU(2)-invariant Lagrangian (12.87) becomes locally SU(2)-invariant if we replaced ∂^μ by D^μ of (13.10), with $\hat{W}^\mu$ now a quantum field:

$$\begin{aligned} \hat{\mathcal{L}}_{\mathrm{D,local\ SU(2)}} &= \bar{\hat{q}}(\mathrm{i}\,\hat{\slashed{D}} - m)\hat{q} \\ &= \bar{\hat{q}}(\mathrm{i}\,\slashed{\partial} - m)\hat{q} - g\bar{\hat{q}}\gamma^\mu \boldsymbol{\tau}/2\hat{q} \cdot \hat{\boldsymbol{W}}_\mu \end{aligned} \tag{13.66}$$

with an interaction of the form 'symmetry current (12.109) dotted into the gauge field'. To this we must add the SU(2) Yang-Mills term

$$\mathcal{L}_{\mathrm{Y-M,SU(2)}} = -\frac{1}{4}\hat{\boldsymbol{F}}_{\mu\nu} \cdot \hat{\boldsymbol{F}}^{\mu\nu} \tag{13.67}$$

to get the local SU(2) analogue of $\mathcal{L}_{\mathrm{QED}}$. It is *not* possible to add a mass term for the gauge fields of the form $\frac{1}{2}\hat{\boldsymbol{W}}^\mu \cdot \hat{\boldsymbol{W}}_\mu$, since such a term would not be invariant under the gauge transformations (13.26) or (13.34) of the W-fields. Thus, just as in the U(1) (electromagnetic) case, the W-quanta of this theory are *massless*. We presumably also need a

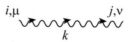

FIGURE 13.3
SU(2) gauge-boson propagator.

gauge-fixing term for the gauge fields, as in section 7.3.2, which we can take to be[1]

$$\mathcal{L}_{\text{gf}} = -\frac{1}{2\xi}\left(\partial_\mu \hat{\boldsymbol{W}}^\mu \cdot \partial_\nu \hat{\boldsymbol{W}}^\nu\right). \tag{13.68}$$

The Feynman rule for the fermion-W vertex is then the same as already given in (13.41), while the W-propagator is (figure 13.3)

$$\frac{\text{i}\left[-g^{\mu\nu} + (1-\xi)k^\mu k^\nu/k^2\right]}{k^2 + \text{i}\epsilon}\,\delta^{ij}. \tag{13.69}$$

Before proceeding to the SU(3) case, we must now emphasize three respects in which our local SU(2) Lagrangian is not suitable (yet) for describing weak interactions. First, weak interactions violate parity, in fact 'maximally', by which is meant that only the 'left handed' part $\hat{\psi}_\text{L}$ of the fermion field enters the interactions with the $\boldsymbol{W}^\mu$ fields, where $\hat{\psi}_\text{L} \equiv \left(\frac{1-\gamma_5}{2}\right)\hat{\psi}$; for this reason the weak isospin group is called SU(2)$_\text{L}$. Secondly, the physical W$^\pm$ are of course not massless, and therefore cannot be described by propagators of the form (13.69). And thirdly, the *fermion* mass term violates the 'left-handed' SU(2) gauge symmetry, as the discussion in section 12.3.2 shows. In this case, however, the chiral symmetry which is broken by fermion masses in the Lagrangian is a local, or gauge, symmetry (in section 12.3.2 the chiral flavour symmetry was a global symmetry). If we want to preserve the chiral gauge symmetry SU(2)$_\text{L}$—and it necessary for renormalizability—then we shall have to replace the simple fermion mass term in (13.66) by something else, as will be explained in chapter 22.

The locally SU(3)$_\text{c}$-invariant Lagrangian for one quark triplet (cf (12.137))

$$\hat{q}_\text{f} = \begin{pmatrix} \hat{f}_\text{r} \\ \hat{f}_\text{b} \\ \hat{f}_\text{g} \end{pmatrix}, \tag{13.70}$$

where 'f' stands for 'flavour', and 'r, b, and g' for 'red, blue, and green', is

$$\bar{\hat{q}}_\text{f}(\text{i}\hat{\not{D}} - m_\text{f})\hat{q}_\text{f} - \frac{1}{4}\hat{F}_{a\mu\nu}\hat{F}_a^{\mu\nu} - \frac{1}{2\xi}(\partial_\mu \hat{A}_a^\mu)(\partial_\nu \hat{A}_a^\nu) \tag{13.71}$$

where $\hat{D}^\mu$ is given by (13.53) with $\boldsymbol{A}^\mu$ replaced by $\hat{\boldsymbol{A}}^\mu$, and the footnote before equation (13.68) also applies here. This leads to the interaction term (cf (13.59))

$$-g_\text{s}\bar{\hat{q}}_\text{f}\gamma^\mu\boldsymbol{\lambda}/2\hat{q}_\text{f}\cdot\hat{\boldsymbol{A}}_\mu \tag{13.72}$$

---

[1]We shall see in section 13.5.3 that in the non-Abelian case this gauge-fixing term does *not* completely solve the problem of quantizing such gauge fields; however, it is adequate for tree graphs.

and the Feynman rule (13.60) for figure 13.2. Once again, the gluon quanta must be *massless*, and their propagator is the same as (13.69), with $\delta_{ij} \to \delta_{ab}$ $(a, b = 1, 2, \ldots 8)$. The different quark flavours are included by simply repeating the first term of (13.71) for all flavours:

$$\sum_{\text{f}} \bar{\hat{q}}_{\text{f}}(i\hat{\slashed{D}} - m_{\text{f}})\hat{q}_{\text{f}}, \qquad (13.73)$$

which incorporates the hypothesis that the $SU(3)_{\text{c}}$-gauge interaction is 'flavour-blind', i.e. exactly the same for each flavour. Note that although the flavour masses are different, the masses of different 'coloured' quarks of the same flavour are the same ($m_{\text{u}} \neq m_{\text{d}}, m_{\text{u,r}} = m_{\text{u,b}} = m_{\text{u,g}}$).

The Lagrangians (13.66) – (13.68), and (13.71), though easily written down after all this preparation, are unfortunately not adequate for anything but tree graphs. We shall indicate why this is so in section 13.3.3. Before that, we want to discuss in more detail the nature of the gauge-field self-interactions contained in the Yang-Mills pieces.

13.3.2   Gauge field self-interactions

We start by pointing out an interesting ambiguity in the prescription for 'covariantizing' wave equations which we have followed, namely 'replace ∂^μ by D^μ'. Suppose we wished to consider the electromagnetic interactions of charged massless spin-1 particles, call them X's, carrying charge e. The standard wave equation for such free massless vector particles would be the same as for A^μ, namely

$$\Box X^\mu - \partial^\mu \partial^\nu X_\nu = 0. \qquad (13.74)$$

To 'covariantize' this (i.e. introduce the electromagnetic coupling) we would replace ∂^μ by $D^\mu = \partial^\mu + ieA^\mu$ so as to obtain

$$D^2 X^\mu - D^\mu D^\nu X_\nu = 0. \qquad (13.75)$$

But this procedure is not unique. If we had started from the perfectly equivalent wave equation

$$\Box X^\mu - \partial^\nu \partial^\mu X_\nu = 0 \qquad (13.76)$$

we would have arrived at

$$D^2 X^\mu - D^\nu D^\mu X_\nu = 0 \qquad (13.77)$$

which is not the same as (13.75), since (cf (13.45))

$$[D^\mu, D^\nu] = ieF^{\mu\nu}. \qquad (13.78)$$

The simple prescription $\partial^\mu \to D^\mu$ has, in this case, failed to produce a unique wave equation. We can allow for this ambiguity by introducing an arbitrary parameter δ in the wave equation, which we write as

$$D^2 X^\mu - D^\nu D^\mu X_\nu + ie\delta F^{\mu\nu} X_\nu = 0. \qquad (13.79)$$

The δ term in (13.79) contributes to the magnetic moment coupling of the X-particle to the electromagnetic field, and is called the 'ambiguous magnetic moment'. Just such an ambiguity would seem to arise in the case of the charged weak interaction quanta $W^\pm$ (their masses do not affect this argument). For the photon itself, of course, $e = 0$ and there is no such ambiguity.

It is important to be clear that (13.79) is fully U(1) gauge-covariant, so that δ cannot be fixed by further appeal to the local U(1) symmetry. Moreover, it turns out that the theory for arbitrary δ is *not renormalizable* (though we shall not show this here). Thus the quantum electrodynamics of charged massless vector bosons is in general non-renormalizable.

However, the theory *is* renormalizable if—to continue with the present terminology—the photon, the X-particle, and its antiparticle the $\bar{X}$ are the members of an SU(2) gauge triplet (like the W's), with gauge coupling constant e. This is, indeed, very much how the photon and the $W^{\pm}$ are 'unified', but there is a complication (as always!) in that case, having to do with the necessity for finding room in the scheme for the neutral weak boson Z^0 as well. We shall see how this works in chapter 19; meanwhile we continue with this $X - \gamma$ model. We shall show that when the $X - \gamma$ interaction contained in (13.79) is regarded as a $3 - X$ vertex in a local SU(2) gauge theory, the value of δ has to equal 1; for this value the theory is renormalizable. In this interpretation, the X^{μ} wave function is identified with '$\frac{1}{\sqrt{2}}(X_1^{\mu} + iX_2^{\mu})$' and $\bar{X}^{\mu}$ with '$\frac{1}{\sqrt{2}}(X_1^{\mu} - iX_2^{\mu})$' in terms of components of the SU(2) triplet X_i^{μ}, while A^{μ} is identified with X_3^{μ}.

Consider then equation (13.79) written in the form[2]

$$\Box X^{\mu} - \partial^{\nu}\partial^{\mu}X_{\nu} = \hat{V}X^{\mu} \tag{13.80}$$

where

$$\begin{aligned}
\hat{V}X^{\mu} &= -ie\{[\partial^{\nu}(A_{\nu}X^{\mu}) + A^{\nu}\partial_{\nu}X^{\mu}] \\
&\quad - (1+\delta)[\partial^{\nu}(A^{\mu}X_{\nu}) + A^{\nu}\partial^{\mu}X_{\nu}] \\
&\quad + \delta[\partial^{\mu}(A^{\nu}X_{\nu}) + A^{\mu}\partial^{\nu}X_{\nu}]\},
\end{aligned} \tag{13.81}$$

and we have dropped terms of $O(e^2)$ which appear in the 'D^2' term; we shall come back to them later. The terms inside the $\{\ \}$ brackets have been written in such a way that each $[\ ]$ bracket has the structure

$$\partial(AX) + A(\partial X) \tag{13.82}$$

which will be convenient for the following evaluation.

The lowest-order $(O(e))$ perturbation theory amplitude for '$X \to X$' under the potential $\hat{V}$ is then

$$-i\int X_{\mu}^{*}(f)\hat{V}X^{\mu}(i)d^4x. \tag{13.83}$$

Inserting (13.81) into (13.83) clearly gives something involving two 'X'-wavefunctions and one 'A' one, i.e. a triple-X vertex (with $A^{\mu} \equiv X_3^{\mu}$), shown in figure 13.4 . To obtain the rule for this vertex from (13.83), consider the first $[\ ]$ bracket in (13.81). It contributes

$$-i(-ie)\int X_{\mu}^{*}(2)\{\partial^{\nu}(X_{3\nu}(3)X^{\mu}(1)) + X_3^{\nu}(3)\partial_{\nu}X^{\mu}(1)\}d^4x \tag{13.84}$$

where the (1), (2), (3) refer to the momenta as shown in figure 13.4, and for reasons of symmetry are all taken to be ingoing; thus

$$X_3^{\mu}(3) = \epsilon_3^{\mu}\exp(-ik_3 \cdot x) \tag{13.85}$$

for example. The first term in (13.84) can be easily evaluated by a partial integration to turn the ∂^{ν} onto the $X_{\mu}^{*}(2)$, while in the second term ∂_{ν} acts straightforwardly on $X^{\mu}(1)$.

---

[2]The sign chosen for $\hat{V}$ here apparently *differs* from that in the KG case (3.101), but it does agree when allowance is made, in the amplitude (13.83), for the fact that the dot product of the polarization vectors is negative (cf (7.87)).

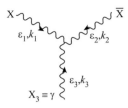

FIGURE 13.4
Triple-X vertex.

Omitting the usual $(2\pi)^4\,\delta^4$ energy-momentum conserving factor, we find (problem 13.8) that (13.84) leads to the amplitude

$$\mathrm{i}e\epsilon_1\cdot\epsilon_2\,(k_1-k_2)\cdot\epsilon_3. \tag{13.86}$$

In a similar way, the other terms in (13.83) give

$$-\mathrm{i}e\delta(\epsilon_1\cdot\epsilon_3\,\epsilon_2\cdot k_2-\epsilon_2\cdot\epsilon_3\,\epsilon_1\cdot k_1) \tag{13.87}$$

and

$$+\mathrm{i}e(1+\delta)(\epsilon_2\cdot\epsilon_3\,\epsilon_1\cdot k_2-\epsilon_1\cdot\epsilon_3\,\epsilon_2\cdot k_1). \tag{13.88}$$

Adding all the terms up and using the 4-momentum conservation condition

$$k_1+k_2+k_3=0 \tag{13.89}$$

we obtain the vertex

$$+\mathrm{i}e\{\epsilon_1\cdot\epsilon_2\,(k_1-k_2)\cdot\epsilon_3+\epsilon_2\cdot\epsilon_3\,(\delta k_2-k_3)\cdot\epsilon_1+\epsilon_3\cdot\epsilon_1\,(k_3-\delta k_1)\cdot\epsilon_2\}. \tag{13.90}$$

It is quite evident from (13.90) that the value $\delta=1$ has a privileged role, and we strongly suspect that this will be the value selected by the proposed SU(2) gauge symmetry of this model. We shall check this in two ways. In the first, we consider a 'physical' process involving the vertex (13.90), and show how requiring it to be SU(2)-gauge invariant fixes δ to be 1; in the second, we 'unpack' the relevant vertex from the compact Yang-Mills Lagrangian $-\frac{1}{4}\hat{\boldsymbol{X}}_{\mu\nu}\cdot\hat{\boldsymbol{X}}^{\mu\nu}$.

The process we shall choose is $X+d\rightarrow X+d$ where d is a fermion (which we call a quark) transforming as the $T_3=-\frac{1}{2}$ component of a doublet under the SU(2) gauge group, its $T_3=+\frac{1}{2}$ partner being the u. There are two contributing Feynman graphs, shown in figure 13.5(a) and (b). Consider first the amplitude for figure 13.5(a). We use the rule of figure 13.1, with the τ-matrix combination $\tau_+=(\tau_1+\mathrm{i}\tau_2)/\sqrt{2}$ corresponding to the absorption of the positively charged X, and $\tau_-=(\tau_1-\mathrm{i}\tau_2)/\sqrt{2}$ for the emission of the X. Then figure 13.5(a) is

$$(-\mathrm{i}e)^2\bar{\psi}^{(\frac{1}{2})}(p_2)\frac{\tau_-}{2}\,\not{\epsilon}_2\frac{\mathrm{i}}{\not{p}_1+\not{k}_1-m}\frac{\tau_+}{2}\,\not{\epsilon}_1\psi^{(\frac{1}{2})}(p_1) \tag{13.91}$$

where

$$\psi^{(\frac{1}{2})}=\begin{pmatrix}u\\d\end{pmatrix}, \tag{13.92}$$

and we have chosen real polarization vectors. Using the explicit forms (12.25) for the τ-matrices, (13.91) becomes

$$(-\mathrm{i}e)^2\bar{d}(p_2)\frac{1}{\sqrt{2}}\,\not{\epsilon}_2\frac{\mathrm{i}}{\not{p}_1+\not{k}_1-m}\frac{1}{\sqrt{2}}\,\not{\epsilon}_1 d(p_1). \tag{13.93}$$

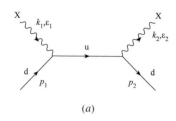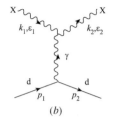

(a) (b)

FIGURE 13.5
Tree graphs contributing to X + d → X + d.

We must now discuss how to implement gauge invariance. In the QED case of electron Compton scattering (section 8.6.2), the test of gauge invariance was that the amplitude should vanish if any photon polarization vector $\epsilon^\mu(k)$ was replaced by k^μ—see (8.165). This requirement was derived from the fact that a gauge transformation on the photon A^μ took the form $A^\mu \to A'^\mu = A^\mu - \partial^\mu \chi$, so that, consistently with the Lorentz condition, ϵ^μ could be replaced by $\epsilon'^\mu = \epsilon^\mu + \beta k^\mu$ (cf 8.163) without changing the physics. But the SU(2) analogue of the U(1) gauge transformation is given by (13.26), for infinitesimal ϵ's, and although there is indeed an analogous '$-\partial^\mu \epsilon$' part, there is also an additional part (with $g \to e$ in our case) expressing the fact that the X's carry SU(2) charge. However this extra part does involve the coupling e. Hence, if we were to make the *full* change corresponding to (13.26) in a tree graph of order e^2, the extra part would produce a term of order e^3. We shall take the view that gauge invariance should hold at each order of perturbation theory separately; thus we shall demand that the tree graphs for X-d scattering, for example, should be invariant under $\epsilon^\mu \to k^\mu$ for any ϵ.

The replacement $\epsilon_1 \to k_1$ in (13.93) produces the result (problem 13.9)

$$(-\mathrm{i}e)^2 \frac{\mathrm{i}}{2} \bar{d}(p_2) \,\not{\epsilon_2} d(p_1) \tag{13.94}$$

where we have used the Dirac equation for the quark spinors of mass m. The term (13.94) is certainly not zero, but we must of course also include the amplitude for figure 13.5(b). Using the vertex of (13.90) with suitable sign changes of momenta, and the photon propagator of (7.119), and remembering that d has $\tau_3 = -1$, the amplitude for figure 13.5(b) is

$$\mathrm{i}e[\epsilon_1 \cdot \epsilon_2 \,(k_1 + k_2)_\mu \quad + \quad \epsilon_{2\mu}\epsilon_1 \cdot (-\delta k_2 - k_2 + k_1) + \epsilon_{1\mu}\epsilon_2 \cdot (k_2 - k_1 - \delta k_1)]$$

$$\times \quad \frac{-\mathrm{i}g^{\mu\nu}}{q^2} \times [-\mathrm{i}e\bar{d}(p_2)\left(-\frac{1}{2}\right)\gamma_\nu d(p_1)], \tag{13.95}$$

where $q^2 = (k_1 - k_2)^2 = -2k_1 \cdot k_2$ using $k_1^2 = k_2^2 = 0$, and where the ξ-dependent part of the γ-propagator vanishes since $\bar{d}(p_2) \,\not{q}d(p_1) = 0$. We now leave it as an exercise (problem 13.10) to verify that, when $\epsilon_1 \to k_1$ in (13.95), the resulting amplitude does exactly cancel the contribution (13.94), *provided that* $\delta = 1$. Thus the $X - \bar{X} - \gamma$ vertex is, assuming the SU(2) gauge symmetry,

$$\mathrm{i}e[\epsilon_1 \cdot \epsilon_2 \,(k_1 - k_2) \cdot \epsilon_3 + \epsilon_2 \cdot \epsilon_3 \,(k_2 - k_3) \cdot \epsilon_1 + \epsilon_3 \cdot \epsilon_1 \,(k_3 - k_1) \cdot \epsilon_2]. \tag{13.96}$$

The verification of this non-Abelian gauge invariance to order e^2 is, of course, not a proof that the entire theory of massless X quanta, γ's and quark isospinors will be gauge invariant if $\delta = 1$. Indeed, having obtained the $X - X - \gamma$ vertex, we immediately have something new to check: we can see if the lowest order $\gamma - X$ scattering amplitude is gauge invariant. The

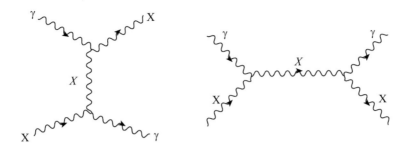

FIGURE 13.6
Tree graphs contributing to $\gamma + X \to \gamma + X$.

$X - X - \gamma$ vertex will generate the $O(e^2)$ graphs shown in figure 13.6, and the dedicated reader may check that the sum of these amplitudes is *not* gauge invariant, again in the (tree-graph) sense of not vanishing when any ϵ is replaced by the corresponding k. But this is actually correct. In obtaining the $X - X - \gamma$ vertex we dropped an $O(e^2)$ term involving the three fields $A, A,$ and X, in going from (13.81) to (13.90). This will generate an $O(e^2)$ $\gamma - \gamma - X - X$ interaction, figure 13.7, when used in lowest order perturbation theory. One can find the amplitude for figure 13.7 by the gauge invariance requirement applied to figures 13.6 and 13.7, but it has to be admitted that this approach is becoming laborious. It is, of course, far more efficient to deduce the vertices from the compact Yang-Mills Lagrangian $-\frac{1}{4}\hat{X}_{\mu\nu} \cdot \hat{X}^{\mu\nu}$, which we shall now do; nevertheless, some of the physical implications of those couplings, such as we have discussed above, are worth exposing.

The SU(2) Yang-Mills Lagrangian for the SU(2) triplet of gauge fields $\hat{X}^{\mu}$ is

$$\hat{\mathcal{L}}_{2,\mathrm{YM}} = -\frac{1}{4}\hat{X}_{\mu\nu} \cdot \hat{X}^{\mu\nu}, \tag{13.97}$$

where

$$\hat{X}^{\mu\nu} = \partial^\mu \hat{X}^\nu - \partial^\nu \hat{X}^\mu - e\hat{X}^\mu \times \hat{X}^\nu. \tag{13.98}$$

$\hat{\mathcal{L}}_{2,\mathrm{YM}}$ can be unpacked a bit into

$$
\begin{aligned}
&- \frac{1}{2}(\partial_\mu \hat{X}_\nu - \partial_\nu \hat{X}_\mu) \cdot (\partial^\mu \hat{X}^\nu) \\
&+ e(\hat{X}_\mu \times \hat{X}_\nu) \cdot \partial^\mu \hat{X}^\nu \\
&- \frac{1}{4}e^2 \left[(\hat{X}^\mu \cdot \hat{X}_\mu)^2 - (\hat{X}^\mu \cdot \hat{X}^\nu)(\hat{X}_\mu \cdot \hat{X}_\nu) \right].
\end{aligned} \tag{13.99}
$$

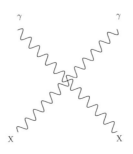

FIGURE 13.7
$\gamma - \gamma - X - X$ vertex.

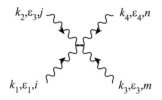

FIGURE 13.8
4-X vertex.

The $X - X - \gamma$ vertex is in the 'e' term, the $X - X - \gamma - \gamma$ one in the 'e^2' term. We give the form of the latter using $SU(2)$ 'i, j, k' labels, as shown in figure 13.8:

$$-\mathrm{i}e^2[\epsilon_{ij\ell}\epsilon_{mn\ell}(\epsilon_1 \cdot \epsilon_3\, \epsilon_2 \cdot \epsilon_4 - \epsilon_1 \cdot \epsilon_4\, \epsilon_2 \cdot \epsilon_3)$$
$$+\epsilon_{in\ell}\epsilon_{jm\ell}(\epsilon_1 \cdot \epsilon_2\, \epsilon_3 \cdot \epsilon_4 - \epsilon_1 \cdot \epsilon_3\, \epsilon_2 \cdot \epsilon_4)$$
$$+\epsilon_{im\ell}\epsilon_{nj\ell}(\epsilon_1 \cdot \epsilon_4\, \epsilon_2 \cdot \epsilon_3 - \epsilon_1 \cdot \epsilon_2\, \epsilon_3 \cdot \epsilon_4)] \tag{13.100}$$

The reason for the collection of terms seen in (13.96) and (13.100) can be understood as follows. Consider the 3 - X vertex

$$\langle k_2, \epsilon_2, j; k_3, \epsilon_3, k \mid e(\hat{\boldsymbol{X}}_\mu \times \hat{\boldsymbol{X}}_\nu) \cdot \partial^\mu \hat{\boldsymbol{X}}^\nu \mid k_1, \epsilon_1, i\rangle \tag{13.101}$$

for example. When each $\hat{\boldsymbol{X}}$ is expressed as a mode expansion, and the initial and final states are also written in terms of appropriate $\hat{a}$'s and $\hat{a}^\dagger$'s, the amplitude will be a vacuum expectation value (vev) of six $\hat{a}$'s and $\hat{a}^\dagger$'s; the different terms in (13.96) arise from the different ways of getting a non-zero value for this vev, by manipulations similar to those in section 6.3.

   We end this chapter by presenting an introduction to the problem of quantizing non-Abelian gauge field theories. Our aim will be, first, to indicate where the approach followed for the Abelian gauge field $\hat{A}^\mu$ in section 7.3.2 fails; and then to show how the assumption (nevertheless) that the Feynman rules we have established for tree graphs work for loops as well, leads to violations of unitarity. This calculation will indicate a very curious way of remedying the situation 'by hand', through the introduction of *ghost particles*, only present in loops.

13.3.3   Quantizing non-Abelian gauge fields

We consider for definiteness the $SU(2)$ gauge theory with massless gauge fields $\hat{\boldsymbol{W}}^\mu(x)$, which we shall call gluons, by a slight abuse of language. We try to carry through for the Yang-Mills Lagrangian

$$\hat{\mathcal{L}}_2 = -\frac{1}{4}\hat{\boldsymbol{F}}_{\mu\nu} \cdot \hat{\boldsymbol{F}}^{\mu\nu}, \tag{13.102}$$

where

$$\hat{\boldsymbol{F}}_{\mu\nu} = \partial_\mu\hat{\boldsymbol{W}}_\nu - \partial_\nu\hat{\boldsymbol{W}}_\mu - g\hat{\boldsymbol{W}}_\mu \times \hat{\boldsymbol{W}}_\nu, \tag{13.103}$$

the same steps we followed for the Maxwell one in section 7.3.2.

   We begin by re-formulating the prescription arrived at in (7.119), which we reproduce again here for convenience:

$$\hat{\mathcal{L}}_\xi = -\frac{1}{4}\hat{F}_{\mu\nu}\hat{F}^{\mu\nu} - \frac{1}{2\xi}(\partial_\mu\hat{A}^\mu)^2. \tag{13.104}$$

$\hat{\mathcal{L}}_{\xi}$ leads to the equation of motion

$$\Box \hat{A}^{\mu} - \partial^{\mu}\partial_{\nu}\hat{A}^{\nu} + \frac{1}{\xi}\partial^{\mu}\partial_{\nu}\hat{A}^{\nu} = 0. \tag{13.105}$$

This has the drawback that the limit $\xi \to 0$ appears to be singular (though the propagator (7.122) is well-behaved as $\xi \to 0$). To avoid this unpleasantness, consider the Lagrangian (Lautrup 1967)

$$\hat{\mathcal{L}}_{\xi B} = -\frac{1}{4}\hat{F}_{\mu\nu}\hat{F}^{\mu\nu} + \hat{B}\partial_{\mu}\hat{A}^{\mu} + \frac{1}{2}\xi\hat{B}^{2} \tag{13.106}$$

where $\hat{B}$ is a scalar field. We may think of the '$\hat{B}\partial \cdot \hat{A}$' term as a field theory analogue of the procedure followed in classical Lagrangian mechanics, whereby a constraint (in this case the gauge-fixing one $\partial \cdot \hat{A} = 0$) is brought into the Lagrangian with a 'Lagrange multiplier' (here the field $\hat{B}$). The momentum conjugate to $\hat{A}^{0}$ is now

$$\hat{\pi}^{0} = \hat{B} \tag{13.107}$$

while the Euler-Lagrange equations for $\hat{A}^{\mu\nu}$ read

$$\Box \hat{A}^{\mu} - \partial^{\mu}\partial_{\nu}\hat{A}^{\nu} = \partial^{\mu}\hat{B}, \tag{13.108}$$

and for $\hat{B}$ yield

$$\partial_{\mu}\hat{A}^{\mu} + \xi\hat{B} = 0. \tag{13.109}$$

Eliminating $\hat{B}$ from (13.106) by means of (13.109) we recover (13.104). Taking ∂_{μ} of (13.108) we learn that $\Box\hat{B} = 0$, so that $\hat{B}$ is a free massless field. Applying $\Box$ to (13.109) then shows that $\Box\partial_{\mu}\hat{A}^{\mu} = 0$, so that $\partial_{\mu}\hat{A}^{\mu}$ is also a free massless field.

In this formulation, the appropriate subsidiary condition for getting rid of the unphysical (non-transverse) degrees of freedom is (cf (7.111))

$$\hat{B}^{(+)}(x) \mid \Psi \rangle = 0. \tag{13.110}$$

Kugo and Ojima (1979) have shown that (13.110) provides a satisfactory definition of the Hilbert space of states. In addition to this it is also essential to prove that all physical results are independent of the gauge parameter ξ.

We now try to generalize the foregoing in a straightforward way to (13.102). The obvious analogue of (13.106) would be to consider

$$\hat{\mathcal{L}}_{2,\xi\,B} = -\frac{1}{4}\hat{\boldsymbol{F}}_{\mu\nu} \cdot \hat{\boldsymbol{F}}^{\mu\nu} + \hat{\boldsymbol{B}} \cdot (\partial_{\mu}\hat{\boldsymbol{W}}^{\mu}) + \frac{1}{2}\xi\hat{\boldsymbol{B}} \cdot \hat{\boldsymbol{B}} \tag{13.111}$$

where $\hat{\boldsymbol{B}}$ is an SU(2) triplet of scalar fields. Equation (13.111) gives (cf (13.108))

$$(\hat{D}^{\nu})_{ij}\hat{F}_{j\mu\nu} + \partial_{\mu}\hat{B}_{i} = 0 \tag{13.112}$$

where the covariant derivative is now the one appropriate to the SU(2) triplet $\boldsymbol{F}_{\mu\nu}$ (see (13.44) with $t = 1$, and (12.48)), and i, j are the SU(2) labels. Similarly, (13.109) becomes

$$\partial_{\mu}\hat{\boldsymbol{W}}^{\mu} + \xi\hat{\boldsymbol{B}} = \mathbf{0}. \tag{13.113}$$

It is possible to verify that

$$(\hat{D}^{\mu})_{ki}(\hat{D}^{\nu})_{ij}\hat{F}_{j\mu\nu} = 0 \tag{13.114}$$

where i, j, k are the SU(2) matrix indices, which implies that

$$(\hat{D}^{\mu})_{ki}\partial_{\mu}\hat{B}_{i} = 0. \tag{13.115}$$

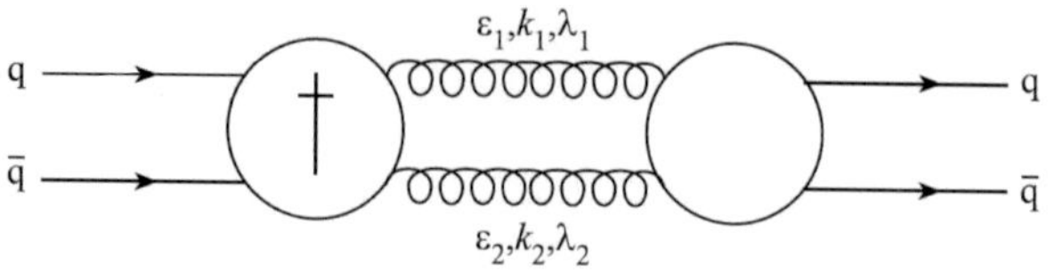

FIGURE 13.9

Two-gluon intermediate state in the unitarity relation for the amplitude for $q\bar{q} \to q\bar{q}$.

This is the crucial result. It implies that the auxiliary field $\hat{B}$ is *not* a free field in this non-Abelian case, and so neither (from (13.113)) is $\partial_\mu \hat{W}^\mu$. In consequence, the obvious generalizations of (7.108) or (13.110) cannot be used to define the physical (transverse) states. The reason is that a condition like (13.110) must hold for all times, and only if the field is free is its time variation known (and essentially trivial).

Let us press ahead nevertheless, and assume that the rules we have derived so far are the correct Feynman rules for this gauge theory. We will see that this leads to physically unacceptable consequences, namely to the *violation of unitarity*.

In fact, this is a problem which threatens all gauge theories if the gauge field is treated covariantly, i.e. as a 4-vector. As we saw in section 7.3.2, this introduces *unphysical degrees of freedom* which must somehow be eliminated from the theory, or at least prevented from affecting physical processes. In QED we do this by imposing the condition (7.111), or (13.110), but as we have seen the analogous conditions will not work in the non-Abelian case, and so unphysical states may make their presence felt, for example in the 'sum over intermediate states' which arises in the unitarity relation. This relation determines the imaginary part of an amplitude via an equation of the form (cf (11.63))

$$2\, \mathrm{Im}\, \langle \mathrm{f} \mid \mathcal{M} \mid \mathrm{i} \rangle = \int \sum_n \langle \mathrm{f} \mid \mathcal{M} \mid \mathrm{n} \rangle \langle \mathrm{n} \mid \mathcal{M}^\dagger \mid \mathrm{i} \rangle \mathrm{d}\rho_n \qquad (13.116)$$

where $\langle \mathrm{f} \mid \mathcal{M} \mid \mathrm{i} \rangle$ is the (Feynman) amplitude for the process $\mathrm{i} \to \mathrm{f}$, and the sum is over a complete set of physical intermediate states $\mid \mathrm{n} \rangle$, which can enter at the given energy; $\mathrm{d}\rho_n$ represents the phase space element for the general intermediate state $\mid \mathrm{n} \rangle$. Consider now the possibility of gauge quanta appearing in the states $\mid \mathrm{n} \rangle$. Since unitarity deals only with physical states, such quanta can have only the two degrees of freedom (polarizations) allowed for a physical massless gauge field (cf section 7.3.1). Now part of the power of the 'Feynman rules' approach to perturbation theory is that it is manifestly covariant. But there is no completely covariant way of selecting out just the two physical components of a massless polarization vector ϵ_μ, from the four originally introduced precisely for reasons of covariance. In fact, when gauge quanta appear as virtual particles in *intermediate* states in Feynman graphs, they will not be restricted to having only two polarization states (as we shall see explicitly in a moment). Hence there is a real chance that when the imaginary part of such graphs is calculated, a contribution from the unphysical polarization states will be found, which has no counterpart at all in the physical unitarity relation, so that unitarity will not be satisfied. Since unitarity is an expression of conservation of probability, its violation is a serious disease indeed.

Consider, for example, the process $q\bar{q} \to q\bar{q}$ (where the 'quarks' are an SU(2) doublet), whose imaginary part has a contribution from a state containing two gluons (figure 13.9):

$$2\, \mathrm{Im}\, \langle q\bar{q} \mid \mathcal{M} \mid q\bar{q} \rangle = \int \sum \langle q\bar{q} \mid \mathcal{M} \mid gg \rangle \langle gg \mid \mathcal{M}^\dagger \mid q\bar{q} \rangle \mathrm{d}\rho_2 \qquad (13.117)$$

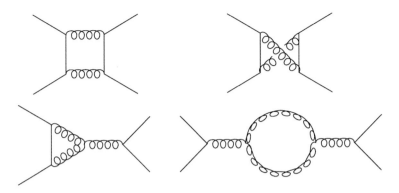

FIGURE 13.10
Some $O(g^4)$ contributions to $q\bar{q} \to q\bar{q}$.

where $d\rho_2$ is the 2-body phase space for the g-g state. The 2-gluon amplitudes in (13.117) must have the form

$$\mathcal{M}_{\mu_1\nu_1}\epsilon_1^{\mu_1}(k_1,\lambda_1)\epsilon_2^{\nu_1}(k_2,\lambda_2) \qquad (13.118)$$

where $\epsilon^\mu(k,\lambda)$ is the polarization vector for the gluon with polarization λ and 4-momentum k. The sum in (13.117) is then to be performed over $\lambda_1 = 1, 2$ and $\lambda_2 = 1, 2$ which are the physical polarization states (cf section 7.3.1). Thus (13.117) becomes

$$2\,\text{Im}\,\mathcal{M}_{q\bar{q}\to q\bar{q}} = \int \sum_{\lambda_1=1,2;\lambda_2=1,2} \mathcal{M}_{\mu_1\nu_1}\epsilon_1^{\mu_1}(k_1,\lambda_1)\epsilon_2^{\nu_1}(k_2,\lambda_2)$$
$$\times \mathcal{M}^*_{\mu_2\nu_2}\epsilon_1^{\mu_2}(k_1,\lambda_1)\epsilon_2^{\nu_2}(k_2,\lambda_2)\mathrm{d}\rho_2. \qquad (13.119)$$

For later convenience we are using real polarization vectors as in (7.81) and (7.82): $\epsilon(k_i,\lambda_i = +1) = (0,1,0,0), \epsilon(k_i,\lambda_i = -1) = (0,0,1,0)$; and of course $k_1^2 = k_2^2 = 0$.

We now wish to find out whether or not a result of the form (13.119) will hold when the $\mathcal{M}$'s represent some suitable Feynman graphs. We first note that we want the unitarity relation (13.119) to be satisfied order by order in perturbation theory: that is to say, when the $\mathcal{M}$'s on both sides are expanded in powers of the coupling strengths (as in the usual Feynman graph expansion), the coefficients of corresponding powers on each side should be equal. Since each emission or absorption of a gluon produces one power of the SU(2) coupling g, the right-hand side of (13.119) involves at least the power g^4. Thus the lowest-order process in which (13.119) may be tested is for the fourth-order amplitude $\mathcal{M}^{(4)}_{q\bar{q}\to q\bar{q}}$. There are quite a number of contributions to $\mathcal{M}^{(4)}_{q\bar{q}\to q\bar{q}}$, some of which are shown in figure 13.10; all contain a loop. On the right-hand side of (13.119), each $\mathcal{M}$ involves two polarization vectors, and so each must represent the $0(g^2)$ contribution to $q\bar{q} \to gg$, which we call $\mathcal{M}^{(2)}_{\mu\nu}$; thus both sides are consistently of order g^4. There are three contributions to $\mathcal{M}^{(2)}_{\mu\nu}$ shown in figure 13.11; when these are placed in (13.119), contributions to the imaginary part of $\mathcal{M}^{(4)}_{q\bar{q}\to q\bar{q}}$ are generated, which should agree with the imaginary part of the total $0(g^4)$ loop-graph contribution. Let us see if this works out. We choose to work in the gauge $\xi = 1$, so that the gluon propagator takes the familiar form $-ig^{\mu\nu}\delta_{ij}/k^2$. According to the rules for propagators and vertices already given, each of the loop amplitudes $\mathcal{M}^{(4)}_{q\bar{q}\to q\bar{q}}$ (e.g. those of figure 13.10) will be proportional to the product of the propagators for the quarks and the gluons, together with appropriate 'γ' and 'τ' vertex factors, the whole being integrated

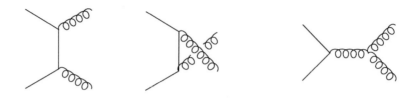

FIGURE 13.11
$O(g^2)$ contributions to $q\bar{q} \to gg$.

over the loop momentum. The extraction of the imaginary part of a Feynman diagram is a technical matter, having to do with careful consideration of the 'iϵ' in the propagators. Rules for doing this exist (Eden *et al.* 1966, section 2.9), and in the present case the result is that, to compute the imaginary part of the amplitudes of figure 13.10, one replaces each gluon propagator of momentum k by

$$\pi(-g^{\mu\nu})\delta(k^2)\theta(k_0)\delta_{ij}. \tag{13.120}$$

That is, the propagator is replaced by a condition stating that, in evaluating the imaginary part of the diagram, the gluon's mass is constrained to have the physical (free-field) value of zero, instead of varying freely as the loop momentum varies, and its energy is positive. These conditions (one for each gluon) have the effect of converting the loop integral with a standard two-body phase space integral for the gg intermediate state, so that eventually

$$2\text{Im } \mathcal{M}^{(4)}_{q\bar{q}\to q\bar{q}} = \int \mathcal{M}^{(2)}_{\mu_1\nu_1}(-g^{\mu_1\mu_2})\mathcal{M}^{(2)}_{\mu_2\nu_2}(-g^{\nu_1\nu_2})d\rho_2 \tag{13.121}$$

where $\mathcal{M}^{(2)}_{\mu_1\nu_1}$ is the sum of the three $O(g^2)$ tree graphs shown in figure 13.11, with all external legs satisfying the 'mass-shell' conditions.

So, the imaginary part of the loop contribution to $\mathcal{M}^{(4)}_{q\bar{q}\to q\bar{q}}$ does seem to have the form (13.116) as required by unitarity, with $|n\rangle$ the gg intermediate state as in (13.119). But there is one essential difference between (13.121) and (13.119): the place of the factor $-g^{\mu\nu}$ in (13.121) is taken in (13.119) by the gluon polarization sum

$$P^{\mu\nu}(k) \equiv \sum_{\lambda=1,2} \epsilon^\mu(k,\lambda)\epsilon^\nu(k,\lambda) \tag{13.122}$$

for $k = k_1, k_2$ and $\lambda = \lambda_1, \lambda_2$, respectively. Thus we have to investigate whether this difference matters.

To proceed further, it is helpful to have an explicit expression for $P^{\mu\nu}$. We might think of calculating the necessary sum over λ by brute force, using two ϵ's specified by the conditions (cf (7.87))

$$\epsilon^\mu(k,\lambda)\epsilon_\mu(k,\lambda') = -\delta_{\lambda\lambda'}, \qquad \epsilon \cdot k = 0. \tag{13.123}$$

The trouble is that conditions (13.123) *do not fix the ϵ's uniquely if $k^2 = 0$*. (Note the $\delta(k^2)$ in (13.120)). Indeed, it is precisely the fact that any given ϵ_μ satisfying (13.123) can be replaced by $\epsilon_\mu + \lambda k_\mu$ that both reduces the degrees of freedom to two (as we saw in section 7.3.1), and evinces the essential arbitrariness in the ϵ_μ specified only by (13.123). In order to calculate (13.122), we need to put another condition on ϵ_μ, so as to fix it uniquely. A standard choice (see e.g. Taylor 1976, pp 14–15) is to supplement (13.123) with the further condition

$$t \cdot \epsilon = 0 \tag{13.124}$$

where t is some 4-vector. This certainly fixes ϵ_μ, and enables us to calculate (13.122), but of course now two further difficulties have appeared, namely, the physical results seem to depend on t_μ; and have we not lost Lorentz covariance, because the theory involves a special 4-vector t_μ?

Setting these questions aside for the moment, we can calculate (13.122) using the conditions (13.123) and (13.124), finding (problem 13.11)

$$P_{\mu\nu} = -g_{\mu\nu} - [t^2 k_\mu k_\nu - k \cdot t(k_\mu t_\nu + k_\nu t_\mu)]/(k \cdot t)^2. \tag{13.125}$$

But only the *first* term on the right-hand side of (13.125) is to be seen in (13.121). A crucial quantity is clearly

$$
\begin{aligned}
U_{\mu\nu}(k,t) &\equiv -g_{\mu\nu} - P_{\mu\nu} \\
&= [t^2 k_\mu k_\nu - k \cdot t(k_\mu t_\nu + k_\nu t_\mu)]/(k \cdot t)^2.
\end{aligned}
\tag{13.126}
$$

We note that whereas

$$k^\mu P_{\mu\nu} = k^\nu P_{\mu\nu} = 0 \tag{13.127}$$

(from the condition $k \cdot \epsilon = 0$), the same is *not* true of $k^\mu U_{\mu\nu}$ - in fact,

$$k^\mu U_{\mu\nu} = -k_\nu \tag{13.128}$$

where we have used $k^2 = 0$. It follows that $U_{\mu\nu}$ may be regarded as including polarization states for which $\epsilon \cdot k \neq 0$. In physical terms, therefore, a gluon appearing internally in a Feynman graph has to be regarded as existing in more than just the two polarization states available to an external gluon (cf section 7.3.1). $U_{\mu\nu}$ characterizes the contribution of these unphysical polarization states.

The discrepancy between (13.121) and (13.119) is then

$$2\mathrm{Im}\,\mathcal{M}^{(4)}_{q\bar{q}\to q\bar{q}} = \int \mathcal{M}^{(2)}_{\mu_1\nu_1}[U^{\mu_1\mu_2}(k_1,t_1)]\mathcal{M}^{(2)}_{\mu_2\nu_2}[U^{\nu_1\nu_2}(k_2,t_2)]\mathrm{d}\rho_2, \tag{13.129}$$

together with similar terms involving one P and one U. It follows that these unwanted contributions will, in fact, vanish if

$$k_1^{\mu_1}\mathcal{M}^{(2)}_{\mu_1\nu_1} = 0, \tag{13.130}$$

and similarly for k_2. This will also ensure that amplitudes are independent of t_μ.

Condition (13.130) is apparently the same as the U(1) gauge invariance requirement of (8.165), already recalled in the previous section. As discussed there, it can be interpreted here also as expressing gauge invariance in the non-Abelian case, working to this given order in perturbation theory. Indeed, the diagrams of figure 13.11 are essentially 'crossed' versions of those in figure 13.5. However, there is one crucial difference here. In figure 13.5, both the X's were physical, their polarizations satisfying the condition $\epsilon \cdot k = 0$. In figure 13.11, by contrast, neither of the gluons, in the discrepant contribution (13.129), satisfies $\epsilon \cdot k = 0$—see the sentence following (13.128). Thus the crucial point is that (13.130) must be true for each gluon, *even when the other gluon has $\epsilon \cdot k \neq 0$*. And, in fact, we shall now see that whereas the (crossed) version of (13.130) did hold for our $\mathrm{dX} \to \mathrm{dX}$ amplitudes of section 13.3.2, (13.130) *fails* for states with $\epsilon \cdot k \neq 0$.

The three graphs of figure 13.11 together yield

$$\mathcal{M}^{(2)}_{\mu_1\nu_1}\epsilon_1^\mu(k_1,\lambda_1)\epsilon_2^\nu(k_2,\lambda_2) = g^2\bar{v}(p_2)\frac{\tau_j}{2}\,\slashed{\epsilon}_2 a_{2j}\frac{1}{\slashed{p}_1-\slashed{k}_1-m}\frac{\tau_i}{2}a_{1i}\,\slashed{\epsilon}_1 u(p_1)$$

$$+ \quad g^2\bar{v}(p_2)\frac{\tau_i}{2}a_{1i}\,\slashed{\epsilon}_1\frac{1}{\slashed{p}_1-\slashed{k}_2-m}\frac{\tau_j}{2}a_{2j}\,\slashed{\epsilon}_2 u(p_1)$$

$$+ \quad (-\mathrm{i})g^2\epsilon_{kij}[(p_1 + p_2 + k_1)^{\nu_1}g^{\mu_1\rho} + (-k_2 - p_1 - p_2)^{\mu_1}g^{\rho\nu_1}$$

$$+ \quad (-k_1 + k_2)^\rho g^{\mu_1\nu_1}]\epsilon_{1\mu_1}a_{1i}a_{2j}\epsilon_{2\nu_1}\frac{-1}{(p_1 + p_2)^2}\bar{v}(p_2)\frac{\tau_k}{2}\gamma_\rho u(p_1) \tag{13.131}$$

where we have written the gluon polarization vectors as a product of a Lorentz 4-vector ϵ_μ and an 'SU(2) polarization vector' a_i to specify the triplet state label. Now replace ϵ_1, say, by k_1. Using the Dirac equation for $u(p_1)$ and $\bar{v}(p_2)$ the first two terms reduce to (cf (13.94))

$$g^2\bar{v}(p_2)\; \not{\epsilon}_2[\tau_i/2, \tau_j/2]u(p_1)a_{1i}a_{2j}$$

$$= \quad \mathrm{i}g^2\bar{v}(p_2)\; \not{\epsilon}_2\epsilon_{ijk}(\tau_k/2)u(p_1)a_{1i}a_{2j} \tag{13.132}$$

using the SU(2) algebra of the τ's. The third term in (13.131) gives

$$-\mathrm{i}g^2\epsilon_{ijk}\bar{v}(p_2)\; \not{\epsilon}_2(\tau_k/2)u(p_1)a_{1i}a_{2j} \tag{13.133}$$

$$+\mathrm{i}g^2\frac{\epsilon_{ijk}}{2k_1 \cdot k_2}\bar{v}(p_2)\; \not{k}_1(\tau_k/2)u(p_1)k_2 \cdot \epsilon_2 a_{1i}a_{2j}. \tag{13.134}$$

We see that the first part (13.133) certainly does cancel (13.132), but there remains the second piece (13.134), *which only vanishes if $k_2 \cdot \epsilon_2 = 0$*. This is not sufficient to guarantee the absence of all unphysical contributions to the imaginary part of the 2-gluon graphs, as the preceding discussion shows. *We conclude that loop diagrams involving two (or, in fact, more) gluons, if constructed according to the simple rules for tree diagrams, will violate unitarity.*

The correct rule for such loops must be as to satisfy unitarity. Since there seems no other way in which the offending piece in (13.134) can be removed, we must infer that the rule for loops will have to involve some extra term, or terms, over and above the simple tree-type constructions, which will cancel the contributions of unphysical polarization states. To get an intuitive idea of what such extra terms might be, we return to expression (13.126) for the sum over unphysical polarization states $U_{\mu\nu}$, and make a specific choice for t. We take $t_\mu = \bar{k}_\mu$, where the 4-vector $\bar{k}$ is defined by $\bar{k} = (-\mid \boldsymbol{k} \mid, \boldsymbol{k})$, and $\boldsymbol{k} = (0, 0, \mid \boldsymbol{k} \mid)$. This choice obviously satisfies (13.124). Then

$$U_{\mu\nu}(k, \bar{k}) = (k_\mu\bar{k}_\nu + k_\nu\bar{k}_\mu)/(2 \mid \boldsymbol{k} \mid^2) \tag{13.135}$$

and unitarity (cf (13.129)) requires

$$\int \mathcal{M}^{(2)}_{\mu_1\nu_1}\mathcal{M}^{(2)}_{\mu_2\nu_2}\frac{(k_1^{\mu_1}\bar{k}_1^{\mu_2} + k_1^{\mu_2}\bar{k}_1^{\mu_1})}{2 \mid \boldsymbol{k}_1 \mid^2}\frac{(k_2^{\nu_1}\bar{k}_2^{\nu_2} + k_2^{\nu_2}\bar{k}_2^{\nu_1})}{2 \mid \boldsymbol{k}_2 \mid^2}\mathrm{d}\rho_2 \tag{13.136}$$

to vanish, but it does not. Let us work in the centre of momentum (CM) frame of the two gluons, with $k_1 = (\mid \boldsymbol{k} \mid, 0, 0, \mid \boldsymbol{k} \mid), k_2 = (\mid \boldsymbol{k} \mid, 0, 0, - \mid \boldsymbol{k} \mid), \bar{k}_1 = (- \mid \boldsymbol{k} \mid, 0, 0, \mid \boldsymbol{k} \mid)$, $\bar{k}_2 = (- \mid \boldsymbol{k} \mid, 0, 0, - \mid \boldsymbol{k} \mid)$, and consider for definiteness the contractions with the $\mathcal{M}^{(2)}_{\mu_1\nu_1}$ term. These are $\mathcal{M}^{(2)}_{\mu_1\nu_1}k_1^{\mu_1}k_2^{\nu_1}, \mathcal{M}^{(2)}_{\mu_1\nu_1}k_1^{\mu_1}\bar{k}_2^{\nu_1}$ etc. Such quantities can be calculated from expression (13.131) by setting $\epsilon_1 = k_1, \epsilon_2 = k_2$ for the first, $\epsilon_1 = k_1, \epsilon_2 = \bar{k}_2$ for the second, and so on. We have already obtained the result of putting $\epsilon_1 = k_1$. From (13.134) it is clear that a term in which ϵ_2 is replaced by k_2 as well as ϵ_1 by k_1 will vanish, since $k_2^2 = 0$. A typical non-vanishing term is of the form $\mathcal{M}^{(2)}_{\mu_1\nu_1}k_1^{\mu_1}\bar{k}_2^{\nu_1}/2 \mid \boldsymbol{k} \mid^2$. From (13.134) this reduces to

$$-\mathrm{i}g^2\frac{\epsilon_{ijk}}{2k_1 \cdot k_2}\bar{v}(p_2)\; \not{k}_1(\tau_k/2)u(p_1)a_{1i}a_{2j} \tag{13.137}$$

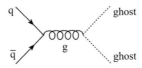

FIGURE 13.12

Tree graph interpretation of the expression (13.138).

using $k_2 \cdot \bar{k}_2 / 2 \mid \mathbf{k} \mid^2 = -1$. We may rewrite (13.137) as

$$j_{\mu k} \frac{-g^{\mu\nu}\delta_{k\ell}}{(k_1 + k_2)^2} ig\epsilon_{ij\ell}a_{1i}a_{2j}k_{1\nu} \tag{13.138}$$

where

$$j_{\mu k} = g\bar{v}(p_2)\gamma_\mu(\tau_k/2)u(p_1) \tag{13.139}$$

is the SU(2) current associated with the q$\bar{\mathrm{q}}$ pair.

The unwanted terms of the form (13.138) can be eliminated of we adopt the following rule (on the grounds of 'forcing the theory to make sense'). In addition to the fourth-order diagrams of the type shown in figure 13.10, constructed according to the simple 'tree' prescriptions, there must exist a previously unknown fourth-order contribution, *only present in loops*, such that it has an imaginary part which is non-zero in the same physical region as the two-gluon intermediate state, and moreover is of just the right magnitude to cancel all the contributions to (13.136) from terms like (13.138). Now (13.138) has the appearance of a one-gluon intermediate state amplitude. The q$\bar{\mathrm{q}} \to$ g vertex is represented by the current (13.139), the gluon propagator appears in Feynman gauge $\xi = 1$, and the rest of the expression would have the interpretation of a coupling between the intermediate gluon and two scalar particles with SU(2) polarizations a_{1i}, a_{2j}. Thus (13.138) can be interpreted as the amplitude for the tree graph shown in figure 13.12, where the dotted lines represent the scalar particles. It seems plausible, therefore, that the fourth-order graph we are looking for has the form shown in figure 13.13. The new scalar particles must be massless, so that this new amplitude has an imaginary part in the same physical region as the gg state. When the imaginary part of figure 13.13 is calculated in the usual way, it will involve contributions from the tree graph of figure 13.12, and these can be arranged to cancel the unphysical polarization pieces like (13.138).

For this cancellation to work, the scalar particle loop graph of figure 13.13 must enter with the opposite sign from the 3-gluon loop graph of figure 13.10, which in retrospect was the cause of all the trouble. Such a relative minus sign between single closed loop graphs would be expected if the scalar particles in figure 13.13 were in fact fermions! (Recall the rule given in section 11.3 and problem 11.2). Thus we appear to need *scalar* particles obeying

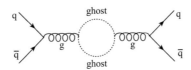

FIGURE 13.13

Ghost loop diagram contributing in fourth order to q$\bar{\mathrm{q}} \to$ q$\bar{\mathrm{q}}$.

Fermi statistics. Such particles are called 'ghosts'. We must emphasize that although we have introduced the tree graph of figure 13.12, which apparently involves ghosts as external lines, in reality the ghosts are always confined to loops, their function being to cancel unphysical contributions from intermediate gluons.

The preceding discussion has, of course, been entirely heuristic. It can be followed through so as to yield the correct prescription for eliminating unphysical contributions from a single closed gluon loop. But, as Feynman recognized (1963, 1977), unitarity alone is not a sufficient constraint to provide the prescription for more than one closed gluon loop. Clearly what is required is some additional term in the Lagrangian, which will do the job in general. Such a term indeed exists, and was first derived using the path integral form of quantum field theory (see chapter 16) by Faddeev and Popov (1967). The result is that the covariant gauge-fixing term (13.68) must be supplemented by the 'ghost Lagrangian'

$$\hat{\mathcal{L}}_g = \partial_\mu \hat{\eta}_i^\dagger \hat{D}_{ij}^\mu \hat{\eta}_j \tag{13.140}$$

where the η field is an SU(2) triplet, and spinless, but obeying *anti*-commutation relations; the covariant derivative is the one appropriate for an SU(2) triplet, namely (from (13.44) and (12.48))

$$\hat{D}_{ij}^\mu = \partial^\mu \delta_{ij} + g\epsilon_{kij} \hat{W}_k^\mu, \tag{13.141}$$

in this case. The result (13.140) is derived in standard books of quantum field theory, for example Cheng and Li (1984), Peskin and Schroeder (1995) or Ryder (1996). We should add the caution that the form of the ghost Lagrangian depends on the choice of the gauge-fixing term; there are gauges in which the ghosts are absent. Feynman rules for non-Abelian gauge field theories are given in Cheng and Li (1984), for example. We give the rules for tree diagrams, for which there are no problems with ghosts, in Appendix Q.

Problems

13.1 Verify that (13.34) reduces to (13.26) in the infinitesimal case.

13.2 Verify equation (13.45).

13.3 Using the expression for D^μ in (13.47), verify (13.48).

13.4 Verify the transformation law (13.51) of $\boldsymbol{F}^{\mu\nu}$ under local SU(2) transformations.

13.5 Verify that $\boldsymbol{F}_{\mu\nu} \cdot \boldsymbol{F}^{\mu\nu}$ is invariant under local SU(2) transformations.

13.6 Verify that the (infinitesimal) transformation law (13.56) for the SU(3) gauge field A_a^μ is consistent with (13.55).

13.7 By considering the commutator of two D^μ's of the form (13.53), verify (13.61).

13.8 Verify that (13.84) reduces to (13.86) (omitting the $(2\pi)^4 \delta^4$ factors).

13.9 Verify that the replacement of ϵ_1 by k_1 in (13.93) leads to (13.94).

13.10 Verify that when ϵ_1 is replaced by k_1 in (13.95), the resulting amplitude cancels the contribution (13.94), provided that $\delta = 1$.

13.11 Show that $P^{\mu\nu}$ of (13.122), with the ϵ's specified by the conditions (13.123) and (13.124), is given by (13.125).

Part VI

QCD and the Renormalization Group

14

QCD I: Introduction, Tree Graph Predictions, and Jets

In the previous chapter we have introduced the elementary concepts and formalism associated with non-Abelian quantum gauge field theories. It is now well established that the strong interactions between quarks are described by a theory of this type, in which the gauge group is an $SU(3)_c$, acting on a degree of freedom called 'colour' (indicated by the subscript c). This theory is called Quantum Chromodynamics, or QCD for short. QCD will be our first application of the theory developed in chapter 13, and we shall devote the next two chapters, and much of chapter 16, to it.

In the present chapter we introduce QCD and discuss some of its simpler experimental consequences. We briefly recall the evidence for the 'colour' degree of freedom in section 14.1, and then proceed to the dynamics of colour, and the QCD Lagrangian, in section 14.2. Perhaps the most remarkable thing about the dynamics of QCD is that, despite its being a theory of the *strong* interactions, there are certain kinematic regimes—roughly speaking, short distances or high energies—in which it is effectively a quite *weakly* interacting theory. This is a consequence of a fundamental property, possessed only by non-Abelian gauge theories, whereby the effective interaction strength becomes progressively smaller in such regimes. This property is called 'asymptotic freedom', and was already mentioned in section 11.5.3 of volume 1. In appropriate cases, therefore, the lowest-order perturbation theory amplitudes (tree graphs) provide a very convincing qualitative, or even 'semi-quantitative', orientation to the data. In sections 14.3 and 14.4 we shall see how the tree graph techniques acquired for QED in volume 1 produce more useful physics when applied to QCD.

However, most of the quantitative experimental support for QCD has come from comparison with predictions which include higher order QCD corrections; indeed, the asymptotic freedom property itself emerges from summing a whole class of higher-order contributions, as we shall indicate at the beginning of chapter 15. This immediately involves all the apparatus of *renormalization*. The necessary calculations quite rapidly become too technical for the intended scope of this book, but in chapter 15 we shall try to provide an elementary introduction to the issues involved, and to the necessary techniques, by building on the discussion of renormalization given in chapters 10 and 11 of volume 1. The main new concept will be the *renormalization group* (and related ideas), which is an essential tool in the modern confrontation of perturbative QCD with data. Some of the simpler predictions of the renormalization group technique will be compared with experimental data in the last part of chapter 15.

In chapter 16 we work towards understanding some non-perturbative aspects of QCD. As a natural concomitant of asymptotic freedom, it is to be expected that the effective coupling strength becomes progressively larger at longer distances or lower energies, ultimately being strong enough to lead (presumably) to the confinement of quarks and gluons; this is sometimes referred to as 'infrared slavery'. In this regime perturbation theory clearly fails. An alternative, purely numerical, approach is available however, namely the method of 'lattice' QCD, which involves replacing the space-time continuum by a *discrete lattice* of points. At first sight, this may seem a topic rather disconnected from everything that has preceded

DOI: 10.1201/9781003411666-14

it. But we shall see that in fact it provides some powerful new insights into several aspects of quantum field theory in general, and in particular of renormalization, by revisiting it in coordinate (rather than momentum) space. Quite apart from this, however, results from lattice QCD now provide independent confirmation of the theory, in the non-perturbative regime.

14.1   The colour degree of freedom

The first intimation of a new, unrevealed degree of freedom of matter came from baryon spectroscopy (Greenberg 1964; see also Han and Nambu 1965, and Tavkhelidze 1965). For a baryon made of three spin-$\frac{1}{2}$ quarks, the original non-relativistic quark model wave-function took the form

$$\psi_{3\mathrm{q}} = \psi_{3\mathrm{q,space}}\psi_{3\mathrm{q,spin}}\psi_{3\mathrm{q,flavour}}. \tag{14.1}$$

It was soon realized (e.g. Dalitz 1965) that the product of these space, spin, and flavour wavefunctions for the ground state baryons was *symmetric* under interchange of any two quarks. For example, the Δ^{++} state mentioned in section 12.2.3 is made of three u quarks (flavour symmetric) in the $J^P = \frac{3}{2}^+$ state, which has zero orbital angular momentum and is hence spatially symmetric, and a symmetric $S = \frac{3}{2}$ spin wavefunction. But we saw in section 7.2 that quantum field theory requires fermions to obey the exclusion principle—i.e. the wavefunction $\psi_{3\mathrm{q}}$ should be *anti*-symmetric with respect to quark interchange. A simple way of implementing this requirement is to suppose that the quarks carry a further degree of freedom, called colour, with respect to which the 3q wavefunction can be antisymmetrized, as follows (Fritzsch and Gell-Mann 1972, Bardeen, Fritzsch and Gell-Mann 1973). We introduce (Gell-Mann 1972) a *colour wavefunction* with colour index α:

$$\psi_\alpha \quad (\alpha = 1, 2, 3).$$

We are here writing the three labels as '1, 2, 3', but they are often referred to by colour names such as 'red, blue, green'; it should be understood that this is merely a picturesque way of referring to the three basic states of this degree of freedom, and has nothing to do with real colour! With the addition of this degree of freedom we can certainly form a three-quark wavefunction which is antisymmetric in colour by using the antisymmetric symbol $\epsilon_{\alpha\beta\gamma}$, namely[1]

$$\psi_{3\mathrm{q, \,colour}} = \epsilon_{\alpha\beta\gamma}\psi_\alpha\psi_\beta\psi_\gamma \tag{14.2}$$

and this must then be multiplied into (14.1) to give the full 3q wavefunction. To date, *all* known baryon states can be described this way, i.e. the symmetry of the 'traditional' space-spin-flavour wavefunction (14.1) is symmetric overall, while the required antisymmetry is restored by the additional factor (14.2). As far as meson ($\bar{\mathrm{q}}\mathrm{q}$) states are concerned, what was previously a π^+ wavefunction d^*u is now

$$\frac{1}{\sqrt{3}}(d_1^* u_1 + d_2^* u_2 + d_3^* u_3) \tag{14.3}$$

which we write in general as $(1/\sqrt{3})d_\alpha^\dagger u_\alpha$. We shall shortly see the group theoretical significance of this 'neutral superposition', and of (14.2). Meanwhile, we note that (14.2) is actually the *only* way of making an antisymmetric combination of the three ψ's; it is therefore called

---

[1] In (14.2) each ψ refers to a different quark, but we have not indicated the quark labels explicitly.

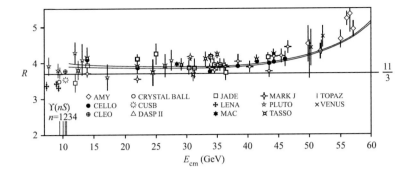

FIGURE 14.1

The ratio R (see (14.4)). Figure reprinted with permission from L. Montanet *et al. Physical Review* D **50** 1173 (1994). Copyright 1994 by the American Physical Society.

a (colour) *singlet*. It is reassuring that there is only one way of doing this—otherwise, we would have obtained more baryon states than are physically observed. As we shall see in section 14.2.1, (14.3) is also a colour singlet combination.

The above would seem a somewhat artificial device unless there were some physical consequences of this increase in the number of quark types—and there are. In any process which we can describe in terms of creation or annihilation of quarks, the *multiplicity* of quark types will enter into the relevant observable cross section or decay rate. For example, at high energies the ratio

$$R = \frac{\sigma(e^+e^- \to \text{hadrons})}{\sigma(e^+e^- \to \mu^+\mu^-)} \tag{14.4}$$

will, in the quark parton model (see section 9.5), reflect the magnitudes of the individual quark couplings to the photon:

$$R = \sum_a e_a^2 \tag{14.5}$$

where a runs over all quark types. For five quarks u, d, s, c, b with respective charges $\frac{2}{3}, -\frac{1}{3}, -\frac{1}{3}, \frac{2}{3}, -\frac{1}{3}$, this yields

$$R_{\text{no colour}} = \frac{11}{9} \tag{14.6}$$

and

$$R_{\text{colour}} = \frac{11}{3} \tag{14.7}$$

for the two cases, as we saw in section 9.5. (The values $R = 2$ below the charm threshold, and $R = 10/3$ below the b threshold, were predicted by Bardeen *et al.* 1973). The data (figure 14.1, see also figure 9.16) rule out (14.6), and are in good agreement with (14.7) at energies well above the b threshold, and well below the Z^0 resonance peak. There is an indication that the data tend to lie *above* the parton model prediction; this is actually predicted by QCD via higher-order corrections, as will be discussed in section 15.1.

A number of branching fractions also provide simple ways of measuring the number of colours N_c. For example, consider the branching fraction for $\tau^- \to e^- \bar{\nu}_e \nu_\tau$ (i.e. the ratio of the rate for $\tau^- \to e^- \bar{\nu}_e \nu_\tau$ to that for all other decays). τ^- decays proceed via the weak process shown in figure 14.2, where the final fermions can be $e^- \bar{\nu}_e$, $\mu^- \bar{\nu}_\mu$, or $\bar{u}d$, the last with multiplicity N_c. Thus

$$B(\tau^- \to e^- \bar{\nu}_e \nu_\tau) \approx \frac{1}{2 + N_c}. \tag{14.8}$$

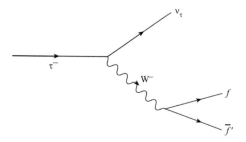

FIGURE 14.2

τ decay.

Experiments give $B \approx 18\%$ and hence $N_c \approx 3$.

Similarly, the branching fraction $B(W^- \to e^- \bar{\nu}_e)$ is $\sim \frac{1}{3+2N_c}$ (from $f = e, \mu, \tau, u,$ and c). Experiment gives a value of 10.7%, so again $N_c \approx 3$.

In chapter 9 we also discussed the Drell-Yan process in the quark parton model; it involves the subprocess $q\bar{q} \to l\bar{l}$ which is the inverse of the one in (14.4). We mentioned that a factor of $\frac{1}{3}$ appears in this case: it arises because we must average over the nine possible initial $q\bar{q}$ combinations (factor $\frac{1}{9}$) and then sum over the number of such states that lead to the colour neutral photon, which is 3 ($\bar{q}_1 q_1, \bar{q}_2 q_2,$ and $\bar{q}_3 q_3$). With this factor, and using quark distribution functions consistent with deep inelastic scattering, the parton model gives a good first approximation to the data.

Finally, we mention the rate for $\pi^0 \to \gamma\gamma$. As will be discussed in section 18.4, this process is entirely calculable from the graph shown in figure 14.3 (and the one with the γ's 'crossed'), where 'q' is u or d. The amplitude is proportional to the square of the quark charges, but because the π^0 is an isovector, the contributions from the $u\bar{u}$ and $d\bar{d}$ states have opposite signs (see section 12.1.3). Thus the rate contains a factor

$$((2/3)^2 - (1/3)^2)^2 = \frac{1}{9}. \tag{14.9}$$

However, the original calculation of this rate by Steinberger (1949) used a model in which the proton and neutron replaced the u and d in the loop, in which case the factor corresponding to (14.9) is just 1 (since the n has zero charge). Experimentally the rate agrees well with Steinberger's calculation, indicating that (14.9) needs to be multiplied by 9, which corresponds to $N_c = 3$ identical amplitudes of the form shown in figure 14.3, as was noted by Bardeen, Fritzsch, and Gell-Mann (1973).

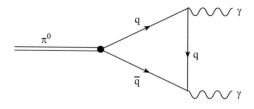

FIGURE 14.3

Triangle graph for π^0 decay.

14.2   The dynamics of colour

14.2.1   Colour as an SU(3) group

We now want to consider the possible dynamical role of colour—in other words, the way in which the forces between quarks depend on their colours. We have seen that we seem to need three different quark types for each given flavour. They must all have the same mass, or else we would observe some 'fine structure' in the hadronic levels. Furthermore, and for the same reason, 'colour' must be an exact symmetry of the Hamiltonian governing the quark dynamics. What symmetry group is involved? We shall consider how some empirical facts suggest that the answer is $SU(3)_c$.

To begin with, it is certainly clear that the interquark force must depend on colour, since we do *not* observe 'colour multiplicity' of hadronic states. For example, we do not see eight other coloured π^+'s ($d_1^* u_2, d_3^* u_1, \dots$) degenerate with the one 'colourless' physical π^+ whose wavefunction was given previously. The observed hadronic states are all *colour singlets*, and the force must somehow be responsible for this. More particularly, the force has to produce only those very restricted *types* of quark configuration which are observed in the hadron spectrum. Consider again the isospin multiplets in nuclear physics discussed in section 12.1.2. There is one very striking difference in the particle physics case: for mesons *only* $T = 0, \frac{1}{2}$, and 1 occur, and for baryons *only* $T = 0, \frac{1}{2}, 1$ and $\frac{3}{2}$, while in nuclei there is nothing in principle to stop us finding $T = \frac{5}{2}, 3, \dots$ states. (In fact such nuclear states are hard to identify experimentally, because they occur at high excitation energy for some of the isobars—cf figure 1.8(c)—where the levels are very dense). The same restriction holds for $SU(3)_f$ also—only **1**'s and **8**'s occur for mesons; and only **1**'s, **8**'s, and **10**'s for baryons. In quark terms, this of course is what is translated into the recipe: 'mesons are $\bar{q}q$, baryons are qqq'. It is as if we said, in nuclear physics, that only $A = 2$ and $A = 3$ nuclei exist! Thus the quark forces must have a dramatic saturation property: apparently no $\bar{q}qq$, no qqqq, qqqqq, $\dots$ states exist. Furthermore, no qq or $\bar{q}\bar{q}$ states exist either — nor, for that matter, do single q's or $\bar{q}$'s. All this can be summarized by saying that the quark colour degree of freedom must be *confined*, a property we shall now assume and return to in chapter 16.

If we assume that only colour singlet states exist (Gell-Mann 1972, Fritzsch and Gell-Mann 1972, Bardeen, Fritzsch and Gell-Mann 1973), and that the strong interquark force depends only on colour, the fact that $\bar{q}q$ states are seen but qq and $\bar{q}\bar{q}$ are not gives us an important clue as to what group to associate with colour. One simple possibility might be that the three colours correspond to the components of an $SU(2)_c$ triplet '$\boldsymbol{\psi}$'. The antisymmetric, colour singlet, three-quark baryon wavefunction of (14.2) is then just the triple scalar product $\boldsymbol{\psi}_1 \cdot \boldsymbol{\psi}_2 \times \boldsymbol{\psi}_3$, which seems satisfactory. But what about the meson wavefunction? Mesons are formed of quarks and antiquarks, and we recall from sections 12.1.3 and 12.2 that antiquarks belong to the complex conjugate of the representation (or multiplet) to which quarks belong. Thus if a quark colour triplet wavefunction ψ_α transforms under a colour transformation as

$$\psi_\alpha \rightarrow \psi_\alpha' = V_{\alpha\beta}^{(1)} \psi_\beta \tag{14.10}$$

where $\mathbf{V}^{(1)}$ is a 3×3 unitary matrix appropriate to the $T = 1$ representation of SU(2) (cf (12.48) and (12.49)), then the wavefunction for the 'anti'-triplet is ψ_α^*, which transforms as

$$\psi_\alpha^* \rightarrow \psi_\alpha^{*'} = V_{\alpha\beta}^{(1)*} \psi_\beta^*. \tag{14.11}$$

Given this information, we can now construct colour singlet wavefunctions for mesons, built from $\bar{q}q$. Consider the quantity (cf (14.3)) $\sum_\alpha \psi_\alpha^* \psi_\alpha$ where ψ^* represents the antiquark and

ψ the quark. This may be written in matrix notation as $\psi^\dagger \psi$ where the $\psi^\dagger$ as usual denotes the transpose of the complex conjugate of the column vector ψ. Then, taking the transpose of (14.11), we find that $\psi^\dagger$ transforms by

$$\psi^\dagger \to \psi^{\dagger\prime} = \psi^\dagger \mathbf{V}^{(1)\dagger} \tag{14.12}$$

so that the combination $\psi^\dagger \psi$ transforms as

$$\psi^\dagger \psi \to \psi^{\dagger\prime} \psi' = \psi^\dagger \mathbf{V}^{(1)\dagger} \mathbf{V}^{(1)} \psi = \psi^\dagger \psi \tag{14.13}$$

where the last step follows since $\mathbf{V}^{(1)}$ is unitary (compare (12.58)). Thus the product is *invariant* under (14.10) and (14.11)—that is, it is a colour singlet, as required. This is the meaning of the superposition (14.3).

All this may seem fine, but there is a problem. The three-dimensional representation of $SU(2)_c$ which we are using here has a very special nature: the matrix $\mathbf{V}^{(1)}$ can be chosen to be *real*. This can be understood 'physically' if we make use of the great similarity between $SU(2)$ and the group of rotations in three dimensions (which is the reason for the geometrical language of isospin 'rotations', and so on). We know very well how real three-dimensional vectors transform, namely by an orthogonal 3×3 matrix. It is the same in $SU(2)$. It is always possible to choose the wavefunctions ψ to be real, and the transformation matrix $\mathbf{V}^{(1)}$ to be real also. Since $\mathbf{V}^{(1)}$ is, in general, unitary, this means that it must be orthogonal. But now the basic difficulty appears: there is no distinction between ψ and ψ^*! They both transform by the real matrix $\mathbf{V}^{(1)}$. This means that we can make $SU(2)$ invariant (colour singlet) combinations for $\bar{q}q$ states, and for qq states, just as well as for $\bar{q}q$ states—indeed they are formally identical. But such 'diquark' (or 'antidiquark') states are not found, and hence—by assumption—should *not* be colour singlets.

The next simplest possibility seems to be that the three colours correspond to the components of an $SU(3)_c$ triplet. In this case the quark colour wavefunction ψ_α transforms as (cf (12.74))

$$\psi \to \psi' = \mathbf{W}\psi \tag{14.14}$$

where $\mathbf{W}$ is a special unitary 3×3 matrix parametrized as

$$\mathbf{W} = \exp(i\boldsymbol{\alpha} \cdot \boldsymbol{\lambda}/2), \tag{14.15}$$

and $\psi^\dagger$ transforms as

$$\psi^\dagger \to \psi^{\dagger\prime} = \psi^\dagger \mathbf{W}^\dagger. \tag{14.16}$$

The proof of the invariance of $\psi^\dagger \psi$ goes through as in (14.13), and it can be shown (problem 14.1(a)) that the antisymmetric 3q combination (14.2) is also an $SU(3)_c$ invariant. Thus both the proposed meson and baryon states are colour singlets. It is *not* possible to choose the $\boldsymbol{\lambda}$'s to be pure imaginary in (14.15), and thus the 3×3 $\mathbf{W}$ matrices of $SU(3)_c$ cannot be real, so that there is a distinction between ψ and ψ^*, as we learned in section 12.2. Indeed, it can be shown (Carruthers (1966) chapter 3, Jones (1990) chapter 8, and see also problem 14.1(b)) that, unlike the case of $SU(2)_c$ triplets, it is not possible to form an $SU(3)_c$ colour singlet combination out of two colour triplets qq or anti-triplets $\bar{q}\bar{q}$. Thus $SU(3)_c$ seems to be a possible and economical choice for the colour group.

14.2.2   Global $SU(3)_c$ invariance, and 'scalar gluons'

As stated above, we are assuming, on empirical grounds, that the only physically observed hadronic states are colour singlets—and this now means singlets under $SU(3)_c$. What sort of interquark force could produce this dramatic result? Consider an $SU(2)$ analogy again, the interaction of two nucleons belonging to the lowest (doublet) representation of SU(2). Labelling the states by an isospin T, the possible T values for two nucleons are $T = 1$ (triplet) and $T = 0$ (singlet). We know of an isospin-dependent force which can produce a splitting between these states, namely $V\boldsymbol{\tau}_1 \cdot \boldsymbol{\tau}_2$, where the '1' and '2' refer to the two nucleons. The total isospin is $\boldsymbol{T} = \frac{1}{2}(\boldsymbol{\tau}_1 + \boldsymbol{\tau}_2)$, and we have

$$\boldsymbol{T}^2 = \frac{1}{4}(\boldsymbol{\tau}_1^2 + 2\boldsymbol{\tau}_1 \cdot \boldsymbol{\tau}_2 + \boldsymbol{\tau}_2^2) = \frac{1}{4}(3 + 2\boldsymbol{\tau}_1 \cdot \boldsymbol{\tau}_2 + 3) \quad (14.17)$$

whence

$$\boldsymbol{\tau}_1 \cdot \boldsymbol{\tau}_2 = 2\boldsymbol{T}^2 - 3. \quad (14.18)$$

In the triplet state $\boldsymbol{T}^2 = 2$, and in the singlet state $\boldsymbol{T}^2 = 0$. Thus

$$\begin{aligned} (\boldsymbol{\tau}_1 \cdot \boldsymbol{\tau}_2)_{T=1} &= 1 & (14.19) \\ (\boldsymbol{\tau}_1 \cdot \boldsymbol{\tau}_2)_{T=0} &= -3 & (14.20) \end{aligned}$$

and if V is positive the $T = 0$ state is pulled down. A similar thing happens in SU(3)$_c$. Suppose this interquark force depended on the quark colours via a term proportional to

$$\boldsymbol{\lambda}_1 \cdot \boldsymbol{\lambda}_2. \quad (14.21)$$

Then, in just the same way, we can introduce the total colour operator

$$\boldsymbol{F} = \frac{1}{2}(\boldsymbol{\lambda}_1 + \boldsymbol{\lambda}_2), \quad (14.22)$$

so that

$$\boldsymbol{F}^2 = \frac{1}{4}(\boldsymbol{\lambda}_1^2 + 2\boldsymbol{\lambda}_1 \cdot \boldsymbol{\lambda}_2 + \boldsymbol{\lambda}_2^2) \quad (14.23)$$

and

$$\boldsymbol{\lambda}_1 \cdot \boldsymbol{\lambda}_2 = 2\boldsymbol{F}^2 - \boldsymbol{\lambda}^2, \quad (14.24)$$

where $\boldsymbol{\lambda}_1^2 = \boldsymbol{\lambda}_2^2 = \boldsymbol{\lambda}^2$, say. Here $\boldsymbol{\lambda}^2 \equiv \sum_{a=1}^{8}(\lambda_a)^2$ is found (see (12.75)) to have the value 16/3 (the unit matrix being understood). The operator $\boldsymbol{F}^2$ commutes with all components of $\boldsymbol{\lambda}_1$ and $\boldsymbol{\lambda}_2$ (as $\boldsymbol{T}^2$ does with $\boldsymbol{\tau}_1$ and $\boldsymbol{\tau}_2$) and represents the quadratic Casimir operator $\hat{C}_2$ of SU(3)$_c$ (see section M.5 of Appendix M), in the colour space of the two quarks considered here. The eigenvalues of $\hat{C}_2$ play a very important role in SU(3)$_c$, analogous to that of the total spin/angular momentum in SU(2). They depend on the SU(3)$_c$ representation. Indeed, they are one of the defining labels of SU(3) representations in general (see section M.5). Two quarks, each in the representation $\boldsymbol{3}_c$, combine to give a $\boldsymbol{6}_c$-dimensional representation and a $\boldsymbol{3}_c^*$ (see problem 14.1(b), and Jones (1990) chapter 8). The value of $\hat{C}_2$ for the singlet $\boldsymbol{6}_c$ representation is 10/3, and for the $\boldsymbol{3}_c^*$ representation is 4/3. Thus the '$\boldsymbol{\lambda}_1 \cdot \boldsymbol{\lambda}_2$' interaction will produce a negative (attractive) eigenvalue $-8/3$ in the $\boldsymbol{3}_c^*$ states, but a repulsive eigenvalue $+4/3$ in the $\boldsymbol{6}_c$ states, for two quarks.

The maximum attraction will clearly be for states in which $\boldsymbol{F}^2$ is zero. This is the singlet representation $\boldsymbol{1}_c$. Two quarks cannot combine to give a colour singlet state, but we have seen in section 12.2 that a quark and an antiquark can: they combine to give $\boldsymbol{1}_c$ and $\boldsymbol{8}_c$. In this case (14.24) is replaced by

$$\boldsymbol{\lambda}_1 \cdot \boldsymbol{\lambda}_2 = 2\boldsymbol{F}^2 - \frac{1}{2}(\boldsymbol{\lambda}_1^2 + \boldsymbol{\lambda}_2^2), \quad (14.25)$$

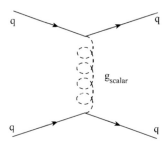

FIGURE 14.4
Scalar gluon exchange between two quarks.

where '1' refers to the quark and '2' to the antiquark. Thus the '$\boldsymbol{\lambda}_1 \cdot \boldsymbol{\lambda}_2$' interaction will give a repulsive eigenvalue $+2/3$ in the $\mathbf{8}_\mathrm{c}$ channel, for which $\hat{C}_2 = 3$, and a 'maximally attractive' eigenvalue $-16/3$ in the $\mathbf{1}_\mathrm{c}$ channel, for a quark and an antiquark.

In the case of baryons, built from three quarks, we have seen that when two of them are coupled to the $\mathbf{3}_\mathrm{c}^*$ state, the eigenvalue of $\boldsymbol{\lambda}_1 \cdot \boldsymbol{\lambda}_2$ is $-8/3$, one half of the attraction in the $\bar{q}q$ colour singlet state, but still strongly attractive. The (qq) pair in the $\mathbf{3}_\mathrm{c}^*$ state can then couple to the remaining third quark to make the overall colour singlet state (14.2), with maximum binding.

Of course, such a simple potential model does not imply that the energy difference between the $\mathbf{1}_\mathrm{c}$ states and all coloured states is *infinite*, as our strict 'colour singlets only' hypothesis would demand, and which would be one (rather crude) way of interpreting confinement. Nevertheless, we can ask: what single particle exchange process between quark (or antiquark) colour triplets produces a $\boldsymbol{\lambda}_1 \cdot \boldsymbol{\lambda}_2$ type of term? The answer is the exchange of an $\mathrm{SU}(3)_\mathrm{c}$ octet ($\mathbf{8}_\mathrm{c}$) of particles, which (anticipating somewhat) we shall call gluons. Since colour is an exact symmetry, the quark wave equation describing the colour interactions must be $\mathrm{SU}(3)_\mathrm{c}$ covariant. A simple such equation is

$$(\mathrm{i}\,\not{\partial} - m)\psi = g_\mathrm{s} \frac{\lambda_a}{2} A_a \psi \tag{14.26}$$

where g_s is a 'strong charge' and A_a ($a = 1, 2, \ldots, 8$) is an octet of *scalar* 'gluon potentials'. Equation (14.26) may be compared with (13.58). In the latter, $\boldsymbol{A}_a$ appears on the right-hand side, because the gauge field quanta are vectors rather than scalars. In (14.26), we are dealing at this stage only with a *global* $\mathrm{SU}(3)$ symmetry, not a local $\mathrm{SU}(3)$ gauge symmetry, and so the potentials may be taken to be scalars, for simplicity. As in (13.60), the vertex corresponding to (14.26) is

$$-\mathrm{i}g_\mathrm{s}\lambda_a/2. \tag{14.27}$$

(14.27) differs from (13.60) simply in the absence of the γ^μ factor, due to the assumed scalar, rather than vector, nature of the 'gluon' here. When we put two such vertices together and join them with a gluon propagator (figure 14.4), the $\mathrm{SU}(3)_\mathrm{c}$ structure of the amplitude will be

$$\frac{\lambda_{1a}}{2} \delta_{ab} \frac{\lambda_{2b}}{2} = \frac{\boldsymbol{\lambda}_1}{2} \cdot \frac{\boldsymbol{\lambda}_2}{2} \tag{14.28}$$

the δ_{ab} arising from the fact that the freely propagating gluon does not change its colour. This interaction has exactly the required '$\boldsymbol{\lambda}_1 \cdot \boldsymbol{\lambda}_2$' character in the colour space.

14.2.3 Local SU(3)$_c$ invariance: the QCD Lagrangian

It is tempting to suppose (Fritzsch and Gell-Mann 1972, Fritzsch, Gell-Mann and Leutwyler 1973) that the 'scalar gluons' introduced in (14.26) are, in fact, vector particles, like the photons of QED. Equation (14.26) then becomes

$$(\mathrm{i}\,\partial\!\!\!/ - m)\psi = g_\mathrm{s}\frac{\lambda_a}{2}\,A\!\!\!/_a\psi \tag{14.29}$$

as in (13.58), and the vertex (14.27) becomes

$$-\mathrm{i}g_\mathrm{s}\frac{\lambda_a}{2}\gamma^\mu \tag{14.30}$$

as in (13.60). One motivation for this is the desire to make the colour dynamics as much as possible like the highly successful theory of QED, and to derive the dynamics from a gauge principle. As we have seen in the last chapter, this involves the simple but deep step of supposing that the quark wave equation is covariant under *local* SU(3)$_c$ transformations of the form

$$\psi \to \psi' = \exp(\mathrm{i}g_\mathbf{s}\boldsymbol{\alpha}(x)\cdot\boldsymbol{\lambda}/2)\psi. \tag{14.31}$$

This is implemented by the replacement

$$\partial_\mu \to \partial_\mu + \mathrm{i}g_\mathrm{s}\frac{\lambda_a}{2}A_{a\mu}(x) \tag{14.32}$$

in the Dirac equation for the quarks, which leads immediately to (14.29) and the vertex (14.30).

Of course, the assumption of local SU(3)$_c$ covariance leads to a great deal more. For example, it implies that the gluons are *massless vector* (spin 1) particles, and that they interact with themselves via *three-gluon* and *four-gluon* vertices, which are the SU(3)$_c$ analogues of the SU(2) vertices discussed in section 13.3.2. The most compact way of summarizing all this structure is via the Lagrangian, most of which we have already introduced in chapter 13. Gathering together (13.71) and (13.140) (adapted to SU(3)$_c$), we write it out here for convenience:

$$\begin{aligned}\mathcal{L}_\mathrm{QCD} &= \sum_{\mathrm{flavours\,f}} \bar{\hat{q}}_{\mathrm{f},\alpha}(\mathrm{i}\hat{D}\!\!\!/ - m_\mathrm{f})_{\alpha\beta}\hat{q}_{\mathrm{f},\beta} - \frac{1}{4}\hat{F}_{a\mu\nu}\hat{F}_a^{\mu\nu}\\ &\quad -\frac{1}{2\xi}(\partial_\mu\hat{A}_a^\mu)(\partial_\nu\hat{A}_a^\nu) + \partial_\mu\hat{\eta}_a^\dagger\hat{D}_{ab}^\mu\hat{\eta}_b.\end{aligned} \tag{14.33}$$

In (14.33), repeated indices are as usual summed over: α and β are SU(3)$_c$-triplet indices running from 1 to 3, and a, b are SU(3)$_c$-octet indices running from 1 to 8. The covariant derivatives are defined by

$$(\hat{D}_\mu)_{\alpha\beta} = \partial_\mu\delta_{\alpha\beta} + \mathrm{i}g_\mathrm{s}\frac{1}{2}(\lambda_a)_{\alpha\beta}\hat{A}_{a\mu} \tag{14.34}$$

when acting on the quark SU(3)$_c$ triplet, as in (13.53), and by

$$(\hat{D}_\mu)_{ab} = \partial_\mu\delta_{ab} + g_s f_{cab}\hat{A}_{c\mu} \tag{14.35}$$

when acting on the octet of ghost fields. For the second of these, note that the matrices representing the SU(3) generators in the octet representation are as given in (12.84), and these take the place of the '$\lambda/2$' in (14.34) (compare (13.141) in the SU(2) case). We remind the reader that the last two terms in (14.33) are the gauge-fixing and ghost terms, respectively, appropriate to a gauge field propagator of the form (13.69) (with δ_{ij} replaced by δ_{ab} here). The Feynman rules following from (14.33) are given in Appendix Q.

As remarked in section 12.3.2, the fact that the QCD interactions (14.33) are 'flavour-blind' implies that the global flavour symmetries discussed in chapter 12 are all preserved by QCD. These include the conservation of each quark flavour (for example, the number of strange quarks minus the number of strange antiquarks is conserved); and the symmetries $SU(2)_f$ and $SU(3)_f$, and the chiral symmetries $SU(2)_{5f}$ and $SU(3)_{5f}$, to the extent that these latter are good symmetries. Further, (14.33) conserves the discrete symmetries **P**, **C**, and **T**, in a manner quite analogous to QED, already covered in section 7.5. In the case of **P** and **T**, the gluon fields $\hat{A}_{a\mu}$ have the same transformation properties as the photon field $\hat{A}_\mu$, and the (normally-ordered) $SU(3)_c$ currents $\hat{j}^\mu_{fa} = \bar{\hat{q}}_f \gamma^\mu \frac{1}{2}\lambda_a \hat{q}_f$ transform in the same way as the electromagnetic current $\bar{\hat{q}}\gamma^\mu \hat{q}$, ensuring **P** and **T** invariance. Under **C**, the quark fields transform as usual according to (7.151). Charge conjugation for the gluon field needs a little more care. The required rule is

$$\hat{\mathbf{C}}\lambda_a \hat{A}_{a\mu} \hat{\mathbf{C}}^{-1} = -\lambda_a^* \hat{A}_{a\mu}. \tag{14.36}$$

The overall minus sign in (14.36) is analogous to that for the photon field (cf (7.152)). To understand the complex conjugate on the right-hand side of (14.36), recall from (7.153) that the complex scalar field $\hat{\phi} = \frac{1}{\sqrt{2}}(\hat{\phi}_1 - i\hat{\phi}_2)$ transforms according to

$$\hat{\mathbf{C}}(\hat{\phi}_1 - i\hat{\phi}_2)\hat{\mathbf{C}}^{-1} = \hat{\phi}_1 + i\hat{\phi}_2. \tag{14.37}$$

Problem 14.2(a) verifies that the (normally ordered) interaction $\hat{j}^\mu_{fa}\hat{A}_{a\mu}$ is then **C**-invariant. As regards the term $\hat{F}_{a\mu\nu}\hat{F}^{\mu\nu}_a$, we can write it as

$$\frac{1}{2}\mathrm{Tr}(\lambda_a \hat{F}_{a\mu\nu}\lambda_b \hat{F}^{\mu\nu}_b) \tag{14.38}$$

using the relation

$$\mathrm{Tr}(\lambda_a \lambda_b) = 2\delta_{ab}. \tag{14.39}$$

A short calculation (problem 14.2(b)) shows that $\lambda_a \hat{F}_{a\mu\nu}$ transforms under **C** the same way as $\lambda_a \hat{A}_{a\mu}$ (i.e. according to (14.36)). Using the complex conjugate of (14.39), it then follows that (14.38) is invariant under **C**.

14.2.4   The θ-term

In arriving at (14.33) we have relied essentially on the 'gauge principle' (invariance under a local symmetry) and the requirement of renormalizability (to forbid the presence of terms with mass dimension higher than 4). The renormalizability of such a theory was proved by 't Hooft (1971a, b). However, there is in fact one more gauge invariant term of mass dimension 4 which can be written down, namely

$$\hat{\mathcal{L}}_\theta = \frac{\theta g_s^2}{64\pi^2}\epsilon_{\mu\nu\rho\sigma}\hat{F}^{\mu\nu}_a \hat{F}^{\rho\sigma}_a; \tag{14.40}$$

this is the 'θ-term' of QCD. A full discussion of this term (see for example Weinberg 1996, section 23.6) is beyond our scope, but we shall give a brief introduction to the main ideas.

The reader may wonder, first of all, whether the θ-term should give rise to a new Feynman rule. The answer to this begins by noting that (14.40) can actually be written as a total divergence:

$$\epsilon_{\mu\nu\rho\sigma}\hat{F}_a^{\mu\nu}\hat{F}_a^{\rho\sigma} = \partial_\mu \hat{K}^\mu. \tag{14.41}$$

This is more easily seen in the analogous term for QED, namely $\epsilon_{\mu\nu\rho\sigma}\hat{F}^{\mu\nu}\hat{F}^{\rho\sigma}$. We have

$$\epsilon_{\mu\nu\rho\sigma}\hat{F}^{\mu\nu}\hat{F}^{\rho\sigma} = \epsilon_{\mu\nu\rho\sigma}(\partial^\mu\hat{A}^\nu - \partial^\nu\hat{A}^\mu)(\partial^\rho\hat{A}^\sigma - \partial^\sigma\hat{A}^\rho) \tag{14.42}$$

$$= 4\epsilon_{\mu\nu\rho\sigma}\partial^\mu\hat{A}^\nu\partial^\rho\hat{A}^\sigma \tag{14.43}$$

$$= \partial^\mu(4\epsilon_{\mu\nu\rho\sigma}\hat{A}^\nu\partial^\rho\hat{A}^\sigma), \tag{14.44}$$

where we have used the antisymmetry of the ϵ symbol in (14.43), and also in (14.44) since the contraction of ϵ with the symmetric tensor $\partial^\mu\partial^\rho$ vanishes. We shall not need the explicit form of $\hat{K}^\mu$.

Any total divergence in a Lagrangian can be integrated to give only a 'surface' term in the action, which can usually be discarded, making conventional assumptions about the vanishing of the fields at spatial infinity. There are, however, field configurations ('instantons') which do contribute to the θ-term. Such configurations are not reachable in perturbation theory, and so no perturbative Feynman rules are associated with (14.40). They approach a pure gauge form at spatial infinity, and are therefore associated with the QCD vacuum state; their effect is equivalent to including the term (14.40) in the QCD Lagrangian (see for example Rajaraman 1982).

The term (14.40) has potentially important phenomenological implications, since it conserves **C** but violates both **P** and **T** (and hence also **CP**). Again, this is easy to see in the QED analogue term (14.42), which equals $8\hat{\boldsymbol{E}} \cdot \hat{\boldsymbol{B}}$ (problem 14.3): we recall that under **P**, $\hat{\boldsymbol{E}} \to -\hat{\boldsymbol{E}}$ and $\hat{\boldsymbol{B}} \to \hat{\boldsymbol{B}}$, while under **T**, $\hat{\boldsymbol{E}} \to \hat{\boldsymbol{E}}$ and $\hat{\boldsymbol{B}} \to -\hat{\boldsymbol{B}}$. But we know (section 4.2) that strong interactions conserve both **P** and **T** to a high degree of accuracy. In particular, the neutron electric dipole moment d_n, which would violate both **P** and **T**, is extremely small (see (4.134)). A very crude estimate of the size of d_n, induced by the θ-term, is given by dimensional analysis as

$$d_n \sim \frac{e}{M_n}\theta, \tag{14.45}$$

where M_n is the neutron mass. This would imply $\theta < 10^{-12}$. In fact, this estimate is too restrictive, since it turns out (Weinberg 1996, section 23.6) that if any quark has zero mass, θ can be reduced to zero by a global chiral U(1) transformation on that quark field. Although neither of the u and d quark masses are zero, they are small on a hadronic scale, and a suppression of (14.45) is expected, increasing the bound on theta. Estimates suggest $\theta < 10^{-9} - 10^{-10}$.

This may seem an unsatisfactorily special value to force on a dimensionless Lagrangian parameter, when there is nothing in the theory, *a priori*, to prevent something of order unity. This perceived difficulty is referred to as the 'strong **CP** problem'. A possible solution to the problem, in which a very small value of θ could arise naturally was suggested by Peccei and Quinn (1977a, 1977b). Their idea goes beyond the Standard Model, and involves the existence of a new very light pseudoscalar particle, the *axion* (Wilczek 1978, Winberg 1978). Axions might be a possible component of Dark Matter. A review of axions may be found (Ringwald, Rosenberg and Rybka 2022) in section 90 of Workman *et al.* 2022.

We proceed now with the main topic of this chapter, which is the application of perturbative QCD.

14.3   Hard scattering processes, QCD tree graphs, and jets

14.3.1   Introduction

The fundamental distinctive feature of non-Abelian gauge theories is that they are 'asymptotically free', meaning that the effective coupling strength becomes progressively smaller at short distances, or high energies (Gross and Wilczek (1973), Politzer (1973)). This property is the most compelling theoretical motivation for choosing a non-Abelian gauge theory for the strong interactions, and it enables a quantitative perturbative approach to be followed (in appropriate circumstances) even in strong interaction physics. This programme has indeed been phenomenally successful, firmly establishing QCD as the theory of strong interactions, and now—in the era of the LHC—serving as a precision tool to guide searches for new physics.

A proper understanding of how this works necessitates a considerable detour, however, into the physics of renormalization. In particular, we need to understand the important cluster of ideas going under the general heading of the 'renormalization group', and this will be the topic of chapter 15. For the moment we proceed with a discussion of some simple tree-level applications of QCD, which provided early confrontation of QCD with experiment.

Let us begin by recapitulating, from a QCD-informed viewpoint, how the parton model successfully interpreted deep inelastic and large-Q^2 data in terms of almost free point-like partons—now to be identified with the QCD quanta: quarks, antiquarks, and gluons.

In section 9.5 we briefly introduced the idea of *jets* in e^+e^- physics: two well collimated sprays of hadrons, apparently created as a quark-antiquark pair separate from each other at high speed. The angular distribution of the two jets followed closely the distribution expected from the parton-level process $e^+e^- \to \bar{q}q$. The dynamics at the parton level was governed by QED, but QCD is responsible for the way the emerging q and $\bar{q}$ turn themselves into hadrons, a process called parton fragmentation (it occurs for gluons too). We may think of it as proceeding in two stages. First, as the rapidly moving q and $\bar{q}$ begin to separate, they develop perturbative showers of narrowly collimated gluons and quark-antiquark pairs. Then, as the partons separate further, the strength of the forces between them increases, becoming strongly non-perturbative at a separation of about 1 fm, and ensuring that the coloured quanta are all confined into hadrons. As yet we do not have a completely quantitative dynamical understanding of the second, hadronization, stage: it is implemented by means of a model. Nevertheless, we can argue that for the forces to be strong enough to produce the observed hadrons, the dominant processes in hadronization must involve small momentum transfers—that is, the exchange of 'soft' quanta. Thus the emerging hadrons are also well collimated into two jets, whose energy and angular distributions reflect the short-distance physics at the parton level. This simple 2-jet picture will be extended in section 14.4, where we consider $e^+e^- \to 3$ jets. .

A somewhat different aspect of parton physics arose in sections 9.2 – 9.3, where we considered deep inelastic electron scattering from nucleons. There the initial state contained one hadron. Correspondingly, one parton appeared in the *initial* state of the parton-level interaction, and the analysis required new functions measuring the probabilities of finding a particular parton in the parent hadron—the parton distribution functions. These too are beyond the reach of perturbation theory.

We may also consider, finally, hadron-hadron collisions. In this case, we need all three of the features we have been discussing: the parton distribution functions, to provide the intial parton-parton state from the two-hadron state; the perturbative short-distance parton-parton interaction; and the parton fragmentation process in the final state. These three parts to the process are pictured in figure 14.5. The identification and analysis of short

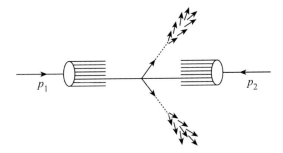

FIGURE 14.5
Hadron-hadron collision involving parton-parton interaction followed by parton fragmentation.

distance parton-parton interactions provide direct tests of the tree-graph structure of QCD, and perturbative corrections to it.

This three-part schematization of certain features of hadronic interactions is useful, because although we cannot yet calculate from first principles either the parton distribution functions or the fragmentation process, both are *universal*. The quark and gluon composition of hadrons is the same for all processes, and so measurements in one experiment can be used to predict the results of others. We saw an example of this in the Drell-Yan process of section 9.4. As regards the fragmentation stage, this too will be universal, provided one is interested in sufficiently inclusive aspects of the final state. The three-part scheme is called *factorization*, and it has been rigorously proved for some cases. We shall return to factorization in section 15.6.2.

Let us turn now to some of the early data on parton-parton interactions in hadron-hadron collisions.

14.3.2 Two-jet events in p̄p collisions

How are short-distance parton-parton interactions to be identified experimentally? The answer is: in just the same way as Rutherford distinguished the presence of a small heavy scattering centre (the nucleus) in the atom: by looking at secondary particles emerging at large angles with respect to the beam direction. For each secondary particle we can define a transverse momentum $p_T = p \sin\theta$ where p is the particle momentum and θ is the emission angle with respect to the beam axis. If hadronic matter were smooth and uniform (cf the Thomson atom), the distribution of events in p_T would be expected to fall off very rapidly at large p_T values—perhaps exponentially. This is just what is observed in the vast majority of events: the average value of p_T measured for charged particles is very low ($\langle p_T \rangle \sim 0.4$ GeV), but in a small fraction of collisions the emission of high-p_T secondaries is observed. They were first seen (Büsser *et al.* 1972, 1973, Alper *et al.* 1973, Banner *et al.* 1982) at the CERN ISR (CMS energies 30-62 GeV), and were interpreted in parton terms as previously indicated. Referring to figure 14.5, a parton from one hadron undergoes a short-distance 'hard scattering' interaction with a parton from the other, leading in lowest-order perturbation theory to two wide-angle partons, which then fragment into two jets.

We now face the experimental problem of picking out, from the enormous multiplicity of total events, just these hard scattering ones, in order to analyze them further. Early experiments used a trigger based on the detection of a single high-p_T particle. But it turns out that such triggering really reduces the probability of observing jets, since the probability

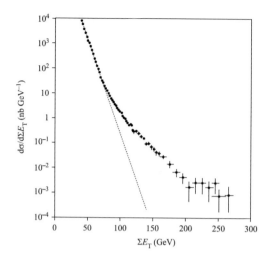

FIGURE 14.6

Distribution of the total transverse energy $\sum E_T$ observed in the UA2 central calorimeter (DiLella 1985).

that a single hadron in a jet will actually carry most of the jet's total transverse momentum is quite small (Jacob and Landshoff 1978; Collins and Martin 1984, Chapter 5). It is much better to surround the collision volume with an array of calorimeters which measure the total energy deposited. *Wide-angle jets* can then be identified by the occurrence of a large amount of total transverse energy deposited in a number of adjacent calorimeter cells: this is then a 'jet trigger'. The importance of calorimetric triggers was first emphasized by Bjorken (1973), following earlier work by Berman, Bjorken, and Kogut (1971). The application of this method to the detection and analysis of wide-angle jets was first reported by the UA2 collaboration at the CERN $\bar{p}p$ collider (Banner *et al.* 1982). An impressive body of quite remarkably clean jet data was subsequently accumulated by both the UA1 and UA2 collaborations (at $\sqrt{s} = 546$ GeV and 630 GeV), and by the CDF and D0 collaborations at the FNAL Tevatron collider ($\sqrt{s} = 1.8$ TeV).

For each event the total transverse energy $\sum E_T$ is measured where

$$\sum E_T = \sum_i E_i \sin \theta_i. \tag{14.46}$$

E_i is the energy deposited in the ith calorimeter cell and θ_i is the polar angle of the cell centre; the sum extends over all cells. Figure 14.6 shows the $\sum E_T$ distribution observed by UA2. It follows the 'soft' exponential form for $\sum E_T \leq 60$ GeV, but thereafter departs from it, showing clear evidence of the wide-angle collisions characteristic of hard processes.

As we shall see shortly, the majority of 'hard' events are of two-jet type, with the jets sharing the $\sum E_T$ approximately equally. Thus a 'local' trigger set to select events with localized transverse energy ≥ 30 GeV and/or a 'global' trigger set at ≥ 60 GeV can be used. At $\sqrt{s} \geq 500 - 600$ GeV there is plenty of energy available to produce such events.

The total $\sqrt{s}$ value is important for another reason. Consider the kinematics of the two-parton collision (figure 14.5) in the $\bar{p}p$ CMS. As in the Drell-Yan process of section 9.4, the right-moving parton has 4-momentum

$$x_1 p_1 = x_1(P, 0, 0, P) \tag{14.47}$$

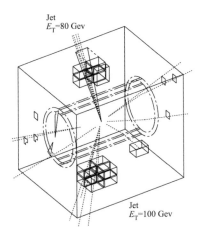

FIGURE 14.7

Two-jet event. Two tightly collimated groups of reconstructed charged tracks can be seen in the cylindrical central detector of UA1, associated with two large clusters of calorimeter energy depositions. Figure reprinted with permission from S Geer in *High Energy Physics 1985, Proc. Yale Advanced Study Institute* eds M J Bowick and F Gursey ; copyright 1986 World Scientific Publishing Company.

and the left-moving one

$$x_2 p_2 = x_2(P, 0, 0, -P) \tag{14.48}$$

where $P = \sqrt{s}/2$ and we are neglecting parton transverse momenta, which are approximately limited by the observed $\langle p_T \rangle$ value (~ 0.4 GeV, and thus negligible on this energy scale). Consider the simple case of 90^0 scattering, which requires (for massless partons) $x_1 = x_2$, equal to x say. The total outgoing transverse energy is then $2xP = x\sqrt{s}$. If this is to be greater than 50 GeV, then partons with $x \geq 0.1$ will contribute to the process. The parton distribution functions are large at these relatively small x values, due to sea quarks and gluons (section 9.3 and figure 9.9), and thus we expect to obtain a reasonable cross section.

What are the characteristics of jet events? When $\sum E_T$ is large enough (≥ 150 GeV), it is found that essentially all of the transverse energy is indeed split roughly equally between two approximately back-to-back jets. A typical such event is shown in figure 14.7. Returning to the kinematics of (14.47) and (14.48), x_1 will not in general be equal to x_2, so that—as is apparent in figure 14.7—the jets will not be collinear. However, to the extent that the transverse parton momenta can be neglected, the jets will be coplanar with the beam direction, i.e. their relative azimuthal angle will be 180^0. Figure 14.8 shows a number of examples in which the distribution of the transverse energy over the calorimeter cells is analyzed as a function of the jet opening angle θ and the azimuthal angle ϕ. It is strikingly evident that we are seeing precisely a kind of 'Rutherford' process, or—to vary the analogy—we might say that hadronic jets are acting as the modern counterpart of Faraday's iron filings, in rendering visible the underlying field dynamics!

We may now consider more detailed features of these two-jet events—in particular, the expectations based on QCD tree graphs. The initial hadrons provide wide-band beams

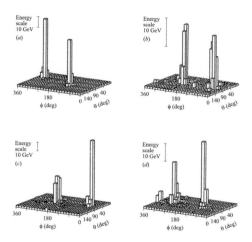

FIGURE 14.8

Four transverse energy distributions for events with $\sum E_T > 100$ GeV, in the θ, ϕ plane (UA2, DiLella 1985). Each bin represents a cell of the UA2 calorimeter. Note that the sum of the ϕ's equals 180^0 (mod 360^0).

of quarks, antiquarks and gluons[2]; thus we shall have many parton subprocesses, such as $qq \to qq$, $q\bar{q} \to q\bar{q}$, $q\bar{q} \to gg$, $gg \to gg$, etc. The most important, numerically, for a $p\bar{p}$ collider are $q\bar{q} \to q\bar{q}$, $gq \to gq$, and $gg \to gg$. The cross section will be given, in the parton model, by a formula of the Drell-Yan type, except that the electromagnetic annihilation cross section

$$\sigma(q\bar{q} \to \mu^+\mu^-) = 4\pi\alpha^2/3q^2 \qquad (14.49)$$

is replaced by the various QCD subprocess cross sections, each one being weighted by the appropriate distribution functions. At first sight this seems to be a very complicated story, with so many contributing parton processes. But a significant simplification comes from the fact that in the CMS of the parton collision, all processes involving one gluon exchange will lead to essentially the same dominant angular distribution of Rutherford-type, $\sim \sin^{-4}\theta/2$, where θ is the parton CMS scattering angle (recall section 1.3.6). This is illustrated in table 14.1 (taken from Combridge *et al.* (1977)), which lists the different relevant spin averaged, squared, one-gluon-exchange matrix elements $|\mathcal{M}|^2$, where the parton differential cross section is given by (cf (6.129))

$$\frac{\mathrm{d}\sigma}{\mathrm{d}\cos\theta} = \frac{\pi\alpha_s^2}{2\hat{s}}|\mathcal{M}|^2 . \qquad (14.50)$$

Here $\alpha_s = g_s^2/4\pi$, and $\hat{s}$, $\hat{t}$, and $\hat{u}$ are the subprocess invariants, so that

$$\hat{s} = (x_1 p_1 + x_2 p_2)^2 = x_1 x_2 s \qquad (\text{cf } (9.85)). \qquad (14.51)$$

Continuing to neglect the parton transverse momenta, the initial parton configuration shown in figure 14.5 can be brought to the parton CMS by a Lorentz transformation along the beam direction, the outgoing partons then emerging back-to-back at an angle θ to the beam axis, so $\hat{t} \propto (1 - \cos\theta) \propto \sin^2\theta/2$. Only the terms in $(\hat{t})^{-2} \sim \sin^{-4}\theta/2$ are given in table 14.1. We note that the $\hat{s}, \hat{t}, \hat{u}$ dependence of these terms is the same for the three types of process (and is in fact the same as that found for the 1γ exchange process $e^-\mu^- \to e^-\mu^-$:

---

[2]In the sense that the partons in hadrons have momentum or energy distributions, which are characteristic of their localization to hadronic dimensions.

TABLE 14.1
Spin-averaged squared matrix elements for one-gluon exchange ($\hat{t}$-channel) processes.

Subprocess	$|\mathcal{M}|^2$
$\left.\begin{array}{l} qq \to qq \\ q\bar{q} \to q\bar{q} \end{array}\right\}$	$\frac{4}{9}\left(\frac{\hat{s}^2+\hat{u}^2}{\hat{t}^2}\right)$
$qg \to qg$	$\frac{\hat{s}^2+\hat{u}^2}{\hat{t}^2} + \cdots$
$gg \to gg$	$\frac{9}{4}\left(\frac{\hat{s}^2+\hat{u}^2}{\hat{t}^2}\right) + \cdots$

see problem 8.17, converting $d\sigma/d\hat{t}$ into $d\sigma/d\cos\theta$). Figure 14.9 shows the two jet angular distribution measured by UA1 (Arnison *et al.* 1985). The broken curve is the exact angular distribution predicted by all the QCD tree graphs — it actually follows the $\sin^{-4}\theta/2$ shape quite closely.

It is interesting to compare this angular distribution with the one predicted on the assumption that the exchanged gluon is a spinless particle, so that the vertices have the form '$\bar{u}u$' rather than '$\bar{u}\gamma_\mu u$'. Problem 14.4 shows that in this case the $1/\hat{t}^2$ factor in the cross section is completely cancelled, thus ruling out such a model.

This analysis provides compelling evidence for elementary hard scattering events proceeding via the exchange of a massless vector quantum. It is possible to go much further. Anticipating our later discussion, the small discrepancy between 'tree graph' theory (which is labelled 'leading order QCD scaling curve' in figure 14.9) and experiment can be accounted for by including corrections which are of higher order in α_s. The solid curve in figure 14.9 includes QCD corrections beyond the tree level, involving the 'running' of the coupling constant α_s and 'scaling violation' in the effective parton distribution functions, both of which effects will be discussed in the following chapter. The corrections lead to good agreement with experiment.

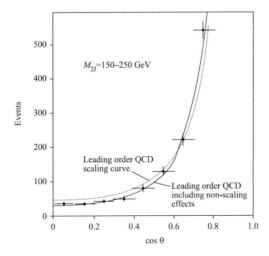

FIGURE 14.9
Two-jet angular distribution plotted against $\cos\theta$ (Arnison *et al.* 1985).

The fact that the angular distributions of all the subprocesses are so similar allows further information to be extracted from these two-jet data. In general, the parton model cross section will have the form (cf (9.91))

$$\frac{\mathrm{d}^3\sigma}{\mathrm{d}x_1\mathrm{d}x_2\mathrm{d}\cos\theta} = \sum_{a,b}\frac{F_a(x_1)}{x_1}\frac{F_b(x_2)}{x_2}\sum_{c,d}\frac{\mathrm{d}\sigma_{ab\to cd}}{\mathrm{d}\cos\theta} \tag{14.52}$$

where $F_a(x_1)/x_1$ is the distribution function for partons of type 'a' (q, $\bar{\mathrm{q}}$ or g), and similarly for $F_b(x_2)/x_2$. Using the near identity of all $\mathrm{d}\sigma/\mathrm{d}\cos\theta$'s, and noting the numerical factors in table 14.1, the sums over parton types reduce to

$$\frac{9}{4}\{g(x_1) + \frac{4}{9}[q(x_1) + \bar{q}(x_1)]\}\{g(x_2) + \frac{4}{9}[q(x_2) + \bar{q}(x_2)]\} \tag{14.53}$$

where $g(x)$, $q(x)$, and $\bar{q}(x)$ are the gluon, quark, and antiquark distribution functions, respectively. Thus effectively the weighted distribution function[3]

$$\frac{F(x)}{x} = g(x) + \frac{4}{9}[q(x) + \bar{q}(x)] \tag{14.54}$$

is measured (Combridge and Maxwell, 1984); in fact, with the weights as in (14.53),

$$\frac{\mathrm{d}^3\sigma}{\mathrm{d}x_1\mathrm{d}x_2\mathrm{d}\cos\theta} = \frac{F(x_1)}{x_1}\cdot\frac{F(x_2)}{x_2}\cdot\frac{\mathrm{d}\sigma_{gg\to gg}}{\mathrm{d}\cos\theta}. \tag{14.55}$$

x_1 and x_2 are kinematically determined from the measured jet variables: from (14.51),

$$x_1 x_2 = \hat{s}/s \tag{14.56}$$

where $\hat{s}$ is the invariant [mass]^2 of the two-jet system and

$$x_1 - x_2 = 2P_{\mathrm{L}}/\sqrt{s} \qquad (\text{cf } (9.83)) \tag{14.57}$$

with P_{L} the total two-jet longitudinal momentum. Figure 14.10 shows $F(x)/x$ obtained in the UA1 (Arnison *et al.* 1984) and UA(2) (Bagnaia *et al.* 1984) experiments. Also shown in this figure is the expected $F(x)/x$ based on contemporary fits to the deep inelastic neutrino scattering data at $Q^2 = 20$ GeV^2 and 2000 GeV^2 (Abramovicz *et al.* 1982a,b, 1983); the reason for the change with Q^2 will be discussed in section 15.6. The agreement is qualitatively very satisfactory. Subtracting the distributions for quarks and antiquarks as found in deep inelastic lepton scattering, UA1 were able to deduce the gluon structure function $g(x)$ shown in figure 14.11. It is clear that gluon processes will dominate at small x—and even at larger x will be important because of the colour factors in table 14.1.

14.3.3 Three-jet events in $\bar{\mathrm{p}}$p collisions

Although most of the high $-\sum E_{\mathrm{T}}$ events at the CERN SPS were two-jet events, in some 10-30% of the cases the energy is shared between three jets. An example is included as (*d*) in the collection of figure 14.8; a clearer one is shown in figure 14.12. In QCD such events are interpreted as arising from a 2 parton → 2 parton + 1 gluon process of the type gg → ggg, gq → ggq, etc. Once again, one can calculate (Kunszt and Piétarinen 1980, Gottschalk and Sivers 1980, Berends *et al.* 1981) all possible contributing tree graphs, of the kind shown in figure 14.13, which should dominate at small α_{s}. They are collectively known as QCD single-bremsstrahlung diagrams. Analysis of triple jets which are well separated both from

---

[3]The $\frac{4}{9}$ reflects the relative strengths of the quark-gluon and gluon-gluon couplings in QCD; see problem 14.5.

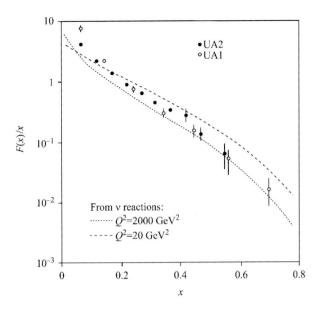

FIGURE 14.10
Effective distribution function measured from two-jet events (Arnison *et al.* 1984 and Bagnaia *et al.* 1984). The broken and chain curves are obtained from deep inelastic neutrino scattering. Taken from DiLella (1985).

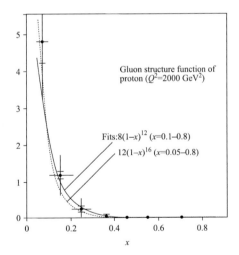

FIGURE 14.11
The gluon distribution function $g(x)$ extracted from the effective distribution function $F(x)$ by subtracting the expected contribution from the quarks and antiquarks. Figure reprinted with permission from S Geer in *High Energy Physics 1985, Proc. Yale Theoretical Advanced Study Institute*, eds M J Bowick and F Gursey; copyright 1986 World Scientific Publishing Company.

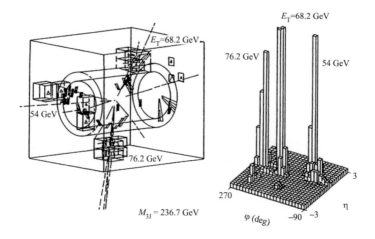

FIGURE 14.12

Three-jet event in the UA1 detector, and the associated transverse energy flow plot. Figure reprinted with permission from S Geer in *High Energy Physics 1985, Proc. Yale Theoretical Advanced Study Institute*, eds M J Bowick and F Gursey; copyright 1986 World Scientific Publishing Company.

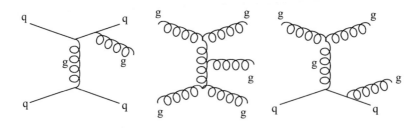

FIGURE 14.13

Some tree graphs associated with three-jet events.

each other and from the beam directions shows that the data are in good agreement with these lowest-order QCD predictions. For example, figure 14.14 shows the production angular distribution of UA2 (Appel *et al.* 1986) as a function of $\cos\theta^*$, where θ^* is the angle between the leading (most energetic) jet momentum and the beam axis, in the three-jet CMS. It follows just the same $\sin^{-4}\theta^*/2$ curve as in the two-jet case (the data for which are also shown in the figure), as expected for massless quantum exchange; the particular curve is for the representative process $gg \to ggg$.

Another qualitative feature is that the ratio of three-jet to two-jet events is controlled, roughly, by α_s (compare figure 14.13 with the graphs in table 14.1). Thus an estimate of α_s can be obtained by comparing the rates of 3-jet to 2-jet events in $\bar{p}p$ collisions. Other interesting predictions concern the characteristics of the 3-jet final state (for example, the distributions in the jet energy variables). At this point, however, it is convenient to leave

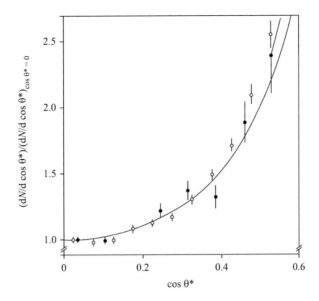

FIGURE 14.14
The distribution of $\cos\theta^*$ ($\bullet$), the angle of the leading jet with respect to the beam line (normalized to unity at $\cos\theta^* = 0$), for three-jet events in $\bar{\text{p}}\text{p}$ collisions (Appel *et al.* 1986). The distribution for two-jet events is also shown ($\circ$). The full curve is a parton model calculation using the tree graph amplitudes for gg $\to$ ggg, and cut-offs in transverse momentum and angular separation to eliminate divergences (see remarks following equation (14.73)).

$\bar{\text{p}}\text{p}$ collisions and consider instead 3-jet events in e⁺e⁻ collisions, for which the complications associated with the initial state hadrons are absent.

14.4   3-jet events in e⁺e⁻ annihilation

Three-jet events in e⁺e⁻ collisions originate, according to QCD, from gluon bremsstrahlung corrections to the two-jet parton level process e⁺e⁻ $\to \gamma^* \to q\bar{q}$, as shown in figure 14.15.[4] This phenomenon was predicted by Ellis *et al.* (1976) and subsequently observed by Brandelik *et al.* (1979) with the TASSO detector at PETRA, and Barber *et al.* (1979) with MARK-J at PETRA, thus providing early encouragement for QCD. The situation here is in many ways simpler and cleaner than in the $\bar{\text{p}}\text{p}$ case; the initial state 'partons' are perfectly physical QED quanta, and their total 4-momentum is zero, so that the three jets have to be coplanar; further, there is only one type of diagram compared to the large number in the $\bar{\text{p}}\text{p}$ case, and much of that diagram involves the easier vertices of QED. Since the calculation of the cross section predicted from figure 14.15 is relevant not only to three-jet production in e⁺e⁻ collisions, but also to a satisfactory definition of the two-jet production cross section,

---

[4]This is assuming that the total e⁺e⁻ energy is far from the Z^0 mass; if not, the contribution from the intermediate Z^0 must be added to that from the photon.

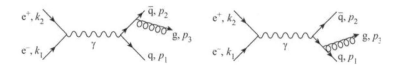

FIGURE 14.15

Gluon brehmsstrahlung corrections to two-jet parton level process.

to QCD corrections to the *total* e^+e^- annihilation cross section, and to scaling violations in deep inelastic scattering as well, we shall now consider it in some detail. It is important to emphasize at the outset that *quark masses will be neglected* in this calculation.

14.4.1   Calculation of the parton-level cross section

The quark, antiquark, and gluon 4-momenta are p_1, p_2, and p_3 respectively, as shown in figure 14.15; the e^- and e^+ 4-momenta are k_1 and k_2. The cross section is then (cf (6.110) and (6.112))

$$d\sigma = \frac{1}{(2\pi)^5}\delta^4(k_1 + k_2 - p_1 - p_2 - p_3)\frac{|\mathcal{M}_{q\bar{q}g}|^2}{2Q^2}\frac{d^3p_1}{2E_1}\frac{d^3p_2}{2E_2}\frac{d^3p_3}{2E_3} \tag{14.58}$$

where (neglecting all masses)

$$\begin{aligned}
\mathcal{M}_{q\bar{q}g} &= \frac{e_a e^2 g_s}{Q^2}\bar{v}(k_2)\gamma^\mu u(k_1)\left(\bar{u}(p_1)\gamma_\nu\frac{\lambda_c}{2}\cdot\frac{(\not{p_1}+\not{p_3})}{2p_1\cdot p_3}\cdot\gamma_\mu v(p_2)\right.\\
&\left.\quad -\bar{u}(p_1)\gamma_\mu\frac{\lambda_c}{2}\cdot\frac{(\not{p_2}+\not{p_3})}{2p_2\cdot p_3}\cdot\gamma_\nu v(p_2)\right)\epsilon^{*\nu}(\lambda)a_c
\end{aligned} \tag{14.59}$$

and $Q^2 = 4E^2$ is the square of the total e^+e^- energy, and also the square of the virtual photon's 4-momentum Q, and e_a (in units of e) is the charge of a quark of type 'a'. Note the minus sign in (14.59): the antiquark coupling is $-g_s$. In (14.59), $\epsilon^{*\nu}(\lambda)$ is the polarization vector of the outgoing gluon with polarization λ; a_c is the colour wavefunction of the gluon $(c = 1 \ldots 8)$, and λ_c is the corresponding Gell-Mann matrix introduced in section 12.2; the colour parts of the q and $\bar{q}$ wavefunctions are understood to be included in the u and v factors; and $(\not{p_1}+\not{p_3})/2p_1\cdot p_3$ is the virtual quark propagator (cf (L.6) in Appendix L of volume 1) before gluon radiation, and similarly for the antiquark. Since the colour parts separate from the Dirac trace parts, we shall ignore them to begin with, and reinstate the result of the colour sum (via problem (14.5)) in the final answer (14.73).

Averaging over $e^\pm$ spins and summing over final state quark spins and gluon polarization λ (using (8.171), and noting the discussion after (13.93)), we obtain (problem 14.6)

$$\frac{1}{4}\sum_{\text{spins},\lambda}|\mathcal{M}_{q\bar{q}g}|^2 = \frac{e^4 e_a^2 g_s^2}{Q^4}L^{\mu\nu}(k_1, k_2)H_{\mu\nu}(p_1, p_2, p_3) \tag{14.60}$$

where the lepton tensor is, as usual (equation (8.119)),

$$L^{\mu\nu}(k_1, k_2) = 2(k_1^\mu k_2^\nu + k_1^\nu k_2^\mu - k_1\cdot k_2 g^{\mu\nu}) \tag{14.61}$$

and the hadron tensor is

$$
\begin{aligned}
H_{\mu\nu}(p_1, p_2, p_3) &= \frac{1}{p_1 \cdot p_3}[L_{\mu\nu}(p_2, p_3) - L_{\mu\nu}(p_1, p_1) + L_{\mu\nu}(p_1, p_2)] \\
&+ \frac{1}{p_2 \cdot p_3}[L_{\mu\nu}(p_1, p_3) - L_{\mu\nu}(p_2, p_2) \\
&+ L_{\mu\nu}(p_1, p_2)] \\
&+ \frac{p_1 \cdot p_2}{(p_1 \cdot p_3)(p_2 \cdot p_3)}[2L_{\mu\nu}(p_1, p_2) + L_{\mu\nu}(p_1, p_3) \\
&+ L_{\mu\nu}(p_2, p_3)]
\end{aligned}
\tag{14.62}
$$

Combining (14.61) and (14.62) allows complete expressions for the five-fold differential cross section to be obtained (Ellis *et al.* 1976).

For the subsequent discussion it will be useful to integrate over the three angles describing the orientation (relative to the beam axis) of the production plane containing the three partons. After this integration, the (doubly differential) cross section is a function of two independent Lorentz invariant variables, which are conveniently taken to be two of the three s_{ij} defined by

$$
s_{ij} = (p_i + p_j)^2.
\tag{14.63}
$$

Since we are considering the massless case $p_i^2 = 0$ throughout, we may also write

$$
s_{ij} = 2p_i \cdot p_j.
\tag{14.64}
$$

These variables are linearly related by

$$
2(p_1 \cdot p_2 + p_2 \cdot p_3 + p_3 \cdot p_1) = Q^2
\tag{14.65}
$$

as follows from

$$
(p_1 + p_2 + p_3)^2 = Q^2
\tag{14.66}
$$

and $p_i^2 = 0$. The integration yields (Ellis *et al.* 1976, 1977)

$$
\frac{\mathrm{d}^2\sigma}{\mathrm{d}s_{13}\mathrm{d}s_{23}} = \frac{2}{3}\alpha^2 e_a^2 \alpha_{\mathrm{s}} \frac{1}{(Q^2)^3}\left(\frac{s_{13}}{s_{23}} + \frac{s_{23}}{s_{13}} + \frac{2Q^2 s_{12}}{s_{13}s_{23}}\right)
\tag{14.67}
$$

where $\alpha_{\mathrm{s}} = g_{\mathrm{s}}^2/4\pi$.

We may understand the form of this result in a simple way, as follows. It seems plausible that after integrating over the production angles, the lepton tensor will be proportional to $Q^2 g^{\mu\nu}$, all directional knowledge of the k_1 having been lost. Indeed, if we use $-g^{\mu\nu}L_{\mu\nu}(p, p) = 4p \cdot p'$ together with (14.62) we easily find that

$$
-\frac{1}{4}g^{\mu\nu}H_{\mu\nu} = \frac{p_1 \cdot p_3}{p_2 \cdot p_3} + \frac{p_2 \cdot p_3}{p_1 \cdot p_3} + \frac{p_1 \cdot p_2 Q^2}{(p_1 \cdot p_3)(p_2 \cdot p_3)} = \frac{s_{13}}{s_{23}} + \frac{s_{23}}{s_{13}} + \frac{2Q^2 s_{12}}{s_{13}s_{23}}
\tag{14.68}
$$

exactly the factor appearing in (14.67). In turn, the result may be given a simple physical interpretation. From (7.118) we note that we can replace $-g^{\mu\nu}$ by $\sum_{\lambda'} \epsilon^{\mu}(\lambda')\epsilon^{\nu*}(\lambda')$ for a virtual photon of polarization λ', the $\lambda' = 0$ state contributing negatively. Thus effectively the result of doing the angular integration is (up to constants and Q^2 factors) to replace the lepton factor $\bar{v}(k_2)\gamma^{\mu}u(k_1)$ by $-i\epsilon^{\mu}(\lambda')$, so that $\mathcal{M}_{\mathrm{q\bar{q}g}}$ is proportional to the $\gamma* \to \mathrm{q\bar{q}g}$ processes shown in figure 14.16. But these are basically the same amplitudes as the ones we already met in Compton scattering (section 8.6). To compare with section 8.6.3, we convert

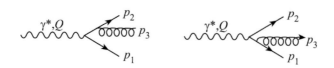

FIGURE 14.16
Virtual photon decaying to q$\bar{\text{q}}$g.

the initial state fermion (electron/quark) into a final state antifermion (positron/antiquark) by $p \to -p$, and then identify the variables of figure 14.16 with those of figure 8.14 (a) by

$$p' \to p_1 \quad k' \to p_3 \quad -p \to p_2 \quad s \to 2p_1 \cdot p_3 = s_{13}$$
$$t \to 2p_1 \cdot p_2 = s_{12} \quad u \to 2p_2 \cdot p_3 = s_{23}. \tag{14.69}$$

Remembering that in (8.181) the virtual γ had squared 4-momentum $-Q^2$, we see that the Compton '$\sum |\mathcal{M}|^2$' of (8.181) indeed becomes proportional to the factor (14.68), as expected.

14.4.2   Soft and collinear divergences

In three-body final states of the type under discussion here it is often convenient to preserve the symmetry between the s_{ij}'s and use *three* (dimensionless) variables x_i defined by

$$s_{23} = Q^2(1 - x_1) \text{ and cyclic permutations.} \tag{14.70}$$

These are related by (14.65), which becomes

$$x_1 + x_2 + x_3 = 2. \tag{14.71}$$

An event with a given value of the set x_i can then be plotted as a point in an equilateral triangle of height 1, as shown in figure 14.17. In order to find the limits of the allowed

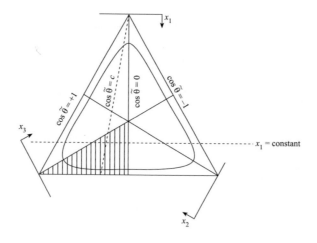

FIGURE 14.17
The kinematically allowed region in (x_i) is the interior of the equilateral triangle.

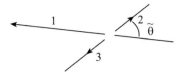

FIGURE 14.18
Definition of $\tilde{\theta}$.

physical region in this x_i space, we now transform from the overall three-body Centre of Mass (CMS) to the CMS of 2 and 3 (figure 14.18). If $\tilde{\theta}$ is the angle between 1 and 3 in this system, then (problem 14.7)

$$
\begin{aligned}
x_2 &= (1 - x_1/2) + (x_1/2)\cos\tilde{\theta} \\
x_3 &= (1 - x_1/2) - (x_1/2)\cos\tilde{\theta}.
\end{aligned}
\tag{14.72}
$$

The limits of the physical region are then clearly $\cos\tilde{\theta} = \pm 1$, which correspond to $x_2 = 1$ and $x_3 = 1$. By symmetry, we see that the entire perimeter of the triangle in figure 14.17 is the required boundary: physical events fall anywhere inside the triangle. (This is the massless limit of the classic Dalitz plot, first introduced by Dalitz (1953) for the analysis of $K \to 3\pi$.) Lines of constant $\tilde{\theta}$ are shown in figure 14.17.

Now consider the distribution provided by the QCD bremsstrahlung process, equation (14.67), which can be written equivalently as

$$
\frac{\mathrm{d}^2\sigma}{\mathrm{d}x_1\mathrm{d}x_2} = \sigma_{\mathrm{pt}} \frac{2\alpha_s}{3\pi} \left(\frac{x_1^2 + x_2^2}{(1 - x_1)(1 - x_2)} \right)
\tag{14.73}
$$

where σ_{pt} is the pointlike e⁺e⁻ → hadrons total cross section of (9.99), and a factor of 4 has been introduced from the colour sum (problem 14.5). The factor in large parentheses is (14.68) written in terms of the x_i (problem 14.8). The most striking feature of (14.73) is that it is *infinite* as x_1 or x_2, or both, tend to 1—and in such a way that the cross section integrated over x_1 and x_2 diverges logarithmically.

This is a quite different infinity from the ones encountered in the loop integrals of chapters 10 and 11. No integral over an arbitrarily large internal momentum is involved here—the tree amplitude itself is becoming singular on the phase space boundary. We can trace the origin of the singularity back to the denominator factors $(p_1 \cdot p_3)^{-1} \sim (1 - x_2)^{-1}$ and $(p_2 \cdot p_3)^{-1} \sim (1 - x_1)^{-1}$ in (14.59). These become zero in two distinct configurations of the gluon momentum:

$$
\begin{aligned}
(a) &\quad p_3 \propto p_1 \text{ or } p_3 \propto p_2 \text{ (using } p_i^2 = 0) &\tag{14.74} \\
(b) &\quad p_3 \to 0 &\tag{14.75}
\end{aligned}
$$

which are easily interpreted physically. Condition (a) corresponds to a situation in which the 4-momentum of the gluon is parallel to that of either the quark of the antiquark; this is called a *collinear divergence* and the configuration is pictured in figure 14.19(a). If we restore the quark masses, $p_1^2 = m_1^2 \neq 0$ and $p_2^2 = m_2^2 \neq 0$, then the factor $(2p_1 \cdot p_3)^{-1}$, for example, becomes $((p_1 + p_3)^2 - m_1^2)^{-1}$ which only vanishes as $p_3 \to 0$, which is condition (b). The divergence of type (a) is therefore also termed a 'mass singularity', as it would be absent if the quarks had mass. Condition (b) corresponds to the emission of a very 'soft' (infrared) gluon (figure 14.19(b)) and is called a *soft divergence*. In contrast to this, the gluon momentum p_3 in type (a) does *not* have to be vanishingly small.

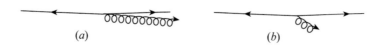

FIGURE 14.19

Gluon configurations leading to divergences of equation (14.73): (a) gluon emitted approximately collinear with quark (or antiquark); (b) soft gluon emission. The events are viewed in the overall CMS.

It is apparent from these figures that in either of these two cases the observed final state hadrons, after the fragmentation process, will in fact resemble a *two*-jet configuration. Such events will be found in the regions $x_1 \approx 1$ and/or $x_2 \approx 1$ of the kinematical plot shown in figure 14.17, which correspond to strips adjacent to two of the boundaries of the triangle. Events outside these strips should be essentially three-jet events, corresponding to the emission of a hard, non-collinear gluon. To isolate such events, we must keep away from the boundaries of the triangle (the strip along the third boundary $x_3 = 1$ will not contain a divergence, but will be included in a physical jet measure—see the following section). Thus to order $\alpha^2 \alpha_s$ the total annihilation cross section to three jets is given by the integral of (14.73) over a suitably defined inner triangular region in figure 14.17.

Assuming such a separation of three- and two-jet events can be done satisfactorily (see the next section), their ratio carries important information—namely, it should be proportional to α_s. This follows simply from the extra factor of g_s associated with the gluon emissions in figure 14.15. Glossing over a number of technicalities (for which the reader is referred to Ellis, Stirling, and Webber 1996, section 3.3), we show in figure 14.20 a compilation of data on the fraction of three-jet events at different e^+e^- annihilation energies. The most remarkable feature of this figure is, of course, that this fraction—and hence α_s—*changes with energy, decreasing as the energy increases*. This is, in fact, direct evidence for asymptotic freedom. A more recent comparison between theory and experiment (the agreement is remarkable) will be presented in the following chapter, section 15.4, after we have introduced the theoretical framework for calculating the energy dependence of α_s.

14.5   Definition of the two-jet cross section in e^+e^- annihilation

As just noted, the integral of (14.73) over the remaining regions of figure 14.17, near the phase-space boundaries, will contribute to the two-jet annihilation cross section—and it is divergent. Clearly this is not a physically acceptable result; we want a finite two-jet cross section. The cure lies in recognizing that at the order to which we are working, namely $\alpha^2 \alpha_s$, other parton-level graphs can contribute. These are the one-gluon loop graphs shown in figure 14.21, which are of order $\alpha \alpha_s$. They turn out to contain exactly the same soft and collinear divergences, this time associated with configurations of virtual momenta inside the loops. In a carefully defined two-jet cross section, these two classes of divergences (one from real gluon emission, the other from virtual gluons) actually cancel.

Let us call the amplitude for the sum of these three graphs F_{vg}, where 'vg' stands for virtual gluon. F_{vg} is the order α_s correction to the original order α parton-level graph of figure 9.17, shown here again in figure 14.22, with amplitude F_γ. The cross section from these contributions is proportional to $|F_\gamma + F_{vg}|^2$. There are three terms in this expression: one of order α^2, from $|F_\gamma|^2$; another of order $\alpha^2 \alpha_s^2$, from $|F_{vg}|^2$, which we drop since it is

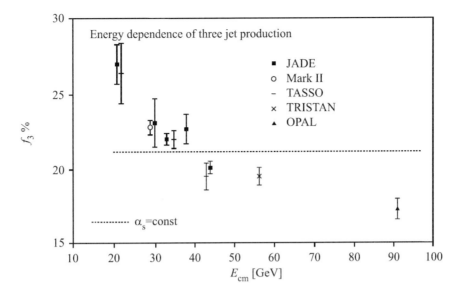

FIGURE 14.20

A compilation of three-jet fractions at different e$^+$e$^-$ annihilation energies. Adapted from Akrawy *et al.* (OPAL) (1990); figure from R K Ellis, W J Stirling and B R Webber (1996) *QCD and Collider Physics*, courtesy Cambridge University Press.

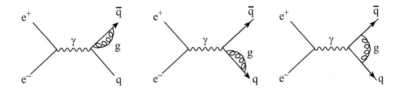

FIGURE 14.21

Virtual gluon corrections to figure 14.22.

of higher order in α_s; and an *interference* term of order $\alpha^2\alpha_s$, the same as (14.73). Thus the interference term must be included in calculating the two-jet cross section to this order. When it is, the soft and collinear divergences cancel[5]: the resulting two-jet cross section is IRC (infrared and collinear) 'safe'.

This result was first shown by Sterman and Weinberg (1977), in a paper which initiated the modern treatment of jets within the framework of QCD. They defined the two-jet differential cross section to include those events in which all but a fraction ϵ of the total e$^+$e$^-$ energy E $(= \sqrt{Q^2})$ is emitted within some pair of oppositely directed cones of half-angle $\delta \ll 1$, lying at an angle θ to the e$^+$e$^-$ beam line. Including the contributions of real and virtual gluons up to order $\alpha\alpha_s$, the result is (Muta 2010, section 5.4.1)

$$\left(\frac{d\sigma}{d\Omega}\right)_{2-jet} = \left(\frac{d\sigma}{d\Omega}\right)_{pt} [1 - \frac{4}{3}\frac{\alpha_s}{\pi}(3\ln\delta + 4\ln\delta\ln 2\epsilon + \frac{\pi^2}{3} - \frac{5}{2})], \qquad (14.76)$$

---

[5]The usual ultraviolet divergences in the loop graphs are removed by conventional renormalization.

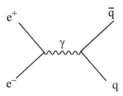

FIGURE 14.22

One-photon annihilation amplitude in $e^+e^- \to \bar{q}q$.

where $(\frac{d\sigma}{d\Omega})_{\mathrm{pt}}$ is the contribution of the lowest order graph, figure 14.22, which is given by equation (9.102) summed over quark colours and charges; terms of order δ and ϵ, and higher powers, are neglected. It is evident from (14.76) that the jet parameters ϵ and δ serve to control the soft and collinear divergences, which reappear as ϵ and δ tend to zero; they are 'resolution parameters'.

The remarkable cancellation of the soft and collinear divergences between the real and virtual emission processes is actually a general result in QED (recall that in chapter 11 we declined to pursue the problem of such infrared divergences). The Bloch-Nordsieck (1937) theorem states that 'soft' singularities cancel between real and virtual processes when one adds up all final states which are indistinguishable by virtue of the energy resolution of the apparatus. The Kinoshita (1962) Lee and Nauenberg (1964) theorem states, roughly speaking, that mass singularities are absent if one adds up all indistinguishable mass-degenerate states. This is the reason for the finiteness of the Sterman-Weinberg 2-jet cross section, in an analogous QCD case.

Returning to (14.76), it is important to note that the angular distribution of this well-defined two-jet process is given precisely by the lowest order expression (9.102), just as was hoped in the simple parton model of section 9.5. Of course, the cross section depends on the jet parameters δ and ϵ. The formula (14.76) can be used, for example, to estimate the angular radius of the jets, as a function of E.

Although the Sterman-Weinberg jet definition was historically the first, it is not the only possible one. Another, in some ways simpler, definition (Kramer and Lampe 1987) is directly phrased in terms of the offending denominators s_{13}^{-1} and s_{23}^{-1} in (14.67). Let us introduce the dimensionless jet mass variables

$$y_{ij} = s_{ij}/Q^2 = 2E_iE_j(1 - \cos\theta_{ij})/Q^2 \tag{14.77}$$

for any two partons i and j; s_{12} will be included, though no singularity is involved. Here E_i and E_j are the (massless) parton energies, and θ_{ij} is the angle between their 3-momenta, in the overall CMS. Then i and j are defined to be in one jet if y_{ij} is less than some given number y. Note that for small θ_{ij}, $s_{ij} \approx E_iE_j\theta_{ij}^2/Q^2$, so the single parameter y provides effectively both an energy and an angle cut. Clearly this definition is equivalent to a formulation in terms of strips $1 \le x_k < 1 - y$ on figure 14.17, as discussed earlier. Including contributions, as before, from figures 14.22, 14.21, and 14.16, the resulting 2-jet cross section is found to be (Kramer and Lampe 1987)

$$\sigma_{2-\mathrm{jet}} = \sigma_{\mathrm{pt}}[1 - \frac{2}{3}\frac{\alpha_{\mathrm{s}}}{\pi}(2\ln^2 y + 3\ln y - 4y\ln y + 1 - \pi^2/3)]. \tag{14.78}$$

Terms of order y were calculated numerically. These include the contribution from the (non-singular) region $y_{12} < y$, where the two quarks are in one jet and the other jet is a pure gluon jet. Plainly the IRC singularities have been eliminated from (14.78), at the cost of the jet mass resolution parameter y. Kramer and Lampe also calculated the order α_s^2 corrections to (14.78).

These two ways of regulating the IRC divergences in the 2-jet partonic cross section have each been extensively developed into *jet algorithms*, as we shall discuss in section 14.6.2.

14.6 Further developments

14.6.1 Test of non-Abelian nature of QCD in $e^+e^- \to$ 4 jets

We have seen in section 14.3.1 how the colour factors associated with different QCD vertices (problem 14.5) play an important part in determining the relative weights of different parton-level processes. The quark-gluon colour factor C_F enters into the parton-level three-jet amplitude (14.67), but the triple-gluon vertex is not involved at order α_s. This vertex is an essential feature of non-Abelian gauge theories, being absent in Abelian theories such as QED. A direct measurement of the triple-gluon vertex colour factor, C_A, can be made in the process $e^+e^- \to$ 4 jets.

4-jet events originate from the parton-level process $e^+e^- \to q\bar{q}g$ via three mechanisms: the emission of a second bremsstrahlung gluon, splitting of the first gluon into two gluons, and splitting of the first gluon into n_f quark pairs. As problem 14.5 shows, these three types of splitting vertices are characterized in cross sections by the colour factors C_F, C_A, and $n_f T_R$, so that the cross section can be written as (Ali and Kramer 2011)

$$\sigma_{4-\mathrm{jet}} = \left(\frac{\alpha_s}{\pi}\right) C_F [C_F \sigma_{\mathrm{bb}} + C_A \sigma_{\mathrm{gg}} + n_f T_R \sigma_{\mathrm{q\bar{q}}}]. \tag{14.79}$$

Measurements yield (Abbiendi *et al.* 2001)

$$\begin{aligned} C_A/C_F &= 2.29 \pm 0.06[\mathrm{stat.}] \pm 0.14[\mathrm{syst.}] \\ T_R/C_F &= 0.38 \pm 0.03[\mathrm{stat.}] \pm 0.06[\mathrm{syst.}], \end{aligned} \tag{14.80}$$

in good agreement with the theoretical predictions $C_A/C_F = 9/4$ and $T_R/C_F = 3/8$ in QCD.

14.6.2 Jet algorithms

From the examples already discussed in this chapter, it is clear that jets are an essential element in making comparisons between experimental measurements involving final state particles in detectors, and theoretical calculations at the parton level using perturbative QCD. Conceptually, jets provide a common representation for these two classes of event— those at the detector level, and those at the parton level. For detailed comparisons, it is necessary to have a practical definition of a jet — a *jet algorithm* — which should be equally applicable at the detector and at the parton level. By a jet algorithm is meant a set of rules, possibly containing a number of parameters, for grouping the particles produced in a final state of a high energy collision process into suitable bunches, which will be jets. Ideally, no prior assumptions need be made regarding the number of jets in a given event: the algorithm is run over all the particles in the event, and returns the jets it finds.

Since Sterman and Weinberg's 1977 paper, many jet algorithms have been developed and applied. All involve the basic notion of grouping together objects that are 'near' to each other, in some sense. Two main classes of algorithm may be distinguished: 'cone' algorithms based on proximity in position space, as in the Sterman-Weinberg approach, and used extensively at hadron colliders before the LHC; and sequential recombination — or 'clustering' — algorithms, based on proximity in momentum space, as in the Kramer and Lampe (1987) definition, which was widely used in e^+e^- and e^-p colliders. The main reason for preferring cone-type algorithms at pre-LHC hadron colliders (such as the Fermilab Tevatron) was that they could be more efficiently implemented, a clear advantage when faced with high final state multiplicities. On the other hand, the available cone algorithms tended not to be IRC safe. By contrast, the cluster algorithms were safe, but faced a computational barrier, because their algorithmic complexity rises as N^3, where N is the number of particles that have to be clustered, which could be thousands. This difficulty was overcome when a fast implementation of cluster algorithms was introduced (Cacciari and Salam 2006). This opened the way for the practical use of the IRC safe cluster algorithms at the LHC, and we shall now discuss them further. We should note, however, that a satisfactory cone type algorithm was developed (Cacciari, Salam and Soyez 2008), called SISCone, which is IRC safe and fast enough for use at the LHC. A useful general review of jet physics is provided by Cacciari (2015)

The JADE algorithm (Bartel *et al.* 1986, Bethke *et al.* 1988) is a prominent early example of a cluster type algorithm applied in e^+e^- annihilation reactions. Particles are clustered in a jet iteratively as long as the quantity y_{ij} of (14.77) is less than some prescribed value y_c. If for some pair (i, j), $y_{ij} < y_c$, particles i and j are combined into a compound object (with the resultant 4-momentum, typically), and the process continues by pairing the compound with a new particle k. The procedure stops when all y_{ij} distances are greater than y_c, and the compounds that remain at this stage are the jets, by definition.

One drawback with this scheme is that in higher orders of perturbation theory one meets terms of the form $\alpha_s^2 \ln^{2n} y$ (generalizations of the $\alpha_s \ln^2 y$ term in (14.78)). Such terms can be large enough to invalidate a perturbative approach. Also, it is possible for two soft particles moving in opposite directions to get combined in the early stages of clustering, which runs counter to the intuitive notion of a jet being restricted in angular radius. The k_t-algorithm (Catani *et al.* 1991) avoids these problems by replacing the y_{ij} of (14.77) by

$$y_{ij} = 2\min.[E_i^2, E_j^2](1 - \cos\theta_{ij})/Q^2. \qquad (14.81)$$

This amounts to defining 'distance' by the minimum transverse momentum k_t of the particles in nearby collinear pairs. The use of the minimum energy ensures that the distance between two soft, back-to-back particles is larger than that between a soft particle and a hard one that is close to it in angle. The k_t algorithm was widely used at LEP.

The basic idea of the k_t algorithm was extended to hadron colliders (Ellis and Soper 1993, Catani *et al.* 1993), where the total energy of the hard scattering particles is not well defined experimentally. Subsequently, a generalization of the k_t algorithm was proposed (Cacciari *et al.* 2008), which encompasses the algorithms used at the LHC. The distance measure y_{ij} is replaced by

$$d_{ij} = \min.[p_{ti}^{2p}, p_{tj}^{2p}][(y_i - y_j)^2 + (\phi_i - \phi_j)^2]/R^2 \qquad (14.82)$$

where, for particle i, p_{ti} is the transverse momentum with respect to the (beam) z-axis, y_i is the rapidity along the beam axis (defined by $y_i = \frac{1}{2}\ln[(E_i + p_{zi})/(E_i - p_{zi})]$), ϕ_i is the azimuthal angle in the plane transverse to the beam, and R is a 'jet radius', which can range from about 0.5 fm to 1.0 fm at the LHC. The variables y_i, ϕ_i have the property that they are invariant under boosts along the beam direction. In addition, recombination with the beam jets is controlled by by the quantity $d_{iB} = k_{ti}^{2p}$, which is included along with the d_{ij}'s when recombining all the particles into (i) jets with non-zero transverse momentum and (ii) beam jets.

The algorithm works by calculating the d_{ij}'s and the d_{iB}'s for all particles in the event, and finding the smallest one. If this smallest distance is interpreted as an interparticle one, the two four-momenta are combined according to a given recombination scheme. If it is a beam distance, the four-momentum concerned is called a jet, and excluded from further processing. The algorithm continues iteratively until no particles in the event remain uncombined. Then all jets with transverse momentum above a specified cut are retained.

The power parameter p can take the value 1, 0 or -1. When $p = 1$, one recovers the basic k_t algorithm. The choice $p = 0$ gives the 'Cambridge-Aachen' algorithm, in which clustering is based simply on the angular separation of the particles. In both these cases the algorithm is expected to roughly correspond to tracing back through the parton branching process induced by QCD emissions, and therefore to produce a jet which approximately reconstructs the original parton (quark or gluon). The case $p = 1$ is called the 'anti-k_t' algorithm. It is IRC safe, and it produces jets with regular borders which rarely extend more than a distance of order R from the hard parton generating event. These algorithms, and the SISCone one, are all implemented in the FatJet package (Cacciari and Salam 2006; Cacciari, Salam, and Soyez 2012); however, the anti-k_t algorithm has become the *de facto* standard for LHC experiments.

In addition to their use in comparisons between experimental measurements and QCD predictions, jets are also used for identifying the hard partonic structure of decays of massive particles such as the top quark, the W and Z bosons, and the Higgs boson. One particularly important example concerns the search at the LHC for the $\bar{\text{b}}$ decay of a Higgs boson, which faces a huge background from standard QCD jets. However, if the Higgs boson is boosted (i.e. has a large transverse momentum $p_t \gg m_H$) the signal to background ratio is substantially improved (Butterworth *et al.* 2008). Such a boosted Higgs boson tends to decay in a collimated way in the laboratory frame, with all decay products within a 'fat' jet. One then looks for such 'fat' jets, and analyzes their substructure to tag those that show a two-prong structure, indicative of a $b\bar{b}$ decay.

In concluding this section, we remark that neural network and deep learning techniques are being applied to jet physics and jet substructure (Louppe *et al.* 2019, Guest *et al.* 2018, Larkoski *et al.* 2020).

.

Problems

14.1

(a) Show that the antisymmetric 3q combination of equation (14.2) is (i) a determinant, and (ii) invariant under the transformation (14.14) for each colour wavefunction.

(b) Suppose that p_α and q_α stand for two $SU(3)_c$ colour wavefunctions, transforming under an infinitesimal $SU(3)_c$ transformation via

$$p' = (1 + i\boldsymbol{\eta} \cdot \boldsymbol{\lambda}/2)p,$$

and similarly for q. Consider the antisymmetric combination of their components, given by

$$\begin{pmatrix} p_2 q_3 - p_3 q_2 \\ p_3 q_1 - p_1 q_3 \\ p_1 q_2 - p_2 q_1 \end{pmatrix} \equiv \begin{pmatrix} Q_1 \\ Q_2 \\ Q_3 \end{pmatrix};$$

that is, $Q_\alpha = \epsilon_{\alpha\beta\gamma} p_\beta q_\gamma$. Check that the three components Q_α transform as a $\mathbf{3}_c^*$, in the particular case for which only the parameters η_1, η_2, η_3, and η_8 are non-zero. [Note: you will need the explicit forms of the $\boldsymbol{\lambda}$ matrices (Appendix M); you need to verify the transformation law

$$Q' = (1 - i\boldsymbol{\eta} \cdot \boldsymbol{\lambda}^*/2)Q.]$$

14.2

(a) Verify that the normally ordered QCD interaction $\bar{\hat{q}}_f \gamma^\mu \frac{1}{2}\lambda_a \hat{q}_f \hat{A}_{a\mu}$ is **C**-invariant.

(b) Show that $\lambda_a \hat{F}_{a\mu\nu}$ transforms under **C** according to (14.36).

14.3 Verify that the Lorentz-invariant 'contraction' $\epsilon_{\mu\nu\rho\sigma} \hat{F}^{\mu\nu} \hat{F}^{\rho\sigma}$ of two U(1) (Maxwell) field strength tensors is equal to $8\boldsymbol{E} \cdot \boldsymbol{B}$.

14.4 Verify that the cross section for the exchange of a single massless scalar gluon between two quarks (or between a quark and an antiquark) contains no '$1/\hat{t}^2$' factor.

14.5 This problem is concerned with the evaluation of various *colour factors*.

(a) Consider first the colour factor needed for equation (14.73). The 'colour wavefunction' part of the amplitude (14.59) is

$$\sum_c a_c(c_3)\chi^\dagger(c_1)\frac{\lambda_c}{2}\chi(c_2) \tag{14.83}$$

where c_1, c_2, and c_3 label the colour degree of freedom of the quark, antiquark, and gluon, respectively, and the sum on the index c has been indicated explicitly. The χ's are the colour wavefunctions of the quark and antiquark, and are represented by 3-component column vectors; a convenient choice is

$$\chi(r) = \begin{pmatrix} 1 \\ 0 \\ 0 \end{pmatrix}, \quad \chi(b) = \begin{pmatrix} 0 \\ 1 \\ 0 \end{pmatrix}, \quad \chi(g) = \begin{pmatrix} 0 \\ 0 \\ 1 \end{pmatrix} \tag{14.84}$$

by analogy with the spin wavefunctions of SU(2). The cross section is obtained by forming the modulus squared of (14.83) and summing over the colour labels c_i:

$$\sum_{c, c_1, c_2, c_3} a_c(c_3)\chi_r^*(c_1)\frac{(\lambda_c)_{rs}}{2}\chi_s(c_2)\chi_l^*(c_2)\frac{(\lambda_d)_{lm}}{2}\chi_m(c_1)a_d^*(c_3) \tag{14.85}$$

where summation is understood on the matrix indices on the χ's and λ's, which have been indicated explicitly. In this form, the expression is very similar to the *spin* summations considered in chapter 8 (cf equation (8.62)). We proceed to evaluate it as follows:

(i) Show that

$$\sum_{c_2} \chi_s(c_2)\chi_l^*(c_2) = \delta_{sl}.$$

(ii) Assuming the analogous result

$$\sum_{c_3} a_c(c_3)a_d^*(c_3) = \delta_{cd}$$

show that (14.85) becomes

$$\sum_{c=1}^{8} \left(\frac{\lambda_c}{2} \frac{\lambda_c}{2} \right)_{rr},$$

where the (implied) sum on r runs from 1 to 3.

(iii) The expression $\sum_c \frac{\lambda_c}{2} \frac{\lambda_c}{2}$ is just the Casimir operator $\hat{C}_2$ (see section M.5 in appendix M) for SU(3) in the fundamental representation **3**, which from (M.67) has the value $C_F \mathbf{1}_3$, where $\mathbf{1}_3$ is the unit 3×3 matrix, and $C_F = 4/3$. Hence show that the colour factor for (14.73) is 4.

Note that if we averaged over the colours of the initial quark, or considered one particular colour, the colour factor would be C_F.

(b) The colour part for the triple gluon vertex $g_1 \to g_2 + g_3$ is

$$\sum_{c,d,e} a_d^*(c_2) a_e^*(c_3) f_{dec} a_c(c_1).$$

Show that the modulus squared of this, averaged over the initial gluon colours and summed over the final gluon colours, is

$$\frac{1}{8} \sum_{c,d,e} f_{dec} f_{dec},$$

where each of c, d, e runs from 1 to 8. Deduce using (12.84) that this expression can be written as

$$\frac{1}{8} \sum_e \left(\sum_d G_d^{(8)} G_d^{(8)} \right)_{ee},$$

where $G_d^{(8)}$, $(d = 1 \ldots 8)$ are the 8×8 matrices representing the generators of SU(3) in the **8**-dimensional (adjoint) representation (see section 12.2). The expression $(\sum_d G_d^{(8)} G_d^{(8)})$ is the SU(3) Casimir operator $\hat{C}_2$ in the adjoint representation, which from (M.67) has the value $C_A \mathbf{1}_8$, where $\mathbf{1}_8$ is the 8×8 unit matrix, and $C_A = 3$. Hence show that the (averaged, summed) triple gluon vertex colour factor is $C_A = 3$.

(c) The colour part of the $g \to q + \bar{q}$ vertex is

$$\chi_r^*(c_3) \left(\frac{\lambda_c}{2} \right)_{rs} \chi_s(c_2) a_c(c_1).$$

Show that the modulus squared of this, averaged over the initial gluon colours and summed over the final quark colours is

$$\frac{1}{8} \sum_c \left(\frac{\lambda_c}{2} \frac{\lambda_c}{2} \right)_{rr} = \frac{1}{2}.$$

This number is usually denoted by T_R.

14.6 Verify equation (14.60).

14.7 Verify equation (14.72).

14.8 Verify that expression (14.68) becomes the factor in large parentheses in equation (14.73), when expressed in terms of the x_i's.

15

QCD II: Asymptotic Freedom, The Renormalization Group, and Scaling Violations

In the previous chapter we learned that QCD amplitudes contributing to $e^+e^- \to$ jets generally have IRC singularities, but that finite physical cross sections can be obtained by including together kinematically indistinguishable final states. The partial cross sections (for example $\sigma(e^+e^- \to 2$ jets$)$) will depend on the IRC cut-off parameter(s). What about the *fully inclusive* process $e^+e^- \to$ hadrons, where all final states are summed over? At order $\alpha\alpha_s$, the parton-level diagrams contributing to this process are the same ones we considered in section 14.5, namely figures 14.16, 14.21, and 14.22. If we denote the amplitudes for these contributions by F_{rg} (for real gluon emission), F_{vg} (for virtual gluon emission), and F_γ for the Born graph, then the partial cross section $\sigma(e^+e^- \to 2$ jets$)$ includes $|F_\gamma|^2$, the interference term $2\text{Re}(F_\gamma F_{vg}^*)$, and the integral of $|F_{rg}|^2$ over strips near the boundaries of figure 14.17. At this order, the partial cross section $\sigma(e^+e^- \to 3$ jets$)$ is given by the integral of $|F_{rg}|^2$ over the remaining (interior) region of figure 14.17. The corresponding total cross section is thus simply the sum of $|F_\gamma|^2$, $2\text{Re}(F_\gamma F_{vg}^*)$, and the integral of $|F_{rg}|^2$ over the whole of the $x_1 - x_2$ phase space. Clearly the IRC singularities will cancel, as in the 2-jet cross section, and the result will not depend on any IRC cut-off parameter. Indeed, the result is (see for example Muta 2010, section 5.1.2)

$$\sigma(e^+e^- \to \text{hadrons}) = \sigma_{pt}(Q^2)(1 + \alpha_s/\pi). \tag{15.1}$$

This fully inclusive cross section is finite and free of IRC cut-offs.

At first sight, this result might appear satisfactory. It predicts a cross section somewhat greater than σ_{pt}, as is observed in figure 14.1—from which we might infer that $\alpha_s \sim 0.5$ or less. Assuming the expansion parameter is α_s/π, the implied perturbation series in powers of α_s would seem to be rapidly convergent. However, this is an illusion, which is dispelled as soon as we go to the next order in α_s (i.e. to the order $\alpha^2\alpha_s^2$ in the cross section).

15.1 Higher-order QCD corrections to $\sigma(e^+e^- \to$ hadrons): large logarithms

Some typical graphs contributing to this order of the cross section are shown in figure 15.1 (note that, as with the $O(\alpha^2\alpha_s)$ terms, some graphs will contribute via their modulus squared and some via interference terms). The result was obtained numerically by Dine and Saperstein (1979), and analytically by Chetyrkin *et al.* (1979) and by Celmaster and Gonsalves (1980). For our present purposes, the crucial feature of the answer is the appearance of a term

$$\sigma_{pt}\left[-\beta_0 \frac{\alpha_s^2}{\pi} \ln(Q^2/\mu^2)\right]. \tag{15.2}$$

DOI: 10.1201/9781003411666-15

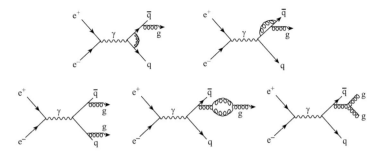

FIGURE 15.1
Some higher-order processes contributing to $e^+e^- \to$ hadrons at the parton level.

where μ is a mass scale (about which we shall shortly have a lot more to say, but which for the moment may be thought of as related in some way to an average quark mass), and the coefficient β_0 is given by

$$\beta_0 = \left(\frac{33 - 2N_f}{12\pi}\right) \tag{15.3}$$

where N_f is the number of 'active' flavours (e.g. $N_f = 5$ above the $b\bar{b}$ threshold). The term (15.2) raises the following problem. The ratio between it and the $O(\alpha\alpha_s)$ term is clearly

$$-\beta_0\alpha_s \ln(Q^2/\mu^2). \tag{15.4}$$

If we take $N_f = 5, \alpha_s \approx 0.4, \mu \sim 1$ GeV, and $Q^2 \sim (10 \text{ GeV})^2$, (15.4) is of order 1, and can in no sense be regarded as a small perturbation. Furthermore, the correction (15.4), by itself, would predict large *scaling violations* in this cross section—that is, large Q^2-dependent departures from the point-like Born cross section, $\sigma_{pt}(Q^2)$. But the data actually follow the point-like prediction very well.

Suppose that, nevertheless, we consider the sum of (15.1) and (15.2), which is

$$\sigma_{pt}[1 + \frac{\alpha_s}{\pi}\{1 - \beta_0\alpha_s \ln(Q^2/\mu^2)\}]. \tag{15.5}$$

This suggests that one effect, at least, of these higher-order corrections is to convert α_s to a Q^2-dependent quantity, namely $\alpha_s\{1 - \beta_0\alpha_s \ln(Q^2/\mu^2)\}$. We have seen something very like this before, in equation (11.56), for the case of QED. There is, however, one remarkable difference: here the coefficient of the ln is *negative*, whereas that in (11.56) is *positive*. Apart from this (vital!) difference, however, we can reasonably begin to think in terms of an effective 'Q^2-dependent strong coupling constant $\alpha_s(Q^2)$'.

Pressing on with the next order $(\alpha^2\alpha_s^3)$ terms, we encounter a term (Samuel and Surguladze 1991, Gorishnii *et al.* 1991)

$$\sigma_{pt}\left[\alpha_s\beta_0 \ln(Q^2/\mu^2)\right]^2 \frac{\alpha_s}{\pi}, \tag{15.6}$$

and the ratio between this and (15.2) is precisely (15.4) once again! We are now strongly inclined to suspect that we are seeing, in this class of terms, an expansion of the form $(1 + x)^{-1} = 1 - x + x^2 - x^3 \ldots$. If true, this would imply that all terms of the form (15.2) and (15.6), and higher, *sum up* to give (cf (11.63))

$$\sigma_{pt}\left[1 + \frac{\alpha_s/\pi}{1 + \alpha_s\beta_0 \ln(Q^2/\mu^2)}\right]. \tag{15.7}$$

The 're-summation' effected by (15.7) has a remarkable effect: the 'dangerous' large logarithms in (15.2) and (15.6) are now effectively in the *denominator* (cf (11.56)), and their effect is such as to *reduce* the effective value of α_s as Q^2 increases—exactly the property of *asymptotic freedom*.

We hasten to say that of course this is not how the property was discovered—which was, rather, through the calculations of Politzer (1973) and Gross and Wilczek (1973). Prior to their work, it was widely believed that any quantum field theory would have a running coupling which behaved like that of QED which, as we saw in section 11.5.3, increases for large Q^2 (short distances). Such behaviour would make the scaling violations due to a term like (15.7) even worse. It was therefore a mystery how quantum field theory could account for the small scaling violations seen in the data. The discovery that the running couplings of non-Abelian gauge theories became weaker at large Q^2 opened the way for a quantitative understanding of parton-model scaling, and perturbative QCD corrections to it.

To place the asymptotic freedom calculation in its proper context requires a considerable detour. Referring to our previous discussion, we may ask: are we guaranteed that still higher order terms will indeed continue to contain pieces corresponding to the expression of (15.7)? And what exactly is the mass parameter μ? Answering these questions will lead to the important body of ideas going under the name of the 'renormalization group'.

15.2   The renormalization group and related ideas in QED

15.2.1   Where do the large logs come from?

We have taken the title of this section from that of section 18.1 in Weinberg (1996), which we have found very illuminating, and to which we refer for a more detailed discussion.

As we have just mentioned, the phenomenon of 'large logarithms' arises also in the simpler case of QED. There, however, the factor corresponding to $\alpha_s \beta_0 \sim \frac{1}{4}$ is $\alpha/3\pi \sim 10^{-3}$, so that it is only at quite unrealistically enormous $|q^2|$ values that the corresponding factor $(\alpha/3\pi)\ln(|q^2|/m_e^2)$ (where m_e is the electron mass) becomes of order unity. Nevertheless, the origin of the logarithmic term is essentially the same in both cases, and the technicalities are much simpler for QED (no photon self-interactions, no ghosts). We shall therefore forget about QCD for a while, and concentrate on QED. Indeed, the discussion of renormalization of QED given in chapter 11 will be sufficient to answer the question in the title of this subsection.

For the answer does, in fact, fundamentally have to do with renormalization. Let us go back to the renormalization of the charge in QED. We learned in chapter 11 that the renormalized charge e was given in terms of the 'bare' charge e_0 by the relation $e = e_0(Z_2/Z_1)Z_3^{\frac{1}{2}}$ (see (11.6)), where in fact due to the Ward identity Z_1 and Z_2 are equal (section 11.6), so that only $Z_3^{\frac{1}{2}}$ is needed. To order e^2 in renormalized perturbation theory, including only the e^+e^- loop of figure 15.2, Z_3 is given by (cf (11.31))

$$Z_3^{[2]} = 1 + \Pi_\gamma^{[2]}(0) \tag{15.8}$$

where, from (11.23) and (11.24),

$$\Pi_\gamma^{[2]}(q^2) = 8e^2\mathrm{i} \int_0^1 \mathrm{d}x \int \frac{\mathrm{d}^4k'}{(2\pi)^4} \frac{x(1-x)}{(k'^2 - \Delta_\gamma + \mathrm{i}\epsilon)^2} \tag{15.9}$$

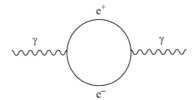

FIGURE 15.2
One-loop vacuum polarization contribution to Z_3.

and $\Delta_\gamma = m_{\mathrm{e}}^2 - x(1-x)q^2$ with $q^2 < 0$. We regularize the k' integral by a cut-off Λ as explained in sections 10.3.1 and 10.3.2, obtaining (problem 15.1)

$$\Pi_\gamma^{[2]}(q^2) = -\frac{e^2}{\pi^2} \int_0^1 \mathrm{d}x \; x(1-x) \left\{ \ln\left(\frac{\Lambda + \sqrt{\Lambda^2 + \Delta_\gamma}}{\Delta_\gamma^{\frac{1}{2}}}\right) - \frac{\Lambda}{(\Lambda^2 + \Delta_\gamma)^{1/2}} \right\}. \quad (15.10)$$

Setting $q^2 = 0$ and retaining the dominant $\ln \Lambda$ term, we find that

$$\left(Z_3^{[2]}\right)^{\frac{1}{2}} = 1 - \left(\frac{\alpha}{3\pi}\right) \ln(\Lambda/m_{\mathrm{e}}). \quad (15.11)$$

It is not a coincidence that the coefficient $\alpha/3\pi$ of the ultraviolet divergence is also the coefficient of the $\ln(|q^2|/m_{\mathrm{e}}^2)$ term in (11.55)—(11.57); we need to understand why.

We first recall how (11.55) was arrived at. It refers to the *renormalized* self-energy part, which is defined by the 'subtracted' form

$$\bar{\Pi}_\gamma^{[2]}(q^2) = \Pi_\gamma^{[2]}(q^2) - \Pi_\gamma^{[2]}(0). \quad (15.12)$$

In the process of subtraction, the dependence on the cut-off Λ disappears and we are left with

$$\bar{\Pi}_\gamma^{[2]}(q^2) = -\frac{2\alpha}{\pi} \int_0^1 \mathrm{d}x \; x(1-x) \ln\left[\frac{m_{\mathrm{e}}^2}{m_{\mathrm{e}}^2 - q^2 x(1-x)}\right] \quad (15.13)$$

as in (11.34). For large values of $|q^2|$ this leads to the 'large log' term
$(\alpha/3\pi) \ln(|q^2|/m_{\mathrm{e}}^2)$. Now, in order to form such a term, it is obviously not possible to have just '$\ln|q^2|$' appearing: the argument of the logarithm must be dimensionless, so that some mass scale must be present, to which $|q^2|$ can be compared. In the present case, that mass scale is evidently m_{e}, which entered via the quantity $\Pi_\gamma^{[2]}(0)$, or equivalently via the renormalization constant $Z_3^{[2]}$ (cf (15.11)). This is the beginning of the answer to our questions.

Why is it m_{e} that enters into $\Pi_\gamma^{[2]}(0)$ or Z_3? Part of the answer—once again—is of course that a '$\ln \Lambda$' cannot appear in that form, but must be '$\ln(\Lambda/\text{some mass})$'. So we must enquire: what determines the 'some mass'? With this question we have reached the heart of the problem (for the moment). The answer is, in fact, not immediately obvious. It lies in the *prescription used to define the renormalized coupling constant*; this prescription, whatever it is, determines Z_3.

The value (15.8) (or (11.31)) was determined from the requirement that the $O(e^2)$ corrected photon propagator (in $\xi = 1$ gauge) had the simple form $-\mathrm{i}g_{\mu\nu}/q^2$ as $q^2 \to 0$; that is, as the photon goes on-shell. Now, this is a perfectly 'natural' definition of the renormalized charge—but it is by no means forced upon us. In fact the appearance of a singularity in

$Z_3^{[2]}$ as $m_e \to 0$ suggests that it is inappropriate to the case in which fermion masses are neglected. We could in principle choose a different value of q^2, say $q^2 = -\mu^2$, at which to 'subtract'. Certainly the difference between $\Pi_\gamma^{[2]}(q^2 = 0)$ and $\Pi_\gamma^{[2]}(q^2 = -\mu^2)$ is finite as $\Lambda \to \infty$, so such a redefinition of 'the' renormalized charge would only amount to a finite shift. Nevertheless, even a finite shift is alarming, to those accustomed to a certain 'sanctity' in the value $\alpha = \frac{1}{137}$! We have to concede, however, that if the point of renormalization is to render amplitudes finite by taking certain constants from experiment, then any choice of such constants should be as good as any other—for example, the 'charge' defined at $q^2 = -\mu^2$ rather than at $q^2 = 0$.

Thus there is, actually, a considerable *arbitrariness* in the way renormalization can be done—a fact to which we did not draw attention in our earlier discussions in chapters 10 and 11. Nevertheless, it must somehow be the case that, despite this arbitrariness, *physical results remain the same*. We shall come back to this important point shortly.

15.2.2   Changing the renormalization scale

The recognition that the *renormalization scale* ($-\mu^2$ in this case) is arbitrary suggests a way in which we might exploit the situation, so as to avoid large '$\ln(|q^2|/m_e^2)$' terms: we renormalize at a *large* value of μ^2! Consider what happens if we define a new $Z_3^{[2]}$ by

$$Z_3^{[2]}(\mu) = 1 + \Pi_\gamma^{[2]}(q^2 = -\mu^2). \tag{15.14}$$

Then for $\mu^2 \gg m_e^2$, but $\mu^2 \ll \Lambda^2$, we have

$$\left(Z_3^{[2]}(\mu)\right)^{\frac{1}{2}} = 1 - \left(\frac{\alpha}{3\pi}\right)\ln\left(\Lambda/\mu\right), \tag{15.15}$$

and a new renormalized self-energy

$$\begin{aligned}\bar{\Pi}_\gamma^{[2]}(q^2, \mu) &= \Pi_\gamma^{[2]}(q^2) - \Pi_\gamma^{[2]}(q^2 = -\mu^2) \\ &= -\frac{e^2}{2\pi^2}\int_0^1 dx\; x(1-x)\ln\left[\frac{m_e^2 + \mu^2 x(1-x)}{m_e^2 - q^2 x(1-x)}\right].\end{aligned} \tag{15.16}$$

For μ^2 and $-q^2$ both $\gg m_e^2$, the logarithm is now $\ln(|q^2|/\mu^2)$ which is small when $|q^2|$ is of order μ^2. It seems, therefore, that with this different renormalization prescription we have 'tamed' the large logarithms.

However, we have forgotten that, for consistency, the 'e' we should now be using is the one defined, in terms of e_0, via

$$e_\mu = \left(Z_3^{[2]}(\mu)\right)^{\frac{1}{2}} e_0 = \left(1 - \frac{\alpha}{3\pi}\ln(\Lambda/\mu)\right)e_0 \tag{15.17}$$

rather than

$$e = \left(Z_3^{[2]}\right)^{\frac{1}{2}} e_0 = \left(1 - \frac{\alpha}{3\pi}\ln(\Lambda/m_e)\right)e_0, \tag{15.18}$$

working always to one-loop order with an e^+e^- loop. The relation between e_μ and e is then

$$e_\mu = \frac{\left(1 - \frac{\alpha}{3\pi}\ln(\Lambda/\mu)\right)}{\left(1 - \frac{\alpha}{3\pi}\ln(\Lambda/m_e)\right)}e \approx \left(1 + \frac{\alpha}{3\pi}\ln(\mu/m_e)\right)e \tag{15.19}$$

to leading order in α. Equation (15.19) indeed represents, as anticipated, a finite shift from 'e' to 'e_μ', but the problem with it is that a 'large log' has resurfaced in the form of $\ln(\mu/m_e)$

(remember that our idea was to take $\mu^2 \gg m_e^2$). Although the numerical coefficient of the log in (15.19) is certainly small, a similar procedure applied to QCD will involve the larger coefficient $\beta_0 \alpha_s$ as in (15.5), and the correction analogous to (15.19) will be of order 1, invalidating the approach.

We have to be more subtle. Instead of making one jump from m_e^2 to a large value μ^2, we need to proceed in stages. We can calculate e_μ from e as long as μ is not too different from m_e. Then we can proceed to $e_{\mu'}$ for μ' not too different from μ, and so on. Rather than thinking of such a process in discrete stages $m_e \to \mu \to \mu' \to \ldots$, it is more convenient to consider infinitesimal steps — that is, we regard $e_{\mu'}$ at the scale μ' as being a continuous function of e_μ at scale μ, and of whatever other dimensionless variables exist in the problem (since the e's are themselves dimensionless). In the present case, these other variables are μ'/μ and m_e/μ, so that $e_{\mu'}$ must have the form

$$e_{\mu'} = E(e_\mu, \mu'/\mu, m_e/\mu). \tag{15.20}$$

Differentiating (15.20) with respect to μ' and letting $\mu' = \mu$ we obtain

$$\mu \frac{\mathrm{d}e_\mu}{\mathrm{d}\mu} = \beta_{\mathrm{em}}(e_\mu, m_e/\mu) \tag{15.21}$$

where

$$\beta_{\mathrm{em}}(e_\mu, m_e/\mu) = \left[\frac{\partial}{\partial z} E(e_\mu, z, m_e/\mu) \right]_{z=1}. \tag{15.22}$$

For $\mu \gg m_e$ equation (15.21) reduces to

$$\mu \frac{\mathrm{d}e_\mu}{\mathrm{d}\mu} = \beta_{\mathrm{em}}(e_\mu, 0) \equiv \beta_{\mathrm{em}}(e_\mu), \tag{15.23}$$

which is a form of *Callan-Symanzik equation* (Callan 1970, Symanzik 1970); it governs the change of the coupling constant e_μ as the renormalization scale μ changes.

To this one-loop order, it is easy to calculate the crucial quantity $\beta_{\mathrm{em}}(e_\mu)$. Returning to (15.17), we may write the bare coupling e_0 as

$$\begin{aligned} e_0 &= e_\mu \left(1 - \frac{\alpha}{3\pi} \ln(\Lambda/\mu) \right)^{-1} \\ &\approx e_\mu \left(1 + \frac{\alpha}{3\pi} \ln(\Lambda/\mu) \right) \\ &\approx e_\mu \left(1 + \frac{\alpha_\mu}{3\pi} \ln(\Lambda/\mu) \right) \end{aligned} \tag{15.24}$$

where the last step follows from the fact that e and e_μ differ by $O(e^3)$, which would be a higher-order correction to (15.24). Now the unrenormalized coupling is certainly independent of μ. Hence, differentiating (15.24) with respect to μ at fixed e_0, we find

$$\frac{\mathrm{d}e_\mu}{\mathrm{d}\mu}\bigg|_{e_0} - \frac{e_\mu \alpha_\mu}{3\pi\mu} - \ln(\Lambda/\mu) \cdot \frac{e_\mu^2}{4\pi^2} \frac{\mathrm{d}e_\mu}{\mathrm{d}\mu}\bigg|_{e_0} = 0. \tag{15.25}$$

Working to order e_μ^3 we can drop the last term in (15.25), obtaining finally (to one-loop order)

$$\mu \frac{\mathrm{d}e_\mu}{\mathrm{d}\mu}\bigg|_{e_0} = \frac{e_\mu^3}{12\pi^2} \quad \left(\equiv \beta_{\mathrm{em}}^{[2]}(e_\mu) \right). \tag{15.26}$$

We can now integrate equation (15.26) to obtain e_μ at an arbitrary scale μ, in terms of its value at some scale $\mu = M$, chosen in practice large enough so that for variable scales μ

greater than M we can neglect m_e compared with μ, but small enough so that $\ln(M/m_e)$ terms do not invalidate the perturbation theory calculation of e_M from e. The solution of (15.26) is then (problem 15.2)

$$\ln(\mu/M) = 6\pi^2 \left(\frac{1}{e_M^2} - \frac{1}{e_\mu^2} \right) \tag{15.27}$$

or equivalently

$$e_\mu^2 = \frac{e_M^2}{1 - \frac{e_M^2}{12\pi^2} \ln(\mu^2/M^2)}, \tag{15.28}$$

which is

$$\alpha_\mu = \frac{\alpha_M}{1 - \frac{\alpha_M}{3\pi} \ln(\mu^2/M^2)} \tag{15.29}$$

where $\alpha = e^2/4\pi$. The crucial point is that the 'large log' is now in the *denominator* (and has coefficient $\alpha_M/3\pi$!). We note that the general solution of (15.23) may be written as

$$\ln(\mu/M) = \int_{e_M}^{e_\mu} \frac{de}{\beta_{em}(e)}. \tag{15.30}$$

We have made progress in understanding how the coupling changes as the renormalization scale changes, and how 'large logarithmic' change as in (15.19) can be brought under control via (15.29). The final piece in the puzzle is to understand how this can help us with the large $-q^2$ behaviour of our cross section, the problem we originally started from.

15.2.3   The RGE and large $-q^2$ behaviour in QED

To see the connection, we need to implement the fundamental requirement, stated at the end of section 15.2.2, that predictions for physically measurable quantities must *not* depend on the renormalization scale μ. Consider, for example, our annihilation cross section σ for $e^+e^- \to$ hadrons, pretending that the one-loop corrections we are interested in are those due to QED rather than QCD. We need to work in the spacelike region, so as to be consistent with all the foregoing discussion. To make this clear, we shall now denote the 4-momentum of the virtual photon by q rather than Q, and take $q^2 < 0$ as in sections 15.2.1 and 15.2.2. Bearing in mind the way we used the 'dimensionless-ness' of the e's in (15.20), let us focus on the dimensionless ratio $\sigma/\sigma_{pt} \equiv S$. Neglecting all masses, S can only be a function of the dimensionless ratio $|q^2|/\mu^2$ and of e_μ:

$$S = S(|q^2|/\mu^2, e_\mu). \tag{15.31}$$

But S must ultimately have no μ dependence. It follows that *the μ^2 dependence arising via the $|q^2|/\mu^2$ argument must cancel that associated with e_μ*. This is why the μ^2-dependence of e_μ controls the $|q^2|$ dependence of S, and hence of σ. In symbols, this condition is represented by the equation

$$\left(\left. \frac{\partial}{\partial \mu} \right|_{e_\mu} + \left. \frac{de_\mu}{d\mu} \right|_{e_0} \frac{\partial}{\partial e_\mu} \right) S\left(|q^2|/\mu^2, e_\mu\right) = 0, \tag{15.32}$$

or

$$\left(\left. \mu \frac{\partial}{\partial \mu} \right|_{e_\mu} + \beta_{em}(e_\mu) \frac{\partial}{\partial e_\mu} \right) S\left(|q|^2/\mu^2, e_\mu\right) = 0. \tag{15.33}$$

Equation (15.33) is referred to as 'the renormalization group equation (RGE) for S'. The terminology goes back to Stueckelberg and Peterman (1953), who were the first to discuss the freedom associated with the choice of renormalization scale. The 'group' connotation is a trifle obscure—but all it really amounts to is the idea that if we do one infinitesimal shift in μ^2, and then another, the result will be a third such shift; in other words, it is a kind of 'translation group'. It was, however, Gell-Mann and Low (1954) who realized how equation (15.33) could be used to calculate the large $|q^2|$ behaviour of S, as we now explain.

It is convenient to work in terms of μ^2 and α rather than μ and e. Equation (15.33) is then

$$\left(\mu^2 \frac{\partial}{\partial \mu^2}\bigg|_{\alpha_\mu} + \beta_{\mathrm{em}}(\alpha_\mu) \frac{\partial}{\partial \alpha_\mu}\right) S\left(|q^2|/\mu^2, \alpha_\mu\right) = 0, \tag{15.34}$$

where $\beta_{\mathrm{em}}(\alpha_\mu)$ is defined by

$$\beta_{\mathrm{em}}(\alpha_\mu) \equiv \mu^2 \frac{\partial \alpha_\mu}{\partial \mu^2}\bigg|_{e_0}. \tag{15.35}$$

From (15.35) and (15.26) we deduce that, to the one-loop order to which we are working,

$$\beta_{\mathrm{em}}^{[2]}(\alpha_\mu) = \frac{e_\mu}{4\pi} \beta_{\mathrm{em}}^{[2]}(e_\mu) = \frac{\alpha_\mu^2}{3\pi}. \tag{15.36}$$

Now introduce the important variable

$$t = \ln(|q^2|/\mu^2). \tag{15.37}$$

Equation (15.34) then becomes

$$\left[-\frac{\partial}{\partial t} + \beta_{\mathrm{em}}(\alpha_\mu) \frac{\partial}{\partial \alpha_\mu}\right] S\left(e^t, \alpha_\mu\right) = 0. \tag{15.38}$$

This is a first-order differential equation which can be solved by implicitly defining a new function—the *running coupling* $\alpha(|q^2|)$—as follows (compare (15.30):

$$t = \int_{\alpha_\mu}^{\alpha(|q^2|)} \frac{\mathrm{d}\alpha}{\beta_{\mathrm{em}}(\alpha)}. \tag{15.39}$$

To see how this helps, we have to recall how to differentiate an integral with respect to one of its limits—or, more generally, the formulae

$$\frac{\partial}{\partial a} \int^{f(a)} g(x)\mathrm{d}x = g\left(f(a)\right) \frac{\partial f}{\partial a}. \tag{15.40}$$

First, let us differentiate (15.39) with respect to t at fixed α_μ; we obtain

$$1 = \frac{1}{\beta_{\mathrm{em}}(\alpha(|q^2|))} \frac{\partial \alpha(|q^2|)}{\partial t}. \tag{15.41}$$

Next, differentiate (15.39) with respect to α_μ at fixed t (note that $\alpha(|q^2|)$ will depend on μ and hence on α_μ); we obtain

$$0 = \frac{\partial \alpha(|q^2|)}{\partial \alpha_\mu} \frac{1}{\beta_{\mathrm{em}}(\alpha(|q^2|))} - \frac{1}{\beta_{\mathrm{em}}(\alpha_\mu)} \tag{15.42}$$

the minus sign coming from the fact that α_μ is the lower limit in (15.39). From (15.41) and (15.42) we find

$$\left[-\frac{\partial}{\partial t} + \beta_{\mathrm{em}}(\alpha_\mu) \frac{\partial}{\partial \alpha_\mu}\right] \alpha(|q^2|) = 0. \tag{15.43}$$

It follows that $S(1, \alpha(|q^2|))$ is a solution of (15.38).

This is a remarkable result. It shows that all the dependence of S on the (momentum)^2 variable $|q^2|$ enters through that of the running coupling $\alpha(|q^2|)$. Of course, this result is only valid in a regime of $-q^2$ which is much greater than all quantities with dimension (mass)^2—for example the squares of all particle masses, which do not appear in (15.31). This is why the technique applies only at 'high' $-q^2$. The result implies that if we can calculate $S(1, \alpha_\mu)$ (i.e. S at the point $q^2 = -\mu^2$) at some definite order in perturbation theory, then replacing α_μ by $\alpha(|q^2|)$ will allow us to predict the q^2-dependence (at large $-q^2$). All we need to do is solve (15.39). Indeed, for QED with one e$^+$e$^-$ loop we have seen that $\beta_{\text{em}}^{[2]}(\alpha) = \alpha^2/3\pi$. Hence integrating (15.39) we obtain

$$\alpha(|q^2|) = \frac{\alpha_\mu}{1 - \frac{\alpha_\mu}{3\pi}t} = \frac{\alpha_\mu}{1 - \frac{\alpha_\mu}{3\pi}\ln(|q^2|/\mu^2)}. \tag{15.44}$$

This is almost exactly the formula we proposed in (11.57), on plausibility grounds.[1]

Suppose now that the leading QED perturbative contribution to $S(1, \alpha_\mu)$ is $S_1\alpha_\mu$. Then the terms contained in $S(1, \alpha(|q^2|))$ in this approximation can be found by expanding in powers of α_μ:

$$\begin{aligned} S(1, \alpha(|q^2|)) &\approx 1 + S_1\alpha(|q^2|) = 1 + S_1\alpha_\mu\left[1 - \frac{\alpha_\mu}{3\pi}t\right]^{-1} \\ &= 1 + S_1\alpha_\mu\left[1 + \frac{\alpha_\mu t}{3\pi} + \left(\frac{\alpha_\mu t}{3\pi}\right)^2 + \dots\right], \end{aligned} \tag{15.45}$$

where $t = \ln(|q^2|/\mu^2)$. The next higher-order calculation of $S(1, \alpha_\mu)$ would be $S_2\alpha_\mu^2$, say, which generates the terms

$$S_2\alpha^2(|q^2|) = S_2\alpha_\mu^2\left[1 + \frac{2\alpha_\mu t}{3\pi} + \dots\right]. \tag{15.46}$$

Comparing (15.45) and (15.46) we see that each power of the large log factor appearing in (15.46) comes with one more power of α_μ than in (15.45). Provided α_μ is small, then, the *leading* terms in $t, t^2, \dots$ are contained in (15.45). It is in this sense that replacing $S(1, \alpha_\mu)$ by $S(1, \alpha(|q^2|))$ sums all 'leading log terms'.

In fact, of course, the one-loop (and higher) corrections to S in which we are really interested are those due to QCD, rather than QED, corrections. But the logic is exactly the same. The leading (O(α_s)) perturbative contribution to $S = \sigma/\sigma_{\text{pt}}$ at $q^2 = -\mu^2$ is given in (15.1) as $\alpha_s(\mu^2)/\pi$. It follows that the 'leading log corrections' at high $-q^2$ are summed up by replacing this expression by $\alpha_s(|q^2|)/\pi$, where the running $\alpha_s(|q^2|)$ is determined by solving (15.39) with the QCD analogue of (15.36)—to which we now turn.

15.3 Back to QCD: asymptotic freedom

15.3.1 One loop calculation

The reader will of course have realized, some time back, that the quantity β_0 introduced in (15.3) must be precisely the coefficient of α_s^2 in the one-loop contribution to the β-function

---

[1] The difference has to do, of course, with the different renormalization prescriptions. Eq (11.57) is written in terms of an 'α' defined at $q^2 = 0$, and without neglect of m_e.

of QCD defined by

$$\beta_s = \mu^2 \frac{\partial \alpha_s}{\partial \mu^2}\bigg|_{\text{fixed bare } \alpha_s} ; \qquad (15.47)$$

that is to say,

$$\beta_s(\text{one loop}) = -\beta_0 \alpha_s^2 \qquad (15.48)$$

with

$$\beta_0 = \frac{33 - 2N_f}{12\pi}. \qquad (15.49)$$

For $N_f \leq 16$ the quantity β_0 is *positive*, so that the sign of (15.48)) is opposite to that of the QED analogue, equation (15.36). Correspondingly, (15.44) is replaced by

$$\alpha_s(|q^2|) = \frac{\alpha_s(\mu^2)}{[1 + \alpha_s(\mu^2)\beta_0 \ln(Q^2/\mu^2)]}, \qquad (15.50)$$

where $Q^2 = |q^2|.[2]$ Then replacing α_s in (15.1) by (15.50) leads to (15.7).

Thus in QCD the strong coupling runs in the opposite way to QED, becoming smaller at large values of Q^2 (or small distances)—the property of asymptotic freedom. The justly famous result (15.49) was first obtained by Politzer (1973), Gross and Wilczek (1973), and 't Hooft. 't Hooft's result, announced at a conference in Marseilles in 1972, was not published. The published calculation of Politzer and of Gross and Wilczek quickly attracted enormous interest, because it immediately offered a way to understand how the successful parton model could be reconciled with the undoubtedly very strong binding forces between quarks. The resolution, we now understand, lies in quite subtle properties of renormalized quantum field theory, involving first the exposure of 'large logarithms', then their re-summation in terms of the running coupling, and of course the crucial sign of the β-function. Not only did the result (15.49) explain the success of the parton model: it also, we repeat, opened the prospect of performing reliable perturbative calculations in a *strongly* interacting theory, at least at high Q^2. For example, at sufficiently high Q^2, we can reliably compute the β function in perturbation theory. The result of Politzer and of Gross and Wilczek, when combined with the motivations for a colour SU(3) group discussed in the previous chapter, led rapidly to the general acceptance of QCD as the theory of strong interactions, a conclusion reinforced by the demonstration by Coleman and Gross (1973) that no theory without Yang-Mills fields possessed the property of asymptotic freedom.

In section 11.5.3 we gave the conventional physical interpretation of the way in which the running of the QED coupling tends to *increase* its value at distances short enough to probe inside the screening provided by e^+e^- pairs ($|q|^{-1} \ll m_e^{-1}$). This vacuum polarization screening effect is also present in (15.49) via the term $-\frac{2N_f}{12\pi}$, the value of which can be quite easily understood. It arises from the '$q\bar{q}$' vacuum polarization diagram of figure 15.3, which is precisely analogous to the e^+e^- diagram used to calculate $\bar{\Pi}_\gamma^{[2]}(q^2)$ in QED. The only new feature in figure 15.3 is the presence of the $\frac{\lambda}{2}$-matrices at each vertex. If 'a' and 'b' are the colour labels of the ingoing and outgoing gluons, the $\frac{\lambda}{2}$-matrix factors must be

$$\sum_{\alpha,\beta=1}^{3} \left(\frac{\lambda_a}{2}\right)_{\alpha\beta} \left(\frac{\lambda_b}{2}\right)_{\beta\alpha} \qquad (15.51)$$

since there are no free quark indices (of type α, β) on the external legs of the diagram.

---

[2]Except that in (15.50) α_s is evaluated at large *spacelike* values of q^2, whereas in (15.7) it is wanted at large *timelike* values. Readers troubled by this may consult Peskin and Schroeder (1995) section 18.5. The difficulty is evaded in the approach of section 15.6.

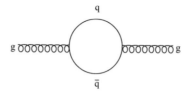

FIGURE 15.3

q$\bar{q}$ vacuum polarization correction to the gluon propagator.

It is simple to check that (15.51) has the value $\frac{1}{2}\delta_{ab}$ (this is, in fact, the way the λ's are conventionally normalized). Hence for one quark flavour we expect '$\alpha/3\pi$' to be replaced by '$\alpha_s/6\pi$', in agreement with the second term in (15.49).

The all-important, positive, first term must therefore be due to the gluons. The one-loop graphs contributing to the calculation of β_0 are shown in figure 15.4. They include figure 15.3, of course, but there are also, characteristically, graphs involving the gluon self-coupling which is absent in QED, and also (in covariant gauges) ghost loops. We do not want to enter into the details of the calculation of $\beta(\alpha_s)$ here (they are given in Peskin and Schroeder (1995) chapter 16, for example), but it would be nice to have a simple intuitive picture of the 'antiscreening' result in terms of the gluon interactions, say. Unfortunately no fully satisfactory simple explanation exists, though the reader may be interested to consult Hughes (1980, 1981) and Nielsen (1981) for a 'paramagnetic' type of explanation, rather than a 'dielectric' one.

Returning to (15.50), we note that the equation effectively provides a prediction of α_s at any scale Q^2, given its value at a particular scale $Q^2 = \mu^2$, which must be taken from experiment. The reference scale is now normally taken to be the Z^0 mass; the value $\alpha_s(m_Z^2)$ then plays the role in QCD that $\alpha \sim 1/137$ does in QED.

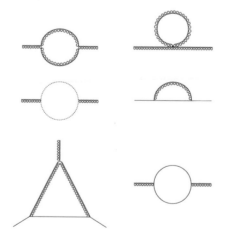

FIGURE 15.4

Graphs contributing to the one-loop β function in QCD. The curly line represents a gluon, a dotted line a ghost (see section 13.3.3) and a straight line a quark.

Despite appearances, equation (15.50) does not really involve two parameters—after all, (15.47) is only a first-order differential equation. By introducing

$$\ln \Lambda_{\text{QCD}}^2 = \ln \mu^2 - 1/(\beta_0 \alpha_{\text{s}}(\mu^2)), \tag{15.52}$$

equation (15.50) can be rewritten (problem 15.3) as

$$\alpha_{\text{s}}(Q^2) = \frac{1}{\beta_0 \ln(Q^2/\Lambda_{\text{QCD}}^2)}. \tag{15.53}$$

Equation (15.53) is equivalent to (cf (15.30))

$$\ln\left(Q^2/\Lambda_{\text{QCD}}^2\right) = \int_{\alpha_{\text{s}}(Q^2)}^{\infty} \frac{\mathrm{d}\alpha_{\text{s}}}{\beta_{\text{s}}(\text{one loop})} \tag{15.54}$$

with $\beta_{\text{s}}(\text{one loop}) = -\beta_0 \alpha_{\text{s}}^2$. Λ_{QCD} is therefore an integration constant, representing the scale at which α_{s} would diverge to infinity (if we extended our calculation beyond its perturbative domain of validity). More usefully, Λ_{QCD} is a measure of the scale at which α_{s} really does become 'strong'. The extraction of a value of Λ_{QCD} is a somewhat complicated matter, as we shall briefly indicate in the following section, but a typical value is in the region of 200 MeV. Note that this is a distance scale of order $(200 \text{ MeV})^{-1} \sim 1$ fm, just about the size of a hadron — a satisfactory connection.

15.3.2 Higher-order calculations and experimental comparison

So far we have discussed only the 'one-loop' calculation of $\beta(\alpha_{\text{s}})$. The general perturbative expansion for β_{s} can be written as

$$\beta_{\text{s}}(\alpha_{\text{s}}) = -\beta_0 \alpha_{\text{s}}^2 - \beta_1 \alpha_{\text{s}}^3 - \beta_2 \alpha_{\text{s}}^4 + \dots \tag{15.55}$$

where β_0 is the one-loop coefficient given in (15.49), β_1 is the two-loop coefficient, and so on. β_1 was calculated by Caswell (1974) and Jones (1974), and has the value

$$\beta_1 = \frac{153 - 19N_{\text{f}}}{24\pi^2}. \tag{15.56}$$

The three-loop coefficient β_2, obtained by Tarasov *et al.* (1980) and by Larin and Vermaseren (1993), is

$$\beta_2 = \frac{77139 - 15099N_{\text{f}} + 325N_{\text{f}}^2}{3456\pi^2}. \tag{15.57}$$

The four-loop coefficient β_3 was calculated by van Ritbergen *et al.* (1997) and by Czakon (2005); the five-loop coefficient β_4 was calculated by Baikov, Chetyrkin and Kühn (2017), Luthe *et al.* (2016), Herzog *et al.* (2017), Luthe *et al.* (2017), and Chetyrkin *et al.* (2017). A technical point to note is that while β_0 and β_1 are independent of the scheme adopted for renormalization (see appendix O), the higher-order coefficients do depend on it; the value (15.57) is in the widely used $\overline{\text{MS}}$ scheme. Likewise, Λ_{QCD} will be scheme-dependent (see appendix O), and the value $\Lambda_{\overline{\text{MS}}}$ will be used here (the 'QCD' now being understood).

The β-function coefficients β_i are given for a theory in which there are N_{f} 'active' or 'light' quark flavours, with masses $m_{\text{q}} \ll Q$, where Q is the running energy scale. For $Q^2 = M_{\text{Z}}^2$, for example, well above the beauty threshold, N_{f} will be 5. As Q^2 runs to smaller values, and a flavour threshold is crossed, N_{f} changes by one unit. Physical quantities must, however, be continuous across such thresholds. This requires that the values of α_{s} above and below that threshold satisfy certain matching conditions (Rodrigo and Santamaria 1993, Bernreuther and Wetzel 1982, Chetyrkin *et al.* 1997).

Only in the one-loop approximation for β_s can an analytic solution of (15.47) be obtained. However, a useful approximate solution can be found iteratively, as follows. Consider the two-loop version of (15.54), namely

$$\ln(Q^2/\Lambda_{\overline{\mathrm{MS}}}^2) = -\int \frac{\mathrm{d}\alpha_s}{\beta_0\alpha_s^2 + \beta_1\alpha_s^3}. \tag{15.58}$$

Expanding the denominator and integrating gives

$$\ln(Q^2/\Lambda_{\overline{\mathrm{MS}}}^2) = \frac{1}{\beta_0\alpha_s} + \frac{b_1}{\beta_0}\ln\alpha_s + C, \tag{15.59}$$

where $b_1 = \beta_1/\beta_0$ and C is a constant. In the $\overline{\mathrm{MS}}$ scheme, C is given by $C = (b_1/\beta_0)\ln\beta_0$. Then the equation for α_s is

$$L = \frac{1}{\beta_0\alpha_s} + \frac{b_1}{\beta_0}\ln\beta_0\alpha_s, \tag{15.60}$$

where we have defined $L = \ln(Q^2/\Lambda_{\overline{\mathrm{MS}}}^2)$. In first approximation, one sets b_1 to zero and finds $\alpha_s = (1/\beta_0 L)$ as before. To obtain the next approximation, we set $\alpha_s = (1/\beta_0 L)$ in the b_1 term of (15.60), and solve for α_s to first order in b_1. This gives (problem 15.4 (a))

$$\alpha_s = \frac{1}{\beta_0 L} - \frac{1}{\beta_0^3 L^2}\beta_1 \ln L. \tag{15.61}$$

Problem 15.4 (b) carries the calculation to the three-loop stage.

This iteration solution of the RGE (15.47) can be carried out to higher loop orders, but an alternative is to solve the RGE equation numerically, including the matching procedure at the flavour thresholds. In this case Λ does not arise directly, and instead the value of $\alpha_s(Q^2)$ at a definite value of Q^2 has to be specified as an integration constant. It is conventional to quote the value at $Q^2 = M_Z^2$.

On the experimental side, α_s has been determined in many different ways, and at different energies. We saw an early example of this in figure 14.20, showing the ratio of 3-jet to 2-jet events in $\mathrm{e^+e^-}$ annihilation to hadrons, which is proportional to α_s in lowest order of QCD perturbation theory. To extract an accurate value of α_s, higher orders will of course have to be included, as will be discussed in the following section. Another example of an α_s determination is provided by corrections to the parton model in deep inelastic scattering, to be discussed in section 15.6. In addition to these two sub-fields, it is customary to list four more experimental sub-fields in which α_s has been determined: hadronic τ decays, heavy quarkonia decays, observables measured in hadron colliders, and electroweak precision fits. To these may also be added the theoretical calculation of α_s in lattice QCD, to be discussed in chapter 16.

These determinations of α_s are regularly reviewed, most recently by Huston, Rabbertz, and Zanderighi (2022), section 9 of Workman *et al.* (2022). These authors give the world average value of $\alpha_s(M_Z^2)$ as

$$\alpha_s(M_Z^2) = 0.1179 \pm 0.0009.. \tag{15.62}$$

which includes the averaged lattice result (Aoki *et al.* 2020). Figure 15.5 (from Huston, Rabbertz and Zanderighi (2022)) shows a summary of measurements of α_s as a function of the energy scale Q. The respective degree of QCD perturbation theory used in the extraction of α_s is shown in brackets: NLO means next-to-leading order, NNLO means next-to-next-to-leading order, and so on. The measurements may be compared to the theoretical prediction,

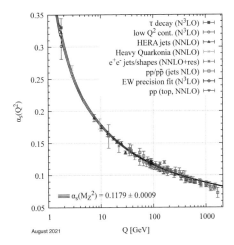

FIGURE 15.5
Comparison between measurements of α_s and the theoretical prediction, as a function of the energy scale Q (Huston, Rabbertz and Zanderighi 2022).

which is the solid line with the quoted value of $\alpha_\mathrm{s}(M_Z^2)$, and evaluated in 4-loop approximation, using 3-loop threshold matching at the charm and beauty thresholds. The agreement is excellent, a triumph for both experiment and theory.

15.4 $\sigma(\mathrm{e^+e^-} \to$ hadrons$)$ revisited

We may now return to the physical process which originally motivated this extensive detour. The perturbative corrections to $\sigma_\mathrm{pt}(Q^2)$ are expressed as a power series in α_s,

$$\sigma(\mathrm{e^+e^-} \to \text{hadrons}) = \sigma_\mathrm{pt}(Q^2)\left[1 + \sum_{n=1}^{\infty} c_n(Q^2/\mu^2)\left(\frac{\alpha_\mathrm{s}(\mu^2)}{\pi}\right)^n\right], \qquad (15.63)$$

where μ is the renormalization scale. (A similar expansion can be written for many other physical quantities too.) The coefficients from c_2 onwards depend on the renormalization scheme (see appendix O), and are usually quoted in the $\overline{\mathrm{MS}}$ scheme. c_1 is the leading order (LO) coefficient, and we already know that $c_1 = 1$ from (15.1). c_2 is the next-to-leading (NLO) coefficient; $c_2(1)$ was calculated by Dine and Sapirstein (1979), Chetyrkin *et al.* (1979) and by Celmaster and Gonsalves (1980), and has the value $1.9857 - 0.1152 N_\mathrm{f}$. The next-to-next-to-leading (NNLO) coefficient $c_3(1)$ was calculated by Samuel and Surguladze (1991) and by Gorishnii *et al.* (1991), and is equal to -12.8 for five flavours. The N3LO coefficient $c_4(1)$ (which requires the evaluation of some 20,000 diagrams) may be found in Baikov *et al.* (2008) and Baikov *et al.* (2009).

The physical cross section $\sigma(\mathrm{e^+e^-} \to$ hadrons$)$ must be independent of the renormalization scale μ^2, and this would also be true of the series in (15.63) if an infinite number of terms were kept: the μ^2-dependence of the coefficients $c_n(Q^2/\mu^2)$ would cancel that of $\alpha_\mathrm{s}(\mu^2)$. This requirement can be imposed order by order in α_s to fix the μ^2-dependence of the coefficients, and is a direct way of applying the RGE idea. Consider, for example, truncating the series at the $n = 2$ stage:

$$\sigma(\mathrm{e^+e^-} \to \text{hadrons}) \approx \sigma_{\mathrm{pt}}(Q^2)\left(1 + \frac{\alpha_{\mathrm{s}}(\mu^2)}{\pi} + c_2(Q^2/\mu^2)(\alpha_{\mathrm{s}}(\mu^2)/\pi)^2\right). \qquad (15.64)$$

Differentiating with respect to μ^2 and setting the result to zero we obtain

$$\mu^2 \frac{\mathrm{d}c_2}{\mathrm{d}\mu^2} = -\frac{\pi\beta(\alpha_{\mathrm{s}}(\mu^2))}{(\alpha_{\mathrm{s}}(\mu^2))^2} \qquad (15.65)$$

where an $O(\alpha_{\mathrm{s}}^3)$ term has been dropped. Substituting the one-loop result (15.48)—as is consistent to this order—we find

$$c_2(Q^2/\mu^2) = c_2(1) - \pi\beta_0 \ln(Q^2/\mu^2). \qquad (15.66)$$

The second term on the right-hand side of (15.66) gives the contribution identified in (15.2).

In practice only a finite number of terms $n = N$ will be available, and a μ^2-dependence will remain, which implies an uncertainty in the prediction of the cross section (and similar physical observables), due to the arbitrariness of the scale choice. This uncertainty will be of the same order as the neglected terms, i.e. of order $\alpha_{\mathrm{s}}^{N+1}$. Thus the scale dependence of a QCD prediction gives a measure of the uncertainties due to neglected terms. For $\sigma(\mathrm{e^+e^-} \to \text{hadrons})$ the choice of scale $\mu^2 = Q^2$ is usually made, so as to avoid large logarithms in relations such as (15.66).

Before proceeding to our second main application of the RGE, scaling violations in deep inelastic scattering, it is necessary to take another detour, to enlarge our understanding of the scope of the RGE.

15.5   A more general form of the RGE: anomalous dimensions and running masses

The reader may have wondered why, for QCD, all the graphs of figure 15.4 are needed, whereas for QED we got away with only figure 11.3. The reason for the simplification in QED was the equality between the renormalization constants Z_1 and Z_2, which therefore cancelled out in the relation between the renormalized and bare charges e and e_0, as briefly stated before equation (15.8) (this equality was discussed in section 11.6). We recall that Z_2 is the field strength renormalization factor for the charged fermion in QED, and Z_1 is the vertex part renormalization constant; their relation to the counter terms was given in equation (11.7). For QCD, although gauge invariance does imply generalizations of the Ward identity used to prove $Z_1 = Z_2$ (Taylor 1971, Slavnov 1972), the consequence is no longer the simple relation '$Z_1 = Z_2$' in this case, due essentially to the ghost contributions. In order to see what change $Z_1 \neq Z_2$ would make, let us return to the one-loop calculation of β for QED, pretending that $Z_1 \neq Z_2$. We have

$$e_0 = \frac{Z_1}{Z_2}Z_3^{-\frac{1}{2}}e_\mu \qquad (15.67)$$

where, because we are renormalizing at scale μ, all the Z_i's depend on μ (as in (15.15)), but we shall now not indicate this explicitly. Taking logs and differentiating with respect to μ at constant e_0, we obtain

$$\mu\frac{\mathrm{d}}{\mathrm{d}\mu}\bigg|_{e_0} \ln Z_1 - \mu\frac{\mathrm{d}}{\mathrm{d}\mu}\bigg|_{e_0} \ln Z_2 - \frac{1}{2}\mu\frac{\mathrm{d}}{\mathrm{d}\mu}\bigg|_{e_0} \ln Z_3 + \frac{\mu}{e_\mu}\frac{\mathrm{d}e_\mu}{\mathrm{d}\mu}\bigg|_{e_0} = 0. \qquad (15.68)$$

Hence

$$\beta(e_\mu) \equiv \mu \frac{\mathrm{d}e_\mu}{\mathrm{d}\mu}\bigg|_{e_0} = e_\mu\gamma_3 + 2e_\mu\gamma_2 - e_\mu\mu\frac{\mathrm{d}}{\mathrm{d}\mu}\ln Z_1, \tag{15.69}$$

where

$$\gamma_2 \equiv \frac{1}{2}\mu\frac{\mathrm{d}}{\mathrm{d}\mu}\bigg|_{e_0}\ln Z_2, \qquad \gamma_3 = \frac{1}{2}\mu\frac{\mathrm{d}}{\mathrm{d}\mu}\bigg|_{e_0}\ln Z_3. \tag{15.70}$$

To leading order in e_μ, the γ_3 term in (15.70) reproduces (15.26) when (15.15) is used for Z_3, the other two terms in (15.68) cancelling via $Z_1 = Z_2$. So if, as in the case of QCD, Z_1 is not equal to Z_2, we need to introduce the contributions from loops determining the fermion field strength renormalization factor, as well as those related to the vertex parts (together with appropriate ghost loops), in addition to the vacuum polarization loop associated in the Z_3.

Quantities such as γ_2 and γ_3 have an interesting and important significance, which we shall illustrate in the case of γ_2 for QED. Z_2 enters into the relation between the propagator of the bare fermion $\langle\Omega|T(\hat{\psi}_0(x)\hat{\psi}_0(0))|\Omega\rangle$ and the renormalized one, via (cf (11.2))

$$\langle\Omega|T(\hat{\psi}(x)\hat{\psi}(0))|\Omega\rangle = \frac{1}{Z_2}\langle\Omega|T(\hat{\psi}_0(x)\hat{\psi}_0(0))|\Omega\rangle, \tag{15.71}$$

where (cf section 10.1.3) $|\Omega\rangle$ is the vacuum of the interacting theory. The Fourier transform of (15.71) is, of course, the Feynman propagator:

$$\tilde{S}'_F(q^2) = \int \mathrm{d}^4x\, e^{iq\cdot x}\langle\Omega|T(\hat{\psi}(x)\hat{\psi}(0))|\Omega\rangle. \tag{15.72}$$

Suppose we now ask: what is the large $-q^2$ behaviour of (15.72) for space-like q^2, with $-q^2 \gg m^2$ where m is the fermion mass? This sounds very similar to the question answered in 15.2.3 for the quantity $S(|q^2|/\mu^2, e_\mu)$. However, the latter was dimensionless whereas (recalling that $\hat{\psi}$ has mass dimension $\frac{3}{2}$) $\tilde{S}'_F(q^2)$ has dimension M^{-1}. This dimensionality is, of course, just what a propagator of the free-field form $i/(\slashed{q} - m)$ would provide.

Accordingly, we extract this $(\slashed{q})^{-1}$ factor (compare $\sigma/\sigma_{\mathrm{pt}}$) and consider the dimensionless ratio $\tilde{R}'_F(|q^2|/\mu^2, \alpha_\mu) = \slashed{q}\tilde{S}'_F(q^2)$. We might guess that, just as for $S(|q^2|/\mu^2, \alpha_\mu)$, to get the leading large $|q^2|$ behaviour we will need to calculate $\tilde{R}'_F$ to some order in α_μ, and then replace α_μ by $\alpha(|q^2|/\mu^2)$. But this is not quite all. The factor Z_2 in (15.71) will—as noted above—depend on the renormalization scale μ, just as Z_3 of (15.15) did. Thus when we change μ, the normalization of the $\hat{\psi}$'s will change via the $Z_2^{\frac{1}{2}}$ factors—of course by a finite amount here—and we must include this change when writing down the analogue of (15.33) for this case (i.e. the condition that the 'total change, on changing μ, is zero'). The required equation is

$$\left[\mu^2\frac{\partial}{\partial\mu^2}\bigg|_{\alpha_\mu} + \beta(\alpha_\mu)\frac{\partial}{\partial\alpha_\mu} + \gamma_2(\alpha_\mu)\right]\tilde{R}'_F(|q^2|/\mu^2, \alpha_\mu) = 0. \tag{15.73}$$

The solution of (15.73) is somewhat more complicated than that of (15.33). We can gain insight into the essential difference caused by the presence of γ_2 by considering the special case $\beta(\alpha_\mu) = 0$. In this case, we easily find

$$\tilde{R}'_F(|q^2|/\mu^2, \alpha_\mu) \propto (\mu^2)^{-\gamma_2(\alpha_\mu)}. \tag{15.74}$$

But since $\tilde{R}'_F$ can only depend on μ via $|q^2|/\mu^2$, we learn that if $\beta = 0$ then the large $|q^2|$ behaviour of $\tilde{R}'_F$ is given by $(|q^2|/\mu^2)^{\gamma_2}$—or, in other words, that at large $|q^2|$

$$\tilde{S}'_F(|q^2|/\mu^2, \alpha_\mu) \propto \frac{1}{\slashed{q}}\left(\frac{|q^2|}{\mu^2}\right)^{\gamma_2(\alpha_\mu)}. \tag{15.75}$$

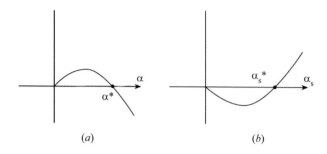

(a) (b)

FIGURE 15.6
Possible behaviour of β functions. (a) The slope is positive near the origin (as in QED), and negative near $\alpha = \alpha^*$. (b) The slope is negative at the origin (as in QCD) and positive near $\alpha_{\rm s} = \alpha_{\rm s}^*$.

Thus, *at a zero of the β-function*, $\tilde{S}'_F$ has an 'anomalous' power law dependence on $|q^2|$ (i.e. in addition to the obvious $\not{q}^{-1}$ factor), which is controlled by the parameter γ_2. The latter is called the 'anomalous dimension' of the fermion field, since its presence effectively means that the $|q^2|$ behaviour of $\tilde{S}'_F$ is not determined by its 'normal' dimensionality M^{-1}. The behaviour (15.75) is often referred to as 'scaling with anomalous dimension', meaning that if we multiply $|q^2|$ by a scale factor λ, then $\tilde{S}'_F$ is multiplied by $\lambda^{\gamma_2(\alpha_\mu)-1}$ rather than just λ^{-1}. Anomalous dimensions turn out to play a vital role in the theory of critical phenomena— they are, in fact, closely related to 'critical exponents' (see section 16.4.3, and Peskin and Schroeder 1995, chapter 13). Scaling with anomalous dimensions is also exactly what occurs in deep inelastic scattering of leptons from nucleons, as we shall see in section 15.6.

The full solution of (15.73) for $\beta \neq 0$ is elegantly discussed in Coleman (1985), chapter 3; see also Peskin and Schroeder (1995) section 12.3. We quote it here:

$$\tilde{R}'_F(|q^2|/\mu^2), \alpha_\mu), = \tilde{R}'_F(1, \alpha(|q^2|/\mu^2)) \exp\left\{ \int_0^t \mathrm{d}t'\, \gamma_2(\alpha(t')) \right\}. \tag{15.76}$$

The first factor is the expected one from section 15.2.3; the second results from the addition of the γ_2 term in (15.73). Suppose now that $\beta(\alpha)$ has a zero at some point α^*, in the vicinity of which $\beta(\alpha) \approx -B(\alpha - \alpha^*)$ with $B > 0$. Then, near this point the evolution of α is given by (cf (15.39))

$$\ln(|q^2|/\mu^2) = \int_{\alpha_\mu}^{\alpha(|q^2|)} = \frac{\mathrm{d}\alpha}{-B(\alpha - \alpha^*)}, \tag{15.77}$$

which implies

$$\alpha(|q^2|) = \alpha^* + \text{constant} \times (\mu^2/|q^2|)^B. \tag{15.78}$$

Thus asymptotically for large $|q^2|$, the coupling will evolve to the *fixed point* α^*. In this case, at sufficiently large $-q^2$, the integral in (15.76) can be evaluated by setting $\alpha(t') = \alpha^*$, and $\tilde{R}'_F$ will scale with an anomalous dimension $\gamma_2(\alpha^*)$ determined by the fixed point value of α. The behaviour of such an α is shown in figure 15.6(a). We emphasize that there is no reason to believe that the QED β function actually does behave like this.

The point α^* in figure 15.6(a) is called an ultraviolet-stable fixed point: α 'flows' to-wards it at large $|q^2|$. In the case of QCD, the β function starts out negative, so that the corresponding behaviour (with a zero at a $\alpha_{\rm s}^* \neq 0$) would look like that shown in figure 15.6(b). In this case, the reader can check (problem 15.5) that $\alpha_{\rm s}^*$ is reached in the infrared limit $q^2 \to 0$, and so α_s^* is called an infrared-stable fixed point. Clearly it is the slope of β near the fixed point that determines whether it is u-v or i-r stable. This applies equally to a fixed point at the origin, so that QED is i-r stable at $\alpha = 0$ while QCD is u-v stable at $\alpha_s = 0$.

We must now point out to the reader an error in the foregoing analysis, in the case of a gauge theory. The quantity Z_2 is not gauge invariant in QED (or QCD), and hence γ_2 depends on the choice of gauge. This is really no surprise, because the full fermion propagator itself is not gauge invariant (the free-field propagator is gauge invariant, of course). What ultimately matters is that the complete physical amplitude for any process, at a given order of α, be gauge invariant. Thus the analysis given above really only applies—in this simple form—to non-gauge theories, such as the ABC model of chapter 6, or to gauge-invariant quantities.

This is an appropriate point at which to consider the treatment of quark masses in the RGE-based approach. Up to now we have simply assumed that the relevant $|q^2|$ is very much greater than all quark masses, the latter therefore being neglected. While this may be adequate for the light quarks u, d, s, it seems surely a progressively worse assumption for c, b, and t. However, in thinking about how to re-introduce the quark masses into our formalism, we are at once faced with a difficulty: how are they to be defined? For an unconfined particle, such as a lepton, it seems natural to define 'the' mass as the position of the pole of the propagator (i.e. the 'on-shell' value $p^2 = m^2$), a definition we followed in chapters 10 and 11. Significantly, renormalization is required (in the shape of a mass counter-term) to achieve a pole at the 'right' physical mass m, in this sense. But this prescription is inherently perturbative, and cannot be used for a confined particle, which never 'escapes' beyond the range of the non-perturbative confining forces, and whose propagator can therefore never approach the form $\sim (\not{p} - m)^{-1}$ of a free particle.

Our present perspective on renormalization suggests an obvious way forward. Just as there was, in principle, no *necessity* to define the QED coupling parameter e via an on-shell prescription, so here a mass parameter in the Lagrangian can be defined in any way we find convenient; all that is necessary is that it should be possible to determine its value from some measurable quantity (for example, quark masses from lattice QCD predictions of hadron masses). Effectively, we are regarding the 'm' in a term such as $-m\bar{\hat{\psi}}(x)\hat{\psi}(x)$ as a 'coupling constant' having mass dimension 1 (and, after all, the ABC coupling itself had mass dimension 1). Incidentally, the operator $\bar{\hat{\psi}}(x)\hat{\psi}(x)$ *is* gauge invariant, as is any such *local* operator. Taking this point of view, it is clear that a renormalization scale will be involved in such a general definition of mass, and we must expect to see our mass parameters 'evolve' with this scale, just as the gauge (or other) couplings do. In turn, this will get translated into a $|q^2|$-dependence of the mass parameters, just as for $\alpha(|q^2|)$ and $\alpha_s(|q^2|)$.

The RGE in such a scheme now takes the form

$$\left[\mu^2 \frac{\partial}{\partial \mu^2} + \beta(\alpha_s)\frac{\partial}{\partial \alpha_s} + \sum_i \gamma_i(\alpha_s) + \gamma_m(\alpha_s)m\frac{\partial}{\partial m}\right] R(|q^2|/\mu^2, \alpha_s, m/|q|) = 0 \qquad (15.79)$$

where the partial derivatives are taken at fixed values of the other two variables. Here the γ_i are the anomalous dimensions relevant to the quantity R, and γ_m is an analogous 'anomalous mass dimension', arising from finite shifts in the mass parameter when the scale μ^2 is changed. Just as with the solution (15.76) of (15.73), the solution of (15.79) is given in terms of a 'running mass' $m(|q^2|)$. Formally, we can think of γ_m in (15.79) as analogous to $\beta(\alpha_s)$ and $\ln m$ as analogous to α_s. Then equation (15.41) for the running α_s,

$$\frac{\partial \alpha_s(|q^2|)}{\partial t} = \beta(\alpha_s(|q^2|)) \qquad (15.80)$$

where $t = \ln(|q^2|/\mu^2)$, becomes

$$\frac{\partial(\ln m(|q^2|))}{\partial t} = \gamma_m(\alpha_{\rm s}(|q|^2)). \tag{15.81}$$

Equation (15.81) has the solution

$$m(|q^2|) = m(\mu^2) \exp \int_{\ln \mu^2}^{\ln |q^2|} {\rm d}\ln |q'^2| \; \gamma_m(\alpha_{\rm s}(|q'^2|)) \tag{15.82}$$

To one-loop order in QCD, $\gamma_m(\alpha_{\rm s})$ turns out to be $-\frac{1}{\pi}\alpha_{\rm s}$ (Peskin and Schroeder 1995, section 18.1). Inserting the one-loop solution for $\alpha_{\rm s}$ in the form (15.53), we find

$$m(|q^2|) = m(\mu^2) \left[\frac{\ln(\mu^2/\Lambda^2)}{\ln(|q^2|/\Lambda^2)} \right]^{\frac{1}{\pi\beta_0}}, \tag{15.83}$$

where $(\pi\beta_0)^{-1} = 12/(33 - 2N_{\rm f})$. Thus the quark masses decrease logarithmically as $|q^2|$ increases, rather like $\alpha_{\rm s}(|q^2|)$. It follows that, in general, quark mass effects are suppressed both by explicit $m^2/|q^2|$ factors, and by the logarithmic decrease given by (15.83). Further discussion of the treatment of quark masses is contained in Ellis, Stirling, and Webber (1996), section 2.4; see also the review by Barnett, Lellouch, and Manohar (2022).

15.6 QCD corrections to the parton model predictions for deep inelastic scattering: scaling violations

As we saw in section 9.2, the parton model provides a simple intuitive explanation for the experimental observation that the nucleon structure functions in deep inelastic scattering depend, to a good first approximation, only on the dimensionless ratio $x = Q^2/2M\nu$, rather than on Q^2 and ν separately; this behaviour is referred to as 'scaling'. Here M is the nucleon mass, and Q^2 and ν are defined in (9.7) and (9.8). In this section we shall show how QCD corrections to the simple parton model, calculated using RGE techniques, predict observable violations of scaling in deep inelastic scattering. As we shall see, comparison between the theoretical predictions and experimental measurements provides strong evidence for the correctness of QCD as the theory of nucleonic constituents.

15.6.1 Uncancelled mass singularities at order $\alpha_{\rm s}$

The free parton model amplitudes we considered in chapter 9 for deep inelastic lepton-nucleon scattering were of the form shown in figure 15.7 (cf figure 9.4). The obvious first QCD corrections will be due to real gluon emission by either the initial or final quark, as shown in figure 15.8, but to these we must add the one-loop virtual gluon processes of figure 15.9 in order (see below) to get rid of infrared divergences similar to those encountered in section 14.4.2, and also the diagram of figure 15.10, corresponding to the presence of gluons in the nucleon. To simplify matters, we shall consider what is called a 'non-singlet structure function' $F_2^{\rm NS}$, such as $F_2^{\rm ep} - F_2^{\rm en}$ in which the (flavour) singlet gluon contribution cancels out, leaving only the diagrams of figures 15.8 and 15.9.

We now want to perform, for these diagrams, calculations analogous to those of section 9.2, which enabled us to find the e-N structure functions νW_2 and $M W_1$ from the simple

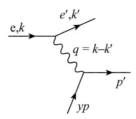

FIGURE 15.7
Electron-quark scattering via one-photon exchange.

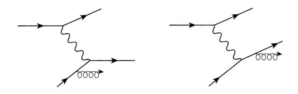

FIGURE 15.8
Electron-quark scattering with single-gluon emission

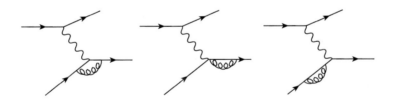

FIGURE 15.9
Virtual single-gluon corrections to figure 15.7.

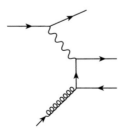

FIGURE 15.10
Electron-gluon scattering with q̄q production.

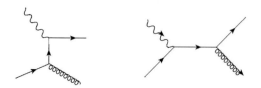

FIGURE 15.11

Virtual photon processes entering into figure 15.8.

parton process of figure 15.7. There are two problems here. One is to find the parton level W's corresponding to   figure 15.8 (leaving aside figure 15.9 for the moment)—cf equations (9.29) and (9.30) in the case of the free parton diagram figure 15.7; the other is to relate these parton W's to observed nucleon W's via an integration over momentum fractions. In section 9.2 we solved the first problem by explicitly calculating the parton level $\mathrm{d}^2\sigma^i/\mathrm{d}Q^2\mathrm{d}\nu$ and picking off the associated νW_2^i, W_1^i. In principle, the same can be done here, starting from the five-fold differential cross section for our $e^- + q \to e^- + q + g$ process. However, a simpler—if somewhat heuristic—way is available. We note from (9.46) that in general $F_1 = MW_1$ is given by the transverse virtual photon cross section

$$W_1 = \sigma_T/(4\pi\alpha^2/K) = \frac{1}{2}\sum_{\lambda=\pm 1}\epsilon_\mu^*(\lambda)\epsilon_\nu(\lambda)W^{\mu\nu} \tag{15.84}$$

where $W^{\mu\nu}$ was defined in (9.3). Further, the Callan-Gross relation is still true (the photon only interacts with the charged partons, which are quarks with spin $\frac{1}{2}$ and charge e_i), and so

$$F_2/x = 2F_1 = 2MW_1 = \sigma_T/(4\pi\alpha^2/2MK). \tag{15.85}$$

These formulae are valid for both parton and proton W_1's and $W^{\mu\nu}$'s, with appropriate changes for parton masses $\hat{M}$. Hence the parton level $2\hat{F}_1$ for figure 15.8 is just the transverse photon cross section as calculated from the graphs of figure 15.11, divided by the factor $4\pi^2\alpha/2\hat{M}\hat{K}$, where as usual ' ^ ' denotes kinematic quantities in the corresponding parton process. This cross section, however, is—apart from a colour factor—just the virtual Compton cross section calculated in section 8.6. Also, taking the same (Hand) convention for the individual photon flux factors,

$$2\hat{M}\hat{K} = \hat{s}. \tag{15.86}$$

Thus for the parton processes of figure 15.9,

$$\begin{aligned}
2\hat{F}_1 &= \hat{\sigma}_T/(4\pi^2\alpha/2\hat{M}\hat{K}) \\
&= \frac{\hat{s}}{4\pi^2\alpha}\int_{-1}^{1}\mathrm{d}\cos\theta \cdot \frac{4}{3}\cdot\frac{\pi e_i^2\alpha\alpha_s(\mu^2)}{\hat{s}}\left(-\frac{\hat{t}}{\hat{s}} - \frac{\hat{s}}{\hat{t}} + \frac{2\hat{u}Q^2}{\hat{s}\hat{t}}\right) \tag{15.87}
\end{aligned}$$

where, in going from (8.180) to (15.87), we have inserted a colour factor $\frac{4}{3}$ (problem 14.5 (a)), renamed the variables $\hat{t} \to \hat{u}, \hat{u} \to \hat{t}$ in accordance with figure 15.11, and replaced α^2 by $e_i^2\alpha\alpha_s(\mu^2)$.

Before proceeding with (15.87), it is helpful to consider the other part of the calculation—namely the relation between the nucleon F_1 and the parton $\hat{F}_1$. We mimic the discussion of section 9.2, but with one significant difference: the quark 'taken' from the proton still has momentum fraction y (momentum yp), but now its longitudinal momentum must be degraded in the final state due to the gluon bremsstrahlung process we are calculating. Let

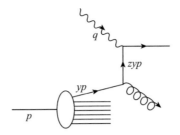

FIGURE 15.12
The first process of figure 15.11, viewed as a contribution to e^--nucleon scattering.

us call the quark momentum after gluon emission zyp (figure 15.12). Then, assuming as in section 9.2 that it stays on-shell, we have

$$q^2 + 2zyq \cdot p = 0 \tag{15.88}$$

or

$$x = yz, \quad x = Q^2/2q \cdot p, \quad q^2 = -Q^2 \tag{15.89}$$

and we can write (cf (9.31))

$$\frac{F_2}{x} = 2F_1 = \sum_i \int_0^1 \mathrm{d}y f_i(y) \int_0^1 \mathrm{d}z \, 2\hat{F}_1^i \delta(x - yz) \tag{15.90}$$

where the $f_i(y)$ are the parton distribution functions introduced in section 9.2 (we often call them $q(x)$ or $g(x)$ as the case may be) for parton type i, and the sum is over contributing partons. The reader may enjoy checking that (15.90) does reduce to (9.34) for free partons by showing that in that case $2\hat{F}_1^i = e_i^2 \delta(1 - z)$ (see Halzen and Martin 1984, section 10.3, for help), so that $2F_1^{\text{free}} = \sum_i e_i^2 f_i(x)$.

To proceed further with the calculation (i.e. of (15.87) inserted into (15.90)), we need to look at the kinematics of the $\gamma q \to qg$ process, in the CMS. Referring to figure 15.13, we let k, k' be the magnitudes of the CMS momenta $\mathbf{k}, \mathbf{k}'$. Then

$$
\begin{aligned}
\hat{s} &= 4k'^2 = (yp + q)^2 = Q^2(1 - z)/z, \quad z = Q^2/(\hat{s} + Q^2) \\
\hat{t} &= (q - p')^2 = -2kk'(1 - \cos\theta) = -Q^2(1 - c)/2z, \quad c = \cos\theta \\
\hat{u} &= (q - q')^2 = -2kk'(1 + \cos\theta) = -Q^2(1 + c)/2z.
\end{aligned}
\tag{15.91}
$$

We now note that in the integral (15.87) for $\hat{F}_1$, when we integrate over $c = \cos\theta$, we shall obtain an infinite result

$$\sim \int^1 \frac{\mathrm{d}c}{1 - c} \tag{15.92}$$

associated with the vanishing of $\hat{t}$ in the 'forward' direction (i.e. when q and p' are parallel). This is a divergence of the 'collinear' type, in the terminology of section 14.4.2—or, as there, a 'mass singularity', occurring in the zero quark mass limit. If we simply replace the propagator factor $\hat{t}^{-1} = [(q - p')^2]^{-1}$ by $[(q - p')^2 - m^2]^{-1}$, where m is a quark mass, then (15.92) becomes

$$\sim \int^1 \frac{\mathrm{d}c}{(1 + 2m^2z/Q^2) - c} \tag{15.93}$$

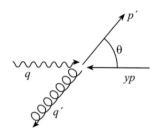

FIGURE 15.13
Kinematics for the parton process of figure 15.12.

which will produce a factor of the form $\ln(Q^2/m^2)$ as $m^2 \to 0$. Thus m regulates the divergence. We have here an uncancelled mass singularity, and it *violates scaling*. This crucial physical result is present in the lowest order QCD correction to the parton model, in this case. As we are learning, such logarithmic violations of scaling are a characteristic feature of all QCD corrections to the free (scaling) parton model.

We may calculate the coefficient of the $\ln Q^2$ term by retaining in (15.87) only the terms proportional to $\hat{t}^{-1}$:

$$2\hat{F}_1^i \approx e_i^2 \int_{-1}^{1} \frac{dc}{1-c} \left(\frac{\alpha_s(\mu^2)}{2\pi} \cdot \frac{4}{3} \cdot \frac{1+z^2}{1-z} \right) \tag{15.94}$$

and so, for just one quark species, this QCD correction contributes (from (15.90)) a term

$$\frac{e_i^2 \alpha_s(\mu^2)}{2\pi} \int_x^1 \frac{dy}{y} q(y) \left\{ \hat{P}_{qq}(x/y) \ln(Q^2/m^2) + C(x/y) \right\} \tag{15.95}$$

to $2F_1$, where

$$\hat{P}_{qq}(z) = \frac{4}{3} \left(\frac{1+z^2}{1-z} \right), \tag{15.96}$$

and $C(x/y)$ has no mass singularity.

Our result so far is therefore that the 'free' quark distribution function $q(x)$, which depended only on the scaling variable x, becomes modified to

$$q(x) + \frac{\alpha_s(\mu^2)}{2\pi} \int_x^1 \frac{dy}{y} q(y) \left\{ \hat{P}_{qq}(x/y) \ln\left(Q^2/m^2\right) + C(x/y) \right\} \tag{15.97}$$

$$= q(x) + \frac{\alpha_s(\mu^2)}{2\pi} \int_0^1 dy \int_0^1 dz \delta(zy - x) q(y) \{ \hat{P}_{qq}(z) \ln(Q^2/m^2)$$

$$+ C(z)\} \tag{15.98}$$

due to lowest-order gluon radiation. Clearly, this corrected distribution function violates scaling because of the $\ln Q^2$ term. But the result as it stands cannot represent a well-controlled approximation, since it contains divergences as $z \to 1$ and as $m^2 \to 0$.

We postpone discussion of the mass divergence until the next section. The divergence as $z \to 1$ is a standard infrared divergence (the quark momentum yzp after gluon emission becomes equal to the quark momentum yp before emission), and we expect that it can be cured by including the virtual gluon diagrams of figure 15.9, as indicated at the start of the section (and as was done analogously in the case of e^+e^- annihilation). This has been verified explicitly by Kim and Schilcher (1978) and by Altarelli *et al.* (1978 a, b; 1979). Alternatively, we follow the procedure of Altarelli and Parisi (1977). First we regulate the divergence as $z \to 1$ by defining a regulated function $1/(1-z)_+$ such that

$$\int_0^1 \frac{f(z)}{(1-z)_+} \mathrm{d}z = \int_0^1 \frac{f(z) - f(1)}{(1-z)} \mathrm{d}z = \int_0^1 \ln(1-z) \frac{\mathrm{d}f(z)}{\mathrm{d}z} \mathrm{d}z, \qquad (15.99)$$

where $f(z)$ is any test function sufficiently regular at the end points. Now the gluon loops which will cancel the i-r divergence only contribute at $z \to 1$, in leading log approximation. Thus the i-r finite version of $\hat{P}_{\mathrm{qq}}$ has the form

$$P_{\mathrm{qq}}(z) = \frac{4}{3} \frac{1 + z^2}{(1-z)_+} + A\delta(1 - z). \qquad (15.100)$$

The coefficient A is determined by the physical requirement that the net number of quarks (i.e. the number of quarks minus the number of antiquarks) does not vary with Q^2. From (15.98) this implies

$$\int_0^1 P_{\mathrm{qq}}(z) \mathrm{d}z = 0. \qquad (15.101)$$

Inserting (15.100) into (15.101), and using (15.99), we find (problem 15.6)

$$A = 2, \qquad (15.102)$$

so that

$$P_{\mathrm{qq}}(z) = \frac{4}{3} \frac{(1 + z^2)}{(1-z)_+} + 2\delta(1 - z). \qquad (15.103)$$

The function P_{qq} is called a 'splitting function', and it has an important physical interpretation. The quantity $\alpha_{\mathrm{s}}(\mu^2)/(2\pi) \, P_{\mathrm{qq}}(z)$ is, for $z < 1$, the probability that, to first order in α_{s}, a quark having radiated a gluon is left with a fraction z of its original momentum. Similar functions arise in QED in connection with what is called the 'equivalent photon approximation' (Weizsäcker 1934, Williams 1934, Chen and Zerwas 1975). The application of these techniques to QCD corrections to the free parton model is due to Altarelli and Parisi (1977), who thereby opened the way to this simpler and more physical way of understanding scaling violations, which had previously been discussed mainly within the rather technical operator product formalism (Wilson 1969).

We must now find some way of making sense, physically, of the uncancelled mass divergence in (15.97).

15.6.2 Factorization and the order α_{s} DGLAP equation

The key is to realize that when two partons are in the collinear configuration their relative momentum is very small, and hence the interaction between them is very strong, beyond the reach of a perturbative calculation. This suggests that we should absorb such uncalculable effects into a modified distribution function $q(x, \mu_{\mathrm{F}}^2)$ given by

$$q(x, \mu_{\mathrm{F}}^2) = q(x) + \frac{\alpha_{\mathrm{s}}(\mu^2)}{2\pi} \int_x^1 \frac{\mathrm{d}y}{y} q(y) P_{\mathrm{qq}}(x/y) \left\{ \ln(\mu_{\mathrm{F}}^2/m^2) + C(x/y) \right\} \qquad (15.104)$$

which we have to take from experiment. Note that we have also absorbed the non-singular term $C(x/y)$ into $q(x, \mu_{\mathrm{F}}^2)$. In terms of this quantity, then, we have

$$F_2(x, Q^2) \equiv e_i^2 x q(x, Q^2) \qquad (15.105)$$

$$= x e_i^2 \int_x^1 \frac{\mathrm{d}y}{y} q(y, \mu_{\mathrm{F}}^2) \left\{ \delta(1 - x/y) + \frac{\alpha_{\mathrm{s}}(\mu^2)}{2\pi} P_{\mathrm{qq}}(x/y) \ln(Q^2/\mu_{\mathrm{F}}^2) \right\}$$

$$(15.106)$$

to this order in α_s, and for one quark type.

This procedure is, of course, very reminiscent of ultraviolet renormalization, in which u-v divergences are controlled by similarly importing some quantities from experiment. In this example, we have essentially made use of the simple fact that

$$\ln(Q^2/m^2) = \ln(Q^2/\mu_F^2) + \ln(\mu_F^2/m^2). \tag{15.107}$$

The arbitrary scale μ_F is analogous to renormalization scale μ (which we have retained in $\alpha_s(\mu^2)$), and is here referred to as a 'factorization scale'. It is the scale entering into the separation in (15.107), between one (uncalculable) factor which depends on the i-r parameter m but not on Q^2, and the other (calculable) factor which depends on Q^2. The scale μ_F can be thought of as one which separates the perturbative short-distance physics from the non-perturbative long-distance physics. Thus partons emitted at small transverse momenta $< \mu_F$ (i.e. approximately collinear processes) should be considered as part of the hadron structure, and are absorbed into $q(x, \mu_F^2)$. Partons emitted at large transverse momenta contribute to the short-distance (calculable) part of the cross section. Just as for the renormalization scale, the more terms that can be included in the perturbative contributions to the mass-singular terms, the weaker the dependence on μ_F will be. We have demonstrated the possibility of factorization only to $O(\alpha_s)$, but proofs to all orders in perturbation theory exist; reviews are provided by Collins and Soper (1987, 1988).

Returning now to (15.106), the reader can guess what is coming next: we shall impose the condition that the physical quantity $F_2(x, Q^2)$ must be independent of the choice of factorization scale μ_F^2. Differentiating (15.106) partially with respect to μ_F^2, and setting the result to zero, we obtain (to order α_s on the right hand side)

$$\mu_F^2 \frac{\partial q(x, \mu_F^2)}{\partial \mu_F^2} = \frac{\alpha_s(\mu^2)}{2\pi} \int_x^1 \frac{dy}{y} P_{qq}(x/y) q(y, \mu_F^2). \tag{15.108}$$

This equation is the analogue of equation (15.35) describing the running of the coupling α_s with μ^2, and is a fundamental equation in the theory of perturbative applications of QCD. It is called the DGLAP equation, after Dokshitzer (1977), Gribov and Lipatov (1972), and Altarelli and Parisi (1977). The above derivation is not rigorous: a more sophisticated treatment (Georgi and Politzer (1974), Gross and Wilczek (1974)) confirms the result and extends it to higher orders.

Equation (15.108) shows that, although perturbation theory cannot be used to calculate the distribution function $q(x, \mu_F^2)$ at any particular value $\mu_F^2 = \mu_0^2$, it can be used to predict how the distribution *changes* (or 'evolves') as μ_F^2 varies. (We recall from (15.105) that $q(x, \mu_0^2)$ can be found experimentally via $xq(x, \mu_0^2) = 2F_2(x, Q^2 = \mu_0^2)/e_i^2$.) As in the case of $\sigma(e^+e^- \rightarrow$ hadrons) and the scale μ^2, the choice of factorization scale is arbitrary, and would cancel from physical quantities if all powers in the perturbation series were included. Truncating at N terms results in an ambiguity of order $\alpha_s^{(N+1)}$. In deep inelastic predictions, the standard choice for scales is $\mu^2 = \mu_F^2 = Q^2$.

The way the non-singlet distribution changes can be understood qualitatively as follows. The change in the distribution for a quark with momentum fraction x, which absorbs the virtual photon, is given by the integral over y of the corresponding distribution for a quark with momentum fraction y, which radiated away (via a gluon) a fraction x/y of its momentum with probability $(\alpha_s/2\pi)P_{qq}(x/y)$. This probability is high for large momentum fractions: high-momentum quarks lose momentum by radiating gluons. Thus there is a predicted tendency for the distribution function $q(x, \mu^2)$ to get smaller at large x as μ^2 increases, and larger at small x (due to the build-up of slower partons), while maintaining the integral of the distribution over x as a constant. The effect is illustrated qualitatively in figure 15.14. In addition, the radiated gluons produce more $q\bar{q}$ pairs at small x. Thus the

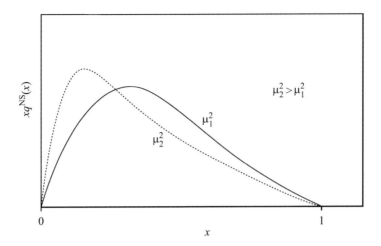

FIGURE 15.14
Evolution of the distribution function with μ^2.

nucleon may be pictured as having more and more constituents, all contributing to its total momentum, as its structure is probed on ever smaller distance (larger μ^2) scales.

In general, the right-hand side of (15.108) will have to be supplemented by terms (calculable from figure 15.10) in which quarks are generated from the gluon distribution; the equations must then be closed by a corresponding one describing the evolution of the gluon distributions (Altarelli 1982). In the now commonly used notation, this generalization of (15.108) reads

$$\mu_{\mathrm{F}}^2 \frac{\partial f_{i/\mathrm{p}}(x, \mu_{\mathrm{F}}^2)}{\partial \mu_{\mathrm{F}}^2} = \sum_{j=\mathrm{q,g}} \frac{\alpha_{\mathrm{s}}(\mu_{\mathrm{F}}^2)}{2\pi} \int_x^1 \frac{\mathrm{d}y}{y} P_{i \leftarrow j}^{(1)}(x/y) f_{j/\mathrm{p}}(y, \mu_{\mathrm{F}}^2), \qquad (15.109)$$

where the sum is over quark types q and gluons g, $P_{i \leftarrow j}^{(1)}$ is the $j \to i$ splitting function to this order, and $f_{i/\mathrm{p}}$ is the parton distribution function for partons of type i in the proton. In our previous notation, $P_{\mathrm{q} \leftarrow \mathrm{q}}^{(1)}(x/y) = P_{\mathrm{qq}}(x/y)$, and $f_{\mathrm{q/p}}(x, \mu_{\mathrm{F}}^2) = q(x, \mu_{\mathrm{F}}^2)$. The other splitting functions may be found in Altarelli (1982).

Both the splitting functions and expression (15.106) for $F_2(x, Q^2)$ can be extended to higher orders in α_{s}. Thus the perturbative expansion (15.106) becomes

$$F_2(x, Q^2) = x \sum_{n=0}^{\infty} \frac{\alpha_{\mathrm{s}}^n(\mu_{\mathrm{F}}^2)}{(2\pi)^n} \sum_{i=\mathrm{q,g}} \int_x^1 \frac{\mathrm{d}z}{z} C_{2,i}^{(2)}(z, Q^2, \mu_{\mathrm{F}}^2) f_{i/\mathrm{p}}(x/z, \mu_{\mathrm{F}}^2), \qquad (15.110)$$

where we have chosen $\mu = \mu_{\mathrm{F}}$. The expansion (15.110) is analogous to (15.63), and as in that case the coefficient functions will depend on μ_{F}^2 in such a way that, order by order, the μ_{F}^2 dependence will cancel. At zeroth order the coefficients are the μ_{F}^2-independent free parton ones, $C_{2,\mathrm{q}}^{(0)} = e_{\mathrm{q}}^2 \delta(1-z)$ and $C_{2,\mathrm{g}}^{(0)} = 0$. Corrections up to order α_{s}^3 have been calculated for electromagnetic (Vermaseren, Vogt and Moch 2005) and weak currents (Moch, Vermaseren and Vogt 2009, Davies *et al.* 2016).

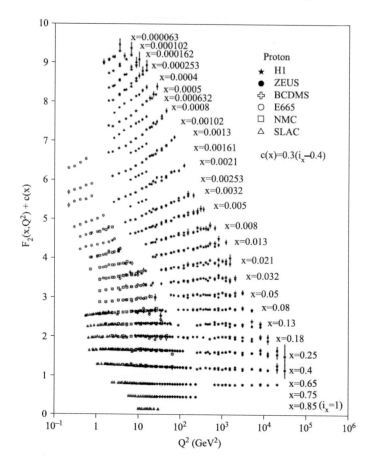

FIGURE 15.15

Q^2-dependence of the proton structure function F_2^{p} for various fixed x values (Hagiwara *et al.* 2002). i_x is a number depending on the x-bin, ranging from $i_x = 1$ ($x = 0.85$) to $i_x = 28$ ($x = 0.000063$). Figure reprinted with permission from K Hagiwara *et al. Phys.Rev.* D **66** 010001 (2002). Copyright 2002 by the American Physical Society.

We ought also to mention that there are in principle non-perturbative corrections to both (15.63) and (15.110), which are of order $(\Lambda^2_{\overline{\mathrm{MS}}}/Q^2)^2$ and $(\Lambda^2_{\overline{\mathrm{MS}}}/Q^2)$ respectively.

15.6.3 Comparison with experiment

Data on nucleon structure functions do indeed show the trend described in the previous section. Figure 15.15 shows the Q^2-dependence of the proton structure function $F_2^{\mathrm{p}}(x, Q^2) = \sum e_i^2 x f_{i/\mathrm{p}}(x, Q^2)$ for various fixed x values, as compiled by B. Foster, A.D. Martin, and M.G. Vincter for the 2002 Particle Data Group review (Hagiwara *et al.* 2002). Clearly at larger x ($x \geq 0.13$) the function gets smaller as Q^2 increases, while at smaller x it increases.

Fits to the data have been made in various ways. One (theoretically convenient) way is to consider 'moments' (Mellin transforms) of the structure functions, defined by

$$M_{\mathrm{q}}^n(t) = \int_0^1 \mathrm{d}x\, x^{n-1} q(x, t), \tag{15.111}$$

where we have taken $\mu^2 = \mu_F^2$ and introduced the variable $t = \ln \mu^2$. Taking moments of both sides of (15.108) and interchanging the order of the x and y integrations, we find

$$\frac{\mathrm{d}M_q^n(t)}{\mathrm{d}t} = \frac{\alpha_s(t)}{2\pi} \int_0^1 \mathrm{d}y\, y^{n-1} q(y,t) \int_0^y \frac{\mathrm{d}x}{y} (x/y)^{n-1} P_{qq}(x/y). \tag{15.112}$$

Changing the variable to $z = x/y$ in the second integral, and defining[3]

$$\gamma_{qq}^n = 4 \int_0^1 \mathrm{d}z\, z^{n-1} P_{qq}(z), \tag{15.113}$$

we obtain

$$\frac{\mathrm{d}M_q^n(t)}{\mathrm{d}t} = \frac{\alpha_s(t)}{8\pi} \gamma_{qq}^n M_q^n(t). \tag{15.114}$$

Thus the integral in (15.108)—which is of convolution type—has been reduced to product form by this transformation. Now we also know from (15.47) and (15.48) that

$$\frac{\mathrm{d}\alpha_s}{\mathrm{d}t} = -\beta_0 \alpha_s^2 \tag{15.115}$$

with $\beta_0 = (33 - 2N_f)/12\pi$ as usual, to this (one-loop) order. Thus (15.114) becomes

$$\frac{\mathrm{d}\ln M_q^n}{\mathrm{d}\ln \alpha_s} = -\frac{\gamma_{qq}^n}{8\pi\beta_0} = -d_{qq}^n, \text{ say.} \tag{15.116}$$

The solution to (15.116) is easily found to be

$$M_q^n(t) = M_q^n(t_0) \left(\frac{\alpha_s(t_0)}{\alpha_s(t)} \right)^{d_{qq}^n}. \tag{15.117}$$

Applying the prescription (15.99) to γ_n, we find (problem 15.9)

$$\gamma_{qq}^n = -\frac{8}{3} \left[1 - \frac{2}{n(n+1)} + 4 \sum_{j=2}^n \frac{1}{j} \right] \tag{15.118}$$

and then

$$d_{qq}^n = \frac{4}{33 - 2N_f} \left[1 - \frac{2}{n(n+1)} + 4 \sum_{j=2}^n \frac{1}{j} \right]. \tag{15.119}$$

We emphasize again that all the foregoing analysis is directly relevant only to distributions in which the flavour singlet gluon distributions do not contribute to the evolution equations. In the more general case, analogous splitting functions P_{qg}, P_{gq}, and P_{gg} will enter, folded appropriately with the gluon distribution function $g(x,t)$, together with the related quantities $\gamma_{qg}^n, \gamma_{gq}^n$, and γ_{gg}^n. Equation (15.108) is then replaced by a 2×2 matrix equation for the evolution of the quark and gluon moments M_q^n and M_g^n.

Returning to (15.117), one way of testing it is to plot the logarithm of one moment, $\ln M_q^n$, versus the logarithm of another, $\ln M_q^m$, for different n, m values. A more direct procedure, applicable to the non-singlet case too of course, is to choose a reference point μ_0^2 and parametrize the parton distribution functions $f_i(x, t_0)$ in some way. These may then

---

[3]The notation is not chosen accidentally: the γ's are indeed anomalous dimensions of certain operators which appear in Wilson's operator product approach to scaling violations (Wilson 1969); interested readers may pursue this with Peskin and Schroeder 1995, chapter 18.

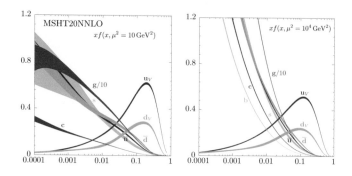

FIGURE 15.16

Distributions of x times the unpolarized parton distributions $f(x, \mu^2)$ (where $f = u_v, d_v, \bar{u}, \bar{d}, s, c, g$) using the NNLO MSHT analysis of Bailey *et al.* (2021) at scales $\mu^2 = 10 \, \text{GeV}^2$ and $\mu^2 = 10^4 \, \text{GeV}^2$.

be evolved numerically, via the DGLAP equations, to the desired scale. As a representative example, figure 15.16 shows the parton distribution functions obtained in the NNLO MSHT analysis (already shown in figure 9.9 of volume 1) at scales $\mu^2 = 10 \, \text{GeV}^2$ and $\mu^2 = 10^4 \, \text{GeV}^2$ (Bailey *et al.* 2021). The value of $\alpha_s(M_Z^2)$ extracted using the NNLO corrections was

$$\alpha_s(M_Z^2) = 0.1174 \pm 0.0013. \tag{15.120}$$

This value contributed to the world average value quoted in (15.62).

It may be worth pausing to reflect on how far our understanding of *structure* has developed, via quantum field theory, from the simple 'fixed number of constituents' models which are useful in atomic and nuclear physics. When nucleons are probed on finer and finer scales, more and more partons (gluons, $q\bar{q}$ pairs) appear, in a way quantitatively predicted by QCD. The precise experimental confirmation of QCD predictions (as discussed by Huston, Rabbertz and Zanderighi (2022) for example) constitutes a remarkable vote of confidence, by Nature, in relativistic quantum field theory.

Problems

15.1 Verify equation (15.10).

15.2 Verify equation (15.27).

15.3 Check that (15.50) can be rewritten as (15.53).

15.4 (a) Verify (15.61). (b) Show that the next term in the expansion (15.60) is

$$\frac{(b_2 - b_1^2)}{\beta_0} \alpha_s$$

where $b_2 = \beta_2/\beta_0$. By iteratively solving the resulting modified equation (15.60), show that the corresponding correction to (15.61) is

$$+\frac{1}{\beta_0^3 L^3}[b_1^2(\ln^2 L^2 - \ln L - 1) + b_2].$$

15.5 Verify that for the type of behaviour of the β function shown in figure 15.7(b), α_s^* is reached as $q^2 \to 0$.

15.6 Verify equation (15.102).

15.7 Check that the electromagnetic charge e has dimension $(\text{mass})^{\epsilon/2}$ in $d = 4 - \epsilon$ dimensions.

15.8 Verify equation (O.20).

15.9 Verify equation (15.118).

16

Lattice Field Theory and the Renormalization Group Revisited

16.1 Introduction

Throughout this book, thus far, we have relied on perturbation theory as the calculational tool, justifying its use in the case of QCD by the smallness of the coupling constant at short distances; note, however, that this result itself required the summation of an infinite series of perturbative terms. As remarked in section 15.3, the concomitant of asymptotic freedom is that α_s really does become strong at small Q^2, or at long distances of order $\Lambda_{\overline{\rm MS}}^{-1} \sim 1$ fm. Here we have no prospect of getting useful results from perturbation theory: it is the *non-perturbative regime*. But this is precisely the regime in which quarks bind together to form hadrons. If QCD is indeed the true theory of the interaction between quarks, then it should be able to explain, ultimately, the vast amount of data that exists in low energy hadronic physics. For example: what are the masses of mesons and baryons? Are there novel colourless states such as glueballs? Is $SU(2)_f$ or $SU(3)_f$ chiral symmetry spontaneously broken? What is the form of the effective interquark potential? What are the hadronic form factors, in electromagnetic (chapter 9) or weak (chapter 20) processes?

After more than 50 years of theoretical development, and machine advances, numerical simulations of *lattice QCD* are now yielding precise answers to many of these questions, thereby helping to establish QCD as the correct theory of the strong interactions of quarks, and also providing reliable input needed for the discovery of new physics. Lattice QCD is a highly mature field, and many technical details are beyond our scope. Rather, in this chapter we aim to give an elementary introduction to lattice field theory in general, including some important insights that it generates concerning the renormalization group. We return to lattice QCD in the final section, with some illustrative results.

In thinking about how to formulate a non-perturbative approach to quantum field theory, several questions immediately arise. First of all, how can we regulate the ultraviolet divergences, and thus define the theory, if we cannot get to grips with them via the specific divergent integrals supplied by perturbation theory? We need to be able to regulate the divergences in a way which does not rely on their appearance in the Feynman graphs of perturbation theory. As Wilson (1974, 1975) was the first to propose, one quite natural non-perturbative way of regulating ultraviolet divergences is to approximate continuous space-time by a discrete lattice of points. Such a lattice will introduce a minimum distance—namely the lattice spacing 'a' between neighbouring points. Since no two points can ever be closer than a, there is now a corresponding maximum momentum $\Lambda = \pi/a$ (see following equation (16.6)) in the lattice version of the theory. Thus the theory is automatically ultraviolet finite from the start, without presupposing the existence of any perturbative expansion; renormalization questions will, however, enter when we consider the a dependence of our parameters. As long as the lattice spacing is much smaller than the physical size of the hadrons one is studying, the lattice version of the theory should be a good approximation. Nevertheless, for finite a there will always be systematic discretization errors, which have to be controlled. In addition, Lorentz invariance is of course sacrificed in such an approach, and replaced by some form of hypercubic symmetry; we must hope that for small enough a this will not matter. We shall discuss how simple field theories are 'discretized' in the next section; scalar fields, gauge fields, and fermion fields each require their own prescriptions.

But how can a quantum field theory can be formulated in a way suitable for numerical computation? Any formalism based on non-commuting *operators* seems to be ruled out, since it is hard to see how they could be numerically simulated. Indeed, the same would be true of ordinary quantum mechanics. Fortunately a formulation does exist which avoids operators: Feynman's *sum over paths* approach, which was briefly mentioned in section 5.2.2. This method is the essential starting point for the lattice approach to quantum field theory, and it will be introduced in section 16.3. The sum over paths approach does not involve quantum operators, but fermions still have to be accommodated somehow. The way this is done is briefly described in section 16.3; see also Appendix P.

It turns out that this formulation enables direct contact to be made between quantum field theory and *statistical mechanics*, as we shall discuss in section 16.3.3. This relationship has proved to be extremely fruitful, allowing physical insights and numerical techniques to pass from one subject to the other, in a way that has been very beneficial to both. In section 16.4 we make a worthwhile detour to explore the physics of renormalization and of the RGE from a lattice/statistical mechanics perspective, before returning to lattice QCD in section 16.5.

16.2 Discretization

16.2.1 Scalar fields

We start by considering a simple classical field theory involving a scalar field ϕ. Postponing until section 16.3 the question of exactly how we shall use it in a quantum context, we assume that we shall still want to formulate the theory in terms of an action of the form

$$S = \int \mathrm{d}^4x \; \mathcal{L}(\phi, \boldsymbol{\nabla}\phi, \dot{\phi}). \tag{16.1}$$

It seems plausible that it might be advantageous to treat space and time as symmetrically as possible, from the start, by formulating the theory in 'Euclidean' space, instead of Minkowskian, by introducing $t = -\mathrm{i}\tau$; further motivation for doing this will be provided in section 16.3. In that case, the action (16.1) becomes

$$S \quad \to \quad -\mathrm{i} \int \mathrm{d}^3\boldsymbol{x}\mathrm{d}\tau \; \mathcal{L}(\phi, \boldsymbol{\nabla}\phi, \mathrm{i}\frac{\partial\phi}{\partial\tau}) \tag{16.2}$$

$$\equiv \quad \mathrm{i} \int \mathrm{d}^3\boldsymbol{x}\mathrm{d}\tau \; \mathcal{L}_\mathrm{E} \equiv \mathrm{i}S_\mathrm{E}. \tag{16.3}$$

A typical free scalar action is then

$$S_\mathrm{E}(\phi) = \frac{1}{2} \int \mathrm{d}^3\boldsymbol{x}\mathrm{d}\tau \left[(\partial_\tau \phi)^2 + (\boldsymbol{\nabla}\phi)^2 + m^2\phi^2 \right]. \tag{16.4}$$

We now represent all of space-time by a finite-volume 'hypercube'. For example, we may have N_1 lattice points along the x-axis, so that a field $\phi(x)$ is replaced by the N_1 numbers

$\phi(n_1 a)$ with $n_1 = 0, 1, \ldots N_1 - 1$. We write $L = N_1 a$ for the length of the cube side. In this notation, integrals and differentials are replaced by the finite sums and difference expressions

$$\int dx \to a \sum_{n_1} , \qquad \frac{\partial \phi}{\partial x} \to \frac{1}{a}[\phi(n_1 + 1) - \phi(n_1)], \qquad (16.5)$$

so that a typical integral (in one dimension) becomes

$$\int dx \left(\frac{\partial \phi}{\partial x}\right)^2 \to a \sum_{n_1} \frac{1}{a^2} [\phi(n_1 + 1) - \phi(n_1)]^2 . \qquad (16.6)$$

As in all our previous work, we can alternatively consider a formulation in momentum space, which will also be discretized. It is convenient to impose periodic boundary conditions such that $\phi(x) = \phi(x + L)$. Then the allowed k-values may be taken to be $k_{\nu_1} = 2\pi\nu_1/L$ with $\nu_1 = -N_1/2 + 1, \ldots 0, \ldots N_1/2$ (we take N_1 to be even). It follows that the maximum allowed magnitude of the momentum is then π/a, indicating that a^{-1} is (as anticipated) playing the role of our earlier momentum cut-off Λ. We then write

$$\phi(n_1) = \sum_{\nu_1} \frac{1}{(N_1 a)^{\frac{1}{2}}} e^{i2\pi\nu_1 n_1/N_1} \tilde{\phi}(\nu_1), \qquad (16.7)$$

which has the inverse

$$\tilde{\phi}(\nu_1) = \left(\frac{a}{N_1}\right)^{\frac{1}{2}} \sum_{n_1} e^{-i2\pi\nu_1 n_1/N_1} \phi(n_1), \qquad (16.8)$$

since (problem 16.1)

$$\frac{1}{N_1} \sum_{n_1=0}^{N_1-1} e^{i2\pi n_1(\nu_1 - \nu_2)/N_1} = \delta_{\nu_1, \nu_2}. \qquad (16.9)$$

Equation (16.9) is a discrete version of the δ-function relation given in (E.25) of volume 1. A one-dimensional version of the mass term in (16.4) then becomes (problem 16.2)

$$\frac{1}{2} \int dx \, m^2 \phi(x)^2 \to \frac{1}{2} m^2 \sum_{\nu_1} \tilde{\phi}(\nu_1) \tilde{\phi}(-\nu_1), \qquad (16.10)$$

while

$$\frac{1}{2} \int dx \left(\frac{\partial \phi}{dx}\right)^2 \to \frac{2}{a^2} \sum_{\nu_1} \tilde{\phi}(\nu_1) \sin^2\left(\frac{\pi\nu_1}{N_1}\right) \tilde{\phi}(-\nu_1) \qquad (16.11)$$

$$= \frac{1}{2a^2} \sum_{k_{\nu_1}} \tilde{\phi}(k_{\nu_1}) 4 \sin^2\left(\frac{k_{\nu_1} a}{2}\right) \tilde{\phi}(-k_{\nu_1}). \qquad (16.12)$$

Thus a one-dimensional version of the free action (16.4) is

$$\frac{1}{2} \sum_{k_{\nu_1}} \tilde{\phi}(k_{\nu_1}) \left[\frac{4 \sin^2(k_{\nu_1} a/2)}{a^2} + m^2\right] \tilde{\phi}(-k_{\nu_1}). \qquad (16.13)$$

In the continuum case, (16.13) would be replaced by the $a \to 0$ limit of (16.13), which is

$$\frac{1}{2} \int \frac{dk}{2\pi} \tilde{\phi}(k) \left[k^2 + m^2\right] \tilde{\phi}(-k) \qquad (16.14)$$

as usual, implying that the propagator in the discrete case is proportional to

$$\left[\frac{4\sin^2(k_{\nu_1}a/2)}{a^2} + m^2\right]^{-1} \tag{16.15}$$

rather than to $\left[k^2 + m^2\right]^{-1}$ (remember we are in one-dimensional Euclidean space). The manipulations we have been going through will be easily recognized by readers familiar with the theory of lattice vibrations and phonons, and lead to a s imple discretization of scalar fields.

For finite a, *discretization errors* will be introduced. In passing from (16.13) to (16.14) the leading order (in powers of a) correction to the continuum result will be a term of order k^4a^2. Such a contribution would be generated by a term of the form $a^2A(\nabla^2\phi)^2$ in the Lagrangian density of (16.4), where A is some constant. This suggests that we might arrange to eliminate such a correction by modifying the original action (16.4) to include just such a term in the first place, and then (in the course of the numerical simulation) tune the coefficient A so as to eliminate the total contribution of order k^4a^2. This is the basic idea of the important 'Symanzik improvement' programme (Symanzik 1983). Note that a term $(\nabla^2\phi)^2$ has mass dimension six, and so (referring to section 11.8 of volume 1) in quantum field theory it would be a non-renormalizable interaction, suppressed by two powers of the cut-off $\Lambda^{-2} \sim a^2$. We shall return to Symanzik improvement later, in section 16.4.3, when we consider renormalization from a lattice viewpoint.

16.2.2 Gauge fields

Having explored the discretization of actions for free scalars, we must now think about how to implement gauge invariance on the lattice. In the usual (continuum) case, we saw in chapter 13 how this was implemented by replacing ordinary derivatives by *covariant derivatives*, the geometrical significance of which (in terms of parallel transport) is discussed in appendix N. It is very instructive to see how the same ideas arise naturally in the lattice case.

We illustrate the idea in the simple case of the Abelian U(1) theory, QED. Consider, for example, a charged scalar field $\phi(x)$, with charge e. To construct a gauge-invariant current, for example, we replaced $\phi^\dagger\partial_\mu\phi$ by $\phi^\dagger(\partial_\mu + ieA_\mu)\phi$, so we ask: what is the discrete analogue of this? The term $\phi^\dagger(x)\frac{\partial}{\partial x}\phi(x)$ becomes, as we have seen,

$$\phi^\dagger(n_1)\frac{1}{a}[\phi(n_1+1) - \phi(n_1)a] \tag{16.16}$$

in one dimension. We do *not* expect (16.16) by itself to be gauge invariant, and it is easy to check that it is not. Under a gauge transformation for the continuous case, we have

$$\phi(x) \to e^{ie\theta(x)}\phi(x), \, A(x) \to A(x) + \frac{d\theta(x)}{dx}; \tag{16.17}$$

then $\phi^\dagger(x)\phi(y)$ transforms by

$$\phi^\dagger(x)\phi(y) \to e^{-ie[\theta(x)-\theta(y)]}\phi^\dagger(x)\phi(y), \tag{16.18}$$

and is clearly not invariant. The essential reason is that this operator involves the fields at two *different* points, and so the term $\phi^\dagger(n_1)\phi(n_1+1)$ in (16.16) will not be gauge invariant either. The discussion in appendix N prepares us for this: we are trying to compare two 'vectors' (here, fields) at two different points, when the 'coordinate axes' are changing as we move about. We need to parallel transport one field to the same point as the other, before

FIGURE 16.1
Link variable $U(n_2; n_1)$ in one dimension.

they can be properly compared. The solution (N.18) shows us how to do this. Consider the quantity

$$\mathcal{O}(x, y) = \phi^\dagger(x)\exp[ie \int_y^x A dx'] \phi(y). \tag{16.19}$$

Under the gauge transformation (16.17), $\mathcal{O}(x, y)$ transforms by

$$\mathcal{O}(x, y) \to \phi^\dagger(x) e^{-ie\theta(x)} e^{\{ie \int_y^x A dx' + ie[\theta(x) - \theta(y)]\}} e^{ie\theta(y)} \phi(y) = \mathcal{O}(x, y), \tag{16.20}$$

and it is therefore gauge invariant. The familiar 'covariant derivative' rule can be recovered by letting $y = x + dx$ for infinitesimal dx, and by considering the gauge-invariant quantity

$$\lim_{dx \to 0} \left[\frac{\mathcal{O}(x, x + dx) - \mathcal{O}(x, x)}{dx} \right]. \tag{16.21}$$

Evaluating (16.21) one finds (problem 16.4) the result

$$\phi^\dagger(x) \left(\frac{d}{dx} - ieA \right) \phi(x) \tag{16.22}$$

$$\equiv \phi^\dagger(x) D_x \phi(x) \tag{16.23}$$

with the usual definition of the covariant derivative. In the discrete case, we merely keep the finite version of (16.19), and replace $\phi^\dagger(n_1)\phi(n_1 + 1)$ in (16.16) by the gauge invariant quantity

$$\phi^\dagger(n_1) U(n_1, n_1 + 1) \phi(n_1 + 1), \tag{16.24}$$

where the *link variable* U is defined by

$$U(n_1, n_1 + 1) = \exp[ie \int_{(n_1+1)a}^{n_1 a} A dx'] \tag{16.25}$$

Note that

$$U(n_1, n_1 + 1) \to \exp[-ieA(n_1)a] \tag{16.26}$$

in the small a limit.

The generalization to more dimensions is straightforward. In the non-Abelian $SU(2)$ or $SU(3)$ case, 'eA' in (16.26) is replaced by $gt^a A^a(n_1)$ where the t's are the appropriate matrices, as in the continuum form of the covariant derivative. A link variable $U(n_2, n_1)$ may be drawn as in figure 16.1. Note that the order of the arguments is significant: $U(n_2, n_1) = U^{-1}(n_1, n_2) = U^\dagger(n_1, n_2)$ from (16.26), which is why the link carries an arrow.

Thus gauge invariant discretized derivatives of charged fields can be constructed. What about the Maxwell action for the U(1) gauge field? This does not exist in only one dimension ($\partial_\mu A_\nu - \partial_\nu A_\mu$ cannot be formed), so let us move into two. Again, our discussion of the

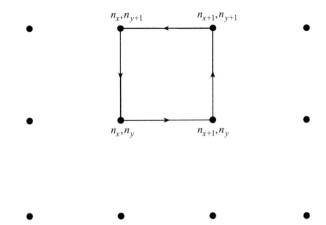

FIGURE 16.2
A simple plaquette in two dimensions.

geometrical significance of $F_{\mu\nu}$ as a curvature guides us to the answer. Consider the product $U_\square$ of link variables around a square path (figure 16.2) of side a (reading from the right):

$$
\begin{aligned}
U_\square \;=\; & U(n_x, n_y; n_x, n_{y+1}) U(n_x, n_{y+1}; n_{x+1}, n_{y+1}) \\
& \times U(n_{x+1}, n_{y+1}; n_{x+1}, n_y) U(n_{x+1}, n_y; n_x, n_y).
\end{aligned}
\tag{16.27}
$$

It is straightforward to verify, first, that $U_\square$ is gauge invariant. Under a gauge transformation, the link $U(n_{x+1}, n_y; n_x, n_y)$, for example, transforms by a factor (cf equation (16.20))

$$
\exp\{ie[\theta(n_{x+1}, n_y) - \theta(n_x, n_y)]\},
\tag{16.28}
$$

and similarly for the three other links in $U_\square$. In this Abelian case the exponentials contain no matrices, and the accumulated phase factors cancel out, verifying the gauge invariance. Next, let us see how to recover the Maxwell action. Adding the exponentials again, we can write

$$
\begin{aligned}
U_\square \;\equiv\; & \exp\{-ieaA_y(n_x, n_y) - ieaA_x(n_x, n_y + 1) \\
& \quad + ieaA_y(n_x + 1, n_y) + ieaA_x(n_x, n_y)\}
\end{aligned}
\tag{16.29}
$$

$$
\begin{aligned}
\;=\; & \exp\left\{ -iea^2 \left[\frac{A_x(n_x, n_y + 1) - A_x(n_x, n_y)}{a} \right] \right. \\
& \left. \quad + iea^2 \left[\frac{A_y(n_x + 1, n_y) - A_y(n_x, n_y)}{a} \right] \right\}
\end{aligned}
\tag{16.30}
$$

$$
\;=\; \exp\left\{ +iea^2 \left(\frac{\partial A_y}{\partial x} - \frac{\partial A_x}{\partial y} \right) \right\},
\tag{16.31}
$$

using the derivative definition of (16.5). For small 'a' we may expand the exponential in (16.31). We also take the real part to remove the imaginary terms, leading to

$$
\sum_\square (1 - \mathrm{Re}\, U_\square) \to \frac{1}{2} \sum_\square e^2 a^4 (F_{xy})^2,
\tag{16.32}
$$

where $F_{xy} = \frac{\partial A_y}{\partial x} - \frac{\partial A_x}{\partial y}$ as usual. To relate this to the continuum limit we must note that we sum over each such *plaquette* with only one definite orientation, so that the sum over plaquettes is equivalent to half of the entire sum. Thus

$$\sum_{\square}(1 - \mathrm{Re}\, U_{\square}) \quad \to \quad \frac{1}{4}\sum_{n_1,n_2} e^2 a^4 F_{xy}^2$$

$$\to \quad e^2 a^2 \int\int \frac{1}{4}F_{xy}^2 \mathrm{d}x\mathrm{d}y. \tag{16.33}$$

(Note that in two dimensions 'e' has dimensions of mass). In four dimensions similar manipulations lead to the form

$$S_{\mathrm{E}} = \frac{1}{e^2}\sum_{\square}(1 - \mathrm{Re}\, U_{\square}) \to \frac{1}{4}\int \mathrm{d}^3\boldsymbol{x}\mathrm{d}\tau F_{\mu\nu}^2 \tag{16.34}$$

for the lattice action, as required. In the non-Abelian case, as noted above, 'eA' is replaced by '$gt \cdot A$'; for SU(3), the analogue of (16.34) is

$$S_g = \frac{6}{g^2}\sum_{\square}\left(1 - \frac{1}{3}\mathrm{TrRe}\, U_{\square}\right), \tag{16.35}$$

where the trace is over the SU(3) matrices.

Once again, the leading discretization errors will be of order a^2, and a 'Symanzik improved' action can be defined (Lüscher and Weisz 1985a, b) by adding loops with six gauge fields (instead of the four in figure 16.2)).

16.2.3   Dirac fields

In the case of fermions, the first obvious problem has already been mentioned: how are we to represent such entirely non-classical objects, which obey anticommutation relations? This is part of the wider problem of representing field operators in a form suitable for numerical simulation, which we defer until section 16.3. There are, however, quite separate and challenging problems which arise when we try to repeat for the Dirac field the discretization used for the scalar field.

First note that the Euclidean Dirac matrices γ_μ^{E} are related to the usual Minkowski ones γ_μ^{M} by $\gamma_{1,2,3}^{\mathrm{E}} \equiv -\mathrm{i}\gamma_{1,2,3}^{\mathrm{M}}, \gamma_4^{\mathrm{E}} \equiv -\mathrm{i}\gamma_4^{\mathrm{M}} \equiv \gamma_0^{\mathrm{M}}$. They satisfy $\{\gamma_\mu^{\mathrm{E}}, \gamma_\nu^{\mathrm{E}}\} = 2\delta_{\mu\nu}$ for $\mu = 1, 2, 3, 4$. The Euclidean Dirac Lagrangian is then $\bar{\psi}(x)\left[\gamma_\mu^{\mathrm{E}}\partial_\mu + m\right]\psi(x)$, which should be written now in Hermitean form

$$m\bar{\psi}(x)\psi(x) + \frac{1}{2}\left\{\bar{\psi}(x)\gamma_\mu^{\mathrm{E}}\partial_\mu\psi(x) - (\partial_\mu\bar{\psi}(x))\gamma_\mu^{\mathrm{E}}\psi(x)\right\}. \tag{16.36}$$

The corresponding 'one-dimensional' discretized action is then

$$a\sum_{n_1} m\bar{\psi}(n_1)\psi(n_1) \quad + \quad \frac{a}{2}\left\{\sum_{n_1}\bar{\psi}(n_1)\gamma_1^{\mathrm{E}}\left[\frac{\psi(n_1+1)-\psi(n_1)}{a}\right]\right.$$

$$\left. - \sum_{n_1}\left(\frac{\bar{\psi}(n_1+1)-\bar{\psi}(n_1)}{a}\right)\gamma_1^{\mathrm{E}}\psi(n_1)\right\} \tag{16.37}$$

$$= a\sum_{n_1}\left\{m\bar{\psi}(n_1)\psi(n_1) + \frac{1}{2a}\left[\bar{\psi}(n_1)\gamma_1^{\mathrm{E}}\psi(n_1+1) - \bar{\psi}(n_1+1)\gamma_1^{\mathrm{E}}\psi(n_1)\right]\right\}, \tag{16.38}$$

which may be called the 'naive' prescription for a free fermion. This action may be made gauge invariant by using the lattice version of the gauge covariant derivative, as in equations (16.16) to (16.24) for the charged scalar field. The Dirac term [.....] in (16.38) is replaced by the gauge-invariant form

$$\bar{\psi}(n_1)\gamma_1^E U(n_1, n_1 + 1)\psi(n_1 + 1) - \bar{\psi}(n_1 + 1)\gamma_1^E U(n_1 + 1, n_1)\psi(n_1). \tag{16.39}$$

In momentum space (16.38) becomes (problem 16.3)

$$\sum_{k_{\nu_1}} \bar{\tilde{\psi}}(k_{\nu_1}) \left[i\gamma_1^E \frac{\sin(k_{\nu_1}a)}{a} + m \right] \tilde{\psi}(k_{\nu_1}), \tag{16.40}$$

and the inverse propagator is $\left[i\gamma_1^E \frac{\sin(k_{\nu_1}a)}{a} + m \right]$. Thus the propagator itself is

$$\left[m - i\gamma_1^E \frac{\sin(k_{\nu_1}a)}{a} \right] \bigg/ \left[m^2 + \frac{\sin^2(k_{\nu_1}a)}{a^2} \right]. \tag{16.41}$$

But here there is a problem: in addition to the correct continuum limit ($a \to 0$) found at $k_{\nu_1} \to 0$, an alternative finite $a \to 0$ limit is found at $k_{\nu_1} \to \pi/a$ (consider expanding $a^{-1}\sin[(\pi/a - \delta)a]$ for small δ). Thus two modes survive as $a \to 0$, a phenomenon known as the 'fermion doubling problem'. Actually in four dimensions there are *sixteen* such corners of the hypercube, so we have far too many degenerate lattice copies (which are called 'doublers').

Various solutions to this problem have been proposed. Wilson (1975), for example, suggested adding the extra term

$$-\frac{1}{2a}r \sum_{n_1} \bar{\psi}(n_1)[\psi(n_1 + 1) + \psi(n_1 - 1) - 2\psi(n_1)] \tag{16.42}$$

to the fermion action in this one-dimensional case, where r is dimensionless. Evidently this is a *second* difference, and it would correspond to the term

$$-\frac{1}{2}ra \int d^3x d\tau \; \bar{\psi}(x)(\partial_\tau^2 + \boldsymbol{\nabla}^2)\psi(x) \tag{16.43}$$

in the four-dimensional continuum action. Note the presence of the lattice spacing 'a' in (16.43), which ensures its disappearance as $a \to 0$. The higher-derivative term $\bar{\psi}(\partial_\tau^2 + \boldsymbol{\nabla}^2)\psi$ has mass dimension 5, and therefore requires a coupling constant with mass dimension -1, i.e. a length in units $\hbar = c = 1$; it is, in fact, a non-renormalizable term, with a coupling proportional to a.

How does the extra term (16.42) help the doubling problem? One easily finds that it changes the (one-dimensional) inverse propagator to

$$\left[i\gamma_1^E \frac{\sin(k_{\nu_1}a)}{a} + m \right] + \frac{r}{a}(1 - \cos(k_{\nu_1}a)). \tag{16.44}$$

By considering the expansion of the cosine near $k_{\nu_1} \approx 0$ it can be seen that the second term disappears in the continuum limit, as expected. However, for $k_{\nu_1} \approx \pi/a$ it gives a large term of order $\frac{1}{a}$ which adds to the mass m, effectively banishing the 'doubled' state to a very high mass, far from the physical spectrum.

Unfortunately there is a price to pay. First of all, the Wilson term introduces discretization errors which are linear in a. They can be eliminated by a Symanzik-type improvement (the 'clover' action of Sheikholeslami and Wohlert 1985), which has been implemented non-perturbatively (Jansen *et al.* 1996). The second disadvantage of Wilson-type fermions is that they break chiral symmetry. This symmetry was introduced in section 12.3.2, as an exact symmetry of QCD with massless quarks. To the extent that m_u and m_d (and m_s, but less so) are small on a hadronic scale such as $\Lambda_{\overline{MS}}$, we expect chiral symmetry to have important physical consequences. These will indeed be explored in chapter 18. For the moment, we note merely that it is important for lattice-based QCD calculations to be able to deal correctly with the light quarks. Now we cannot simply choose the bare Lagrangian mass parameters to be small, and leave it at that. In any interacting theory, renormalization effects will cause shifts in these masses. In a chirally symmetric theory, or one which is chirally symmetric as a fermion mass goes to zero, such a mass shift is proportional to the fermion mass itself; in particular it does not simply add to the mass. We drew attention to this fact in the case of the electron mass renormalization in QED, in section 11.2. So in chirally symmetric theories, mass renormalizations are 'protected', in this sense. But the modification (16.42), while avoiding physical fermion doublers, breaks chiral symmetry badly. This can easily be seen by noting (see (12.154) for example) that the crucial property required for chiral symmetry to hold is

$$\gamma_5 \, \slashed{D} + \slashed{D} \gamma_5 = 0, \qquad\qquad (16.45)$$

where $\slashed{D}$ is the $SU(3)_c$-covariant Dirac derivative. Any addition to $\slashed{D}$ which is proportional to the unit 4×4 matrix will violate (16.45), and hence break chiral symmetry. The Lagrangian mass m itself is of this form, and it breaks chiral symmetry, but 'softly'—i.e. in a way that disappears as m goes to zero (thereby preserving the symmetry in this limit). The Wilson addition (16.42) also breaks chiral symmetry, but it remains there even as $m \to 0$: it is a 'hard' breaking.

This means that in the theory with Wilson fermions, fermion mass renormalization will not be protected by the chiral symmetry, so that large additive renormalizations are possible. This will require repeated fine-tunings of the bare mass parameters, to bring them down to the desired small values. And it turns out that this seriously lengthens the computing time.

Another approach ('staggered fermions') was suggested by Kogut and Susskind (1975), Banks *et al.* (1976), and Susskind (1977). This essentially involves distributing the 4 spin degrees of freedom of the Dirac field across different lattice sites (we shall not need the details). At each site there is now a 1-component fermion, with the colour degrees of freedom, which speeds the calculations. The 16-fold 'doubling' degeneracy can be re-arranged as four different 'tastes' of 4-component fermions, while retaining enough chiral symmetry to forbid additive mass renormalizations.

Since the different components of the staggered Dirac field now live on different sites, they will experience slightly different gauge field interactions. (These are of course local in the continuum limit, but the point remains true after discretization, as we shall see in the following section.) These interactions will mix fields of different tastes, causing new problems, but they can be suppressed by adding further terms to the action. There is still the 4-fold degeneracy to get rid of, but a trick ('rooting') is available for that, as we shall explain in section 16.3.2.

As usual, the staggered fermion action can be improved to reduce discretization errors. Recent calculations use the HISQ (highly improved staggered quark) action, introduced by Follana *et al.* (2007). It can deal with light quarks, and has been successfully used for processes with charm and bottom quarks (Davies *et al.* 2010, Donald *et al.* 2012).

One might wonder if a lattice theory with fermions could be formulated such that it both avoids doublers and preserves chiral symmetry. For quite a long time it was believed that this was not possible—a conclusion which was essentially the content of the Nielsen-Ninomaya theorem (Nielsen and Ninomaya 1981a, b, c). But a way was found to formulate chiral gauge theories with fermions satisfactorily on the lattice at finite spacing a. The key is to replace the condition (16.45) by the Ginsparg-Wilson (1982) relation

$$\gamma_5 \,\slashed{D} + \slashed{D}\gamma_5 = a\, \slashed{D}\gamma_5\, \slashed{D} \tag{16.46}$$

which reduces to (16.45) in the $a \to 0$ continuum limit. This relation implies (Lüscher 1998) that the associated action has an exact symmetry, with infinitesimal variations proportional to

$$\delta\psi = \gamma_5 \left(1 - \frac{1}{2}a\,\slashed{D}\right)\psi \tag{16.47}$$

$$\delta\overline{\psi} = \overline{\psi}\left(1 - \frac{1}{2}a\,\slashed{D}\right)\gamma_5 . \tag{16.48}$$

The symmetry under (16.47)–(16.48), which is proportional to the infinitesimal version of (12.152) as $a \to 0$, provides a lattice theory with all the fundamental symmetry properties of continuum chiral gauge theories (Hasenfratz *et al.* 1998).

Two types of Ginsparg-Wilson fermions are currently used in large-scale lattice QCD numerical simulations: 'domain wall' fermions (Kaplan 1992, Shamir 1993), and 'overlap' fermions (Narayanan and Neuberger 1993a, 1993b, 1994, 1995, Neuberger 1998a, 1998b), which are, in fact, essentially equivalent (Borici 2000). Unfortunately both these actions are computationally more expensive than the Wilson or staggered fermion alternatives.,

Finally, in this rather lengthy discussion of lattice fermions, in particular for QCD calculations, there is the problem of 'heavy' quarks. It turns out that there are discretization errors proportional to powers of am_q, and if m_q is greater than, or of order, $1/a$ these errors are large and uncontrolled. Lattice QCD simulations currently have $a^{-1} \sim 2-4\text{GeV}$, which means that c quarks are borderline and an alternative approach has to be found for b quarks. The possibilities are summarized in the review by Hashimoto and Sharpe (2022).

16.3   Representation of quantum amplitudes

So (with some suitable fermionic action) we have a gauge-invariant 'classical' field theory defined on a lattice, with a suitable continuum limit. (Actually, the $a \to 0$ limit of the quantum theory is, as we shall see in section 16.5, more subtle than the naive replacements (16.5) because of renormalization issues, as should be no surprise to the reader by now). However, we have not yet considered how we are going to turn this classical lattice theory into a quantum one. The fact that the calculations are mostly going to have to be done numerically seems at once to require a formulation that avoids non-commuting operators. This is precisely what is provided by Feynman's *sum over paths* formulation of quantum mechanics and of quantum field theory, and it is therefore an essential element in the lattice approach to quantum field theory. In this section we give a brief introduction to this formalism, starting with quantum mechanics.

16.3.1   Quantum mechanics

In section 5.2.2 we stated that in this approach the amplitude for a quantum system, described by a Lagrangian L depending on one degree of freedom $q(t)$, to pass from a state in which $q = q^i$ at $t = t_i$ to a state in which $q = q^f$ at time $t = t_f$, is proportional to (with $\hbar = 1$)

$$\sum_{\text{all paths } q(t)} \exp\left(i \int_{t_i}^{t_f} L(q(t), \dot{q}(t)) dt\right), \tag{16.49}$$

where $q(t_i) = q^i$, and $q(t_f) = q^f$. We shall now provide some justification for this assertion.

We begin by recalling how, in ordinary quantum mechanics, state vectors and observables are related in the Schrödinger and Heisenberg pictures (see Appendix I of volume 1). Let $\hat{q}$ be the canonical coordinate operator in the Schrödinger picture, with an associated complete set of eigenvectors $|q\rangle$ such that

$$\hat{q}|q\rangle = q|q\rangle . \tag{16.50}$$

The corresponding Heisenberg operator $\hat{q}_H(t)$ is defined by

$$\hat{q}_H(t) = e^{i\hat{H}(t-t_0)} \hat{q} e^{-i\hat{H}(t-t_0)} \tag{16.51}$$

where $\hat{H}$ is the Hamiltonian, and t_0 is the (arbitrary) time at which the two pictures coincide. Now define the Heisenberg picture state $|q_t\rangle_H$ by

$$|q_t\rangle_H = e^{i\hat{H}(t-t_0)}|q\rangle . \tag{16.52}$$

We then easily obtain from (16.50)–(16.52) the result

$$\hat{q}_H(t)|q_t\rangle_H = q|q_t\rangle_H , \tag{16.53}$$

which shows that $|q_t\rangle_H$ is the (Heisenberg picture) state which at time t is an eigenstate of $\hat{q}_H(t)$ with eigenvalue q. Consider now the quantity

$$_H\langle q^f_{t_f}|q^i_{t_i}\rangle_H \tag{16.54}$$

which is, indeed, the amplitude for the system described by $\hat{H}$ to go from q^i at t_i to q^f at t_f. Using (16.52) we can write

$$_H\langle q^f_{t_f}|q^i_{t_i}\rangle_H = \langle q^f|e^{-i\hat{H}(t_f-t_i)}|q^i\rangle ; \tag{16.55}$$

we want to understand how (16.55) can be represented as (16.49).

We shall demonstrate the connection explicitly for the special case of a free particle, for which

$$\hat{H} = \frac{\hat{p}^2}{2m} . \tag{16.56}$$

For this case, we can evaluate (16.55) directly as follows. Inserting a complete set of momentum eigenstates, we obtain[1]

$$\begin{aligned}
\langle q^f|e^{-i\hat{H}(t_f-t_i)}|q^i\rangle &= \int_{-\infty}^{\infty} \langle q^f|p\rangle\langle p|e^{-i\hat{H}(t_f-t_i)}|q^i\rangle \, dp \\
&= \frac{1}{2\pi} \int_{-\infty}^{\infty} e^{ipq^f} e^{-ip^2(t_f-t_i)/2m} e^{-ipq^i} \, dp \\
&= \frac{1}{2\pi} \int_{-\infty}^{\infty} \exp\left\{-i\left[\frac{p^2(t_f-t_i)}{2m} - p(q^f-q^i)\right]\right\} \, dp .
\end{aligned} \tag{16.57}$$

---

[1] Remember that $\langle q|p\rangle$ is the q-space wavefunction of a state with definite momentum p, and is therefore a plane wave; we are using the normalization of equation (E.26) in volume 1.

To evaluate the integral, we complete the square via the steps

$$
\begin{aligned}
\frac{p^2(t_f - t_i)}{2m} - p(q^f - q^i) &= \left(\frac{t_f - t_i}{2m}\right)\left[p^2 - \frac{2mp(q^f - q^i)}{t_f - t_i}\right] \\
&= \left(\frac{t_f - t_i}{2m}\right)\left\{\left[p - \frac{m(q^f - q^i)}{t_f - t_i}\right]^2 - \frac{m^2(q^f - q^i)^2}{(t_f - t_i)^2}\right\} \\
&= \left(\frac{t_f - t_i}{2m}\right)p'^2 - \frac{m(q^f - q^i)^2}{2(t_f - t_i)} \,,
\end{aligned}
\tag{16.58}
$$

where

$$
p' = p - \frac{m(q^f - q^i)}{t_f - t_i} \,.
\tag{16.59}
$$

We then shift the integration variable in (16.57) to p', and obtain

$$
\langle q^f|e^{-i\hat{H}(t_f - t_i)}|q^i\rangle = \frac{1}{2\pi}\exp\left[i\frac{m(q^f - q^i)^2}{2(t_f - t_i)}\right]\int_{-\infty}^{\infty}dp'\exp\left[-\frac{i(t_f - t_i)p'^2}{2m}\right] \,.
\tag{16.60}
$$

As it stands, the integral in (16.60) is not well-defined, being rapidly oscillatory for large p'. However, it is at this point that the motivation for passing to 'Euclidean' spacetime arises. If we make the replacement $t \to -i\tau$, (16.60) becomes

$$
\langle q^f|e^{-\hat{H}(\tau_f - \tau_i)}|q^i\rangle = \frac{1}{2\pi}\exp\left[-\frac{m(q^f - q^i)^2}{2(\tau_f - \tau_i)}\right]\int_{-\infty}^{\infty}dp'\exp\left[-\frac{(\tau_f - \tau_i)p'^2}{2m}\right]
\tag{16.61}
$$

and the integral is a simple convergent Gaussian. Using the result

$$
\int_{-\infty}^{\infty}d\xi e^{-b\xi^2} = \sqrt{\frac{\pi}{b}}
\tag{16.62}
$$

we finally obtain

$$
\langle q^f|e^{-\hat{H}(\tau_f - \tau_i)}|q^i\rangle = \left[\frac{m}{2\pi(\tau_f - \tau_i)}\right]^{\frac{1}{2}}\exp\left[-\frac{m(q^f - q^i)^2}{2(\tau_f - \tau_i)}\right] \,.
\tag{16.63}
$$

We must now understand how the result (16.63) can be represented in the form (16.49). In Euclidean space, (16.49) is

$$
\sum_{\text{paths}}\exp\left(-\int_{\tau_i}^{\tau_f}\frac{1}{2}m\left(\frac{dq}{d\tau}\right)^2 d\tau\right)
\tag{16.64}
$$

in the free-particle case. We interpret the τ integral in terms of a discretization procedure, similar to that introduced in section 16.2. We split the interval $\tau_f - \tau_i$ into N segments each of size ϵ, as shown in figure 16.3. The τ-integral in (16.64) becomes the sum

$$
m\sum_{j=1}^{N}\frac{(q^j - q^{j-1})^2}{2\epsilon} \,,
\tag{16.65}
$$

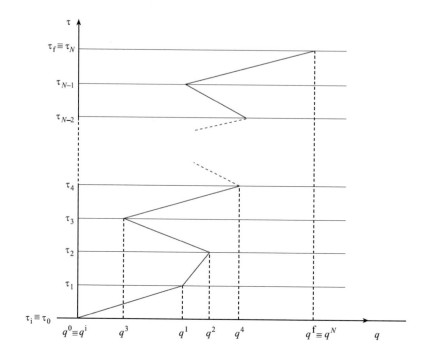

FIGURE 16.3
A 'path' from $q^0 \equiv q^i$ at τ_i to $q^N \equiv q^f$ at τ_f, via the intermediate positions $q^1, q^2, \ldots, q^{N-1}$ at $\tau_1, \tau_2, \ldots, \tau_{N-1}$.

and the 'sum over paths', in going from $q^0 \equiv q^i$ at τ_i to $q^N \equiv q^f$ at τ_f, is now interpreted as a multiple integral over all the intermediate positions $q^1, q^2, \ldots, q^{N-1}$ which paths can pass through at 'times' $\tau_1, \tau_2, \ldots, \tau_{N-1}$:

$$\frac{1}{A(\epsilon)} \int \int \cdots \int \exp\left[-m \sum_{j=1}^{N} \frac{(q^j - q^{j-1})^2}{2\epsilon}\right] \frac{dq^1}{A(\epsilon)} \frac{dq^2}{A(\epsilon)} \cdots \frac{dq^{N-1}}{A(\epsilon)}, \tag{16.66}$$

where $A(\epsilon)$ is a normalizing factor, depending on ϵ, which is to be determined.

The integrals in (16.66) are all of Gaussian form, and since the integral of a Gaussian is again a Gaussian (cf the manipulations leading from (16.57) to (16.60), but without the 'i' in the exponents), we may perform all the integrations analytically. We follow the method of Feynman and Hibbs (1965), section 3.1. Consider the integral over q^1:

$$I^1 \equiv \int \exp\left\{-\frac{m}{2\epsilon}\left[(q^2 - q^1)^2 + (q^1 - q^i)^2\right]\right\} dq^1 . \tag{16.67}$$

This can be evaluated by completing the square, shifting the integration variable, and using (16.62), to obtain (problem 16.5)

$$I^1 = \left(\frac{\pi\epsilon}{m}\right)^{\frac{1}{2}} \exp\left[\frac{-m}{4\epsilon}(q^2 - q^i)^2\right] . \tag{16.68}$$

Now the procedure may be repeated for the q^2 integral

$$I^2 \equiv \int \exp\left\{-\frac{m}{4\epsilon}(q^2 - q^i)^2 - \frac{m}{2\epsilon}(q^3 - q^2)^2\right\} dq^2 , \tag{16.69}$$

which yields (problem 16.5)

$$I^2 = \left(\frac{4\pi\epsilon}{3m}\right)^{\frac{1}{2}} \exp\left[\frac{-m}{6\epsilon}(q^3 - q^{\mathrm{i}})^2\right] . \tag{16.70}$$

As far as the exponential factors in (16.63) in (16.64) are concerned, the pattern is now clear. After $n - 1$ steps we shall have an exponential factor

$$\exp\left[-m(q^n - q^{\mathrm{i}})^2/(2n\epsilon)\right] . \tag{16.71}$$

Hence, after $N - 1$ steps we shall have a factor

$$\exp\left[-m(q^{\mathrm{f}} - q^{\mathrm{i}})^2/2(\tau_{\mathrm{f}} - \tau_{\mathrm{i}})\right] , \tag{16.72}$$

remembering that $q^N \equiv q^{\mathrm{f}}$ and that $\tau_{\mathrm{f}} - \tau_{\mathrm{i}} = N\epsilon$. So we have recovered the correct exponential factor of (16.63), and all that remains is to choose $A(\epsilon)$ in (16.66) so as to produce the same normalization as (16.63).

The required $A(\epsilon)$ is

$$A(\epsilon) = \sqrt{\frac{2\pi\epsilon}{m}} , \tag{16.73}$$

as we now verify. For the first (q^1) integration, the formula (16.66) contains two factors of $A^{-1}(\epsilon)$, so that the result (16.68) becomes

$$\frac{1}{[A(\epsilon)]^2}I^1 = \frac{m}{2\pi\epsilon}\left(\frac{\pi\epsilon}{m}\right)^{\frac{1}{2}}\exp\left[-\frac{m}{4\epsilon}(q^2 - q^{\mathrm{i}})^2\right]$$

$$= \left(\frac{m}{2\pi 2\epsilon}\right)^{\frac{1}{2}}\exp\left[-\frac{m}{4\epsilon}(q^2 - q^{\mathrm{i}})^2\right] . \tag{16.74}$$

For the second (q^2) integration, the accumulated constant factor is

$$\frac{1}{A(\epsilon)}\left(\frac{m}{2\pi 2\epsilon}\right)^{\frac{1}{2}}\left(\frac{4\pi\epsilon}{3m}\right)^{\frac{1}{2}} = \left(\frac{m}{2\pi 3\epsilon}\right)^{\frac{1}{2}} . \tag{16.75}$$

Proceeding in this way, one can convince oneself that after $N - 1$ steps, the accumulated constant is

$$\left(\frac{m}{2\pi N\epsilon}\right)^{\frac{1}{2}} = \left[\frac{m}{2\pi(\tau_{\mathrm{f}} - \tau_{\mathrm{i}})}\right]^{\frac{1}{2}} , \tag{16.76}$$

as in (16.63).

The equivalence of (16.63) and (16.64) (in the sense $\epsilon \to 0$) is therefore established for the free-particle case. More general cases are discussed in Feynman and Hibbs (1965) chapter 5, and in Peskin and Schroeder (1995) chapter 9. The conventional notation for the path-integral amplitude is

$$\langle q^{\mathrm{f}}|\mathrm{e}^{-\hat{H}(\tau_{\mathrm{f}} - \tau_{\mathrm{i}})}|q^{\mathrm{i}}\rangle = \int \mathcal{D}q(\tau)\mathrm{e}^{-\int_{\tau_{\mathrm{i}}}^{\tau_{\mathrm{f}}} L \, \mathrm{d}\tau} , \tag{16.77}$$

where the right-hand side of (16.77) is interpreted in the sense of (16.66).

We now proceed to discuss further aspects of the path-integral formulation. Consider the (Euclideanized) amplitude $\langle q^{\mathrm{f}}|\mathrm{e}^{-\hat{H}(\tau_{\mathrm{f}} - \tau_{\mathrm{i}})}|q^{\mathrm{i}}\rangle$, and insert a complete set of energy eigenstates $|n\rangle$ such that $\hat{H}|n\rangle = E_n|n\rangle$:

$$\langle q^{\mathrm{f}}|\mathrm{e}^{-\hat{H}(\tau_{\mathrm{f}} - \tau_{\mathrm{i}})}|q^{\mathrm{i}}\rangle = \sum_n \langle q^{\mathrm{f}}|n\rangle\langle n|q^{\mathrm{i}}\rangle\mathrm{e}^{-E_n(\tau_{\mathrm{f}} - \tau_{\mathrm{i}})} . \tag{16.78}$$

Equation (16.78) shows that if we take the limits $\tau_i \to -\infty$, $\tau_f \to \infty$, then the state of lowest energy E_0 (the ground state) provides the dominant contribution. Thus, in this limit, our amplitude will represent the process in which the system begins in its ground state $|\Omega\rangle$ at $\tau_i \to -\infty$, with $q = q^i$, and ends in $|\Omega\rangle$ at $\tau_f \to \infty$, with $q = q^f$.

How do we represent propagators in this formalism? Consider the expression (somewhat analogous to a field theory propagator)

$$G_{fi}(t_a, t_b) \equiv \langle q^f_{t_f} | T\{\hat{q}_H(t_a)\hat{q}_H(t_b)\} | q^i_{t_i}\rangle , \qquad (16.79)$$

where T is the usual time-ordering operator. Using (16.51) and (16.52), (16.79) can be written, for $t_b > t_a$, as

$$G_{fi}(t_a, t_b) = \langle q^f | e^{-i\hat{H}(t_f - t_b)} \hat{q} e^{-i\hat{H}(t_b - t_a)} \hat{q} e^{-i\hat{H}(t_a - t_i)} | q^i \rangle . \qquad (16.80)$$

Inserting a complete set of states and Euclideanizing, (16.80) becomes

$$
\begin{aligned}
G_{fi}(t_a, t_b) = \int \mathrm{d}q^a \mathrm{d}q^b \, q^a q^b \langle q^f | e^{-\hat{H}(\tau_f - \tau_b)} | q^b \rangle \\
\times \langle q^b | e^{-\hat{H}(\tau_b - \tau_a)} | q^a \rangle \langle q^a | e^{-\hat{H}(\tau_a - \tau_i)} | q^i \rangle .
\end{aligned}
\qquad (16.81)
$$

Now, each of the three matrix elements has a discretized representation of the form (16.63), with say $N_1 - 1$ variables in the interval (τ_a, τ_i), $N_2 - 1$ in (τ_b, τ_a) and $N_3 - 1$ in (τ_f, τ_b). Each such representation carries one 'surplus' factor of $[A(\epsilon)]^{-1}$, making an overall factor of $[A(\epsilon)]^{-3}$. Two of these factors can be associated with the $\mathrm{d}q^a \mathrm{d}q^b$ integration in (16.81), so that we have a total of $N_1 + N_2 + N_3 - 1$ properly normalized integrations, and one 'surplus' factor $[A(\epsilon)]^{-1}$ as in (16.66). If we now identify $q(\tau_a) \equiv q^a$, $q(\tau_n) \equiv q^b$, it follows that (16.81) is simply

$$\int \mathcal{D}q(\tau) q(\tau_a) q(\tau_b) e^{-\int_{\tau_i}^{\tau_f} L \, \mathrm{d}\tau} . \qquad (16.82)$$

In obtaining (16.82), we took the case $\tau_b > \tau_a$. Suppose alternatively that $\tau_a > \tau_b$. Then the order of τ_a and τ_b inside the interval (τ_i, τ_f) is simply reversed, but since q^a and q^b in (16.81), or $q(\tau_a)$ and $q(\tau_b)$ in (16.82), are ordinary (commuting) numbers, the formula (16.82) is unaltered, and actually does represent the matrix element (16.79 of the time-ordered product.

16.3.2 Quantum field theory

The generalizations of these results to the field theory case are intuitively clear. For example, in the case of a single scalar field $\phi(\mathbf{x})$, we expect the analogue of (16.82) to be (cf (16.4))

$$\int \mathcal{D}\phi(x) \, \phi(x_a)\phi(x_b) \exp\left[-\int_{\tau_i}^{\tau_f} \mathcal{L}_E(\phi, \nabla\phi, \partial_\tau\phi) \, \mathrm{d}^4 x_E\right] , \qquad (16.83)$$

where

$$\mathrm{d}^4 x_E = \mathrm{d}^3 \mathbf{x} \mathrm{d}\tau, \qquad (16.84)$$

and the boundary conditions are given by $\phi(\mathbf{x}, \tau_i) = \phi^i(x)$, $\phi(\mathbf{x}, \tau_f) = \phi^f(x)$, $\phi(\mathbf{x}, \tau_a) = \phi^a(x)$ and $\phi(\mathbf{x}, \tau_b) = \phi^b(x)$, say. In (16.83), we have to understand that a *four*-dimensional discretization of Euclidean spacetime is implied, the fields being Fourier-analyzed by four-dimensional generalizations of expressions such as (16.7). Just as in (16.79)–(16.82), (16.83) is equal to

$$\langle \phi^f(x) | e^{-\hat{H}\tau_f} T\{\hat{\phi}_H(x_a)\hat{\phi}_H(x_b)\} e^{-\hat{H}\tau_i} | \phi^i(x) \rangle . \qquad (16.85)$$

Taking the limits $\tau_i \to -\infty$, $\tau_f \to \infty$ will project out the configuration of lowest energy, as discussed after (16.78), which in this case is the (interacting) vacuum state $|\Omega\rangle$. Thus in this limit the surviving part of (16.85) is

$$\langle \phi^f(x)|\Omega\rangle e^{-E_\Omega \tau} \langle \Omega|T\left\{\hat{\phi}_H(x_a)\hat{\phi}_H(x_b)\right\}|\Omega\rangle e^{-E_\Omega \tau}\langle \Omega|\phi^i(x)\rangle \qquad (16.86)$$

with $\tau \to \infty$. The exponential and overlap factors can be removed by dividing by the same quantity as (16.85) but without the additional fields $\phi(x_a)$ and $\phi(x_b)$. In this way, we obtain the formula for the *field theory propagator* in four-dimensional Euclidean space:

$$\langle \Omega|T\left\{\hat{\phi}_H(x_a)\hat{\phi}_H(x_b)\right\}|\Omega\rangle = \lim_{\tau \to \infty} \frac{\int \mathcal{D}\phi \, \phi(x_a)\phi(x_b)\exp[-\int_{-\tau}^{\tau}\mathcal{L}_E d^4 x_E]}{\int \mathcal{D}\phi \exp[-\int_{-\tau}^{\tau}\mathcal{L}_E d^4 x_E]}. \qquad (16.87)$$

Vacuum expectation values of time-ordered products of more fields will simply have more factors of ϕ on both sides.

Perturbation theory can be developed in this formalism also. Suppose $\mathcal{L}_E = \mathcal{L}_E^0 + \mathcal{L}_E^{int}$, where $\mathcal{L}_E^0$ describes a free scalar field and $\mathcal{L}_E^{int}$ is an interaction, for example $\lambda \phi^4$. Then, assuming λ is small, the exponential in (16.87) can be expressed as

$$\exp\left[-\int d^4 x_E \, (\mathcal{L}_E^0 + \mathcal{L}_E^{int})\right] = \left(\exp - \int d^4 x_E \, \mathcal{L}_E^0\right)\left(1 - \lambda \int d^4 x_E \phi^4 + \ldots\right) \qquad (16.88)$$

and both numerator and denominator of (16.87) may be expressed as vevs of products of free fields. Compact techniques exist for analyzing this formulation of perturbation theory (Ryder (1996), chapter 6, Peskin & Schroeder (1995) chapter 9), and one finds exactly the same 'Feynman rules' as in the canonical (operator) approach.

In the case of gauge theories, we can easily imagine a formula similar to (16.87) for the gauge field propagator, in which the integral is carried out over all gauge fields $A_\mu(x)$ (in the $U(1)$ case, for example). But we already know from chapter 7 (or from chapter 13 in the non-Abelian case) that we shall not be able to construct a well-defined perturbation theory in this way, since the gauge field propagator will not exist unless we 'fix the gauge' by imposing some constraint, such as the Lorentz gauge condition. Such constraints can be imposed on the corresponding path integral, and indeed this was the route followed by Faddeev and Popov (1967) in first obtaining the Feynman rules for non-Abelian gauge theories, as mentioned in section 13.5.3.

In the discrete case, the appropriate integration variables are the link variables $U(l_i)$ where l_i is the ith link. They are elements of the relevant gauge group—for example $U(n_1, n_1 + 1)$ of (16.3.1) is an element of $U(1)$. In the case of the unitary groups, such elements typically have the form (c.f. 12.3.5) $\sim \exp$ (i Hermitian matrix), where the 'Hermitean matrix' can be parametrized in some convenient way—for example, as in (12.31) for $SU(2)$. In all these cases, the variables in the parametrization of U vary over some bounded domain (they are essentially 'angle-type' variables, as in the simple $U(1)$ case), and so, with a finite number of lattice points, the integral over the link variables is well-defined without gauge-fixing. The integration measure for the link variables can be chosen so as to be gauge invariant, and hence provided the action is gauge invariant, the formalism provides well-defined expressions, independently of perturbation theory, for vevs of gauge invariant quantities. This is an important advantage of the lattice approach.

There remains one more conceptual problem to be addressed in this approach: namely, how are we to deal with fermions? It seems that we must introduce new variables which, though not quantum field operators, must nevertheless *anti*-commute with each other. Such 'classical' anti-commuting variables are called *Grassmann variables*, and are briefly described in appendix O. Further details are contained in Ryder (1996) and in Peskin and

Schroeder (1995) section 9.5). For our purposes, the important point is that the fermion Lagrangian is *bilinear* in the (Grassmann) fermion fields ψ, the fermionic action for one flavour having the form

$$S_{\psi_{\mathrm{f}}} = \int \mathrm{d}^4 x_{\mathrm{E}} \, \bar{\psi}_{\mathrm{f}} M_{\mathrm{f}}(U) \psi_{\mathrm{f}}, \tag{16.89}$$

where M_{f} is a matrix representing the Dirac operator $\mathrm{i} \not{D} - m_{\mathrm{f}}$ in its discretized and Euclideanized form. This means that in a typical fermionic amplitude of the form (cf the denominator of (16.87))

$$Z_{\psi_{\mathrm{f}}} = \int \mathcal{D}\bar{\psi}_{\mathrm{f}} \mathcal{D}\psi_{\mathrm{f}} \exp[-S_{\psi_{\mathrm{f}}}], \tag{16.90}$$

one has essentially an integral of Gaussian type (albeit with Grassmann variables), which can actually be performed analytically[2]. The result is simply $\det[M_{\mathrm{f}}(U)]$, the determinant of the Dirac operator matrix. For N_{f} flavours, this easily generalizes to

$$\prod_{\mathrm{f}=1}^{N_{\mathrm{f}}} \det M_{\mathrm{f}}(U). \tag{16.91}$$

Now may write

$$\prod_{\mathrm{f}=1}^{N_{\mathrm{f}}} \det M_{\mathrm{f}}(U) = \exp\left[\sum_{\mathrm{f}} \ln \det M_{\mathrm{f}}(U)\right], \tag{16.92}$$

so that the effect of N_{f} fermions is to contribute an additional term

$$S_{\mathrm{eff}}(U) = -\sum_{\mathrm{f}} \ln \det[M_{\mathrm{f}}(U)] \tag{16.93}$$

to the gluonic action. But although formally correct, this fermionic contribution is computationally very time-consuming to include. Until the mid-1990s it could not be done, and instead calculations were made using the *quenched approximation*, in which the determinant is set equal to a constant independent of the link variables U. This is equivalent to the neglect of closed fermion loops in a Feynman graph approach, i.e. no vacuum polarization insertions on virtual gluon lines. Vacuum polarization amplitudes typically behave as q^2/m_{f}^2 for $q^2 \ll m_{\mathrm{f}}^2$, where q is the momentum flowing into the loop (see equation (11.39), for example, in the case of QED). The quenched approximation is therefore poorer for the light quarks u, d and s.

By the later 1990s it was possible to include the determinant provided the quark masses were not too small. The computation slowed down seriously for light quark masses. So calculations were done for unphysically large values of $m_{\mathrm{u}}, m_{\mathrm{d}}$, and m_{s}, and the results extrapolated towards the physical values. This was a source of large systematic errors.

Beginning in the early 2000s, however, more precise calculations with substantially lighter quark masses became possible, using the (variously improved) staggered fermion formulations discussed in section 16.2.2. It will be recalled that this saves a factor of four in the number of degrees of freedom. But there is still the remaining problem of the four unwanted additional 'tastes'. If these tastes are degenerate, as they would be in the continuum limit, then we can use the simple trick of replacing $S_{\mathrm{eff}}(U)$ by $\frac{1}{4}S_{\mathrm{eff}}(U)$, which means that we take the fourth root of the staggered fermion determinant (the 'rooting' procedure). The true physical (non-degenerate) quark flavour multiplicity still remains, of course, and we arrive at

$$S_{\mathrm{eff,stag.}} = -\ln \det\{M_{\mathrm{stag.\ u}}(U) M_{\mathrm{stag.\ d}}(U) M_{\mathrm{stag.\ s}}(U)\}^{1/4}. \tag{16.94}$$

---

[2]See Appendix P.

Unfortunately, things are not so simple away from the continuum limit, at finite lattice spacing a. Bernard, Golterman, and Shamir (2006) pointed out that the quantity

$$\{\det M_{\text{stag.}}(U)\}^{1/4} \tag{16.95}$$

cannot be represented by a local single-taste theory except in the continuum limit: at finite a, it represents a non-local single-taste action. Locality is a very fundamental property of all successful quantum field theories, and its recovery from (16.95) in the limit $a \to 0$ is not obvious. We refer to Sharpe (2006) for a full discussion, and further references; see also Golterman (2008).

In the last decade, it has become possible to perform simulations with physical isospin-symmetric light quark masses ($m_{\text{u}} = m_{\text{d}}$), removing the need for 'rooting' and for extrapolation in the masses, thereby significantly reducing the overall error. Calculations have controlled errors of the order of a few percent, so that further progress will have to include isospin breaking and electromagnetic effects. Such calculations are referred to as '$N_{\text{f}} = 2+1$' simulations for example, meaning that they treat three light quarks, the mass-degenerate u and d quarks, and the s quark. Some simulations also now include the c quark, in a '$N_{\text{f}} = 2+1+1$' framework.

16.3.3   Connection with statistical mechanics

Not the least advantage of the path integral formulation of quantum field theory (especially in its lattice form) is that it enables a highly suggestive connection to be set up between quantum field theory and statistical mechanics. We introduce this connection now, by way of a preliminary to the discussion of renormalization in the following section.

The connection is made via the fundamental quantity of equilibrium statistical mechanics, the *partition function* Z defined by

$$Z = \sum_{\text{configurations}} \exp\left(-\frac{H}{k_{\text{B}}T}\right), \tag{16.96}$$

which is simply the 'sum over states' (or configurations) of the relevant degrees of freedom, with the Boltzmann weighting factor. H is the classical Hamiltonian evaluated for each configuration. Consider, for comparison, the denominator in (16.87), namely

$$Z_\phi = \int \mathcal{D}\phi \, \exp(-S_{\text{E}}), \tag{16.97}$$

where

$$S_{\text{E}} = \int \mathrm{d}^4 x_{\text{E}} \mathcal{L}_{\text{E}} = \int \mathrm{d}^4 x_{\text{E}} \left\{\frac{1}{2}(\partial_\tau \phi)^2 + \frac{1}{2}(\boldsymbol{\nabla}\phi)^2 + \frac{1}{2}m^2\phi^2 + \lambda\phi^4\right\} \tag{16.98}$$

in the case of a single scalar field with mass m and self-interaction $\lambda\phi^4$. The Euclideanized Lagrangian density $\mathcal{L}_{\text{E}}$ is like an energy density: it is bounded from below, and increases when the field has large magnitude or has large gradients in τ or $\boldsymbol{x}$. The factor $\exp(-S_{\text{E}})$ is then a sensible statistical weight for the fluctuations in ϕ, and Z_ϕ may be interpreted as the partition function for a system described by the field degree of freedom ϕ, but of course in *four* 'spatial' dimensions.

The parallel becomes perhaps even stronger when we discretize spacetime. In an Ising model (see the following section), the Hamiltonian has the form

$$H = -J\sum_n s_n s_{n+1}, \tag{16.99}$$

where J is a constant, and the sum is over lattice sites n, the system variables taking the values ± 1. When (16.99) is inserted into (16.96), we arrive at something very reminiscent of the $\phi(n_1)\phi(n_1 + 1)$ term in (16.6). Naturally, the effective 'Hamiltonian' is not quite the same—though we may note that Wilson (1971b) argued that in the case of a ϕ^4 interaction the parameters can be chosen so as to make the values $\phi = \pm 1$ the most heavily weighted in S_E. Statistical mechanics does, of course, deal in three spatial dimensions, not the four of our Euclideanized spacetime. Nevertheless, it is remarkable that quantum field theory in three spatial dimensions appears to have such a close relationship to equilibrium statistical mechanics in four spatial dimensions.

One insight we may draw from this connection is that, in the case of pure gauge actions (16.34) or (16.35), the gauge coupling is seen to be analogous to an inverse temperature, by comparison with (16.96). One is led to wonder whether something like *transitions between different 'phases'* exist, as coupling constants (or other parameters) vary—and, indeed, such changes of 'phase' can occur.

A second point is somewhat related to this. In statistical mechanics, an important quantity is the *correlation length* ξ, which for a spin system may be defined via the *spin-spin correlation function*

$$G(\boldsymbol{x}) = \langle s(\boldsymbol{x})s(\boldsymbol{0}) \rangle = \sum_{\text{all } s(\boldsymbol{x})} s(\boldsymbol{x})s(\boldsymbol{0}) e^{-H/k_B T} , \qquad (16.100)$$

where we are once more reverting to a continuous $\boldsymbol{x}$ variable. For large $|\boldsymbol{x}|$, this takes the form

$$G(\boldsymbol{x}) \propto \frac{1}{|\boldsymbol{x}|} \exp\left(\frac{-|\boldsymbol{x}|}{\xi(T)}\right) . \qquad (16.101)$$

The Fourier transform of this (in the continuum limit) is

$$\tilde{G}(\boldsymbol{k}^2) \propto \left(\boldsymbol{k}^2 + \xi^{-2}(T)\right)^{-1} , \qquad (16.102)$$

as we learned in section 1.3.3. Comparing (16.100) with (16.87), it is clear that (16.100) is proportional to the propagator (or Green function) for the field $s(\boldsymbol{x})$; (16.102) then shows that $\xi^{-1}(T)$ is playing the role of a mass term m. Now, near a critical point for a statistical system, correlations exist over very large scales ξ compared to the inter-atomic spacing a; in fact, at the critical point $\xi(T_c) \sim L$, where L is the size of the system. In the quantum field theory, as indicated earlier, we may regard a^{-1} as playing a role analogous to a momentum cut-off Λ, so the regime $\xi \gg a$ is equivalent to $m \ll \Lambda$, as was indeed always our assumption. Thus studying a quantum field theory this way is analogous to studying a four-dimensional statistical system near a critical point. This shows rather clearly why it is not going to be easy: correlations over all scales will have to be included. At this point, we are naturally led to the consideration of *renormalization* in the lattice formulation.

16.4   Renormalization, and the renormalization group, on the lattice

16.4.1   Introduction

In the continuum formulation which we have used elsewhere in this book, fluctuations over short distances of order Λ^{-1} generally lead to divergences in the limit $\Lambda \to \infty$, which are controlled (in a renormalizable theory) by the procedure of renormalization. Such divergent fluctuations turn out, in fact, to affect a renormalizable theory only through the values of some of its parameters and, if these parameters are taken from experiment, all other quantities become finite, even as $\Lambda \to \infty$. This latter assertion is not easy to prove, and indeed is quite surprising. However, this is by no means all there is to renormalization theory: we have seen the power of 'renormalization group' ideas in making testable predictions for QCD. Nevertheless, the methods of chapter 15 were rather formal, and the reader may well feel the need of a more physical picture of what is going on. Such a picture was provided by Wilson (1971a) (see also Wilson and Kogut 1974), using the 'lattice + path integral' approach. Another important advantage of this formalism is, therefore, precisely the way in which, thanks to Wilson's work, it provides access to a more intuitive way of understanding renormalization theory. The aim of this section is to give a brief introduction to Wilson's ideas, so as to illuminate the formal treatment of the previous chapter.

In the 'lattice + path integral' approach to quantum field theory, the degrees of freedom involved are the values of the field(s) at each lattice site, as we have seen. Quantum amplitudes are formed by integrating suitable quantities over all values of these degrees of freedom, as in (16.87) for example. From this point of view, it should be possible to examine specifically how the 'short distance' or 'high momentum' degrees of freedom affect the result. In fact, the idea suggests itself that we might be able to perform explicitly the integration (or summation) over those degrees of freedom located near the cutoff Λ in momentum space, or separated by only a lattice site or two in co-ordinate space. If we can do this, the result may be compared with the theory as originally formulated, to see how this 'integration over short-distance degrees of freedom' affects the physical predictions of the theory. Having done this once, we can imagine doing it again—and indeed *iterating* the process, until eventually we arrive at some kind of 'effective theory' describing physics in terms of 'long-distance' degrees of freedom.

There are several aspects of such a programme which invite comment. First, the process of 'integrating out' short-distance degrees of freedom will obviously *reduce* the number of effective degrees of freedom, which is necessarily very large in the case $\xi \gg a$, as envisaged above. Thus it must be a step in the right direction. Secondly, the above sketch of the 'integrating out' procedure suggests that, at any given stage of the integration, we shall be considering the system as described by parameters (including masses and couplings) *appropriate to that scale*, which is of course strongly reminiscent of RGE ideas. And thirdly, we may perhaps anticipate that the result of this 'integrating out' will be not only to render the parameters of the theory scale-dependent, but also, in general, to introduce new kinds of *effective interactions* into the theory. We now consider some simple examples which we hope will illustrate these points.

16.4.2   Two one-dimensional examples

Consider first a simple one-dimensional Ising model with Hamiltonian (16.99) and partition function

$$Z = \sum_{\{s_n\}} \exp \left[K \sum_{n=0}^{N-1} s_n s_{n+1} \right], \tag{16.103}$$

where $K = J/(k_B T) > 0$. In (16.103) all the s_n variables take the values ± 1 and the 'sum over $\{s_n\}$' means that all possible configurations of the N variables $s_0, s_1, s_2, \ldots, s_{N-1}$ are to be included. The spin s_n is located at the lattice site na, and we shall (implicitly) be assuming the periodic boundary condition $s_n = s_{N+n}$. Figure 16.4 shows a portion of the one-dimensional lattice with the spins on the sites, each site being separated by the

FIGURE 16.4
A portion of the one-dimensional lattice of spins in the Ising model.

lattice constant a. Thus, for the portion $\{s_{N-1}, s_0, \ldots s_4\}$ we are evaluating

$$\sum_{s_{N-1}, s_0, s_1, s_2, s_3, s_4} \exp[K(s_{N-1}s_0 + s_0s_1 + s_1s_2 + s_2s_3 + s_3s_4)]. \tag{16.104}$$

Now suppose we want to describe the system in terms of a 'coarser' lattice, with lattice spacing $2a$, and corresponding new spin variables s_n'. There are many ways we could choose to describe the s_n', but here we shall only consider a very simple one (Kadanoff 1977) in which each s_n' is simply identified with the s_n at the corresponding site (see figure 16.5). For the portion of the lattice under consideration, then, (16.104) becomes

$$\sum_{s_{N-1}, s_0', s_1, s_1', s_3, s_2'} \exp\left[K\left(s_{N-1}s_0' + s_0's_1 + s_1s_1' + s_1's_3 + s_3s_2'\right)\right]. \tag{16.105}$$

If we can now perform the sums over s_1 and s_3 in (16.105), we shall end up (for this portion) with an expression involving the 'effective' spin variables s_0', s_1', and s_2', situated twice as far apart as the original ones, and therefore providing a more 'coarse grained' description of the system. Summing over s_1 and s_3 corresponds to 'integrating out' two short-distance degrees of freedom as discussed earlier.

In fact, these sums are easy to do. Consider the quantity $\exp(Ks_0's_1)$, expanded as a power series:

$$\exp(Ks_0's_1) = 1 + Ks_0's_1 + \frac{K^2}{2!} + \frac{K^3}{3!}(s_0's_1) + \ldots \tag{16.106}$$

where we have used $(s_0's_1)^2 = 1$. It follows that

$$\exp(Ks_0's_1) = \cosh K \left(1 + s_0's_1 \tanh K\right), \tag{16.107}$$

and similarly

$$\exp(Ks_1s_1') = \cosh K \left(1 + s_1s_1' \tanh K\right). \tag{16.108}$$

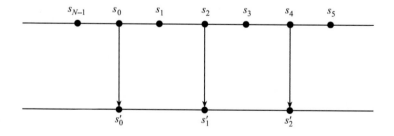

FIGURE 16.5
A 'coarsening' transformation applied to the lattice portion shown in figure 16.4. The new (primed) spin variables are situated twice as far apart as the original (unprimed) ones.

Thus the sum over s_1 is

$$\sum_{s_1=\pm 1} \cosh^2 K \left(1 + s_0' s_1 \tanh K + s_1 s_1' \tanh K + s_0' s_1' \tanh^2 K\right). \tag{16.109}$$

Clearly, the terms linear in s_1 vanish after summing, and the s_1 sum becomes just

$$2 \cosh^2 K \left(1 + s_0' s_1' \tanh^2 K\right). \tag{16.110}$$

Remarkably, (16.110) contains a new 'nearest-neighbour' interaction, $s_0' s_1'$, just like the original one in (16.103), but with an *altered coupling* (and a different spin-independent piece). In fact, we can write (16.110) in the standard form

$$\exp\left[g_1(K) + K' s_0' s_1'\right] \tag{16.111}$$

and then use (16.107) to set

$$\tanh K' = \tanh^2 K \tag{16.112}$$

and identify

$$g_1(K) = \ln\left(\frac{2 \cosh^2 K}{\cosh K'}\right). \tag{16.113}$$

Exactly the same steps can be followed through for the sum on s_3 in (16.105), and indeed for *all* the sums over the 'integrated out' spins. The upshot is that, apart from the accumulated spin-independent part, the new partition function, defined on a lattice of size $2a$, has the same form as the old one, but with a new coupling K' related to the old one K by (16.112).

Equation (16.112) is an example of a *renormalization transformation*: the number of degrees of freedom has been halved, the lattice spacing has doubled, and the coupling K has been renormalized to K'.

It is clear that we could apply the same procedure to the new Hamiltonian, introducing a coupling K'' which is related to K', and thence to K by

$$\tanh K'' = (\tanh K')^2 = (\tanh K)^4. \tag{16.114}$$

This is equivalent to *iterating* the renormalization transformation; after n iterations, the effective lattice constant is $2^n a$, and the effective coupling is given by

$$\tanh K^{(n)} = (\tanh K)^n. \tag{16.115}$$

The successive values $K', K'', \ldots$ of the coupling under these iterations can be regarded as a *flow* in the (one-dimensional) space of K-values: a *renormalization flow*.

Of particular interest is a point (or points) K^* such that

$$\tanh K^* = \tanh^2 K^*. \tag{16.116}$$

This is called a *fixed point* of the renormalization tranformation. At such a point in K-space, changing the scale by a factor of 2 (or 2^n for that matter) will make no difference, which means that the system must be in some sense ordered. Remembering that $K = J/(k_B T)$, we see that $K = K^*$ when the temperature is 'tuned' to the value $T = T^* = J/(k_B K^*)$. Such a T^* would be the temperature of a *critical point* for the thermodynamics of the system, corresponding to the onset of ordering. In the present case, the only fixed points are $K^* = \infty$ and $K^* = 0$. Thus there is no critical point at a non-zero T^*, and hence no transition to an ordered phase. However, we may describe the behaviour as $T \to 0$ as 'quasi-critical'. For large K, we may use

$$\tanh K \simeq 1 - 2e^{-2K} \tag{16.117}$$

FIGURE 16.6

'Renormalization flow': the arrows show the direction of flow of the coupling K as the lattice constant is increased.
The starred values are fixed points.

to write (16.115) as

$$K^{(n)} = K - \frac{1}{2}\ln n, \tag{16.118}$$

which shows that K^n changes only very slowly (logarithmically) under iterations when in the vicinity of a very large value of K, so that this is 'almost' a fixed point.

We may represent the flow of K, under the renormalization transformation (16.115), as in figure 16.6. Note that the flow is away from the quasi-fixed point at $K^* = \infty$ ($T = 0$) and towards the (non-interacting) fixed point at $K^* = 0$.

A renormalization transformation which has a fixed point at a finite (neither zero nor infinite) value of the coupling is clearly of greater interest, since this will correspond to a critical point at a finite temperature. A simple such example given by Kadanoff (1977) is the transformation

$$K' = \frac{1}{2}(2K)^2 \tag{16.119}$$

for a doubling of the effective lattice size, or

$$K^{(n)} = \frac{1}{2}(2K)^n \tag{16.120}$$

for n such iterations. The model leading to (16.120) involves fermions in one dimension, but the details are irrelevant to our purpose here. The renormalization transformation (16.120) has three fixed points: $K^* = 0$, $K^* = \infty$, and the finite point $K^* = \frac{1}{2}$. The renormalization flow is shown in figure 16.7.

The striking feature of this flow is that the motion is always away from the finite fixed point, under successive iterations. This may be understood by recalling that at the fixed point (which is a critical point for the statistical system) the correlation length ξ must be infinite (as $L \to \infty$). As we iterate away from this point, ξ decreases and we leave the fixed (or critical) point. For this model, ξ is given by Kadanoff (1977) as

$$\xi = \frac{a}{|\ln 2K|} \tag{16.121}$$

which indeed goes to infinity at $K = \frac{1}{2}$.

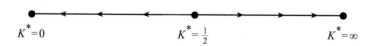

FIGURE 16.7

The renormalization flow for the transformation (16.120).

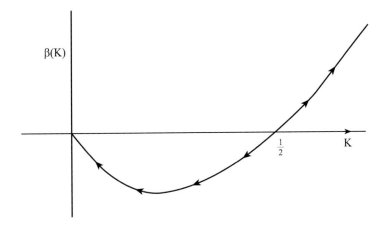

FIGURE 16.8
The β-function of (16.124); the arrows indicate increasing f.

16.4.3   Connections with particle physics

Let us now begin to think about how all this may relate to the treatment of the renormalization group in particle physics, as given in the previous chapter. First, we need to consider a continuous change of scale, say by a factor of f. In the present model, the transformation (16.120) then becomes

$$K(fa) = \frac{1}{2}(2K(a))^f. \tag{16.122}$$

Differentiating (16.122) with respect to f, we find

$$f\frac{\mathrm{d}K(fa)}{\mathrm{d}f} = K(fa)\ln\left[2K(fa)\right]. \tag{16.123}$$

We may reasonably call (16.123) a renormalization group equation, describing the 'running' of $K(fa)$ with the scale f, analogous to the RGE's for α and α_s considered in chapter 15. In this case, the β-function is

$$\beta(K) = K\ln(2K), \tag{16.124}$$

which is sketched in figure 16.8. The zero of β is indeed at the fixed (critical) point $K = \frac{1}{2}$, and this is an infrared unstable fixed point, the flow being away from it as f increases.

The foregoing is exactly analogous to the discussion in section 15.5: see in particular figure 15.6 and the related discussion. Note, however, that in the present case we are considering rescalings in *position* space, not momentum space. Since momenta are measured in units of a^{-1}, it is clear that scaling a by f is the same as scaling k by $f^{-1} = t$, say. This will produce a change in sign in $\mathrm{d}K/\mathrm{d}t$ relative to $\mathrm{d}K/\mathrm{d}f$, and accounts for the fact that $K = \frac{1}{2}$ is an infrared unstable fixed point in figure 16.8, while α_s^* is an infrared stable fixed point in figure 15.6(b). Allowing for the change in sign, figure 16.8 is quite analogous to figure 15.6(a).

We have emphasized that, at a critical point, and in the continuum limit, the correlation length $\xi \to \infty$, or equivalently the mass parameter (cf (16.102)) $m = \xi^{-1} \to 0$. In this case, the Fourier transform of the spin-spin correlation function should behave as

$$\tilde{G}(\mathbf{k}^2) \propto \frac{1}{\mathbf{k}^2}. \tag{16.125}$$

This is indeed the $\mathbf{k}^2$-dependence of the propagator of a free, massless scalar particle, but—as we learned for the fermion propagator in section 15.5—it is no longer true in an interacting theory. In the interacting case, (16.125) generally becomes modified to

$$\tilde{G}(\mathbf{k}^2) \propto \frac{1}{(\mathbf{k}^2)^{1-\frac{\eta}{2}}}, \tag{16.126}$$

or equivalently

$$G(\mathbf{x}) \propto \frac{1}{|\mathbf{x}|^{1+\eta}} \tag{16.127}$$

in three spatial dimensions, and in the continuum limit. Thus, at a critical point, the spin-spin correlation function exhibits scaling under the transformation $\mathbf{x}' = f\mathbf{x}$, but it is not free-field scaling. Comparing (16.126) with (15.75), we see that $\eta/2$ is precisely the *anomalous dimension* of the field $s(\mathbf{x})$, so—just as in section 15.5—we have an example of scaling with anomalous dimension. In the statistical mechanics case, η is a *critical exponent*, one of a number of such quantities characterizing the critical behaviour of a system. In general, η will depend on the coupling constant $\eta(K)$: at a non-trivial fixed point, η will be evaluated at the fixed point value K^*, $\eta(K^*)$. Enormous progress was made in the theory of critical phenomena when the powerful methods of quantum field theory were applied to calculate critical exponents (see for example Peskin & Schroeder (1995) chapter 13, and Binney *et al.* (1992)).

In our discussion so far, we have only considered simple models with just one 'coupling constant', so that diagrams of renormalization flow were one-dimensional. Generally, of course, Hamiltonians will consist of several terms, and the behaviour of all their coefficients will need to be considered under a renormalization transformation. The general analysis of renormalization flow in multi-dimensional coupling space was given by Wegner (1972). In simple terms, the coefficients show one of three types of behaviour under renormalization transformations such that $a \to fa$, characterized by their behaviour in the vicinity of a fixed point: (i) the difference from the fixed point value grows as f increases, so that the system moves away from the fixed point (as in the single-coupling examples considered earlier); (ii) the difference decreases as f increases, so the system moves towards the fixed point; (iii) there is no change in the value of the coupling as f changes. The corresponding coefficients are called, respectively, (i) *relevant*, (ii) *irrelevant*, and (iii) *marginal* couplings; the terminology is also frequently applied to the operators in the Hamiltonians themselves. The intuitive meaning of 'irrelevant' is clear enough: the system will head towards a fixed point as $f \to \infty$ whatever the initial values of the irrelevant couplings. The critical behaviour of the system will therefore be independent of the number and type of all irrelevant couplings, and will be determined by the relatively few (in general) marginal and relevant couplings. Thus *all* systems which flow close to the fixed point will display the same critical exponents determined by the dynamics of these few couplings. This explains the property of *universality* observed in the physics of phase transitions, whereby many apparently quite different physical systems are described (in the vicinity of their critical points) by the same critical exponents.

Additional terms in the Hamiltonian are, in fact, generally introduced following a renormalization transformation. In the quantum field case, we may expect that renormalization transformations associated with $a \to fa$, and iterations thereof, will in general lead to an effective theory involving all possible couplings allowed by whatever symmetries are assumed to be relevant. Thus, if we start with a typical 'ϕ^4' scalar theory as given by (16.98), we shall expect to generate all possible couplings involving ϕ and its derivatives. At first sight, this may seem disturbing: after all, the original theory (in four dimensions) is a renormalizable one, but an interaction such as $A\phi^6$ is *not* renormalizable according to the criterion given in section 11.8 (in four dimensions ϕ has mass dimension unity, so that A must have mass dimension -2). It is, however, essential to remember that in this 'Wilsonian' approach to renormalization, summations over momenta appearing in loops do not, after one iteration $a \to fa$, run up to the original cut-off value π/a, but only up to the lower cut-off π/fa. The additional interactions compensate for this change.

In fact, we shall now see how the coefficients of non-renormalizable interactions correspond precisely to *irrelevant* couplings in Wilson's approach, so that their effect becomes negligible as we iterate to scales much larger than a. We consider continuous changes of scale characterized by a factor f, and we discuss a theory with only a single scalar field ϕ for simplicity. Imagine, therefore, that we have integrated out, in (16.97), those components of $\phi(\mathbf{x})$ with $a < |\mathbf{x}| < fa$. We will be left with a functional integral of the form (16.97), but with $\phi(\mathbf{x})$ restricted to $|\mathbf{x}| > fa$, and with additional interaction terms in the action. In order to interpret the result in Wilson's terms, we must rewrite it so that it has the same general form as the original Z_ϕ of (16.97). A simple way to do this is to rescale distances by

$$\mathbf{x}' = \frac{\mathbf{x}}{f} \tag{16.128}$$

so that the functional integral is now over $\phi(\mathbf{x}')$ with $|\mathbf{x}'| > a$, as in (16.97). We now *define* the fixed point of the renormalization transformation to be that in which all the terms in the action are *zero*, except the 'kinetic' piece; this is the 'free-field' fixed point. Thus, we require the kinetic action to be unchanged:

$$\int d^4 x_E \, (\partial_\mu \phi)^2 \;=\; \int d^4 x'_E (\partial'_\mu \phi')^2$$

$$=\; \int \frac{1}{f^2} d^4 x_E (\partial_\mu \phi')^2 \;, \tag{16.129}$$

from which it follows that $\phi' = f\phi$. Consider now a term of the form $A\phi^6$:

$$A \int d^4 x_E \, \phi^6 = \frac{A}{f^2} \int d^4 x'_E \, \phi'^6 \;. \tag{16.130}$$

(16.130) shows that the 'new' A' is related to the old one by $A' = \frac{A}{f^2}$, and in particular that, as f increases, A' decreases and is therefore an *irrelevant* coupling, tending to zero as we reach large scales. But such an interaction is precisely a non-renormalizable one (in four dimensions), according to the criterion of section 11.8. The mass dimension of ϕ is unity, and hence that of A must be -2 so that the action is dimensionless; couplings with negative mass dimensions correspond to non-renormalizable interactions. The reader may verify the generality of this result for any interaction with p powers of ϕ, and q derivatives of ϕ.

However, the mass term $m^2 \phi^2$ behaves differently:

$$m^2 \int d^4 x_E \, \phi^2 = m^2 f^2 \int d^4 x'_E \, \phi'^2 \tag{16.131}$$

showing that $m'^2 = m^2 f^2$ and the 'coupling' m^2 is *relevant*, since it grows with f^2. Such a term has positive mass dimension, and corresponds to a 'super-renormalizable' interaction. Finally, the $\lambda \phi^4$ interaction transforms as

$$\lambda \int d^4 x_E \, \phi^4 = \lambda \int d^4 x'_E \, \phi'^4 \tag{16.132}$$

and so $\lambda' = \lambda$. The coupling is *marginal*, which may correspond (though not necessarily) to a renormalizable interaction. To find out if such couplings increase or decrease with f, we have to include higher-order loop corrections. The foregoing analysis in terms of the suppression of non-renormalizable interactions by powers of f^{-1} parallels precisely the similar one in section 11.8. We saw that such terms were suppressed at low energies by factors of E/Λ, where Λ is the cut-off scale beyond which the theory is supposed to fail on physical grounds. The result is that as we renormalize, in Wilson's sense, down to much lower energy scales, the non-renormalizable terms disappear and we are left with an effective renormalizable theory. This is the field theory analogue of 'universality'.

These ideas have an important application in lattice QCD. One of the reasons for systematic inaccuracies in lattice computations is that the continuum is being simulated by a lattice of finite spacing. As already mentioned, Symanzik (1983) showed that corrections to continuum theory results stemming from finite lattice spacing could be diminished systematically by the use of lattice actions that also include suitable irrelevant terms. This procedure (Lüscher and Weisz 1985a, b) is routinely adopted in accurate lattice calculations with 'Symanzik-improved' actions.

One further word should be said about terms such as '$m^2\phi^2$' (which arise in the Higgs sector of the standard model, for instance). As we have seen, m^2 scales by $m'^2 = m^2 f^2$, which is a rapid growth with f. If we imagine starting at a very high scale, such as 10^{15} TeV and flowing down to 1 TeV, then the 'initial' value of m will have to be very finely 'tuned' in order to end up with a mass of order 1 TeV. Thus, in this picture, it seems unnatural to have scalar particles with masses much less than the physical cut-off scale, unless some symmetry principle 'protects' their light masses. We shall return to this problem in sections 22.8.1 and 23.3.

We now return to lattice QCD, with a brief survey of some of the impressive results which have been obtained numerically.

16.5 Lattice QCD

16.5.1 Introduction and the continuum limit

The basic inputs for lattice QCD calculations are the same as for continuum QCD, namely the coupling $\alpha_s = g_s^2/4\pi$ at some scale, the quark masses, and the θ parameter. The last one (see section 14.2.4)) is usually set to zero. The others must be determined from experiment.

Let us begin by considering some numbers. The lattice must be large enough so that the spatial dimension R of the object we wish to describe—say the size of a hadron—fits comfortably inside it, otherwise the result will be subject to 'finite size effects' as the hypercube side length L is varied. We also need $R \gg a$, or else the granularity of the lattice resolution will become apparent. Further, as indicated earlier, we expect the mass m (which is of order R^{-1}) to be very much less than a^{-1}. Thus ideally we need

$$a \ll R \sim 1/m \ll L = Na \tag{16.133}$$

so that N must be large. For example, if $N = 64$ and $a \sim 0.1\,\text{fm}$ the condition (16.133) would be reasonably satisfied by a light hadron mass. But remember that each field at each lattice point is an independent degree of freedom and dealing with integrals such as (16.87) presents a formidable numerical challenge.

Ignoring any statistical inaccuracy, the results will depend on the parameters g_L and N, where g_L is the bare lattice gauge coupling (we assume for simplicity that the quarks are massless). Despite the fact that g_L is dimensionless, we shall now see that its value actually controls the physical size of the lattice spacing a, as a result of renormalization effects. The computed mass of a hadron M, say, must be related to the only quantity with mass dimension, a^{-1}, by a relation of the form

$$M = \frac{1}{a} f(g_{\mathrm{L}}). \tag{16.134}$$

Thus in approaching the continuum limit $a \to 0$, we shall also have to change g_{L} suitably, so as to ensure that M remains finite. This is, of course, quite analogous to saying that, in a renormalizable theory, the bare parameters of the theory depend on the momentum cut-off Λ in such a way that, as $\Lambda \to \infty$, finite values are obtained for the corresponding physical parameters (see the last paragraph of section 10.1.2, for example). In practice, of course, the extent to which the lattice 'a' can really be taken to be very small is severely limited by the computational resources available—that is, essentially, by the number of mesh points N.

Equation (16.134) should therefore really read

$$M = \frac{1}{a} f\left(g_{\mathrm{L}}(a)\right). \tag{16.135}$$

As $a \to 0$, M should be finite and independent of a. However, we know that the behaviour of $g_{\mathrm{L}}(a)$ at small scales is in fact calculable in perturbation theory, thanks to the asymptotic freedom of QCD. This will allow us to determine the form of $f(g_{\mathrm{L}})$, up to a constant, and lead to an interesting prediction for M (equations (16.141)–(16.142)).

Differentiating (16.135) we find

$$0 = \frac{\mathrm{d}M}{\mathrm{d}a} = -\frac{1}{a^2} f\left(g_{\mathrm{L}}(a)\right) + \frac{1}{a} \frac{\mathrm{d}f}{\mathrm{d}g_{\mathrm{L}}} \frac{\mathrm{d}g_{\mathrm{L}}(a)}{\mathrm{d}a}, \tag{16.136}$$

so that

$$\left(a \frac{\mathrm{d}g_{\mathrm{L}}(a)}{\mathrm{d}a}\right) \frac{\mathrm{d}f}{\mathrm{d}g_{\mathrm{L}}} = f\left(g_{\mathrm{L}}(a)\right). \tag{16.137}$$

Meanwhile, the scale dependence of g_{L} is given (to one loop order) by

$$a \frac{\mathrm{d}g_{\mathrm{L}}(a)}{\mathrm{d}a} = \frac{\beta_0}{4\pi} g_{\mathrm{L}}^3(a), \tag{16.138}$$

where the sign is the opposite of (15.47) since $a \sim \mu^{-1}$ is the relevant scale parameter here (compare the comments after equation (16.124)). The integration of (16.138) requires, as usual, a dimensionful constant of integration (c.f. (15.53)):

$$\frac{g_{\mathrm{L}}^2(a)}{4\pi} = \frac{1}{\beta_0 \ln(1/a^2 \Lambda_{\mathrm{L}}^2)}. \tag{16.139}$$

Equation (16.139) shows that $g_{\mathrm{L}}(a)$ tends logarithmically to zero as $a \to 0$, as we expect from asymptotic freedom. Λ_{L} can be regarded as a lattice equivalent of the continuum $\Lambda_{\overline{\mathrm{MS}}}$, and it is defined (at one loop order) by

$$\Lambda_{\mathrm{L}} \equiv \lim_{g_{\mathrm{L}} \to 0} \frac{1}{a} \exp\left(-\frac{2\pi}{\beta_0 g_{\mathrm{L}}^2}\right). \tag{16.140}$$

Equation (16.140) may also be read as showing that the lattice spacing a must go exponentially to zero as g_{L} tends to zero. Higher order corrections can of course be included.

In a similar way, integrating (16.137) using (16.138) gives, in (16.134),

$$M = \text{constant} \times \left[\frac{1}{a} \exp\left(-\frac{2\pi}{\beta_0 g_{\mathrm{L}}^2}\right)\right] \tag{16.141}$$

$$= \text{constant} \times \Lambda_{\mathrm{L}}. \tag{16.142}$$

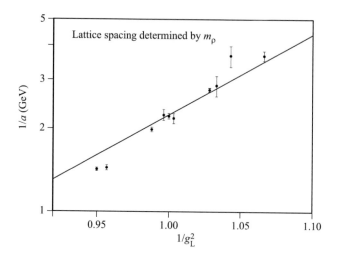

FIGURE 16.9

$\ln(a^{-1}$ in GeV$)$ plotted against $1/g_{\rm L}^2$; figure from R K Ellis, W J Stirling, and B R Webber (1996) *QCD and Collider Physics*, courtesy Cambridge University Press, as adapted from Allton (1995).

Equation (16.141) is known as *asymptotic scaling*. It predicts how any physical mass, expressed in lattice units a^{-1}, should vary as a function of $g_{\rm L}$. The form (16.142) is remarkable, as it implies that all calculated masses must be proportional, in the continuum limit $a \to 0$, to the same universal scale factor $\Lambda_{\rm L}$.

How are masses calculated on the lattice? The principle is very similar to the way in which the ground state was selected out as $\tau_{\rm i} \to -\infty$, $\tau_{\rm f} \to +\infty$ in (16.78). Consider a correlation function for a scalar field, for simplicity:

$$
\begin{aligned}
C(\tau) &= \langle \Omega | \phi(\mathbf{x} = 0, \tau) \phi(0) | \Omega \rangle \\
&= \sum_n |\langle \Omega | \phi(0) | n \rangle|^2 \, {\rm e}^{-E_n \tau} .
\end{aligned}
\tag{16.143}
$$

As $\tau \to \infty$, the term with the minimum value of E_n, namely $E_n = M_\phi$, will survive; M_ϕ can be measured from a fit to the exponential fall-off as a function of τ.

The behaviour predicted by (16.141) and (16.142) can be tested in actual calculations. A quantity such as the ρ meson mass is calculated (via a correlation function of the form (16.143), the result being expressed in terms of a certain number of lattice units a^{-1} at a certain value of $g_{\rm L}$. By comparison with the known ρ mass, a^{-1} can be converted to GeV. Then the calculation is repeated for a different $g_{\rm L}$ value and the new a^{-1} (GeV) extracted. A plot of $\ln[a^{-1}({\rm GeV})]$ versus $1/g_{\rm L}^2$ should then give a straight line with slope $2\pi/\beta_0$ and intercept $\ln \Lambda_{\rm L}$. Figure 16.9 shows such a plot, taken from Ellis, Stirling, and Webber (1996), from which it appears that the calculations are indeed being performed close to the continuum limit. The value of $\Lambda_{\rm L}$ has been adjusted to fit the numerical data, and has the value $\Lambda_{\rm L} = 1.74$ MeV in this case. This may seem alarmingly far from the kind of value expected for $\Lambda_{\rm QCD}$, but we must remember that the renormalization schemes involved in the two cases are quite different. In fact, we may expect $\Lambda_{\rm QCD} \approx 50\Lambda_{\rm L}$ (Montvay and Munster (1994), section 5.1.6).

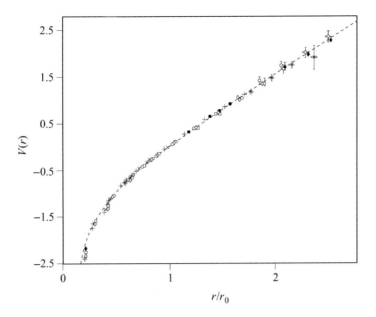

FIGURE 16.10
The static QCD potential, expressed in units of r_0. The broken curve is the functional form
(16.147). Figure reprinted with permission from C R Allton *et al.* (UKQCD Collaboration)
Phys. Rev. D **65** 054502 (2002). Copyright 2002 by the American Physical Society.

16.5.2 The static q$\bar{\text{q}}$ potential

The calculations of m_ρ represented in figure 16.9 were done in the quenched approximation.
As a first example of a calculation with dynamical (unquenched) fermions we show in
figure 16.10 a lattice calculation of the static q$\bar{\text{q}}$ potential (Allton *et al.* 2002, UKQCD
Collaboration), using two degenerate flavours of dynamical quarks[3] on a $16^3 \times 32$ lattice.
As usual, one dimensionful quantity has to be fixed in order to set the scale. In the present
case this has been done via the scale parameter r_0 of Sommer (1994), defined by

$$r_0^2 \left.\frac{\mathrm{d}V}{\mathrm{d}r}\right|_{r=r_0} = 1.65 \,. \tag{16.144}$$

Applying (16.144) to the Cornell (Eichten *et al.* 1980) or Richardson (1979) phenomenological potentials gives $r_0 \simeq 0.49$ fm, conveniently in the range which is well-determined by c$\bar{\text{c}}$
and b$\bar{\text{b}}$ data. The data are well described by the expression

$$V(r) = V_0 + \sigma r - \frac{A}{r} \,, \tag{16.145}$$

where in accordance with (16.144)

$$\sigma = \frac{(1.65 - A)}{r_0^2} \,, \tag{16.146}$$

---

[3]Comparison with matched data in the quenched approximation revealed very little difference, in this
case.

and where V_0 has been chosen such that $V(r_0) = 0$. Thus (16.145) becomes

$$r_0 V(r) = (1.65 - A)\left(\frac{r}{r_0} - 1\right) - A\left(\frac{r_0}{r} - 1\right) . \qquad (16.147)$$

This is—up to a constant—exactly the functional form mentioned in chapter 1, equation (1.33). The quantity $\sqrt{\sigma}$ (there called b) is referred to as the 'string tension', and has a value of about 465 MeV in the present calculations. Phenomenological models suggest a value of around 440 MeV (Eichten *et al.* 1980). The parameter A is found to have a value of about 0.3. In lowest order perturbation theory, and in the continuum limit, A would be given by one-gluon exchange as

$$A = \frac{4}{3}\alpha_{\rm s}(\mu) \qquad (16.148)$$

where μ is some energy scale. This would give $\alpha_{\rm s} \simeq 0.22$, a reasonable value for $\mu \simeq 3$ GeV. Interestingly, the form (16.147) is predicted by the 'universal bosonic string model' (Lüscher *et al.* 1980, Lüscher 1981), in which A has the 'universal' value $\frac{\pi}{12} \simeq 0.26$.

The existence of the linearly rising term with $\sigma > 0$ is a signal for confinement, since—if the potential maintained this form—it would cost an infinite amount of energy to separate a quark and an antiquark. But at some point, enough energy will be stored in the 'string' to create a q$\bar{\rm q}$ pair from the vacuum: the string then breaks, and the two q$\bar{\rm q}$ pairs form mesons. There is no evidence for string breaking in figure 16.10, but we must note that the largest distance probed is only about 1.3 fm.

16.5.3   The results of 'mature' lattice QCD calculations

Earlier sections have described how there are, in fact, many different choices for lattice QCD actions, all of which have the same continuum limit, which is, of course, the physical case. Results of different lattice calculations, with different actions and different systematic errors, must all be consistent (within errors) in their predictions for a given physical quantity, just as different experimental determinations must. Indeed, such consistency has now been achieved for many physical quantities. In short, the field of lattice QCD is now 'mature'. The number, variety, and precision of calculations is now such that an international collaboration, the Flavour Averaging Group (FLAG), has been formed, which provides averages of those results which pass certain quality criteria. A recent FLAG review is by Aoki *et al.* (2020), where the reader will find a very extensive discussion of the details of the calculations, and of the treatment of the systematic errors. In the following we shall consider just two particular areas - the calculation of $\alpha_{\rm s}(M_{\rm Z})$ and the hadron spectrum.

1. Calculation of $\alpha(M_{\rm Z)}$

The general strategy for calculating the value of $\alpha_{\rm s}(\mu)$ at some scale μ is to calculate a short distance quantity A say non-perturbatively on the lattice, and equate the result to a perturbative calculation of the same quantity:

$$A({\rm Lattice}) = \sum_{n=1}^{M} c_n \alpha_{\rm s}^n(\mu). \qquad (16.149)$$

Solving this equation gives $\alpha_{\rm s}$ at the scale of the computed quantity. To define the calculation on the lattice, the values of the quark masses $m_{\rm u} = m_{\rm d}, m_{\rm s}, m_{\rm c}$, and $m_{\rm b}$ have to be determined by tuning them to four experimentally measured quantities, and the lattice spacing is adjusted to fit a fifth measured quantity.

Various short-distance quantities have been considered in this context. In the work of the HPQCD Collaboration (Davies *et al.* 2008, McNeile *et al.* 2010) the quantities calculated were vacuum expectation values of small Wilson loop operators W_{mn} (and related quantities) where

$$W_{mn} \equiv \frac{1}{3}\langle 0|\mathrm{Re\, Tr\, P}\exp[-ig_{\mathrm{L}}\int_{nm} A\cdot \mathrm{d}x]|0\rangle, \qquad (16.150)$$

where P denotes path ordering, $A_\mu = \boldsymbol{\lambda}/2\cdot \boldsymbol{A}_\mu$ is the QCD (matrix-valued) vector potential, and the integral is over a closed $ma \times na$ rectangular path, not necessarily planar. The 1×1 Wilson loop is just the vev of the simple plaquette operator $U_\square$ of section 16.2.3.

In order to use the expansion (16.149), one has to decide what renormalization scheme to employ in defining $\alpha_{\mathrm{s}}(\mu)$. Davies *et al.* used the 'V-scheme' defined through the heavy quark potential. This can be converted to the $\overline{\mathrm{MS}}$ scheme of continuum QCD, and the resultant $\alpha_{\mathrm{s}}(\mu)$ evolved to $\mu = M_{\mathrm{Z}}$. This gave $\alpha_{\mathrm{s}}(M_{\mathrm{Z}}) = 0.1183(8)$.

The various methods for calculating α_{s} are reviewed in Aoki *et al.* (2020), which reports the averaged lattice value

$$\alpha_{\mathrm{s}}(M_{\mathrm{Z}}) = 0.1182(8). \qquad (16.151)$$

It is noteworthy that the error here is closely similar to that of the averaged experimental value quoted earlier, $\alpha_{\mathrm{s}}(M_{\mathrm{Z}}) = 0.1176(10)$ (Huston, Rabbertz, and Zanderighi 2022). Combining this figure with (16.151) gives the current world average figure quoted earlier, in chapter 15.

2. Hadronic spectroscopy

For our second example of precise lattice QCD calculations, it is appropriate to consider the mass spectrum of hadrons. After all, protons and neutrons account for nearly all the mass of ordinary matter, and 95% of their mass is the result of QCD interactions. It has long been a fundamental challenge to predict hadron masses accurately from QCD, thus showing that it does accurately describe strong interactions in the low-energy domain.

As our first illustration of such calculations, we show in figure 16.11 the light hadron spectrum of QCD as reported by Dürr *et al.* (2008). Horizontal lines and bands are the experimental values (which have been isospin-averaged) with their decay widths. The solid circles are the predicted values. Vertical error bars represent combined statistical and systematic error estimates. The masses of the π, K, and Ξ have no error bars, because they have been used to set the values of $m_{\mathrm{u}} = m_{\mathrm{d}}$, m_{s}, and the overall scale, respectively. Once again, the agreement with experiment is very impressive.

These calculations used a Symanzik-improved gauge action (Lüscher and Weisz 1985a, b), and 2+1 flavours of light dynamical Wilson fermions, with various improvements (Morningstar and Peardon (2004)). The physical scale was set either by fitting to the mass of the Ξ, or to the mass of the Ω; the two ways gave consistent results. Pion masses in the range (approximately) 800 MeV to 190 MeV were used to extrapolate to the physical value, with lattice sizes approximately four times the inverse pion mass. A particular type of finite-volume effect arises in the case of strongly decaying resonant states: a procedure for reconstructing the infinite-volume resonance mass, given by Lüscher (1986, 1991a, 1991b), was followed here. This was satisfactory, except for the ρ and Δ at the lightest pion mass point, which was omitted from the extrapolation for these two channels. For further details, and additional references, we refer the reader to the supplementary material to Dürr *et al.* (2008) provided online.

A number of groups have reported results on the masses of charm baryons, among them Alexandrou *et al.* (2014). They used physical values for light quark masses, thereby avoiding

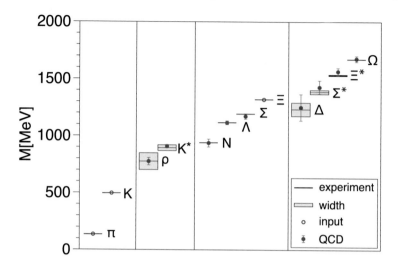

FIGURE 16.11
The light hadron spectrum of QCD, from Dürr *et al.* (2008).

chiral extrapolations. They found perfect agreement with experiment for the masses of the non-charm baryon octet and decuplet, and for the masses of the known spin-1/2 singly-charmed baryons. They predicted a doubly-charmed spin-1/2 state Ξ_{cc}^* at 3.682(10)(26) GeV. This state was subsequently confirmed by LHCb (Aaij *et al.* 2017) at 3.62155 GeV. Predictions were also made for doubly- and triply-charmed baryons.

The spectrum of mesons with one heavy and one light quark has also been well explored. Dowdall *et al.* (2012) presented results for the B meson spectrum (B, B_s, and B_c) with very high statistics, and including $N_f = 2 + 1 + 1$ quark flavours with the HISQ action. All parameters were fixed from earlier calculations, so that this was a parameter-free test of lattice QCD. Agreement with experiment, where data existed, was excellent, and predictions were made for new states. High precision results are also available for charmonium and bottomonium states. DeTar *et al.* (2019) used $N_f = 2 + 1$ fermions with an improved staggered fermion action, obtaining essentially perfect agreement with experiment.

These calculations all refer to low-lying states, in the sense that they do not deal with states that can decay by strong interactions (or such decays are ignored and the widths are not calculated). More recently, however, this situation has changed, and methods have been found for extracting scattering amplitudes (which pertain to infinite volume) from the discrete spectra provided by finite-volume QCD. The idea builds on seminal work by Lüscher (1986, 1991a, b). Following Briceño *et al.* (2018), it can be illustrated by considering a very simple potential model. Two spinless particles, separated by a distance $|x|$, interact via a finite-range potential. Outside the potential, the wavefunction has the form $\psi_V(x) \sim \cos(p|x| + \delta(p))$, where p is a continuous variable, and δ is the elastic scattering phase shift, accounting for the effect of the potential. Now imagine putting the system in a periodic one-dimensional box of side L with periodic boundary conditions (as on the lattice), so that the free-particle momenta are now discrete, $p_n = 2\pi n/L$. To include the effect of the potential, we apply periodic boundary conditions at $x = \pm L/2$ to ψ_V and its derivative, which leads to the following condition on the momentum:

$$p = \frac{2\pi}{L}n - \frac{2}{L}\delta(p). \tag{16.152}$$

The crucial point is that there will only be certain *discrete* values of p (and hence of the energy E) which solve this equation, and they depend both on the size of the box L and on the infinite-volume phase-shift δ. Put differently, the infinite-volume scattering phase-shift controls the discrete spectrum of eigenstates in a finite periodic box. In principle, then, it should be possible to reverse the situation, and determine the phase shift from the calculated energy levels in the box. How this can be done is well explained in the review article by Briceño *et al.* (2018).

At present, the approach is fully established for decays to two-particle states. Initial applications concentrated on the $\pi - \pi$ system, below the $K\overline{K}$ threshold. Dudek, Edwards and Thomas (2013), with $m_\pi = 391$ MeV, extracted the $I = 1$, P-wave phase shift, which closely followed the Breit-Wigner resonance parametrization, with a resonant mass of 854.1 $\pm$ 1.1 MeV ($m_\rho/m_\pi \sim 2.2$), and a width of 12.4 ± 0.6 MeV. Buleva *et al.* (2016), using $N_{\mathrm{f}-} = 2+1$ clover improved fermions with $m_\pi = 230$ MeV, obtained the $I = 1$ P-wave and $I = 2$ S-wave phase shifts. They quote results in terms of the ratio $m_\rho/m_\pi = 3.350(24)$. The baryon $\Delta(1236)$ state has also been studied (Andersen *et al.* 2018). In general, the computed masses and widths are still some distance from the observed values. The formalism is being extended to include decays to three-particle states — as described, for example, by Briceño *et al.* (2017). This is an active area of research, in which more progress can be expected.

We have been able to give only a brief introduction into what is now, some fifty years after its initial inception by Wilson (1974), the highly mature field of lattice QCD. There are now many physics applications, including pseudoscalar meson decay constants, form factors in semi-leptonic decays, hadronic vacuum polarization contributions to the muon $g - 2$ value, and the non-perturbative (long-distance) contribution to K -$\overline{\mathrm{K}}$ mixing. A great deal of effort has gone into ingenious and subtle improvements to the lattice action, to the numerical algorithms, and to the treatment of fermions—to name a few of the issues. Lattice QCD is now a major part of particle physics. From the perspective of this chapter and the previous one, we can confidently say that, both in the short-distance (perturbative) regime, and in the long-distance (non-perturbative) regime, QCD is established as the correct theory of the strong interactions of quarks, beyond reasonable doubt.

Problems

16.1 Verify equation (16.9).

16.2 Verify equation (16.10).

16.3 Show that the momentum space version of (16.38) is (16.40).

16.4 Use (16.19) in (16.21) to verify (16.22).

16.5 Verify (16.68) and (16.70).

16.6 In a modified one-dimensional Ising model, spin variables s_n at sites labelled by $n = 1, 2, 3, \ldots N$ take the values $s_n = \pm 1$, and the energy of each spin configuration is

$$E = -\sum_{n=1}^{N-1} J_n s_n s_{n+1} ,$$

where all the constants J_n are positive. Show that the partition function Z_N is given by

$$Z_N = 2 \prod_{n=1}^{N-1} \left(2\cosh K_n\right) ,$$

where $K_n = J_n/k_BT$. Hence calculate the entropy for the particular case in which all the J_n's are equal to J and $N \gg 1$, and discuss the behaviour of the entropy in the limits $T \to \infty$ and $T \to 0$.

Let 'p' denote a particular site such that $1 \ll p \ll N$. Show that the average value $\langle s_p s_{p+1} \rangle$ of the product $s_p s_{p+1}$ is given by

$$\langle s_p s_{p+1} \rangle = \frac{1}{Z_N} \frac{\partial Z_N}{\partial K_p}.$$

Show further that

$$\langle s_p s_{p+j} \rangle = \frac{1}{Z_N} \frac{\partial^j Z_N}{\partial K_p \partial K_{p+1} \ldots \partial K_{p+j-1}}.$$

Hence show that in the case $J_1 = J_2 = \ldots = J_N = J$,

$$\langle s_p s_{p+j} \rangle = e^{-ja/\xi} ,$$

where

$$\xi = -a/[\ln(\tanh K)] ,$$

and $K = J/k_BT$. Discuss the physical meaning of ξ, considering the $T \to \infty$ and $T \to 0$ limits explicitly.

Part VII

Spontaneously Broken Symmetry

17

Spontaneously Broken Global Symmetry

Previous chapters have introduced the non-Abelian symmetries SU(2) and SU(3) in both global and local forms, and we have seen how they may be applied to describe such typical physical phenomena as particle multiplets, and massless gauge fields. Remarkably enough, however, these symmetries are also applied, in the standard model (SM), in two cases where the physical phenomena appear to be very different. Consider the following two questions: (i) Why are there no signs in the baryonic spectrum, such as parity doublets in particular, of the global chiral symmetry introduced in section 12.3.2? (ii) How can weak interactions be described by a local non-Abelian gauge theory when we know the mediating gauge field quanta are not massless? The answers to these questions each involve the same fundamental idea, which is a crucial component of the standard model, and perhaps also of theories which go beyond it. This is the idea that a symmetry can be 'spontaneously broken', or 'hidden'. By contrast, the symmetries considered hitherto may be termed 'manifest symmetries'.

The physical consequences of spontaneous symmetry breaking turn out to be rather different in the global and local cases. However, the essentials for a theoretical understanding of the phenomenon are contained in the simpler global case, which we consider in this chapter. The application to spontaneously broken chiral symmetry will be treated in chapter 18, and spontaneously broken local symmetry will be discussed in chapter 19, and applied in chapter 22.

17.1 Introduction

We begin by considering, in response to question (i) above, what could go wrong with the argument for symmetry multiplets that we gave in chapter 12. To understand this, we must use the field theory formulation of section 12.3, in which the generators of the symmetry are Hermitian field operators, and the states are created by operators acting on the vacuum. Thus consider two states $|A\rangle$, $|B\rangle$[1] :

$$|A\rangle = \hat{\phi}_A^\dagger |0\rangle, \qquad |B\rangle = \hat{\phi}_B^\dagger |0\rangle \tag{17.1}$$

where $\hat{\phi}_A^\dagger$ and $\hat{\phi}_B^\dagger$ are related to each other by (cf (12.100))

$$[\hat{Q}, \hat{\phi}_A^\dagger] = \hat{\phi}_B^\dagger \tag{17.2}$$

for some generator $\hat{Q}$ of a symmetry group, such that

$$[\hat{Q}, \hat{H}] = 0. \tag{17.3}$$

(17.2) is equivalent to

$$\hat{U}\hat{\phi}_A^\dagger \hat{U}^{-1} \approx \hat{\phi}_A^\dagger + i\epsilon\hat{\phi}_B^\dagger \tag{17.4}$$

---

[1]We now revert to the ordinary notation $|0\rangle$ for the vacuum state, rather than $|\Omega\rangle$, but it must be borne in mind that $|0\rangle$ is the full (interacting) vacuum.

DOI: 10.1201/9781003411666-17

for an infinitesimal transformation $\hat{U} \approx 1 + i\epsilon\hat{Q}$. Thus $\hat{\phi}_A^\dagger$ is 'rotated' into $\hat{\phi}_B^\dagger$ by $\hat{U}$, and the operators will create states related by the symmetry transformation. We want to see what are the assumptions necessary to prove that

$$E_A = E_B, \quad \text{where} \quad \hat{H}|A\rangle = E_A|A\rangle \quad \text{and} \quad \hat{H}|B\rangle = E_B|B\rangle. \tag{17.5}$$

We have

$$E_B|B\rangle = \hat{H}|B\rangle = \hat{H}\hat{\phi}_B^\dagger|0\rangle = \hat{H}(\hat{Q}\hat{\phi}_A^\dagger - \hat{\phi}_A^\dagger\hat{Q})|0\rangle. \tag{17.6}$$

Now if

$$\hat{Q}|0\rangle = 0 \tag{17.7}$$

we can rewrite the right hand side of (17.6) as

$$\begin{aligned}
\hat{H}\hat{Q}\hat{\phi}_A^\dagger|0\rangle &= \hat{Q}\hat{H}\hat{\phi}_A^\dagger|0\rangle \quad \text{using (17.3)} \quad = \hat{Q}\hat{H}|A\rangle = E_A\hat{Q}|A\rangle \\
&= E_A\hat{Q}\hat{\phi}_A^\dagger|0\rangle = E_A(\hat{\phi}_B^\dagger + \hat{\phi}_A^\dagger\hat{Q})|0\rangle \quad \text{using (17.2)} \\
&= E_A|B\rangle \quad \text{if (17.7) holds;}
\end{aligned} \tag{17.8}$$

whence, comparing (17.8) with (17.6), we see that

$$E_A = E_B \quad \text{if (17.7) holds.} \tag{17.9}$$

Remembering that $\hat{U} = \exp(i\alpha\hat{Q})$, we see that (17.7) is equivalent to

$$|0\rangle' \equiv \hat{U}|0\rangle = |0\rangle. \tag{17.10}$$

Thus a multiplet structure will emerge provided that the vacuum is left invariant under the symmetry transformation. *The 'spontaneously broken symmetry' situation arises in the contrary case—that is, when the vacuum is not invariant under the symmetry*, which is to say when

$$\hat{Q}|0\rangle \neq 0. \tag{17.11}$$

In this case, the argument for the existence of symmetry multiplets breaks down, and although the Hamiltonian or Lagrangian may exhibit a non-Abelian symmetry, this will not be manifested in the form of multiplets of mass-degenerate particles.

The preceding italicized sentence does correctly define what is meant by a spontaneously broken symmetry in field theory, but there is another way of thinking about it which is somewhat less abstract though also less rigorous. The basic condition is $\hat{Q}|0\rangle \neq 0$, and it seems tempting to infer that, in this case, the application of $\hat{Q}$ to the vacuum gives, not zero, but *another possible vacuum*, $|0\rangle'$. Thus we have the physically suggestive idea of 'degenerate vacua' (they must be degenerate since $[\hat{Q}, H] = 0$). We shall see in a moment why this notion, though intuitively helpful, is not rigorous.

It would seem, in any case, that the properties of the *vacuum* are all-important, so we begin our discussion with a somewhat formal, but nonetheless fundamental, theorem about the quantum field vacuum.

17.2   The Fabri–Picasso theorem

Suppose that a given Lagrangian $\hat{\mathcal{L}}$ is invariant under some one-parameter continuous global internal symmetry with a conserved Noether current $\hat{j}^\mu$, such that $\partial_\mu\hat{j}^\mu = 0$. The associated 'charge' is the Hermitian operator $\hat{Q} = \int \hat{j}^0 \mathrm{d}^3\boldsymbol{x}$, and $\dot{\hat{Q}} = 0$. We have hitherto assumed that the transformations of such a U(1) group are representable in the space of physical states by unitary operations $\hat{U}(\lambda) = \exp \mathrm{i}\lambda\hat{Q}$ for arbitrary λ, with the vacuum invariant under $\hat{U}$, so that $\hat{Q}|0\rangle = 0$. Fabri and Picasso (1966) showed that there are actually *two* possibilities:

(a) $\hat{Q}|0\rangle = 0$, and $|0\rangle$ is an eigenstate of $\hat{Q}$ with eigenvalue 0, so that $|0\rangle$ is invariant under $\hat{U}$(i.e. $\hat{U}|0\rangle = |0\rangle$);

or

(b) $\hat{Q}|0\rangle$ does not exist in the space (its norm is infinite).

The statement (b) is technically more correct than the more intuitive statements '$\hat{Q}|0\rangle \neq 0$' or '$\hat{U}|0\rangle = |0\rangle$'', suggested above.

To prove this result, consider the vacuum matrix element $\langle 0|\hat{j}^0(x)\hat{Q}|0\rangle$. From translation invariance, implemented by the unitary operator[2] $\hat{U}(x) = \exp \mathrm{i}\hat{P} \cdot x$ (where $\hat{P}^\mu$ is the 4-momentum operator) we obtain

$$
\begin{aligned}
\langle 0|\hat{j}^0(x)\hat{Q}|0\rangle &= \langle 0|\mathrm{e}^{\mathrm{i}\hat{P}\cdot x}\hat{j}^0(0)\mathrm{e}^{-\mathrm{i}\hat{P}\cdot x}\hat{Q}|0\rangle \\
&= \langle 0|\mathrm{e}^{\mathrm{i}\hat{P}\cdot x}\hat{j}^0(0)\hat{Q}\mathrm{e}^{-\mathrm{i}\hat{P}\cdot x}|0\rangle
\end{aligned}
$$

where the second line follows from

$$[\hat{P}^\mu, \hat{Q}] = 0 \tag{17.12}$$

since $\hat{Q}$ is an *internal* symmetry. But the vacuum is an eigenstate of $\hat{P}^\mu$ with eigenvalue zero, and so

$$\langle 0|\hat{j}^0(x)\hat{Q}|0\rangle = \langle 0|\hat{j}^0(0)\hat{Q}|0\rangle \tag{17.13}$$

which states that the matrix element we started from is in fact independent of x. Now consider the norm of $\hat{Q}|0\rangle$:

$$
\begin{aligned}
\langle 0|\hat{Q}\hat{Q}|0\rangle &= \int \mathrm{d}^3\boldsymbol{x}\langle 0|\hat{j}^0(x)\hat{Q}|0\rangle \tag{17.14} \\
&= \int \mathrm{d}^3\boldsymbol{x}\langle 0|\hat{j}^0(0)\hat{Q}|0\rangle, \tag{17.15}
\end{aligned}
$$

which must diverge in the infinite volume limit, unless $\hat{Q}|0\rangle = 0$. Thus either $\hat{Q}|0\rangle = 0$ or $\hat{Q}|0\rangle$ has infinite norm. The foregoing can be easily generalized to non-Abelian symmetry operators $\hat{T}_i$.

Remarkably enough, the argument can also, in a sense, be reversed. Coleman (1966) proved that if an operator

$$\hat{Q}(t) = \int \mathrm{d}^3\boldsymbol{x}\hat{j}^0(x) \tag{17.16}$$

is the spatial integral of the $\mu = 0$ component of a 4-vector (but *not assumed* to be conserved), and if it annihilates the vacuum

$$\hat{Q}(t)|0\rangle = 0, \tag{17.17}$$

then in fact $\partial_\mu\hat{j}^\mu = 0$, $\hat{Q}$ is independent of t, and the symmetry is unitarily implementable by operators $\hat{U} = \exp(\mathrm{i}\lambda\hat{Q})$.

We might now simply proceed to the chiral symmetry application. We believe, however, that the concept of spontaneous symmetry breaking is so important to particle physics that

---

[2]If this seems unfamiliar, it may be regarded as the 4-dimensional generalization of the transformation (I.7) in Appendix I of volume 1, from Schrödinger picture operators at $t = 0$ to Heisenberg operators at $t \neq 0$.

a more extended discussion is amply justified. In particular, there are crucial insights to be gained by considering the analogous phenomenon in condensed matter physics. After a brief look at the ferromagnet, we shall describe the Bogoliubov model for the ground state of a superfluid, which provides an important physical example of a spontaneously broken global Abelian U(1) symmetry. We shall see that the excitations away from the ground state are *massless modes* and we shall learn, via Goldstone's theorem, that such modes are an inevitable result of spontaneously breaking a global symmetry. Next, we shall introduce the 'Goldstone model' which is the simplest example of a spontaneously broken global U(1) symmetry, involving just one complex scalar field. The generalization of this to the non-Abelian case will draw us in the direction of the Higgs sector of the standard model. Returning to condensed matter systems, we introduce the BCS ground state for a superconductor, in a way which builds on the Bogoliubov model of a superfluid. We are then prepared for the application, in chapter 18, to spontaneous chiral symmetry breaking (question (i) above), following Nambu's profound analogy with one aspect of superconductivity. In chapter 19 we shall see how a different aspect of superconductivity provides a model for the answer to question (ii) above.

17.3 Spontaneously broken symmetry in condensed matter physics

17.3.1 The ferromagnet

We have seen that everything depends on the properties of the vacuum state. An essential aid to understanding hidden symmetry in quantum field theory is provided by Nambu's (1960) remarkable insight that the *vacuum* state of a quantum field theory is analogous to the ground state of an interacting many-body system. It is the state of lowest energy—the equilibrium state, given the kinetic and potential energies as specified in the Hamiltonian. Now the ground state of a complicated system (for example, one involving interacting fields) may well have unsuspected properties—which may, indeed, be very hard to predict from the Hamiltonian. But we can postulate (even if we cannot yet prove) properties of the quantum field theory vacuum $|0\rangle$ which are analogous to those of the ground states of many physically interesting many-body systems—such as superfluids and superconductors, to name two with which we shall be principally concerned.

Now it is generally the case, in quantum mechanics, that the ground state of any system described by a Hamiltonian is non-degenerate. Sometimes we may meet systems in which apparently more than one state has the same lowest energy eigenvalue. Yet in fact none of these states will be the true ground state: tunnelling will take place between the various degenerate states, and the true ground state will turn out to be a unique linear superposition of them. This is, in fact, the only possibility for systems of finite spatial extent, though in practice a state which is not the true ground state may have an extremely long lifetime. However, in the case of fields (extending presumably throughout all space), the Fabri–Picasso theorem shows that there is an alternative possibility, which is often described as involving a 'degenerate ground state'—a term we shall now elucidate. In case (a) of the theorem, the ground state is unique. For, suppose that several ground states $|0, a\rangle, |0, b\rangle, \ldots$ existed, with the symmetry unitarily implemented. Then one ground state will be related to another by

$$|0, a\rangle = e^{i\lambda \hat{Q}}|0, b\rangle \tag{17.18}$$

for some λ. However, in case (a) the charge annihilates a ground state, and so all of them are really identical. In case (b), on the other hand, we cannot write (17.18)—since $\hat{Q}|0\rangle$ does not exist—and we do have the possibility of many degenerate ground states. In simple models one can verify that these alternative ground states are all orthogonal to each other, in the infinite volume limit—or perhaps more physically, the limit in which the number of degrees of freedom becomes infinite. And each member of every 'tower' of excited states, built on these alternative ground states, is also orthogonal to all the members of other towers. But any single tower must constitute a complete space of states. It follows that states in different towers belong to *different* complete spaces of states, that is to different—and inequivalent— 'worlds', each one built on one of the possible orthogonal ground states.

At first sight, a familiar example of these ideas seems to be that of a ferromagnet, below its Curie temperature T_C. Consider an 'ideal Heisenberg ferromagnet' with N atoms each of spin 1/2, described by a Hamiltonian of Heisenberg exchange form $H_S = -J \sum \hat{\boldsymbol{S}}_i \cdot \hat{\boldsymbol{S}}_j$, where i and j label the atomic sites. This Hamiltonian is invariant under spatial rotations, since it only depends on the dot product of the spin operators. Such rotations are implemented by unitary operators $\exp(i\hat{\boldsymbol{S}} \cdot \boldsymbol{\alpha})$ where $\hat{\boldsymbol{S}} = \sum_i \hat{\boldsymbol{S}}_i$, and spins at different sites are assumed to commute. As usual with angular momentum in quantum mechanics, the eigenstates of H_S are labelled by the eigenvalues of total squared spin, and of one component of spin, say of $\hat{S}_z = \sum_i \hat{S}_{iz}$. The quantum mechanical ground state of H_S is an eigenstate with total spin quantum number $S = N/2$, and this state is $(2 \cdot N/2 + 1) = (N+1)-$ fold degenerate, allowing for all the possible eigenvalues $(N/2, N/2 - 1, \ldots - N/2)$ of $\hat{S}_z$ for this value of S. We are free to choose any one of these degenerate states as 'the' ground state, say the state with eigenvalue $S_z = N/2$.

It is clear that the ground state is not invariant under the spin-rotation symmetry of H_S, which would require the eigenvalues $S = S_z = 0$. Furthermore, this ground state is degenerate. So two important features of what we have so far learned to expect of a spontaneously broken symmetry are present—namely, 'the ground state is not invariant under the symmetry of the Hamiltonian', and 'the ground state is degenerate'. However, it has to be emphasized that this ferromagnetic ground state does, in fact, respect the symmetry of H_S, in the sense that it belongs to an irreducible representation of the symmetry group. The unusual feature is that it is not the 'trivial' (singlet) representation, as would be the case for an invariant ground state. The spontaneous symmetry breaking which is the true model for particle physics is that in which a many-body ground state is *not* an eigenstate (trivial or otherwise) of the symmetry operators of the Hamiltonian; rather it is a superposition of such eigenstates. We shall explore this for the superfluid and the superconductor in due course.

Nevertheless, there are some useful insights to be gained from the ferromagnet. First, consider two ground states differing by a spin rotation. In the first, the spins are all aligned along the 3-axis, say, and in the second along the axis $\hat{\boldsymbol{n}} = (0, \sin\alpha, \cos\alpha)$. Thus the first ground state is

$$\chi_0 = \begin{pmatrix} 1 \\ 0 \end{pmatrix}_1 \begin{pmatrix} 1 \\ 0 \end{pmatrix}_2 \cdots \begin{pmatrix} 1 \\ 0 \end{pmatrix}_N \qquad \text{(N products)} \qquad (17.19)$$

while the second is (cf (4.74))

$$\chi_0^{(\alpha)} = \begin{pmatrix} \cos\alpha/2 \\ i\sin\alpha/2 \end{pmatrix}_1 \cdots \begin{pmatrix} \cos\alpha/2 \\ i\sin\alpha/2 \end{pmatrix}_N. \qquad (17.20)$$

The scalar product of (17.19) and (17.20) is $(\cos\alpha/2)^N$, which goes to zero as $N \to \infty$. Thus any two such 'rotated ground states' are indeed orthogonal in the infinite volume (or infinite number of degrees of freedom) limit.

We may also enquire about the excited states built on one such ground state, say the one with $\hat{S}_z$ eigenvalue $N/2$. Suppose for simplicity that the magnet is one-dimensional (but the spins have all three components). Consider the state $\chi_n = \hat{S}_{n-}\chi_0$ where $\hat{S}_{n-}$ is the spin lowering operator $\hat{S}_{n-} = (\hat{S}_{nx} - i\hat{S}_{ny})$ at site n, such that

$$\hat{S}_{n-} \begin{pmatrix} 1 \\ 0 \end{pmatrix}_n = \begin{pmatrix} 0 \\ 1 \end{pmatrix}_n ; \tag{17.21}$$

so $\hat{S}_{n-}\chi_0$ differs from the ground state χ_0 by having the spin at site n flipped. The action of $\hat{H}_S$ on χ_n can be found by writing

$$\sum_{i \neq j} \hat{\boldsymbol{S}}_i \cdot \hat{\boldsymbol{S}}_j = \sum_{i \neq j} \frac{1}{2}(\hat{S}_{i-}\hat{S}_{j+} + \hat{S}_{j-}\hat{S}_{i+}) + \hat{S}_{iz}\hat{S}_{jz} \tag{17.22}$$

(remembering that spins on different sites commute), where $\hat{S}_{i+} = \hat{S}_{ix} + i\hat{S}_{iy}$. Since all $\hat{S}_{i+}$ operators give zero on a spin 'up' state, the only non-zero contributions from the first (bracketed) term in (17.22) come from terms in which either $\hat{S}_{i+}$ or $\hat{S}_{j+}$ act on the 'down' spin at n, so as to restore it to 'up'. The 'partner' operator $\hat{S}_{i-}$ (or $\hat{S}_{j-}$) then simply lowers the spin at i (or j), leading to the result

$$\sum_{i \neq j} \frac{1}{2}(\hat{S}_{i-}\hat{S}_{j+} + \hat{S}_{j-}\hat{S}_{i+})\chi_n = \sum_{i \neq n} \chi_i. \tag{17.23}$$

Thus the state χ_n is not an eigenstate of $\hat{H}_S$. However, a little more work shows that the superpostitions

$$\tilde{\chi}_q = \frac{1}{\sqrt{N}} \sum_n e^{iqna}\chi_n \tag{17.24}$$

are eigenstates; here q is one of the discretized wavenumbers produced by appropriate boundary conditions, as is usual in one-dimensional 'chain' problems. The states (17.24) represent *spin waves*, and they have the important feature that for low q (long wavelength) their frequency ω tends to zero with q (actually $\omega \propto q^2$). In this respect, therefore, they behave like massless particles when quantized—and this is another feature we should expect when a symmetry is spontaneously broken.

The ferromagnet gives us one more useful insight. We have been assuming that one particular ground state (e.g. the one with $S_z = N/2$) has been somehow 'chosen'. But what does the choosing? The answer to this is clear enough in the (perfectly realistic) case in which the Hamiltonian $\hat{H}_S$ is supplemented by a term $-g\mu B \sum_i \hat{S}_{iz}$, representing the effect of an applied field B directed along the z-axis. This term will indeed ensure that the ground state is unique, and has $S_z = N/2$. Consider now the two limits $B \to 0$ and $N \to \infty$, both at finite temperature. When $B \to 0$ at finite N, the $N + 1$ different S_z eigenstates become degenerate, and we have an ensemble in which each enters with an equal weight; there is therefore no loss of symmetry, even as $N \to \infty$ (but only *after* $B \to 0$). On the other hand, if $N \to \infty$ at finite $B \neq 0$, the single state with $S_z = N/2$ will be selected out as the unique ground state and this asymmetric situation will persist even in the limit $B \to 0$. In a (classical) mean field theory approximation we suppose that an 'internal field' is 'spontaneously generated', which is aligned with the external B and survives even as $B \to 0$, thus 'spontaneously' breaking the symmetry.

The ferromagnet therefore provides an easily pictured system exhibiting many of the features associated with spontaneous symmetry breaking; most importantly, it strongly suggests that what is really characteristic about the phenonenon is that it entails 'spontaneous ordering'.[3] Generally such ordering occurs below some characteristic 'critical temperature',

---

[3]It is worth pausing to reflect on the idea that *ordering* is associated with *symmetry breaking*.

T_C. The field which develops a non-zero equilibrium value below T_C is called an 'order parameter'. This concept forms the basis of Landau's theory of second-order phase transitions (see for example chapter XIV of Landau and Lifshitz 1980).

We now turn to an example much more closely analogous to the particle physics applications: the superfluid.

17.3.2   The Bogoliubov superfluid

Consider the non-relativistic Hamiltonian (in the Schrödinger picture)

$$
\begin{aligned}
\hat{H} &= \frac{1}{2m} \int \mathrm{d}^3 \boldsymbol{x} \boldsymbol{\nabla} \hat{\phi}^\dagger \cdot \boldsymbol{\nabla} \hat{\phi} \\
&+ \frac{1}{2} \int \int \mathrm{d}^3 \boldsymbol{x} \, \mathrm{d}^3 \boldsymbol{y} \, v(|\boldsymbol{x} - \boldsymbol{y}|) \hat{\phi}^\dagger(\boldsymbol{x}) \hat{\phi}^\dagger(\boldsymbol{y}) \hat{\phi}(\boldsymbol{y}) \hat{\phi}(\boldsymbol{x})
\end{aligned} \tag{17.25}
$$

where $\hat{\phi}^\dagger(\boldsymbol{x})$ creates a boson of mass m at position $\boldsymbol{x}$. This $\hat{H}$ describes identical bosons interacting via a potential v, which is assumed to be weak (see, for example, Schiff 1968 section 55, or Parry 1973 chapter 1). We note at once that $\hat{H}$ is invariant under the global U(1) symmetry

$$
\hat{\phi}(\boldsymbol{x}) \to \hat{\phi}'(\boldsymbol{x}) = \mathrm{e}^{-\mathrm{i}\alpha} \hat{\phi}(\boldsymbol{x}), \tag{17.26}
$$

the generator being the conserved number operator

$$
\hat{N} = \int \hat{\phi}^\dagger \hat{\phi} \, \mathrm{d}^3 \boldsymbol{x} \tag{17.27}
$$

which obeys $[\hat{N}, \hat{H}] = 0$. Our ultimate concern will be with the way this symmetry is 'spontaneously broken' in the superfluid ground state. Naturally, since this is an Abelian, rather than a non-Abelian, symmetry the physics will not involve any (hidden) multiplet structure. But the nature of the 'symmetry breaking ground state' in this U(1) case (and in the BCS model of section 17.7) will serve as a physical model for non-Abelian cases also.

We begin by re-writing $\hat{H}$ in terms of mode creation and annihilation operators in the usual way. We expand $\hat{\phi}(\boldsymbol{x})$ as a superposition of solutions of the $v = 0$ problem, which are plane waves quantized in a large cube of volume Ω:

$$
\hat{\phi}(\boldsymbol{x}) = \frac{1}{\Omega^{\frac{1}{2}}} \sum_{\boldsymbol{k}} \hat{a}_{\boldsymbol{k}} \mathrm{e}^{\mathrm{i}\boldsymbol{k} \cdot \boldsymbol{x}} \tag{17.28}
$$

where $\hat{a}_{\boldsymbol{k}}|0\rangle = 0$, $\hat{a}_{\boldsymbol{k}}^\dagger|0\rangle$ is a one-particle state, and $[\hat{a}_{\boldsymbol{k}}, \hat{a}_{\boldsymbol{k}'}^\dagger] = \delta_{\boldsymbol{k},\boldsymbol{k}'}$, with all other commutators vanishing. We impose periodic boundary conditions at the cube faces, and the free particle energies are $\epsilon_k = \boldsymbol{k}^2/2m$. Inserting (17.28) into (17.25) leads (problem 17.1) to

$$
\hat{H} = \sum_{\boldsymbol{k}} \epsilon_k \hat{a}_{\boldsymbol{k}}^\dagger a_{\boldsymbol{k}} + \frac{1}{2\Omega} \sum_\Delta \bar{v}(|\boldsymbol{k}_1 - \boldsymbol{k}_1'|) \hat{a}_{\boldsymbol{k}_1}^\dagger \hat{a}_{\boldsymbol{k}_2}^\dagger \hat{a}_{\boldsymbol{k}_2'} \hat{a}_{\boldsymbol{k}_1'} \Delta(\boldsymbol{k}_1 + \boldsymbol{k}_2 - \boldsymbol{k}_1' - \boldsymbol{k}_2') \tag{17.29}
$$

where the sum is over all momenta $\boldsymbol{k}_1, \boldsymbol{k}_2, \boldsymbol{k}_1', \boldsymbol{k}_2'$ subject to the conservation law imposed by the Δ function:

$$
\begin{aligned}
\Delta(\boldsymbol{k}) &= 1 \qquad \text{if } \boldsymbol{k} = \boldsymbol{0} \tag{17.30} \\
&= 0 \qquad \text{if } \boldsymbol{k} \neq \boldsymbol{0}. \tag{17.31}
\end{aligned}
$$

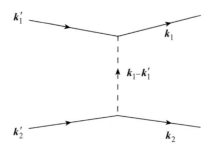

FIGURE 17.1
The interaction term in (17.29).

The interaction term in (17.29) is easily visualized as in figure 17.1. A pair of particles in states k_1', k_2' is scattered (conserving momentum) to a pair in states k_1, k_2 via the Fourier transform of v:

$$\bar{v}(|\boldsymbol{k}|) = \int v(\boldsymbol{r})\mathrm{e}^{-\mathrm{i}\boldsymbol{k}\cdot\boldsymbol{r}}\mathrm{d}^3\boldsymbol{r}. \tag{17.32}$$

Now, below the superfluid transition temperature T_{S}, we know that in the limit as $v \to 0$ the ground state has all the particles 'condensed' into the lowest energy state, which has $\boldsymbol{k} = \boldsymbol{0}$. Thus the ground state will be proportional to

$$|N,0\rangle = (\hat{a}_0^\dagger)^N |0\rangle. \tag{17.33}$$

When a weak repulsive v is included, it is reasonable to hope that most of the particles remain in the condensate, only relatively few being excited to states with $\boldsymbol{k} \neq \boldsymbol{0}$. Let N_0 be the number of particles with $\boldsymbol{k} = \boldsymbol{0}$, where by assumption $N_0 \approx N$. We now consider the limit N (and N_0) $\to \infty$ and $\Omega \to \infty$ such that the density $\rho = N/\Omega$ (and $\rho_0 = N_0/\Omega$) stays constant. Bogoliubov (1947) argued that in this limit we may effectively replace both $\hat{a}_0$ and $\hat{a}_0^\dagger$ in the second term in (17.29) by the number $N_0^{1/2}$. This amounts to saying that in the commutator

$$\frac{\hat{a}_0}{\Omega^{1/2}}\frac{\hat{a}_0^\dagger}{\Omega^{1/2}} - \frac{\hat{a}^\dagger}{\Omega^{1/2}}\frac{\hat{a}_0}{\Omega^{1/2}} = \frac{1}{\Omega} \tag{17.34}$$

the two terms on the left-hand side are each of order N_0/Ω and hence finite, while their difference may be neglected as $\Omega \to \infty$. Replacing $\hat{a}_0$ and $\hat{a}_0^\dagger$ by $N_0^{1/2}$ leads (problem 17.2) to the following approximate form for $\hat{H}$:

$$\hat{H} \approx \hat{H}_{\mathrm{B}} \equiv \sum_{\boldsymbol{k}}' \hat{a}_{\boldsymbol{k}}^\dagger \hat{a}_{\boldsymbol{k}} E_k + \frac{1}{2}\frac{N^2}{\Omega}\bar{v}(0)$$

$$+ \frac{1}{2}\sum_{\boldsymbol{k}}' \frac{N}{\Omega}\bar{v}(|\boldsymbol{k}|)[\hat{a}_{\boldsymbol{k}}^\dagger \hat{a}_{-\boldsymbol{k}}^\dagger + \hat{a}_{\boldsymbol{k}}\hat{a}_{-\boldsymbol{k}}], \tag{17.35}$$

where

$$E_k = \epsilon_k + \frac{N}{\Omega}\bar{v}(|\boldsymbol{k}|), \tag{17.36}$$

primed summations do not include $\boldsymbol{k} = \boldsymbol{0}$, and terms which tend to zero as $\Omega \to \infty$ have been dropped (thus, N_0 has been replaced by N).

The most immediately striking feature of (17.35), as compared with $\hat{H}$ of (17.29), is that $\hat{H}_B$ does not conserve the U(1) (number) symmetry (17.26) while $\hat{H}$ does: it is easy to see that for (17.26) to be a good symmetry, the number of $\hat{a}$'s must equal the number of $\hat{a}^\dagger$'s in every term. Thus the ground state of $\hat{H}_B$, $|\text{ground}\rangle_B$, cannot be expected to be an eigenstate of the number operator. However, it is important to be clear that the number non-conserving aspect of (17.35) is of a completely different kind, conceptually, from that which would be associated with a (hypothetical) *explicit* number violating term in the original Hamiltonian—for example, the addition of a term of the form '$\hat{a}^\dagger\hat{a}\hat{a}$'. In arriving at (17.35), we effectively replaced (17.28) by

$$\hat{\phi}_B(\boldsymbol{x}) = \rho_0^{1/2} + \frac{1}{\Omega^{1/2}} \sum_{\boldsymbol{k}\neq\boldsymbol{0}} \hat{a}_{\boldsymbol{k}} e^{i\boldsymbol{k}\cdot\boldsymbol{x}} \tag{17.37}$$

where $\rho_0 = N_0/\Omega, N_0 \approx N$, and N_0/Ω remains finite as $\Omega \to \infty$. The limit is crucial here, it enables us to picture the condensate N_0 as providing an infinite reservoir of particles, with which excitations away from the ground state can exchange particle number. From this point of view, a number non-conserving ground state may appear more reasonable. The ultimate test, of course, is whether such a state is a good approximation to the true ground state, for a large but finite system.

What is $|\text{ground}\rangle_B$? Remarkably, $\hat{H}_B$ can be exactly diagonalized by means of the *Bogoliubov quasiparticle operators* (for $\boldsymbol{k} \neq \boldsymbol{0}$)

$$\hat{\alpha}_{\boldsymbol{k}} = f_k \hat{a}_{\boldsymbol{k}} + g_k \hat{a}^\dagger_{-\boldsymbol{k}}, \quad \hat{\alpha}^\dagger_{\boldsymbol{k}} = f_k \hat{a}^\dagger_{\boldsymbol{k}} + g_k \hat{a}_{-\boldsymbol{k}} \tag{17.38}$$

where f_k and g_k are real functions of $k = |\boldsymbol{k}|$. We must again at once draw attention to the fact that this transformation does not respect the symmetry (17.26) either, since $\hat{a}_{\boldsymbol{k}} \to e^{-i\alpha}\hat{a}_{\boldsymbol{k}}$ while $\hat{a}^\dagger_{-\boldsymbol{k}} \to e^{+i\alpha}\hat{a}^\dagger_{-\boldsymbol{k}}$. In fact, the operators $\hat{\alpha}^\dagger_{\boldsymbol{k}}$ will turn out to be precisely *creation operators for quasiparticles* which exchange particle number with the ground state.

The commutator of $\hat{\alpha}_{\boldsymbol{k}}$ and $\hat{\alpha}^\dagger_{\boldsymbol{k}}$ is easily evaluated:

$$[\hat{\alpha}_{\boldsymbol{k}}, \hat{\alpha}^\dagger_{\boldsymbol{k}}] = f_k^2 - g_k^2, \tag{17.39}$$

while two $\hat{\alpha}$'s or two $\hat{\alpha}^\dagger$'s commute. We choose f_k and g_k such that $f_k^2 - g_k^2 = 1$, so that the $\hat{a}$'s and the $\hat{\alpha}$'s have the same (bosonic) commutation relations, and the transformation (17.38) is 'canonical'. A convenient choice is $f_k = \cosh\theta_k, g_k = \sinh\theta_k$. We now assert that $\hat{H}_B$ can be written in the form

$$\hat{H}_B = \sum_{\boldsymbol{k}}{}' \omega_k \hat{\alpha}^\dagger_{\boldsymbol{k}} \hat{\alpha}_{\boldsymbol{k}} + \beta \tag{17.40}$$

for certain ω_k and β. Equation (17.40) implies, of course, that the eigenvalues of $\hat{H}_B$ are $\beta + \sum_{\boldsymbol{k}}(n+1/2)\omega_k$, and that $\hat{\alpha}^\dagger_{\boldsymbol{k}}$ acts as the creation operator for the quasiparticle of energy ω_k, as just anticipated.

We verify (17.40) slightly indirectly. We note first that it implies that

$$[\hat{H}_B, \hat{\alpha}^\dagger_{\boldsymbol{l}}] = \omega_l \hat{\alpha}^\dagger_{\boldsymbol{l}}. \tag{17.41}$$

Substituting for $\hat{\alpha}^\dagger_{\boldsymbol{l}}$ from (17.38), we require

$$[\hat{H}_B, \cosh\theta_l \, \hat{a}^\dagger_{\boldsymbol{l}} + \sinh\theta_l \, \hat{a}_{-\boldsymbol{l}}] = \omega_l(\cosh\theta_l \, \hat{a}^\dagger_{\boldsymbol{l}} + \sinh\theta_l \, \hat{a}_{-\boldsymbol{l}}), \tag{17.42}$$

which must hold as an identity in the $\hat{a}$'s and $\hat{a}^\dagger$'s. Using the expression (17.35) for $\hat{H}_B$, and some patient work with the commutation relations (problem 17.3), one finds

$$(\omega_l - E_l) \cosh \theta_l \quad + \quad \frac{N}{\Omega} \bar{v}(|\boldsymbol{l}|) \sinh \theta_l = 0 \tag{17.43}$$

$$\frac{N}{\Omega} \bar{v}(|\boldsymbol{l}|) \cosh \theta_l \quad - \quad (\omega_l + E_l) \sinh \theta_l = 0. \tag{17.44}$$

For consistency, therefore, we require

$$E_l^2 - \omega_l^2 - \left(\frac{N}{\Omega^2}\right)^2 (\bar{v}(|\boldsymbol{l}|))^2 = 0, \tag{17.45}$$

or (recalling the definitions of E_l and ϵ_l)

$$\omega_l = \left[\frac{\boldsymbol{l}^2}{2m}\left(\frac{\boldsymbol{l}^2}{2m} + 2\rho\bar{v}(|\boldsymbol{l}|)\right)\right]^{1/2} \tag{17.46}$$

where $\rho = N/\Omega$. The value of $\tanh \theta_l$ is then determined via either of (17.43) and (17.44).

Equation (17.46) is an important result, giving the frequency as a function of the momentum (or wavenumber); it is an example of a 'dispersion relation'. At the risk of stating the obvious, let us emphasize that equation (17.40) tells us that the original system of interacting bosons is equivalent (under the approximations made) to a system of non-interacting quasiparticles, whose frequency ω_l is related to wavenumber by (17.46). These are the true modes of the system. Let us consider this dispersion relation.

First of all, in the non-interacting case $\bar{v} = 0$, we recover the usual frequency-wavenumber relation for a massive non-relativistic particle, $\omega_l = \boldsymbol{l}^2/2m$. But if $\bar{v}(0) \neq 0$, the behaviour at small $\boldsymbol{l}$ is very different: $\omega_l \approx c_s|\boldsymbol{l}|$, where $c_s = (\rho\bar{v}(0)/m)^{1/2}$. This dispersion relation is characteristic of a massless mode, but in this case it is sound rather than light, with speed of sound c_s. The spectrum is therefore phonon-like, not (non-relativistic) particle-like. The two behaviours can be easily distinguished experimentally, by measuring the low-temperature specific heat: in three dimensions, for $\omega_l \sim \boldsymbol{l}^2$ it goes to zero as $T^{3/2}$, whereas for $\omega_l \sim |\boldsymbol{l}|$ it goes as T^3. The latter behaviour is observed in superfluids. At large values of $|\boldsymbol{l}|$, however, ω_l behaves essentially like $\boldsymbol{l}^2/2m$ and the spectrum returns to the 'particle-like' one of massive bosons. Thus (17.46) interpolates between phonon-like behaviour at small $|\boldsymbol{l}|$ and particle-like behaviour at large $|\boldsymbol{l}|$.

There is still more to be learned from (17.46). If, in fact, $\bar{v}(|\boldsymbol{l}|) \sim 1/\boldsymbol{l}^2$, then $\omega_l \to$ constant as $|\boldsymbol{l}| \to 0$, and the spectrum would *not* be phonon-like. Indeed, if $\bar{v}(|\boldsymbol{l}|) \sim e^2/\boldsymbol{l}^2$, then $\omega_l \sim |e|(\rho/m)^{1/2}$ for small $|\boldsymbol{l}|$, which is just the 'plasma frequency' ω_p. In particle physics terms, this would be analogous to a dispersion relation of the form $\omega_l \sim (\omega_p^2 + \boldsymbol{l}^2)^{1/2}$, which describes a particle with mass ω_p. Such a $\bar{v}$ is, of course, Colombic (the Fourier transform of $e^2/|\boldsymbol{x}|$), indicating that *in the case of such a long-range force the frequency spectrum acquires a mass-gap*. This will be the topic of chapter 19.

Having discussed the spectrum of quasiparticle excitations, let us now concentrate on the ground state in this model. From (17.40), it is clear that it is defined as the state $|\text{ground}\rangle_B$ such that

$$\hat{\alpha}_{\boldsymbol{k}}|\text{ground}\rangle_B = 0 \qquad \text{for all } \boldsymbol{k} \neq \boldsymbol{0}; \tag{17.47}$$

i.e. as the state with no non-zero-momentum quasiparticles in it. This is a complicated state in terms of the original $\hat{a}_{\boldsymbol{k}}$ and $\hat{a}_{\boldsymbol{k}}^\dagger$ operators, but we can give a formal expression for it, as follows. Since the $\hat{\alpha}$'s and $\hat{a}$'s are related by a canonical transformation, there must exist a unitary operator $\hat{U}_B$ such that

$$\hat{\alpha}_{\boldsymbol{k}} = \hat{U}_B \hat{a}_{\boldsymbol{k}} \hat{U}_B^{-1}, \qquad \hat{a}_{\boldsymbol{k}} = \hat{U}_B^{-1} \hat{\alpha}_{\boldsymbol{k}} \hat{U}_B. \tag{17.48}$$

Now we know that $\hat{a}_{\boldsymbol{k}}|0\rangle = 0$. Hence it follows that

$$\hat{\alpha}_{\boldsymbol{k}}\hat{U}_{\mathrm{B}}|0\rangle = 0, \tag{17.49}$$

and we can identify $|\text{ground}\rangle_{\mathrm{B}}$ with $\hat{U}_{\mathrm{B}}|0\rangle$. In problem 17.4, $\hat{U}_{\mathrm{B}}$ is evaluated for an $\hat{H}_{\mathrm{B}}$ consisting of a single $\boldsymbol{k}$-mode only, in which case the operator effecting the transformation analogous to (17.48) is $\hat{U}_1 = \exp[\theta(\hat{a}\hat{a} - \hat{a}^\dagger\hat{a}^\dagger)/2]$ where θ replaces θ_k in this case. This generalizes (in the form of products of such operators) to the full $\hat{H}_{\mathrm{B}}$ case, but we shall not need the detailed result; an analogous result for the BCS ground state is discussed more fully in section 17.7. The important point is the following. It is clear from expanding the exponentials that $\hat{U}_{\mathrm{B}}$ creates a state in which the number of a-quanta (i.e. the original bosons) *is not fixed*. Thus unlike the simple non-interacting ground state $|N, 0\rangle$ of (17.33), $|\text{ground}\rangle_{\mathrm{B}} = \hat{U}_{\mathrm{B}}|0\rangle$ does not have a fixed number of particles in it: that is to say, it is *not* an eigenstate of the symmetry operator $\hat{N}$, as anticipated in the comment following (17.36). This is just the situation alluded to in the paragraph before equation (17.19), in our discussion of the ferromagnet.

Consider now the expectation value of $\hat{\phi}(\boldsymbol{x})$ in any state of definite particle number—that is, in an eigenstate of the symmetry operator $\hat{N}$. It is easy to see that this must vanish (remember that $\hat{\phi}$ destroys a boson, and so $\hat{\phi}|N\rangle$ is proportional to $|N-1\rangle$, which is orthogonal to $|N\rangle$). On the other hand, this is *not* true of $\hat{\phi}_{\mathrm{B}}(\boldsymbol{x})$: for example, in the non-interacting ground state (17.33), we have

$$\langle N, 0|\hat{\phi}_{\mathrm{B}}(\boldsymbol{x})|N, 0\rangle = \rho_0^{1/2}. \tag{17.50}$$

Furthermore, using the inverse of (17.38)

$$\hat{a}_{\boldsymbol{k}} = \cosh\theta_k\hat{\alpha}_{\boldsymbol{k}} - \sinh\theta_k\hat{\alpha}^\dagger_{-\boldsymbol{k}} \tag{17.51}$$

together with (17.47), we find the similar result:

$$_{\mathrm{B}}\langle\text{ground}|\hat{\phi}_{\mathrm{B}}(\boldsymbol{x})|\text{ground}\rangle_{\mathrm{B}} = \rho_0^{1/2}. \tag{17.52}$$

The question is now how to generalize (17.50) or (17.52) to the complete $\hat{\phi}(\boldsymbol{x})$ and the true ground state $|\text{ground}\rangle$, in the limit $N, \Omega \to \infty$ with fixed N/Ω. We make the *assumption* that

$$\langle\text{ground}|\hat{\phi}(\boldsymbol{x})|\text{ground}\rangle \neq 0; \tag{17.53}$$

that is, we abstract from the Bogoliubov model the crucial feature that *the field acquires a non-zero expectation value in the ground state*, in the infinite volume limit.

We are now at the heart of spontaneous symmetry breaking in field theory. Condition (17.53) has the form of an 'ordering' condition: it is analogous to the non-zero value of the total spin in the ferromagnetic case, but in (17.53)—we must again emphasize—$|\text{ground}\rangle$ is *not* an eigenstate of the symmetry operator $\hat{N}$; if it were, (17.53) would vanish, as we have just seen. Recalling the association 'quantum vacuum $\leftrightarrow$ many body ground state' we expect that the occurrence of a non-zero vacuum expectation value (vev) for an operator transforming non-trivially under a symmetry operator will be the key requirement for spontaneous symmetry breaking in field theory. Such operators are generically called *order parameters*. In the next section we show how this requirement necessitates one (or more) massless modes, via Goldstone's theorem (1961).

Before leaving the superfluid, we examine (17.37) and (17.52) in another way, which is only rigorous for a finite system but is nevertheless very suggestive. Since the original $\hat{H}$

has a U(1) symmetry under which $\hat{\phi}$ transforms to $\hat{\phi}' = \exp(-i\alpha)\hat{\phi}$, we should be at liberty to replace (17.37) by

$$\hat{\phi}'_{\mathrm{B}} = \mathrm{e}^{-i\alpha}\rho_0^{1/2} + \frac{1}{\Omega^{1/2}} \sum_{\boldsymbol{k}\neq\boldsymbol{0}} \hat{a}_{\boldsymbol{k}} \mathrm{e}^{-i\alpha} \mathrm{e}^{i\boldsymbol{k}\cdot\boldsymbol{x}}. \tag{17.54}$$

But in that case our condition (17.52) becomes

$$_{\mathrm{B}}\langle\mathrm{ground}|\hat{\phi}'_{\mathrm{B}}|\mathrm{ground}\rangle_{\mathrm{B}} = \mathrm{e}^{-i\alpha}{}_{\mathrm{B}}\langle\mathrm{ground}|\hat{\phi}_{\mathrm{B}}|\mathrm{ground}\rangle_{\mathrm{B}}. \tag{17.55}$$

Now $\hat{\phi}' = \hat{U}_\alpha \hat{\phi} \hat{U}_\alpha^{-1}$ where $\hat{U}_\alpha = \exp(i\alpha\hat{N})$. Hence (17.55) may be written as

$$_{\mathrm{B}}\langle\mathrm{ground}|\hat{U}_\alpha \hat{\phi} \hat{U}_\alpha^{-1}|\mathrm{ground}\rangle_{\mathrm{B}} = \mathrm{e}^{-i\alpha}{}_{\mathrm{B}}\langle\mathrm{ground}|\hat{\phi}_{\mathrm{B}}|\mathrm{ground}\rangle_{\mathrm{B}}. \tag{17.56}$$

If $|\mathrm{ground}\rangle_{\mathrm{B}}$ were an eigenstate of $\hat{N}$ with eigenvalue N, say, then the $\hat{U}_\alpha$ factors in (17.56) would become just $\mathrm{e}^{i\alpha N} \cdot \mathrm{e}^{-i\alpha N}$ and would cancel out, leaving a contradiction. Instead, however, knowing that $|\mathrm{ground}\rangle_{\mathrm{B}}$ is not an eigenstate of $\hat{N}$, we can regard $\hat{U}_\alpha^{-1}|\mathrm{ground}\rangle_{\mathrm{B}}$ as an 'alternative ground state' $|\mathrm{ground}, \alpha\rangle_{\mathrm{B}}$ such that

$$_{\mathrm{B}}\langle\mathrm{ground}, \alpha|\hat{\phi}|\mathrm{ground}, \alpha\rangle_{\mathrm{B}} = \mathrm{e}^{-i\alpha}{}_{\mathrm{B}}\langle\mathrm{ground}|\hat{\phi}_{\mathrm{B}}|\mathrm{ground}\rangle_{\mathrm{B}}, \tag{17.57}$$

the original choice (17.52) corresponding to $\alpha = 0$. There are infinitely many such ground states since α is a continuous parameter. No physical consequence follows from choosing one rather than another, but we do have to choose one, thus 'spontaneously' breaking the symmetry. In choosing say $\alpha = 0$, we are deciding (arbitrarily) to pick the ground state such that $_{\mathrm{B}}\langle\mathrm{ground}|\hat{\phi}|\mathrm{ground}\rangle_{\mathrm{B}}$ is aligned in the 'real' direction. By hypothesis, a similar situation obtains for the true ground state. None of the states $|\mathrm{ground}, \alpha\rangle$ is an eigenstate for $\hat{N}$: instead, they are certain coherent superpositions of states with different eigenvalues N, such that the expectation value of $\hat{\phi}$ has a definite phase.

17.4   Goldstone's theorem

We return to quantum field theory proper, and show following Goldstone (1961) (see also Goldstone, Salam and Weinberg (1962)) how in case (b) of the Fabri–Picasso theorem massless particles will necessarily be present. Whether these particles will actually be observable depends, however, on whether the theory also contains gauge fields. In this chapter we are concerned solely with global symmetries, and gauge fields are absent; the local symmetry case is treated in chapter 19.

Suppose, then, that we have a Lagrangian $\hat{\mathcal{L}}$ with a continuous symmetry generated by a charge $\hat{Q}$, which is independent of time, and is the space integral of the $\mu = 0$ component of a conserved Noether current:

$$\hat{Q} = \int \hat{j}_0(x)\, \mathrm{d}^3\boldsymbol{x}. \tag{17.58}$$

We consider the case in which the vacuum of this theory is not invariant, i.e. is not annihilated by $\hat{Q}$.

Suppose $\hat{\phi}(y)$ is some field operator which is not invariant under the continuous symmetry in question, and consider the vacuum expectation value

$$\langle 0|[\hat{Q}, \hat{\phi}(y)]|0\rangle. \tag{17.59}$$

Just as in equation (17.13), translation invariance implies that this vev is, in fact, independent of y, and we may set $y = 0$. If $\hat{Q}$ were to annihilate $|0\rangle$, the expression (17.18) would clearly vanish; we investigate the consequences of it *not* vanishing. Since $\hat{\phi}$ is not invariant under $\hat{Q}$, the commutator in (17.59) will give some other field, call it $\hat{\phi}'(y)$; thus the hallmark of the hidden symmetry situation is the existence of some field (here $\hat{\phi}'(y)$) with *non-vanishing vacuum expectation value*, just as in (17.53).

From (17.58), we can write (17.59) as

$$0 \neq \langle 0|\hat{\phi}'(y)|0\rangle \tag{17.60}$$

$$= \langle 0|[\int \mathrm{d}^3\boldsymbol{x}\,\hat{\jmath}_0(x), \hat{\phi}(y)]|0\rangle. \tag{17.61}$$

Since, by assumption, $\partial_\mu \hat{\jmath}^\mu = 0$, we have as usual

$$\frac{\partial}{\partial x^0}\int \mathrm{d}^3\boldsymbol{x}\,\hat{\jmath}_0(x) + \int \mathrm{d}^3\boldsymbol{x}\,\mathrm{div}\hat{\boldsymbol{\jmath}}(\mathrm{x}) = 0, \tag{17.62}$$

whence

$$\frac{\partial}{\partial x^0}\int \mathrm{d}^3\boldsymbol{x}\,\langle 0|[\hat{\jmath}_0(x), \hat{\phi}(y)]|0\rangle = -\int \mathrm{d}^3\boldsymbol{x}\,\langle 0|[\mathrm{div}\hat{\boldsymbol{\jmath}}(x), \hat{\phi}(y)]|0\rangle \tag{17.63}$$

$$= -\int \mathrm{d}\boldsymbol{S}\cdot\langle 0|[\hat{\boldsymbol{\jmath}}(x), \hat{\phi}(y)]|0\rangle. \tag{17.64}$$

If the surface integral vanishes in (17.64), (17.61) will be independent of x_0. The commutator in (17.64) involves local operators separated by a very large space-like interval, and therefore the vanishing of (17.64) would seem to be unproblematic. Indeed so it is—with the exception of the case in which the symmetry is local and gauge fields are present. A detailed analysis of exactly how this changes the argument being presented here will take us too far afield at this point, and the reader is referred to Guralnik *et al.* (1968) and Bernstein (1974). We shall treat the 'spontaneously broken' gauge theory case in chapter 19, but in less formal terms.

Let us now see how the independence of (17.61) on x_0 leads to the necessity for a massless particle in the spectrum. Inserting a complete set of states in (17.61), we obtain

$$0 \neq \int \mathrm{d}^3\boldsymbol{x}\sum_n\{\langle 0|\hat{\jmath}_0(x)|n\rangle\langle n|\hat{\phi}(y)|0\rangle - \langle 0|\hat{\phi}(y)|n\rangle\langle n|\hat{\jmath}_0(x)|0\rangle\} \tag{17.65}$$

$$= \int \mathrm{d}^3\boldsymbol{x}\sum_n\{\langle 0|\hat{\jmath}_0(0)|n\rangle\langle n|\hat{\phi}(y)|0\rangle\mathrm{e}^{-\mathrm{i}p_n\cdot x} - \langle 0|\hat{\phi}(y)|n\rangle\langle n|\hat{\jmath}_0(0)|0\rangle\mathrm{e}^{\mathrm{i}p_n\cdot x}\} \tag{17.66}$$

using translation invariance, with p_n the four-momentum eigenvalue of the state $|n\rangle$. Performing the spatial integral on the right-hand side we find (omitting the irrelevant $(2\pi)^3$)

$$0 \neq \sum_n\delta^3(\boldsymbol{p}_n)[\langle 0|\hat{\jmath}_0(0)|n\rangle\langle n|\hat{\phi}(y)|0\rangle\mathrm{e}^{\mathrm{i}p_{n0}x_0} - \langle 0|\hat{\phi}(y)|n\rangle\langle n|\hat{\jmath}_0(0)|0\rangle\mathrm{e}^{-\mathrm{i}p_{n0}x_0}]. \tag{17.67}$$

But this expression is independent of x_0. *Massive* states $|n\rangle$ will produce explicit x_0-dependent factors $\mathrm{e}^{\pm\mathrm{i}M_nx_0}$ ($p_{n0} \to M_n$ as the δ-function constrains $\boldsymbol{p}_n = \boldsymbol{0}$), hence the matrix elements of $\hat{\jmath}_0$ between $|0\rangle$ and such a massive state must *vanish*, and such states contribute zero to (17.67). Equally, if we take $|n\rangle = |0\rangle$, (17.67) vanishes identically. But it has been assumed to be not zero. Hence *some* state or states must exist among $|n\rangle$ such that $\langle 0|j_0|n\rangle \neq 0$ and yet (17.67) is independent of x_0. The only possibility is states whose energy p_{n0} goes to zero as their three-momentum does (from $\delta^3(\boldsymbol{p}_n)$). Such states are, of course, massless; they are called generically *Goldstone modes*. Thus the existence of a non-vanishing vacuum expectation value for a field, in a theory with a continuous symmetry, appears to lead inevitably to the necessity of having a massless particle, or particles, in the theory. This is the Goldstone result.

The superfluid provided us with an explicit model exhibiting the crucial non-zero expectation value $\langle \text{ground}|\hat{\phi}|\text{ground}\rangle \neq 0$, in which the now expected massless mode emerged dynamically. We now discuss a simpler, relativistic model, in which the symmetry breaking is brought about more 'by hand'—that is, by choosing a parameter in the Lagrangian appropriately. Although in a sense less 'dynamical' than the Bogoliubov superfluid (or the BCS superconductor, to be discussed shortly) this *Goldstone model* does provide a very simple example of the phenomenon of spontaneous symmetry breaking in field theory.

17.5 Spontaneously broken global U(1) symmetry: the Goldstone model

We consider, following Goldstone (1961), a complex scalar field $\hat{\phi}$ as in section 7.1, with

$$\hat{\phi} = \frac{1}{\sqrt{2}}(\hat{\phi}_1 - i\hat{\phi}_2), \qquad \hat{\phi}^\dagger = \frac{1}{\sqrt{2}}(\hat{\phi}_1 + i\hat{\phi}_2), \tag{17.68}$$

described by the Lagrangian

$$\hat{\mathcal{L}}_{\mathrm{G}} = (\partial_\mu \hat{\phi}^\dagger)(\partial^\mu \hat{\phi}) - \hat{V}(\hat{\phi}). \tag{17.69}$$

We begin by considering the 'normal' case in which the potential has the form

$$\hat{V} = \hat{V}_{\mathrm{S}} \equiv \frac{1}{4}\lambda(\hat{\phi}^\dagger\hat{\phi})^2 + \mu^2\hat{\phi}^\dagger\hat{\phi} \tag{17.70}$$

with $\mu^2, \lambda > 0$. The Hamiltonian density is then

$$\hat{\mathcal{H}}_{\mathrm{G}} = \dot{\hat{\phi}}^\dagger \dot{\hat{\phi}} + \boldsymbol{\nabla}\hat{\phi}^\dagger \cdot \boldsymbol{\nabla}\hat{\phi} + \hat{V}(\hat{\phi}). \tag{17.71}$$

Clearly $\hat{\mathcal{L}}_{\mathrm{G}}$ is invariant under the global U(1) symmetry

$$\hat{\phi} \to \hat{\phi}' = e^{-i\alpha}\hat{\phi}, \tag{17.72}$$

the generator being $\hat{N}_\phi$ of (7.23). We shall see how this symmetry may be 'spontaneously broken'.

We know that everything depends on the nature of the ground state of this field system—that is, the vacuum of the quantum field theory. In general, it is a difficult, non-perturbative, problem to find the ground state (or a good approximation to it—witness the superfluid). But we can make some progress by first considering the theory *classically*. It is clear that the absolute minimum of the classical Hamiltonian $\mathcal{H}_{\mathrm{G}}$ is reached for

(i) $\phi = \text{constant}$, which reduces the $\dot{\phi}$ and $\boldsymbol{\nabla}\phi$ terms to zero;

(ii) $\phi = \phi_0$ where ϕ_0 is the minimum of the classical version of the potential, V.

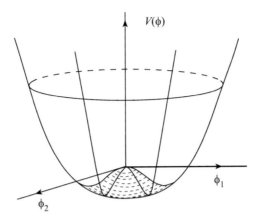

FIGURE 17.2
The classical potential V_{SB} of (17.77).

For $V = V_S$ as in (17.70) but without the hats, and with λ and μ^2 both positive, the minimum of V_S is clearly at $\phi = 0$, and is unique. In the quantum theory, we expect to treat small oscillations of the field about this minimum as approximately harmonic, leading to the usual quantized modes. To implement this, we expand $\hat{\phi}$ about the classical minimum at $\phi = 0$, writing as usual

$$\hat{\phi} = \int \frac{\mathrm{d}^3 \boldsymbol{k}}{(2\pi)^3 \sqrt{2\omega}} [\hat{a}(k)\mathrm{e}^{-\mathrm{i}k \cdot x} + b^\dagger(k)\mathrm{e}^{\mathrm{i}k \cdot x}] \tag{17.73}$$

where the plane waves are solutions of the 'free' ($\lambda = 0$) problem. For $\lambda = 0$ the Lagrangian is simply

$$\hat{\mathcal{L}}_{\text{free}} = \partial_\mu \hat{\phi}^\dagger \partial^\mu \hat{\phi} - \mu^2 \hat{\phi}^\dagger \hat{\phi}, \tag{17.74}$$

which represents a complex scalar field, consisting of two degrees of freedom, each with the same mass μ (see section 7.1). Thus in (17.73) $\omega = (\boldsymbol{k}^2 + \mu^2)^{1/2}$, and the vacuum is defined by

$$\hat{a}(k)|0\rangle = \hat{b}(k)|0\rangle = 0, \tag{17.75}$$

and so clearly

$$\langle 0|\hat{\phi}|0\rangle = 0. \tag{17.76}$$

It seems reasonable to interpret quantum field average values as corresponding to classical field values, and on this interpretation (17.76) is consistent with the fact that the classical minimum energy configuration has $\phi = 0$.

Consider now the case in which the classical minimum is *not* at $\phi = 0$. This can be achieved by altering the sign of μ^2 in (17.70) 'by hand', so that the classical potential is now the 'symmetry breaking' one

$$V = V_{SB} \equiv \frac{1}{4}\lambda(\phi^\dagger \phi)^2 - \mu^2 \phi^\dagger \phi. \tag{17.77}$$

This is sketched versus ϕ_1 and ϕ_2 in figure 17.2. This time, although the origin $\phi_1 = \phi_2 = 0$ is a stationary point, it is an (unstable) maximum rather than a minimum. The minimum of V_{SB} occurs when

$$(\phi^\dagger \phi) = \frac{2\mu^2}{\lambda}, \tag{17.78}$$

or alternatively when

$$\phi_1^2 + \phi_2^2 = \frac{4\mu^2}{\lambda} \equiv v^2 \tag{17.79}$$

where

$$v = \frac{2|\mu|}{\lambda^{1/2}}. \tag{17.80}$$

The condition (17.79) can also be written as

$$|\phi| = v/\sqrt{2}. \tag{17.81}$$

To have a clearer picture, it is helpful to introduce the 'polar' variables $\rho(x)$ and $\theta(x)$ via

$$\phi(x) = (\rho(x)/\sqrt{2}) \exp(i\theta(x)/v) \tag{17.82}$$

where for convenience the v is inserted so that θ has the same dimension (mass) as ρ and ϕ. The minimum condition (17.81) therefore represents the circle $\rho = v$; any point on this circle, at any value of θ, represents a possible classical ground state—and it is clear that they are (infinitely) degenerate.

Before proceeding further, we briefly outline a condensed matter analogue of (17.77) and (17.81) which may help in understanding the change in sign of the parameter μ^2. Consider the free energy F of a ferromagnet as a function of the magnetization $\boldsymbol{M}$ at temperature T, and make an expansion of the form

$$F \approx F_0(T) + \mu^2(T)\boldsymbol{M}^2 + \frac{\lambda}{4}\boldsymbol{M}^4 + \cdots \tag{17.83}$$

valid for weak and slowly varying magnetization. If the parameter μ^2 is positive, it is clear that F has a simple 'bowl' shape as a function of $|\boldsymbol{M}|$, with a minimum at $|\boldsymbol{M}| = 0$. This is the case for T greater than the ferromagnetic transition temperature T_C. However, if one assumes that $\mu^2(T)$ changes sign at T_C, becoming negative for $T < T_C$, then F will now resemble a vertical section of figure 17.2, the minimum being at $|\boldsymbol{M}| \neq 0$. Any direction of $\boldsymbol{M}$ is possible (only $|\boldsymbol{M}|$ is specified); but the system must choose one particular direction (e.g. via the influence of a very weak external field, as discussed in section 17.3.1), and when it does so the rotational invariance exhibited by F of (17.83) is lost. This symmetry has been broken 'spontaneously'—though this is still only a classical analogue. Nevertheless, the model is essentially the Landau mean field theory of ferromagnetism, and suggests that we should think of the 'symmetric' and 'broken symmetry' situations as different phases of the same system. It may also be the case in particle physics, that parameters such as μ^2 change sign as a function of T, or some other variable, thereby effectively precipitating a phase change.

If we maintain the idea that the vacuum expectation value of the quantum field should equal the ground state value of the classical field, the vacuum in this $\mu^2 < 0$ case must therefore be $|0\rangle_\mathrm{B}$ such that $_\mathrm{B}\langle 0|\hat{\phi}|0\rangle_\mathrm{B}$ does not vanish, in contrast to (17.76). It is clear that this is exactly the situation met in the superfluid (but 'B' here will stand for 'broken symmetry'), and is moreover the condition for the existence of massless (Goldstone) modes. Let us see how they emerge in this model.

In quantum field theory, particles are thought of as excitations from a ground state, which is the vacuum. Figure 17.2 strongly suggests that if we want a sensible quantum interpretation of a theory with the potential (17.77), we had better expand the fields about a point on the circle of minima, about which stable oscillations are likely, rather than about the obviously unstable point $\hat{\phi} = 0$. Let us pick the point $\rho = v$, $\theta = 0$ in the classical case. We might well guess that 'radial' oscillations in $\hat{\rho}$ would correspond to a conventional massive field (having a parabolic restoring potential), while 'angle' oscillations in $\hat{\theta}$—which which pass through all the degenerate vacua—have no restoring force and are massless. Accordingly, we set

$$\hat{\phi}(x) = \frac{1}{\sqrt{2}}(v + \hat{h}(x)) \exp(-i\hat{\theta}(x)/v) \tag{17.84}$$

and find (problem 17.5) that $\hat{\mathcal{L}}_G$ (with $\hat{V} = \hat{V}_{SB}$ of (17.77) with hats on) becomes

$$
\begin{aligned}
\hat{\mathcal{L}}_G &= \frac{1}{2}\partial_\mu\hat{h}\partial^\mu\hat{h} - \mu^2\hat{h}^2 + \frac{1}{2}\partial_\mu\hat{\theta}\partial^\mu\hat{\theta} + \mu^4/\lambda \\
&\quad + \frac{\hat{h}}{v}\partial_\mu\hat{\theta}\partial^\mu\hat{\theta} + \frac{1}{2}\frac{\hat{h}^2}{v^2}\partial_\mu\hat{\theta}\partial^\mu\hat{\theta} - \frac{\lambda}{16}v\hat{h}^3 - \frac{\lambda}{16}\hat{h}^4,
\end{aligned}
\tag{17.85}
$$

Equation (17.85) is very important. First of all, the first line shows that the particle spectrum in the 'spontaneously broken' case is dramatically different from that in the normal case: instead of two degrees of freedom with the same mass μ, one (the θ-mode) is massless, and the other (the h-mode) has a mass of $\sqrt{2}\mu$. We expect the vacuum $|0\rangle_B$ to be annihilated by the mode operators $\hat{a}_h$ and $\hat{a}_\theta$ for these fields. This implies, however, that

$$_B\langle 0|\hat{\phi}|0\rangle_B = v/\sqrt{2} \tag{17.86}$$

which is consistent with our interpretation of the vacuum expectation value (vev) as the classical minimum, and with the occurrence of massless modes. (The constant term in (17.85), which does not affect equations of motion, merely reflects the fact that the minimum value of V_{SB} is $-\mu^4/\lambda$.) The ansatz (17.84) and the non-zero vev (17.86) may be compared with (17.37) and (17.52), respectively, in the superfluid case.

Secondly, the second line of equation (17.85) shows that only the *derivative* of the $\hat{\theta}$ field appears in the interaction terms, whereas this is not true of the $\hat{h}$ field. Indeed, the Lagrangian for the θ-mode cannot have any dependence on a *constant* value of $\hat{\theta}$, since this could be transformed away by a global U(1) transformation (17.72), which is a symmetry of the theory, and under which $\hat{\theta} \to \hat{\theta} + v\alpha$. This will be an important point to remember when we consider effective Lagrangians for Goldstone modes in section 18.3.

Goldstone's model, then, contains much of the essence of spontaneous symmetry breaking in field theory: a non-zero vacuum value of a field which is not an invariant under the symmetry group, zero mass bosons, and massive excitations in a direction in field space which is 'orthogonal' to the degenerate ground states. However, it has to be noted that the triggering mechanism for the symmetry breaking ($\mu^2 \to -\mu^2$) has to be put in by hand, in contrast to the—admittedly approximate, but more 'dynamical'—Bogoliubov approach. The Goldstone model, in short, is essentially phenomenological.

As in the case of the superfluid, we may perfectly well choose a vacuum corresponding to a classical ground state with non-zero θ, say $\theta = -v\alpha$. Then

$$_B\langle 0, \alpha|\hat{\phi}|0, \alpha\rangle_B = e^{-i\alpha}\frac{v}{\sqrt{2}} \tag{17.87}$$

$$= e^{-i\alpha}{}_B\langle 0|\hat{\phi}|0\rangle_B, \tag{17.88}$$

as in (17.57). But we know (see (7.27) and (7.28)) that

$$e^{-i\alpha}\hat{\phi} = \hat{\phi}' = \hat{U}_\alpha\hat{\phi}\hat{U}_\alpha^{-1} \tag{17.89}$$

where

$$\hat{U}_\alpha = e^{i\alpha\hat{N}_\phi}. \tag{17.90}$$

So (17.88) becomes

$$_B\langle 0, \alpha|\hat{\phi}|0, \alpha\rangle_B = {}_B\langle 0|\hat{U}_\alpha \hat{\phi} \hat{U}_\alpha^{-1}|0\rangle_B \tag{17.91}$$

and we may interpret $\hat{U}_\alpha^{-1}|0\rangle_B$ as the 'alternative vacuum' $|0, \alpha\rangle_B$ (this argument is, as usual, not valid in the infinite volume limit where $\hat{N}_\phi$ fails to exist).

It is interesting to find out what happens to the symmetry current corresponding to the invariance (17.72), in the 'broken symmetry' case. This current is given in (7.23) which we write again here in slightly different notation:

$$\hat{j}_\phi^\mu = i(\hat{\phi}^\dagger \partial^\mu \hat{\phi} - (\partial^\mu \hat{\phi})^\dagger \hat{\phi}), \tag{17.92}$$

normal ordering being understood. Written in terms of the $\hat{h}$ and $\hat{\theta}$ of (17.84), $\hat{j}_\phi^\mu$ becomes

$$\hat{j}_\phi^\mu = v\partial^\mu \hat{\theta} + 2\hat{h}\partial^\mu \hat{\theta} + \hat{h}^2 \partial^\mu \hat{\theta}/v. \tag{17.93}$$

The term involving just the *single* field $\hat{\theta}$ is very remarkable: it tells us that there is a non-zero matrix element of the form

$$_B\langle 0|\hat{j}_\phi^\mu(x)|\theta, p\rangle = -ip^\mu v e^{-ip \cdot x} \tag{17.94}$$

where $|\theta, p\rangle$ stands for the state with one θ-quantum (Goldstone boson), with momentum p^μ. This is easily seen by writing the usual normal mode expansion for $\hat{\theta}$, and using the standard bosonic commutation relations for $\hat{a}_\theta(k), \hat{a}_\theta^\dagger(k')$. In words, (17.94) asserts that, *when the symmetry is spontaneously broken, the symmetry current connects the vacuum to a state with one Goldstone quantum, with an amplitude which is proportional to the symmetry breaking vacuum expectation value v, and which vanishes as the 4-momentum goes to zero.* The matrix element (17.94), with $x = 0$, is precisely of the type that was shown to be non-zero in the proof of the Goldstone theorem, after (17.67). Note also that (17.94) is consistent with $\partial_\mu \hat{j}_\phi^\mu = 0$ only if $p^2 = 0$, as is required for the massless θ.

The expression (17.93) for the symmetry current allows us to write the θ-dependent part of (17.85) as

$$\hat{\mathcal{L}}_G(\hat{\theta}) = \frac{1}{2v}\hat{j}^\mu \partial_\mu \hat{\theta} \tag{17.95}$$

which is the effective interaction between the symmetry current and the Goldstone field. The current of (17.93) has mass dimension 3, so a quantity of mass dimension 1 is required in the denominator of (17.95). In this case that is the symmetry breaking scale, v. It follows that the interaction will be 'very weak' if v is 'very large', relative to other normal scales; this may be the case for the axion (for a review, see Ringwald, Rosenberg, and Rybka (2022) in Workman *et al.* (2022)).

We are now ready to generalize the Abelian U(1) model to the (global) non-Abelian case.

17.6 Spontaneously broken global non-Abelian symmetry

We can illustrate the essential features by considering a particular example, which in fact forms part of the Higgs sector of the SM. We consider an SU(2) doublet, but this time not of fermions as in section 12.3, but of bosons:

$$\hat{\phi} = \begin{pmatrix} \hat{\phi}^+ \\ \hat{\phi}^0 \end{pmatrix} \equiv \begin{pmatrix} \frac{1}{\sqrt{2}}(\phi_1 + i\phi_2) \\ \frac{1}{\sqrt{2}}(\phi_3 + i\phi_4) \end{pmatrix} \tag{17.96}$$

where the complex scalar field $\hat{\phi}^+$ destroys positively charged particles and creates negatively charged ones, and the complex scalar field $\hat{\phi}^0$ destroys neutral particles and creates neutral antiparticles. As we shall see in a moment, the Lagrangian we shall use has an additional U(1) symmetry, so that the full symmetry is SU(2)×U(1). This U(1) symmetry leads to a conserved quantum number which we call y. We associate the physical charge Q with the eigenvalue t_3 of the SU(2) generator $\hat{t}_3$, and with y, via

$$Q = e(t_3 + y/2) \tag{17.97}$$

so that $y(\phi^+) = 1 = y(\phi^0)$. Thus ϕ^+ and ϕ^0 can be thought of as analogous to the hadronic iso-doublet (K$^+$, K^0).

The Lagrangian we choose is a simple generalization of (17.69) and (17.77):

$$\hat{\mathcal{L}}_\Phi = (\partial_\mu \hat{\phi}^\dagger)(\partial^\mu \hat{\phi}) + \mu^2 \hat{\phi}^\dagger \hat{\phi} - \frac{\lambda}{4}(\hat{\phi}^\dagger \hat{\phi})^2 \tag{17.98}$$

which has the 'spontaneous symmetry breaking' choice of sign for the parameter μ^2. Plainly, for the 'normal' sign of μ^2, in which '$+\mu^2 \hat{\phi}^\dagger \hat{\phi}$' is replaced by '$-\mu^2 \hat{\phi}^\dagger \hat{\phi}$', with μ^2 positive in both cases, the free ($\lambda = 0$) part would describe a complex doublet, with four degrees of freedom, each with the same mass μ. Let us see what happens in the broken symmetry case.

For the Lagrangian (17.98) with $\mu^2 > 0$, the minimum of the classical potential is at the point

$$(\phi^\dagger \phi)_{\text{min}} = 2\mu^2/\lambda \equiv v^2/2. \tag{17.99}$$

As in the U(1) case, we interpret (17.99) as a condition on the vev of $\hat{\phi}^\dagger \hat{\phi}$,

$$\langle 0|\hat{\phi}^\dagger \hat{\phi}|0\rangle = v^2/2. \tag{17.100}$$

Before proceeding we note that (17.98) is invariant under global SU(2) transformations

$$\hat{\phi} \to \hat{\phi}' = \exp(-i\boldsymbol{\alpha} \cdot \boldsymbol{\tau}/2)\hat{\phi} \tag{17.101}$$

but also under a separate global U(1) transformation

$$\hat{\phi} \to \hat{\phi}' = \exp(-i\alpha)\hat{\phi} \tag{17.102}$$

where α is to be distinguished from $\boldsymbol{\alpha} \equiv (\alpha_1, \alpha_2, \alpha_3)$. The symmetry is then referred to as SU(2)×U(1), which is the symmetry of the electroweak sector of the SM, except that in that case it is a *local* symmetry.

As before, in order to get a sensible particle spectrum we must expand the fields $\hat{\phi}$ not about $\hat{\phi} = 0$ but about a point satisfying the stable ground state (vacuum) condition (17.99). That is, we need to define '$\langle 0|\hat{\phi}|0\rangle$' and expand about it, as in (17.84). In the present case, however, the situation is more complicated than (17.84) since the complex doublet (17.96) contains four real fields as indicated in (17.96), and (17.99) becomes

$$\langle 0|\hat{\phi}_1^2 + \hat{\phi}_2^2 + \hat{\phi}_3^2 + \hat{\phi}_4^2|0\rangle = v^2. \tag{17.103}$$

It is evident that we have a lot of freedom in choosing the $\langle 0|\hat{\phi}_i|0\rangle$ so that (17.103) holds, and it is not at first obvious what an appropriate generalization of (17.84) and (17.85) might be.

Furthermore, in this more complicated (non-Abelian) situation a qualitatively new feature can arise: it may happen that the chosen condition $\langle 0|\hat{\phi}_i|0\rangle \neq 0$ is *invariant* under some subset of the allowed symmetry transformations. This would effectively mean that this particular choice of the vacuum state respected that subset of symmetries, which would therefore not be 'spontaneously broken' after all. Since each broken symmetry is associated with a massless Goldstone boson, we would then get fewer of these bosons than expected. Just this happens (by design) in the present case.

Suppose, then, that we could choose the $\langle 0|\hat{\phi}_i|0\rangle$ so as to break this SU(2)×U(1) symmetry completely: we would then expect four massless fields. Actually, however, it is not possible to make such a choice. An analogy may make this point clearer. Suppose we were considering just SU(2), and the field '$\hat{\phi}$' was an SU(2)-triplet, $\hat{\phi}$. Then we could always write $\langle 0|\hat{\phi}|0\rangle = v\boldsymbol{n}$ where $\boldsymbol{n}$ is a unit vector; but this form is invariant under rotations about the $\boldsymbol{n}$-axis, irrespective of where that points. In the present case, by using the freedom of global SU(2)×U(1) phase changes, an arbitrary $\langle 0|\hat{\phi}|0\rangle$ can be brought to the form

$$\langle 0|\hat{\phi}|0\rangle = \begin{pmatrix} 0 \\ v/\sqrt{2} \end{pmatrix}. \tag{17.104}$$

In considering what symmetries are respected or broken by (17.104), it is easiest to look at infinitesimal transformations. It is then clear that the particular transformation

$$\delta\hat{\phi} = -\mathrm{i}\epsilon(1 + \tau_3)\hat{\phi} \tag{17.105}$$

(which is a combination of (17.102) and the 'third component' of (17.101)) is still a symmetry of (17.104) since

$$(1 + \tau_3)\begin{pmatrix} 0 \\ v/\sqrt{2} \end{pmatrix} = \begin{pmatrix} 0 \\ 0 \end{pmatrix}, \tag{17.106}$$

so that

$$\langle 0|\phi|0\rangle = \langle 0|\phi + \delta\phi|0\rangle; \tag{17.107}$$

we say that 'the vacuum in invariant under (17.105)', and when we look at the spectrum of oscillations about that vacuum we expect to find only three massless bosons, not four.

Oscillations about (17.104) are conveniently parametrized by

$$\hat{\phi} = \exp(-\mathrm{i}\hat{\boldsymbol{\theta}}(x) \cdot \boldsymbol{\tau}/2v)\begin{pmatrix} 0 \\ \frac{1}{\sqrt{2}}(v + \hat{H}(x)) \end{pmatrix}, \tag{17.108}$$

which is to be compared with (17.84). Inserting (23.29) into (17.98) (see problem 17.6) we easily find that no mass term is generated for the θ fields, while the H field piece is

$$\hat{\mathcal{L}}_H = \frac{1}{2}\partial_\mu\hat{H}\partial^\mu\hat{H} - \mu^2\hat{H}^2 + \text{interactions} \tag{17.109}$$

just as in (17.85), showing that $m_H = \sqrt{2}\mu$.

Let us now note carefully that whereas in the 'normal symmetry' case with the opposite sign for the μ^2 term in (17.98), the free-particle spectrum consisted of a degenerate doublet of four degrees of freedom all with the same mass μ, in the 'spontaneously broken' case no such doublet structure is seen: instead, there is one massive scalar field, and three massless scalar fields. The number of degrees of freedom is the same in each case, but the physical spectrum is completely different.

In the application of this to the electroweak sector of the SM, the SU(2)×U(1) symmetry will be 'gauged' (i.e. made local), which is easily done by replacing the ordinary derivatives in (17.98) by suitable covariant ones. We shall see in chapter 19 that the result, with the choice (23.29), will be to end up with three *massive* gauge fields (those mediating the weak interactions) and one *massless* gauge field (the photon). We may summarize this (anticipated) result by saying, then, that when a spontaneously broken non-Abelian symmetry is gauged, those gauge fields corresponding to symmetries that are broken by the choice of $\langle 0|\hat{\phi}|0\rangle$ acquire a mass, while those that correspond to symmetries that are respected by $\langle 0|\hat{\phi}|0\rangle$ do not. Exactly how this happens will be the subject of chapter 19.

We end this chapter by considering a second important example of spontaneous symmetry breaking in condensed matter physics, as a preliminary to our discussion of chiral symmetry breaking in the following chapter.

17.7   The BCS superconducting ground state

We shall not attempt to provide a self-contained treatment of the Bardeen–Cooper–Schrieffer (1957)—or BCS—theory; rather, we wish simply to focus on one aspect of the theory, namely the occurrence of an *energy gap* separating the ground state from the lowest excited levels of the fermionic energy spectrum. The existence of such a gap is a fundamental ingredient of the theory of superconductivity; in the following chapter we shall see how Nambu (1960) interpreted a chiral symmetry breaking fermionic mass term as an analogous 'gap'. We emphasize at the outset that we shall here not treat electromagnetic interactions in the superconducting state, leaving that topic for chapter 19.

Our discussion will deliberately have some similarity to that of section 17.3.2. In the present case, of course, we shall be dealing with fermions—namely electrons—rather than the bosons of a superfluid. Nevertheless, we shall see that a similar kind of 'condensation' occurs in the superconductor too. Naturally, such a phenomenon can only occur for bosons. Thus an essential element in the BCS theory is the identification of a mechanism whereby pairs of electrons become correlated, the behaviour of which may have some similarity to that of bosons. Now, direct Coulomb interaction between a pair of electrons is repulsive, and it remains so despite the screening that occurs in a solid. But the positively charged ions do provide sources of attraction for the electrons, and may be used as intermediaries (via 'electron-phonon interactions') to promote an effective attraction between electrons in certain circumstances. At this point we recall the characteristic feature of a weakly interacting gas of electrons at zero temperature: thanks to the Exclusion Principle, the electrons populate single particle energy levels up to some maximum energy E_F (the Fermi energy), whose value is fixed by the electron density. It turns out (see for example Kittel (1987) chapter 8) that electron–electron scattering, mediated by phonon exchange, leads to an effective attraction between two electrons whose energies ϵ_k lie in a thin band $E_F - \omega_D < \epsilon_k < E_F + \omega_D$ around E_F, where ω_D is the Debye frequency associated with lattice vibrations. Cooper (1956) was the first to observe that the Fermi 'sea' was unstable with respect to the formation of bound pairs, in the presence of an attractive interaction. What this means is that the energy of the system can be lowered by exciting a pair of electrons above E_F, which then become bound to a state with a total energy less than $2E_F$. This instability modifies the Fermi sea in a fundamental way: a sort of 'condensate' of pairs is created around the Fermi energy, and we need a many-body formalism to handle the situation.

For simplicity we shall consider pairs of equal and opposite momentum $\boldsymbol{k}$, so their total momentum is zero. It can also be argued that the effective attraction will be greater when the spins are antiparallel, but the spin will not be indicated explicitly in what follows: '$\boldsymbol{k}$' will stand for '$\boldsymbol{k}$ with spin up', and '$-\boldsymbol{k}$' for '$-\boldsymbol{k}$ with spin down'. With this by way of motivation, we thus arrive at the *BCS reduced Hamiltonian*

$$\hat{H}_{\mathrm{BCS}} = \sum_{\boldsymbol{k}} \epsilon_k \hat{c}^\dagger_{\boldsymbol{k}} \hat{c}_{\boldsymbol{k}} - V \sum_{\boldsymbol{k},\boldsymbol{k}'} \hat{c}^\dagger_{\boldsymbol{k}'} \hat{c}^\dagger_{-\boldsymbol{k}'} \hat{c}_{-\boldsymbol{k}} \hat{c}_{\boldsymbol{k}} \tag{17.110}$$

which is the starting point of our discussion. In (17.110), the $\hat{c}$'s are fermionic operators obeying the usual anticommutation relations, and the ground state is such that $\hat{c}_{\boldsymbol{k}}|0\rangle = 0$. The sum is over states lying near E_{F}, as above, and the single particle energies ϵ_k are measured relative to E_{F}. The constant V (with the minus sign in front) represents a simplified form of the effective electron–electron attraction. Note that, in the non-interacting ($V = 0$) part, $\hat{c}^\dagger_{\boldsymbol{k}} \hat{c}_{\boldsymbol{k}}$ is the number operator for the electrons, which because of the Pauli Principle has eigenvalues 0 or 1; this term is of course completely analogous to (7.55), and sums the single particle energies ϵ_k for each occupied level.

We immediately note that $\hat{H}_{\mathrm{BCS}}$ is invariant under the global U(1) transformation

$$\hat{c}_{\boldsymbol{k}} \rightarrow \hat{c}'_{\boldsymbol{k}} = e^{-i\alpha} \hat{c}_{\boldsymbol{k}} \tag{17.111}$$

for all $\boldsymbol{k}$, which is equivalent to $\hat{\psi}'(\boldsymbol{x}) = e^{-i\alpha} \hat{\psi}(\boldsymbol{x})$ for the electron field operator at $\boldsymbol{x}$. Thus fermion number is conserved by $\hat{H}_{\mathrm{BCS}}$. However, just as for the superfluid, we shall see that the BCS ground state does not respect the symmetry.

We follow Bogoliubov (1958) and Bogoliubov *et al.* (1959) (see also Valatin 1958), and make a canonical transformation on the operators $\hat{c}_{\boldsymbol{k}}$, $\hat{c}^\dagger_{-\boldsymbol{k}}$ similar to the one employed for the superfluid problem in (17.38), as motivated by the 'pair condensate' picture. We set

$$\begin{aligned}
\hat{\beta}_{\boldsymbol{k}} &= u_k \hat{c}_{\boldsymbol{k}} - v_k \hat{c}^\dagger_{-\boldsymbol{k}}, \qquad & \hat{\beta}^\dagger_{\boldsymbol{k}} &= u_k \hat{c}^\dagger_{\boldsymbol{k}} - v_k \hat{c}_{-\boldsymbol{k}} \\
\hat{\beta}_{-\boldsymbol{k}} &= u_k \hat{c}_{-\boldsymbol{k}} + v_k \hat{c}^\dagger_{\boldsymbol{k}}, \qquad & \hat{\beta}^\dagger_{-\boldsymbol{k}} &= u_k \hat{c}^\dagger_{-\boldsymbol{k}} + v_k \hat{c}_{\boldsymbol{k}}
\end{aligned} \tag{17.112}$$

where u_k and v_k are real, depend only on $k = |\boldsymbol{k}|$, and are chosen so as to preserve *anti*commutation relations for the β's. This last condition implies (problem 17.7)

$$u_k^2 + v_k^2 = 1 \tag{17.113}$$

so that we may conveniently set

$$u_k = \cos\theta_k, v_k = \sin\theta_k. \tag{17.114}$$

Just as in the superfluid case, the transformations (17.112) only make sense in the context of a number non-conserving ground state, since they do not respect the symmetry (17.111). Although $\hat{H}_{\mathrm{BCS}}$ of (17.110) is number conserving, we shall shortly make a crucial number non-conserving approximation.

We seek a diagonalization of (17.110), analogous to (17.40), in terms of the mode operators $\hat{\beta}$ and $\hat{\beta}^\dagger$:

$$\hat{H}_{\mathrm{BCS}} = \sum_{\boldsymbol{k}} \omega_k (\hat{\beta}^\dagger_{\boldsymbol{k}} \hat{\beta}_{\boldsymbol{k}} + \hat{\beta}^\dagger_{-\boldsymbol{k}} \hat{\beta}_{-\boldsymbol{k}}) + \gamma. \tag{17.115}$$

It is easy to check (problem 17.8) that the form (17.115) implies

$$[\hat{H}_{\mathrm{BCS}}, \hat{\beta}^\dagger_{\boldsymbol{l}}] = \omega_l \hat{\beta}^\dagger_{\boldsymbol{l}} \tag{17.116}$$

as in (17.41), despite the fact that the operators obey *anti*-commution relations. Equation (17.116) then implies that the ω_k are the energies of states created by the *quasiparticle operators* $\hat{\beta}^\dagger_{\boldsymbol{k}}$ and $\hat{\beta}^\dagger_{-\boldsymbol{k}}$, the ground state being defined by

$$\hat{\beta}_{\boldsymbol{k}}|\text{ground}\rangle_{\mathrm{BCS}} = \hat{\beta}_{-\boldsymbol{k}}|\text{ground}\rangle_{\mathrm{BCS}} = 0. \tag{17.117}$$

Substituting for $\hat{\beta}_l^\dagger$ in (17.116) from (17.112) we therefore require

$$[\hat{H}_{\mathrm{BCS}}, \cos\theta_l \,\hat{c}_l^\dagger - \sin\theta_l \,\hat{c}_{-l}] = \omega_l(\cos\theta_l \,\hat{c}_l^\dagger - \sin\theta_l \,\hat{c}_{-l}), \qquad (17.118)$$

which must hold as an identity in the $\hat{c}_l$'s and $\hat{c}_l^\dagger$'s. Evaluating (17.118) one obtains (problem 17.9)

$$(\omega_l - \epsilon_l)\cos\theta_l - V\sin\theta_l \sum_{\boldsymbol{k}} \hat{c}_{-\boldsymbol{k}}\hat{c}_{\boldsymbol{k}} = 0 \qquad (17.119)$$

$$-V\cos\theta_l \sum_{\boldsymbol{k}} \hat{c}_{\boldsymbol{k}}^\dagger \hat{c}_{-\boldsymbol{k}}^\dagger + (\omega_l + \epsilon_l)\sin\theta_l = 0. \qquad (17.120)$$

It is at this point that we make the crucial 'condensate' assumption: we replace the *operator* expressions $\sum_{\boldsymbol{k}} \hat{c}_{-\boldsymbol{k}}\hat{c}_{\boldsymbol{k}}$ and $\sum_{\boldsymbol{k}} \hat{c}_{\boldsymbol{k}}^\dagger \hat{c}_{-\boldsymbol{k}}^\dagger$ by their average values, which are *assumed to be non-zero in the ground state*. Since these operators carry fermion number ± 2, it is clear that this assumption is only valid if the ground state does not, in fact, have a definitive number of particles—just as in the superfluid case. We accordingly make the replacements

$$V\sum_{\boldsymbol{k}} \hat{c}_{-\boldsymbol{k}}\hat{c}_{\boldsymbol{k}} \to V\,_{\mathrm{BCS}}\langle\mathrm{ground}| \sum_{\boldsymbol{k}} \hat{c}_{-\boldsymbol{k}}\hat{c}_{\boldsymbol{k}} |\mathrm{ground}\rangle_{\mathrm{BCS}} \equiv \Delta \qquad (17.121)$$

$$V\sum_{\boldsymbol{k}} \hat{c}_{\boldsymbol{k}}^\dagger \hat{c}_{-\boldsymbol{k}}^\dagger \to V\,_{\mathrm{BCS}}\langle\mathrm{ground}| \sum_{\boldsymbol{k}} \hat{c}_{\boldsymbol{k}}^\dagger \hat{c}_{-\boldsymbol{k}}^\dagger |\mathrm{ground}\rangle_{\mathrm{BCS}} \equiv \Delta^*. \qquad (17.122)$$

In that case, equations (17.119) and (17.120) become

$$\omega_l \cos\theta_l = \epsilon_l \cos\theta_l + \Delta\sin\theta_l \qquad (17.123)$$

$$\omega_l \sin\theta_l = -\epsilon_l \sin\theta_l + \Delta^*\cos\theta_l \qquad (17.124)$$

which are consistent if

$$\omega_l = \pm[\epsilon_l^2 + |\Delta|^2]^{1/2}. \qquad (17.125)$$

Equation (17.125) is the fundamental result at this stage. Recalling that ϵ_l is measured relative to E_{F}, we see that it implies that all excited states are separated from E_{F} by a finite amount, namely $|\Delta|$.

In interpreting (17.125) we must however be careful to reckon energies for an excited state as relative to a BCS state having the same number of pairs, if we consider experimental probes which do not inject or remove electrons. Thus relative to a component of $|\mathrm{ground}\rangle_{\mathrm{BCS}}$ with N pairs, we may consider the excitation of two particles above a BCS state with $N-1$ pairs. The minimum energy for this to be possible is $2|\Delta|$. It is this quantity which is usually called the *energy gap*. Such an exited state is represented by $\beta_{\boldsymbol{k}}^\dagger \beta_{-\boldsymbol{k}}^\dagger |\mathrm{ground}\rangle_{\mathrm{BCS}}$.

We shall need the expressions for $\cos\theta_l$ and $\sin\theta_l$ which may be obtained as follows. Squaring (17.123), and taking Δ now to be real and equal to $|\Delta|$, we obtain

$$|\Delta|^2(\cos^2\theta_l - \sin^2\theta_l) = 2\epsilon_l|\Delta|\cos\theta_l \sin\theta_l, \qquad (17.126)$$

which leads to

$$\tan 2\theta_l = |\Delta|/\epsilon_l \qquad (17.127)$$

and then

$$\cos\theta_l = \left[\frac{1}{2}\left(1 + \frac{\epsilon_l}{\omega_l}\right)\right]^{1/2}, \qquad \sin\theta_l = \left[\frac{1}{2}\left(1 - \frac{\epsilon_l}{\omega_l}\right)\right]^{1/2}. \qquad (17.128)$$

All our experience to date indicates that the choice '$\Delta = $ real' amounts to a choice of phase for the ground state value:

$$V\,_{\mathrm{BCS}}\langle\mathrm{ground}| \sum_{\boldsymbol{k}} \hat{c}_{-\boldsymbol{k}}c_{\boldsymbol{k}} |\mathrm{ground}\rangle_{\mathrm{BCS}} = |\Delta|. \qquad (17.129)$$

By making use of the U(1) symmetry (17.111), other phases for Δ are equally possible.

The condition (17.129) has, of course, the by now anticipated form for a spontaneously broken U(1) symmetry, and we must therefore expect the occurrence of a massless mode (which we do not demonstrate here). However, we may now recall that the electrons are charged, so that when electromagnetic interactions are included in the superconducting state, we have to allow the α in (17.111) to become a local function of x. At the same time, the massless photon field will enter. Remarkably, we shall learn in chapter 19 that the expected massless (Goldstone) mode is, in this case, not observed. Instead, that degree of freedom is incorporated into the gauge field, rendering it massive. As we shall see, this is the physics of the Meissner effect in a superconductor, and that of the 'Higgs mechanism' in the SM. Thus in the (charged) BCS model, both a fermion mass and a gauge boson mass are dynamically generated.

An explicit formula for Δ can be found by using the definition (17.121), together with the expression for $\hat{c}_{\boldsymbol{k}}$ found by inverting (17.112):

$$\hat{c}_{\boldsymbol{k}} = \cos\theta_k\,\hat{\beta}_{\boldsymbol{k}} + \sin\theta_k\,\hat{\beta}^{\dagger}_{-\boldsymbol{k}}. \tag{17.130}$$

This gives, using (17.121) and (17.130),

$$
\begin{aligned}
|\Delta| &= V_{\text{BCS}}\langle\text{ground}|\sum_{\boldsymbol{k}}(\cos\theta_k\hat{\beta}_{-\boldsymbol{k}} + \sin\theta_k\,\hat{\beta}^{\dagger}_{\boldsymbol{k}}) \\
&\quad \times (\cos\theta_k\,\hat{\beta}_{\boldsymbol{k}} + \sin\theta_k\,\hat{\beta}^{\dagger}_{-\boldsymbol{k}})|\text{ground}\rangle_{\text{BCS}} \\
&= V_{\text{BCS}}\langle\text{ground}|\sum_{\boldsymbol{k}}\cos\theta_k\sin\theta_k\hat{\beta}_{-\boldsymbol{k}}\hat{\beta}^{\dagger}_{-\boldsymbol{k}}|\text{ground}\rangle_{\text{BCS}}, \\
&= V\sum_{\boldsymbol{k}}\frac{|\Delta|}{2[\epsilon_k^2 + |\Delta|^2]^{1/2}}.
\end{aligned}
\tag{17.131}
$$

The sum in (17.131) is only over the small band $E_{\text{F}} - \omega_{\text{D}} < \epsilon_k < E_{\text{F}} + \omega_{\text{D}}$ over which the effective electron–electron attraction operates. Replacing the sum by an integral, we obtain the *gap equation*

$$
\begin{aligned}
1 &= \frac{1}{2}V \cdot N_{\text{F}}\int_{-\omega_{\text{D}}}^{\omega_{\text{D}}}\frac{\mathrm{d}\epsilon}{[\epsilon^2 + |\Delta|^2]^{\frac{1}{2}}} \\
&= VN_{\text{F}}\sinh^{-1}(\omega_{\text{D}}/|\Delta|)
\end{aligned}
\tag{17.132}
$$

where N_{F} is the density of states at the Fermi level. Equation (17.132) yields

$$|\Delta| = \frac{\omega_{\text{D}}}{\sinh(1/VN_{\text{F}})} \approx 2\omega_{\text{D}}\mathrm{e}^{-1/VN_{\text{F}}} \tag{17.133}$$

for $VN_{\text{F}} \ll 1$. This is the celebrated BCS solution for the gap parameter $|\Delta|$. Perhaps the most significant thing to note about it, for our purpose, is that the expression for $|\Delta|$ is not an analytic function of the dimensionless interaction parameter VN_{F} (it cannot be expanded as a power series in this quantity), and so no perturbative treatment starting from a normal ground state could reach this result. The estimate (17.133) is in reasonably good agreement with experiment, and may be refined.

The explicit form of the ground state in this model can be found by a method similar to the one indicated in section 17.3.2 for the superfluid. Since the transformation from the $\hat{c}$'s to the $\hat{\beta}$'s is canonical, there must exist a unitary operator which effects it via (compare (17.48))

$$\hat{U}_{\text{BCS}}\,\hat{c}_{\boldsymbol{k}}\,\hat{U}^{\dagger}_{\text{BCS}} = \hat{\beta}_{\boldsymbol{k}}, \quad \hat{U}_{\text{BCS}}\,\hat{c}^{\dagger}_{-\boldsymbol{k}}\,\hat{U}^{\dagger}_{\text{BCS}} = \hat{\beta}^{\dagger}_{-\boldsymbol{k}}. \tag{17.134}$$

The operator $\hat{U}_{\mathrm{BCS}}$ is (Blatt 1964 section V.4, Yosida 1958, and compare problem 17.4))

$$\hat{U}_{\mathrm{BCS}} = \prod_{\boldsymbol{k}} \exp[\theta_k(\hat{c}_{\boldsymbol{k}}^{\dagger}\hat{c}_{-\boldsymbol{k}}^{\dagger} - \hat{c}_{\boldsymbol{k}}\hat{c}_{-\boldsymbol{k}})]. \tag{17.135}$$

Then, since $\hat{c}_{\boldsymbol{k}}|0\rangle = 0$, we have

$$\hat{U}_{\mathrm{BCS}}^{\dagger}\hat{\beta}_{\boldsymbol{k}}\hat{U}_{\mathrm{BCS}}|0\rangle = 0 \tag{17.136}$$

showing that we may identify

$$|\mathrm{ground}\rangle_{\mathrm{BCS}} = \hat{U}_{\mathrm{BCS}}|0\rangle \tag{17.137}$$

via the condition (17.117). When the exponential in $\hat{U}_{\mathrm{BCS}}$ is expanded out, and applied to the vacuum state $|0\rangle$, great simplifications occur. Consider the operator

$$\hat{s}_{\boldsymbol{k}} = \hat{c}_{\boldsymbol{k}}^{\dagger}\hat{c}_{-\boldsymbol{k}}^{\dagger} - \hat{c}_{\boldsymbol{k}}\hat{c}_{-\boldsymbol{k}}. \tag{17.138}$$

We have

$$\hat{s}_{\boldsymbol{k}}^2 = -\hat{c}_{\boldsymbol{k}}^{\dagger}\hat{c}_{-\boldsymbol{k}}^{\dagger}\hat{c}_{\boldsymbol{k}}\hat{c}_{-\boldsymbol{k}} - \hat{c}_{\boldsymbol{k}}\hat{c}_{-\boldsymbol{k}}\hat{c}_{\boldsymbol{k}}^{\dagger}\hat{c}_{-\boldsymbol{k}}^{\dagger} \tag{17.139}$$

so that $\hat{s}_{\boldsymbol{k}}^2|0\rangle = -|0\rangle$. It follows that

$$
\begin{aligned}
\exp(\theta_k\hat{s}_{\boldsymbol{k}})|0\rangle &= (1 + \theta_k\hat{s}_{\boldsymbol{k}} - \frac{\theta_k^2}{2} - \frac{\theta_k^3}{3}\hat{s}_{\boldsymbol{k}}\ldots)|0\rangle \\
&= (\cos\theta_k + \sin\theta_k\,\hat{s}_{\boldsymbol{k}})|0\rangle \\
&= (\cos\theta_k + \sin\theta_k\,\hat{c}_{\boldsymbol{k}}^{\dagger}\hat{c}_{-\boldsymbol{k}}^{\dagger})|0\rangle
\end{aligned} \tag{17.140}
$$

and hence

$$|\mathrm{ground}\rangle_{\mathrm{BCS}} = \prod_{\boldsymbol{k}}(\cos\theta_k + \sin\theta_k\,\hat{c}_{\boldsymbol{k}}^{\dagger}\hat{c}_{-\boldsymbol{k}}^{\dagger})|0\rangle. \tag{17.141}$$

As for the superfluid, (17.141) represents a coherent superposition of correlated pairs, with no restraint on the particle number.

We should emphasize that the above is only the barest outline of a simple version of BCS theory, with no electromagnetic interactions, from which many subtleties have been omitted. Consider, for example, the binding energy E_{b} of a pair, to calculate which one needs to evaluate the constant γ in (17.115). To a good approximation one finds (see for example Enz 1992) $E_{\mathrm{b}} \approx 3\Delta^2/E_{\mathrm{F}}$. One can also calculate the approximate spatial extension of a pair, which is denoted by the *coherence length* ξ and is of order $v_{\mathrm{F}}/\pi\Delta$ where $k_{\mathrm{F}} = mv_{\mathrm{F}}$ is the Fermi momentum. If we compare E_{b} to the Coulomb repulsion at a distance ξ we find

$$E_{\mathrm{b}}/(\alpha/\xi) \sim a_0/\xi \tag{17.142}$$

where a_0 is the Bohr radius. Numerical values show that the right-hand side of (17.142), in conventional superconductors, is of order 10^{-3}. Hence the pairs are not really bound, only correlated, and as many as 10^6 pairs may have their centres of mass within one coherence length of each other. Nevertheless, the simple theory presented here contains the essential features which underlie all attempts to understand the dynamical occurrence of spontaneous symmetry breaking in fermionic systems.

We now proceed to an important application in particle physics.

Problems

17.1 Verify (17.29).

17.2 Verify (17.35).

17.3 Derive (17.43) and (17.44).

17.4 Let

$$\hat{U}_\lambda = \exp[\frac{1}{2}\lambda\theta(\hat{a}^2 - \hat{a}^{\dagger 2})]$$

where $[\hat{a}, \hat{a}^\dagger] = 1$ and λ, θ are real parameters.
(a) Show that $\hat{U}_\lambda$ is unitary.
(b) Let

$$\hat{I}_\lambda = \hat{U}_\lambda \hat{a} \hat{U}_\lambda^{-1} \quad \text{and} \quad \hat{J}_\lambda = \hat{U}_\lambda \hat{a}^\dagger \hat{U}_\lambda^{-1}.$$

Show that

$$\frac{\mathrm{d}\hat{I}_\lambda}{\mathrm{d}\lambda} = \theta\hat{J}_\lambda$$

and

$$\frac{\mathrm{d}^2\hat{I}_\lambda}{\mathrm{d}\lambda^2} = \theta^2\hat{I}_\lambda.$$

(c) Hence show that

$$\hat{I}_\lambda = \cosh(\lambda\theta)\,\hat{a} + \sinh(\lambda\theta)\,\hat{a}^\dagger,$$

and thus finally (compare (17.38) and (17.48)) that

$$\hat{U}_1 \hat{a} \hat{U}_1^{-1} = \cosh\theta\,\hat{a} + \sinh\theta\,\hat{a}^\dagger \equiv \hat{\alpha}$$

and

$$\hat{U}_1 \hat{a}^\dagger \hat{U}_1^{-1} = \sinh\theta\,\hat{a} + \cosh\theta\,\hat{a}^\dagger \equiv \hat{\alpha}^\dagger,$$

where

$$\hat{U}_1 \equiv \hat{U}_{\lambda=1} = \exp[\frac{1}{2}\theta(\hat{a}^2 - \hat{a}^{\dagger 2})].$$

17.5 Insert the ansatz (17.84) for $\hat{\phi}$ into $\hat{\mathcal{L}}_\mathrm{G}$ of (17.69), with $\hat{V} = \hat{V}_\mathrm{SB}$ of (17.77), and show that the result for the constant term, and the quadratic terms in $\hat{h}$ and $\hat{\theta}$, is as given in (17.85).

17.6 Verify that when (23.29) is inserted in (17.98), the terms quadratic in the fields $\hat{H}$ and $\hat{\theta}$ reveal that $\hat{\theta}$ is a massless field, while the quanta of the $\hat{H}$ field have mass $\sqrt{2}\mu$.

17.7 Verify that the $\hat{\beta}$'s of (17.112) satisfy the required anticommutation relations if (17.113) holds.

17.8 Verify (17.116).

17.9 Derive (17.119) and (17.120).

18

Chiral Symmetry Breaking

In section 12.4.2 we arrived at a puzzle: there seemed good reason to think that a world consisting of u and d quarks and their antiparticles, interacting via the colour gauge fields of QCD, should exhibit signs of the non-Abelian *chiral symmetry* $SU(2)_{f5}$, which was exact in the massless limit $m_u, m_d \to 0$. But, as we showed, one of the simplest consequences of such a symmetry should be the existence of nucleon parity doublets, which are not observed. We can now resolve this puzzle by making the hypothesis (section 18.1) first articulated by Nambu (1960) and Nambu and Jona-Lasinio (1961a), that this chiral symmetry is *spontaneously broken* as a dynamical effect—presumably, from today's perspective, as a property of the QCD interactions, as discussed in section 18.2. If this is so, an immediate physical consequence should be the appearance of massless (Goldstone) bosons, one for every symmetry not respected by the vacuum. Indeed, returning to (12.168) which we repeat here for convenience,

$$\hat{T}^{(\frac{1}{2})}_{+5}|d\rangle = |\tilde{u}\rangle, \tag{18.1}$$

we now interpret the state $|\tilde{u}\rangle$ (which is degenerate with $|d\rangle$) as $|d + `\pi^+`\rangle$ where 'π^+' is a massless particle of positive charge, but a *pseudo*scalar (0^-) rather than a scalar (0^+) since, as we saw, $|\tilde{u}\rangle$ has opposite parity to $|u\rangle$. In the same way, 'π^-' and 'π^0' will be associated with $\hat{T}^{(\frac{1}{2})}_{-5}$ and $\hat{T}^{(\frac{1}{2})}_{35}$. Of course, no such *massless* pseudoscalar particles are observed. But it is natural to hope that when the small up and down quark masses are included, the real pions (π^+, π^-, π^0) will emerge as 'anomalously light', rather than strictly massless. This is indeed how they do appear, particularly with respect to the octet of mesons, which differ only in q$\bar{\text{q}}$ spin alignment from the 0^- octet. As Nambu and Jona-Lasinio (1961a) said, 'it is perhaps not a coincidence that there exists such an entity [i.e. the Goldstone state(s)] in the form of the pion'.

The pion is often cited as an example of a 'pseudo-Goldstne boson', meaning that it is an anomalously light state (relative to some appropriate scale), and would be massless if the symmetry which is spontaneously broken were exact, but which is in fact explicitly broken by a certain parameter in the Lagrangian (here the u and d quark masses).

If this was the only observable consequence of spontaneously breaking chiral symmetry, it would perhaps hardly be sufficient grounds for accepting the hypothesis. But there are two circumstances which greatly increase the phenomenological implications of the idea. First, the vector and axial vector symmetry currents $\hat{\boldsymbol{T}}^{(\frac{1}{2})\mu}$ and $\hat{\boldsymbol{T}}^{(\frac{1}{2})\mu}_5$ of the u-d strong interaction SU(2) symmetries (see (12.109) and (12.165)) happen to be the very same currents which enter into strangeness-conserving semileptonic weak interactions (such as n $\to$ pe$^-\bar{\nu}_e$ and $\pi^- \to \mu^-\bar{\nu}_\mu$), as we shall see in chapter 20. Thus some remarkable connections between weak- and strong-interaction parameters can be established, such as the Goldberger–Treiman (1958) relation (see section 18.3) and the Adler–Weisberger (Adler 1965, Weisberger 1965) relation. Second, it turns out that the dynamics of the Goldstone modes, and their interactions with other hadrons such as nucleons, are strongly constrained by the underlying chiral symmetry of QCD; indeed, surprisingly detailed *effective theories* (see section 18.3) have been developed, which provide a very successful description of the

DOI: 10.1201/9781003411666-18

low energy dynamics of the Goldstone degrees of freedom. Finally we shall introduce the subject of chiral anomalies in section 18.4.

It would take us too far from our main focus on gauge theories to pursue these interesting avenues in any detail. But we hope to convince the reader, in this chapter, that chiral symmetry breaking is an integral part of the SM, being a fundamental property of QCD.

18.1 The Nambu analogy

We recall from section 12.4.2 that for 'almost massless' fermions it is natural to use the representation (3.40) for the Dirac matrices, in terms of which the Dirac equation reads

$$E\phi \;=\; \boldsymbol{\sigma}\cdot\boldsymbol{p}\phi + m\chi \qquad (18.2)$$

$$E\chi \;=\; -\boldsymbol{\sigma}\cdot\boldsymbol{p}\chi + m\phi. \qquad (18.3)$$

Nambu (1960) and Nambu and Jona-Lasinio (1961a) pointed out a remarkable analogy between (18.2) and (18.3) and equations (17.123) and (17.124) which describe the elementary excitations in a superconductor (in the case Δ is real), and which we repeat here for convenience:

$$\omega_l\cos\theta_l \;=\; \epsilon_l\cos\theta_l + \Delta\sin\theta_l \qquad (18.4)$$

$$\omega_l\sin\theta_l \;=\; -\epsilon_l\sin\theta_l + \Delta\cos\theta_l. \qquad (18.5)$$

In (18.4) and (18.5), $\cos\theta_l$ and $\sin\theta_l$ are respectively the components of the electron destruction operator $\hat{c}_{\boldsymbol{l}}$ and the electron creation operator $\hat{c}^{\dagger}_{-\boldsymbol{l}}$ in the quasiparticle operator $\hat{\beta}_{\boldsymbol{l}}$ (see (17.112)):

$$\hat{\beta}_{\boldsymbol{l}} = \cos\theta_l\,\hat{c}_{\boldsymbol{l}} - \sin\theta_l\,\hat{c}^{\dagger}_{-\boldsymbol{l}}. \qquad (18.6)$$

The superposition in $\hat{\beta}_{\boldsymbol{l}}$ combines operators which transform differently under the U(1) (number) symmetry. The result of this spontaneous breaking of the U(1) symmetry is the creation of the gap Δ (or 2Δ for a number-conserving excitation), and the appearance of a massless mode. If Δ vanishes, (17.127) implies that $\theta_l = 0$, and we revert to the symmetry-respecting operators $\hat{c}_{\boldsymbol{l}}, \hat{c}^{\dagger}_{-\boldsymbol{l}}$. Consider now (18.2) and (18.3). Here ϕ and χ are the components of definite chirality in the Dirac spinor ω (compare (12.149)), which is itself not a chirality eigenstate when $m \neq 0$. When m vanishes, the Dirac equation for ω decouples into two separate ones for the chirality eigenstates $\phi_{\mathrm{R}} \equiv \begin{pmatrix} \phi \\ 0 \end{pmatrix}$ and $\phi_{\mathrm{L}} \equiv \begin{pmatrix} 0 \\ \chi \end{pmatrix}$. Nambu therefore made the following analogy:

$$\begin{aligned}
\text{Superconducting gap parameter } \Delta &\leftrightarrow \text{Dirac mass } m \\
\text{quasiparticle excitation} &\leftrightarrow \text{massive Dirac particle} \\
\text{U(1) number symmetry} &\leftrightarrow \text{U(1)}_5 \text{ chirality symmetry} \\
\text{Goldstone mode} &\leftrightarrow \text{massless boson.}
\end{aligned}$$

In short, the mass of a Dirac particle arises from the (presumed) spontaneous breaking of a chiral (or γ_5) symmetry, and this will be accompanied by a massless boson.

FIGURE 18.1
The type of fermion-antifermion in the Nambu 'chiral condensate'.

Before proceeding we should note that there are features of the analogy, on both sides, which need qualification. First, the particle symmetry we want to interpret this way is $SU(2)_{f5}$ not $U(1)_5$, so the appropriate generalization (Nambu and Jona-Lasinio (1961b)) has to be understood. Second, we must again note that the BCS electrons are charged, so that in the real superconducting case we are dealing with a spontaneously broken *local* $U(1)$ symmetry, not a global one. By contrast, the $SU(2)_{f5}$ chiral symmetry is not gauged.

As usual, the quantum field theory vacuum is analogous to the many-body ground state. According to Nambu's analogy, therefore, the vacuum for a massive Dirac particle is to be pictured as a condensate of correlated pairs of massive fermions. Since the vacuum carries neither linear nor angular momentum, the members of a pair must have equal and opposite spin: they therefore have the same helicity. However, since the vacuum does *not* violate fermion number conservation, one has to be a fermion and the other an antifermion. This means (recalling the discussion after (12.147)) that they have opposite chirality. Thus a typical pair in the Nambu vacuum is as shown in figure 18.1. We may easily write down an expression for the Nambu vacuum, analogous to (17.141) for the BCS ground state. Consider solutions ϕ_+ and χ_+ of positive helicity in (18.2) and (18.3); then

$$E\phi_+ = |\boldsymbol{p}|\phi_+ + m\chi_+ \tag{18.7}$$

$$E\chi_+ = -|\boldsymbol{p}|\chi_+ + m\phi_+. \tag{18.8}$$

Comparing (18.7) and (18.8) with (18.4) and (18.5), we can read off the mixing coefficients $\cos\theta_p$ and $\sin\theta_p$ as (cf (17.128))

$$\cos\theta_p = \left[\frac{1}{2}\left(1 + \frac{|\boldsymbol{p}|}{E}\right)\right]^{1/2} \tag{18.9}$$

$$\sin\theta_p = \left[\frac{1}{2}\left(1 - \frac{|\boldsymbol{p}|}{E}\right)\right]^{1/2} \tag{18.10}$$

where $E = (m^2 + \boldsymbol{p}^2)^{1/2}$. The Nambu vacuum is then given by[1]

$$|0\rangle_N = \prod_{\boldsymbol{p},s}(\cos\theta_p - \sin\theta_p \hat{c}_s^\dagger(\boldsymbol{p})\hat{d}_s^\dagger(-\boldsymbol{p}))|0\rangle_{m=0}, \tag{18.11}$$

where $\hat{c}_s^\dagger$'s and $\hat{d}_s^\dagger$'s are the operators in *massless* Dirac fields. Depending on the sign of the helicity s, each pair in (18.11) carries ± 2 units of chirality. We may check this by noting that in the mode expansion of the Dirac field $\hat{\psi}$, $\hat{c}_s$ $(\boldsymbol{p})$ operators go with u-spinors for which the γ_5 eigenvalue equals the helicity, while $\hat{d}_s^\dagger(-\boldsymbol{p})$ operators accompany v-spinors for which the γ_5 eigenvalue equals minus the helicity. Thus under a chiral transformation $\hat{\psi}' = e^{-i\beta\gamma_5}\hat{\psi}$, $\hat{c}_s \to e^{-i\beta s}\hat{c}_s$ and $\hat{d}_s^\dagger \to e^{i\beta s}\hat{d}_s^\dagger$, for a given s. Hence $\hat{c}_s^\dagger\hat{d}_s^\dagger$ acquires a factor $e^{2i\beta s}$. Thus the

---

[1]A different phase convention is used for $\hat{d}_s^\dagger(-\boldsymbol{p})$ as compared to that for $\hat{c}_{-\boldsymbol{k}}^\dagger$ in (17.112).

Nambu vacuum does not have a definite chirality, and operators carrying non-zero chirality can have non-vanishing vacuum expectation values. A mass term $\hat{\bar{\psi}}\hat{\psi}$ is of just this kind, since under $\hat{\psi} = e^{-i\beta\gamma_5}\hat{\psi}$ we find $\hat{\psi}^\dagger\gamma^0\hat{\psi} \to \hat{\psi}^\dagger e^{i\beta\gamma_5}\gamma^0 e^{-i\beta\gamma_5}\hat{\psi} = \hat{\bar{\psi}}e^{-2i\beta\gamma_5}\hat{\psi}$. Thus, in analogy with (17.121), a Dirac mass is associated with a non-zero value for $_N\langle 0|\hat{\bar{\psi}}\hat{\psi} = \hat{\bar{\psi}}_L\hat{\psi}_R + \hat{\bar{\psi}}_R\hat{\psi}_L|0\rangle_N$.

In the original conception by Nambu and co-workers, the fermion under discussion was taken to be the nucleon, with 'm' the (spontaneously generated) nucleon mass. The fermion–fermion interaction—necessarily invariant under chiral transformations—was taken to be of the four-fermion type. As we have seen in volume 1, this is actually a non-renormalizable theory, but a physical cut-off was employed, somewhat analogous to the Fermi energy E_F. Thus the nucleon mass could not be dynamically predicted, unlike the analogous gap parameter Δ in BCS theory. Nevertheless, a gap equation similar to (17.132) could be formulated, and it was possible to show that when it had a non-trivial solution, a massless bound state automatically appeared in the $\bar{f}f$ channel (Nambu and Jona-Lasinio 1961a). This work was generalized to the $SU(2)_{f5}$ case by Nambu and Jona-Lasinio (1961b), who showed that if the chiral symmetry was broken explicitly by the introduction of a small nucleon mass (~ 5 MeV), then the Goldstone pions would have their observed non-zero (but small) mass; they would be pseudo-Goldstone bosons. In addition, the Goldberger-Treiman (1958) relation was derived, and a number of other applications were suggested. Subsequently, Nambu with other collaborators (Nambu and Lurie 1962, Nambu and Schrauner 1962) showed how the amplitudes for the emission of a single 'soft' (nearly massless, low momentum) pion could be calculated, for various processes. These developments culminated in the Adler-Weisberger relation (Adler 1965, Weisberger 1965) which involves *two* soft pions.

This work was all done in the absence of an agreed theory of the strong interactions (the NJ-L theory was an illustrative working model of dynamically-generated spontaneous symmetry breaking, but not a complete theory of strong interactions). QCD became widely accepted as that theory around 1973. In this case, of course, the 'fermions in question' are quarks, and the interactions between them are gluon exchanges, which conserve chirality as noted in section 12.4.2. The bulk of the masses of the qqq bound states which form baryons is then interpreted as being spontaneously generated, while a small explicit quark mass term in the Lagrangian is responsible for the non-zero pion mass. Let us therefore now turn to two-flavour QCD.

18.1.1 Two flavour QCD and $SU(2)_{fL} \times SU(2)_{fR}$

Let us begin with the massless case, for which the fermionic part of the Lagrangian is

$$\hat{\mathcal{L}}_q = \hat{\bar{u}}\,i\hat{\slashed{D}}\hat{u} + \hat{\bar{d}}\,i\hat{\slashed{D}}\hat{d} \tag{18.12}$$

where $\hat{u}$ and $\hat{d}$ now stand for the field operators,

$$\hat{D}^\mu = \partial^\mu + ig_s\boldsymbol{\lambda}/2 \cdot \boldsymbol{A}^\mu, \tag{18.13}$$

and the $\boldsymbol{\lambda}$ matrices act on the colour (r,b,g) degree of freedom of the u and d quarks. This Lagrangian is invariant under

   (i) $U(1)_f$ 'quark number' transformations

$$\hat{q} \to e^{-i\alpha}\hat{q}; \tag{18.14}$$

   (ii) $SU(2)_f$ 'flavour isospin' transformations

$$\hat{q} \to \exp(-i\boldsymbol{\alpha}\cdot\boldsymbol{\tau}/2)\,\hat{q}; \tag{18.15}$$

(iii) U(1)$_{\mathrm{f}\,5}$ 'axial quark number' transformations

$$\hat{q} \to \mathrm{e}^{-\mathrm{i}\beta\gamma_5}\hat{q};\tag{18.16}$$

(iv) SU(2)$_{\mathrm{f}\,5}$ 'axial flavour isospin' transformations

$$\hat{q} \to \exp(-\mathrm{i}\boldsymbol{\beta}\cdot\boldsymbol{\tau}/2\gamma_5)\,\hat{q},\tag{18.17}$$

where

$$\hat{q} = \left(\begin{array}{c}\hat{u}\\\hat{d}\end{array}\right).\tag{18.18}$$

Symmetry (i) is unbroken, and its associated 'charge' operator (the quark number operator) commutes with all other symmetry operators, so it need not concern us further. Symmetry (ii) is the standard isospin symmetry of chapter 12, explicitly broken by the electromagnetic interactions (and by the difference in the masses m_{u} and m_{d}, when included). Symmetry (iii) does not correspond to any known conservation law; on the other hand, there are not any near-massless isoscalar 0^- mesons, either, such as must be present if the symmetry is spontaneously broken. The η meson is an isoscalar 0^- meson, but with a mass of 547 MeV it is considerably heavier than the pion. In fact, it can be understood as one of the Goldstone bosons associated with the spontaneous breaking of the larger group SU(3)$_{\mathrm{f}\,5}$, which includes the s quark (see section 18.3.3). In that case, the symmetry (iii) becomes extended to

$$\hat{u} \to \mathrm{e}^{-\mathrm{i}\beta\gamma_5}\hat{u}, \quad \hat{d} \to \mathrm{e}^{-\mathrm{i}\beta\gamma_5}\hat{d}, \quad \hat{s} \to \mathrm{e}^{-\mathrm{i}\beta\gamma_5}\hat{s},\tag{18.19}$$

but there is still a missing light isoscalar 0^- meson. It can be shown that its mass must be less than or equal to $\sqrt{3}\,m_\pi$ (Weinberg 1975), but no such particle exists. This is the famous 'U(1) problem': it was resolved by 't Hooft (1976a, 1986), by showing that the inclusion of instanton configurations (Belavin *et al.* 1975) in path integrals leads to violations of symmetry (iii)—see, for example, Weinberg (1996) section 23.5. Finally, symmetry (iv) is the one with which we are presently concerned.

The symmetry currents associated with (iv) are those already given in (12.165), but we give them again here in a slightly different notation which will be similar to the one used for weak interactions:

$$\hat{j}^{\mu}_{i,5} = \bar{\hat{q}}\gamma^{\mu}\gamma_5\frac{\tau_i}{2}\hat{q} \qquad i = 1, 2, 3.\tag{18.20}$$

Similarly the currents associated with (ii) are

$$\hat{j}^{\mu}_{i} = \bar{\hat{q}}\gamma^{\mu}\frac{\tau_i}{2}\hat{q} \qquad i = 1, 2, 3.\tag{18.21}$$

The corresponding 'charges' are (compare (12.166))

$$\hat{Q}_{i,5} \equiv \int \hat{j}^{0}_{i,5}\,\mathrm{d}^3\boldsymbol{x} = \int \hat{q}^{\dagger}\gamma_5\frac{\tau_i}{2}\hat{q}\,\mathrm{d}^3\boldsymbol{x},\tag{18.22}$$

previously denoted by $\hat{T}^{(\frac{1}{2})}_{i,5}$, and (compare (12.101)),

$$\hat{Q}_i = \int \hat{q}^{\dagger}\frac{\tau_i}{2}\hat{q}\,\mathrm{d}^3\boldsymbol{x},\tag{18.23}$$

previously denoted by $\hat{\boldsymbol{T}}^{(\frac{1}{2})}_5$. As with all symmetries, it is interesting to discover the *algebra* of the generators, which are the six charges $\hat{Q}_i, \hat{Q}_{i,5}$ in this case. Patient work with the anticommutation relations for the operators in $\hat{q}(x)$ and $\hat{q}^\dagger(x)$ gives the results (problem 18.1)

$$[\hat{Q}_i, \hat{Q}_j] = \mathrm{i}\epsilon_{ijk}\hat{Q}_k \qquad (18.24)$$

$$[\hat{Q}_i, \hat{Q}_{j,5}] = \mathrm{i}\epsilon_{ijk}\hat{Q}_{k,5} \qquad (18.25)$$

$$[\hat{Q}_{i,5}, \hat{Q}_{j,5}] = \mathrm{i}\epsilon_{ijk}\hat{Q}_k. \qquad (18.26)$$

Relation (18.24) has been seen before in (12.101), and simply says that the $\hat{Q}_i$'s obey a SU(2) algebra. A simple trick reduces the rather complicated algebra of (18.24)–(18.26) to something much simpler. Defining

$$\hat{Q}_{i,\mathrm{R}} = \frac{1}{2}(\hat{Q}_i + \hat{Q}_{i,5}) \qquad \hat{Q}_{i,\mathrm{L}} = \frac{1}{2}(\hat{Q}_i - \hat{Q}_{i,5}) \qquad (18.27)$$

we find (problem 18.2)

$$[\hat{Q}_{i,\mathrm{R}}, \hat{Q}_{j,\mathrm{R}}] = \mathrm{i}\epsilon_{ijk}\hat{Q}_{k,\mathrm{R}} \qquad (18.28)$$

$$[\hat{Q}_{i,\mathrm{L}}, \hat{Q}_{j,\mathrm{L}}] = \mathrm{i}\epsilon_{ijk}\hat{Q}_{k,\mathrm{L}} \qquad (18.29)$$

$$[\hat{Q}_{i,\mathrm{R}}, \hat{Q}_{j,\mathrm{L}}] = 0. \qquad (18.30)$$

The operators $\hat{Q}_{i,\mathrm{R}}, \hat{Q}_{i,\mathrm{L}}$ therefore behave like *two commuting (independent) angular momentum operators*, each obeying the algebra of SU(2). For this reason, the symmetry group of the combined symmetries (ii) and (iv) is called $SU(2)_{f\,\mathrm{L}} \times SU(2)_{f\,\mathrm{R}}$.

The decoupling effected by (18.27) has a simple interpretation. Referring to (18.22) and (18.23), we see that

$$\hat{Q}_{i,\mathrm{R}} = \int \hat{q}^\dagger \left(\frac{1+\gamma_5}{2}\right)\frac{\tau_i}{2}\hat{q}\,\mathrm{d}^3\boldsymbol{x} \qquad (18.31)$$

and similarly for $\hat{Q}_{i,\mathrm{L}}$. But $((1\pm\gamma_5)/2)$ are just the projection operators $P_{\mathrm{R,L}}$ introduced in section 12.3.2, which project out the chiral parts of any fermion field. Furthermore, it is easy to see that $P_{\mathrm{R}}^2 = P_{\mathrm{R}}$ and $P_{\mathrm{L}}^2 = P_{\mathrm{L}}$, so that $\hat{Q}_{i,\mathrm{R}}$ and $\hat{Q}_{i,\mathrm{L}}$ can also be written as

$$\hat{Q}_{i,\mathrm{R}} = \int \hat{q}_{\mathrm{R}}^\dagger \frac{\tau_i}{2}\hat{q}_{\mathrm{R}}\,\mathrm{d}^3\boldsymbol{x} \qquad \hat{Q}_{i,\mathrm{L}} = \int \hat{q}_{\mathrm{L}}^\dagger \frac{\tau_i}{2}\hat{q}_{\mathrm{L}}\,\mathrm{d}^3\boldsymbol{x}, \qquad (18.32)$$

where $\hat{q}_{\mathrm{R}} = ((1+\gamma_5)/2)\hat{q}, \hat{q}_{\mathrm{L}} = ((1-\gamma_5)/2)\hat{q}$. In a similar way, the currents (18.20) and (18.21) can be written as

$$\hat{\jmath}_i^\mu = \hat{\jmath}_{i,\mathrm{R}}^\mu + \hat{\jmath}_{i,\mathrm{L}}^\mu \qquad \hat{\jmath}_{i,5}^\mu = \hat{\jmath}_{i,\mathrm{R}}^\mu - \hat{\jmath}_{i,\mathrm{L}}^\mu, \qquad (18.33)$$

where

$$\hat{\jmath}_{i,\mathrm{R}} = \bar{\hat{q}}_{\mathrm{R}}\gamma^\mu\frac{\tau_i}{2}\hat{q}_{\mathrm{R}} \qquad \hat{\jmath}_{i,\mathrm{L}}^\mu = \bar{\hat{q}}_{\mathrm{L}}\gamma^\mu\frac{\tau_i}{2}\hat{q}_{\mathrm{L}}. \qquad (18.34)$$

Thus the $SU(2)_{\mathrm{L}}$ and $SU(2)_{\mathrm{R}}$ refer to the two chiral components of the fermion fields, which is why it is called *chiral symmetry*.

Under infinitesimal SU(2) isospin and axial isospin transformations, $\hat{q}$ transforms by

$$\hat{q} \to \hat{q}' = (1 - \mathrm{i}\boldsymbol{\epsilon}\cdot\boldsymbol{\tau}/2 - \mathrm{i}\boldsymbol{\eta}\cdot\boldsymbol{\tau}/2\,\gamma_5)\hat{q}. \qquad (18.35)$$

This can be rewritten in terms of $\hat{q}_{\mathrm{R}}$ and $\hat{q}_{\mathrm{R}}$, using

$$\hat{q} = \hat{q}_{\mathrm{R}} + \hat{q}_{\mathrm{L}}, \quad \gamma_5\hat{q}_{\mathrm{R}} = \hat{q}_{\mathrm{R}}, \quad \gamma_5\hat{q}_{\mathrm{L}} = -\hat{q}_{\mathrm{L}}. \qquad (18.36)$$

We find that

$$\hat{q}'_{\mathrm{R}} = (1 - \mathrm{i}(\boldsymbol{\epsilon} + \boldsymbol{\eta}) \cdot \boldsymbol{\tau}/2)\hat{q}_{\mathrm{R}} \tag{18.37}$$

and similarly

$$\hat{q}'_{\mathrm{L}} = (1 - \mathrm{i}(\boldsymbol{\epsilon} - \boldsymbol{\eta}) \cdot \boldsymbol{\tau}/2)\hat{q}_{\mathrm{L}}. \tag{18.38}$$

Hence $\hat{q}_{\mathrm{R}}$ and $\hat{q}_{\mathrm{L}}$ transform quite independently[2], which is why $[\hat{Q}_{i,\mathrm{R}}, \hat{Q}_{j,\mathrm{L}}] = 0$.

This formalism allows us to see immediately why (18.12) is chirally invariant: problem 18.3 verifies that $\hat{\mathcal{L}}_q$ can be written as

$$\hat{\mathcal{L}}_q = \bar{\hat{q}}_{\mathrm{R}} \mathrm{i}\, \hat{\not{D}} q_{\mathrm{R}} + \bar{\hat{q}}_{\mathrm{L}} \mathrm{i}\, \hat{\not{D}} \hat{q}_{\mathrm{L}} \tag{18.39}$$

which is plainly invariant under (18.37) and (18.38), since $\hat{D}$ is flavour-blind.

There is as yet no formal proof that this $\mathrm{SU}(2)_{\mathrm{L}} \times \mathrm{SU}(2)_{\mathrm{R}}$ chiral symmetry is spontaneously broken in QCD, though it can be argued that the larger symmetry $\mathrm{SU}(3)_{\mathrm{L}} \times \mathrm{SU}(3)_{\mathrm{R}}$—appropriate to three massless flavours—must be spontaneously broken (see Weinberg (1996) section 22.5). This is, of course, an issue that cannot be settled within perturbation theory (compare the comments after (17.133)). Numerical solutions of QCD on a lattice (see chapter 16) do provide strong evidence that baryons acquire large dynamical $(\mathrm{SU}(2)_{\mathrm{f}5}\text{-breaking})$ mass.

Even granted that chiral symmetry is spontaneously broken in massless two-flavour QCD, how do we know that it breaks in such a way as to leave the isospin ('R+L') symmetry unbroken? A plausible answer can be given if we restore the quark mass terms via

$$\hat{\mathcal{L}}_m = -m_{\mathrm{u}}\bar{\hat{u}}\hat{u} - m_{\mathrm{d}}\bar{\hat{d}}\hat{d} = -\frac{1}{2}(m_{\mathrm{u}} + m_{\mathrm{d}})\bar{\hat{q}}\hat{q} - \frac{1}{2}(m_{\mathrm{u}} - m_{\mathrm{d}})\bar{\hat{q}}\tau_3\hat{q}. \tag{18.40}$$

Now

$$\bar{\hat{q}}\hat{q} = \bar{\hat{q}}_{\mathrm{L}}\hat{q}_{\mathrm{R}} + \bar{\hat{q}}_{\mathrm{R}}\hat{q}_{\mathrm{L}} \tag{18.41}$$

and

$$\bar{\hat{q}}\tau_3\hat{q} = \bar{\hat{q}}_{\mathrm{L}}\tau_3 q_{\mathrm{R}} + \bar{\hat{q}}_{\mathrm{R}}\tau_3\hat{q}_{\mathrm{L}}. \tag{18.42}$$

Including these extra terms is somewhat analogous to switching on an external field in the ferromagnetic problem, which determines a preferred direction for the symmetry breaking. It is clear that neither of (18.41) and (18.42) preserves $\mathrm{SU}(2)_{\mathrm{L}} \times \mathrm{SU}(2)_{\mathrm{R}}$ since they treat the L and R parts differently. Indeed from (18.37) and (18.38) we find

$$\begin{aligned}
\bar{\hat{q}}_{\mathrm{L}}\hat{q}_{\mathrm{R}} \to \bar{\hat{q}}'_{\mathrm{L}}q'_{\mathrm{R}} &= \bar{\hat{q}}_{\mathrm{L}}(1 + \mathrm{i}(\boldsymbol{\epsilon} - \boldsymbol{\eta}) \cdot \boldsymbol{\tau}/2)(1 - \mathrm{i}(\boldsymbol{\epsilon} + \boldsymbol{\eta}) \cdot \boldsymbol{\tau}/2)\hat{q}_{\mathrm{R}} \tag{18.43} \\
&= \bar{\hat{q}}_{\mathrm{L}}\hat{q}_{\mathrm{R}} - \mathrm{i}\boldsymbol{\eta} \cdot \bar{\hat{q}}_{\mathrm{L}}\boldsymbol{\tau}\hat{q}_{\mathrm{R}} \tag{18.44}
\end{aligned}$$

and

$$\bar{\hat{q}}_{\mathrm{R}}\hat{q}_{\mathrm{L}} \to \bar{\hat{q}}_{\mathrm{R}}\hat{q}_{\mathrm{L}} + \mathrm{i}\boldsymbol{\eta} \cdot \bar{\hat{q}}_{\mathrm{R}}\boldsymbol{\tau}\hat{q}_{\mathrm{L}}. \tag{18.45}$$

Equations (18.44) and (18.45) confirm that the term $\bar{\hat{q}}\hat{q}$ in (18.40) is invariant under the isospin part of $\mathrm{SU}(2)_{\mathrm{L}} \times \mathrm{SU}(2)_{\mathrm{R}}$ (since $\boldsymbol{\epsilon}$ is not involved), but not invariant under the axial isospin transformations parametrized by $\boldsymbol{\eta}$. The $\bar{\hat{q}}\tau_3\hat{q}$ term explicitly breaks the third component of isospin (resembling an electromagnetic effect), but its magnitude may be expected to be smaller than that of the $\bar{\hat{q}}\hat{q}$ term, being proportional to the difference of the masses, rather than their sum. This suggests that the vacuum will 'align' in such a way as to preserve isospin, but break axial isospin.

---

[2]We may set $\boldsymbol{\gamma} = \boldsymbol{\epsilon} + \boldsymbol{\eta}$, and $\boldsymbol{\delta} = \boldsymbol{\epsilon} - \boldsymbol{\eta}$.

18.2   Pion decay and the Goldberger-Treiman relation

We now discuss some of the rather surprising phenomenological implications of spontaneously broken chiral symmetry—specifically, the spontaneous breaking of the axial isospin symmetry. We start by ignoring any 'explicit' quark masses, so that the axial isospin current is conserved, $\partial_\mu \hat{j}^\mu_{i,5} = 0$. From sections 17.4 and 17.5 (suitably generalized) we know that this current has non-zero matrix elements between the vacuum and a 'Goldstone' state, which in our case is the pion. We therefore set (cf (17.94))

$$\langle 0|\hat{j}^\mu_{i,5}(x)|\pi_j, p\rangle = -\mathrm{i}p^\mu f_\pi \mathrm{e}^{-\mathrm{i}p\cdot x}\delta_{ij} \tag{18.46}$$

where f_π is a constant with dimensions of mass, and which we expect to be related to a symmetry breaking vev. This is just what we shall find in section 18.3.1. Note that (18.46) is consistent with $\partial_\mu \hat{j}^\mu_{i,5} = 0$ if $p^2 = 0$, i.e. if the pion is massless.

We treat f_π as a phenomenological parameter. Its value can be determined from the rate for the decay $\pi^+ \to \mu^+ \nu_\mu$ by the following reasoning. In chapter 20 we shall learn that the effective weak Hamiltonian density for this low energy *strangeness non-changing semileptonic transition* is

$$\hat{\mathcal{H}}_\mathrm{W}(x) \;=\; \frac{G_\mathrm{F}}{\sqrt{2}}V_\mathrm{ud}\bar{\hat{\psi}}_\mathrm{d}(x)\gamma^\mu(1-\gamma_5)\hat{\psi}_\mathrm{u}(x)$$
$$\times[\bar{\hat{\psi}}_{\nu_\mathrm{e}}(x)\gamma_\mu(1-\gamma_5)\hat{\psi}_\mathrm{e}(x) + \bar{\hat{\psi}}_{\nu_\mu}(x)\gamma_\mu(1-\gamma_5)\hat{\psi}_\mu(x)] \tag{18.47}$$

where G_F is Fermi constant and V_ud is an element of the Cabibbo-Kobayashi-Maskawa (CKM) matrix (see section 20.7.3). Thus the lowest order contribution to the S-matrix is

$$-\mathrm{i}\langle\mu^+, p_1; \nu_\mu, p_2|\int \mathrm{d}^4x \hat{\mathcal{H}}_\mathrm{W}(x)|\pi^+, p\rangle$$
$$= -\mathrm{i}\frac{G_\mathrm{F}}{\sqrt{2}}V_\mathrm{ud}\int \mathrm{d}^4x\langle\mu^+, p_1; \nu_\mu, p_2|\bar{\hat{\psi}}_{\nu_\mu}(x)\gamma_\mu(1-\gamma_5)\hat{\psi}_\mu(x)|0\rangle$$
$$\times\langle 0|\bar{\hat{\psi}}_\mathrm{d}(x)\gamma^\mu(1-\gamma_5)\hat{\psi}_\mathrm{u}(x)|\pi^+, p\rangle. \tag{18.48}$$

The leptonic matrix element gives $\bar{u}_\nu(p_2)\gamma_\mu(1-\gamma_5)v_\mu(p_1)\mathrm{e}^{\mathrm{i}(p_1+p_2)\cdot x}$. For the pionic one, we note that

$$\bar{\hat{\psi}}_\mathrm{d}(x)\gamma^\mu(1-\gamma_5)\hat{\psi}_\mathrm{u}(x) = \hat{j}^\mu_1(x) - \mathrm{i}\hat{j}^\mu_2(x) - \hat{j}^\mu_{1,5}(x) + \mathrm{i}\hat{j}^\mu_{2,5}(x) \tag{18.49}$$

from (18.20) and (18.21). Further, the currents $\hat{j}^\mu_i$ can have no matrix elements between the vacuum (which is a 0^+ state) and the π (which is 0^-), by the following argument. From Lorentz invariance such a matrix element has to be a 4-vector. But since the initial and final parities are different, it would have to be an axial 4-vector[3]. However, the only 4-vector available is the pion's momentum p^μ which is an ordinary (not an axial) 4-vector. On the other hand, precisely for this reason the axial currents $\hat{j}^\mu_{i,5}$ do have a non-zero matrix element, as in (18.46). Noting that $|\pi^+\rangle = \frac{1}{\sqrt{2}}|\pi_1 + \mathrm{i}\pi_2\rangle$, we find that

$$\langle 0|\bar{\hat{\psi}}_\mathrm{d}(x)\gamma^\mu(1-\gamma_5)\hat{\psi}_\mathrm{u}(x)|\pi^+, p\rangle \;=\; -\frac{\mathrm{i}}{\sqrt{2}}\langle 0|\hat{j}^\mu_{1,5} - \mathrm{i}\hat{j}^\mu_{2,5}|\pi_1 + \mathrm{i}\pi_2\rangle \tag{18.50}$$

$$=\; -\sqrt{2}p^\mu f_\pi \mathrm{e}^{-\mathrm{i}p\cdot x} \tag{18.51}$$

---

[3]See chapter 4 of volume 1.

FIGURE 18.2
Helicities of *massless* leptons in $\pi^+ \to \mu^+ \nu_\mu$ due to the 'V-A' interaction.

so that (18.48) becomes

$$i(2\pi)^4 \delta^4(p_1 + p_2 - p)[G_F V_{ud} \bar{u}_\nu(p_2) \gamma_\mu (1 - \gamma_5) v(p_1) p^\mu f_\pi]. \tag{18.52}$$

The quantity in brackets is, therefore, the invariant amplitude for the process, $\mathcal{M}$. Using $p = p_1 + p_2$, we may replace $\not{p}$ in (18.52) by m_μ, neglecting the neutrino mass.

Before proceeding, we comment on the physics of (18.52). The $(1 - \gamma_5)$ factor acting on a v spinor selects out the $\gamma_5 = -1$ eigenvalue which, *if the muon was massless*, would correspond to positive helicity for the μ^+ (compare the discussion in section 12.4.2). Likewise, taking the $(1 - \gamma_5)$ through the $\gamma^0 \gamma^\mu$ factor to act on $u_\nu^\dagger$, it selects the negative helicity neutrino state. Hence the configuration is as shown in figure 18.2, so that the leptons carry off a net spin angular momentum. But this is forbidden, since the pion spin is zero. Hence the amplitude vanishes for massless muons and neutrinos. Now the muon, at least, is not massless, and some 'wrong' helicity is present in its wavefunction, in an amount proportional to m_μ. This is why, as we have just remarked after (18.52), the amplitude is proportional to m_μ. The rate is therefore proportional to m_μ^2. This is a very important conclusion, because it implies that the rate to muons is $\sim (m_\mu/m_e)^2 \sim (400)^2$ times greater than the rate to electrons—a result which agrees with experiment, while grossly contradicting the naive expectation that the rate with the larger energy release should dominate. This, in fact, is one of the main indications for the 'vector-axial vector', or 'V-A', structure of (18.47), as we shall see in more detail in section 20.2.

Problem 18.4 shows that the rate computed from (18.52) is

$$\Gamma_{\pi \to \mu\nu} = \frac{G_F^2 m_\mu^2 f_\pi^2 (m_\pi^2 - m_\mu^2)^2}{4\pi m_\pi^3} |V_{ud}|^2. \tag{18.53}$$

Including radiative corrections, the value

$$f_\pi \simeq 92 \text{ MeV} \tag{18.54}$$

can be extracted.

Consider now another matrix element of $\hat{j}_{i,5}^\mu$, this time between nucleon states. Following an analysis similar to that in section 8.8 for the matrix elements of the electromagnetic current operator between nucleon states, we write

$$\langle N, p' | \hat{j}_{i,5}^\mu(0) | N, p \rangle$$
$$= \bar{u}(p') \left[\gamma^\mu \gamma_5 F_1^5(q^2) + \frac{i\sigma^{\mu\nu}}{2M} q_\nu \gamma_5 F_2^5(q^2) + q^\mu \gamma_5 F_3^5(q^2) \right] \frac{\tau_i}{2} u(p), \tag{18.55}$$

where the F_i^5's are certain form factors, M is the nucleon mass, and $q = p - p'$. The spinors in (18.55) are understood to be written in flavour and Dirac space. Since (with massless

FIGURE 18.3
One pion intermediate state contribution to F_3^5.

quarks) $\hat{j}_{i,5}^\mu$ is conserved—that is $q_\mu \hat{j}_{i,5}^\mu(0) = 0$—we find

$$
\begin{aligned}
0 &= \bar{u}(p')[\not{q}\gamma_5 F_1^5(q^2) + q^2\gamma_5 F_3^5(q^2)]\frac{\tau_i}{2}u(p) \\
&= \bar{u}(p')[(\not{p}-\not{p}')\gamma_5 F_1^5(q^2) + q^2\gamma_5 F_3^5(q^2)]\frac{\tau_i}{2}u(p) \\
&= \bar{u}(p')[-2M\gamma_5 F_1^5(q^2) + q^2\gamma_5 F_3^5(q^2)]\frac{\tau_i}{2}u(p),
\end{aligned}
\tag{18.56}
$$

using $\not{p}\gamma_5 = -\gamma_5 \not{p}$ and the Dirac equations for $u(p), \bar{u}(p')$. Hence the form factors F_1^5 and F_3^5 must satisfy

$$
2M F_1^5(q^2) = q^2 F_3^5(q^2).
\tag{18.57}
$$

Now the matrix element (18.55) enters into neutron β-decay (as does the matrix element of $\hat{j}_i^\mu(0)$). Here, $q^2 \simeq 0$ and (18.57) appears to predict, therefore, that either $M = 0$ (which is certainly not so) or $F_1^5(0) = 0$. But $F_1^5(0)$ can be measured in β decay, and is found to be approximately equal to 1.26; it is conventionally called g_A. The only possible conclusion is that F_3^5 *must contain a part proportional to* $1/q^2$. Such a contribution can only arise from the propagator of a massless particle—which, of course, is the pion. This elegant physical argument, first given by Nambu (1960), sheds a revealing new light on the phenomenon of spontaneous symmetry breaking: the existence of the massless particle coupled to the symmetry current $\hat{j}_{i,5}^\mu$ 'saves' the conservation of the current.

We calculate the pion contribution to F_3^5 as follows. The process is pictured in figure 18.3. The pion-current matrix element is given by (18.46), and the (massless) propagator is i/q^2. For the $\pi - N$ vertex, the conventional Lagrangian is

$$
ig_{\pi NN}\hat{\pi}_i \bar{\hat{N}}\gamma_5\tau_i\hat{N},
\tag{18.58}
$$

which is $SU(2)_f$-invariant and parity conserving since the pion field is a pseudoscalar, and so is $\bar{N}\gamma_5 N$. Putting these pieces together, the contribution of figure 18.3 to the current matrix element is

$$
2g_{\pi NN}\bar{u}(p')\gamma_5\frac{\tau_i}{2}u(p)\frac{i}{q^2}(-iq^\mu f_\pi),
\tag{18.59}
$$

and so

$$
F_3^5(q^2) = \frac{1}{q^2}2g_{\pi NN}f_\pi
\tag{18.60}
$$

from this contribution. Combining (18.57) with (18.60) we deduce

$$
g_A \equiv \lim_{q^2 \to 0} F_1^5(q^2) = \frac{g_{\pi NN}f_\pi}{M},
\tag{18.61}
$$

the Goldberger-Treiman (1958) relation. Taking $M = 939$ MeV, $g_A = 1.26$, and $f_\pi = 92$ MeV one obtains $g_{\pi NN} \approx 12.9$, which is only 5% below the experimental value of this effective pion-nucleon coupling constant.

We can repeat the argument leading to the G-T relation but retaining $m_\pi^2 \neq 0$. Equation (18.46) tells us that $\partial_\mu \hat{j}_{i,5}^\mu / (m_\pi^2 f_\pi)$ behaves like a properly normalized pion field, at least when operating on a near mass-shell pion state. This means that the one-nucleon matrix element of $\partial_\mu \hat{j}_{i,5}^\mu$ is (cf (18.59))

$$2g_{\pi NN} \bar{u}(p') \gamma_5 \frac{\tau_i}{2} u(p) \frac{\mathrm{i}}{q^2 - m_\pi^2} m_\pi^2 f_\pi, \tag{18.62}$$

while from (18.55) it is given by

$$\mathrm{i}\bar{u}(p')[-2M\gamma_5 F_1^5(q^2) + q^2 \gamma_5 F_3^5(q^2)]\frac{\tau_i}{2} u(p). \tag{18.63}$$

Hence

$$-2MF_1^5(q^2) + q^2 F_3^5(q^2) = \frac{2g_{\pi NN} m_\pi^2 f_\pi}{q^2 - m_\pi^2}. \tag{18.64}$$

Also, in place of (18.60) we now have

$$F_3^5(q^2) = \frac{1}{q^2 - m_\pi^2} 2g_{\pi NN} f_\pi. \tag{18.65}$$

Equations (18.64) and (18.65) are consistent for $q^2 = m_\pi^2$ if

$$F_1^5(q^2 = m_\pi^2) = g_{\pi NN} f_\pi / M. \tag{18.66}$$

$F_1^5(q^2)$ varies only slowly from $q^2 = 0$ to $q^2 = m_\pi^2$, since it contains no rapidly varying pion pole contribution, and so we recover the G-T relation again.

Amplitudes involving *two* Goldstone pions can be calculated by an extension of these techniques. However, a much more efficient method is available, through the use of *effective chiral Lagrangians*, which capture the low-energy dynamics of the Goldstone modes.

18.3 Effective Chiral Lagrangians

18.3.1 The linear and non-linear σ-models

We begin by considering the linear σ-model, which has the same Lagrangian as the one considered in section 17.6,

$$\hat{\mathcal{L}}_\Phi = (\partial_\mu \hat{\phi}^\dagger)(\partial^\mu \hat{\phi}) + \mu^2 \hat{\phi}^\dagger \hat{\phi} - \frac{\lambda}{4}(\hat{\phi}^\dagger \hat{\phi})^4, \tag{18.67}$$

but we shall interpret it differently here. The sign of the μ^2 term has been chosen to induce spontaneous symmetry breaking. In section 17.6, $\hat{\phi}$ was the SU(2) doublet

$$\hat{\phi} = \begin{pmatrix} \frac{1}{\sqrt{2}}(\hat{\phi}_1 + \mathrm{i}\hat{\phi}_2) \\ \frac{1}{\sqrt{2}}(\hat{\phi}_3 + \mathrm{i}\hat{\phi}_4) \end{pmatrix}, \tag{18.68}$$

in terms of which (18.67) becomes

$$\hat{\mathcal{L}}_\Phi = \frac{1}{2} \partial_\mu \hat{\phi}_a \partial^\mu \hat{\phi}_a + \frac{1}{2}\mu^2 \hat{\phi}_a \hat{\phi}_a - \frac{\lambda}{16}(\hat{\phi}_a \hat{\phi}_a)^2, \tag{18.69}$$

where the sum on $a = 1$ to 4 is understood. Evidently (18.69) is invariant under transformations which preserve the 'dot product' $\hat{\phi}_a \hat{\phi}_a$, namely the transformations of SO(4). This group is discussed in appendix M, section M.4.3. We note there that the algebra of the generators of SO(4) is the same as that of SU(2)$\times$ SU(2), which is the algebra of the chiral charges in (18.28)–(18.30). This suggests that we should rewrite (18.69) in such a way as to reveal its SU(2)$_L\times$SU(2)$_R$ symmetry, rather than its O(4) symmetry. Three of the four fields will then be identified with the Goldstone bosons associated with the spontaneous breaking of the 'R-L' part; they will in turn be identified with the (massless) pions.

One way to bring out the chiral symmetry of (18.69) is to write

$$\hat{\phi} = \left(\begin{array}{c} (\hat{\pi}_2 + i\hat{\pi}_1)/\sqrt{2} \\ (\hat{\sigma} - i\hat{\pi}_3)/\sqrt{2} \end{array} \right) = \frac{1}{\sqrt{2}}\hat{\Sigma} \left(\begin{array}{c} 0 \\ 1 \end{array} \right), \tag{18.70}$$

where

$$\hat{\Sigma} = \hat{\sigma} + i\boldsymbol{\tau} \cdot \hat{\boldsymbol{\pi}}. \tag{18.71}$$

Then

$$\hat{\phi}^\dagger \hat{\phi} = \frac{1}{4}\mathrm{Tr}(\hat{\Sigma}^\dagger\hat{\Sigma}), \tag{18.72}$$

and (18.69) becomes

$$\hat{\mathcal{L}}_\Sigma = \frac{1}{4}\mathrm{Tr}(\partial_\mu\hat{\Sigma}^\dagger\partial^\mu\hat{\Sigma}) + \frac{\mu^2}{4}\mathrm{Tr}(\hat{\Sigma}^\dagger\hat{\Sigma}) - \frac{\lambda}{64}\mathrm{Tr}(\hat{\Sigma}^\dagger\hat{\Sigma})^2. \tag{18.73}$$

This Lagrangian is invariant under the SU(2)$_L\times$ SU(2)$_R$ transformation

$$\hat{\Sigma} \to U_L\hat{\Sigma}U_R^\dagger \tag{18.74}$$

where

$$U_L = \exp(-i\boldsymbol{\alpha}_L \cdot \boldsymbol{\tau}/2), \quad U_R = \exp(-i\boldsymbol{\alpha}_R \cdot \boldsymbol{\tau}/2) \tag{18.75}$$

are two independent SU(2) transformations (remember that $\mathrm{Tr}AB = \mathrm{Tr}BA$). For the case of infinitesimal transformations, we find (problem 18.5)

$$\hat{\sigma} \to \hat{\sigma} - \boldsymbol{\eta} \cdot \hat{\boldsymbol{\pi}} \tag{18.76}$$

$$\hat{\boldsymbol{\pi}} \to \hat{\boldsymbol{\pi}} + \boldsymbol{\eta}\hat{\sigma} + \boldsymbol{\epsilon} \times \hat{\boldsymbol{\pi}}, \tag{18.77}$$

where

$$\boldsymbol{\eta} = (\boldsymbol{\epsilon}_R - \boldsymbol{\epsilon}_L)/2, \quad \boldsymbol{\epsilon} = (\boldsymbol{\epsilon}_R + \boldsymbol{\epsilon}_L)/2. \tag{18.78}$$

Evidently $\boldsymbol{\epsilon}_R = \boldsymbol{\eta}+\boldsymbol{\epsilon}$ and $\boldsymbol{\epsilon}_L = \boldsymbol{\epsilon}-\boldsymbol{\eta}$, which we may compare with the L and R transformation of the quark fields in (18.37), (18.38).

With the sign of μ^2 as in (18.73), the classical potential has a minimum at

$$\hat{\sigma}^2 + \hat{\boldsymbol{\pi}}^2 = 4\mu^2/\lambda \equiv v^2, \tag{18.79}$$

which we interpret as the symmetry breaking condition

$$\langle 0|\hat{\sigma}^2 + \hat{\boldsymbol{\pi}}^2|0\rangle = v^2. \tag{18.80}$$

Let us choose the particular ground state

$$\langle 0|\hat{\sigma}|0\rangle = v, \quad \langle 0|\hat{\boldsymbol{\pi}}|0\rangle = 0, \tag{18.81}$$

which is actually the same as (17.104). Referring back to (18.76) and (18.77) we see that this vacuum is invariant under 'L + R' transformations with parameters $\boldsymbol{\epsilon}$, but not under

'L - R' transformations with parameters $\boldsymbol{\eta}$. These correspond respectively to the $SU(2)_f$ flavour isospin, and $SU(2)_{f5}$ axial flavour isospin, transformations on the quark fields. So this vacuum spontaneously breaks the axial isospin symmetry. Fluctuations away from this minimum are described by fields $\hat{\boldsymbol{\pi}}$ and $\hat{s} = \hat{\sigma} - v$. Placing this shift into (18.73) we find that $\hat{\mathcal{L}}_\Sigma$ becomes $\hat{\mathcal{L}}_s$ where

$$\hat{\mathcal{L}}_s = \frac{1}{2}\partial_\mu \hat{s} \partial^\mu \hat{s} - \mu^2 \hat{s}^2 + \frac{1}{2}\partial_\mu \hat{\boldsymbol{\pi}} \cdot \partial^\mu \hat{\boldsymbol{\pi}} - \frac{\lambda}{4}v\hat{s}(\hat{s}^2 + \hat{\boldsymbol{\pi}}^2) - \frac{\lambda}{16}(\hat{s}^2 + \hat{\boldsymbol{\pi}}^2)^2, \tag{18.82}$$

discarding an irrelevant constant. As expected, the field $\hat{s}$ is massive (with mass $\sqrt{2}\mu$), while the fields $\hat{\boldsymbol{\pi}}$ are massless, and may be identified with the Goldstone modes associated with the spontaneous breaking of the axial isospin symmetry.

The Lagrangian $\hat{\mathcal{L}}_s$ incorporates the correct symmetries, and can be used to calculate $\pi-\pi$ scattering, for example (in the massless limit). But it is not the most efficient Lagrangian to use, as we can see from the following considerations. Consider the amplitude for $\pi^+ - \pi^0$ scattering, in tree approximation (Donoghue *et al.* 1992). The contributing terms in $\hat{\mathcal{L}}_s$ are

$$\hat{\mathcal{L}}_{\pi-\pi} = -\frac{\lambda}{16}(\hat{\boldsymbol{\pi}}^2)^2 - \frac{\lambda}{4}v\hat{s}\hat{\boldsymbol{\pi}}^2, \tag{18.83}$$

which we can rewrite in terms of the charged and neutral fields as

$$\hat{\mathcal{L}}_{\pi-\pi} = -\frac{\lambda}{16}(2\hat{\pi}_+^\dagger \hat{\pi}_+ + \hat{\pi}^{0\,2})^2 - \frac{\lambda}{4}v\hat{s}(2\hat{\pi}_+^\dagger \hat{\pi}_+ + \hat{\pi}^{0\,2}). \tag{18.84}$$

Then the terms responsible for $\pi^+ - \pi^0$ scattering at tree level are

$$-\frac{\lambda}{4}\hat{\pi}_+^\dagger \hat{\pi}_+ \hat{\pi}^{0\,2} - \frac{\lambda}{2}v\hat{s}(\hat{\pi}_+^\dagger \hat{\pi}_+ + \frac{1}{2}\hat{\pi}^{0\,2}). \tag{18.85}$$

The first of these represents a four-pion contact interaction with amplitude

$$-i\lambda/2, \tag{18.86}$$

while the second contributes an s-exchange graph in the t-channel with amplitude

$$(-i\lambda v/2)^2 \frac{i}{q^2 - 2\mu^2}, \tag{18.87}$$

where q is the four-momentum transfer $q = p'_+ - p_+ = p_0 - p'_0$. The sum of these is

$$-i\lambda/2 \frac{q^2}{q^2 - 2\mu^2}, \tag{18.88}$$

which reduces to iq^2/v^2 for $q \approx 0$. Thus, despite the apparent constant 4-boson piece (18.86), the total amplitude in fact *vanishes* as $q^2 \to 0$, due to a cancellation.

This cancellation is not an accident. It is generally true that Goldstone fields enter only via their derivatives, which bring factors of momenta into the amplitudes. We drew attention to this following equation (17.85), and the same is true of the $\hat{\boldsymbol{\theta}}$ fields in (23.29). This suggests that it is both possible, and more efficient, to recast $\hat{\mathcal{L}}_s$ into a form in which only the derivatives of the Goldstone fields enter. Equation (23.29) indicates how to do this. We define new pion fields (but call them the same) by

$$\hat{\Sigma} = (v + \hat{S})\hat{U}, \quad \hat{U} = \exp(i\boldsymbol{\tau} \cdot \hat{\boldsymbol{\pi}}/v), \tag{18.89}$$

where $\hat{S}$ is invariant under $SU(2)_L \times SU(2)_R$, and where $\hat{U}$ transforms by

$$\hat{U} \to U_L \hat{U} U_R^\dagger. \tag{18.90}$$

Now $\hat{\Sigma}^\dagger \hat{\Sigma} = (v + \hat{S})^2$, and the Goldstone modes have been transformed away from the potential terms in $\hat{\mathcal{L}}_\Sigma$, reappearing in the derivative terms instead. We write the transformed Lagrangian as $\hat{\mathcal{L}}_S$ where

$$\hat{\mathcal{L}}_S = \frac{1}{2}\partial_\mu \hat{S} \partial^\mu \hat{S} - \mu^2 \hat{S}^2 + \frac{1}{4}(v + \hat{S})^2 \text{Tr}(\partial_\mu \hat{U} \partial^\mu \hat{U}^\dagger) - \frac{1}{4}\lambda v \hat{S}^3 - \frac{\lambda}{16}\hat{S}^4, \tag{18.91}$$

where we have used

$$\hat{U}^\dagger \partial^\mu \hat{u} + \partial^\mu \hat{U}^\dagger \hat{U} = 0, \tag{18.92}$$

which follows from the unitary condition $\hat{U}^\dagger \hat{U} = 1$.

When $\partial_\mu \hat{U}$ is expanded in powers of $\hat{\boldsymbol{\pi}}$, we recover a kinetic energy piece

$$\frac{1}{2}\partial_\mu \hat{\boldsymbol{\pi}} \cdot \partial^\mu \hat{\boldsymbol{\pi}}, \tag{18.93}$$

and all other terms involve derivatives of $\hat{\boldsymbol{\pi}}$. In particular, the term with the lowest number of derivatives which contributes to the $\pi - \pi$ scattering amplitude is

$$\frac{1}{6v^2}[(\hat{\boldsymbol{\pi}} \cdot \partial_\mu \hat{\boldsymbol{\pi}})(\hat{\boldsymbol{\pi}} \cdot \partial^\mu \hat{\boldsymbol{\pi}}) - \hat{\boldsymbol{\pi}}^2 \partial_\mu \hat{\boldsymbol{\pi}} \cdot \partial^\mu \hat{\boldsymbol{\pi}}], \tag{18.94}$$

since the $\hat{S} - \hat{\boldsymbol{\pi}} - \hat{\boldsymbol{\pi}}$ vertex already has two derivatives. The reader may check that the amplitude for $\pi^+\pi^0 \to \pi^+\pi^0$ calculated from (18.94) is iq^2/v^2, exactly as before, but this time without having to go through the cancellation argument.

The fields in $\hat{\Sigma}$ on the one hand, and in $\hat{S}$ and $\hat{U}$ on the other, are related non-linearly, but a physical amplitude calculated with either representation has turned out to be the same, in this simple case. It is in fact generally true that such non-linear field redefinitions lead to the same physics (Haag 1958, Coleman, Wess and Zumino 1969, Callan, Coleman, Wess and Zumino 1969). It is clearly advantageous to work with $\hat{\mathcal{L}}_S$, which builds in the desired derivatives of the Goldstone modes.

Indeed, we can simplify matters even further. Since $\hat{S}$ is invariant under $SU(2)_L \times SU(2)_R$, the full symmetry of the Lagrangian is maintained with only the field $\hat{U}$, transforming by (18.90), and we may as well discard $\hat{S}$ altogether. The dynamics of the Goldstone sector is then described by the *non-linear σ-model*, with Lagrangian

$$\hat{\mathcal{L}}_2 = \frac{v^2}{4}\text{Tr}(\partial_\mu \hat{U} \partial^\mu \hat{U}^\dagger). \tag{18.95}$$

This is the most general Lagrangian that involves the Goldstone fields, exhibits the desired symmetry, and contains only two derivatives.

Since $\hat{\mathcal{L}}_2$ is invariant under the $SU(2)_L \times SU(2)_R$ transformations (18.75), we can calculate the associated Noether currents (problem 18.6), obtaining

$$\hat{j}_{i,L}^\mu(\hat{U}) = \frac{-iv^2}{8}\text{Tr}[\tau_i \hat{U}\partial^\mu \hat{U}^\dagger - \tau_i(\partial^\mu \hat{U})\hat{U}^\dagger] = \frac{-iv^2}{4}\text{Tr}(\tau_i \hat{U}\partial^\mu \hat{U}^\dagger), \tag{18.96}$$

and

$$\hat{j}_{i,R}^\mu(\hat{U}) = \frac{iv^2}{8}\text{Tr}[\tau_i(\partial^\mu \hat{U}^\dagger)\hat{U} - \tau_i \hat{U}^\dagger \partial^\mu \hat{U}] = \frac{-iv^2}{4}\text{Tr}(\tau_i \hat{U}^\dagger \partial^\mu \hat{U}). \tag{18.97}$$

The axial 'R-L' current is then

$$\hat{j}_{i5}^\mu(\hat{U}) = \frac{iv^2}{4}\text{Tr}[\tau_i(\hat{U}\partial^\mu \hat{U}^\dagger - \hat{U}^\dagger \partial^\mu \hat{U})], \tag{18.98}$$

and the vector 'R+L' current is

$$\hat{j}_i^\mu(\hat{U}) = \frac{iv^2}{4}\text{Tr}[\tau_i(\hat{U}\partial^\mu\hat{U}^\dagger + \hat{U}^\dagger\partial^\mu\hat{U})]. \tag{18.99}$$

Expanding (18.98) in powers of the pion field, we find

$$\hat{j}_{i,5}^\mu(\hat{U}) = v\partial^\mu\hat{\pi}_i + \dots, \tag{18.100}$$

which we may compare with (17.93). Just as in equation (17.94), (18.100) implies that this axial current has a matrix element between the vacuum and the one-Goldstone state:

$$\langle 0|\hat{j}_{i,5}^\mu(\hat{U})|\pi_j,p\rangle = -ip^\mu v e^{-ip\cdot x}\delta_{ij}. \tag{18.101}$$

Now comes the pay-off: this is the *same* symmetry current which enters into weak interactions, for which we already defined the vacuum-to-one-particle matrix element in terms of the pion decay constant f_π, via equation (18.46). Comparing (18.101) with (18.46) we identify

$$v = f_\pi. \tag{18.102}$$

Thus, finally, the dynamics of our massless pions, to lowest order in an expansion in powers of momenta, is given by the Lagrangian

$$\hat{\mathcal{L}}_2 = \frac{1}{4}f_\pi^2\text{Tr}(\partial_\mu\hat{U}\partial^\mu\hat{U}^\dagger). \tag{18.103}$$

It is quite remarkable that the low-energy dynamics of the (massless) Goldstone modes is completely determined in terms of one constant, measurable in π decay.

The Lagrangian of (18.103) is an example of an *effective Lagrangian*. By this is meant, broadly, any Lagrangian which involves the presumed relevant degrees of freedom (here the Goldstone modes), and respects desired symmetries of the theory. Evidently it is implied that there is some 'underlying theory', couched in terms of different degrees of freedom (here quarks and gluons), from which the symmetries have been abstracted. It is important to realize that an effective Lagrangian may or may not be renormalizable. Whereas our starting Lagrangian $\hat{\mathcal{L}}_s$ is renormalizable, $\hat{\mathcal{L}}_2$ is not. Clearly the latter contains terms with arbitrarily many pion fields, which are operators of arbitrarily high dimension, compensated by negative powers of f_π^2. As it stands, $\hat{\mathcal{L}}_2$ can only be used at tree level—as, for example, in the calculation of $\pi - \pi$ scattering using the interaction(18.94), for which the amplitude has an energy dependence of the form E^2/f_π^2, where E is the order of magnitude of the particles' energy or momentum. This interaction has mass dimension 6, and its coupling $1/f_\pi^2$ has dimension $(\text{mass})^{-2}$, like the 4-fermion interaction considered in section 11.8. It is therefore not renormalizable. However, the argument of section 11.8 suggests that a loop-by-loop renormalization programme is possible, and this was shown to be the case by Weinberg (1979a). Each loop built from the interaction $(\hat{18.94})$ will carry an extra two powers of energy, to compensate the $1/f_\pi^2$ in the coupling. Thus f_π (or perhaps this multiplied by factors like 4 and π, if we are lucky) provides the energy scale characteristic of a non-renormalizable theory: as we go up in energy, we need more loops. But, at each loop order new divergences appear, which require additional counter terms for renormalization. Thus at any given order in E^2/f_π^2, we must ensure that our effective Lagrangian contains all the appropriate counter terms which are allowed by the symmetry. For example, at one-loop order for $\hat{\mathcal{L}}_2$, we need to include the 4-derivative terms

$$\hat{\mathcal{L}}_4 = c_1\text{Tr}(\partial_\mu\hat{U}\partial^\mu\hat{U}^\dagger\partial_\nu\hat{U}\partial^\nu\hat{U}^\dagger) + c_2\text{Tr}(\partial_\mu\hat{U}\partial_\nu\hat{U}^\dagger\partial^\mu\hat{U}\partial^\nu\hat{U}^\dagger). \tag{18.104}$$

To perform a one-loop calculation, one uses $\hat{\mathcal{L}}_2$ at tree-level and in one-loop diagrams, and $\hat{\mathcal{L}}_4$ at tree-level only.

Real pions, however, are not massless, nor are real quarks. We need to extend our effective Lagrangian to include *explicit* chiral symmetry breaking mass terms.

18.3.2 Inclusion of explicit symmetry breaking: masses for pions and quarks

Consider the term

$$\hat{\mathcal{L}}_{m_\pi} = \frac{m_\pi^2}{4} \text{Tr}(\hat{U} + \hat{U}^\dagger). \tag{18.105}$$

This is invariant only under the restricted set of transformations with $\boldsymbol{\alpha}_L = \boldsymbol{\alpha}_R$, that is transformations such that $U_R = U_L$, for then $\text{Tr}\hat{U} \to \text{Tr}(U_R\hat{U}U_R^\dagger) = \text{Tr}\hat{U}$. Such transformations form the SU(2) flavour isospin group. The term (18.105) breaks the axial isospin group explicitly, which would correspond to transformations with $\boldsymbol{\alpha}_L = -\boldsymbol{\alpha}_R$, or equivalently $U_L = U_R^\dagger$, under which $\hat{U} \to U_L\hat{U}U_L$. Expanding (18.105) to second order in the pion fields, we find the term

$$\hat{\mathcal{L}}_{\text{quad},m_\pi} = -\frac{1}{2}m_\pi^2\boldsymbol{\pi}^2 \tag{18.106}$$

which, together with (18.93), shows that the pion field now has mass m_π. Higher-order terms can be added, m_π^2 counting as equivalent to two derivatives. The low energy expansion is now an expansion in both the energy E and the pion mass m_π. This is called *chiral perturbation theory*, or ChPT for short.

For example, to calculate $\pi - \pi$ scattering to order E^2, we use $\hat{\mathcal{L}}_2 + \hat{\mathcal{L}}_{m_\pi}$ at tree-level, expanded up to fourth power in the pion fields. The result is to change the amplitude for $\pi^+\pi^0 \to \pi^+\pi^0$ from $\text{i}(p'_+ - p_+)^2/f_\pi^2$ to $\text{i}[(p'_+ - p_+)^2 - m_\pi^2]/f_\pi^2$. By considering the general $\pi - \pi$ amplitude, predictions for the scattering lengths can be made for low energy observables, for example the s-wave scattering lengths a_0 and a_2 in the isospin 0 and 2 channels. The results (first calculated by Weinberg 1966 using current algebra techniques) are

$$a_0 = \frac{7m_\pi^2}{32\pi f_\pi^2} = 0.16\,m_\pi^{-1}, \quad a_2 = -\frac{m_\pi^2}{16\pi f_\pi^2} = -0.045\,m_\pi^{-1} \tag{18.107}$$

The experimental values are $a_0 = 0.26 \pm 0.05\,m_\pi^{-1}$ and $a_2 = -0.028 \pm 0.012\,m_\pi^{-1}$, as given by Donoghue *et al.*(1992). The next order in ChPT improves upon these results.

A systematic exposition of ChPT at the one-loop level was given by Gasser and Leutwyler (1984). Bijnens *et al.* (1996) carried the $\pi - \pi$ calculation to two-loop order; see also Colangelo *et al.* (2001).

It is clear that there must be some relation between the masses of the u and d quarks (in the SU(2) flavour case) and the pion mass, since the latter must vanish in the limit $m_u = m_d = 0$. To see this connection, we consider the quark mass term in the 2-flavour QCD Lagrangian, which is

$$-\bar{\hat{q}}\,\boldsymbol{m}_2\hat{q}, \quad \boldsymbol{m}_2 = \begin{pmatrix} m_u & 0 \\ 0 & m_d \end{pmatrix}. \tag{18.108}$$

Let us now redefine the quark fields (compare (23.29) and (18.17)) by

$$\hat{q} = \exp[-\text{i}\boldsymbol{\tau} \cdot \hat{\boldsymbol{\pi}}\gamma_5/(2f_\pi)]\,\hat{f}. \tag{18.109}$$

This transformation is a perfectly good parametrization of the Goldstone fields associated with the axial symmetry (18.17), and effectively removes them from the new fermion fields $\hat{f}$. The quark mass term now becomes

$$-\bar{\hat{f}}\exp[-\text{i}\boldsymbol{\tau} \cdot \hat{\boldsymbol{\pi}}\gamma_5/(2f_\pi)]\,\boldsymbol{m}_2\exp[-\text{i}\boldsymbol{\tau} \cdot \hat{\boldsymbol{\pi}}/(2f_\pi)]\hat{f}. \tag{18.110}$$

We now make the assumption that the axial SU(2) is spontaneously broken in QCD, by imposing a non-zero vev on the symmetry-breaking operator $\bar{\hat{f}}\hat{f}$:

$$\langle 0|\bar{\hat{f}}_i\hat{f}_j|0\rangle = -f_\pi^2 B\delta_{ij} \quad (i,j = 1,2). \tag{18.111}$$

Expanding (18.110) up to second order in the pion fields, retaining just the expectation value of the fermion bilinear[4], we find a mass term

$$-\frac{1}{2}B(m_{\mathrm{u}}+m_{\mathrm{d}})\hat{\boldsymbol{\pi}}^2,$$ (18.112)

from which the relation (Gasser and Leutwyler 1982)

$$m_\pi^2 = -\frac{(m_{\mathrm{u}}+m_{\mathrm{d}})}{f_\pi^2}\langle 0|\bar{\hat{f}}\hat{f}|0\rangle$$ (18.113)

follows, where $\bar{\hat{f}}\hat{f}$ represents either $\bar{\hat{f}}_{\mathrm{u}}\hat{f}_{\mathrm{u}}$ or $\bar{\hat{f}}_{\mathrm{d}}\hat{f}_{\mathrm{d}}$. From (18.113) we can see that the *square* of the pion mass is proportional to the average u-d quark mass (provided of course that B does not accidentally vanish), and goes to zero as they do. $\langle 0|\bar{\hat{f}}\hat{f}|0\rangle$ is the 'chiral condensate' (cf figure 18.1), with mass dimension 3.

Lattice QCD (see chapter 16) can be used to test equation (18.113), since simulations can be done for a range of quark masses, and the relation between m_π^2 and $m_{\mathrm{u,d}}$ can be checked. Conversely, ChPT was able to assist lattice QCD calculations by guiding the extrapolation of the calculated results to quark mass values lighter than could be simulated. For example, Noaki *et al.* (2008) have reported the results of such a calculation, using 2 light dynamical quark flavours, in the overlap fermion formalism (Neuberger 1998a, 1998b), which preserves chiral symmetry at finite lattice spacing. Their pion masses ranged from 290 MeV to 750 MeV, and they compared their results with the predictions of ChPT at one-loop (Gasser and Leutwyler 1984) and two-loop (Colangelo *et al.* 2001). They found good fits to the ChPT formulae, and extracted quark masses (in the $\overline{\mathrm{MS}}$ scheme at the scale 2 GeV) of about 4.5 MeV; they also found $|\langle 0|\bar{\hat{f}}\hat{f}|0\rangle| \sim (235\,\mathrm{GeV})^3$, in the $\overline{\mathrm{MS}}$ scheme at 2 GeV scale. Studies by this and other groups are continuing, with 3 light flavours, lighter pion masses, and other lattice fermion formalisms, as described in section 16.5.3.

18.3.3   Extension to $\mathbf{SU(3)_{fL}\times SU(3)_{fR}}$

To the extent that the strange quark is also 'light' on hadronic scales, the QCD Lagrangian has the larger symmetry of $\mathrm{SU(3)_{fL}\times SU(2)_{fR}}$, which breaks spontaneously so as to preserve the flavour symmetry $\mathrm{SU(3)_f}$, and produce an SU(3) octet of pseudoscalar Goldstone bosons: $\pi^\pm, \pi^0, \mathrm{K}^\pm, \mathrm{K}^0, \bar{\mathrm{K}}^0$ and η_8 (see figure 12.4). The effective Lagrangian approach to the dynamics of the Goldstone fields can be easily extended to chiral SU(3). One simply replaces $\hat{U} = \exp(\mathrm{i}\boldsymbol{\tau}\cdot\hat{\boldsymbol{\pi}}/f_\pi)$ by $\hat{V} = \exp(\mathrm{i}\boldsymbol{\lambda}\cdot\hat{\boldsymbol{\phi}}/f_\pi)$ where

$$\frac{1}{\sqrt{2}}\sum_{a=1}^{8}\lambda_a\hat{\phi}_a = \begin{pmatrix} \frac{1}{\sqrt{2}}\hat{\pi}^0 + \frac{1}{\sqrt{6}}\hat{\eta}_8 & \hat{\pi}^+ & \hat{K}^+ \\ \hat{\pi}^- & -\frac{1}{\sqrt{2}}\hat{\pi}^0 + \frac{1}{\sqrt{6}}\hat{\eta}_8 & \hat{K}^0 \\ \hat{K}^- & \bar{K}^0 & -\frac{2}{\sqrt{6}}\hat{\eta}_8 \end{pmatrix}.$$ (18.114)

One easily verifies that the kinetic terms in

$$\hat{\mathcal{L}}_2 = \frac{f_\pi^2}{4}\mathrm{Tr}\partial_\mu\hat{V}\partial^\mu\hat{V}$$ (18.115)

have the correct normalization, using $\mathrm{Tr}\lambda_a\lambda_b = 2\delta_{ab}$. The 3-flavour quark mass term is now

$$-\bar{\hat{f}}\exp[-\mathrm{i}\boldsymbol{\lambda}\cdot\hat{\boldsymbol{\phi}}\gamma_5/(2f_\pi)]\,\boldsymbol{m}_3\exp[-\mathrm{i}\boldsymbol{\lambda}\cdot\hat{\boldsymbol{\phi}}\gamma_5/(2f_\pi)]\,\hat{f}$$ (18.116)

---

[4]A formal justification of this step is provided by Weinberg (1996), section 19.6.

where

$$\boldsymbol{m}_3 = \begin{pmatrix} m_\mathrm{u} & 0 & 0 \\ 0 & m_\mathrm{d} & 0 \\ 0 & 0 & m_\mathrm{s} \end{pmatrix}. \tag{18.117}$$

The axial SU(3) symmetry breaking vev is

$$\langle 0| \bar{\hat{f}}_i \hat{f}_j |0\rangle = -f_\pi^2 B \delta_{ij} \quad (i,j=1,2,3) \tag{18.118}$$

and the meson mass term is

$$-\frac{B}{2}\mathrm{Tr}\left\{(\boldsymbol{\lambda}\cdot\hat{\boldsymbol{\phi}})^2 \boldsymbol{m}_3\right\}. \tag{18.119}$$

This yields (problem 18.7)

$$m_{\pi^+}^2 = m_{\pi^0}^2 = B(m_\mathrm{u} + m_\mathrm{d}), \tag{18.120}$$
$$m_{K^+}^2 = B(m_\mathrm{u} + m_\mathrm{s}), \tag{18.121}$$
$$m_{K^0}^2 = B(m_\mathrm{d} + m_\mathrm{s},) \tag{18.122}$$
$$m_{\eta_8}^2 = \frac{1}{3}B(m_\mathrm{u} + m_\mathrm{d} + 4m_\mathrm{s}), \tag{18.123}$$

and there is also a term which mixes π^0 and η_8:

$$m_{\pi\,\eta}^2 = \frac{B}{\sqrt{3}}(m_\mathrm{u} - m_\mathrm{d}). \tag{18.124}$$

It is interesting that the charged and neutral pions have the same mass, even though we have made no assumption about the ratio of m_u to m_d. The observed pion mass differences arise from electromagnetism.

If we ignore for the moment electromagnetic mass differences, we can deduce from (18.120)–(18.122) the relation

$$\frac{m_\pi^2}{2m_K^2 - m_\pi^2} = \frac{m_\mathrm{u} + m_\mathrm{d}}{m_\mathrm{s}}. \tag{18.125}$$

The left-hand side is approximately equal to 0.04, so we learn that the non-strange quarks are about $1/25$ times as heavy as the strange quark. We also obtain

$$m_{\eta_8}^2 = \frac{1}{3}(4m_K^2 - m_\pi^2), \tag{18.126}$$

which is the Gell-Mann—Okubo formula for the (squared) masses of the pseudoscalar meson octet (Gell-Mann 1961, Okubo 1962). Using average values for the K and π masses, the relation (18.126) predicts $m_{\eta_8}^2 = 566$ MeV, quite close to the η (548 MeV).

Further progress requires the inclusion of electromagnetic effects, since m_u and m_d are themselves comparable to the electromagnetic mass differences. Including these effects, Weinberg (1996) estimates

$$\frac{m_\mathrm{d}}{m_\mathrm{s}} \approx 0.050, \quad \frac{m_\mathrm{u}}{m_\mathrm{s}} \approx 0.027; \tag{18.127}$$

see also Leutwyler (1996). Note that the d quark is almost twice as heavy as the u quark: according to QCD, the origin of SU(2) isospin symmetry is not that $m_\mathrm{u} \approx m_\mathrm{d}$, but that both are very small compared with, say, $\Lambda_{\overline{\mathrm{MS}}}$.

All the results we have given are subject to correction by the inclusion of higher order effects in the ChPT expansion. In the case of chiral SU(3), the fourth order Lagrangian $\hat{\mathcal{L}}_4$ contains 8 terms (Gasser and Leutwyler 1984, 1985). Donoghue *et al.* (1992) give a clear exposition of ChPT to one-loop order.

18.4   Chiral anomalies

In all our discussions of symmetries so far—unbroken, approximate, and spontaneously broken—there is one result on which we have relied, and never queried. We refer to Noether's theorem, as discussed in section 12.3.1. This states that for every continuous symmetry of a Lagrangian, there is a corresponding conserved current. We demonstrated this result in some special cases, but we have now to point out that while it is undoubtedly valid at the level of the *classical* Lagrangian and field equations, we did not investigate whether quantum corrections might violate the classical conservation law. This can, in fact, happen and when it does the afflicted current (or its divergence) is said to be 'anomalous', or to contain an 'anomaly'. General analysis shows that anomalies occur in renormalizable theories of fermions coupled to both vector and axial vector currents. We may therefore expect to find anomalies among the vector and axial vector flavour currents which we have been discussing.

One way of understanding how anomalies arise is through consideration of the renormalization process, which is in general necessary once we get beyond the classical ('tree level') approximation. As we saw in volume 1, this will invariably entail some *regularization* of divergent integrals. But the specific example of the $O(e^2)$ photon self-energy studied in section 11.3 showed that a simple cut-off form of regularization already violated the current conservation (or gauge invariance) condition (11.21). In that case, it was possible to find alternative regularizations which respected electromagnetic current conservation, and were satisfactory. Anomalies arise when *both* axial and vector symmetry currents are present, since it is not possible to find a regularization scheme which preserves both vector and axial vector current conservation (Adler 1970, Jackiw 1972, Adler and Bardeen 1969).

We shall not attempt an extended discussion of this technical subject. But we do want to alert the reader to the existence of these anomalies; to indicate how they arise in one simple model; and to explain why, in some cases, they are to be welcomed, while in others they must be eliminated.

We consider the classic case of $\pi^0 \to \gamma\gamma$, in the context of spontaneously broken chiral flavour symmetry, with massless quarks and pions. The axial isospin current $\hat{j}^\mu_{i,5}(x)$ should then be conserved, but we shall see that this implies that the amplitude for $\pi^0 \to \gamma\gamma$ must vanish, as first pointed out by Veltman (1967) and Sutherland (1967). We begin by writing the matrix element of $\hat{j}^\mu_{3,5}(x)$ between the vacuum and a 2γ state, in momentum space, as

$$\int \mathrm{d}^4 x \mathrm{e}^{-\mathrm{i}q\cdot x} \langle \gamma, k_1, \epsilon_1; \gamma, k_2, \epsilon_2 | \hat{j}^\mu_{3,5}(x) | 0 \rangle$$
$$= (2\pi)^4 \delta^4(k_1 + k_2 - q)\epsilon^*_{1\nu}(k_1)\epsilon^*_{2\lambda}(k_2)\mathcal{M}^{\mu\nu\lambda}(k_1, k_2). \qquad (18.128)$$

As in figure 18.3, one contribution to $\mathcal{M}^{\mu\nu\lambda}$ has the form $(\text{constant}/q^2)$ due to the massless π^0 propagator, shown in figure 18.4. This is, once again, because when chiral symmetry is spontaneously broken, the axial current connects the pion state to the vacuum, as described by the matrix element (18.46). The contribution of the process shown in figure 18.4 to $\mathcal{M}^{\mu\nu\lambda}$ is then

$$\mathrm{i}q^\mu f_\pi \frac{\mathrm{i}}{q^2} \mathrm{i}A\epsilon^{\nu\lambda\alpha\beta} k_{1\alpha} k_{2\beta} \qquad (18.129)$$

where we have parametrized the $\pi^0 \to \gamma\gamma$ amplitude as $A\epsilon^{\nu\lambda\alpha\beta}\epsilon^*_{1\nu}(k_1)\epsilon^*_{2\lambda}(k_2)k_{1\alpha}k_{2\beta}$. Note that this automatically incorporates electromagnetic gauge invariance (the amplitude vanishes when the polarization vector of either photon is replaced by its 4-momentum, due to the antisymmetry of the ϵ symbol), and it is symmetrical under interchange of the photon

FIGURE 18.4
The amplitude considered in (18.128), and the one pion intermediate state contribution to it.

labels. Now consider replacing $\hat{j}^\mu_{3,5}(x)$ in (18.128) by $\partial_\mu \hat{j}^\mu_{3,5}(x)$, which should be zero. A partial integration then shows that this implies that

$$q_\mu \mathcal{M}^{\mu\nu\lambda} = 0 \tag{18.130}$$

which with (18.129) implies that $A = 0$, and hence that $\pi^0 \to \gamma\gamma$ is forbidden. It is important to realize that all other contributions to $\mathcal{M}^{\mu\nu\lambda}$, apart from the π^0 one shown in figure 18.4, will *not* have the $1/q^2$ factor in (18.129), and will therefore give a vanishing contribution to $q_\mu \mathcal{M}^{\mu\nu\lambda}$ at $q^2 = 0$ which is the on-shell point for the (massless) pion.

It is of course true that $m^2_\pi \neq 0$. But estimates (Adler 1969) of the consequent corrections suggest that the predicted rate of $\pi^0 \to \gamma\gamma$ for real π^0's is far too small. Consequently, there is a problem for the hypothesis of spontaneously broken (approximate) chiral symmetry.

In such a situation it is helpful to consider a detailed calculation performed within a specific model. This is supplied by Itzykson and Zuber (1980), section 11.5.2; in essentials it is the same as the one originally considered by Steinberger (1949) in the first calculation of the $\pi^0 \to \gamma\gamma$ rate, and subsequently by Bell and Jackiw (1969) and by Adler (1969). It employs (scalar) σ and (pseudoscalar) π^0 meson fields, augmented by a fermion of mass m and charge $+e$, representing the proton. To order α, there are two graphs to consider, shown in figure 18.5(a) and (b). It turns out that the fermion loop integral is actually convergent. In the limit $q^2 \to 0$ the result is

$$A = \frac{e^2}{4\pi^2 f_\pi} \tag{18.131}$$

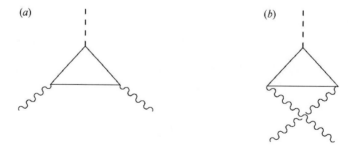

FIGURE 18.5
The two $O(\alpha)$ graphs contributing to $\pi^0 \to \gamma\gamma$ decay.

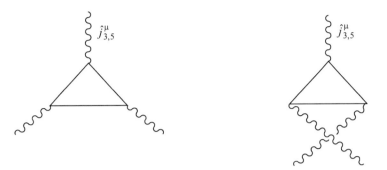

FIGURE 18.6
$O(\alpha)$ contributions to the matrix element in (18.128).

where A is the $\pi^0 \to \gamma\gamma$ amplitude introduced above. Problem 18.8 evaluates the $\pi^0 \to \gamma\gamma$ rate using (18.131), to give

$$\Gamma(\pi^0 \to 2\gamma) = \frac{\alpha^2}{64\pi^3} \frac{m_\pi^3}{f_\pi^2}. \tag{18.132}$$

(18.132) is in very good agreement with experiment.

In principle, various possibilities now exist. But a careful analysis of the 'triangle' graph contributions to the matrix element $\mathcal{M}^{\mu\nu\lambda}$ of (18.128), shown in figure 18.6, reveals that the fault lies in assuming that a regularization exists such that for these amplitudes the conservation equation $q_\mu \langle \gamma\gamma | \hat{j}_{3,5}^\mu(0)|0\rangle = 0$ can be maintained, at the same time as electromagnetic gauge variance. In fact, no such regularization can be found. When the amplitudes of figure 18.6 are calculated using an (electromagnetic) gauge invariant procedure, one finds a non-zero result for $q_\mu \langle \gamma\gamma | \hat{j}_{3,5}^\mu(0)|0\rangle$ (again the details are given in Itzykson and Zuber (1980)). This implies that $\partial_\mu \hat{j}_{3,5}^\mu(x)$ is not zero after all, the calculation producing the specific value

$$\partial_\mu \hat{j}_{3,5}^\mu(x) = -\frac{e^2}{32\pi^2} \epsilon^{\alpha\nu\beta\lambda} \hat{F}_{\alpha\nu} \hat{F}_{\beta\lambda} \tag{18.133}$$

where the F's are the usual electromagnetic field strengths.

Equation (18.133) means that (18.130) is no longer valid, so that A need no longer vanish: indeed, (18.133) predicts a definite value for A, so we need to see if it is consistent with (18.131). Taking the vacuum $\to 2\gamma$ matrix element of (18.133) produces (problem 18.9)

$$iq_\mu \mathcal{M}^{\mu\nu\lambda} = \frac{e^2}{4\pi^2} \epsilon^{\alpha\nu\beta\lambda} k_{1\alpha} k_{2\beta} \tag{18.134}$$

which is indeed consistent with (18.128) and (18.131), after suitably interchanging the labels on the ϵ symbol.

Equation (18.133) is therefore a typical example of 'an anomaly'—the violation, at the quantum level, of a symmetry of the classical Lagrangian. It might be thought that the result (18.133) is only valid to order α (though the $O(\alpha^2)$ correction would presumably be very small). But Adler and Bardeen (1969) showed that such 'triangle' loops give the *only* anomalous contributions to the $\hat{j}_{i,5}^\mu - \gamma - \gamma$ vertex, so that (18.133) is true to all orders in α.

The triangles considered above actually used a fermion with integer charge (the proton). We clearly should use quarks, which carry fractional charge. In this case, the previous numerical value for A is multiplied by the factor $\tau_3 Q^2$ for each contributing quark. For the u and d quarks of chiral SU(2)×SU(2), this gives 1/3. Consequently agreement with experiment is lost unless there exist three replicas of each quark, identical in their electromagnetic and SU(2)×SU(2) properties. Colour supplies just this degeneracy, and thus the $\pi^0 \to \gamma\gamma$ rate is important evidence for such a degree of freedom, as we noted in chapter 14.

In the foregoing discussion, the axial isospin current was associated with a global symmetry; only the electromagnetic currents (in the case of $\pi^0 \to \gamma\gamma$) were associated with a local (gauged) symmetry, and they remained conserved (anomaly free). If, however, we have an anomaly in a current associated with a local symmetry, we will have a serious problem. The whole rather elaborate construction of a quantum gauge field theory relies on current conservation equations such as (11.21) or (13.130) to eliminate unwanted gauge degrees of freedom, and ensure unitarity of the S-matrix. So anomalies in currents coupled to gauge fields cannot be tolerated. As we shall see in chapter 20, and is already evident from (18.48), axial currents are indeed present in weak interactions and they are coupled to the $W^\pm, Z^0$ gauge fields. Hence, if this theory is to be satisfactory at the quantum level, all anomalies must somehow cancel away. That this is possible rests essentially on the observation that the anomaly (18.133) is independent of the mass of the circulating fermion. Thus cancellations are in principle possible between quark and lepton 'triangles' in the weak interaction case. Bouchiat *et al.* (1972) were the first to point out that, for each generation of quarks and leptons, the anomalies will cancel between quarks and leptons if the fractionally charged quarks come in three colours. The condition that anomalies cancel in the gauged currents of the SM is the remarkably simple one (Ryder 1996, p384):

$$N_c(Q_u + Q_d) + Q_e = 0 \qquad (18.135)$$

where N_c is the number of colours and Q_u, Q_d, and Q_e are the charges (in units of e) of the 'u', 'd', and 'e' type fields in each generation. Clearly (18.135) is true for each generation of the SM; the condition indicates a remarkable connection, at some deep level, between the facts that quarks come in three colours and have charges which are 1/3 fractions. The SM provides no explanation for this connection. Anomaly cancellation is a powerful constraint on possible theories ('t Hooft 1980, Weinberg 1996 section 22.4).

Problems

18.1 Verify (18.24) – (18.26).

18.2 Verify (18.28) – (18.30).

18.3 Show that $\mathcal{L}_q$ of (18.12) can be written as (18.39).

18.4 Show that the rate for $\pi^+ \to \mu^+ \nu_\mu$, calculated from the lowest order matrix element (18.52), is given by (18.53).

18.5 Verify the transformation equations (18.76) and (18.77).

18.6

(a) Consider a Lagrangian $\hat{\mathcal{L}}(\hat{\phi}_r, \partial_\mu\hat{\phi}_r)$ where the $\hat{\phi}_r$ could be either bosonic or fermionic fields. Let the fields transform by an infinitesimal local (x-dependent) transformation

$$\hat{\phi}_r \to \hat{\phi}_r - i\epsilon_\alpha(x)T^\alpha_{rs}\hat{\phi}_s \quad \text{(sum on } s\text{)}.$$

Show that the change in $\hat{\mathcal{L}}$ may be written as

$$\delta\hat{\mathcal{L}} = \hat{j}^{\alpha\mu}(x)\partial_\mu\epsilon_\alpha(x) + \epsilon_\alpha(x)\partial_\mu\hat{j}^{\alpha\mu}(x)$$

where

$$\hat{j}^{\alpha\mu}(x) = -iT^\alpha_{rs}\hat{\phi}_s\frac{\partial\hat{\mathcal{L}}}{\partial(\partial_\mu\hat{\phi}_r)} = \frac{\partial(\delta\hat{\mathcal{L}})}{\partial(\partial_\mu\epsilon_\alpha(x)))}$$

and

$$\partial_\mu\hat{j}^{\alpha\mu}(x) = \frac{\partial(\delta\hat{\mathcal{L}})}{\partial\epsilon_\alpha(x)}. \qquad (1)$$

Deduce that if $\hat{\mathcal{L}}$ is invariant under the global form of this transformation (i.e. constant ϵ_α), then the current defined by (1) is conserved. [This procedure for finding conserved currents for global symmetries is due to Gell-Mann and Levy (1960).]

(b) Apply the method of part (a) to verify the form of the currents (18.96) and (18.97).

18.7 Verify equation (18.120)–(18.124).

18.8 Verify (18.132), and calculate the π^0 lifetime in seconds.

18.9 Verify (18.134).

19

Spontaneously Broken Local Symmetry

In earlier parts of this book we have briefly indicated why we might want to search for a *gauge* theory of the weak interactions. The reasons include: (a) the goal of unification (e.g. with the U(1) gauge theory QED), as mentioned in section 1.3..5 and (b) certain 'universality' phenomena (to be discussed more fully in chapter 20), which are reminiscent of a similar situation in QED (see comment (ii) in section 2.6, and also section 11.6), and which are particularly characteristic of a non-Abelian gauge theory, as pointed out in section 13.1 after equation (13.44). However, we also know from section 1.3.5 that weak interactions are short-ranged, so that their mediating quanta must be massive. At first sight, this seems to rule out the possibility of a gauge theory of weak interactions, since a simple gauge boson mass violates gauge invariance, as we pointed out for the photon in section 11.3 and for non-Abelian gauge quanta in section 13.3.1, and will review again in the following section. Nevertheless, there is a way of giving gauge field quanta a mass, which is by *spontaneously breaking* the gauge (i.e. local) symmetry. This is the topic of the present chapter. The detailed application to the electroweak theory will be made in chapter 21.

19.1 Massive and massless vector particles

Let us begin by noting an elementary (classical) argument for why a gauge field quantum cannot have mass. The electromagnetic potential satisfies the Maxwell equation (cf (2.22))

$$\Box A^\nu - \partial^\nu(\partial_\mu A^\mu) = j^\nu_{\text{em}} \tag{19.1}$$

which, as discussed in section 2.3, is invariant under the gauge transformation

$$A^\mu \to A'^\mu = A^\mu - \partial^\mu \chi. \tag{19.2}$$

However, if A^μ were to represent a *massive* field, the relevant wave equation would be

$$(\Box + M^2)A^\nu - \partial^\nu(\partial_\mu A^\mu) = j^\nu_{\text{em}}. \tag{19.3}$$

To get this, we have simply replaced the massless 'Klein–Gordon' operator $\Box$ by the corresponding massive one, $\Box + M^2$ (compare sections 3.1 and 5.3). Equation (19.3) is manifestly *not* invariant under (19.2), and it is precisely the mass term $M^2 A^\nu$ that breaks the gauge invariance. The same conclusion follows in a Lagrangian treatment; to obtain (19.3) as the corresponding Euler-Lagrange equation, one adds a mass term $+\frac{1}{2}M^2 A_\mu A^\mu$ to the Lagrangian of (7.66) (see also sections 11.4 and 13.3.1), and this clearly violates invariance under (19.2). Similar reasoning holds for the non-Abelian case too. Perhaps, then, we must settle for a theory involving massive charged vector bosons, $W^\pm$ for example, without it being a gauge theory.

DOI: 10.1201/9781003411666-19

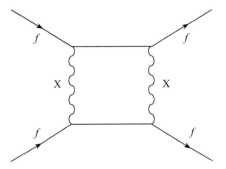

FIGURE 19.1
Fermion-fermion scattering via exchange of two X bosons.

Such a theory is certainly possible, but it will not be *renormalizable*, as we now discuss. Consider figure 19.1, which shows some kind of fermion–fermion scattering (we need not be more specific), proceeding in fourth order of perturbation theory via the exchange of two massive vector bosons, which we will call X-particles. To calculate this amplitude, we need the propagator for the X-particle, which can be found by following the 'heuristic' route outlined in section 7.3.2 for photons. We consider the momentum-space version of (19.3) for the corresponding X^ν field, but without the current on the right-hand side (so as to describe a free field):

$$[(-k^2 + M^2)g^{\nu\mu} + k^\nu k^\mu]\tilde{X}_\mu(k) = 0, \tag{19.4}$$

which should be compared with (7.90). Apart from the 'iϵ', the propagator should be proportional to the inverse of the quantity in the square brackets in (19.4). Problem 19.1 shows that *unlike* the (massless) photon case, this inverse does exist, and is given by

$$\frac{-g^{\mu\nu} + k^\mu k^\nu/M^2}{k^2 - M^2}. \tag{19.5}$$

A proper field-theoretic derivation would yield this result multiplied by an overall factor 'i' as usual, and would also include the 'iϵ' via $k^2 - M^2 \to k^2 - M^2 + i\epsilon$. We remark immediately that (19.5) gives nonsense in the limit $M \to 0$, thus indicating already that a massless vector particle seems to be a very different kind of thing from a massive one (we can't just take the massless limit of the latter).

Now consider the loop integral in figure 19.1. At each vertex we will have a coupling constant g, associated with an interaction Lagrangian having the general form $g\bar{\hat{\psi}}\gamma_\mu\hat{\psi}\hat{X}^\mu$ (a $\gamma_\mu\gamma_5$ coupling could also be present but will not affect the argument). Just as in QED, this 'g' is dimensionless but, as we warned the reader in section 11.8, this may not guarantee renormalizability, and indeed this is a case where it does not. To get an idea of why this might be so, consider the leading divergent behaviour of figure 19.1. This will be associated with the $k^\mu k^\nu$ terms in the numerator of (19.5), so that the leading divergence is effectively

$$\sim \int d^4k \left(\frac{k^\mu k^\nu}{k^2}\right)\left(\frac{k^\rho k^\sigma}{k^2}\right)\frac{1}{\not{k}}\frac{1}{\not{k}} \tag{19.6}$$

for high k-values (we are not troubling to get all the indices right, we are omitting the spinors altogether, and we are looking only at the large k part of the propagators). Now the first two bracketed terms in (19.6) behave like a constant at large k, so that the divergence is effectively

$$\sim \int d^4k \frac{1}{\not{k}}\frac{1}{\not{k}} \tag{19.7}$$

which is quadratically divergent, and indeed exactly what we would get in a 'four-fermion' theory—see (11.98) for example. This strongly suggests that the theory is non-renormalizable.

Where have these dangerous powers of k in the numerator of (19.6) come from? The answer is simple and important. They come from the *longitudinal* polarization state of the massive X-particle, as we shall now explain. The free-particle wave equation is

$$(\Box + M^2)X^\nu - \partial^\nu(\partial_\mu X^\mu) = 0 \tag{19.8}$$

and plane wave solutions have the form

$$X^\nu = \epsilon^\nu e^{-ik\cdot x}. \tag{19.9}$$

Hence the polarization vectors ϵ^ν satisfy the condition

$$(-k^2 + M^2)\epsilon^\nu + k^\nu k_\mu \epsilon^\mu = 0. \tag{19.10}$$

Taking the 'dot' product of (19.10) with k_ν leads to

$$M^2 k \cdot \epsilon = 0, \tag{19.11}$$

which implies (for $M^2 \neq 0$!)

$$k \cdot \epsilon = 0. \tag{19.12}$$

Equation (19.12) is a covariant condition, which has the effect of ensuring that there are just three independent polarization vectors, as we expect for a spin-1 particle. Let us take $k^\mu = (k^0, 0, 0, |\boldsymbol{k}|)$; then the x- and y-directions are 'transverse' while the z-direction is 'longitudinal'. Now, in the rest frame of the X, such that $k_{\text{rest}} = (M, 0, 0, 0)$, (19.12) reduces to $\epsilon^0 = 0$, and we may choose three independent ϵ's as

$$\epsilon^\mu(k_{\text{rest}}, \lambda) = (0, \boldsymbol{\epsilon}(\lambda)) \tag{19.13}$$

with

$$\boldsymbol{\epsilon}(\lambda = \pm 1) = \mp 2^{-1/2}(1, \pm i, 0) \tag{19.14}$$
$$\boldsymbol{\epsilon}(\lambda = 0) = (0, 0, 1). \tag{19.15}$$

The $\boldsymbol{\epsilon}$'s are 'orthonormalized' so that (cf (7.86))

$$\boldsymbol{\epsilon}(\lambda)^* \cdot \boldsymbol{\epsilon}(\lambda') = \delta_{\lambda\lambda'}. \tag{19.16}$$

These states have definite spin projection ($\lambda = \pm 1, 0$) along the z-axis. For the result in a general frame, we can Lorentz transform $\epsilon^\mu(k_{\text{rest}}, \lambda)$ as required. For example, in a frame such that $k^\mu = (k^0, 0, 0, |\boldsymbol{k}|)$, we find

$$\epsilon^\mu(k, \lambda = \pm 1) = \epsilon^\mu(k_{\text{rest}}, \lambda = \pm 1) \tag{19.17}$$

as before, but the longitudinal polarization vector becomes (problem 19.2)

$$\epsilon^\mu(k, \lambda = 0) = M^{-1}(|\boldsymbol{k}|, 0, 0, k^0). \tag{19.18}$$

Note that $k \cdot \epsilon^\mu(k, \lambda = 0) = 0$ as required.

From (19.17) and (19.18) it is straightforward to verify the result (problem 19.3)

$$\sum_{\lambda=0,\pm 1} \epsilon^\mu(k,\lambda)\epsilon^{\nu*}(k,\lambda) = -g^{\mu\nu} + k^\mu k^\nu/M^2. \qquad (19.19)$$

Consider now the propagator for a spin-1/2 particle, given in (7.63):

$$\frac{i(\slashed{k}+m)}{k^2 - m^2 + i\epsilon}. \qquad (19.20)$$

Equation (7.64) shows that the factor in the numerator of (19.20) arises from the spin sum

$$\sum_s u_\alpha(k,s)\bar{u}_\beta(k,s) = (\slashed{k}+m)_{\alpha\beta}. \qquad (19.21)$$

In just the same way, the massive spin-1 propagator is given by

$$\frac{i[-g^{\mu\nu} + k^\mu k^\nu/M^2]}{k^2 - M^2 + i\epsilon}, \qquad (19.22)$$

the numerator in (19.22) arising from the spin sum (19.19). Thus the dangerous factor $k^\mu k^\nu/M^2$ can be traced to the spin sum (19.19): in particular, at large values of k the longitudinal state $\epsilon^\mu(k, \lambda = 0)$ is proportional to k^μ, and this is the origin of the numerator factors $k^\mu k^\nu/M^2$ in (19.22).

We shall not give further details here (see also section 22.1.2), but merely state that theories with massive charged vector bosons are indeed non-renormalizable. Does this matter? In section 11.8 we explained why it is thought that the relevant theories at presently accessible energy scales should be renormalizable theories. And, apart from anything else, they are much more predictive. Is there, then, any way of getting rid of the offending '$k^\mu k^\nu$' terms in the X-propagator, so as (perhaps) to render the theory renormalizable? Consider the photon propagator of chapter 7 repeated here:

$$\frac{i[-g^{\mu\nu} + (1-\xi)k^\mu k^\nu/k^2]}{k^2 + i\epsilon}. \qquad (19.23)$$

This contains somewhat similar factors of $k^\mu k^\nu$ (admittedly divided by k^2 rather than M^2), but they are gauge-dependent, and can in fact be 'gauged away' entirely, by choice of the gauge parameter ξ (namely by taking $\xi = 1$). But, as we have seen, such 'gauging'—essentially the freedom to make gauge transformations—seems to be possible only in a massless vector theory.

A closely related point is that, as section 7.3.1 showed, free photons exist in only two polarization states (electromagnetic waves are purely transverse), instead of the three we might have expected for a vector (spin-1) particle—and as do indeed exist for massive vector particles. This gives another way of seeing in what way a massless vector particle is really very different from a massive one: the former has only two (spin) degrees of freedom, while the latter has three, and it is not at all clear how to 'lose' the offending longitudinal state smoothly (certainly not, as we have seen, by letting $M \to 0$ in (19.5)).

These considerations therefore suggest the following line of thought: is it possible somehow to create a theory involving massive vector bosons, in such a way that the dangerous $k^\mu k^\nu$ term can be 'gauged away', making the theory renormalizable? The answer is yes, via the idea of *spontaneous breaking* of gauge symmetry. This is the natural generalization of the spontaneous global symmetry breaking considered in chapter 17. By way of advance notice, the crucial formula is (19.75) for the propagator in such a theory, which is to be compared with (19.22).

The first serious challenge to the then widely held view that electromagnetic gauge invariance requires the photon to be massless was made by Schwinger (1962), as we pointed out in section 11.4. Soon afterwards, Anderson (1963) argued that several situations in solid state physics could be interpreted in terms of an effectively massive electromagnetic field. He outlined a general framework for treating the phenomenon of the acquisition of mass by a gauge boson, and discussed its possible relevance to contemporary attempts (Sakurai 1960) to interpret the recently discovered vector mesons ($\rho, \omega, \phi \ldots$) as the gauge quanta associated with a local extension of hadronic flavour symmetry. From his discussion, it is clear that Anderson had his doubts about the hadronic application, precisely because, as he remarked, gauge bosons can only acquire a mass if the symmetry is spontaneously broken. This has the consequence, as we saw in chapter 17, that the multiplet structure ordinarily associated with a non-Abelian symmetry would be lost. But we know that flavour symmetry, even if admittedly not exact, certainly leads to identifiable multiplets, which are at least approximately degenerate in mass. It was Weinberg (1967) and Salam (1968) who made the correct application of these ideas, to the generation of mass for the gauge quanta associated with the weak force. There is, however, nothing specifically relativistic about the basic mechanism involved, nor need we start with the non-Abelian case. In fact, the physics is well illustrated by the non-relativistic Abelian (i.e. electromagnetic) case—which is nothing but the physics of superconductivity. Our presentation is influenced by that of Anderson (1963).

19.2   The generation of 'photon mass' in a superconductor: Ginzburg-Landau theory and the Meissner effect

In chapter 17, section 17.7, we gave a brief introduction to some aspects of the BCS theory of superconductivity. We were concerned mainly with the nature of the BCS ground state, and with the non-perturbative origin of the energy gap for elementary excitations. In particular, as noted after (17.129), we omitted completely all electromagnetic couplings of the electrons in the 'microscopic' Hamiltonian. It is certainly possible to complete the BCS theory in this way, so as to include within the same formalism a treatment of electromagnetic effects (e.g. the Meissner effect) in a superconductor. We refer interested readers to the book by Schrieffer (1964), chapter 8. Instead, we shall follow a less 'microscopic' and somewhat more 'phenomenological' approach, which has a long history in theoretical studies of superconductivity, and is in some ways actually closer (at least formally) to our eventual application in particle physics.

In section 17.3.1 we introduced the concept of an 'order parameter', a quantity which was a measure of the 'degree of ordering' of a system below some transition temperature. In the case of superconductivity, the order parameter (in this sense) is taken to be a complex scalar field ψ, as originally proposed by Ginzburg and Landau (1950), well before the appearance of BCS theory. Subsequently, Gorkov (1959) and others showed how the Ginzburg-Landau description could be derived from BCS theory, in certain domains of temperature and magnetic field. This work all relates to static phenomena. More recently, an analogous 'effective theory' for time-dependent phenomena (at zero temperature) has been derived from a BCS-type model (Aitchison *et al.* 1995). For the moment, we shall follow a more qualitative approach.

The Ginzburg–Landau field ψ is commonly referred to as the 'macroscopic wave function'. This terminology originates from the recognition that in the BCS ground state a macroscopic number of Cooper pairs have 'condensed' into the state of lowest energy, a situation similar to that in the Bogoliubov superfluid. Further, this state is highly *coherent*, all pairs having the same total momentum (namely zero, in the case of (17.141)). These considerations suggest that a successful phenomenology can be built by invoking the idea of a macroscopic wavefunction ψ, describing the condensate. Note that ψ is a 'bosonic' quantity, referring essentially to *paired* electrons. Perhaps the single most important property of ψ is that it is assumed to be normalized to the total density of Cooper pairs n_c via the relation

$$|\psi|^2 = n_c = n_s/2 \tag{19.24}$$

where n_s is the density of superconducting electrons. The quantities n_c and n_s will depend on temperature T, tending to zero as T approaches the superconducting transition temperature T_c from below. The precise connection between ψ and the microscopic theory is indirect; in particular, ψ has no knowledge of the coordinates of individual electron pairs. Nevertheless, as an 'empirical' order parameter, it may be thought of as in some way related to the ground state 'pair' expectation value introduced in (17.121): in particular, the charge associated with ψ is taken to be $-2e$, and the mass is $2m_e$.

The Ginzburg-Landau description proceeds by considering the quantum-mechanical electromagnetic current associated with ψ, in the presence of a static external electromagnetic field described by a vector potential $\boldsymbol{A}$. This current was considered in section 2.4, and is given by the gauge-invariant form of (A.7), namely

$$\boldsymbol{j}_{\text{em}} = \frac{-2e}{4m_e\mathrm{i}}[\psi^*(\boldsymbol{\nabla} + 2\mathrm{i}e\boldsymbol{A})\psi - \{(\boldsymbol{\nabla} + 2\mathrm{i}e\boldsymbol{A})\psi\}^*\psi]. \tag{19.25}$$

Note that we have supplied an overall factor of $-2e$ to turn the Schrödinger 'number density' current into the appropriate electromagnetic current. Assuming now that, consistently with (19.24), ψ is varying primarily through its *phase* degree of freedom ϕ, rather than its modulus $|\psi|$, we can rewrite (19.25) as

$$\boldsymbol{j}_{\text{em}} = -\frac{2e^2}{m_e}\left(\boldsymbol{A} + \frac{1}{2e}\boldsymbol{\nabla}\phi\right)|\psi|^2 \tag{19.26}$$

where $\psi = \mathrm{e}^{\mathrm{i}\phi}|\psi|$. We easily verify that (19.26) is invariant under the gauge transformation (2.41), which can be written in this case as

$$\boldsymbol{A} \quad \to \quad \boldsymbol{A} + \boldsymbol{\nabla}\chi \tag{19.27}$$

$$\phi \quad \to \quad \phi - 2e\chi. \tag{19.28}$$

We now replace $|\psi|^2$ in (19.26) by $n_s/2$ in accordance with (19.24), and take the curl of the resulting equation to obtain

$$\boldsymbol{\nabla} \times \boldsymbol{j}_{\text{em}} = -\left(\frac{e^2 n_s}{m_e}\right)\boldsymbol{B}. \tag{19.29}$$

Equation (19.29) is known as the London equation (London 1950) and is one of the fundamental phenomenological relations in superconductivity.

The significance of (19.29) emerges when we combine it with the (static) Maxwell equation

$$\boldsymbol{\nabla} \times \boldsymbol{B} = \boldsymbol{j}_{\text{em}}. \tag{19.30}$$

Taking the curl of (19.30), and using $\boldsymbol{\nabla} \times (\boldsymbol{\nabla} \times \boldsymbol{B}) = \boldsymbol{\nabla}(\boldsymbol{\nabla} \cdot \boldsymbol{B}) - \boldsymbol{\nabla}^2\boldsymbol{B}$ and $\boldsymbol{\nabla} \cdot \boldsymbol{B} = 0$, we find

$$\boldsymbol{\nabla}^2\boldsymbol{B} = \left(\frac{e^2 n_s}{m_e}\right)\boldsymbol{B}. \tag{19.31}$$

The variation of magnetic field described by (19.31) is a very characteristic one encountered in a number of contexts in condensed matter physics. First, we note that the quantity $(e^2 n_s/m_e)$ must—in our units—have the dimensions of (length)$^{-2}$, by comparison with the left-hand side of (19.31). Let us write

$$\left(\frac{e^2 n_s}{m_e}\right) = \frac{1}{\lambda^2}. \tag{19.32}$$

Next, consider for simplicity one-dimensional variation

$$\frac{\mathrm{d}^2 \boldsymbol{B}}{\mathrm{d}x^2} = \frac{1}{\lambda^2}\boldsymbol{B} \tag{19.33}$$

in the half-plane $x \geq 0$, say. Then the solutions of (19.33) have the form

$$\boldsymbol{B}(x) = \boldsymbol{B}_0 \exp{-(x/\lambda)}; \tag{19.34}$$

the exponentially growing solution is rejected as unphysical. The field therefore penetrates only a distance of order λ into the region $x \geq 0$. The range parameter λ is called the *screening length*. This expresses the fact that, in a medium such that (19.29) holds, the magnetic field will be 'screened out' from penetrating further into the medium.

The physical origin of the screening is provided by Lenz's law. When a magnetic field is applied to a system of charged particles, induced EMFs are set up which accelerate the particles, and the magnetic effect of the resulting currents tends to cancel (or screen) the applied field. On the atomic scale this is the cause of atomic diamagnetism. Here the effect is occurring on a macroscopic scale (as mediated by the 'macroscopic wavefunction' ψ), and leads to the Meissner effect—the exclusion of flux from the interior of a superconductor. In this case, screening currents are set up within the superconductor, over distances of order λ from the exterior boundary of the material. These exactly cancel—perfectly screen—the applied flux density in the interior. With $n_s \sim 4 \times 10^{28}$ m^{-3} (roughly one conduction electron per atom) we find

$$\lambda = \left(\frac{m_e}{n_s e^2}\right)^{1/2} \approx 10^{-8} \text{ m}, \tag{19.35}$$

which is the correct order of magnitude for the thickness of the surface layer within which screening currents flow, and over which the applied field falls to zero. As $T \to T_c$, $n_s \to 0$ and λ becomes arbitrarily large, so that flux is no longer screened.

It is quite simple to interpret equation (19.31) in terms of an 'effective non-zero photon mass'. Consider the equation (19.8) for a free massive vector field. Taking the divergence via ∂_ν leads to

$$M^2 \partial_\nu X^\nu = 0 \tag{19.36}$$

(cf (19.11)), and so (19.8) can be written as

$$(\Box + M^2)X^\nu = 0, \tag{19.37}$$

which simply expresses the fact that each component of X^ν has mass M. Now consider the static version of (19.37), in the rest frame of the X-particle in which (see equation (19.13)) the $\nu = 0$ component vanishes. Equation (19.37) reduces to

$$\boldsymbol{\nabla}^2 \boldsymbol{X} = M^2 \boldsymbol{X} \tag{19.38}$$

which is exactly the same in form as (19.31) (if $\boldsymbol{X}$ were the electromagnetic field $\boldsymbol{A}$, we could take the curl of (19.38) to obtain (19.31) via $\boldsymbol{B} = \boldsymbol{\nabla} \times \boldsymbol{A}$). The connection is made precise by making the association

$$M^2 = \left(\frac{e^2 n_{\mathrm{s}}}{m_{\mathrm{e}}}\right) = \frac{1}{\lambda^2}. \tag{19.39}$$

Equation (19.39) shows very directly another way of understanding the 'screening length $\leftrightarrow$ photon mass' connection. In our units $\hbar = c = 1$, a mass has the dimension of an inverse length, and so we naturally expect to be able to interpret λ^{-1} as an equivalent mass (for the photon, in this case).

The above treatment conveys much of the essential physics behind the phenomenon of 'photon mass generation' in a superconductor. In particular, it suggests rather strongly that a *second* field, in addition to the electromagnetic one, is an essential element in the story (here, it is the ψ field). This provides a partial answer to the puzzle about the discontinuous change in the number of spin degrees of freedom in going from a massless to a massive gauge field; actually, some other field has to be supplied. Nevertheless, many questions remain unanswered so far. For example, how is all the foregoing related to what we learned in chapter 17 about spontaneous symmetry breaking? Where is the Goldstone mode? Is it really all gauge invariant? And what about Lorentz invariance? Can we provide a Lagrangian description of the phenomenon? The answers to these questions are mostly contained in the model to which we now turn, which is due to Higgs (1964) and is essentially the *local* version of the U(1) Goldstone model of section 17.5.

19.3 Spontaneously broken local U(1) symmetry: the Abelian Higgs model

This model is just $\hat{\mathcal{L}}_{\mathrm{G}}$ of (17.69) and (17.77), extended so as to be locally, rather than merely globally, U(1) invariant. Due originally to Higgs (1964), it provides a deservedly famous and beautifully simple model for investigating what happens when a *gauge* symmetry is spontaneously broken.

To make (17.69) locally U(1) invariant, we need only replace the ∂'s by $\hat{D}$'s according to the rule (7.123), and add the Maxwell piece. This produces

$$\hat{\mathcal{L}}_{\mathrm{H}} = [(\partial^\mu + iq\hat{A}^\mu)\hat{\phi}]^\dagger [(\partial_\mu + iq\hat{A}_\mu)\hat{\phi}] - \frac{1}{4}\hat{F}_{\mu\nu}\hat{F}^{\mu\nu} - \frac{1}{4}\lambda(\hat{\phi}^\dagger\hat{\phi})^2 + \mu^2(\hat{\phi}^\dagger\hat{\phi}). \tag{19.40}$$

(19.40) is invariant under the local version of (17.72), namely

$$\hat{\phi}(x) \to \hat{\phi}'(x) = \mathrm{e}^{-i\hat{\alpha}(x)}\hat{\phi}(x) \tag{19.41}$$

when accompanied by the gauge transformation on the potentials

$$\hat{A}^\mu(x) \to \hat{A}'^\mu(x) = \hat{A}^\mu(x) + \frac{1}{q}\partial^\mu\hat{\alpha}(x). \tag{19.42}$$

Before proceeding any further, we note at once that this model contains four field degrees of freedom—two in the complex scalar Higgs field $\hat{\phi}$, and two in the massless gauge field $\hat{A}^\mu$.

We learned in section 17.5 that the form of the potential terms in (19.40) (specifically the μ^2 one) does not lend itself to a natural particle interpretation, which only appears after making a 'shift to the classical minimum', as in (17.84). But there is a remarkable difference between the global and local cases. In the present (local) case, the phase of $\hat{\phi}$ is completely arbitrary, since any change in $\hat{\alpha}$ of (19.41) can be compensated by an appropriate transformation (19.42) on $\hat{A}^\mu$, leaving $\hat{\mathcal{L}}_{\mathrm{H}}$ the same as before. Thus the field $\hat{\theta}$ in (17.84) can be 'gauged away' altogether, if we choose! But $\hat{\theta}$ was precisely the Goldstone field, in the global case. This must mean that there is somehow no longer any physical manifestation of the massless mode. This is the first unexpected result in the local case. We may also be reminded of our desire to 'gauge away' the longitudinal polarization states for a 'massive gauge' boson: we shall return to this later.

However, a degree of freedom (the Goldstone mode) cannot simply disappear. Somehow the system must keep track of the fact that we started with four degrees of freedom. To see what is going on, let us study the field equation for $\hat{A}^\nu$, namely

$$\Box \hat{A}^\nu - \partial^\nu(\partial_\mu \hat{A}^\mu) = \hat{j}_{\mathrm{em}}^\nu, \tag{19.43}$$

where $\hat{j}_{\mathrm{em}}^\nu$ is the electromagnetic current contained in (19.40). This current can be obtained just as in (7.141), and is given by

$$\hat{j}_{\mathrm{em}}^\nu = \mathrm{i}q(\hat{\phi}^\dagger \partial^\nu \hat{\phi} - (\partial^\nu \hat{\phi}^\dagger)\hat{\phi}) - 2q^2 \hat{A}^\nu \hat{\phi}^\dagger \hat{\phi}. \tag{19.44}$$

We now insert the field parametrization (cf (17.84))

$$\hat{\phi}(x) = \frac{1}{\sqrt{2}}(v + \hat{h}(x))\exp(-\mathrm{i}\hat{\theta}(x)/v) \tag{19.45}$$

into (19.44) where $v/\sqrt{2} = 2^{1/2}|\mu|/\lambda^{\frac{1}{2}}$ is the position of the minimum of the classical potential as a function of $|\phi|$, as in (17.81). We obtain (problem 19.4)

$$\hat{j}_{\mathrm{em}}^\nu = -v^2 q^2 \left(\hat{A}^\nu - \frac{\partial^\nu \hat{\theta}}{vq} \right) + \text{terms quadratic and cubic in the fields.} \tag{19.46}$$

The linear part of the right-hand side of (19.46) is directly analogous to the non-relativistic current (19.26), interpreting $\hat{\theta}$ as essentially playing the role of ϕ, and $|\psi|^2$ the role of v^2. Retaining just the linear terms in (19.46) (the others would appear on the right-hand side of equation (19.47) following, where they would represent interactions), and placing this $\hat{j}_{\mathrm{em}}^\nu$ in (19.43), we obtain

$$\Box \hat{A}^\nu - \partial^\nu \partial_\mu \hat{A}^\mu = -v^2 q^2 \left(\hat{A}^\nu - \frac{\partial^\nu \hat{\theta}}{vq} \right). \tag{19.47}$$

Now a gauge transformation on $\hat{A}^\nu$ has the form shown in (19.42), for arbitrary $\hat{\alpha}$. So we can certainly regard the whole expression $(\hat{A}^\nu - \partial^\nu \hat{\theta}/vq)$ as a perfectly acceptable gauge field. Let us define

$$\hat{A}'^\nu = \hat{A}^\nu - \frac{\partial^\nu \hat{\theta}}{vq}. \tag{19.48}$$

Then, since we know that the left-hand side of (19.47) is invariant under (19.42), the resulting equation for $\hat{A}'^\nu$ is

$$\Box \hat{A}'^\nu - \partial^\nu \partial_\mu \hat{A}'^\mu = -v^2 q^2 \hat{A}'^\nu, \tag{19.49}$$

or

$$(\Box + v^2 q^2)\hat{A}^{'\nu} - \partial^\nu \partial_\mu \hat{A}^{'\mu} = 0. \tag{19.50}$$

But (19.50) is nothing but the equation (19.8) for a free massive vector field, with mass $M = vq$! This fundamental observation was first made, in the relativistic context, by Englert and Brout (1964), Higgs (1964), and Guralnik *et al.* (1964); for a full account, see Higgs (1966).

The foregoing analysis shows us two things. First, the current (19.46) is indeed a relativistic analogue of (19.26), in that it provides a 'screening' (mass generation) effect on the gauge field. Second, equation (19.48) shows how the *phase* degree of freedom of the Higgs field $\hat{\phi}$ has been incorporated into a new gauge field $\hat{A}^{'\nu}$, which is massive, and therefore has *three* spin degrees of freedom. In fact, we can go further. If we imagine plane wave solutions for $\hat{A}^{'\nu}$, $\hat{A}^\nu$, and $\hat{\theta}$, we see that the $\partial^\nu \hat{\theta}/vq$ part of (19.48) will contribute something proportional to k^ν/M to the polarization vector of $A^{'\nu}$ (recall $M = vq$). But this is exactly the (large k) behaviour of the longitudinal polarization vector of a massive vector particle. We may therefore say that the massless gauge field $\hat{A}^\nu$ has 'swallowed' the Goldstone field $\hat{\theta}$ via (19.48) to make the massive vector field $\hat{A}^{'\nu}$. The Goldstone field disappears as a massless degree of freedom, and reappears, via its gradient, as the longitudinal part of the massive vector field. In this way the four degrees of freedom are all now safely accounted for: three are in the massive vector field $\hat{A}^{'\nu}$, and one is in the real scalar field $\hat{h}$ (to which we shall return shortly).

In this (relativistic) case, we know from Lorentz covariance that all the components (transverse and longitudinal) of the vector field must have the same mass, and this has of course emerged automatically from our covariant treatment. But the transverse and longitudinal degrees of freedom respond differently in the non-relativistic (superconductor) case. There, the longitudinal part of $\boldsymbol{A}$ couples strongly to longitudinal excitations of the electrons: primarily, as Bardeen (1957) first recognized, to the collective density fluctuation mode of the electron system—that is, to plasma oscillations. This is a high frequency mode, and is essentially the one discussed in section 17.3.2, after equation (17.46). When this aspect of the dynamics of the electrons is included, a fully gauge invariant description of the electromagnetic properties of superconductors, within the BCS theory, is obtained (Schrieffer 1964, chapter 8).

We return to equations (19.48)–(19.50). Taking the divergence of (19.50) leads, as we have seen, to the condition

$$\partial_\mu \hat{A}^{'\mu} = 0 \tag{19.51}$$

on $\hat{A}^{'\mu}$. It follows that in order to interpret the relation (19.48) as a gauge transformation on $\hat{A}^\nu$ we must, to be consistent with (19.51), regard $\hat{A}^\mu$ as being in a gauge specified by

$$\partial_\mu \hat{A}^\mu = \frac{1}{vq}\Box\hat{\theta} = \frac{1}{M}\Box\hat{\theta}. \tag{19.52}$$

In going from the situation described by $\hat{A}^\mu$ and $\hat{\theta}$ to one described by $\hat{A}^{'\mu}$ alone via (19.48), we have evidently chosen a gauge function (cf (19.42))

$$\hat{\alpha}(x) = -\hat{\theta}(x)/v. \tag{19.53}$$

Recalling then the form of the associated local phase change on $\hat{\phi}(x)$

$$\hat{\phi}(x) \to \hat{\phi}'(x) = e^{-i\hat{\alpha}(x)}\hat{\phi}(x) \tag{19.54}$$

we see that the phase of $\hat{\phi}$ in (19.45) has been reduced to zero, in this choice of gauge. Thus it is indeed possible to 'gauge $\hat{\theta}$ away' in (19.45), but then the vector field we must use is $\hat{A}^{'\mu}$, satisfying the massive equation (19.50) (ignoring other interactions). In superconductivity, the choice of gauge which takes the macroscopic wavefunction to be real (i.e. $\phi = 0$ in (19.26)) is called the 'London gauge'. In the next section we shall discuss a subtlety in the argument which applies in the case of real superconductors, and which leads to the phenomenon of flux quantization.

The fact that this 'Higgs mechanism' leads to a massive vector field can be seen very economically by working in the particular gauge for which $\hat{\phi}$ is real, and inserting the parametrization (cf (19.45))

$$\hat{\phi} = \frac{1}{\sqrt{2}}(v + \hat{h}) \tag{19.55}$$

into the Lagrangian $\hat{\mathcal{L}}_H$. Retaining only the terms quadratic in the fields one finds (problem 19.5)

$$\begin{aligned} \hat{\mathcal{L}}_H^{\text{quad}} &= -\frac{1}{4}(\partial_\mu \hat{A}_\nu - \partial_\nu \hat{A}_\mu)(\partial^\mu \hat{A}^\nu - \partial^\nu \hat{A}^\mu) + \frac{1}{2}q^2 v^2 \hat{A}_\mu \hat{A}^\mu \\ &\quad + \frac{1}{2}\partial_\mu \hat{h}\partial^\mu \hat{h} - \mu^2 \hat{h}^2. \end{aligned} \tag{19.56}$$

The first line of (19.56) is exactly the Lagrangian for a spin-1 field of mass vq—i.e. the Maxwell part with the addition of a mass term (note that the sign of the mass term is correct for the spatial (physical) degrees of freedom); the second line is the Lagrangian of a scalar particle of mass $\sqrt{2}\mu$. The latter is the mass of excitations of the Higgs field $\hat{h}$ away from its vacuum value (compare the global U(1) case discussed in section 17.5). The necessity for the existence of one or more massive *scalar* particles ('Higgs bosons'), when a gauge symmetry is spontaneously broken in this way, was pointed out by Higgs (1964).

We may now ask: what happens if we start with a certain phase $\hat{\theta}$ for $\hat{\phi}$ but do *not* make use of the gauge freedom in $\hat{A}^\nu$ to reduce $\hat{\theta}$ to zero? We shall see in section 19.5 that the equation of motion, and hence the propagator, for the vector particle *depends on the choice of gauge*; furthermore, Feynman graphs involving quanta corresponding to the degree of freedom associated with the phase field $\hat{\theta}$ will have to be included for a consistent theory, even though this must be an unphysical degree of freedom, as follows from the fact that a gauge can be chosen in which this field vanishes. That the propagator is gauge dependent should, on reflection, come as a relief. After all, if the massive vector boson generated in this way were *simply* described by the wave equation (19.50), all the troubles with massive vector particles outlined in section 19.1 would be completely unresolved. As we shall see, a different choice of gauge from that which renders $\hat{\phi}$ real has precisely the effect of ameliorating the bad high-energy behaviour associated with (19.50). This is ultimately the reason for the following wonderful fact: *massive vector theories, in which the vector particles acquire mass through the spontaneous symmetry breaking mechanism, are renormalizable* ('t Hooft 1971b).

However, before discussing other gauges than the one in which $\hat{\phi}$ is given by (19.55), we first explore another interesting aspect of superconductivity.

19.4   Flux quantization in a superconductor

Though a slight diversion, it is convenient to include a discussion of flux quantization at this point, while we have a number of relevant results assembled. Apart from its intrinsic interest,

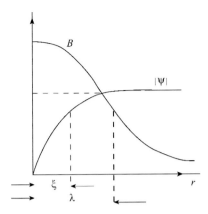

FIGURE 19.2
Magnetic field B and modulus of the macroscopic (pair) wavefunction $|\psi|$ in the neighbourhood of a flux filament.

the phenomenon may also provide a useful physical model for the 'confining' property of QCD, as already discussed in sections 1.3.6 and 16.5.3.

Our discussion of superconductivity so far has dealt, in fact, with only one class of superconductors, called type I; these remain superconducting throughout the bulk of the material (exhibiting a complete Meissner effect), when an external magnetic field of less than a certain critical value is applied. There is a quite separate class—type II superconductors—which allow partial entry of the external field, in the form of thin filaments of flux. Within each filament the field is high, and the material is not superconducting. Outside the core of the filaments, the material is superconducting and the field dies off over the characteristic penetration length λ. Around each filament of magnetic flux there circulates a vortex of screening current; the filaments are often called vortex lines. It is as if numerous thin cylinders, each enclosing flux, had been drilled in a block of type I material, thereby producing a non-simply connected geometry.

In real superconductors, screening currents are associated with the macroscopic pair wavefunction (field) ψ. For type II behaviour to be possible, $|\psi|$ must vanish at the centre of a flux filament, and rise to the constant value appropriate to the superconducting state over a distance $\xi < \lambda$, where ξ is the 'coherence length' of section 17.7. According to the Ginzburg-Landau (GL) theory, a more precise criterion is that type II behaviour holds if $\xi < 2^{1/2}\lambda$; both ξ and λ are, of course, temperature-dependent. The behaviour of $|\psi|$ and B in the vicinity of a flux filament is shown in figure 19.2. Thus, whereas for simple type I superconductivity, $|\psi|$ is simply set equal to a constant, in the type II case $|\psi|$ has the variation shown in this figure. Solutions of the coupled GL equations for $\boldsymbol{A}$ and ψ can be obtained which exhibit this behaviour.

An important result is that the flux through a vortex line is quantized. To see this, we write

$$\psi = \mathrm{e}^{\mathrm{i}\phi}|\psi| \tag{19.57}$$

as before. The expression for the electromagnetic current is

$$\boldsymbol{j}_{\mathrm{em}} = -\frac{q^2}{m}\left(\boldsymbol{A} - \frac{\boldsymbol{\nabla}\phi}{q}\right)|\psi|^2 \tag{19.58}$$

as in (19.26), but in (19.58) we are leaving the charge parameter q undetermined for the moment; the mass parameter m will be unimportant. Rearranging, we have

$$A = -\frac{m}{q^2|\psi|^2}\boldsymbol{j}_{\text{em}} + \frac{\boldsymbol{\nabla}\phi}{q}. \tag{19.59}$$

Let us integrate equation (19.59) around any closed loop $\mathcal{C}$ in the type II superconductor, which encloses a flux (or vortex) line. Far enough away from the vortex, the screening currents $\boldsymbol{j}_{\text{em}}$ will have dropped to zero, and hence

$$\oint_{\mathcal{C}} A \cdot \mathrm{d}s = \frac{1}{q}\oint_{\mathcal{C}} \boldsymbol{\nabla}\phi \cdot \mathrm{d}s = \frac{1}{q}[\phi]_{\mathcal{C}} \tag{19.60}$$

where $[\phi]_{\mathcal{C}}$ is the change in phase around $\mathcal{C}$. If the wavefunction ψ is single-valued, the change in phase $[\phi]_{\mathcal{C}}$ for any closed path can only be zero or an integer multiple of 2π. Transforming the left-hand side of (19.60) by Stokes' Theorem, we obtain the result that the flux Φ through any surface spanning $\mathcal{C}$ is quantized:

$$\Phi = \int B \cdot \mathrm{d}S = \frac{2\pi n}{q} = n\Phi_0 \tag{19.61}$$

where $\Phi_0 = 2\pi/q$ is the flux equation (or $2\pi\hbar/q$ in ordinary units). It is not entirely self-evident why ψ should be single-valued, but experiments do indeed demonstrate the phenomenon of flux quantization, in units of Φ_0 with $|q| = 2e$ (which maybe interpreted as the charge on a Cooper pair, as usual). The phenomenon is seen in non-simply connected specimens of type I superconductors (i.e. ones with holes in them, such as a ring), and in the flux filaments of type II materials; in the latter case each filament carries a single flux quantum Φ_0.

It is interesting to consider now a situation—so far entirely hypothetical—in which a magnetic monopole is placed in a superconductor. Dirac showed (1931) that for consistency with quantum mechanics the monopole strength g_{m} had to satisfy the 'Dirac quantization conduction'

$$qg_{\text{m}} = n/2 \tag{19.62}$$

where q is any electronic charge, and n is an integer. It follows from (19.62) that the flux $4\pi g_{\text{m}}$ out of any closed surface surrounding the monopole is quantized in units of Φ_0. Hence a flux filament in a superconductor can originate from, or be terminated by, a Dirac monopole (with the appropriate sign of g_{m}), as was first pointed out by Nambu (1974).

This is the basic model which, in one way or another, underlies many theoretical attempts to understand confinement. The monopole-antimonopole pair in a type II superconducting vacuum, joined by a quantized magnetic flux filament, provides a model of a meson. As the distance between the pair—the length of the filament—increases, so does the energy of the filament, at a rate proportional to its length, since the flux cannot spread out in directions transverse to the filament. This is exactly the kind of linearly rising potential energy required by hadron spectroscopy (see equations (1.33) and (16.145)). The configuration is stable, because there is no way for the flux to leak away; it is a conserved quantized quantity.

For the eventual application to QCD, one will want (presumably) particles carrying non-zero values of the colour quantum numbers to be confined. These quantum numbers are the analogues of electric charge in the U(1) case, rather than of magnetic charge. We imagine, therefore, interchanging the roles of magnetism and electricity in all of the foregoing. Indeed, the Maxwell equations have such a symmetry when monopoles are present, as well as charges. The essential feature of the superconducting ground state was that it involved the coherent state formed by condensation of electrically charged bosonic fermion pairs. A vacuum which confined filaments of $\boldsymbol{E}$ rather than $\boldsymbol{B}$ may be formed as a coherent state of condensed magnetic monopoles (Mandelstam 1976, 't Hooft 1976b). These $\boldsymbol{E}$ filaments would then terminate on electric charges. Now magnetic monopoles do not occur naturally as solutions of QED: they would have to be introduced by hand. Remarkably enough, however, solutions of the magnetic monopole type do occur in the case of non-Abelian gauge field theories, whose symmetry is spontaneously broken to an electromagnetic $U(1)_{em}$ gauge group. Just this circumstance can arise in a grand unified theory which contains $SU(3)_c$ and a residual $U(1)_{em}$. Incidentally, these monopole solutions provide an illuminating way of thinking about charge quantization. As Dirac (1931) pointed out, the existence of just one monopole implies, from his quantization condition (19.62), that charge is quantized.

When these ideas are applied to QCD, $\boldsymbol{E}$ and $\boldsymbol{B}$ must be understood as the appropriate colour fields (i.e. they carry an $SU(3)_c$ index). The group structure of $SU(3)$ is also quite different from that of $U(1)$ models, and we do not want to be restricted just to static solutions (as in the GL theory, here used as an analogue). Whether in fact the real QCD vacuum (ground state) is formed as some such coherent plasma of monopoles, with confinement of electric charges and flux, is a subject of continuing research; other schemes are also possible. As so often stressed, the difficulty lies in the non-perturbative nature of the confinement problem.

19.5 't Hooft's gauges

We must now at last grasp the nettle and consider what happens if, in the parametrization

$$\hat{\phi} = |\hat{\phi}| \exp(i\hat{\theta}(x)/v) \tag{19.63}$$

we do not choose the gauge (cf (19.52))

$$\partial_\mu \hat{A}^\mu = \Box \hat{\theta}/M. \tag{19.64}$$

This was the gauge that enabled us to transform away the phase degree of freedom and reduce the equation of motion for the electromagnetic field to that of a massive vector boson. Instead of using the modulus and phase as the two independent degrees of freedom for the complex Higgs field $\hat{\phi}$, we now choose to parametrize $\hat{\phi}$, quite generally, by the decomposition

$$\hat{\phi} = 2^{-1/2}[v + \hat{\chi}_1(x) + i\hat{\chi}_2(x)], \tag{19.65}$$

where the vacuum values of $\hat{\chi}_1$ and $\hat{\chi}_2$ are zero. Substituting this form for $\hat{\phi}$ into the master equation for $\hat{A}^\nu$ (obtained from (19.43) and (19.44))

$$\Box \hat{A}^\nu - \partial^\nu(\partial_\mu \hat{A}^\mu) = iq[\hat{\phi}^\dagger \partial^\nu \hat{\phi} - (\partial^\nu \hat{\phi})^\dagger \hat{\phi}] - 2q^2 \hat{A}^\nu \hat{\phi}^\dagger \hat{\phi}, \tag{19.66}$$

leads to the equation of motion

$$\begin{aligned}(\Box + M^2)\hat{A}^\nu - \partial^\nu(\partial_\mu \hat{A}^\mu) &= -M\partial^\nu \hat{\chi}_2 + q(\hat{\chi}_2 \partial^\nu \hat{\chi}_1 - \hat{\chi}_1 \partial^\nu \hat{\chi}_2) \\ &\quad -q^2 \hat{A}^\nu(\hat{\chi}_1^2 + 2v\hat{\chi}_1 + \hat{\chi}_2^2)\end{aligned} \tag{19.67}$$

with $M = qv$. At first sight this just looks like the equation of motion of an ordinary massive vector field $\hat{A}^\nu$ coupled to a rather complicated current. However, this certainly cannot be right, as we can see by a count of the degrees of freedom. In the previous gauge we had four degrees of freedom, counted either as two for the original massless $\hat{A}^\nu$ plus one each

FIGURE 19.3

$\hat{A}^\nu - \hat{\chi}_2$ coupling.

for $\hat{\theta}$ and $\hat{h}$, or as three for the massive $\hat{A}'^\nu$ and one for $\hat{h}$. If we take this new equation at face value, there seem to be three degrees of freedom for the massive field $\hat{A}^\nu$, and one for each of $\hat{\chi}_1$ and $\hat{\chi}_2$, making *five* in all. Actually, we know perfectly well that we can make use of the freedom gauge choice to set $\hat{\chi}_2$ to zero, say, reducing $\hat{\phi}$ to a real quantity and eliminating a spurious degree of freedom: we have then returned to the form (19.55). In terms of (19.67), the consequence of the unwanted degree of freedom is quite subtle, but it is basic to all gauge theories and already appeared in the photon case, in section 7.3.2. The difficulty arises when we try to calculate the propagator for $\hat{A}^\nu$ from equation (19.67).

The operator on the left-hand side can be simply inverted, as was done in section 19.1, to yield (apparently) the standard massive vector boson propagator

$$\mathrm{i}(-g^{\mu\nu} + k^\mu k^\nu/M^2)/(k^2 - M^2). \tag{19.68}$$

However, the current on the right-hand side of (19.67) is rather peculiar. Instead of having only terms corresponding to $\hat{A}^\nu$ coupling to two or three particles, there is also a term involving only one field. This is the term $-M\partial^\nu\hat{\chi}_2$, which tells us that $\hat{A}^\nu$ actually couples directly to the scalar field χ_2 via the gradient coupling $(-M\partial^\nu)$. In momentum space this corresponds to a coupling strength $-\mathrm{i}k^\nu M$ and an associated vertex as shown in figure 19.3. Clearly, for a scalar particle, the momentum 4-vector is the only quantity that can couple to the vector index of the vector boson. The existence of this coupling shows that the propagators of $\hat{A}^\nu$ and $\hat{\chi}_2$ are necessarily mixed: the complete vector propagator must be calculated by summing the infinite series shown diagrammatically in figure 19.4. This complication is, of course, completely eliminated by the gauge choice $\hat{\chi}_2 = 0$. However, we are interested in pursuing the case $\hat{\chi}_2 \neq 0$.

In figure 19.4 the only unknown factor is the propagator for $\hat{\chi}_2$. This can be easily found by substituting (19.65) into $\hat{\mathcal{L}}_\mathrm{H}$ and examining the part which is quadratic in the $\hat{\chi}$'s; we find (problem 19.6)

$$\hat{\mathcal{L}}_\mathrm{H} = \frac{1}{2}\partial_\mu\hat{\chi}_1\partial^\mu\hat{\chi}_1 + \frac{1}{2}\partial_\mu\hat{\chi}_2\partial^\mu\hat{\chi} - \mu^2\hat{\chi}_1^2 + \text{and cubic and quartic terms.} \tag{19.69}$$

Equation (19.69) confirms that $\hat{\chi}_1$ is a massive field with mass $\sqrt{2}\mu$ (like the $\hat{h}$ in (19.56)), while $\hat{\chi}_2$ is massless. The $\hat{\chi}_2$ propagator is therefore i/k^2. Now that all the elements of the diagrams are known, we can formally sum the series by generalizing the well-known result ((cf 10.12)and (11.27))

$$(1-x)^{-1} = 1 + x + x^2 + x^3 + \dots \tag{19.70}$$

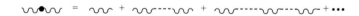

FIGURE 19.4

Series for the full $\hat{A}^\nu$ propagator.

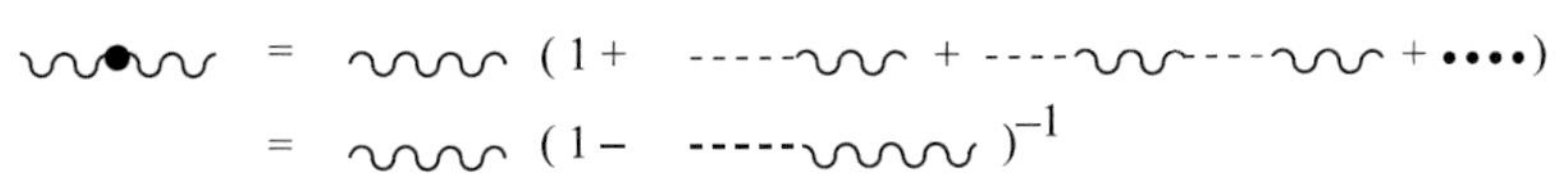

FIGURE 19.5
Formal summation of the series in figure 19.4.

Diagrammatically, we rewrite the propagator of figure 19.4 as in figure 19.5 and perform the sum. Inserting the expressions for the propagators and vector-scalar coupling, and keeping track of the indices, we finally arrive at the result (problem 19.7)

$$\mathrm{i}\left(\frac{-g^{\mu\lambda} + k^\mu k^\lambda/M^2}{k^2 - M^2}\right)(g^\nu_\lambda - k^\nu k_\lambda/k^2)^{-1} \tag{19.71}$$

for the full propagator. But the inverse required in (19.71) is precisely (with a lowered index) the one we needed for the photon propagator in (7.91)—and, as we saw there, it does not exist. At last the fact that we are dealing with a gauge theory has caught up with us!

As we saw in section 7.3.2, to obtain a well-defined gauge field propagator we need to *fix the gauge*. A clever way to do this in the present (spontaneously broken) case was suggested by 't Hooft (1971b). His proposal was to set

$$\partial_\mu \hat{A}^\mu = M\xi\hat{\chi}_2 \tag{19.72}$$

where ξ is an arbitrary gauge parameter[1] (not to be confused with the superconducting coherence length). This condition is manifestly covariant, and moreover it effectively reduces the degrees of freedom by one. Inserting (19.72) into (19.67) we obtain

$$(\Box + M^2)\hat{A}^\nu - \partial^\nu(\partial_\mu\hat{A}^\mu)(1 - 1/\xi) = q(\hat{\chi}_2\partial^\nu\hat{\chi}_1 - \hat{\chi}_1\partial^\nu\hat{\chi}_2) \tag{19.73}$$
$$-q^2\hat{A}^\nu(\hat{\chi}_1^2 + 2v\hat{\chi}_1 + \hat{\chi}_2^2). \tag{19.74}$$

The operator appearing on the left-hand side now *does* have an inverse (see problem 19.8) and yields the general form for the gauge boson propagator

$$\mathrm{i}\left[-g^{\mu\nu} + \frac{(1-\xi)k^\mu k^\nu}{k^2 - \xi M^2}\right](k^2 - M^2)^{-1}. \tag{19.75}$$

This propagator is very remarkable[2]. The standard massive vector boson propagator

$$\mathrm{i}(-g^{\mu\nu} + k^\mu k^\nu/M^2)(k^2 - M^2)^{-1} \tag{19.76}$$

is seen to correspond to the limit $\xi \to \infty$, and in this gauge the high-energy disease outlined in section 19.1 appears to threaten renormalizability (in fact, it can be shown that there is a consistent set of Feynman rules for this gauge, and the theory is renormalizable thanks to many cancellations of divergences). For any finite ξ, however, the high-energy behaviour of the gauge boson propagator is actually $\sim 1/k^2$, which is as good as the *renormalizable* theory of QED (in Lorentz gauge). Note, however, that there seems to be another pole in the

---

[1] We shall not enter here into the full details of quantization in such a gauge. We shall effectively treat (19.72) as a classical field relation.

[2] A vector boson propagator of similar form was first introduced by Lee and Yang (1962), but their discussion was not within the framework of a spontaneously broken theory, so that Higgs particles were not present, and the physical limit was obtained *only* as $\xi \to 0$.

propagator (19.75) at $k^2 = \xi M^2$. This is surely unphysical since it depends on the arbitrary parameter ξ. A full treatment ('t Hooft 1971b) shows that this pole is always cancelled by an exactly similar pole in the propagator for the $\hat{\chi}_2$ field itself. These finite-ξ gauges are called *R gauges* (since they are 'manifestly renormalizable') and typically involve unphysical Higgs fields such as $\hat{\chi}_2$. The infinite-ξ gauge is known as the *U gauge* (U for unitary) since only physical particles appear in this gauge. For tree diagram calculations, of course, it is easiest to use the U gauge Feynman rules. The technical difficulties with this gauge choice only enter in loop calculations, for which the R gauge choice is easier.

Notice that in our master formula (19.75) for the gauge boson propagator the limit $M \to 0$ may be safely taken (compare the remarks about this limit for the 'naive' massive vector boson propagator in section 19.1). This yields the massless vector boson (photon) propagator in a general ξ-gauge, exactly as in equation (7.122) or (19.23).

We now proceed with the generalization of these ideas to the non-Abelian SU(2) case, which is the one relevant to the electroweak theory. The general non-Abelian case was treated by Kibble (1967).

19.6   Spontaneously broken local SU(2)×U(1) symmetry

We shall limit our discussion of the spontaneous breaking of a local non-Abelian symmetry to the particular case needed for the electroweak part of the standard model. This is, in fact, just the local version of the model studied in section 17.6. As noted there, the Lagrangian $\hat{\mathcal{L}}_\Phi$ of (17.98) is invariant under global SU(2) transformations of the form (17.101), and also global U(1) transformations (17.102). Thus in the local version we shall have to introduce three SU(2) gauge fields (as in section 13.1), which we call $\hat{W}_i^\mu(x)(i = 1, 2, 3)$, and one U(1) gauge field $\hat{B}^\mu(x)$. We recall that the scalar field $\hat{\phi}$ is an SU(2)-doublet

$$\hat{\phi} = \begin{pmatrix} \hat{\phi}^+ \\ \hat{\phi}^0 \end{pmatrix}, \tag{19.77}$$

so that the SU(2) covariant derivative acting on $\hat{\phi}$ is as given in (13.10), namely

$$\hat{D}^\mu = \partial^\mu + ig\boldsymbol{\tau} \cdot \hat{\boldsymbol{W}}^\mu/2. \tag{19.78}$$

To this must be added the U(1) piece, which we write as $ig'\hat{B}^\mu/2$, the $\frac{1}{2}$ being for later convenience. The Lagrangian (without gauge-fixing and ghost terms) is therefore

$$\hat{\mathcal{L}}_{G\Phi} = (\hat{D}_\mu\hat{\phi})^\dagger(\hat{D}^\mu\hat{\phi}) + \mu^2\hat{\phi}^\dagger\hat{\phi} - \frac{\lambda}{4}(\hat{\phi}^\dagger\hat{\phi})^2 - \frac{1}{4}\hat{\boldsymbol{F}}_{\mu\nu} \cdot \hat{\boldsymbol{F}}^{\mu\nu} - \frac{1}{4}\hat{G}_{\mu\nu}\hat{G}^{\mu\nu} \tag{19.79}$$

where

$$\hat{D}^\mu\hat{\phi} = (\partial^\mu + ig\boldsymbol{\tau} \cdot \hat{\boldsymbol{W}}^\mu/2 + ig'\hat{B}^\mu/2)\hat{\phi}, \tag{19.80}$$

$$\hat{\boldsymbol{F}}^{\mu\nu} = \partial^\mu\hat{\boldsymbol{W}}^\nu - \partial^\nu\hat{\boldsymbol{W}}^\mu - g\hat{\boldsymbol{W}}^\mu \times \hat{\boldsymbol{W}}^\nu, \tag{19.81}$$

and

$$\hat{G}^{\mu\nu} = \partial^\mu\hat{B}^\nu - \partial^\nu\hat{B}^\mu. \tag{19.82}$$

We must now decide how to choose the non-zero vacuum expectation value that breaks this symmetry. The essential point for the electroweak application is that, after symmetry breaking, we should be left with three massive boson gauge bosons (which will be the $W^\pm$ and Z^0) and one massless gauge boson, the photon. We may reasonably guess that the massless boson will be associated with a symmetry that is *un*broken by the vacuum expectation value. Put differently, we certainly do not want a 'superconducting' massive photon to emerge from the theory in this case, as the physical vacuum is not an electromagnetic superconductor. This means that we do not want to give a vacuum value to a charged field (as is done in the BCS ground state). On the other hand, we do want it to behave as a 'weak' superconductor, generating mass for $W^\pm$ and Z^0. The choice suggested by Weinberg (1967)) was

$$\langle 0|\hat{\phi}|0\rangle = \begin{pmatrix} 0 \\ v/\sqrt{2} \end{pmatrix} \tag{19.83}$$

where $v/\sqrt{2} = \sqrt{2}\mu/\lambda^{1/2}$, which we already considered in the global case in section 17.6. As pointed out there, (19.83) implies that the vacuum remains invariant under the combined transformation of 'U(1) + third component of SU(2) isospin'—that is, (19.83) implies

$$(\frac{1}{2} + t_3^{(\frac{1}{2})})\langle 0|\hat{\phi}|0\rangle = 0 \tag{19.84}$$

and hence

$$\langle 0|\hat{\phi}|0\rangle \rightarrow (\langle 0|\hat{\phi}|0\rangle)' = \exp[i\alpha(\frac{1}{2} + t_3^{(1/2)})]\langle 0|\hat{\phi}|0\rangle = \langle 0|\hat{\phi}|0\rangle, \tag{19.85}$$

where as usual $t_3^{(1/2)} = \tau_3/2$ (we are using lower case t for isospin now, anticipating that it is the *weak*, rather than hadronic, isospin—see chapter 21).

We now need to consider oscillations about (19.83) in order to see the physical particle spectrum. As in (23.29) we parametrize these conveniently as

$$\hat{\phi} = \exp(-i\hat{\boldsymbol{\theta}}(x) \cdot \boldsymbol{\tau}/2v) \begin{pmatrix} 0 \\ \frac{1}{\sqrt{2}}(v + \hat{H}(x))n \end{pmatrix} \tag{19.86}$$

(compare (19.45)). However this time, in contrast to (23.29) but just as in (19.55), we can reduce the phase fields $\hat{\boldsymbol{\theta}}$ to zero by an appropriate gauge transformation, and it is simplest to examine the particle spectrum in this (*unitary*) gauge. Substituting

$$\hat{\phi} = \begin{pmatrix} 0 \\ \frac{1}{\sqrt{2}}(v + \hat{H}(x)) \end{pmatrix} \tag{19.87}$$

into (19.79) and retaining only terms which are second order in the fields (i.e. kinetic energies or mass terms) we find (problem 19.9)

$$\begin{aligned}
\hat{\mathcal{L}}_{G\Phi}^{\text{free}} =\ & \frac{1}{2}\partial_\mu \hat{H}\partial^\mu \hat{H} - \mu^2 \hat{H}^2 \\
& - \frac{1}{4}(\partial_\mu \hat{W}_{1\nu} - \partial_\nu \hat{W}_{1\mu})(\partial^\mu \hat{W}_1^\nu - \partial^\nu \hat{W}_1^\mu) + \frac{1}{8}g^2 v^2 \hat{W}_{1\mu}\hat{W}_1^\mu \\
& - \frac{1}{4}(\partial_\mu \hat{W}_{2\nu} - \partial_\nu \hat{W}_{2\mu})(\partial^\mu \hat{W}_2^\nu - \partial^\nu \hat{W}_1^\mu) + \frac{1}{8}g^2 v^2 \hat{W}_{2\mu}\hat{W}_2^\mu \\
& - \frac{1}{4}(\partial_\mu \hat{W}_{3\nu} - \partial_\nu \hat{W}_{3\mu})(\partial^\mu \hat{W}_3^\nu - \partial^\nu \hat{W}_3^\mu) - \frac{1}{4}\hat{G}_{\mu\nu}\hat{G}^{\mu\nu} \\
& + \frac{1}{8}v^2(g\hat{W}_{3\mu} - g'\hat{B}_\mu)(g\hat{W}_3^\mu - g'\hat{B}^\mu). \tag{19.88}
\end{aligned}$$

The first line of (19.88) tells us that we have a scalar field of mass $\sqrt{2}\mu$ (the Higgs boson, again). The next two lines tell us that the components $\hat{W}_1$ and $\hat{W}_2$ of the triplet ($\hat{W}_1$, $\hat{W}_2$, $\hat{W}_3$) acquire a mass (cf (19.56) in the U(1) case)

$$M_1 = M_2 = gv/2 \equiv M_W. \tag{19.89}$$

The last two lines show us that the fields $\hat{W}_3$ and $\hat{B}$ are mixed. But they can easily be unmixed by noting that the last term in (19.88) involves only the combination $g\hat{W}_3^\mu - g'\hat{B}^\mu$, which evidently acquires a mass. This suggests introducing the normalized linear combination

$$\hat{Z}^\mu = \cos\theta_W \hat{W}_3^\mu - \sin\theta_W \hat{B}^\mu \tag{19.90}$$

where

$$\cos\theta_W = g/(g^2 + g'^2)^{1/2} \qquad \sin\theta_W = g'/(g^2 + g'^2)^{1/2}, \tag{19.91}$$

together with the orthogonal combination

$$\hat{A}^\mu = \sin\theta_W \hat{W}_3^\mu + \cos\theta_W \hat{B}^\mu. \tag{19.92}$$

We then find that the last two lines of (19.88) become

$$-\frac{1}{4}(\partial_\mu \hat{Z}_\nu - \partial_\nu \hat{Z}_\mu)(\partial_\mu \hat{Z}^\nu - \partial^\nu \hat{Z}^\mu) + \frac{1}{8}v^2(g^2 + g'^2)\hat{Z}_\mu \hat{Z}^\mu - \frac{1}{4}\hat{F}_{\mu\nu}\hat{F}^{\mu\nu}, \tag{19.93}$$

where

$$\hat{F}_{\mu\nu} = \partial_\mu \hat{A}_\nu - \partial_\nu \hat{A}_\mu. \tag{19.94}$$

Thus

$$M_Z = \frac{1}{2}v(g^2 + g'^2)^{1/2} = M_W / \cos\theta_W \tag{19.95}$$

and

$$M_A = 0. \tag{19.96}$$

Counting degrees of freedom as in the local U(1) case, we originally had 12 in (19.79)— three massless $\hat{W}$'s and one massless $\hat{B}$, which is 8 degrees of freedom in all, together with 4 $\hat{\phi}$-fields, all with the same mass. After symmetry breaking, we have three massive vector fields $\hat{W}_1$, $\hat{W}_2$, and $\hat{Z}$ with 9 degrees of freedom, one massless vector field $\hat{A}$ with 2, and one massive scalar $\hat{H}$. Of course, the physical application will be to identify the $\hat{W}$ and $\hat{Z}$ fields with those physical particles, the $\hat{A}$ field with the massless photon, and the $\hat{H}$ field with the Higgs boson. In the gauge (19.87), the W and Z particles have propagators of the form (19.22).

The identification of $\hat{A}^\mu$ with the photon field is made clearer if we look at the form of $D_\mu \hat{\phi}$ written in terms of $\hat{A}_\mu$ and $\hat{Z}_\mu$, discarding the $\hat{W}_1$, $\hat{W}_2$ pieces:-

$$\begin{aligned}
D_\mu \hat{\phi} = & \left\{ \partial_\mu + ig\sin\theta_W \left(\frac{1 + \tau_3}{2} \right) \hat{A}_\mu \right. \\
& \left. + \frac{ig}{\cos\theta_W}\left[\frac{\tau_3}{2} - \sin^2\theta_W \left(\frac{1 + \tau_3}{2} \right) \right] \hat{Z}_\mu \right\} \hat{\phi}.
\end{aligned} \tag{19.97}$$

Now the operator $(1 + \tau_3)$ acting on $\langle 0|\hat{\phi}|0\rangle$ gives zero, as observed in (19.84), and this is why $\hat{A}_\mu$ does not acquire a mass when $\langle 0|\hat{\phi}|0\rangle \neq 0$ (gauge fields coupled to *unbroken* symmetries of $\langle 0|\hat{\phi}|0\rangle$ do *not* become massive). Although certainly not unique, this choice of $\hat{\phi}$ and $\langle 0|\hat{\phi}|0\rangle$ is undoubtedly very economical and natural. We are interpreting the zero eigenvalue of $(1 + \tau_3)$ as the electromagnetic charge of the vacuum, which we do not wish to be non-zero. We then make the identification

$$e = g \sin \theta_{\mathrm{W}} \tag{19.98}$$

in order to get the right 'electromagnetic D_μ' in (19.97).

We emphasize once more that the particular form of (19.88) corresponds to a *choice of gauge*, namely the unitary one (cf the discussions in sections 19.3 and 19.5). There is always the possibility of using other gauges, as in the Abelian case, and this will in general be advantageous when doing loop calculations involving renormalization. We would then return to a general parametrization such as (cf (19.65) and (17.96))

$$\hat{\phi} = \left(\begin{array}{c} 0 \\ v/\sqrt{2} \end{array} \right) + \frac{1}{\sqrt{2}} \left(\begin{array}{c} \hat{\phi}_2 - \mathrm{i}\hat{\phi}_1 \\ \hat{\sigma} - \mathrm{i}\hat{\phi}_3 \end{array} \right), \tag{19.99}$$

and add 't Hooft gauge-fixing terms

$$-\frac{1}{2\xi} \left\{ \sum_{i=1,2} (\partial_\mu \hat{W}_i^\mu + \xi M_{\mathrm{W}} \hat{\phi}_i)^2 + (\partial_\mu \hat{Z}^\mu + \xi M_{\mathrm{Z}} \hat{\phi}_3)^2 + (\partial_\mu \hat{A}^\mu)^2 \right\}. \tag{19.100}$$

In this case the gauge boson propagators are all of the form (19.75), and ξ-dependent. In such gauges, the Feynman rules will have to involve graphs corresponding to exchange of quanta of the 'unphysical' fields $\hat{\phi}_i$, as well as those of the physical Higgs scalar $\hat{\sigma}$. These will also have to be suitable ghost interactions in the non-Abelian sector as discussed in section 13.3.3. The complete Feynman rules of the electroweak theory are given in Appendix B of Cheng and Li (1984), for example.

The model introduced here is actually the 'Higgs sector' of the standard model, but without any couplings to fermions. We have seen how, by supposing that the potential in (19.79) has the symmetry-breaking sign of the parameter μ^2, the W$^\pm$ and Z^0 gauge bosons can be given masses. This seems to be an ingenious and even elegant 'mechanism' for arriving at a renormalizable theory of massive vector bosons. One may of course wonder whether this 'mechanism' is after all purely phenomenological, somewhat akin to the GL theory of a superconductor. In the latter case, we know that it can be derived from 'microscopic' BCS theory, and this naturally leads to the question whether there could be a similar underlying 'dynamical' theory, behind the Higgs sector.

It is, in fact, quite straightforward to imagine a theory in which the Higgs fields $\hat{\phi}$ appear as bound, or composite, states of a new sector of strongly interacting fermions, and which serve to give mass to the standard model gauge bosons. Such theories are generically known as technicolour theories, and were pioneered by Weinberg (1976, 1979b) and by Susskind (1979). These theories typically predict many new states, some analogous to the hadrons of QCD, which can be searched for at the LHC. These searches, reviewed by Black, Sakhar Chivukula and Narain (2022), have so far proved unsuccessful. The simplicity of the original models has also been severely challenged by the more recent precision electroweak experiments. We shall discuss such theories further in section 23.4.2.

But generating masses for the gauge bosons is not the only job that the Higgs sector does, in the SM. It also generates masses for all the fermions. As we will see in chapter 22, the gauge symmetry of the weak interactions is a *chiral* one, which requires that there should be no explicit fermion masses in the Lagrangian. We saw in chapter 18 how there is good evidence that the strong QCD interactions break chiral symmetry spontaneously, but that there is also a need for small Lagrangian masses for the quarks, which break chiral symmetry explicitly (so as to give mass to the pions, for example). The leptons are of course not coupled to QCD, and we have to assume Lagrangian masses for them too. Thus for both quarks and leptons chiral-symmetry-breaking mass terms seem to be required. The only way to preserve the weak chiral gauge symmetry is to assume that these fermion masses must, in their turn, be interpreted as arising 'spontaneously' also; that is, *not* via an explicit mass term in the Lagrangian. The dynamical generation of quark and lepton masses would, in fact, be closely analogous to the generation of the energy gap in the BCS theory, as we saw in section 18.1. So we may ask: is it possible to find a dynamical theory which generates masses for *both* the vector bosons, *and* the fermions? Such theories are known as 'extended technicolour models', and were introduced by Dimopoulos and Susskind (1979) and by Eichten and Lane (1980). Again, the fact that the couplings of the Higgs boson to the standard model fermions are now known to differ by no more than 10% from their standard model predictions places severe constraints on the models. Dynamical symmetry breaking will be discussed further in section 23.4.2, but it seems fair to say that no satisfactory dynamical theory of electroweak symmetry breaking has yet been constructed. Within the SM, one proceeds along what seems a more phenomenological route, attributing the masses of fermions to their couplings with the Higgs field, in a way that will be explained in chapter 22: briefly, the couplings have the Yukawa form $g_f \bar{\hat{f}} \hat{f} \hat{\phi}$, so that when $\hat{\phi}$ develops a vev v, the fermions gain a mass $m_f = g_f v$.

We now turn, in the last part of the book, to weak interactions and the electroweak theory.

Problems

19.1 Show that

$$[(-k^2 + M^2) g^{\nu\mu} + k^\nu k^\mu] \left(\frac{-g_{\mu\rho} + k_\mu k_\rho / M^2}{k^2 - M^2} \right) = g^\nu_\rho.$$

19.2 Verify (19.18).

19.3 Verify (19.19).

19.4 Verify (19.46).

19.5 Insert (19.55) into $\hat{\mathcal{L}}_H$ of (19.40) and derive (19.56) for the quadratic terms.

19.6 Insert (19.65) into $\hat{\mathcal{L}}_H$ of (19.40) and derive the quadratic terms of (19.69).

19.7 Derive (19.71).

19.8 Write the left-hand side of (19.74) in momentum space (as in (19.4)), and show that the inverse of the factor multiplying $\tilde{\hat{A}}^\mu$ is (19.75) without the 'i' (cf problem 19.1).

19.9 Verify (19.88).

Part VIII

Weak Interactions and the Electroweak Theory

20

Introduction to the Phenomenology of Weak Interactions

Public letter to the group of the Radioactives at the district society meeting in Tübingen:

Physikalisches Institut
der Eidg. Technischen Hochschule
Gloriastr.
Zürich

Zürich, 4. Dec. 1930

Dear Radioactive Ladies and Gentlemen,

As the bearer of these lines, to whom I graciously ask you to listen, will explain to you in more detail, how because of the "wrong " statistics of the N and ^{6}Li nuclei and the continuous β-spectrum, I have hit upon a desperate remedy to save the "exchange theorem" of statistics and the law of conservation of energy. Namely, the possiblity that there could exist in the nuclei electrically neutral particles, that I wish to call neutrons, which have the spin $\frac{1}{2}$ and obey the exclusion principle and which further differ from light quanta in that they do not travel with the velocity of light. The mass of the neutrons should be of the same order of magnitude as the electron mass and in any event not larger than 0.01 proton masses.—The continuous β-spectrum would then become understandable by the assumption that in β-decay, a neutron is emitted in addition to the electron such that the sum of the energies of the neutron and electron is constant. ...

I admit that on a first look my way out might seem to be quite unlikely, since one would certainly have seen the neutrons by now if they existed. But nothing ventured nothing gained, and the seriousness of the matter with the continuous β-spectrum is illustrated by a quotation of my honoured predecessor in office, Mr. Debye, who recently told me in Brussels: "Oh, it is best not to think about it, like the new taxes." Therefore one should earnestly discuss each way of salvation.—So, dear Radioactives, examine and judge it.— Unfortunately I cannot appear in Tübingen personally, since I am indispensable here in Zürich because of a ball on the night of 6/7 December.—With my best regards to you, and also Mr. Back, your humble servant,

W. Pauli

Quoted from Winter (2000), pages 4–5.

At the end of the previous chapter we arrived at an important part of the Lagrangian of the standard model, namely the terms involving just the gauge and Higgs fields. The full electroweak Lagrangian also includes, of course, the couplings of these fields to the quarks

DOI: 10.1201/9781003411666-20

and leptons. We could at this point simply write these couplings down, with little motivation, and proceed at once to discuss the empirical consequences. But such an approach, though economical, would assume considerable knowledge of weak interaction phenomenology on the reader's part. We prefer to keep this book as self-contained as possible, and so in the present chapter we shall provide an introduction to this phenomenology, following a 'semi-historical' route (for fuller historical treatments we refer the reader to Marshak *et al.* (1969), or to Winter (2000), for example).

Much of what we shall discuss is still, for many purposes, a very useful approximation to the full theory at energies well below the masses of the $W^{\pm}$ ($\sim$80 GeV) and Z^0 ($\sim$90 GeV). The reason for this is that in the electroweak theory (chapter 22), tree-level amplitudes have a structure very similar to that in the purely electromagnetic case, namely (see equation (8.101))

$$j_{\text{wk}}^{\mu} \frac{(-g_{\mu\nu} + q_{\mu}q_{\nu}/M_{\text{W,Z}}^2)}{q^2 - M_{\text{W,Z}}^2} j_{\text{wk}}^{\nu} \tag{20.1}$$

where j_{wk}^{μ} is a weak current, and we are using (19.76) for the propagator of the exchanged W or Z bosons. For $q^2 \ll M_{\text{W,Z}}^2$, (20.1) becomes proportional to the product of two currents; this 'current-current' form was for many years the basis of weak interaction phenomenology, as we now describe.

20.1 Fermi's 'current-current' theory of nuclear β-decay and its generalizations

The first quantum field theory of a weak interaction process was proposed by Fermi (1934a,b) for nuclear β-decay, building on the 'neutrino hypothesis' of Pauli. In 1930, Pauli (in his 'Dear Radioactive Ladies and Gentlemen' letter) had suggested that the continuous e^- spectrum in β-decay could be understood by supposing that, in addition to the e^-, the decaying nucleus also emitted a light, spin-$\frac{1}{2}$, electrically neutral particle, which he called the 'neutron'. In this first version of the proposal, Pauli regarded his hypothetical particle as a constituent of the nucleus. This had the attraction of solving not only the problem with the continuous e^- spectrum, but a second problem as well—what he called the 'wrong' statistics of the ^{14}N and ^{6}Li nuclei. Taking ^{14}N for definiteness, the problem was as follows. Assuming that the nucleus was somehow composed of the only particles (other than the photon) known in 1930, namely electrons and protons, one requires 14 protons and 7 electrons for the known charge of 7. This implies a half-odd integer value for the total nuclear spin. But data from molecular spectra indicated that the nitrogen nuclei obeyed Bose–Einstein, not Fermi–Dirac statistics, so that—if the usual 'spin-statistics' connection were to hold—the spin of the nitrogen nucleus should be an integer, not a half-odd integer. This second part of Pauli's hypothesis was quite soon overtaken by the discovery of the (real) neutron by Chadwick (1932), after which it was rapidly accepted that nuclei consisted of protons and (Chadwick's) neutrons.

However, the β-spectrum problem remained, and at the Solvay Conference in 1933 Pauli restated his hypothesis (Pauli 1934), using now the name 'neutrino' which had meanwhile been suggested by Fermi. Stimulated by the discussions at the Solvay meeting, Fermi then developed his theory of β-decay. In the new picture of the nucleus, neither the electron nor the neutrino were to be thought of as nuclear constituents. Instead, the electron-neutrino pair had somehow to be created and emitted in the transition process of the nuclear decay, much as a photon is created and emitted in nuclear γ-decay. Indeed, Fermi relied heavily

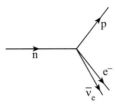

FIGURE 20.1
Four-fermion interaction for neutron β-decay.

on the analogy with electromagnetism. The basic process was assumed to be the transition neutron→proton, with the emission of an $e^-\nu$ pair, as shown in figure 20.1. The n and p were then regarded as 'elementary' and without structure (point-like); the whole process took place at a single space-time point, like the emission of a photon in QED. Further, Fermi conjectured that the nucleons participated via a weak interaction analogue of the electromagnetic transition *currents* frequently encountered in volume 1 for QED. In this case, however, rather than having the 'charge conserving' form of $\bar{u}_p\gamma^\mu u_p$ for instance, the 'weak current' had the form $\bar{u}_p\gamma^\mu u_n$, in which the charge of the nucleon changed. The lepton pair was also charged, obviously. The whole interaction then had to be Lorentz invariant, implying that the $e^-\nu$ pair had also to appear in a similar (4-vector) 'current' form. Thus a 'current–current' amplitude was proposed, of the form

$$A\bar{u}_p\gamma^\mu u_n\bar{u}_{e^-}\gamma_\mu u_\nu, \tag{20.2}$$

where A was a constant. Correspondingly, the process was described field theoretically in terms of the local interaction density

$$A\hat{\bar{\psi}}_p(x)\gamma^\mu\hat{\psi}_n(x)\hat{\bar{\psi}}_e(x)\gamma_\mu\hat{\psi}_\nu(x). \tag{20.3}$$

The discovery of positron β-decay soon followed, and then of electron capture; these processes were easily accommodated by adding to (20.3) its Hermitian conjugate

$$A\hat{\bar{\psi}}_n(x)\gamma^\mu\hat{\psi}_p(x)\hat{\bar{\psi}}_\nu(x)\gamma_\mu\hat{\psi}_e(x), \tag{20.4}$$

taking A to be real. The sum of (20.3) and (20.4) gave a good account of many observed characteristics of β-decay, when used to calculate transition probabilities in first-order perturbation theory.

Soon after Fermi's theory was presented, however, it became clear that the observed selection rules in some nuclear transitions could not be accounted for by the forms (20.3) and (20.4). Specifically, in 'allowed' transitions (where the orbital angular momentum carried by the leptons is zero) it was found that, while for many transitions the nuclear spin did not change ($\Delta J = 0$), for others—of comparable strength—a change of nuclear spin by one unit ($\Delta J = 1$) occurred. Now, in nuclear decays the energy release is very small ($\sim$ few MeV) compared to the mass of a nucleon, and so the non-relativistic limit is an excellent approximation for the nucleon spinors. It is then easy to see (problem 20.1) that, in this limit, the interactions (20.3) and (20.4) imply that the nucleon spins cannot 'flip'. Hence some other interaction(s) must be present. Gamow and Teller (1936) introduced the general four-fermion interaction, constructed from bilinear combinations of the nucleon pair and of the lepton pair, but not their derivatives. For example, the combination

$$\hat{\bar{\psi}}_p(x)\hat{\psi}_n(x)\hat{\bar{\psi}}_e(x)\hat{\psi}_\nu(x) \tag{20.5}$$

could occur, and also

$$\bar{\hat{\psi}}_{\mathrm{p}}(x)\sigma_{\mu\nu}\hat{\psi}_{\mathrm{n}}(x)\bar{\hat{\psi}}_{\mathrm{e}}\sigma^{\mu\nu}\hat{\psi}_{\nu}(x) \tag{20.6}$$

where

$$\sigma_{\mu\nu} = \frac{\mathrm{i}}{2}(\gamma_\mu\gamma_\nu - \gamma_\nu\gamma_\mu). \tag{20.7}$$

The non-relativistic limit of (20.5) gives $\Delta J = 0$, but (20.6) allows $\Delta J = 1$. Other combinations are also possible, as we shall discuss shortly. Note that the interaction must always be Lorentz invariant.

Thus began a long period of difficult experimentation to establish the correct form of the β-decay interaction. With the discovery of the muon (in 1937) and the pion (ten years later) more weak decays became experimentally accessible, for example μ decay

$$\mu^- \to \mathrm{e}^- + \nu + \nu \tag{20.8}$$

and π decay

$$\pi^- \to \mathrm{e}^- + \nu. \tag{20.9}$$

Note that we have deliberately called all the neutrinos just 'ν', without any particle/antiparticle indication, or lepton flavour label; we shall have more to say on these matters in section 20.3. There were hopes that the couplings of the pairs (p,n), (ν, e^-), and (ν, μ^-) might have the same form ('universality') but the data was incomplete, and in part apparently contradictory.

The breakthrough came in 1956, when Lee and Yang (1956) suggested that parity was not conserved in all weak decays. Hitherto, it had always been assumed that any physical interaction had to be such that parity was conserved, and this assumption had been built into the structure of the proposed β-decay interactions, such as (20.3), (20.5), or (20.6). Once it was looked for properly, following the analysis of Lee and Yang, parity violation was indeed found to be a strikingly evident feature of weak interactions.

20.2   Parity violation in weak interactions and V–A theory

20.2.1   Parity violation

In 1957, the experiment of Wu *et al.* (1957) established for the first time that parity was violated in a weak interaction, specifically nuclear β-decay. The experiment involved a sample of $^{60}\mathrm{Co}$ ($J = 5$) cooled to 0.01 K in a solenoid. At this temperature most of the nuclear spins are aligned by the magnetic field, and so there is a net polarization $\langle \boldsymbol{J} \rangle$, which is in the direction opposite to the applied magnetic field. $^{60}\mathrm{Co}$ decays to $^{60}\mathrm{Ni}$ ($J = 4$), a $\Delta J = 1$ transition. The degree of $^{60}\mathrm{Co}$ alignment was measured from observations of the angular distribution of γ-rays from $^{60}\mathrm{Ni}$. The relative intensities of electrons emitted along and against the magnetic field direction were measured, and the results were consistent with a distribution of the form

$$
\begin{aligned}
I(\theta) &= 1 - \langle \boldsymbol{J} \rangle \cdot \boldsymbol{p}/E & (20.10) \\
&= 1 - Pv\cos\theta & (20.11)
\end{aligned}
$$

where $v, \boldsymbol{p}$, and E are respectively the electron speed, momentum, and energy, P is the magnitude of the polarization, and θ is the angle of emission of the electron with respect to $\langle \boldsymbol{J} \rangle$.

Why does this indicate parity violation? To see this, we recall from the discussion of the parity operation $\mathbf{P}$ in section 4.2.1 that the angular momentum $\boldsymbol{J}$ is an axial vector such that $\langle \boldsymbol{J} \rangle \to \langle \boldsymbol{J} \rangle$ under $\mathbf{P}$, while $\boldsymbol{p}$ is a polar vector transforming by $\boldsymbol{p} \to -\boldsymbol{p}$. Hence, in the parity-transformed system, the distribution (20.11) would have the form

$$I_{\mathbf{P}}(\theta) = 1 + Pv\cos\theta \tag{20.12}$$

The difference between (20.12) and (20.11) implies that, by performing the measurement, we can *determine* which of the two coordinate systems we must in fact be using. The two are inequivalent, in contrast to all the other coordinate system equivalences which we have previously studied (e.g. under three-dimensional rotations, and Lorentz transformations). This is an operational consequence of 'parity violation'. The crucial point in this example, evidently, is the appearance of the *pseudoscalar* quantity $\langle \boldsymbol{J} \rangle \cdot \boldsymbol{p}$ in (20.10), alongside the obviously scalar quantity '1'.

The Fermi theory, employing only vector currents, needs a modification to accommodate this result. We saw in section 4.2.1 that a combination of vector ('V') and axial vector ('A') currents would be parity-violating. Indeed, after many years of careful experiments, and many false trails, it was eventually established (always, of course, to within some experimental error) that the currents participating in Fermi's current–current interaction are, in fact, certain combinations of V-type and A-type currents, for both nucleons and leptons.

20.2.2 V-A theory: chirality and helicity

Quite soon after the discovery of parity violation, Sudarshan and Marshak (1958), and then Feynman and Gell-Mann (1958) and Sakurai (1958), proposed a specific form for the current-current interaction, namely the V-A ('V minus A') structure. For example, in place of the leptonic combination $\bar{u}_{e^-}\gamma_\mu u_\nu$, these authors proposed the form $\bar{u}_{e^-}\gamma_\mu(1-\gamma_5)u_\nu$, being the difference (with equal weight) of a V-type and an A-type current. For the part involving the nucleons the proposal was slightly more complicated, having the form $\bar{u}_p\gamma_\mu(1-r\gamma_5)u_n$ where r had the empirical value $r \approx 1.2$. From our present perspective, of course, the hadronic transition is actually occurring at the quark level, so that rather than a transition $n \to p$ we now think in terms of a $d \to u$ one. In this case, the remarkable fact is that the appropriate current to use is, once again, essentially the simple 'V-A' one, $\bar{u}_u\gamma_\mu(1-\gamma_5)u_d$[1]. *This V-A structure for quarks and leptons is fundamental to the standard model.*

We must now at once draw the reader's attention to a rather remarkable feature of this V-A structure, which is that the $(1-\gamma_5)$ factor can be thought of as acting either on the u spinor or on the $\bar{u}$ spinor. Consider, for example, a term $\bar{u}_{e^-}\gamma_\mu(1-\gamma_5)u_\nu$. We have

$$
\begin{aligned}
\bar{u}_{e^-}\gamma_\mu(1-\gamma_5)u_\nu &= u_{e^-}^\dagger \beta\gamma_\mu(1-\gamma_5)u_\nu \\
&= u_{e^-}^\dagger (1-\gamma_5)\beta\gamma_\mu u_\nu \\
&= [(1-\gamma_5)u_{e^-}]^\dagger \beta\gamma_\mu u_\nu \\
&= \overline{[(1-\gamma_5)u_{e^-}]}\,\gamma_\mu u_\nu.
\end{aligned}
\tag{20.13}
$$

To understand the significance of this, it is advantageous to work in the representation (3.40) of the Dirac matrices, in which γ_5 is diagonal, namely

$$
\gamma_5 = \begin{pmatrix} 1 & 0 \\ 0 & -1 \end{pmatrix} \qquad
\boldsymbol{\alpha} = \begin{pmatrix} \boldsymbol{\sigma} & 0 \\ 0 & -\boldsymbol{\sigma} \end{pmatrix} \qquad
\beta = \begin{pmatrix} 0 & 1 \\ 1 & 0 \end{pmatrix} \qquad
\boldsymbol{\gamma} = \begin{pmatrix} 0 & -\boldsymbol{\sigma} \\ \boldsymbol{\sigma} & 0 \end{pmatrix}.
\tag{20.14}
$$

---

[1] We shall see in section 20.7 that a slight modification is necessary.

Readers who have not worked through problem 9.4 might like to do so now; we may also suggest a backward glance at section 12.4.2 and chapter 17.

First of all it is clear that any combination '$(1 - \gamma_5)u$' is an eigenstate of γ_5 with eigenvalue -1:

$$\gamma_5(1 - \gamma_5)u = (\gamma_5 - 1)u = -(1 - \gamma_5)u \tag{20.15}$$

using $\gamma_5^2 = 1$. In the terminology of section 12.4.2, '$(1 - \gamma_5)u$' has definite *chirality*, namely L ('left-handed'), meaning that it belongs to the eigenvalue -1 of γ_5. We may introduce the projection operators P_R, P_L of section 12.4.2,

$$P_L \equiv \left(\frac{1 - \gamma_5}{2}\right) \qquad P_R \equiv \left(\frac{1 + \gamma_5}{2}\right) \tag{20.16}$$

satisfying

$$P_R^2 = P_R \qquad P_L^2 = P_L \qquad P_R P_L = P_L P_R = 0 \qquad P_R + P_L = 1, \tag{20.17}$$

and define

$$u_L \equiv P_L u, \qquad u_R \equiv P_R u \tag{20.18}$$

for any u. Then

$$
\begin{aligned}
\bar{u}_1 \gamma_\mu \left(\frac{1 - \gamma_5}{2}\right) u_2 &= \bar{u}_1 \gamma_\mu P_L u_2 = \bar{u}_1 \gamma_\mu P_L^2 u_2 \\
&= \bar{u}_1 \gamma_\mu P_L u_{2L} = \bar{u}_1 P_R \gamma_\mu u_{2L} \\
&= u_1^\dagger P_L \beta \gamma_\mu u_{2L} = \bar{u}_{1L} \gamma_\mu u_{2L}
\end{aligned} \tag{20.19}
$$

which formalizes (20.13) and emphasizes the fact that *only the chiral L components of the u spinors enter into weak interactions*, a remarkably simple statement.

To see the physical consequences of this, we need the forms of the Dirac spinors in this new representation, which we shall now derive explicitly, for convenience. As usual, positive energy spinors are defined as solutions of $(\not{p} - m)u = 0$, so that writing

$$u = \begin{pmatrix} \phi \\ \chi \end{pmatrix} \tag{20.20}$$

we obtain

$$
\begin{aligned}
(E - \boldsymbol{\sigma} \cdot \boldsymbol{p})\phi &= m\chi \\
(E + \boldsymbol{\sigma} \cdot \boldsymbol{p})\chi &= m\phi.
\end{aligned} \tag{20.21}
$$

A convenient choice of 2-component spinors ϕ, χ is to take them to be *helicity eigenstates* (see section 3.3). For example, the eigenstate ϕ_+ with positive helicity $\lambda = +1$ satisfies

$$\boldsymbol{\sigma} \cdot \boldsymbol{p}\phi_+ = |\boldsymbol{p}|\phi_+ \tag{20.22}$$

while the eigenstate ϕ_- with $\lambda = -1$ satisfies (20.22) with a minus on the right-hand side. Thus the spinor $u(p, \lambda = +1)$ can be written as

$$u(p, \lambda = +1) = N \begin{pmatrix} \phi_+ \\ \frac{(E - |\boldsymbol{p}|)}{m}\phi_+ \end{pmatrix}. \tag{20.23}$$

The normalization N is fixed as usual by requiring $\bar{u}u = 2m$, from which it follows (problem 20.2) that $N = (E + |\boldsymbol{p}|)^{1/2}$. Thus finally we have

$$u(p, \lambda = +1) = \begin{pmatrix} \sqrt{E + |\boldsymbol{p}|}\phi_+ \\ \sqrt{E - |\boldsymbol{p}|}\phi_+ \end{pmatrix}. \tag{20.24}$$

Similarly

$$u(p, \lambda = -1) = \begin{pmatrix} \sqrt{E - |\boldsymbol{p}|}\phi_- \\ \sqrt{E + |\boldsymbol{p}|}\phi_- \end{pmatrix}. \tag{20.25}$$

Now we have agreed that only the chiral 'L' components of all u-spinors enter into weak interactions, in the standard model. But from the explicit form of γ_5 given in (20.14), we see that when acting on any spinor u, the projector P_L 'kills' the top two components:

$$P_L \begin{pmatrix} \phi \\ \chi \end{pmatrix} = \begin{pmatrix} 0 \\ \chi \end{pmatrix}. \tag{20.26}$$

In particular

$$P_L u(p, \lambda = +1) = \begin{pmatrix} 0 \\ \sqrt{E - |\boldsymbol{p}|}\phi_+ \end{pmatrix} \tag{20.27}$$

and

$$P_L u(p, \lambda = -1) = \begin{pmatrix} 0 \\ \sqrt{E + |\boldsymbol{p}|}\phi_- \end{pmatrix}. \tag{20.28}$$

Equations (20.27) and (20.28) are very important. In particular, equation (20.27) implies that in the limit of zero mass m (and hence $E \to |\boldsymbol{p}|$), only the *negative helicity* u-spinor will enter. More quantitatively, using

$$\frac{\sqrt{E - |\boldsymbol{p}|}}{\sqrt{E + |\boldsymbol{p}|}} = \frac{\sqrt{E^2 - \boldsymbol{p}^2}}{(E + |\boldsymbol{p}|)} \approx \frac{m}{2E} \qquad \text{for } m \ll E, \tag{20.29}$$

we can say that *positive helicity components of all fermions are suppressed in V-A matrix elements, relative to the negative helicity components, by factors of order* (m/E). Bearing in mind that the helicity operator $\boldsymbol{\sigma} \cdot \boldsymbol{p}/|\boldsymbol{p}|$ is a pseudoscalar, this 'unequal' treatment for $\lambda = +1$ and $\lambda = -1$ components is, of course, precisely related to the parity violation built in to the V-A structure.

A similar analysis may be done for the v-spinors. They satisfy $(\not{p} + m)v = 0$ and the normalization $\bar{v}v = -2m$. We must however remember the 'small subtlety' to do with the labelling of v-spinors, discussed in section 3.4.3: the 2-component spinors χ_- in $v(p, \lambda = +1)$ actually satisfy $\boldsymbol{\sigma} \cdot \boldsymbol{p}\chi_- = -|\boldsymbol{p}|\chi_-$, and similarly the χ_+'s in $v(p, \lambda = -1)$ satisfy $\boldsymbol{\sigma} \cdot \boldsymbol{p}\chi_+ = |\boldsymbol{p}|\chi_+$. We then find (problem 20.3) the results

$$v(p, \lambda = +1) = \begin{pmatrix} -\sqrt{E - |\boldsymbol{p}|}\chi_- \\ \sqrt{E + |\boldsymbol{p}|}\chi_- \end{pmatrix} \tag{20.30}$$

and

$$v(\lambda = -1) = \begin{pmatrix} \sqrt{E + |\boldsymbol{p}|}\chi_+ \\ -\sqrt{E - |\boldsymbol{p}|}\chi_+ \end{pmatrix}. \tag{20.31}$$

Once again, the action of P_L removes the top two components, leaving the result that, in the massless limit, only the $\lambda = +1$ state survives. Recalling the 'hole theory' interpretation of section 3.4.3, this would mean that *the positive helicity components of all antifermions dominate in V-A interactions*, negative helicity components being suppressed by factors of order m/E. The proportionality of the negative helicity amplitude to the mass of the antifermion is of course exactly as noted for $\pi^+ \to \mu^+ \nu_\mu$ decay in section 18.2.

We should emphasize that although the above results, stated in italics, were derived in the convenient representation (20.14) for the Dirac matrices, they actually hold independently of any choice of representation. This can be shown by using general helicity projection operators.

In Pauli's original letter, he suggested that the mass of the neutrino might be of the same order as the electron mass. Immediately after the discovery of parity violation, it was realized that the result could be elegantly explained by the assumption that the neutrinos were strictly massless particles (Landau (1957), Lee and Yang (1957) and Salam (1957)). In this case, u and v spinors satisfy the same equation $p\!\!\!/(u \text{ or } v) = 0$, which reduces via (20.21) (in the $m = 0$ limit) to the two independent 2-component 'Weyl' equations.

$$E\phi_0 = \boldsymbol{\sigma} \cdot \boldsymbol{p}\, \phi_0 \qquad E\chi_0 = -\boldsymbol{\sigma} \cdot \boldsymbol{p}\, \chi_0. \qquad (20.32)$$

Remembering that $E = |\boldsymbol{p}|$ for a massless particle, we see that ϕ_0 has positive helicity and χ_0 negative helicity. In this strictly massless case, helicity is Lorentz invariant, since the direction of $\boldsymbol{p}$ cannot be reversed by a velocity transformation with $v < c$. Furthermore, each of the equations in (20.32) violates parity, since E is clearly a scalar while $\boldsymbol{\sigma} \cdot \boldsymbol{p}$ is a pseudoscalar (note that when $m \neq 0$ we can infer from (20.21) that, in this representation, $\phi \leftrightarrow \chi$ under **P**, which is consistent with (20.32) and with the form of β in (20.14)). Thus the (massless) neutrino could be 'blamed' for the parity violation. In this model, neutrinos have one definite helicity, either positive or negative. As we have seen, the massless limit of the (4-component) V-A theory leads to the same conclusion.

Which helicity is actually chosen by Nature was determined in a classic experiment by Goldhaber *et al.* (1958), involving the K-capture reaction

$$e^- + {}^{152}\text{Eu} \rightarrow \nu + {}^{152}\text{Sm}^*, \qquad (20.33)$$

as described by Bettini (2008), for example. They found that the helicity of the emitted neutrino was (within errors) 100% *negative*, a result taken as confirming the '2-component' neutrino theory and the V-A theory.

We now know that neutrinos are not massless. This information does not come from studies of nuclear decays, but rather from a completely different phenomenon—that of *neutrino oscillations*, which we shall mention again in the following section, and treat more fully in section 21.4. Neutrino masses are so small that the existence of the 'wrong helicity' component cannot be detected experimentally in processes such as (20.33), or indeed in any of the reactions we shall discuss, apart from neutrino oscillations.

In section 4.2.2 we introduced the charge conjugation operation **C** (see also section 7.5.2). As we noted there, **C** is not a good symmetry in weak interactions. The V-A interaction treats a negative helicity fermion very differently from a negative helicity antifermion, while one is precisely transformed into the other under **C**. However, it is clear that the helicity operator itself is odd under **P**. Thus the **CP** conjugate of a negative helicity fermion is positive helicity antifermion, which is what the V-A interaction selects. It may easily be verified (problem 20.4) that the '2-component' theory of (20.32) automatically incorporates **CP** invariance. Elegance notwithstanding, however, there are **CP**-violating weak interactions, as mentioned in section 4.2.3. How this is accommodated within the SM we shall discuss in section 20.7.3.

For charged fermions the distinction between particle and antiparticle is clear; but is there a conserved quantum number which we can use instead of charge to distinguish a neutrino from an antineutrino? That is the question to which we now turn.

20.3   Lepton number and lepton flavours

In section 1.2.1 of volume 1 we gave a brief discussion of leptonic quantum numbers ('lepton flavours'), adopting a traditional approach in which the data is interpreted in terms of conserved quantum numbers carried by neutrinos, which serve to distinguish neutrinos from antineutrinos. We must now examine the matter more closely, in the light of what we have learned about the helicity properties of the V-A interaction.

In 1995, Davis (1955)—following a suggestion made by Pontecorvo (1946)—argued as follows. Consider the e^- capture reaction $e^- + p \rightarrow \nu + n$, which was of course well established. Then in principle the inverse reaction $\nu + n \rightarrow e^- + p$ should also exist. Of course, the cross section is extremely small, but by using a large enough target volume this might perhaps be compensated. Specifically, the reaction $\nu + {}^{37}_{17}Cl \rightarrow e^- + {}^{37}_{18}Ar$ was proposed, the argon being detected through its radioactive decay. Suppose, however, that the 'neutrinos' actually used are those which accompany electrons in β^--decay. If (as was supposed in section 1.2.1) these are to be regarded as antineutrinos, '$\bar\nu$', carrying a conserved lepton number, then the reaction

$$\text{`}\bar\nu\text{'} + {}^{37}_{17}Cl \rightarrow e^- + {}^{37}_{18}Ar \tag{20.34}$$

should *not* be observed. If, on the other hand, the 'ν' in the capture process and the '$\bar\nu$' in β-decay are not distinguished by the weak interaction, the reaction (20.34) should be observed. Davis found no evidence for reaction (20.34), at the expected level of cross section, a result which could clearly be interpreted as confirming the 'conserved electron number hypothesis'.

However, another interpretation is possible. The e^- in β-decay has predominately negative helicity, and its accompanying '$\bar\nu$' has predominately positive helicity. The fraction of the other helicity present is of the order m/E, where $E \sim$ few Mev, and the neutrino mass is less than 1eV; this is, therefore, an almost undetectable 'contamination' of negative helicity component in the '$\bar\nu$'. Now the property of the V-A interaction is that it conserves helicity in the zero mass limit (in which chirality is the same as helicity). Hence the positive helicity '$\bar\nu$' from β^--decay will (predominately) produce a positive helicity lepton, which must be the e^+ not the e^-. Thus the property of the V-A interaction, together with the very small value of the neutrino mass, conspire effectively to forbid (20.34), independently of any considerations about 'lepton number'.

Indeed, the 'helicity-allowed' reaction

$$\text{`}\bar\nu\text{'} + p \rightarrow e^+ + n \tag{20.35}$$

was observed by Reines and Cowan (1956) (see also Cowan *et al.* (1956)). Reaction (20.35) too, of course, can be interpreted in terms of '$\bar\nu$' carrying a lepton number of -1, equal to that of the e^+. It was also established that only 'ν' produced e^- via (20.34), where 'ν' is the helicity -1 state (or, on the other interpretation, the carrier of lepton number $+1$).

The situation may therefore be summarized as follows. In the case of e^- and e^+, all four 'modes'—$e^-(\lambda = +1), e^-(\lambda = -1), e^+(\lambda = +1), e^+(\lambda = -1)$—are experimentally accessible via electromagnetic interactions, even though only two generally dominate in weak interactions ($e^-(\lambda = -1)$ and $e^+(\lambda = +1)$). Neutrinos, on the other hand, seem to interact only weakly. In their case, we may if we wish say that the participating states are (in association with e^- or e^+) $\bar\nu_e$ $(\lambda = +1)$ and $\nu_e(\lambda = -1)$, to a very good approximation. But we may also regard these two states as simply two different helicity state of one particle, rather than of a particle and its antiparticle. As we have seen, the helicity rules do the job required just as well as the lepton number rules. In short, the question is : are these 'neutrinos' distinguished only by their helicity, or is there an additional distinguishing characteristic ('electron number')? In the latter case we should expect the 'other' two states $\bar\nu_e(\lambda = -1)$ and $\nu_e(\lambda = +1)$ to exist as well as the ones known from weak interactions.

If, in fact, no quantum number—other than the helicity—exists which distinguishes the neutrino states, then we would have to say that the **C**-conjugate of a neutrino state is a neutrino, not an antineutrino—that is, 'neutrinos are their own antiparticles'. A neutrino

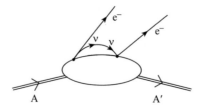

FIGURE 20.2
Double β-decay without emission of a neutrino, a test for Majorana-type neutrinos.

would be a fermionic state somewhat like a photon, which is of course also its own antiparticle. Such '**C**-self-conjugate' fermions are called *Majorana fermions* (Majorana 1937), in contrast to the *Dirac* variety, which have all four possible modes present (2 helicities, 2 particle/antiparticle). We discussed Majorana fermions in sections 4.2.2 and 7.5.2.

The distinction between the 'Dirac' and 'Majorana' neutrino possibilities becomes an essentially 'metaphysical' one in the limit of strictly massless neutrinos, since then (as we have seen) a given helicity state cannot be flipped by going to a suitably moving Lorentz frame, nor by any weak (or electromagnetic) interaction, since they both conserve chirality which is the same as helicity in the massless limit. We would have just the two states $\nu_e(\lambda = -1)$ and $\bar{\nu}_e(\lambda = +1)$, and no way of creating $\nu_e(\lambda = +1)$ or $\bar{\nu}_e(\lambda = -1)$. The '$-$' label then becomes superfluous. Unfortunately, the massless limit is approached smoothly, and neutrino masses are, in fact, so small that the 'wrong helicity' suppression factors will make it very difficult to see the presence of the possible states $\nu_e(\lambda = +1), \bar{\nu}_e(\lambda = -1)$.

One much-discussed experimental test case (see, for example, the review by Gonzalez-Garcia and Yokoyama 2022) concerns 'neutrinoless double β-decay' (often denoted by $0\nu\beta\beta$), which is the process A $\to$ A$' + e^- + e^-$, where A, A$'$ are nuclei. If the neutrino emitted in the first β-decay carries no electron-type conserved quantum number, then in principle it can initiate a second weak interaction, exactly as in Davis' original argument, via the diagram shown in figure 20.2. Note that this is a second-order weak process, so that the amplitude contains the very small factor G_F^2. Furthermore, the ν emitted along with the e$^-$ at the first vertex will be predominately $\lambda = +1$, but in the second vertex the V-A interaction will 'want' it to have $\lambda = -1$, like the outgoing e$^-$. Thus there is bound to be one 'm/E' suppression factor, whichever vertex we choose to make 'easy'. (In the case of 3-state neutrino mixing—see section 21.4—the quantity 'm' will be an appropriately averaged mass.) There is also a complicated nuclear physics overlap factor. The expected half-lives of neutrinoless double β decays depend on the decaying nucleus, but are typically longer than 10^{24}–10^{25} years. The current best limit is set by the GERDA experiment (M. Agostini *et al.* 2020): the half-life of ^{76}Ge for $0\nu\beta\beta$ decay is greater than 1.8×10^{26} at 90% CL.

In the same way, '$\bar{\nu}''$' particles accompanying the μ^-'s in π^- decay

$$\pi^- \to \mu^- + {}^{\prime}\bar{\nu}'' \qquad (20.36)$$

are observed to produce only μ^+'s when they interact with matter, not μ^-'s. Again this can be interpreted either in terms of helicity conservation or in terms of conservation of a leptonic quantum number L_μ. We shall assume the analogous properties are true for the $\bar{\nu}'''$s accompanying τ leptons.

On the other hand, helicity arguments alone would allow the reaction

$$\text{`}\bar{\nu}'' + \text{p} \to \text{e}^+ + \text{n} \qquad (20.37)$$

to proceed, but as we saw in section 1.2.1 the experiment of Danby *et al.* (1962) found no evidence for it. Thus there is evidence, in this type reaction, for a flavour quantum number distinguishing neutrinos which interact in association with one kind of charged lepton from those which interact in association with a different charged lepton. The electroweak sector of the SM was originally formulated on the assumption that the three lepton flavours L_e, L_μ, and L_τ are conserved, and that the neutrinos are massless. It turns out that these two assumptions are related, in the sense that if neutrinos have mass, then (barring degeneracies) 'neutrino oscillations' can occur, in which a state of one lepton flavour can acquire a component of another, as it propagates. Compelling evidence accumulated during the 2000s for oscillations of neutrinos caused by non-zero masses and neutrino mixing. Strictly speaking, neutrino masses and oscillations lie outside the framework of the original SM, and they are sometimes so regarded. Apart from anything else, the phenomenology of massive neutrinos has to allow for the possibility that they are Majorana, rather than Dirac, fermions. For the moment, we shall continue with a semi-historical path, and proceed with weak interaction phenomenology on the basis of the original SM, with massless neutrinos. We return to the question of neutrino mass when we discuss neutrino oscillations (along with analogous oscillations in meson systems) in chapter 21.

20.4 The universal current × current theory for weak interactions of leptons

After the breakthroughs of parity violation and V-A theory, the earlier hopes (Pontecorvo 1947, Klein 1948, Puppi 1948, Lee, Rosenbluth and Yang 1949, Tiomno and Wheeler 1949) were revived of a universal weak interaction among the pairs of particles (p,n), (ν_e, e^-), (ν_μ, μ^-), using the V-A modification to Fermi's theory. From our modern standpoint, this list has to be changed by the replacement of (p,n) by the corresponding quarks (u,d), and by the inclusion of the third lepton pair (ν_τ, τ^-) as well as two other quark pairs (c,s) and (t,b). It is to these pairs that the 'V-A' structure applies, as already indicated in section 20.2.2, and a certain form of 'universality' does hold, as we now describe.

Because of certain complications which arise, we shall postpone the discussion of the quark currents until section 20.7, concentrating here on the leptonic currents[2]. In this case, Fermi's original vector-like current $\bar{\hat{\psi}}_e \gamma^\mu \hat{\psi}_\nu$ becomes modified to a *total leptonic charged current*

$$\hat{j}^\mu_{\text{CC}}(\text{leptons}) = \hat{j}^\mu_{\text{wk}}(e) + \hat{j}^\mu_{\text{wk}}(\mu) + \hat{j}^\mu_{\text{wk}}(\tau) \tag{20.38}$$

where, for example,

$$\hat{j}^\mu_{\text{wk}}(e) = \bar{\hat{\nu}}_e \gamma^\mu (1 - \gamma_5) \hat{e}. \tag{20.39}$$

In (20.39) we are now adopting, for the first time, a useful shorthand whereby the field operator for the electron field, say, is denoted by $\hat{e}(x)$ rather than $\hat{\psi}_e(x)$, and the 'x' argument is suppressed. The 'charged' current terminology refers to the fact that these weak current operators $\hat{j}^\mu_{\text{wk}}$ carry net charge, in contrast to an electromagnetic current operator such as $\bar{\hat{e}} \gamma^\mu \hat{e}$ which is electrically neutral. We shall see in section 20.6 that there are also electrically neutral weak currents.

---

[2]Very much the same complications arise for the leptonic currents too, in the case of massive neutrinos, as we shall see in section 21.4.

The interaction Hamiltonian density accounting for all leptonic weak interactions is then taken to be

$$\hat{\mathcal{H}}_{\rm CC}^{\rm lep} = \frac{G_{\rm F}}{\sqrt{2}} \hat{j}_{\rm CC}^{\mu}(\text{leptons})\hat{j}_{{\rm CC}\mu}^{\dagger}(\text{leptons}). \tag{20.40}$$

Note that

$$(\bar{\hat{\nu}}_{\rm e}\gamma^{\mu}(1-\gamma_5)\hat{e})^{\dagger} = \bar{\hat{e}}\gamma^{\mu}(1-\gamma_5)\hat{\nu}_{\rm e} \tag{20.41}$$

and similarly for the other bilinears. The currents can also be written in terms of the chiral components of the fields (recall section 20.2.2) using

$$2\bar{\hat{\nu}}_{\rm eL}\gamma^{\mu}\hat{e}_{\rm L} = \bar{\hat{\nu}}_{\rm e}\gamma^{\mu}(1-\gamma_5)\hat{e}, \tag{20.42}$$

for example. 'Universality' is manifest in the fact that all the lepton pairs have the same form of the V-A coupling, and the same 'strength parameter' $G_{\rm F}/\sqrt{2}$ multiplies all of the products in (20.40).

The terms in (20.40), when it is multiplied out, describe many physical processes. For example, the term

$$\frac{G_{\rm F}}{\sqrt{2}}\bar{\hat{\nu}}_{\mu}\gamma^{\mu}(1-\gamma_5)\hat{\mu}\,\bar{\hat{e}}\gamma_{\mu}(1-\gamma_5)\hat{\nu}_{\rm e} \tag{20.43}$$

describes μ^- decay:

$$\mu^- \to \nu_{\mu} + {\rm e}^- + \bar{\nu}_{\rm e}, \tag{20.44}$$

as well as all the reactions related by 'crossing' particles from one side to the other, for example

$$\nu_{\mu} + {\rm e}^- \to \mu^- + \nu_{\rm e}. \tag{20.45}$$

The value of $G_{\rm F}$ can be determined from the rate for process (20.44) (see for example Renton (1990) section 6.1.2), and it is found to be

$$G_{\rm F} \simeq 1.166 \times 10^{-5}{\rm GeV}^{-2}. \tag{20.46}$$

This is a convenient moment to notice that the theory is *not renormalizable* according to the criteria discussed in section 11.8 at the end of the previous volume: $G_{\rm F}$ has dimensions $({\rm mass})^{-2}$. We shall return to this aspect of Fermi-type V-A theory in section 22.1.

There are also what we might call 'diagonal' terms in which the same lepton pair is taken from $\hat{j}_{\rm wk}^{\mu}$ and $\hat{j}_{{\rm wk}\mu}^{\dagger}$, for example

$$\frac{G_{\rm F}}{\sqrt{2}}\bar{\hat{\nu}}_{\rm e}\gamma^{\mu}(1-\gamma_5)\hat{e}\,\bar{\hat{e}}\gamma_{\mu}(1-\gamma_5)\hat{\nu}_{\rm e} \tag{20.47}$$

which describes reactions such as

$$\bar{\nu}_{\rm e} + {\rm e}^- \to \bar{\nu}_{\rm e} + {\rm e}^-. \tag{20.48}$$

The cross section for (20.48) was measured by Reines, Gurr, and Sobel (1976) after many years of effort; the value obtained was consistent with the Glashow-Salam-Weinberg theory (see section 22.3), with the parameter $\sin^2\theta_{\rm W} = 0.29 \pm 0.05$.

It is interesting that some seemingly rather similar processes are forbidden to occur, to first order in $\hat{\mathcal{H}}_{\rm wk}^{\rm lep}$, for example

$$\bar{\nu}_{\mu} + {\rm e}^- \to \bar{\nu}_{\mu} + {\rm e}^-. \tag{20.49}$$

For reasons which will become clearer in section 20.6, (20.49) is called a 'neutral current' process, in contrast to all the others (such as β-decay or μ-decay) we have discussed so far, which are called 'charged current' processes. If the lepton pairs are arranged so as to have no net lepton number (for example $\mathrm{e}^-\bar{\nu}_\mathrm{e}, \mu^+\nu_\mu, \nu_\mu\bar{\nu}_\mu$ etc.) then pairs with non-zero charge occur in charged current processes, while those with zero charge participate in neutral current processes. In the case of (20.48), the leptons can be grouped either as $(\bar{\nu}_\mathrm{e}\mathrm{e}^-)$ which is charged, or as $(\bar{\nu}_\mathrm{e}\nu_\mathrm{e})$ or $(\mathrm{e}^+\mathrm{e}^-)$ which are neutral. On the other hand, there is no way of pairing the leptons in (20.49) so as to cancel the lepton number and have non-zero charge. So (20.49) is a *purely* 'neutral current' process, while *some* 'neutral current' contribution could be present in (20.48), in principle. In 1973 such neutral current processes were discovered (Hasert *et al.* 1973), generating a whole new wave of experimental activity. Their existence had, in fact, been *predicted* in the first version of the standard model, due to Glashow (1961). Today we know that charged current processes are mediated by the $\mathrm{W}^\pm$ bosons, and the neutral current ones by the Z^0. We shall discuss the neutral current couplings in section 20.6.

20.5   Calculation of the cross section for $\nu_\mu + \mathrm{e}^- \to \mu^- + \nu_\mathrm{e}$

After so much qualitative discussion it is time to calculate something. We choose the process (20.45), sometimes called inverse muon decay, which is a pure 'charged current' process. The amplitude, in the Fermi-like V-A current theory, is

$$\mathcal{M} = -\mathrm{i}(G_\mathrm{F}/\sqrt{2})\bar{u}(\mu, k')\gamma_\mu(1 - \gamma_5)u(\nu_\mu, k)\bar{u}(\nu_\mathrm{e}, p')\gamma^\mu(1 - \gamma_5)u(\mathrm{e}, p). \tag{20.50}$$

We shall be interested in energies much greater than any of the leptons, and so we shall work in the *massless limit*; this is mainly for ease of calculation—the full expressions for non-zero masses can be obtained with more effort.

From the general formula (6.129) for $2 \to 2$ scattering in the CM system, we have, neglecting all masses,

$$\frac{\mathrm{d}\sigma}{\mathrm{d}\Omega} = \frac{1}{64\pi^2 s}|\overline{\mathcal{M}}|^2 \tag{20.51}$$

where $|\overline{\mathcal{M}}|^2$ is the appropriate spin-averaged matrix element squared, as in (8.183) for example. In the case of neutrino-electron scattering, we must average over initial electron states for unpolarized electrons and sum over the final muon polarization states. For the neutrinos there is no averaging over initial neutrino helicities, since only left-handed (massless) neutrinos participate in the weak interaction. Similarly, there is no sum over final neutrino helicities. However, for convenience of calculation, we can in fact sum over both helicity states of both neutrinos since the $(1 - \gamma_5)$ factors guarantee that right-handed neutrinos contribute nothing to the cross section. As for the $\mathrm{e}\mu$ scattering example in section 8.7, the calculation then reduces to a product of traces:

$$|\overline{\mathcal{M}}|^2 = \left(\frac{G_\mathrm{F}^2}{2}\right)\mathrm{Tr}[\not{k}'\gamma_\mu(1 - \gamma_5)\,\not{k}\gamma_\nu(1 - \gamma_5)]\frac{1}{2}\mathrm{Tr}[\not{p}'\gamma^\mu(1 - \gamma_5)\,\not{p}\gamma^\nu(1 - \gamma_5)], \tag{20.52}$$

all lepton masses being neglected. We define

$$|\overline{\mathcal{M}}|^2 = \left(\frac{G_\mathrm{F}^2}{2}\right)N_{\mu\nu}E^{\mu\nu} \tag{20.53}$$

where the $\nu_\mu \to \mu^-$ tensor $N_{\mu\nu}$ is given by

$$N_{\mu\nu} = \mathrm{Tr}[\not{k}'\gamma_\mu(1 - \gamma_5)\,\not{k}\gamma_\nu(1 - \gamma_5)] \tag{20.54}$$

without a $1/(2s+1)$ factor, and the $e^- \to \nu_e$ tensor is

$$E^{\mu\nu} = \frac{1}{2}\mathrm{Tr}[\not{p}'\gamma^\mu(1-\gamma_5)\not{p}\gamma^\nu(1-\gamma_5)] \qquad (20.55)$$

including a factor of $\frac{1}{2}$ for spin averaging.

Since this calculation involves a couple of new features, let us look at it in some detail. By commuting the $(1-\gamma_5)$ factor through two γ matrices ($\not{p}\gamma^\nu$) and using the result that

$$(1-\gamma_5)^2 = 2(1-\gamma_5) \qquad (20.56)$$

the tensor $N_{\mu\nu}$ may be written as

$$\begin{aligned}
N_{\mu\nu} &= 2\mathrm{Tr}[\not{k}'\gamma_\mu(1-\gamma_5)\,\not{k}\gamma_\nu] \\
&= 2\mathrm{Tr}(\not{k}'\gamma_\mu\,\not{k}\gamma_\nu) - 2\mathrm{Tr}(\gamma_5\,\not{k}\gamma_\nu\,\not{k}'\gamma_\mu).
\end{aligned} \qquad (20.57)$$

The first trace is the same as in our calculation of $e\mu$ scattering (cf (8.186)):

$$\mathrm{Tr}(\not{k}'\gamma_\mu\,\not{k}\gamma_\nu) = 4[k'_\mu k_\nu + k'_\nu k_\mu + (q^2/2)g_{\mu\nu}]. \qquad (20.58)$$

The second trace must be evaluated using the result

$$\mathrm{Tr}(\gamma_5\,\not{a}\,\not{b}\,\not{c}\,\not{d}) = 4i\epsilon_{\alpha\beta\gamma\delta}a^\alpha b^\beta c^\gamma d^\delta \qquad (20.59)$$

(see equation (J.37) in Appendix J of volume 1). The totally antisymmetric tensor $\epsilon_{\alpha\beta\gamma\delta}$ is just the generalization of ϵ_{ijk} to four dimensions, and is defined by

$$\epsilon_{\alpha\beta\gamma\delta} = \begin{cases} +1 & \text{for } \epsilon_{0123} \text{ and all even permutations of } 0,1,2,3 \\ -1 & \text{for } \epsilon_{1023} \text{ and all odd permutations of } 0,1,2,3 \\ 0 & \text{otherwise.} \end{cases} \qquad (20.60)$$

Its appearance here is a direct consequence of parity violation. Notice that this definition has the consequence that

$$\epsilon_{0123} = +1 \qquad (20.61)$$

but

$$\epsilon^{0123} = -1. \qquad (20.62)$$

We will also need to contract two ϵ tensors. By looking at the possible combinations, it should be easy to convince yourself of the result

$$\epsilon_{ijk}\epsilon_{ilm} = \begin{vmatrix} \delta_{jl} & \delta_{jm} \\ \delta_{kl} & \delta_{km} \end{vmatrix} \qquad (20.63)$$

i.e.

$$\epsilon_{ijk}\epsilon_{ilm} = \delta_{jl}\delta_{km} - \delta_{kl}\delta_{jm}. \qquad (20.64)$$

For the four-dimensional ϵ tensor one can show (see problem 20.6)

$$\epsilon_{\mu\nu\alpha\beta}\epsilon^{\mu\nu\gamma\delta} = -2! \begin{vmatrix} \delta_\alpha^\gamma & \delta_\beta^\gamma \\ \delta_\alpha^\delta & \delta_\beta^\delta \end{vmatrix} \qquad (20.65)$$

where the minus sign arises from (20.62) and the 2! from the fact that the two indices are contracted.

We can now evaluate $N_{\mu\nu}$. We obtain, after some rearrangement of indices, the result for the $\nu_\mu \to \mu^-$ tensor:

$$N_{\mu\nu} = 8[(k'_\mu k_\nu + k'_\nu k_\mu + (q^2/2)g_{\mu\nu}) - i\epsilon_{\mu\nu\alpha\beta}k^\alpha k'^\beta]. \qquad (20.66)$$

For the electron tensor $E^{\mu\nu}$ we have a similar result (divided by 2):

$$E^{\mu\nu} = 4[(p'^{\mu}p^{\nu} + p'^{\nu}p^{\mu} + (q^2/2)g^{\mu\nu}) - i\epsilon^{\mu\nu\gamma\delta}p_\gamma p'_\delta]. \tag{20.67}$$

Next, we have to perform the contraction $N_{\mu\nu}E^{\mu\nu}$ in (20.53). In the case of elastic $e^-\mu^-$ scattering considered in section 8.7, the analogous contraction between the tensors $L_{\mu\nu}$ and $M^{\mu\nu}$ was simplified by using the conditions $q^\mu L_{\mu\nu} = q^\nu L_{\mu\nu} = 0$ (see (8.189)), which followed from electromagnetic current conservation at the electron vertex (see (8.188)): $q^\mu \bar{u}(k')\gamma_\mu u(k) = 0$. Here, the analogous vertex is $\bar{u}(\mu, k')\gamma_\mu(1 - \gamma_5)u(\nu_\mu, k)$. In this case, when we contract this with $q^\mu = (k - k')^\mu$ we find a non-zero result:

$$(m_{\nu_\mu} - m_\mu)\bar{u}(\mu, k')u(\nu_\mu, k) + (m_\mu + m_{\nu_\mu})\bar{u}(\mu, k')\gamma_5 u(\nu_\mu, k), \tag{20.68}$$

using the on-shell conditions for the spinors. (In the electromagnetic case, there was no γ_5 term, and the intial and final masses were the same.) The quantity (20.68) vanishes only when the lepton masses vanish, and that is the approximation we shall make: i.e. we shall neglect all lepton masses. Then

$$q^\mu N_{\mu\nu} = q^\nu N_{\mu\nu} = 0, \tag{20.69}$$

and we may write

$$p' = p + q \tag{20.70}$$

and drop all terms involving q in the contraction with $N_{\mu\nu}$. In the antisymmetric term, however, we have

$$\epsilon^{\mu\nu\gamma\delta}p_\gamma(p_\delta + q_\delta) = \epsilon^{\mu\nu\gamma\delta}p_\gamma q_\delta \tag{20.71}$$

since the term with p_δ vanishes because of the antisymmetry of $\epsilon_{\mu\nu\gamma\delta}$. Thus we arrive at

$$E^{\mu\nu}_{\text{eff}} = 8p^\mu p^\nu + 2q^2 g^{\mu\nu} - 4i\epsilon^{\mu\nu\gamma\delta}p_\gamma q_\delta. \tag{20.72}$$

We must now evaluate the '$N \cdot E$' contraction in (20.53). Since we are neglecting all masses, it is easiest to perform the calculation in invariant from before specializing to the 'laboratory' frame. The usual Mandelstam variables are (neglecting all masses)

$$s = 2k \cdot p \tag{20.73}$$
$$u = -2k' \cdot p \tag{20.74}$$
$$t = -2k \cdot k' = q^2 \tag{20.75}$$

satisfying

$$s + t + u = 0. \tag{20.76}$$

The result of performing the contraction

$$N_{\mu\nu}E^{\mu\nu} = N_{\mu\nu}E^{\mu\nu}_{\text{eff}} \tag{20.77}$$

may be found using the result (20.65) for the contraction of two ϵ tensors (see problem 20.6). The answer for $\nu_\mu e^- \rightarrow \mu^- \nu_e$ is

$$N_{\mu\nu}E^{\mu\nu} = 16(s^2 + u^2) + 16(s^2 - u^2) \tag{20.78}$$

where the first term arises from the symmetric part of $N_{\mu\nu}$ similar to $L_{\mu\nu}$, and the second term from the antisymmetric part involving $\epsilon_{\mu\nu\alpha\beta}$. We have also used

$$t = q^2 = -(s + u) \tag{20.79}$$

valid in the approximation in which we are working. Thus for $\nu_\mu e^- \to \mu^- \nu_e$ we have

$$N_{\mu\nu} E^{\mu\nu} = +32 s^2 \tag{20.80}$$

and with

$$\frac{d\sigma}{d\Omega} = \frac{1}{64\pi^2 s} \left(\frac{G_F^2}{2} \right) N_{\mu\nu} E^{\mu\nu} \tag{20.81}$$

we finally obtain the result

$$\frac{d\sigma}{d\Omega} = \frac{G_F^2 s}{4\pi^2}. \tag{20.82}$$

The total cross section is then

$$\sigma = \frac{G_F^2 s}{\pi}. \tag{20.83}$$

Since $t = -2p^2(1 - \cos\theta)$, where p is the CM momentum and θ the CM scattering angle, (20.82) can alternatively be written in invariant form as (problem 20.7)

$$\frac{d\sigma}{dt} = \frac{G_F^2}{\pi}. \tag{20.84}$$

All other purely leptonic processes may be calculated in an analogous fashion (see Bailin (1982) and Renton (1990) for further examples).

When we discuss deep inelastic neutrino scattering in section 20.7.2, we shall be interested in neutrino 'laboratory' cross sections, as in the electron scattering case of chapter 9. A simple calculation gives $s \simeq 2m_e E$ (neglecting squares of lepton masses by comparison with $m_e E$), where E is the 'laboratory' energy of a neutrino incident, in this example, on a stationary electron. It follows that *the total 'laboratory' cross section in this Fermi-like current-current model rises linearly with E*. We shall return to the implications of this in section 20.7.2.

The process (20.45) was measured by Bergsma *et al.* (1983) using the CERN wide-band beam ($E_\nu \sim 20$ GeV). The ratio of the observed number of events to that expected for pure V-A was quoted as 0.98±0.12.

20.6 Leptonic weak neutral currents

The first observations of the weak neutral current process $\bar{\nu}_\mu e^- \to \bar{\nu}_\mu e^-$ were reported by Hasert *et al.* (1973), in a pioneer experiment using the heavy-liquid bubble chamber Gargamelle at CERN, irradiated with a $\bar{\nu}_\mu$ beam. As in the case of the charged currents, much detailed experimental work was necessary to determine the precise form of the neutral current couplings. They are, of course, *predicted* by the Glashow-Salam-Weinberg theory, as we shall explain in chapter 22. For the moment, we continue with the current-current approach, parametrizing the currents in a convenient way.

There are two types of 'neutral current' couplings, those involving neutrinos of the form $\bar{\hat{\nu}}_l \ldots \hat{\nu}_l$, and those involving the charged leptons of the form $\bar{\hat{l}} \ldots \hat{l}$. We shall assume the following form for these currents (with one eye on the GSW theory to come):

(1) neutrino neutral current

$$g_N c^{\nu_l} \bar{\hat{\nu}}_l \gamma^\mu \left(\frac{1 - \gamma_5}{2} \right) \hat{\nu}_l \qquad l = e, \mu, \tau; \tag{20.85}$$

(2) charged lepton neutral current

$$g_N \bar{l} \gamma^\mu \left[c_L^l \frac{(1 - \gamma_5)}{2} + c_R^l \frac{(1 + \gamma_5)}{2} \right] \hat{l} \qquad l = e, \mu, \tau. \qquad (20.86)$$

This is, of course, by no means the most general possible parametrization. The neutrino coupling is retained as pure 'V-A', while the coupling in the charged lepton sector is now a combination of 'V-A' and 'V+A' with certain coefficients c_L^l and c_R^l. We may also write the coupling in terms of 'V' and 'A' coefficients defined by $c_V^l = c_L^l + c_R^l, c_A^l = c_L^l - c_R^l$. An overall factor g_N determines the strength of the neutral currents as compared to the charged ones; the c's determine the relative amplitudes of the various neutral current processes.

As we shall see, an essential feature of the GSW theory is its prediction of weak neutral current processes, with couplings determined in terms of one parameter of the theory called 'θ_W', the 'weak mixing angle' (Glashow (1961), Weinberg(1967)). The GSW predictions for the parameter g_N and the c's are (see equations (22.59 - (22.62))

$$g_N = g / \cos \theta_W, \quad c^{\nu_l} = \frac{1}{2}, \quad c_L^l = -\frac{1}{2} + a, \quad c_R^l = a \qquad (20.87)$$

for $l = e, \mu, \tau$, where $a = \sin^2 \theta_W$ and g is the SU(2) gauge coupling. Note that a strong form of 'universality' is involved here too; the coefficients are independent of the 'flavour' e, μ, or τ, for both neutrinos and charged leptons.

The following reactions are available for experimental measurement (in addition to the charged current process (20.45) already discussed):

$$\nu_\mu e^- \rightarrow \nu_\mu e^-, \quad \bar{\nu}_\mu e^- \rightarrow \bar{\nu}_\mu e^- \text{ (NC)} \qquad (20.88)$$
$$\nu_e e^- \rightarrow \nu_e e^-, \quad \bar{\nu}_e e^- \rightarrow \bar{\nu}_e e^- \text{ (NC + CC)} \qquad (20.89)$$

where 'NC' means neutral current and 'CC' charged current. Formulae for these cross sections are given in section 22.3. The experiments are discussed and reviewed in Commins and Buksbaum (1983), Renton (1990), and by Winter (2000). All observations are in excellent agreement with the GSW predictions, with θ_W determined as $\sin^2 \theta_W \simeq 0.23$. The reader must note, however, that modern precision measurements are sensitive to higher order (loop) corrections, which must be included in comparing the full GSW theory with experiment (see section 22.6). The simultaneous fit of data from all four reactions in terms of the single parameter θ_W provides already strong confirmation of the theory—and indeed such confirmation was already emerging in the late 1970's and early 1980's, before the actual discovery of the $W^\pm$ and Z^0 bosons. It is also interesting to note that the presence of vector (V) interactions in the neutral current processes may suggest the possibility of some kind of link with electromagnetic interactions, which are of course also 'neutral' (in this sense) and vector-like. In the GSW theory, this linkage is provided essentially through the parameter θ_W, as we shall see.

20.7   Quark weak currents

We now turn our attention to the weak interactions of quarks. We shall begin by considering an earlier world, when only two generations (four flavours) were known.

20.7.1   Two generations

The original version of V-A theory was framed in terms of a nucleonic current of the form $\bar{\hat{\psi}}_{\mathrm{p}}\gamma^{\mu}(1-r\gamma_5)\hat{\psi}_{\mathrm{n}}$. With the acceptance of quark substructure it was natural to re-interpret such a hadronic transition by a charged current of the form $\bar{\hat{u}}\gamma^{\mu}(1-\gamma_5)\hat{d}$, very similar to the charged lepton currents; indeed, here was a further example of 'universality', this time between quarks and leptons. Detailed comparison with experiment showed, however, that such d $\rightarrow$ n transitions were very slightly weaker than the analogous leptonic ones; this could be established by comparing the rates for n $\rightarrow$ pe$^-\bar{\nu}_{\mathrm{e}}$ and $\bar{\mu}\rightarrow\nu_{\mu}e^-\bar{\nu}_{\mathrm{e}}$.

But for quarks (or their hadronic composites) there is a further complication, which is the very familiar phenomenon of flavour change in weak hadronic processes (recall the discussion in section 1.2.2). The first step towards the modern theory of quark currents was taken by Cabibbo (1963); in a sense, it restored universality. Cabibbo postulated that the strength of the hadronic weak interaction was *shared* between the $\Delta S = 0$ and $\Delta S = 1$ transitions (where S is the strangeness quantum number), the latter being relatively suppressed as compared to the former. According to Cabibbo's hypothesis, phrased in terms of quarks, the total weak charged current for u, d, and s quarks is

$$\hat{j}^{\mu}_{\mathrm{Cab}}(\mathrm{u,d,s}) = \cos\theta_{\mathrm{C}}\bar{\hat{u}}\gamma^{\mu}\frac{(1-\gamma_5)}{2}\hat{d} + \sin\theta_{\mathrm{C}}\bar{\hat{u}}\gamma^{\mu}\frac{(1-\gamma_5)}{2}\hat{s}, \tag{20.90}$$

where θ_{C} is the 'Cabibbo angle' (not to be confused with θ_{W}). We can now postulate a total weak charged current

$$\hat{j}^{\mu}_{\mathrm{CC}}(\mathrm{total}) = \hat{j}^{\mu}_{\mathrm{CC}}(\mathrm{leptons}) + \hat{j}^{\mu}_{\mathrm{Cab}}(\mathrm{u,d,s}), \tag{20.91}$$

where $\hat{j}^{\mu}_{\mathrm{CC}}(\mathrm{leptons})$ is given by (20.38), and then generalize (20.40) to

$$\hat{\mathcal{H}}^{\mathrm{tot}}_{\mathrm{CC}} = \frac{G_{\mathrm{F}}}{\sqrt{2}}\hat{j}^{\mu}_{\mathrm{CC}}(\mathrm{total})\hat{j}^{\dagger}_{\mathrm{CC}\mu}(\mathrm{total}). \tag{20.92}$$

The effective interaction (20.92) describes a great many processes. The purely leptonic ones discussed previously are, of course, present in the term $\hat{j}^{\mu}_{\mathrm{CC}}(\mathrm{leptons})\hat{j}^{\dagger}_{\mathrm{CC}\mu}(\mathrm{leptons})$. But there are also now all the *semi-leptonic* processes such as the $\Delta S = 0$ (strangeness conserving) one

$$\mathrm{d} \rightarrow \mathrm{u} + \mathrm{e}^- + \bar{\nu}_{\mathrm{e}}, \tag{20.93}$$

and the $\Delta S = 1$ (strangeness changing) one

$$\mathrm{s} \rightarrow \mathrm{u} + \mathrm{e}^- + \bar{\nu}_{\mathrm{e}}. \tag{20.94}$$

The notion that the 'total current' should be the sum of a hadronic and a leptonic part is already familiar from electromagnetism—see, for example, equation (8.91).

The transition (20.94), for example, is the underlying process in semi-leptonic decays such as

$$\Sigma^- \rightarrow \mathrm{n} + \mathrm{e}^- + \bar{\nu}_{\mathrm{e}} \tag{20.95}$$

and

$$\mathrm{K}^- \rightarrow \pi^0 + \mathrm{e}^- + \bar{\nu}_{\mathrm{e}} \tag{20.96}$$

as indicated in figure 20.3.

The 's' quark is assigned $S = -1$ and charge $-\frac{1}{3}e$. The s $\rightarrow$ u transition is then referred to as one with '$\Delta S = \Delta Q$', meaning that the change in the quark (or hadronic) strangeness is equal to the change in the quark (or hadronic) charge: both the strangeness and the charge

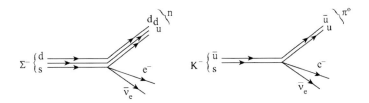

FIGURE 20.3
Strangeness-changing semi-leptonic weak decays.

increase by 1 unit. Prior to the advent of the quark model, and the Cabibbo hypothesis, it had been established empirically that all known strangeness-changing semileptonic decays satisfied the rules $|\Delta S| = 1$ and $\Delta S = \Delta Q$. The u-s current in (20.90) satisfies these rules automatically. Note, for example, that the process apparently similar to (20.95), $\Sigma^+ \rightarrow$ n + e$^+$ + ν_e, is forbidden in the lowest order (it requires a double quark transition from suu to udd). All known data on such decays can be fit with a value $\sin\theta_C \simeq 0.22$ for the Cabibbo angle θ_C. This relatively small angle is therefore a measure of the suppression of $|\Delta S| = 1$ processes relative to $\Delta S = 0$ ones.

The Cabibbo current can be written in a more compact form by introducing the 'mixed' field

$$\hat{d}' \equiv \cos\theta_C \hat{d} + \sin\theta_C \hat{s}. \tag{20.97}$$

Then

$$\hat{j}^\mu_{\text{Cab}}(\text{u},\text{d},\text{s}) = \bar{\hat{u}}\gamma^\mu \frac{(1-\gamma_5)}{2}\hat{d}'. \tag{20.98}$$

In 1970 Glashow, Iliopuolos and Maiani (GIM) (1970) drew attention to a theoretical problem with the interaction (20.92) if used in *second* order. Now it is, of course, the case that this interaction is not renormalizable, as noted previously for the purely leptonic one (20.40), since G_F has dimensions of an inverse mass squared. As we saw in section 11.7, this means that one-loop diagrams will typically diverge quadratically, so that the contribution of such a second-order process will be of order $(G_F.G_F\Lambda^2)$ where Λ is a cut-off, compared to the first-order amplitude G_F. Recalling from (20.46) that $G_F \sim 10^{-5}$ GeV^{-2}, we see that for $\Lambda \sim 10$ GeV such a correction could be significant if accurate enough data exists. GIM pointed out, in particular, that some second-order processes could be found which violated the (hitherto) well-established phenomenological selection rules, such as the $|\Delta S| = 1$ and $\Delta S = \Delta Q$ rules already discussed. For example, there could be $\Delta S = 2$ amplitudes contributing to the $K_L - K_S$ mass difference (see Renton (1990) section 9.1.6, for example), as well as contributions to rare decay modes such as

$$K^+ \rightarrow \pi^+ + \nu + \bar{\nu} \tag{20.99}$$

which has a *neutral* lepton pair in association with a *strangeness change* for the hadron. In fact, experiment then placed very tight limits on the rate for (20.99). [3] This seemed to imply a surprisingly low value of the cut-off, say ~ 3GeV (Mohapatra *et al.* (1968)).

Partly in order to address this problem, and partly as a revival of an earlier lepton-quark symmetry proposal (Bjorken and Glashow 1964), GIM introduced a fourth quark, now called c (the charm quark) with charge $\frac{2}{3}e$. Note that in 1970 the τ-lepton had not been

---

[3]The rate for (20.99) can be calculated in the SM — see for example the review by Lichtenberg and Valencia (2022), section 65 of Workman *et al.* (2022). The theoretical value is consistent with the most recent experimental value from the NA62 experiment (Cortina Gil *et al.* 2021).

discovered, so only two lepton family pairs (ν_e, e), (ν_μ, μ) were known; this fourth quark therefore did restore the balance, via the two quark family pairs (u,d), (c,s). In particular, a second quark current could now be hypothesized, involving the (c,s) pair. GIM postulated that the c-quark was coupled to the 'orthogonal' d-s combination (cf (20.97))

$$\hat{s}' = -\sin\theta_C \hat{d} + \cos\theta_C \hat{s}. \tag{20.100}$$

The complete four-quark charged current is then

$$\hat{j}^\mu_{GIM}(u, d, c, s) = \bar{\hat{u}}\gamma^\mu \frac{(1 - \gamma_5)}{2}\hat{d}' + \bar{\hat{c}}\gamma^\mu \frac{(1 - \gamma_5)}{2}\hat{s}'. \tag{20.101}$$

The form (20.101), had already been suggested by Bjorken and Glashow (1964). The new feature of GIM was the observation that, assuming an exact $SU(4)_f$ symmetry for the four quarks (in particular, equal masses), all second-order contributions which could have violated the $|\Delta S| = 1, \Delta S = \Delta Q$ selection rules now vanished. Further, to the extent that the (unknown) mass of the charm quark functioned as an effective cut-off Λ, due to breaking of the $SU(4)_f$ symmetry, they estimated m_c to lie in the range 3–4 GeV, from the observed $K_L - K_S$ mass difference.

GIM went on to speculate that the non-renormalizability could be overcome if the weak interactions were described by an SU(2) Yang-Mills gauge theory, involving a triplet (W^+, W^-, W^0) of gauge bosons. In this case, it is natural to introduce the idea of (weak) 'isospin', in terms of which the pairs (ν_e, e), (ν_μ, μ), (u,d'), (c, s') are all $t = \frac{1}{2}$ doublets with $t_3 = \pm\frac{1}{2}$. Charge-changing currents then involve the 'raising' matrix

$$\tau_+ \equiv \frac{1}{2}(\tau_1 + i\tau_2) = \begin{pmatrix} 0 & 1 \\ 0 & 0 \end{pmatrix} \tag{20.102}$$

and charge-lowering ones the matrix $\tau_- = (\tau_1 - i\tau_2)/2$. The full symmetry must also involve the matrix τ_3, given by the commutator $[\tau_+, \tau_-] = \tau_3$. Whereas τ_+ and τ_- would (in this model) be associated with transitions mediated by $W^\pm$, transitions involving τ_3 would be mediated by W^0, and would correspond to 'neutral current' transitions for quarks. We now know that things are slightly more complicated than this: the correct symmetry is the SU(2) × U(1) of Glashow (1961), also invoked by GIM. Skipping therefore some historical steps, we parametrize the *weak quark neutral current* as (cf (20.86) for the leptonic analogue)

$$g_N \sum_{q=u,c,d',s'} \bar{\hat{q}}\gamma^\mu[c^q_L \frac{(1 - \gamma_5)}{2} + c^q_R \frac{(1 + \gamma_5)}{2}]\hat{q} \tag{20.103}$$

for the four flavours so far in play. In the GSW theory, the c^q_L's are predicted to be

$$c^{u,c}_L = \frac{1}{2} - \frac{2}{3}a \qquad c^{u,c}_R = -\frac{2}{3}a \tag{20.104}$$

$$c^{d,s}_L = -\frac{1}{2} + \frac{1}{3}a \qquad c^{d,s}_R = \frac{1}{3}a \tag{20.105}$$

where $a = \sin^2\theta_W$ as before, and $g_N = g/\cos\theta_W$.

One feature of (20.103) is very important. Consider the terms

$$\bar{\hat{d}}'\{\ldots\}\hat{d}' + \bar{\hat{s}}'\{\ldots\}\hat{s}'. \tag{20.106}$$

It is simple to verify that, whereas either part of (20.106) alone contains a *strangeness changing neutral* combination such as $\bar{\hat{d}}\{\ldots\}\hat{s}$ or $\bar{\hat{s}}\{\ldots\}\hat{d}$, such combinations vanish in the

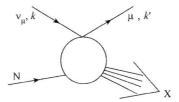

FIGURE 20.4
Inelastic neutrino scattering from a nucleon.

sum, leaving the result *diagonal in quark flavour*. Thus there are no first-order neutral flavour-changing currents in this model, a result which will be extended to three flavours in section 20.7.3.

In 1974, Gaillard and Lee (1974) performed a full one-loop calculation of the $K_L - K_S$ mass difference in the GSW model as extended by GIM to quarks and using the renormalization techniques recently developed by 't Hooft (1971b). They were able to predict $m_c \sim 1.5$ GeV for the charm quark mass, a result spectacularly confirmed by the subsequent discovery of the $c\bar{c}$ states in charmonium, and of charmed mesons and baryons of the appropriate mass.

In summary, then, the essential feature of the quark weak currents in the two-generation model is that they have the universal V-A form, but the participating fields are $(\hat{u}, \hat{d}'), (\hat{c}, \hat{s}')$ where $\hat{d}'$ and $\hat{s}'$ are not the fields $\hat{d}, \hat{s}$ with definite mass, but rather are related to them by an orthogonal transformation:

$$\begin{pmatrix} \hat{d}' \\ \hat{s}' \end{pmatrix} = \begin{pmatrix} \cos\theta_C & \sin\theta_C \\ -\sin\theta_C & \cos\theta_C \end{pmatrix} \begin{pmatrix} \hat{d} \\ \hat{s} \end{pmatrix}. \tag{20.107}$$

In section 20.8 we shall enlarge this picture to three generations, where significant new features occur, specifically **CP** violation. In chapter 22 we shall see how this transformation from the 'mass' basis to the 'weak interaction' basis arises via the gauge-invariant interactions of the Standard Model.

20.7.2   Deep inelastic neutrino scattering

We now have enough theory to present another illustrative calculation within the framework of the 'current-current' model, this time involving neutrinos and quarks. We shall calculate cross sections for deep inelastic neutrino scattering from nucleons, using the parton model introduced (for electromagnetic interactions) in chapter 9. In particular, we shall consider the processes

$$\begin{aligned} \nu_\mu + N &\rightarrow \mu^- + X \tag{20.108} \\ \bar{\nu}_\mu + N &\rightarrow \mu^+ + X \tag{20.109} \end{aligned}$$

which of course involve the charged currents, for both leptons and quarks. Studies of these reactions at Fermilab and CERN in the 1970s and 1980s played a crucial part in establishing the quark structure of the nucleon, in particular the quark distribution functions.

The general process is illustrated in figure 20.4. By now we are becoming accustomed to the idea that such processes are in fact mediated by the W^+, but we shall assume that the momentum transfers are such that the W-propagator is effectively constant. The effective lepton-quark interaction will then take the form

$$\hat{\mathcal{H}}_{\nu q}^{\text{eff}} = \frac{G_F}{\sqrt{2}} \bar{\hat{\mu}} \gamma_\mu (1 - \gamma_5) \hat{\nu}_\mu [\bar{\hat{u}} \gamma^\mu (1 - \gamma_5) \hat{d} + \bar{\hat{c}} \gamma^\mu (1 - \gamma_5) \hat{s}], \tag{20.110}$$

leading to expressions for the parton-level subprocess amplitudes which are exactly similar to that in (20.50) for $\nu_\mu + e^- \to \mu^- + \nu_e$. Note that we are considering only the four flavours u,d,c,s to be 'active', and we have set $\theta_C \approx 0$.

As in (20.53), the ν_μ cross section will have the general form

$$d\sigma^{(\nu)} \propto N_{\mu\nu} W_{(\nu)}^{\mu\nu}(q, p) \tag{20.111}$$

where $N_{\mu\nu}$ is the neutrino tensor of (20.67). The form of the weak hadron tensor $W_{(\nu)}^{\mu\nu}$ is deduced from Lorentz invariance. In the approximation of neglecting lepton masses, we can ignore any dependence on the 4-vector q since

$$q^\mu N_{\mu\nu} = q^\nu N_{\mu\nu} = 0. \tag{20.112}$$

Just as $N_{\mu\nu}$ contains the pseudotensor $\epsilon_{\mu\nu\alpha\beta}$ so too will $W_{(\nu)}^{\mu\nu}$ since parity is not conserved. In a manner similar to equation (9.10) for the case of electron scattering, and following the steps that led from (20.67) to (20.72), we define effective neutrino structure functions by

$$W_{(\nu)}^{\mu\nu} = (-g^{\mu\nu}) W_1^{(\nu)} + \frac{1}{M^2} p^\mu p^\nu W_2^{(\nu)} - \frac{i}{2M^2} \epsilon^{\mu\nu\gamma\delta} p_\gamma q_\delta W_3^{(\nu)}. \tag{20.113}$$

In general, the structure functions depend on two variables, say Q^2 and ν, where $Q^2 = -(k - k')^2$ and $\nu = p \cdot q / M$; but in the Bjorken limit approximate scaling is observed, as in the electron case:

$$\left. \begin{array}{c} Q^2 \to \infty \\ \nu \to \infty \end{array} \right\} \quad x = Q^2 / 2M\nu \ \text{ fixed} \tag{20.114}$$

$$\nu W_2^{(\nu)}(Q^2, \nu) \quad \to \quad F_2^{(\nu)}(x) \tag{20.115}$$

$$M W_1^{(\nu)}(Q^2, \nu) \quad \to \quad F_1^{(\nu)}(x) \tag{20.116}$$

$$\nu W_3^{(\nu)}(Q^2, \nu) \quad \to \quad F_3^{(\nu)}(x) \tag{20.117}$$

where, as with (9.21) and (9.22), the physics lies in the assertion that the F's are finite. This scaling can again be interpreted in terms of pointlike scattering from partons—which we shall take to have quark quantum numbers.

In the 'laboratory' frame (in which the nucleon is at rest) the cross section in terms of W_1, W_2, and W_3 may be derived in the usual way from (cf equation (9.11))

$$d\sigma^{(\nu)} = \left(\frac{G_F}{\sqrt{2}} \right)^2 \frac{1}{4k \cdot p} 4\pi M N_{\mu\nu} W_{(\nu)}^{\mu\nu} \frac{d^3 k'}{2k'(2\pi)^3}. \tag{20.118}$$

In terms of 'laboratory' variables, one obtains (problem 20.9)

$$\frac{d^2 \sigma^{(\nu)}}{dQ^2 d\nu} = \frac{G_F^2}{2\pi} \frac{k'}{k} \left(W_2^{(\nu)} \cos^2(\theta/2) + W_1^{(\nu)} 2 \sin^2(\theta/2) + \frac{k + k'}{M} \sin^2(\theta/2) W_3^{(\nu)} \right). \tag{20.119}$$

For an incoming antineutrino beam, the W_3 term changes sign.

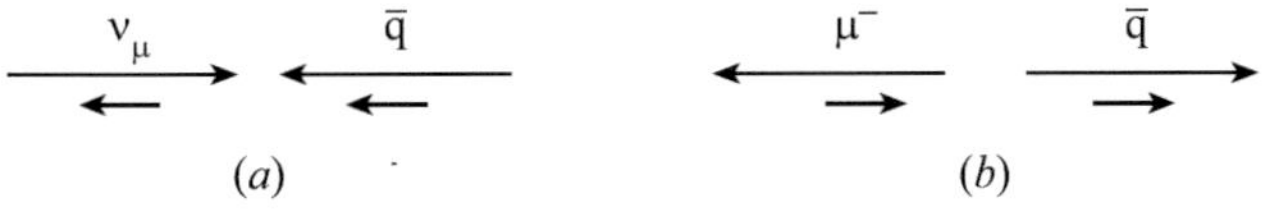

FIGURE 20.5
Suppression of $\nu_\mu \bar{q} \to \mu^- \bar{q}$ for $y = 1$: (a) initial state helicities; (b) final state helicities at $y = 1$.

In neutrino scattering it is common to use the variables x, ν, and the 'inelasticity' y where

$$y = p \cdot q / p \cdot k. \qquad (20.120)$$

In the 'laboratory' frame, $\nu = E - E'$ (the energy transfer to the nucleon) and $y = \nu/E$. The cross section can be written in the form (see problem 20.9)

$$\frac{\mathrm{d}^2\sigma^{(\nu)}}{\mathrm{d}x\mathrm{d}y} = \frac{G_F^2}{2\pi} s \left(F_2^{(\nu)} \frac{1 + (1 - y)^2}{2} + x F_3^{(\nu)} \frac{1 - (1 - y)^2}{2} \right) \qquad (20.121)$$

in terms of the Bjorken scaling functions, and we have assumed the relation

$$2x F_1^{(\nu)} = F_2^{(\nu)} \qquad (20.122)$$

appropriate for spin-$\frac{1}{2}$ constituents.

We now turn to the parton-level subprocesses. Their cross sections can be straightforwardly calculated in the same way as for $\nu_\mu e^-$ scattering in section 20.5. We obtain (problem 20.10)

$$\nu q, \bar{\nu}\bar{q} : \frac{\mathrm{d}^2\sigma}{\mathrm{d}x\mathrm{d}y} = \frac{G_F^2}{\pi} s x \delta \left(x - \frac{Q^2}{2M\nu} \right) \qquad (20.123)$$

$$\nu\bar{q}, \bar{\nu}q : \frac{\mathrm{d}^2\sigma}{\mathrm{d}x\mathrm{d}y} = \frac{G_F^2}{\pi} s x (1 - y)^2 \delta \left(x - \frac{Q^2}{2M\nu} \right). \qquad (20.124)$$

The factor $(1 - y)^2$ in the $\nu\bar{q}, \bar{\nu}q$ cases means that the reaction is forbidden at $y = 1$ (backwards in the CM frame). This follows from the V-A nature of the current, and angular momentum conservation, as a simple helicity argument shows. Consider for example the case $\nu\bar{q}$ shown in figure 20.5, with the helicities marked as shown. In our current-current interaction there are no gradient coupling terms and therefore no momenta in the momentun-space matrix element. This means that no orbital angular momentum is available to account for the reversal of net helicity in the initial and final states in figure 20.5. The lack of orbital angular momentum can also be inferred physically from the 'pointlike' nature of the current-current coupling. For the νq or $\bar{\nu}\bar{q}$ cases, the initial and final helicities add to zero, and backward scattering is allowed.

The contributing processes are

$$\nu d \rightarrow l^- u, \qquad \bar{\nu}d \to l^+ \bar{u} \qquad (20.125)$$

$$\nu\bar{u} \rightarrow l^- \bar{d}, \qquad \bar{\nu}u \to l^+ d, \qquad (20.126)$$

the first pair having the cross section (20.123), the second (20.124). Following the same steps as in the electron scattering case (sections 9.2 and 9.3) we obtain

$$F_2^{\nu p} = F_2^{\bar{\nu}n} = 2x[d(x) + \bar{u}(x)] \qquad (20.127)$$

$$F_3^{\nu p} = F_3^{\bar{\nu} n} = 2[d(x) - \bar{u}(x)] \tag{20.128}$$

$$F_2^{\nu n} = F_2^{\bar{\nu} p} = 2x[u(x) + \bar{d}(x)] \tag{20.129}$$

$$F_3^{\nu n} = F_3^{\bar{\nu} p} = 2[u(x) - \bar{d}(x)]. \tag{20.130}$$

Inserting (20.127) and (20.128) into (20.121), for example, we find

$$\frac{d^2\sigma^{(\nu p)}}{dxdy} = 2\sigma_0 x[d(x) + (1-y)^2 \bar{u}(x)] \tag{20.131}$$

where

$$\sigma_0 = \frac{G_F^2 s}{2\pi} = \frac{G_F^2 ME}{\pi} \simeq 1.5 \times 10^{-42}(E/\text{GeV})\text{m}^2 \tag{20.132}$$

is the basic 'pointlike' total cross section (compare (20.83)). Note the small magnitude of this cross section, as compared with the electromagnetic one of equation (B.18) in volume 1, which was $\sigma \approx \frac{86.8}{(s/\text{GeV}^2)} \times 10^{-37}\text{m}^2$. Similarly, one finds

$$\frac{d^2\sigma^{(\bar{\nu} p)}}{dxdy} = 2\sigma_0 x[(1-y)^2 u(x) + \bar{d}(x)]. \tag{20.133}$$

The corresponding results for νn and $\bar{\nu} n$ are given by interchanging $u(x)$ and $d(x)$, and $\bar{u}(x)$ and $\bar{d}(x)$.

The target nuclei usually have high mass number (in order to increase the cross section), with approximately equal numbers of protons and neutrons; it is then appropriate to average the 'n' and 'p' results to obtain an 'isoscalar' cross section $\sigma^{(\nu N)}$ or $\sigma^{(\bar{\nu} N)}$:

$$\frac{d^2\sigma^{(\nu N)}}{dxdy} = \sigma_0 x[q(x) + (1-y)^2 \bar{q}(x)] \tag{20.134}$$

$$\frac{d^2\sigma^{(\bar{\nu} N)}}{dxdy} = \sigma_0 x[(1-y)^2 q(x) + \bar{q}(x)] \tag{20.135}$$

where $q(x) = u(x) + d(x)$ and $\bar{q}(x) = \bar{u}(x) + \bar{d}(x)$.

Many simple and striking predictions now follow from these quark parton results. For example, by integrating (20.134) and (20.135) over x we can write

$$\frac{d\sigma^{(\nu N)}}{dy} = \sigma_0[Q + (1-y)^2 \bar{Q}] \tag{20.136}$$

$$\frac{d\sigma^{(\bar{\nu} N)}}{dy} = \sigma_0[(1-y)^2 Q + \bar{Q}] \tag{20.137}$$

where $Q = \int xq(x)dx$ is the fraction of the nucleon's momentum carried by quarks, and similarly for $\bar{Q}$. These two distributions in y ('inelasticity distributions') therefore give a direct measure of the quark and antiquark composition of the nucleon. Figure 20.6 shows the inelasticity distributions as reported by the CDHS collaboration (de Groot *et al.* (1979)), from which the authors extracted the ratio

$$\bar{Q}/(Q+\bar{Q}) = 0.15 \pm 0.03 \tag{20.138}$$

after applying radiative corrections. An even more precise value can be obtained by looking at the region near $y = 1$ for $\bar{\nu} N$ which is dominated by $\bar{Q}$, the small Q contribution ($\propto (1-y)^2$) being subtracted out using νN data at the same y. This method yields

$$\bar{Q}/(Q+\bar{Q}) = 0.15 \pm 0.01. \tag{20.139}$$

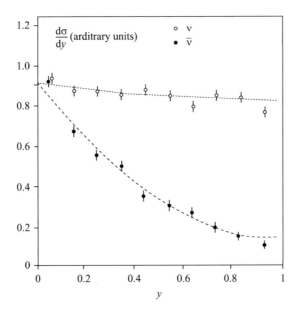

FIGURE 20.6
Charged-current inelasticity (y) distribution as measured by CDHS; figure from K Winter (2000) *Neutrino Physics* 2nd edn, courtesy Cambridge University Press.

Integrating (20.136) and (20.137) over y gives

$$\sigma^{(\nu N)} = \sigma_0(Q + \frac{1}{3}\bar{Q}) \tag{20.140}$$

$$\sigma^{(\bar{\nu} N)} = \sigma_0(\frac{1}{3}Q + \bar{Q}) \tag{20.141}$$

and hence

$$Q + \bar{Q} = 3(\sigma^{(\nu N)} + \sigma^{(\bar{\nu} N)})/4\sigma_0 \tag{20.142}$$

while

$$\bar{Q}/(Q + \bar{Q}) = \frac{1}{2}\left(\frac{3r - 1}{1 + r}\right) \tag{20.143}$$

where $r = \sigma^{(\nu N)}/\sigma^{(\bar{\nu} N)}$. From total cross section measurements, and including c and s contributions, the CHARM collaboration (Allaby *et al.* (1988)) reported

$$Q + \bar{Q} = 0.492 \pm 0.006(\text{stat}) \pm 0.019(\text{syst}) \tag{20.144}$$
$$\bar{Q}/Q + \bar{Q} = 0.154 \pm 0.005(\text{stat}) \pm 0.011(\text{syst}). \tag{20.145}$$

The second figure is in good agreement with (20.139), and the first shows that only about 50% of the nucleon momentum is carried by charged partons, the rest being carried by the gluons, which do not have weak or electromagnetic interactions.

Equations (20.140) and (20.141), together with (20.132), predict that the total cross sections $\sigma^{\nu N}$ and $\sigma^{\bar{\nu} N}$ rise linearly with energy E. This (parton model) prediction was confirmed as early as 1975 (Perkins 1975), soon after the model's success in deep inelastic electron scattering; later data is included in figure 20.7. In fact, both $\sigma^{\nu N}/E$ and $\sigma^{\bar{\nu} N}/E$ are found to be independent of E up to $E \sim 350$ GeV (Nakamura *et al.* 2010).

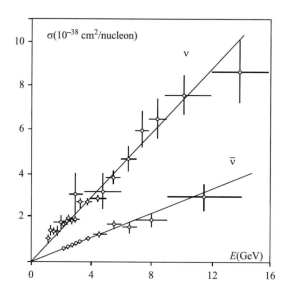

FIGURE 20.7

Low-energy ν and $\bar{\nu}$ cross-sections; figure from K Winter (2000) *Neutrino Physics* 2nd edn, courtesy Cambridge University Press.

Detailed comparison between the data at high energies and the earlier data of figure 20.7 at E_ν up to 15 GeV reveals that the $\bar{Q}$ fraction is increasing with energy. This is in accordance with the expectation of QCD corrections to the parton model (section 15.6): the $\bar{Q}$ distribution is large at small x, and scaling violations embodied in the evolution of the parton distributions predict a rise at small x as the energy scale increases.

Returning now to (20.127) – (20.130), the two sum rules of (9.65) and (9.66) can be combined to give

$$3 = \int_0^1 \mathrm{d}x[u(x) + d(x) - \bar{u}(x) - \bar{d}(x)] \tag{20.146}$$

$$= \frac{1}{2} \int_0^1 \mathrm{d}x(F_3^{\nu \mathrm{p}} + F_3^{\nu \mathrm{n}}) \tag{20.147}$$

$$\equiv \int_0^1 \mathrm{d}x F_3^{\nu \mathrm{N}} \tag{20.148}$$

which is the Gross-Llewellyn Smith sum rule (1969), expressing the fact that the number of valence quarks per nucleon is three. The CDHS collaboration (de Groot *et al.* 1979), quoted

$$I_{\mathrm{GLLS}} \equiv \int_0^1 \mathrm{d}x F_3^{\nu \mathrm{N}} = 3.2 \pm 0.5. \tag{20.149}$$

In perturbative QCD there are corrections expressible as a power series in α_{s}, so that the parton model result is only reached as $Q^2 \to \infty$:

$$I_{\mathrm{GLLS}}(Q^2) = 3[1 + d_1 \alpha_{\mathrm{s}}/\pi + d_2 \alpha_{\mathrm{s}}^2/\pi^2 + \ldots] \tag{20.150}$$

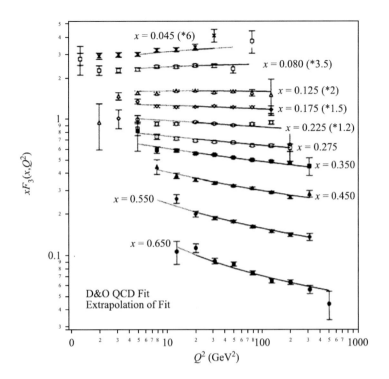

FIGURE 20.8
CCFR neutrino-iron structure functions $xF_3^{(\nu)}$ (Shaevitz *et al.* 1995). The solid line is the next-to-leading order (one-loop) QCD prediction, and the dotted line is an extrapolation to regions outside the kinematic cuts for the fit.

where $d_1 = -1$ (Altarelli *et al.* (1978 a, 1978b)), $d_2 = -55/12 + N_{\rm f}/3$ (Gorishnii and Larin (1986)) where $N_{\rm f}$ is the number of active flavours. The CCFR collaboration (Shaevitz *et al.* 1995) measured $I_{\rm GLLS}$ in antineutrino-nucleon scattering at $\langle Q^2 \rangle \sim 3{\rm GeV}^2$. They obtained

$$I_{\rm GLLS}(\langle Q^2 \rangle = 3 \text{ GeV}^2) = 2.50 \pm 0.02 \pm 0.08 \tag{20.151}$$

in agreement with the $O(\alpha_{\rm s}^3)$ calculation of Larin and Vermaseren (1991) using $\Lambda_{\overline{MS}} = 250 \pm 50{\rm MeV}$.

The predicted Q^2 evolution of xF_3 is particularly simple since it is not coupled to the gluon distribution. To leading order, the xF_3 evolution is given by (cf (15.109))

$$\frac{\rm d}{\rm d\ln Q^2}(xF_3(x,Q^2)) = \frac{\alpha_{\rm s}(Q^2)}{2\pi} \int_x^1 P_{\rm qq}(z) xF_3\left(\frac{x}{z}, Q^2\right) \frac{\rm dz}{z}. \tag{20.152}$$

Figure 20.8, taken from Shaevitz *et al.* (1995) shows a comparison of the CCFR data with the next-to-leading order calculation of Duke and Owens (1984). This fit yields a value of $\alpha_{\rm s}$ at $Q^2 = M_{\rm Z}^2$ given by

$$\alpha_{\rm s}(M_{\rm Z}^2) = 0.111 \pm 0.002 \pm 0.003. \tag{20.153}$$

The Adler sum rule (Adler 1963) involves the functions $F_2^{\bar{\nu}{\rm p}}$ and $F_2^{\nu{\rm p}}$:

$$I_{\rm A} = \int_0^1 \frac{\rm dx}{x}(F_2^{\bar{\nu}{\rm p}} - F_2^{\nu{\rm p}}). \tag{20.154}$$

In the simple model of (20.127) – (20.130), the right hand side of I_A is just

$$2 \int_0^1 \mathrm{d}x (u(x) + \bar{d}(x) - d(x) - \bar{u}(x)) \tag{20.155}$$

which represents four times the average of I_3 (isospin) of the target, which is $\frac{1}{2}$ for the proton. This sum rule follows from the conservation of the charged weak current (as will be true in the standard model, since this is a gauge symmetry current, as we shall see in the following chapter). Its measurement, however, depends precisely on separating the non-isoscalar contribution (I_A vanishes for the isoscalar average 'N'). The BEBC collaboration (Allasia *et al.* 1984, 1985) reported:

$$I_A = 2.02 \pm 0.40; \tag{20.156}$$

in agreement with the expected value 2.

Relations (20.127)–(20.130) allow the F_2 functions for electron (muon) and neutrino scattering to be simply related. From (9.58) and (9.61) we have

$$F_2^{\mathrm{eN}} = \frac{1}{2}(F_2^{\mathrm{ep}} + F_2^{\mathrm{en}}) = \frac{5}{18} x(u + \bar{u} + d + \bar{d}) + \frac{1}{9} x(s + \bar{s}) + \cdots \tag{20.157}$$

while (20.127) and (20.129) give

$$F_2^{\nu \mathrm{N}} \equiv \frac{1}{2}(F_2^{\nu \mathrm{p}} + F_2^{\nu \mathrm{n}}) = x(u + d + \bar{u} + \bar{d}). \tag{20.158}$$

Assuming that the non-strange contributions dominate, the neutrino and charged lepton structure functions should be approximately in the ratio 18/5, which is the reciprocal of the mean squared charged of the u and d quarks in the nucleon. Figure 20.9 shows the neutrino results on F_2 and xF_3 together with those from several μN experiments scaled by the factor 18/5. The agreement is satisfactory for a tree-level parton model calculation.

From (20.127)–(20.130) we see that $F_2^{\nu \mathrm{N}} - xF_3^{\nu \mathrm{N}} = 2x(\bar{u} + \bar{d})$, which is just the sea distribution; figure 20.9 shows that this is concentrated at small x, as we already inferred in section 9.3.

We have mentioned QCD corrections to the simple parton model at several points. Clearly the full machinery introduced in chapter 16, in the context of deep inelastic charged lepton scattering, can be employed for the case of neutrino scattering also. For further access to this area we refer to Ellis, Stirling and Webber (1996), chapter 4, and Winter (2000) chapter 5.

20.7.3   Three generations

We have seen in section 20.2.2 that the V-A interaction violates both **P** and **C**, and that it conserves **CP** in interactions with massless neutrinos. But we know (section 4.2.3) that **CP**-violating transitions occur, among states formed from quarks in the first two generations, albeit at a very slow rate. Is it possible, in fact, to incorporate **CP**-violation with only two generations of quarks?

To answer this question, we need to go back and examine the **CP**-transformation properties of the interactions in more detail. Rather than work with the current-current form, which is after all only an approximation valid for energies much less than $M_{\mathrm{W,Z}}$, we shall look at the actual gauge interactions of the electroweak theory. Given the form of those interactions, we want to know the condition for **CP**-violation to be present.

Consider then the particular interaction involved in u $\leftrightarrow$ d transitions:

$$V_{\mathrm{ud}} \bar{\hat{u}} \gamma^\mu \hat{d} \, \hat{W}_\mu + V_{\mathrm{ud}}^* \bar{\hat{d}} \gamma^\mu \hat{u} \, \hat{W}_\mu^\dagger - V_{\mathrm{ud}} \bar{\hat{u}} \gamma^\mu \gamma_5 \hat{d} \, \hat{W}_\mu - V_{\mathrm{ud}}^* \bar{\hat{d}} \gamma^\mu \gamma_5 \hat{u} \, \hat{W}_\mu^\dagger, \tag{20.159}$$

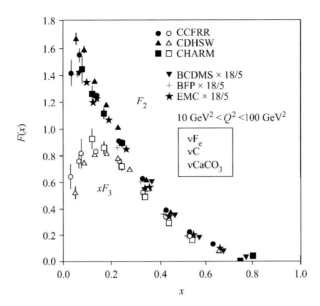

FIGURE 20.9

Comparison of neutrino results (experiments CCFRR, CDHSW and CHARM) on $F_2(x)$ and $xF_3(x)$ with those from muon production (experiments BCDMS, BFP and EMC) properly rescaled by the factor 18/5, for a Q^2 ranging between 10 and 1000 GeV^2; figure from K Winter (2000) *Neutrino Physics* 2nd edn, courtesy Cambridge University Press.

where $\hat{W}_\mu = (\hat{W}_\mu^1 - i\hat{W}_\mu^2)/\sqrt{2}$ destroys the W$^+$ or creates the W$^-$. We have written out the Hermitian conjugate terms explicitly, keeping the coupling $V_{\rm ud}$ complex for the sake of generality, and separating the vector from the axial vector parts. Problem 20.11 shows that the different parts of (20.159) transform under **C** as follows (normal ordering being understood in all cases):

$$\mathbf{C}: \bar{\hat{u}}\gamma^\mu\hat{d} \to -\bar{\hat{d}}\gamma^\mu\hat{u}, \quad \bar{\hat{u}}\gamma^\mu\gamma_5\hat{d} \to +\bar{\hat{d}}\gamma^\mu\gamma_5\hat{u}, \qquad (20.160)$$

and we also know that under **C**, $\hat{W}_\mu \to -\hat{W}_\mu^\dagger$ (the dagger is as in the charged scalar field case, and the minus sign is as in the photon $\hat{A}_\mu$ case). Hence under **C**, (20.159) transforms into

$$V_{\rm ud}\bar{\hat{d}}\gamma^\mu\hat{u}\,\hat{W}_\mu^\dagger + V_{\rm ud}^*\bar{\hat{u}}\gamma^\mu\hat{d}\,\hat{W}_\mu + V_{\rm ud}\bar{\hat{d}}\gamma^\mu\gamma_5\hat{u}\,\hat{W}_\mu^\dagger + V_{\rm ud}^*\bar{\hat{u}}\gamma^\mu\gamma_5\hat{d}\,\hat{W}_\mu. \qquad (20.161)$$

Under **P**, $\hat{W}_\mu$ behaves like an ordinary four-vector, so the 'vector·vector' products in (20.161) are even under **P**, while the 'axial vector·vector' products are odd under **P**. Thus finally, under the combined **CP** transformation (20.159) becomes

$$V_{\rm ud}\bar{\hat{d}}\gamma^\mu\hat{u}\,\hat{W}_\mu^\dagger + V_{\rm ud}^*\bar{\hat{u}}\gamma^\mu\hat{d}\,\hat{W}_\mu - V_{\rm ud}\bar{\hat{d}}\gamma^\mu\gamma_5\hat{u}\,\hat{W}_\mu^\dagger - V_{\rm ud}^*\bar{\hat{u}}\gamma^\mu\gamma_5\hat{d}\,\hat{W}_\mu. \qquad (20.162)$$

Comparing (20.159) with (20.162) we deduce the essential result that this interaction conserves **CP** if and only if

$$V_{\rm ud} = V_{\rm ud}^*, \qquad (20.163)$$

that is, if the coupling is real. The same is true for all the other couplings $V_{\rm ij}$.

The couplings we have introduced in this chapter so far only involve the real Fermi constant G_F, and the elements of the Cabibbo-GIM matrix which enters into the relation between the weakly interacting fields $(\hat{d}', \hat{s}')$ and the fields with definite mass $(\hat{d}, \hat{s})$:

$$\begin{pmatrix} \hat{d}' \\ \hat{s}' \end{pmatrix} = \begin{pmatrix} \cos\theta_C & \sin\theta_C \\ -\sin\theta_C & \cos\theta_C \end{pmatrix} \begin{pmatrix} \hat{d} \\ \hat{s} \end{pmatrix} \equiv \mathbf{V}_{CGIM} \begin{pmatrix} \hat{d} \\ \hat{s} \end{pmatrix}. \qquad (20.164)$$

All these couplings are plainly real. But could we perhaps parametrize the $(\hat{d}', \hat{s}') \leftrightarrow (\hat{d}, \hat{s})$ differently, so as to smuggle in a complex, **CP**-violating, coupling?

This is the question that Kobayashi and Maskawa asked themselves in 1972 (Kobayashi 2009, Maskawa 2009). To answer it is not completely straightforward, because we can always change the phases of the quark fields by independent constant amounts. A rephasing of the quark fields in the transition i ↔ j with coupling V_{ij} changes V_{ij} by the phase factor $\exp i(\alpha_i - \alpha_j)$. We need to know whether, after allowing for this rephasing of the quark fields, an 'irreducible' complex coupling can remain.

First of all, note that the matrix $\mathbf{V}_{CGIM}$ appearing in (20.164) is orthogonal, and this property guaranteed the vanishing of tree-level neutral strangeness-changing transitions, as we saw after (20.106). But this could just as well be achieved if the matrix was unitary. Now a general 2×2 matrix has 8 real parameters; unitarity gives 2 real conditions from the diagonal elements of $\mathbf{V}_{CGIM}\mathbf{V}_{CGIM}^{\dagger} = \mathbf{I}$, and one complex condition from the off-diagonal elements, leaving four real parameters. If all the elements are taken to be real from the beginning, the matrix becomes orthogonal, as in (20.164), and depends on only one real parameter, the 'rotation' in the 2-dimensional $\hat{d} - \hat{s}$ space. So in the general, unitary case, the matrix will have one real angle parameter, and three phase parameters. But we have four quark fields whose phases we can adjust. In fact, since only phase differences enter, we really only have three free phases at our disposal, but that is just enough to transform away the three phases in the unitary version of $\mathbf{V}_{CGIM}$, leaving it in the real orthogonal form (20.164). Kobayashi and Maskawa therefore concluded that the 2-generation GIM-type theory could not accommodate **CP**-violation.

In a step which may seem natural now but was very bold in 1972, they decided to see if there was room for **CP**-violation in a 3-generation model. (Remember that there was no sign of any third generation particles at that time.) The matrix transforming from the mass basis to the weak basis is now a 3×3 unitary matrix $\mathbf{V}$, with 18 real parameters. There are three real diagonal conditions from unitarity, and three complex off-diagonal conditions, leaving 9 real parameters. If the elements of $\mathbf{V}$ are taken to be real, one has an orthogonal (rotation) matrix, which can be parametrized by three real Euler angles. That leaves 6 phase parameters in the general unitary $\mathbf{V}$. We also have 6 quark fields, with 5 phase differences which can be adjusted. Thus just one irreducible phase degree of freedom can remain in $\mathbf{V}$, after quark rephasing. Consequently, the three-generation model naturally accommodates **CP**-violation in the quark sector: this was the great discovery of Kobayashi and Maskawa (1973). It was another four years before the existence of the b quark was established, and more than twenty before the t quark was produced.

The 3-generation matrix $\mathbf{V}$, written out in full, is

$$\mathbf{V} = \begin{pmatrix} V_{ud} & V_{us} & V_{ub} \\ V_{cd} & V_{cs} & V_{cb} \\ V_{td} & V_{ts} & V_{tb} \end{pmatrix}, \qquad (20.165)$$

and is called the CKM matrix, after Cabibbo, Kobayashi, and Maskawa. Clearly, there is no unique parametrization of $\mathbf{V}$. One that has now become standard is (Chau and Keung

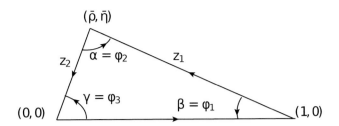

FIGURE 20.10
The unitarity triangle represented by (20.168).

1984)

$$\mathbf{V} = \begin{pmatrix} c_{12}c_{13} & s_{12}c_{13} & s_{13}e^{-i\delta} \\ -s_{12}c_{23} - c_{12}s_{23}s_{13}e^{i\delta} & c_{12}c_{23} - s_{12}s_{23}s_{13}e^{i\delta} & s_{23}c_{13} \\ s_{12}s_{23} - c_{12}c_{23}s_{13}e^{i\delta} & -c_{12}s_{23} - s_{12}c_{23}s_{13}e^{i\delta} & c_{23}c_{13} \end{pmatrix} \qquad (20.166)$$

where $c_{ij} = \cos\theta_{ij}$, $s_{ij} = \sin\theta_{ij}$ with $i, j = 1, 2, 3$; the θ_{ij} may be thought of as the three Euler angles in an orthogonal $\mathbf{V}$, and δ is the remaining irreducible **CP**-violating phase. In the limit $\theta_{13} = \theta_{23} = 0$, this CKM matrix reduces to the Cabibbo-GIM matrix with $\theta_{12} \equiv \theta_C$.

However, it would also be desirable to have a measure of **CP**-violation that was independent of quark rephasing. Consider one of the off-diagonal unitarity conditions,

$$V_{ud}V_{ub}^* + V_{cd}V_{cb}^* + V_{td}V_{tb}^* = 0. \qquad (20.167)$$

(Note that the complex conjugate of this equation gives another, independent, condition.) The best-measured of these products is $V_{cd}V_{cb}^*$; dividing by this quantity, (20.167) can be written as

$$1 + z_1 + z_2 = 0, \qquad (20.168)$$

where

$$z_1 = \frac{V_{td}V_{tb}^*}{V_{cd}V_{cb}^*}, \quad z_2 = \frac{V_{ud}V_{ub}^*}{V_{cd}V_{cb}^*}. \qquad (20.169)$$

When viewed in the complex plane, relation (20.168) is the statement that the vectors $(1,0)$, z_1 and z_2 close to form a triangle as shown in figure 20.10, one of six such *unitarity triangles* that can be formed. The area Δ of this triangle is

$$\Delta = \frac{1}{2}\mathrm{Im}(z_2 z_1^*) = \frac{1}{2}\mathrm{Im}\left(\frac{V_{ud}V_{ub}^* V_{td}^* V_{tb}}{|V_{cd}|^2 |V_{cb}|^2}\right). \qquad (20.170)$$

Recalling that a rephasing multiplies V_{ij} by $\exp i(\alpha_i - \alpha_j)$, we see that Δ is rephasing invariant; in particular, so is the numerator J where

$$J \equiv \mathrm{Im}(V_{ud}V_{tb}V_{ub}^* V_{td}^*) \qquad (20.171)$$

is a *Jarlskog invariant* (Jarlskog 1985). J may be thought of as follows: (i) strike out the 'c' row and 's' column of $\mathbf{V}$; (ii) take the complex conjugate of the off-diagonal elements in the 2×2 matrix that remains; (iii) multiply the four elements and take the imaginary part. There is nothing special about this particular row and column. There are nine different ways of choosing to pair one row with one column, but all such Js are equal up to a sign, because of the unitarity of $\mathbf{V}$. In the parametrization (20.166), J takes the form

$$J = c_{12}s_{12}c_{23}s_{23}c_{13}^2 s_{13}\sin\delta, \tag{20.172}$$

which vanishes if any $\theta_{ij} = 0$, or $\pi/2$, or if $\delta = 0$ or π.

The CKM matrix is an integral part of the SM, and testing its validity is an important experimental goal. Various tests are possible. Consider first the magnitudes of the CKM elements. These must satisfy six relations following from the unitarity of $\mathbf{V}$: namely, the sum of the squares of the absolute values of the elements of each row, and of each column, must add up to unity.

The magnitudes of the six elements of the first two rows have been determined from measurements of semileptonic decay rates. For example, the amplitude for the tree-level process $d \to u + e^- + \bar{\nu}_e$ is proportional to V_{ud}. But non-perturbative strong interaction effects enter into the amplitudes for corresponding measured hadronic transitions, such as $n \to p + e^- + \bar{\nu}_e$ or $\pi^- \to \pi^0 + e^- + \bar{\nu}_e$. In many cases these hadronic factors in the matrix elements can now be calculated by unquenched lattice QCD.

The status of the experimental determination of the moduli $|V_{ij}|$ is regularly reviewed by the Particle Data Group. The current results for the unitarity checks are (Ceccucci, Ligeti and Sakai 2022)

$$|V_{ud}|^2 + |V_{us}|^2 + |V_{ub}|^2 = 0.9985 \pm 0.0007 \tag{20.173}$$
$$|V_{cd}|^2 + |V_{cs}|^2 + |V_{cb}|^2 = 1.001 \pm 0.012 \tag{20.174}$$
$$|V_{ud}|^2 + |V_{cd}|^2 + |V_{td}|^2 = 0.9972 \pm 0.0020 \tag{20.175}$$
$$|V_{us}|^2 + |V_{cs}|^2 + |V_{ts}|^2 = 1.004 \pm 0.012. \tag{20.176}$$

Evidently these results are consistent with the CKM prediction of unitarity, up to a 2.2 σ tension in (20.173).

The most accurate values of the nine magnitudes are obtained by a global fit to all the available measurements, imposing the constraints of 3-generation unitarity. The current result for the magnitudes, imposing these constraints, is (Ceccucci, Ligeti and Sakai 2022)

$$\mathbf{V} = \begin{pmatrix} 0.97435 \pm 0.00016 & 0.22500 \pm 0.00067 & 0.00369 \pm 0.00011 \\ 0.22486 \pm 0.00067 & 0.97349 \pm 0.00016 & 0.04182^{+0.00085}_{-0.00074} \\ 0.00857^{+0.00020}_{-0.00018} & 0.04110^{+0.00083}_{-0.00072} & 0.99918^{+0.000031}_{-0.000036} \end{pmatrix}, \tag{20.177}$$

and the Jarlskog invariant is $J = (3.08^{+0.15}_{-0.13}) \times 10^{-5}$.

From (20.177) it follows that the mixing angles are small, and moreover satisfy a definite hierarchy

$$1 \gg \theta_{12} \gg \theta_{23} \gg \theta_{13}. \tag{20.178}$$

In more physical terms, hadrons evidently prefer to decay semileptonically to the nearest generation. Also, because the elements V_{ub}, V_{cb}, V_{td}, and V_{ts}, which connect the third generation to the first two, are all quite small, the physics of the first two generations is hardly influenced by the presence of the third. This reflects, in quantitative terms, the success of the Cabibbo-GIM description, and the fact that the **CP**-violation seen in the K-meson sector is so weak. **CP**-violation is much more visible in B physics, as Carter and Sanda (1980, 1981) were the first to suggest, and as we shall discuss in the following chapter.

Consider now the complex-valued off-diagonal unitarity conditions, in particular the condition (20.168). Following Wolfenstein (1983), we identify s_{12} as the small parameter

λ, and write $V_{cb} \simeq s_{23} = A\lambda^2$ and $V_{ub} = s_{13}\exp(-i\delta) = A\lambda^3(\rho - i\eta)$ with $A \simeq 1$ and $|\rho - i\eta| < 1$. This gives

$$\mathbf{V} = \begin{pmatrix} 1 - \lambda^2/2 & \lambda & A\lambda^3(\rho - i\eta) \\ -\lambda & 1 - \lambda^2/2 & A\lambda^2 \\ A\lambda^3(1 - \rho - i\eta) & -A\lambda^2 & 1 \end{pmatrix}, \qquad (20.179)$$

neglecting terms of order λ^4 and higher. Then

$$z_1 = \frac{V_{td}V_{tb}^*}{V_{cd}V_{cb}^*} \simeq \rho + i\eta - 1, \quad z_2 = \frac{V_{ud}V_{ub}^*}{V_{cd}V_{cb}^*} \simeq -(\rho + i\eta), \quad J \simeq A^2\lambda^6\eta. \qquad (20.180)$$

The unitarity triangle represented by the condition (20.168), or alternatively $-z_1 - z_2 = 1$, is therefore a triangle on the base (1,0), with sides $\rho + i\eta$ and $1 - (\rho + i\eta)$. Buras *et al.* (1994) showed that including terms up to order λ^5 changes (ρ, η) to $(\bar\rho, \bar\eta)$ where $\bar\rho = (1 - \lambda^2/2)\rho, \bar\eta = (1 - \lambda^2/2)\eta$. The top vertex of the triangle in figure 20.10 is therefore at the point $(\bar\rho, \bar\eta)$. The angles α, β, and γ (also called ϕ_2, ϕ_1, and ϕ_3) are defined by

$$\alpha \equiv \phi_2 \equiv \arg\left(-\frac{V_{td}V_{tb}^*}{V_{ud}V_{ub}^*}\right) \approx \arg - \left(\frac{1 - \bar\rho - i\bar\eta}{\bar\rho + i\bar\eta}\right) \qquad (20.181)$$

$$\beta \equiv \phi_1 \equiv \arg\left(-\frac{V_{cd}V_{cb}^*}{V_{td}V_{tb}^*}\right) \approx \arg\left(\frac{1}{1 - \bar\rho - i\bar\eta}\right) \qquad (20.182)$$

$$\gamma \equiv \phi_3 \equiv \arg\left(-\frac{V_{ud}V_{ub}^*}{V_{cd}V_{cb}^*}\right) \approx \arg(\bar\rho + i\bar\eta) \qquad (20.183)$$

The sides of this triangle are determined by the magnitudes of the CKM elements, and so another check is provided by the condition that the three sides should close to form a triangle. Further independent constraints are provided by measurements of the angles α, β, and γ which are directly related to **CP**-violation effects, as we shall discuss in the following chapter. The fit displayed in (20.177) gave $\alpha + \beta + \gamma = (173 \pm 6)°$. Figure 20.11 shows a plot of all the constraints in the $\bar\rho, \bar\eta$ plane from many different measurements (combined following the approach of Charles *et al.* 2005 and Höcker *et al.* 2001, for example), and the global fit, as presented by Ceccucci, Ligeti, and Sakai (2022). The annular region labelled by $|V_{ub}|$ represents, for example, the uncertainty in the determination of $|z_2| = |V_{ud}V_{ub}^*/V_{cd}V_{cb}^*|$, which is principally due to the uncertainty in $|V_{ub}|$. The region labelled by Δm_d represents the constraint on $|z_1| = |V_{td}V_{tb}^*/V_{cd}V_{cb}^*|$, where $|V_{td}|$ is deduced from the value of the $B^0 - \bar{B}^0$ mass difference Δm_d measured in $B^0 - \bar{B}^0$ oscillations mediated by top-quark dominated box diagrams (see section 21.2.1 in the following chapter); here the uncertainties are dominated by lattice QCD. Figure 20.11 represents an enormous sustained experimental effort. The 95% CL regions all overlap consistently. It is quite remarkable how the single **CP**-violating parameter, three-generation scheme of Kobayashi and Maskawa (1973) has withstood this searching test.

20.8   Non-leptonic weak interactions

The CKM 6-quark charged weak current, which replaces the GIM current (20.101), is

$$\hat{j}_{CKM}^\mu(u, d, s, c, t, b) = \bar{\hat{u}}\gamma^\mu\frac{(1 - \gamma_5)}{2}\hat{d}' + \bar{\hat{c}}\gamma^\mu\frac{(1 - \gamma_5)}{2}\hat{s}' + \bar{\hat{t}}\gamma^\mu\frac{(1 - \gamma_5)}{2}\hat{b}', \qquad (20.184)$$

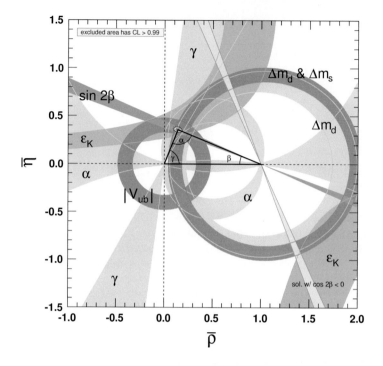

FIGURE 20.11

Constraints in the $\bar{\rho}, \bar{\eta}$ plane. The shaded areas have 95% CL. [Figure reproduced, courtesy of the Particle Data Group, from the review of the CKM Quark-Mixing Matrix by A Ceccucci, Z Ligeti and Y Sakai (2022), section 12 of Workman *et al.* (2022)]

and the effective weak Hamiltonian of (20.92) (as modified by CKM) clearly contains the term

$$\hat{\mathcal{H}}_{\mathrm{CC}}^{q}(x) = \frac{G_{\mathrm{F}}}{\sqrt{2}}\hat{j}_{\mathrm{CKM}}^{\mu}(x)\hat{j}_{\mu\,\mathrm{CKM}}^{\dagger}(x) \tag{20.185}$$

in which no lepton fields are present (just as there are no quarks in (20.40)). This interaction is responsible, at the quark level, for transitions involving four quark (or antiquark) fields at a point. For example, the process shown in figure 20.12 can occur. By 'adding on' another two quark lines u and d, which undergo no weak interaction, we arrive at figure 20.13, which represents the non-leptonic decay $\Lambda^{0} \rightarrow \mathrm{p}\pi^{-}$.

This figure is, of course, rather symbolic since there are strong QCD interactions (not shown) which are responsible for binding the three-quark systems into baryons, and the q$\bar{\mathrm{q}}$ system into a meson. Unlike the case of deep inelastic lepton scattering, these QCD interactions cannot be treated perturbatively, since the distance scales involved are typically those of the hadron sizes (~ 1 fm), where perturbation theory fails. This means that non-leptonic weak interactions among hadrons are difficult to analyze quantitatively, though progress can be made via lattice QCD. Similar difficulties also arise, evidently, in the case of semi-leptonic decays. In general, one has to begin in a phenomenological way, parametrizing the decay amplitudes in terms of appropriate form factors (which are analogous to the electromagnetic form factors introduced in chapter 8). In the case of transitions involving at least one heavy quark Q, Isgur and Wise (1989, 1990) noticed that a considerable simplification occurs in the linit $m_{\mathrm{Q}} \rightarrow \infty$. For example, one universal function (the 'Isgur-Wise form factor') is sufficient to describe a large number of hadronic form factors introduced for

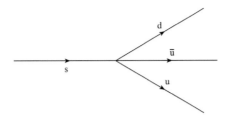

FIGURE 20.12
Effective four-fermion non-leptonic weak transition at the quark level.

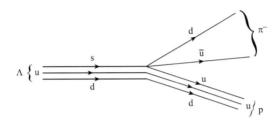

FIGURE 20.13
Non-leptonic weak decay of Λ^0 using the process of figure 20.12, with the addition of two 'spectator' quarks.

semi-leptonic transitions between two heavy pseudoscalar (0^-) or vector (1^-) mesons. For an introduction to the Isgur-Wise theory, we refer to Donoghue *et al.* (1992).

The non-leptonic sector is, however, the scene of some very interesting physics, such as $K^0 - \bar{K}^0$ and $B^0 - \bar{B}^0$ oscillations, and **CP** violation in the $K^0 - \bar{K}^0, D^0 - \bar{D}^0$, and $B^0 - \bar{B}^0$ systems. We shall discuss these phenomena in the following chapter.

Problems

20.1 Show that in the non-relativistic limit $(|\boldsymbol{p}| \ll M)$ the matrix element $\bar{u}_p \gamma^\mu u_n$ of (20.2) vanishes if p and n have different spin states.

20.2 Verify the normalization $N = (E + |\boldsymbol{p}|)^{1/2}$ in (20.23).

20.3 Verify (20.30) and (20.31).

20.4 Verify that equations (20.32) are invariant under **CP**.

20.5 The matrix γ_5 is defined by $\gamma_5 = i\gamma^0\gamma^1\gamma^2\gamma^3$. Prove the following properties:

(a) $\gamma_5^2 = 1$ and hence that
$$(1 + \gamma_5)(1 - \gamma_5) = 0;$$

(b) from the anticommutation relations of the other γ matrices, show that
$$\{\gamma_5, \gamma_\mu\} = 0$$

and hence that

$$(1 + \gamma_5)\gamma_0 = \gamma_0(1 - \gamma_5)$$

and

$$(1 + \gamma_5)\gamma_0\gamma_\mu = \gamma_0\gamma_\mu(1 + \gamma_5).$$

20.6

(a) Consider the two-dimensional antisymmetric tensor ϵ_{ij} defined by

$$\epsilon_{12} = +1, \epsilon_{21} = -1, \qquad \epsilon_{11} = \epsilon_{22} = 0.$$

By explicitly enumerating all the possibilities (if necessary), convince yourself of the result

$$\epsilon_{ij}\epsilon_{kl} = +1(\delta_{ik}\delta_{jl} - \delta_{il}\delta_{jk}).$$

Hence prove that

$$\epsilon_{ij}\epsilon_{il} = \delta_{jl} \qquad \text{and} \qquad \epsilon_{ij}\epsilon_{ij} = 2$$

(remember, in two dimensions, $\sum_i \delta_{ii} = 2$).

(b) By similar reasoning to that in part (a) of this question, it can be shown that the product of two three-dimensional antisymmetric tensors has the form

$$\epsilon_{ijk}\epsilon_{lmn} = \begin{vmatrix} \delta_{il} & \delta_{im} & \delta_{in} \\ \delta_{jl} & \delta_{jm} & \delta_{jn} \\ \delta_{kl} & \delta_{km} & \delta_{kn} \end{vmatrix}.$$

Prove the results

$$\epsilon_{ijk}\epsilon_{imn} = \begin{vmatrix} \delta_{jm} & \delta_{jn} \\ \delta_{km} & \delta_{kn} \end{vmatrix} \qquad \epsilon_{ijk}\epsilon_{ijn} = 2\delta_{kn} \qquad \epsilon_{ijk}\epsilon_{ijk} = 3!$$

(c) Extend these results to the case of the four-dimensional (Lorentz) tensor $\epsilon_{\mu\nu\alpha\beta}$ (remember that a minus sign will appear as a result of $\epsilon_{0123} = +1$ but $\epsilon^{0123} = -1$).

20.7 Starting from the amplitude for the process

$$\nu_\mu + e^- \to \mu^- + \nu_e$$

given by the current-current theory of weak interactions,

$$\mathcal{M} = -i(G_F/\sqrt{2})\bar{u}(\mu)\gamma_\mu(1 - \gamma_5)u(\nu_\mu)g^{\mu\nu}\bar{u}(\nu_e)\gamma_\nu(1 - \gamma_5)u(e),$$

verify the intermediate results given in section 20.5 leading to the result

$$d\sigma/dt = G_F^2/\pi$$

(neglecting all lepton masses). Hence show that the local total cross section for this process rises linearly with s:

$$\sigma = G_F^2 s/\pi.$$

20.8 The invariant amplitude for $\pi^+ \to e^+\nu$ decay may be written as (see (18.52))

$$\mathcal{M} = (G_F V_{ud})f_\pi p^\mu \bar{u}(\nu)\gamma_\mu(1 - \gamma_5)v(e)$$

where p^μ is the 4-momentum of the pion, and the neutrino is taken to be massless. Evaluate the decay rate in the rest frame of the pion using the decay rate formula

$$\Gamma = (1/2m_\pi)|\overline{\mathcal{M}}|^2 \mathrm{dLips}(m_\pi^2; k_e, k_\nu).$$

Show that the ratio of $\pi^+ \to e^+\nu$ and $\pi^+ \to \mu^+\nu$ rates is given by

$$\frac{\Gamma(\pi^+ \to e^+\nu)}{\Gamma(\pi^+ \to \mu^+\nu)} = \left(\frac{m_e}{m_\mu}\right)^2 \left(\frac{m_\pi^2 - m_e^2}{m_\pi^2 - m_\mu^2}\right)^2.$$

Repeat the calculation using the amplitude

$$\mathcal{M}' = (G_F V_{ud}) f_\pi p^\mu \bar{u}(\nu) \gamma_\mu (g_V + g_A \gamma_5) v(e)$$

and retaining a finite neutrino mass. Discuss the e^+/μ^+ ratio in the light of your result.

20.9

(a) Verify that the inclusive inelastic neutrino-proton scattering differential cross section has the form

$$\frac{d^2\sigma^{(\nu)}}{dQ^2 d\nu} = \frac{G_F^2 k'}{2\pi k} \left(W_2^{(\nu)} \cos^2(\theta/2) + W_1^{(\nu)} 2\sin^2(\theta/2)\right.$$
$$\left. + \frac{(k+k')}{M} \sin^2(\theta/2) W_3^{(\nu)}\right)$$

in the notation of section 20.7.2.

(b) Using the Bjorken scaling behaviour

$$\nu W_2^{(\nu)} \to F_2^{(\nu)} \qquad MW_1^{(\nu)} \to F_1^{(\nu)} \qquad \nu W_3^{(\nu)} \to F_3^{(\nu)}$$

rewrite this expression in terms of the scaling functions. In terms of the variables x and y, neglect all masses and show that

$$\frac{d^2\sigma^{(\nu)}}{dxdy} = \frac{G_F^2}{2\pi} s[F_2^{(\nu)}(1-y) + F_1^{(\nu)} xy^2 + F_3^{(\nu)}(1-y/2)yx].$$

Remember that

$$\frac{k' \sin^2(\theta/2)}{M} = \frac{xy}{2}.$$

(c) Insert the Callan-Gross relation

$$2x F_1^{(\nu)} = F_2^{(\nu)}$$

to derive the result quoted in section 20.7.2:

$$\frac{d^2\sigma^{(\nu)}}{dxdy} = \frac{G_F^2}{2\pi} s F_2^{(\nu)} \left(\frac{1 + (1-y)^2}{2} + \frac{x F_3^{(\nu)}}{F_2^{(\nu)}} \frac{1 - (1-y)^2}{2}\right).$$

20.10 The differential cross section for $\nu_\mu q$ scattering by charged currents has the same form (neglecting masses) as the $\nu_\mu e^- \to \mu^- \nu_e$ result of problem 20.7, namely

$$\frac{d\sigma}{dt}(\nu q) = \frac{G_F^2}{\pi}.$$

(a) Show that the cross section for scattering by antiquarks $\nu_\mu \bar{q}$ has the form

$$\frac{d\sigma}{dt}(\nu\bar{q}) = \frac{G_F^2}{\pi}(1-y)^2.$$

(b) Hence prove the results quoted in section 20.7.2:

$$\frac{\mathrm{d}^2\sigma}{\mathrm{d}x\mathrm{d}y}(\nu\mathrm{q}) = \frac{G_{\mathrm{F}}^2}{\pi}sx\delta(x - Q^2/2M\nu)$$

and

$$\frac{\mathrm{d}^2\sigma}{\mathrm{d}x\mathrm{d}y}(\nu\bar{\mathrm{q}}) = \frac{G_{\mathrm{F}}^2}{\pi}sx(1 - y^2)\delta(x - Q^2/2M\nu)$$

(where M is the nucleon mass).

(c) Use the parton model prediction

$$\frac{\mathrm{d}^2\sigma^{(\nu)}}{\mathrm{d}x\mathrm{d}y} = \frac{G_{\mathrm{F}}^2}{\pi}sx[q(x) + \bar{q}(x)(1 - y)^2]$$

to show that

$$F_2^{(\nu)} = 2x[q(x) + \bar{q}(x)]$$

and

$$\frac{xF_3^{(\nu)}(x)}{F_2^{(\nu)}(x)} = \frac{q(x) - \bar{q}(x)}{q(x) + \bar{q}(x)}.$$

20.11 Verify the transformation laws (20.160).

21

CP Violation and Oscillation Phenomena

In this chapter we shall continue with the phenomenology of weak interactions, introducing two topics which have been the focus of intense experimental effort in recent decades : **CP** violation in B meson decays, and oscillations in both neutral meson and neutrino systems. In the following chapter we take up again the gauge theory theme, with the Glashow-Salam-Weinberg electroweak theory.

CP violation was first discovered in the decays of neutral K mesons (Christenson *et al.* 1964), but we shall not follow a historical approach to this subject. Instead we shall concentrate on B-meson decays, where the effects are far larger, and much clearer to interpret theoretically than in the K-meson case. **CP** violation is reviewed in Branco *et al.* (1999), Bigi and Sanda (2000), Harrison and Quinn (1998), and Gershon and Gligorov (2017). We aim simply to illustrate the principles with some particular examples. In particular, we shall generally not discuss theoretical predictions; our main emphasis will be on describing selected experiments which have allowed determinations of the angles α, β, and γ of the unitarity triangle, figure 20.10.

We saw in section 20.7.3 that, in the SM, **CP** violation is attributable solely to one irreducible phase degree of freedom, δ, in the CKM matrix $\mathbf{V}$. Clearly, to measure this phase, it is necessary (as usual in quantum mechanics) to create situations where it enters into the *interference* between two complex amplitudes. Two situations may be distinguished (Carter and Sanda 1980): (a) interference between two decay amplitudes $B^0 \to X$ and $\bar{B}^0 \to X$, where the B^0 and $\bar{B}^0$ have been produced in a coherent state by mixing, and decay to a common hadronic final state X; and (b) interference between two different amplitudes for a single B-meson to decay to a final state X. Method (b) ("direct **CP** violation") can be applied to charged as well as neutral mesons.

The mixing in method (a) is formally similar to that involved in neutrino oscillations, which we treat after the meson case. We shall therefore start in section 21.1 with an example illustrating method (b). We set up the mixing formalism and apply it to **CP** violation in B decays in section 21.2; we discuss K decays in section 21.3. Neutrino oscillations are treated in section 21.4.

21.1   Direct CP violation in B decays

Consider the decays

$$B^0 \to K^+\pi^- \quad \text{and} \quad \bar{B}^0 \to K^-\pi^+ \tag{21.1}$$

which provided the first observations of direct **CP** violation in B decays. The first of these can proceed via the quark transitions shown in figure 21.1, which (in parton-like language) is a "tree-diagram". Of course, long-distance strong interaction effects will come into play in forming the hadronic states B^0, K^+, and π^-, and in final state interactions between the K^+ and π^-; we do not represent these strong interactions in figure 21.1, or in subsequent

DOI: 10.1201/9781003411666-21

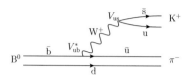

FIGURE 21.1

Tree diagram contribution to $B^0 \to K^+\pi^-$ via the quark transition $\bar{b} \to \bar{s}u\bar{u}$.

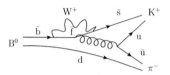

FIGURE 21.2

Penguin diagrams $(\bar{f} = \bar{u}, \bar{c}, \bar{t})$ contributing to $B^0 \to K^+\pi^-$ via the quark transition $\bar{b} \to \bar{s}u\,\bar{u}$.

similar diagrams. We are specifically interested in the *weak phase* of figure 21.1, since it is this quantity which changes sign under the **CP** transformation ($V_{ij} \to V_{ij}^*$), and this phase change will lead to observable **CP** violation effects. By contrast, the strong interaction phases—which will play an important role—will be **CP** invariant, but we do not need to display them yet. So we write the amplitude for figure 21.1 as

$$A_{\mathrm{T}}(B^0 \to K^+\pi^-) = V_{ub}^* V_{us}\, t_{\bar{u}}, \tag{21.2}$$

where the CKM couplings have been displayed.

There are, however, three order-α_s loop corrections to figure 21.1, shown in figure 21.2, where $\bar{f} = \bar{u}, \bar{c}$ and $\bar{t}$. We write the amplitude for the sum of these three "penguin" diagrams[1] as

$$A_{\mathrm{P}}(B^0 \to K^+\pi^-) = V_{us}V_{ub}^*\, p_{\bar{u}} + V_{cs}V_{cb}^*\, p_{\bar{c}} + V_{ts}V_{tb}^*\, p_{\bar{t}}, \tag{21.3}$$

where $p_{\bar{f}}$ is the penguin amplitude with $\bar{f}$ in the loop. It is convenient to use a unitarity relation to rewrite $V_{ts}V_{tb}^*$ in terms of the other two related CKM products:

$$V_{ts}V_{tb}^* = -V_{us}V_{ub}^* - V_{cs}V_{cb}^*, \tag{21.4}$$

so that the total amplitude becomes

$$A(B^0 \to K^+\pi^-) = V_{ub}^* V_{us} T_{K\pi} + V_{cs}V_{cb}^* P_{K\pi}, \tag{21.5}$$

where

$$T_{K\pi} = t_{\bar{u}} + p_{\bar{u}} - p_{\bar{t}}, \qquad P_{K\pi} = p_{\bar{c}} - p_{\bar{t}}. \tag{21.6}$$

In terms of the parametrization (20.179), (21.5) becomes

$$A(B^0 \to K^+\pi^-) = A\lambda^4(\rho + i\eta)T_{K\pi} + A\lambda^2(1 - \lambda^2/2)P_{K\pi} \tag{21.7}$$

where (see equation (20.183)) γ is the phase of $\rho + i\eta$. Similarly, the amplitude for the charge-conjugate reaction is

$$A(\bar{B}^0 \to K^-\pi^+) = A\lambda^4(\rho - i\eta)T_{K\pi} + A\lambda^2(1 - \lambda^2/2)P_{K\pi}. \tag{21.8}$$

---

[1] "Loop" diagrams might be better, but the name (which has historical origins) seems to have stuck.

The interference between the tree and the penguin terms will be sensitive to the phase angle γ (Bander, Silverman and Soni 1979).

We can now calculate the decay-rate asymmetry

$$\mathcal{A}_{K\pi} = \frac{|A(\bar{B}^0 \to K^-\pi^+)|^2 - |A(B^0 \to K^+\pi^-)|^2}{|A(\bar{B}^0 \to K^-\pi^+)|^2 + |A(B^0 \to K^+\pi^-)|^2}. \tag{21.9}$$

To simplify things, let us take a common complex factor K out of the expressions (21.7) and (21.8) and write them as

$$A(B^0 \to K^+\pi^-) = K(e^{i\gamma} + Re^{i(\delta_P - \delta_T)}) \tag{21.10}$$

$$A(\bar{B}^0 \to K^-\pi^+) = K(e^{-i\gamma} + Re^{i(\delta_P - \delta_T)}), \tag{21.11}$$

where R is real, and $\delta_P - \delta_T$ is the difference in (strong) phases between $P_{K\pi}$ and $T_{K\pi}$. Then we easily find

$$\mathcal{A}_{K\pi} = \frac{2R\sin\gamma \sin(\delta_T - \delta_P)}{1 + R^2 + 2R\cos\gamma \cos(\delta_T - \delta_P)}. \tag{21.12}$$

Thus we see that, for a **CP**-violating signal, there must be two interfering amplitudes leading to a common final state, and the amplitudes must have both different weak phases and different strong phases. An order of magnitude estimate of the effect can be made as follows. First, note that $P_{K\pi}$ is not ultraviolet divergent, since it is the difference of two penguin contributions; its magnitude is expected to be of order $\alpha_s/\pi \sim 0.05$. The tree contribution in (21.7) carries an extra factor of $\lambda^2 \sim 0.05$ as compared with the penguin contribution, so that R is of order 1. This indicates that the asymmetry should be significant.

Indeed non-zero values of $\mathcal{A}_{K\pi}$ have been reported by both the BaBar and Belle collaborations:

$$\text{BaBar} \ (\text{B.Aubert } et \ al. \ 2004) : \mathcal{A}_{K\pi} = -0.133 \pm 0.030 \pm 0.009 \tag{21.13}$$

$$\text{Belle} \ (\text{Y.Chao } et \ al. \ 2005) : \mathcal{A}_{K\pi} = -0.113 \pm 0.022 \pm 0.008 \tag{21.14}$$

where the first error is statistical and the second is systematic.

Altough $\mathcal{A}_{K\pi}$ is sensitive to the **CP**-violating angle γ, it is not easy to extract γ cleanly from these measurements. Both the tree and the penguin amplitudes involve non-perturbative factors for producing a particular meson state from the corresponding q$\bar{\text{q}}$ state; the strong phases are also not known.

A decay with no penguin contributions, but still with two interfering channels, would have fewer uncertainties. (It is also less likely to be affected by new physics, which could provide short-distance corrections to penguin loops.) One such example is provided by the decays (a) $B^- \to D^0 K^-$ and (b) $B^- \to \bar{D}^0 K^-$, which can interfere when the $(D^0 K^-)$ and $(\bar{D}^0 K^-)$ states decay to a common final state. Here the quark transition in (a) is b $\to$ c$\bar{\text{u}}$s, and in (b) is b $\to$ u$\bar{\text{c}}$s; in neither case is a penguin contribution possible.

The tree-level diagrams which contribute are shown in the left-hand parts of figure 21.3 (we shall discuss the right-hand parts in a moment). We denote the amplitude for $B^- \to D^0 K^-$ by A_B, and note that $A_B \sim A\lambda^3$. The amplitude for $B^- \to \bar{D}^0 K^-$, $\tilde{A}_B$, differs in three ways from A_B: (i) it is colour-suppressed by a factor $1/3$ since the $\bar{\text{c}}$ and u have to be colour matched; (ii) it contains the factor $V_{ub}V_{cs}^* \sim A\lambda^3(\rho - i\eta)$ and (iii) it will have a different strong interaction phase. With these factors in mind, we write

$$\tilde{A}_B = r_B A_B e^{i(\delta_B - \gamma)} \tag{21.15}$$

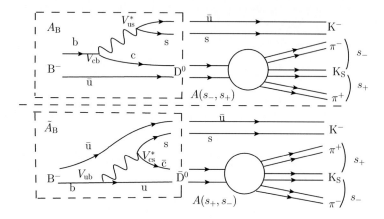

FIGURE 21.3
Left-hand part: tree diagram contributions to $B^- \rightarrow D^0 K^-$ (upper diagram, via quark transition $b \rightarrow c\bar{u}s$), and to $B^- \rightarrow \bar{D}^0 K^-$ (lower diagram, via quark transition $b \rightarrow u\bar{c}s$). Right-hand part: decays of D^0 and $\bar{D}^0$ to the common $\pi^+ \pi^- K_S$ state.

where δ_B is the difference in strong phases between $\tilde{A}_B$ and A_B, and r_B is the magnitude ratio of the amplitudes. Since $|\rho - i\eta| \sim 0.38$, r_B is of order $0.1 - 0.2$, allowing for the colour suppression.

Once again, the asymmetry is proportional to

$$|1 + r_B \exp[i(\delta_B - \gamma)]|^2 - |1 + r_B \exp[i(\delta_B + \gamma)]|^2 \approx 4 r_B \sin \delta_B \sin \gamma. \qquad (21.16)$$

This involves γ, but the relative smallness of r_B tends to reduce the sensitivity to γ. An alternative determination of γ can be made (Attwood *et al.* 2001, Giri *et al.* 2003) by making use of three-body decays (to a common channel) of D^0 and $\bar{D}^0$, such as D^0, $\bar{D}^0 \rightarrow K_S \pi^+ \pi^-$. If we denote the amplitude for $D^0 \rightarrow K_S \pi^+ \pi^-$ by $A(s_-, s_+)$ (see figure 21.3), where $s_- = (p_K + p_{\pi^-})^2$ and $s_+ = (p_K + p_{\pi^+})^2$ are the indicated invariant masses, then the amplitude for the B^- to decay to $K^- K_S \pi^+ \pi^-$ via the two coherent paths is[2]

$$A_- = A_B D[A(s_-, s_+) + r_B e^{i(\delta_B - \gamma)} A(s_+, s_-)], \qquad (21.17)$$

and the amplitude for the charge conjugate reaction $B^+ \rightarrow K^+ K_S \pi^- \pi^+$ is

$$A_+ = A_B D[A(s_+, s_-) + r_B e^{i(\delta_B + \gamma)} A(s_-, s_+)], \qquad (21.18)$$

where D is the D meson propagator. The event rate for the B^- decay is then $\Gamma_-(s_-, s_+)$ where (Aubert *et al.* 2008)

$$\Gamma_-(s_-, s_+) \propto |A(s_-, s_+)|^2 + r_B^2 |A(s_+, s_-)|^2 +$$
$$2[x_- \text{Re}\{A(s_-, s_+)A^*(s_+, s_-)\} + y_- \text{Im}\{A(s_-, s_+)A^*(s_+, s_-)\}] \qquad (21.19)$$

and the rate for B^+ decay is $\Gamma_+(s_-, s_+)$ where

$$\Gamma_+(s_-, s_+) \propto |A(s_+, s_-)|^2 + r_B^2 |A(s_-, s_+)|^2 +$$
$$2[x_+ \text{Re}\{A(s_+, s_-)A^*(s_-, s_+)\} + y_+ \text{Im}\{A(s_+, s_-)A^*(s_-, s_+)\}]. \qquad (21.20)$$

---

[2]We are neglecting $D^0 - \bar{D}^0$ mixing and **CP** asymmetries in D decays, which are at the 1% or less level (Grossman *et al.* 2005).

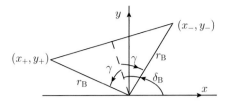

FIGURE 21.4

Geometry of the **CP**-violating parameters $x_\pm, y_\pm$.

Here

$$x_- = r_B \cos(\delta_B - \gamma), \quad y_- = r_B \sin(\delta_B - \gamma) \tag{21.21}$$

$$x_+ = r_B \cos(\delta_B + \gamma), \quad y_+ = r_B \sin(\delta_B + \gamma). \tag{21.22}$$

The geometry of the **CP**-violating parameters is shown in figure 21.4. Note that the separation of the B^- and B^+ positions in the (x, y) plane is equal to $2r_B|\sin\gamma|$, and is a measure of direct **CP** violation. The angle between the lines connecting the B^- and B^+ centres to the origin $(0,0)$ is equal to 2γ.

If the functional dependence of both the modulus and the phase of $A(s_-, s_+)$ were known, then the rates would depend on only three variables, r_B, δ_B, and γ (or equivalently on $x_\pm, y_\pm$). In fact, $A(s_-, s_+)$ can be determined from a Dalitz plot analysis of the decays of D^0 mesons coming from $D^{*+} \to D^0 \pi^+$ decays produced in $e^+ e^- \to c\bar{c}$ events; the charge of the low-momentum π^+ identifies the flavour of the D^0. Such an analysis is a well-established tool in the study of three-hadron final states, originating in the pioneer work of Dalitz (1953), in connection with the decay $K \to 3\pi$. The partial rate for $D^0(\bar{D}^0) \to K_S \pi^+ \pi^-$ is (see the kinematics section of Workman *et al.* 2022)

$$d\Gamma \propto |A(s_-, s_+)|^2 ds_- ds_+. \tag{21.23}$$

The physical region in the s_-, s_+ plane is a bounded oval-like region, which would be uniformly populated if $A(s_-, s_+)$ were a constant. In reality, the decay is dominated by quasi two-body states, in particular

$$
\begin{aligned}
D^0 &\to K^{*-}(s_-)\,\pi^+ \quad \text{(CA)} \\
&\to K^{*+}(s_+)\,\pi^- \quad \text{(DCS)} \\
&\to K_S\,\rho^0(s_0), \quad \text{(CP)}
\end{aligned} \tag{21.24}
$$

where (CA) means CKM-favoured, (DCS) means doubly CKM-suppressed, and (CP) means that it is a **CP** eigenstate. The Dalitz plot shows a dense band of events at $s_- = m^2_{K^{*-}}$ corresponding to the K^{*-} resonance, a band at $s_+ = m^2_{K^{*+}}$, and a band at $s_0 = m^2_\rho$, where $s_0 = (p_{\pi^+} + p_{\pi^-})^2$ and $s_+ + s_- + s_0 = m^2_D + m^2_K + 2m^2_\pi$.

The Dalitz-plot analysis proceeds by writing (Aubert *et al.* 2008) $A(s_-, s_+)$ as a coherent sum of terms representing the quasi two-body modes, together with a non-resonant background. Once $A(s_-, s_+)$ is known, it is inserted into $\Gamma_\mp(s_-, s_+)$ to obtain $(x_\pm, y_\pm)$ from the Dalitz plot distributions of the signal modes of the $B^\mp$ decays. From these, the quantities r_B, δ_B, and finally γ can be inferred.

This method has been applied by both BaBar and Belle to determine γ. Their results were

$$\text{BaBar (del Amo Sanchez } et\ al.\ 2010): \quad \gamma = 68 \pm 14 \pm 4 \pm 3° \tag{21.25}$$

$$\text{Belle (Poluektov } et\ al.\ 2010): \quad \gamma = 78.4^{+10.8}_{-11.6} \pm 3 \pm 8.9° \tag{21.26}$$

where the last uncertainty is due to the D-decay modelling (which ignores, for example, rescattering among the three final state particles). Both these experiments use decays $B^\pm \to DK^\pm, B^\pm \to D^*K^\pm$ with $D^* \to D\pi^0$ and $D^* \to D\gamma$; BaBar in addition uses the decays $D^0 \to K_S K^+ K^-$.

More recently LHCb has significantly reduced the uncertainty in γ (Aaij *et al.* 2021a). The sensitivity to γ comes almost entirely from $B^\pm \to DK^\pm$ decays, where there are approximately 13,600 (1,900) reconstructed events in the $D \to K_S^0 \pi^+ \pi^-$ ($D \to K_S^0 K^+ K^-$) modes. The analysis is performed in bins of the D-decay Dalitz plot, where a combination of measurements by CLEO and BESII (Libby *et al.* (CLEO) 2010, Ablikin *et al.* (BESIII) 2020a, 2020b, 2020c) provided input on the strong interaction phase variation across the Dalitz plot, thus avoiding uncertainties in the modelling of that phase variation. The CKM angle γ was determined to be $\gamma = (68.7^{+5.2}_{-5.1})°$, where the uncertainties are statistical. This is currently the most precise single determination of γ. The current world average is (Ceccucci, Ligeti and Sakai 2022, section 12 in Workman *et al.* 2022)

$$\gamma = (65.9^{+3.3}_{-3.5})°. \tag{21.27}$$

We now turn to the other main method of detecting **CP** violation, through the interference between decays of (for example) B^0 and $\bar{B}^0$ mesons that have been produced in a coherent state by mixing. For this we need to set up the formalism describing time-dependent mixing.

21.2   CP violation in B meson oscillations

B^0 - $\bar{B}^0$ oscillations have been studied by the BaBar and Belle collaborations at the PEP2 and KEKB asymmetric e^+e^- colliders. These machines operate at a centre of mass energy equal to the mass of the $\Upsilon(4S)$ resonance state, which is some 20 MeV above the threshold for $B^0 \bar{B}^0$ production. If produced in a symmetric e^+e^- collider (with equal and opposite momenta for the e^+ and e^-), the produced B mesons would move very slowly, $v/c \sim 0.06$, covering a distance of only some 30 μm before decaying ($c\tau$ for B mesons is about 460 μm). This would make it impossible to resolve the decay vertices of the two Bs, as is required in order to observe B^0 - $\bar{B}^0$ oscillations, since the accuracy of the decay vertex reconstruction is roughly 100 μm. Oddone (1989) suggested making $e^+ e^-$ collisions with asymmetric energy colliding beams, so that the B mesons now move with the motion of the centre of mass, which can be considerable. For example, at PEP2 (e^- 9 GeV, e^+ 3.1 GeV) $\beta_{cm} \sim 0.5$ and $\gamma_{cm} \sim 1.15$, so that the distance travelled in the (asymmetric) lab frame during the lifetime of an average B meson is ~ 250 μm, which is measurable. At KEKB (e^- 8 GeV, e^- 3.5 GeV), $\beta_{cm}\gamma_{cm} \sim 0.425$.

Since the $\Upsilon(4S0)$ state has $J = 1$, the decay $\Upsilon \to BB$ leaves the B mesons in a p wave state, which is forbidden for two identical spinless bosons; therefore one must be a B^0 and the other a $\bar{B}^0$, but we do not know which is which until one has been identified ("tagged") in some way. The flavour of the tagged B may be determined, for example, by the charge of the lepton emitted in the semi-leptonic decays $B^0 \to D^- \ell^+ \nu_e, \bar{B}^0 \to D^+ \ell^- \bar{\nu}_e$. We shall not describe the evolution of the BB coherent state following production; interested readers may consult Cohen *et al.* (2009) for an instructive discussion, which also covers neutrino oscillations. We shall be interested in the time dependence of the state of the meson which partners the tagged meson, once the correlated state has been collapsed by the tagging at time $t = 0$ say; the partner meson will be reconstructed by its decay products. Note that the partner meson can decay earlier or later than the tagged one; its state vector has that time dependence which ensures that it becomes the correlate of the tagged particle at $t = 0$.

21.2.1   Time-dependent mixing formalism

We denote the neutral meson by B (which will usually be B^0, but could also be K^0 or D^0), and its **CP**-conjugate by $\bar{B}$. According to the theory of Weisskopf and Wigner (1930a, 1930b) (see also appendix A of Kabir (1968)) a state that is initially in some superposition of $|B\rangle$ and $|\bar{B}\rangle$, say

$$|\psi(0)\rangle = a(0)|B\rangle + b(0)|\bar{B}\rangle, \tag{21.28}$$

will evolve in time to a general superposition

$$|\psi(t)\rangle = a(t)|B\rangle + b(t)|\bar{B}\rangle \tag{21.29}$$

governed by an effective Hamiltonian $\mathbf{H}$ with matrix elements, in the 2-state subspace,

$$\mathbf{H} = \mathbf{M} - \mathrm{i}\frac{\mathbf{\Gamma}}{2} = \begin{pmatrix} A & p^2 \\ q^2 & A \end{pmatrix} \tag{21.30}$$

where $\mathbf{M}$ and $\mathbf{\Gamma}$ are Hermitian, and the equality $M_{11} - \mathrm{i}\Gamma_{11}/2 = M_{22} - \mathrm{i}\Gamma_{22}/2 = A$ follows from **CPT** invariance, which we shall assume. If **CP** is a good symmetry, then

$$\begin{aligned} \langle\bar{B}|\mathbf{H}|B\rangle &= \langle\bar{B}|(\mathbf{CP})^{-1}(\mathbf{CP})\mathbf{H}(\mathbf{CP})^{-1}\mathbf{CP}|B\rangle \\ &= \langle B|\mathbf{H}|\bar{B}\rangle \end{aligned} \tag{21.31}$$

so that p would equal q. Since $\mathbf{M}$ and $\mathbf{\Gamma}$ are both Hermitian, this would imply that M_{12} and Γ_{12} are both real; in the **CP** non-invariant world, this is not the case.

The eigenvalues of $\mathbf{H}$ are

$$\omega_{\mathrm{L}} \equiv m_{\mathrm{L}} - \mathrm{i}\Gamma_{\mathrm{L}}/2 = A + pq, \quad \omega_{\mathrm{H}} \equiv m_{\mathrm{H}} - \mathrm{i}\Gamma_{\mathrm{H}}/2 = A - pq, \tag{21.32}$$

and the corresponding eigenstates are

$$\begin{aligned} |B_{\mathrm{L}}\rangle &= (p\,|B\rangle + q|\bar{B}\rangle)/(|p|^2 + |q|^2)^{1/2} \\ |B_{\mathrm{H}}\rangle &= (p\,|B\rangle - q|\bar{B}\rangle)/(|p|^2 + |q|^2)^{1/2}. \end{aligned} \tag{21.33}$$

The states $|B_{\mathrm{L}}\rangle, |B_{\mathrm{H}}\rangle$ have definite masses m_{H} and m_{L} and widths Γ_{L} and Γ_{H}. The widths Γ_{L} and Γ_{H} are equal to a very good approximation for B and D mesons, because the Q-values of both are large; in the case of K-mesons (see section 21.3), one state decays predominantly to 2π and the other to 3π, with different Q-values, and the lifetimes are very different.

Suppose now that at time $t = 0$ the 'tag' shows that a B^0 has decayed. Then the partner is a $\bar{B}^0$ at $t = 0$, described by the superposition

$$|\bar{B}^0\rangle = -\frac{\sqrt{|p|^2 + |q|^2}}{2q}(|B_{\mathrm{H}}\rangle - |B_{\mathrm{L}}\rangle). \tag{21.34}$$

At a later time t in the $\bar{B}^0$ rest-frame, this state evolves to (problem 21.1)

$$|\bar{B}^0_t\rangle = g_+(t)\,|\bar{B}^0\rangle + (p/q)g_-(t)|B^0\rangle \tag{21.35}$$

where

$$\begin{aligned} g_+(t) &= \mathrm{e}^{-\mathrm{i}m_{\mathrm{B}}t}\mathrm{e}^{-\Gamma t/2}\cos(\Delta m_{\mathrm{B}}t/2) \end{aligned} \tag{21.36}$$

$$\begin{aligned} g_-(t) &= \mathrm{i}\mathrm{e}^{-\mathrm{i}m_{\mathrm{B}}t}\mathrm{e}^{-\Gamma t/2}\sin(\Delta m_{\mathrm{B}}t/2) \end{aligned} \tag{21.37}$$

with $m_{\mathrm{B}} = \frac{1}{2}(m_{\mathrm{H}} + m_{\mathrm{L}})$ and $\Delta m_{\mathrm{B}} = m_{\mathrm{H}} - m_{\mathrm{L}}$. Note, from (21.35), that the state which started as a $\bar{B}^0$ at $t = 0$ develops also a B^0 component at a later time. Similarly, if the tag shows that a $\bar{B}^0$ has decayed, the partner meson at $t = 0$ is a B^0, and its state evolves to

$$|B^0_t\rangle = (q/p)g_-(t)|\bar{B}^0\rangle + g_+(t)|B^0\rangle. \tag{21.38}$$

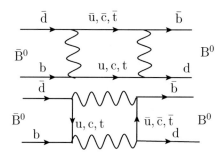

FIGURE 21.5
Box diagram contributions to $B^0 - \bar{B}^0$ mixing.

Consider first the semileptonic decays of B^0 and $\bar{B}^0$, where the only transitions that can occur are

$$B^0 \rightarrow \ell^+ \nu_\ell X, \quad \bar{B}^0 \rightarrow \ell^- \bar{\nu}_\ell X. \tag{21.39}$$

The state $|\bar{B}_t^0\rangle$ of (21.35), however, which was pure $\bar{B}^0$ at $t = 0$, will be able to decay to a positively charged lepton via the admixture of the $|B^0\rangle$ component; similarly negatively charged leptons may appear in the decay of $|B_t^0\rangle$. From (21.35) and (21.38) we obtain directly the amplitudes for these 'wrong sign' transitions:

$$\langle \ell^- \bar{\nu}_\ell X | \hat{\mathcal{H}}_{\mathrm{sl}} | B_t^0 \rangle = (q/p) g_-(t) \langle \ell^- \bar{\nu}_\ell X | \hat{\mathcal{H}}_{\mathrm{sl}} | \bar{B}^0 \rangle \tag{21.40}$$

and

$$\langle \ell^+ \nu_\ell X | \hat{\mathcal{H}}_{\mathrm{sl}} | \bar{B}_t^0 \rangle = (p/q) g_-(t) \langle \ell^+ \nu_\ell X | \hat{\mathcal{H}}_{\mathrm{sl}} | B^0 \rangle, \tag{21.41}$$

where $\hat{\mathcal{H}}_{\mathrm{sl}}$ is the relevant semileptonic part of the complete weak current-current Hamiltonian. Hence the semileptonic asymmetry is

$$\mathcal{A}_{\mathrm{SL}} = \frac{\Gamma(\bar{B}_t^0 \rightarrow \ell^+ \nu_\ell X) - \Gamma(B_t^0 \rightarrow \ell^- \bar{\nu}_\ell X)}{\Gamma(\bar{B}_t^0 \rightarrow \ell^+ \nu_\ell X) + \Gamma(B_t^0 \rightarrow \ell^- \bar{\nu}_\ell X)} = \frac{1 - |q/p|^4}{1 + |q/p|^4}, \tag{21.42}$$

independent of time. In (21.42) we have used the fact that $\langle \ell^- \bar{\nu}_\ell X | \hat{\mathcal{H}}_{\mathrm{sl}} | \bar{B}^0 \rangle = \langle \ell^+ \nu_\ell X | \hat{\mathcal{H}}_{\mathrm{sl}} | B^0 \rangle^*$. The upper bound on $\mathcal{A}_{\mathrm{SL}}$ is of order 10^{-3} (Workman *et al.* 2022). At the present level of experimental precision, it is a very good approximation to take $|q/p| = 1$. Since $q/p = [(M_{12}^* - i\Gamma_{12}^*/2)/(M_{12} - i\Gamma_{12}/2)]^{1/2}$, it follows that in this approximation we can neglect Γ_{12}, and the phase of q/p is just minus the phase of M_{12}.

In the Standard Model, the B^0 - $\bar{B}^0$ mixing amplitude occurs via the box diagrams of figure 21.5. The box amplitude is approximately proportional to the product of the masses of the internal quarks, and in this case the t quark contribution dominates (the magnitudes of the CKM couplings are all comparable). The phase of M_{12} is then that of $(V_{\mathrm{td}}^* V_{\mathrm{tb}})^2$, which is the phase of $((1 - \rho - i\eta)^*)^2$ in the parametrization of (20.179), which in turn is equal to the angle 2β. Hence

$$(q/p) = e^{-2i\beta}, \tag{21.43}$$

neglecting terms of order λ^4. Equation (21.43) will be important in what follows.

From (21.35) we can now read off that the probability that the state $|\bar{B}_t^0\rangle$ (which—we remind the reader—is the partner of the state tagged as a B^0 at $t = 0$, and which is pure $\bar{B}^0$ at $t = 0$) decays as a $\bar{B}^0$ at $t \neq 0$, is $|g_+(t)|^2 = \exp(-\Gamma t) \cos^2 \Delta m_\mathrm{B}^2 t/2$. Similarly, the probability that this state decays as a B^0 at time t is $\exp(-\Gamma t) \sin^2 \Delta m_\mathrm{B} t/2$, taking

$|(p/q)| = 1$. Hence the difference in these probabilities, normalized to their sum, is $\cos \Delta m_{\rm B} t$. Measurements of this *flavour asymmetry* yields the value of $\Delta m_{\rm B}$, currently (Workman *et al.* 2022)[3]

$$\Delta m_{\rm B} = (0.5065 \pm 0.0019) \times 10^{12} \, \hbar \, {\rm s}^{-1} \tag{21.44}$$

More generally, we define decay amplitudes to final states $|f\rangle$ by

$$A_f = \langle f|\hat{\mathcal{H}}_{\rm wk}|{\rm B}^0\rangle \quad , \quad \bar{A}_f = \langle f|\hat{\mathcal{H}}_{\rm wk}|\bar{\rm B}^0\rangle \tag{21.45}$$

$$A_{\bar{f}} = \langle \bar{f}|\hat{\mathcal{H}}_{\rm wk}|{\rm B}^0\rangle \quad , \quad \bar{A}_{\bar{f}} = \langle \bar{f}|\hat{\mathcal{H}}_{\rm wk}|\bar{\rm B}^0\rangle, \tag{21.46}$$

where ${\rm CP}|f\rangle = |\bar{f}\rangle$ and $\hat{\mathcal{H}}_{\rm wk}$ is the weak interaction Hamiltonian responsible for the transition. We can now calculate the rates for $|\bar{\rm B}_t^0\rangle$ to go to $|f\rangle$, and for $|{\rm B}_t^0\rangle$ to go to $|f\rangle$; up to a common normalization factor, which we omit, these rates are (problem 21.2)

$$\Gamma(\bar{\rm B}_t^0 \to f) = \frac{1}{2}{\rm e}^{-\Gamma t}\{|\bar{A}_f|^2 + |(p/q)A_f|^2 + (|\bar{A}_f|^2 - |(p/q)A_f|^2)\cos\Delta m_{\rm B} t$$
$$+2{\rm Im}(\bar{A}_f\frac{q}{p}A_f^*)\sin\Delta m_{\rm B} t\}, \tag{21.47}$$

and

$$\Gamma({\rm B}_t^0 \to f) = \frac{1}{2}{\rm e}^{-\Gamma t}\{|A_f|^2 + |(q/p)\bar{A}_f|^2 + (|A_f|^2 - |(q/p)\bar{A}_f|^2)\cos\Delta m_{\rm B} t$$
$$-2{\rm Im}(\bar{A}_f\frac{q}{p}A_f^*)\sin\Delta m_{\rm B} t\}. \tag{21.48}$$

The rates to $|\bar{f}\rangle$ are obtained by the substitutions $A_f \to A_{\bar{f}}, \bar{A}_f \to \bar{A}_{\bar{f}}$.

We can now derive the basic formulae for the time-dependent **CP** asymmetry of neutral B decays to a final state f common to ${\rm B}^0$ and $\bar{\rm B}^0$ (problem 21.3):

$$\mathcal{A}_f = \frac{\Gamma(\bar{\rm B}_t^0 \to f) - \Gamma({\rm B}_t^0 \to f)}{\Gamma(\bar{\rm B}_t^0 \to f) + \Gamma({\rm B}_t^0 \to f)} = S_f \sin(\Delta m_{\rm B} t) - C_f \cos\Delta m_{\rm B} t) \tag{21.49}$$

where

$$S_f = \frac{2{\rm Im}\lambda_f}{1 + |\lambda_f|^2}, \quad C_f = \frac{1 - |\lambda_f|^2}{1 + |\lambda_f|^2}, \quad \lambda_f = \frac{q}{p}\left(\frac{\bar{A}_f}{A_f}\right). \tag{21.50}$$

21.2.2 Determination of the angles $\alpha(\phi_2)$ and $\beta(\phi_1)$ of the unitarity triangle

A very large number of measurements have been made, constraining the parameters of the CKM matrix, or equivalently the unitarity triangle of figure 20.10. We shall limit our discussion to those measurements which determine the angles $\alpha(\phi_2)$ and $\beta(\phi_1)$ of the triangle.

(1) The angle β (ϕ_1)

One of the cleanest examples is the decay

$${\rm B}^0 \to J/\psi + {\rm K}_{\rm S,L}^0. \tag{21.51}$$

The tree diagram is shown in figure 21.6(a), and the penguins in figure 21.6(b). The tree diagram contributes CKM factors $V_{\rm cb}^* V_{\rm cs} = A\lambda^2(1 - \lambda^2/2)$. The $\bar{\rm f} = \bar{\rm u}$ penguin has factors $V_{\rm ub}^* V_{\rm us} = A\lambda^4(\rho - {\rm i}\eta)$ which is suppressed by two powers of λ; it also carries a loop factor

---

[3]$\Delta m_{\rm B}$ is a measure of 2π times the ${\rm B}^0 \to \bar{\rm B}^0$ oscillation frequency in time-dependent mixing experiments.

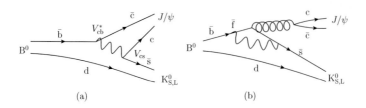

FIGURE 21.6
Tree (a) and penguin (b) contributions to $B^0 \to J/\psi + K^0_{S,L}$ via the quark transitions $\bar{b} \to \bar{c}c\bar{s}$.

$\sim \alpha_s/\pi$, and it may therefore be safely neglected. The other two penguins have the same weak phase as the tree diagram. Hence to a good approximation we can write the amplitude as

$$A_{\psi K} = (V^*_{cb}V_{cs})T_{\psi K}. \tag{21.52}$$

There is one subtlety: to get the two final states from B^0 and $\bar{B}^0$ to interfere, we need $K^0 - \bar{K}^0$ mixing to produce the (very nearly) **CP** eigenstates K^0_S (**CP** =+1) and K^0_L (**CP**=-1). (We shall discuss the $K^0 - \bar{K}^0$ system briefly in section 21.3.) This introduces a factor $(q/p)_K = (V^*_{cd}V_{cs}/V_{cd}V^*_{cs})$, quite analogously to (21.43), but its effect on $\lambda_{\psi K}$ is negligible. So, remembering that the relative orbital angular momentum of the two final state particles is $\ell = 1$, we have $\lambda_{\psi K_S} = -\exp(-2i\beta)$ and $S_f = \sin 2\beta$, while the $J/\psi K^0_L$ state has **CP**=+1 and $S_f = -\sin 2\beta$. Hence $S_{\psi K}$ measures $-\eta_f \sin 2\beta$, where η_f is the **CP** eigenvalue of the $J/\psi K^0_{S,L}$ state. The sinusoidal oscillations in the asymmetry $\mathcal{A}_{\psi K}$ for the two modes S and L will have the same amplitude and opposite phase.

Both BaBar and Belle reported increasingly precise measurements of $\mathcal{A}_{\psi K}$ in these modes. The early results (Abashian *et al.* 2001, Aubert *et al.* 2001) were the first direct measurements of one of the angles of the unitarity triangle, offering a test of the consistency of the CKM mechanism for **CP** violation. Later measurements (Aubert *et al.* 2009, Adachi *et al.* 2012, Aaij *et al.* 2015) have achieved accuracies of about ±5%. The current world average for $\sin 2\beta$ is (Ceccucci, Ligeti and Sakai 2022, section 12 in Workman *et al.* 2022)

$$\sin 2\beta = 0.699 \pm 0.017. \tag{21.53}$$

Figure 21.7 shows the asymmetry (before corrections for experimental effects) for $\eta_f = -1$ and $\eta_f = +1$ candidates as measured by BaBar (Aubert *et al.* 2007a); Belle and LHCb have reported similar results. We should note that a measurement of $\sin 2\beta$ still leaves ambiguities in β (for example, $\beta \to \pi/2 - \beta$), which can be resolved by other measurements (Ceccucci, Ligeti and Sakai, section 12 in Workman *et al.* 2022).

(2) The angle $\alpha(\phi_2)$

The angle α is the phase between $V^*_{tb}V_{td}$ and $V^*_{ub}V_{ud}$. It can be measured in decays dominated by the quark transition b $\to$ u ū d. Consider, for example, the decays $B^0 \to \pi^+\pi^-, \bar{B}^0 \to \pi^+\pi^-$. Figure 21.8 shows the tree graph (a) and penguin (b) contributions to $B^0 \to \pi^+\pi^-$. Exposing the weak phases as before, the amplitude is

$$
\begin{aligned}
A_{+-} &= V^*_{ub}V_{ud}(t + p_{\bar{u}} - p_{\bar{c}}) + V^*_{tb}V_{td}(p_{\bar{t}} - p_{\bar{c}}) \\
&\equiv V^*_{ub}V_{ud}T_{+-} + V^*_{tb}V_{td}P_{+-}.
\end{aligned}
\tag{21.54}
$$

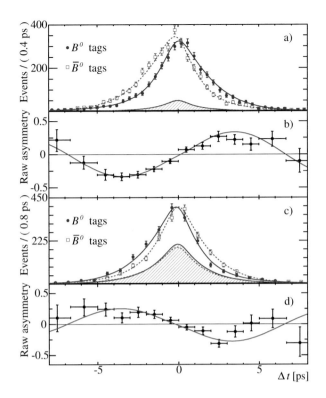

FIGURE 21.7
(a) Number of $\eta_f = -1$ candidates in the signal region with a B^0 tag (N_{B^0}) and with a $\bar{B}^0$ tag ($N_{\bar{B}^0}$), and (b) the measured asymmetry $(N_{B^0} - N_{\bar{B}^0})/(N_{B^0} + N_{\bar{B}^0})$, as functions of t; (c) and (d) are the corresponding distributions for the $\eta_f = +1$ candidates. Figure reprinted with permission from B Aubert *et al.* (BaBar Collaboration) *Phys. Rev. Lett.* **99** 171803 (2007). Copyright 2007 by the American Physical Society.

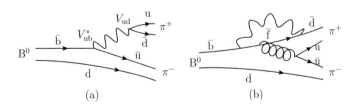

FIGURE 21.8
Tree graph (a) and penguin (b) contributions to $B^0 \to \pi^+\pi^-$, via quark transitions $\bar{b} \to d\bar{u}u$.

Suppose first that the penguin contributions could be neglected. Then the asymmetry $\mathcal{A}_{\pi^+\pi^-}$ would measure

$$
\begin{aligned}
\mathrm{Im}\lambda_{\pi^+\pi^-} &= \mathrm{Im}\left(e^{-2\mathrm{i}\beta}\frac{\bar{A}_{+-}}{A_{+-}}\right) = \mathrm{Im}\left(e^{-2\mathrm{i}\beta}\frac{V_{ub}V_{ud}^*}{V_{ub}^*V_{ud}}\right) \\
&= \mathrm{Im}\,e^{-2\mathrm{i}(\gamma+\beta)} = \sin 2\alpha
\end{aligned}
\tag{21.55}
$$

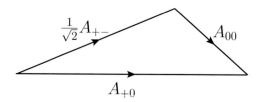

FIGURE 21.9
The triangle formed by the three amplitudes A_{ij} in equation (21.58).

where α is defined as $\pi - \beta - \gamma$. Unfortunately, this simple result is spoiled by the penguin contributions, which there is no good reason to ignore. However, Gronau and London (1990) showed how an isospin analysis could disentangle the tree and penguin parts. The method involves the three amplitudes $A_{+-}, A_{00}(\mathrm{B}^0 \to \pi^0\pi^0)$, and $A_{+0}(\mathrm{B}^+ \to \pi^+\pi^0)$.

First of all, note that Bose statistics for the 2π states requires them to have only the symmetric isospin states $I = 0$ or 2, since the angular momentum is zero. Next, the effective non-leptonic weak Hamiltonian $\hat{\mathcal{H}}_{\mathrm{nl}}$ acting in the tree diagram transition contains both $\Delta I = 1/2$ and $\Delta I = 3/2$ pieces; combining with the initial $I = 1/2$ of the B meson, the first piece will lead only to the $I = 0$ final state, while the second contributes to both $I = 0$ and $I = 2$ final states. However, since the gluon in the penguin diagrams carries no isospin, these diagrams can only change the isospin by $\Delta I = 1/2$, which connects only to the $I = 0$ final state. The conclusion is that the $I = 2$ final state is free of penguins, and carries the pure tree phase.

This information can be exploited as follows (Gronau and London 1990). First, the action of $\hat{\mathcal{H}}_{\mathrm{nl}}$ on the B^0 state can be written as

$$\hat{\mathcal{H}}_{\mathrm{nl}}|\tfrac{1}{2}\ -\tfrac{1}{2}\rangle = \frac{1}{\sqrt{2}}A_{3/2}|20\rangle + \frac{1}{\sqrt{2}}A_{1/2}|00\rangle \tag{21.56}$$

where as usual $|II_3\rangle$ is the state with isospin I and third component I_3. Expanding the states $\pi^+\pi^-, \pi^+\pi^0$, and $\pi^0\pi^0$ in terms of definite isospin states, one finds (problem 21.4)

$$A_{+-} = \frac{1}{\sqrt{6}}A_{3/2} + \frac{1}{\sqrt{3}}A_{1/2}$$

$$A_{+0} = \frac{\sqrt{3}}{2}A_{3/2}$$

$$A_{00} = \frac{1}{\sqrt{3}}A_{3/2} - \frac{1}{\sqrt{6}}A_{1/2} \tag{21.57}$$

where A_{ij} is the amplitude $\langle \pi^i\pi^j|\hat{\mathcal{H}}_{\mathrm{nl}}|B^{i+j}\rangle$. The $\pi^+\pi^0$ state can have only $I = 2$, and arises solely from the tree diagram. Furthermore, the three complex amplitudes A_{+-}, A_{+0}, and A_{00} are expressed in terms of only two reduced amplitudes $A_{3/2}$ and $A_{1/2}$, leading to one relation between them:

$$\frac{1}{\sqrt{2}}A_{+-} + A_{00} = A_{+0}, \tag{21.58}$$

which can be represented as a triangle in the complex plane, as shown in figure 21.9. There is a similar triangle for the charge conjugate processes

$$\frac{1}{\sqrt{2}}\bar{A}_{+-} + \bar{A}_{00} = \bar{A}_{+0}, \tag{21.59}$$

where the $\bar{A}$ amplitudes are obtained from the As by complex conjugating the CKM couplings, the strong phases remaining the same as usual.

Since A_{+0} is pure tree, its weak phase is well defined, namely that of $V_{ub}^* V_{ud}$, which is γ. It is convenient to define (Lipkin *et al.* 1991) $\tilde{A} = \exp(2i\gamma)\bar{A}$, so that $\tilde{A}_{+0} = A_{+0}$. Then the two triangles have a common base, A_{+0}. The failure of the two triangles to overlap exactly is a measure of the penguin contribution. In principle, by measuring the asymmetry coefficients $S_{\pi^+\pi^-}, C_{\pi^+\pi^-}$, the branching fractions of all three modes, and C_{00}, one can construct the triangles. But unfortunately the relative orientation of the triangles is not known, which leads to 16 solutions to α in the range $0 < \alpha < 2\pi$. In addition, the data on $\pi^0\pi^0$ (with a branching ratio of order 10^{-6}) has sizeable experimental errors, and only a relatively loose constraint on α can be obtained.

A much better constraint can be found from the **CP** asymmetries in $B \to \pi\rho$ decays (Snyder and Quinn 1993). The method is essentially a time-dependent version of the Dalitz plot analysis discussed in section 21.1. The available channels are

$$
\begin{aligned}
B^0 &\to \{\rho^+\pi^-, \rho^-\pi^+, \rho^0\pi^0\} \to \pi^+\pi^-\pi^0 \\
\bar{B}^0 &\to \{\rho^-\pi^+, \rho^+\pi^-, \rho^0\pi^0\} \to \pi^+\pi^-\pi^0
\end{aligned}
\tag{21.60}
$$

where all result in the final state $\pi^+\pi^-\pi^0$ after the decay of the ρ mesons, and interferences following $B^0 - \bar{B}^0$ mixing are possible.

Returning then to equations (21.47) and (21.48), the rate for the 3π decay following a B^0 tag at $t = 0$ is

$$
\Gamma(\bar{B}_t^0 \to 3\pi) = \frac{1}{4}\Gamma e^{-\Gamma t}[|A_{3\pi}|^2 + |\bar{A}_{3\pi}|^2 + (|\bar{A}_{3\pi}|^2 - |A_{3\pi}|^2)\cos\Delta m_B t
$$

$$
+ 2\mathrm{Im}\left(\frac{q}{p}\bar{A}_{3\pi}A_{3\pi}^*\right)\sin\Delta m_B t],
\tag{21.61}
$$

and there is a similar formula, with appropriate changes, for the case of a $\bar{B}^0$ tag at $t = 0$. We now write

$$
A_{3\pi} = f_+(s_+)F^+ + f_-(s_-)F^- + f_0(s_0)F^0
\tag{21.62}
$$

and similarly

$$
\bar{A}_{3\pi} = f_+(s_+)\bar{F}^+ + f_-(s_-)\bar{F}^- + f_0(s_0)\bar{F}^0,
\tag{21.63}
$$

where $s_+ = (p_{\pi^+} + p_{\pi^0})^2, s_- = (p_{\pi^-} + p_{\pi^0})^2$, and $s_0 = (p_{\pi^+} + p_{\pi^-})^2$, satisfying $s_+ + s_- + s_0 = m_B^2 + 2m_{\pi^+}^2 + m_{\pi^0}^2$. $f_\kappa(s_\kappa)$ is the sum of three relativistic Breit-Wigner resonance amplitudes, together with appropriate angular momentum and angle factors, corresponding to the $\rho(770), \rho(1450)$, and $\rho(1700)$ resonances. F^κ is the amplitude for the quasi two-body mode $B^0 \to \rho^\kappa \pi^{\bar{\kappa}}$. Here κ takes the values $+, -$, and 0, and correspondingly $\bar{\kappa} = -, +, 0$. The amplitudes F^κ are complex and include the strong and weak transition phases, from tree and penguin diagrams; they are, however, independent of the Dalitz plot variables.

The $\rho\pi$ states have the same decomposition into tree and penguin parts as discussed previously for the $\pi\pi$ states, namely

$$
F^\kappa = e^{i\gamma}T^\kappa + e^{-i\beta}P^\kappa,
\tag{21.64}
$$

where the magnitudes of the weak couplings have been absorbed into T^κ and P^κ. We can rewrite (21.64) as

$$
e^{i\beta}F^\kappa = -e^{-i\alpha}T^\kappa + P^\kappa \equiv A^\kappa,
\tag{21.65}
$$

and similarly

$$\mathrm{e}^{i\beta}(q/p)\bar{F}^\kappa = -\mathrm{e}^{i\alpha}T^{\bar\kappa} + P^{\bar\kappa} \equiv \bar{A}^\kappa. \tag{21.66}$$

Then (21.62) and (21.63) become

$$A_{3\pi} = \sum_\kappa f_\kappa(s_\kappa)A^\kappa \tag{21.67}$$

$$(q/p)\bar{A}_{3\pi} = \sum_\kappa f_\kappa(s_\kappa)\bar{A}^\kappa, \tag{21.68}$$

disregarding a common overall phase $\mathrm{e}^{-i\beta}$. When (21.67) and (21.68) are inserted into (21.61), it is clear that one obtains many terms, for example

$$\mathrm{Re}(f_+f_-^*)\,\mathrm{Im}(\bar{A}^+A^{-*} + \bar{A}^-A^{+*}), \quad \mathrm{Im}(f_+f_-^*)\,\mathrm{Re}(\bar{A}^+A^{-*} - \bar{A}^-A^{+*}), \tag{21.69}$$

and so on, in which different resonances interfere on the Dalitz plot. The strong, and known, rapid phase variation in these interference regions, via factors such as $f_+f_-^*$, is a powerful tool for extracting the complex amplitudes A^κ, $\bar{A}^\kappa$, and hence via (21.65) and (21.66) the phase α. The quantities multiplying the interference terms $\mathrm{Re}(f_\kappa f_\sigma^*)$ and $\mathrm{Im}(f_\kappa f_\sigma^*)$ are the key degrees of freedom which allow this analysis to determine the penguin contributions and the strong phases, and hence α. However, the resonance overlap regions cover a small fraction of the Dalitz plot, so that a substantial data sample (a few thousand events) is needed to constrain all the amplitude parameters.

An isospin analysis similar to that of the $\pi\pi$ states can be done for the $\rho\pi$ states, but now there is no reason to forbid the final state to have $I = 1$. Nevertheless, if charged B decays are also included, there are five physical amplitudes ($\rho^0 \to \pi^+\pi^-, \pi^-\pi^+, \pi^0\pi^0, \rho^+ \to \pi^+\pi^0, \rho^- \to \pi^-\pi^0$) which are expressible in terms of two pure tree transitions ($\Delta I = 3/2$) transitions to $I = 1, 2$ final states. One of the pure tree amplitudes may be written (Gronau 1991) as the sum $A^+ + A^- + 2A^0$, and hence the ratio $(\bar{A}^+ + \bar{A}^- + 2\bar{A}^0)/(A^+ + A^- + 2A^0)$ has the phase 2α.

This approach was followed by both BaBar and Belle, with the results

$$\text{BaBar (Aubert } et\ al.\ 2007b) \qquad \alpha = 87^{+45°}_{-13°} \tag{21.70}$$

$$\text{Belle (Kusaka } et\ al.\ 2007) \qquad 68 < \alpha < 95°; \tag{21.71}$$

see also Lees *et al.* (BaBar Collaboration) (2013). These moderate constraints are consistent with the values of β and γ given in (21.53), (21.25), and (21.26), given the definition $\alpha = \pi - \beta - \gamma$.

There are now very many independent measurements of the magnitudes of the CKM matrix elements, as well as the angles (the determination of which is reviewed by Gershon, Kenzle, and Trabelsi (2022) in section 77 of Workman *et al.* 2022). We shall not describe these here, referring the reader to the regular updates by the Particle Data Group (currently Workman *et al.* 2022). We showed in figure 20.11 the 2022 plot of the constraints in the $\bar\rho, \bar\eta$ plane, presented by Ceccucci, Ligeti, and Sakai, section 12 of Workman *et al.* (2022). They concluded that that the 95% CL regions all overlapped consistently around the global fit region. The consistency represented in figure 20.11 must be counted as a major triumph of the SM, in particular of the original analysis by Kobayashi and Maskawa (1973). It must be remembered, though, that many extensions of the SM allow considerable room for new **CP**-violating effects, which could be revealed by increasingly precise determinations of the CKM parameters.

21.3 CP violation in neutral K-meson and D-meson decays

Although the formalism is similar, the phenomenology of **CP** violation in neutral K-meson decays is very different from that in neutral B-meson decays. In the K case, **CP** violation is a very small effect, typically at the level of parts per thousand or smaller; its observation by Cristenson *et al.* (1964) was a historic achievement. But the neutral K system is most simply (and traditionally) approached by starting with the assumption that **CP** is conserved.

We will define **CP**$|K^0\rangle = -|\bar{K}^0\rangle$; then the **CP** eigenstates are

$$|K_\pm\rangle = \frac{1}{\sqrt{2}}(|K^0\rangle \mp |\bar{K}^0\rangle) \tag{21.72}$$

The **CP** $=1$ state can decay to two pions in an s-state, but not to three pions if (as we are assuming to start with) **CP** is a good symmetry; the situation is the opposite for the **CP** $= -1$ state. The Q-value for the three pion mode is very much smaller than for the two pion mode, with the result that the $|K_+\rangle$ state, decaying to two pions, has a much shorter lifetime than the $|K_-\rangle$ state: $\tau_{2\pi} \sim 0.9 \times 10^{-10}s$, $\tau_{3\pi} \sim 5 \times 10^{-8}s$. Due to **CP** violation, the actual eigenstates $|K_L\rangle$ and $|K_S\rangle$ of the effective Hamiltonian are slightly different from $|K_\pm\rangle$ (see (21.76) and (21.77)), with masses m_S and m_L, and widths Γ_S and Γ_L. At this point, however, we shall associate m_S and Γ_S with $|K_+\rangle$, and m_L and Γ_L with $|K_-\rangle$.

A K^0 is produced in strangeness-conserving reactions such as $K^+n \to K^0p$, and a $\bar{K}^0$ in $K^- + p \to \bar{K}^0 + n$, for example. However, the two states can mix following production, since (as usual) it is the Hamiltonian eigenstates which propagate in free space, and they are the superpositions $|K_\pm\rangle$, assuming **CP** is conserved. Hence, as time proceeds following production, a state produced as a K^0 at time $t = 0$ will evolve into the state

$$|K_t^0\rangle = \frac{1}{2}(e^{-\Gamma_L t/2 - im_L t} + e^{-\Gamma_S t/2 - im_S t})|K^0\rangle + (e^{-\Gamma_L t/2 - im_L t} - e^{-\Gamma_S t/2 - im_S t})|\bar{K}^0\rangle. \tag{21.73}$$

The probability that a $K^0(\bar{K}^0)$ will then be observed at time t following production (in the K-meson rest frame) is

$$P_{+(-)} = \frac{1}{4}[e^{-\Gamma_L t} + e^{-\Gamma_S t} + (-)2e^{-(\Gamma_L + \Gamma_S)t/2} \cos \Delta mt] \tag{21.74}$$

where $\Delta m = m_L - m_S$. This is the famous phenomenon of strangeness oscillations, predicted by Gell-Mann and Pais (1955). Experimentally, the strangeness of the state at time t is defined by the modes $K^0 \to \pi^- \ell^+ \nu_\ell$ and $\bar{K}^0 \to \pi^+ \ell^- \bar{\nu}_\ell$. The difference $P_+(t) - P_-(t)$ is measured, and although the oscillations are heavily damped by $\exp(-\Gamma_S t)$, the mass difference can be determined:

$$\Delta m = (3.483 \pm 0.006) \times 10^{-12} \text{ meV} \tag{21.75}$$

However, this is not the whole story. Christenson *et al.* (1964) found that, after many τ_S lifetimes, some 2π events were observed, indicating that the surviving state K_L was capable of decaying to 2π after all (albeit very rarely). The same conclusion follows from the fact that $P_+(t) - P_-(t)$ does not go to zero at long times, as it should from (21.74). Accordingly, the true Hamiltonian eigenstates are not quite the **CP** eigenstates, but rather

$$|K_L\rangle = [(1 + \bar{\epsilon})|K^0\rangle + (1 - \bar{\epsilon})|\bar{K}^0\rangle]/\sqrt{2(1 + |\bar{\epsilon}|^2)} \tag{21.76}$$

$$|K_S\rangle = [(1 + \bar{\epsilon})|K^0\rangle - (1 - \bar{\epsilon})|\bar{K}^0\rangle]/\sqrt{2(1 + |\bar{\epsilon}|^2)}. \tag{21.77}$$

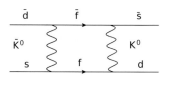

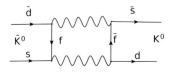

FIGURE 21.10
Box diagrams contributing to $K^0 - \bar{K}^0$ mixing.

This is a traditional parametrization in K-physics, similar to that in (21.33) with $q/p = (1 - \bar{\epsilon})/(1 + \bar{\epsilon})$ (this is why we chose **CP** $|K^0\rangle = -|K^0\rangle$). We now find that a state which starts at $t = 0$ as a K^0 evolves to

$$|K_t^0\rangle = g_+(t)|K^0\rangle + \frac{1 - \bar{\epsilon}}{1 + \bar{\epsilon}} g_-(t)|\bar{K}^0\rangle \tag{21.78}$$

where

$$g_\pm(t) = e^{-\Gamma_L t/2} e^{-i m_L t}[1 \pm e^{-\Delta \Gamma t/2} e^{i\Delta m t}], \tag{21.79}$$

with $\Delta \Gamma = \Gamma_S - \Gamma_L$, $\Delta m = m_L - m_S$, and we have omitted a normalization factor. Similarly, a state tagged as $\bar{K}^0$ at $t = 0$ evolves to

$$|\bar{K}_t^0\rangle = \frac{1 - \bar{\epsilon}}{1 + \bar{\epsilon}} g_-(t)|K^0\rangle + g_+(t)|\bar{K}^0\rangle. \tag{21.80}$$

The $K^0 - \bar{K}^0$ mixing amplitude arises in the SM from the box graphs shown in figure 21.9 (cf figure 21.5). These contain factors of m_f^2, but the magnitude of the four CKM couplings to the t quark are of order λ^{10}, compared with λ^6 for the c quark, so that the c quark diagram dominates, with a CKM factor of $(V_{cs} V_{cd}^*)^2$, which is real to a good approximation. This means that $\mathrm{Im}\,\bar{\epsilon}$ is very small. A comparison of the mass difference Δm predicted from figure 21.10 and the experimental value is complicated by uncertainties in the hadronic matrix element.

The traditional reactions in which **CP** violation is probed in K decays are the 2π modes, where one looks for the existence of $K_L \to 2\pi$. There is also the semileptonic asymmetry. Three common observables are defined by

$$\eta_{00} = \frac{\langle \pi^0 \pi^0 | \hat{\mathcal{H}}_{nl} | K_L \rangle}{\langle \pi^0 \pi^0 | \hat{\mathcal{H}}_{nl} | K_S \rangle}, \quad \eta_{+-} = \frac{\langle \pi^+ \pi^- | \hat{\mathcal{H}}_{nl} | K_L \rangle}{\langle \pi^+ \pi^- | \hat{\mathcal{H}}_{nl} | K_S \rangle} \tag{21.81}$$

and

$$\delta_L = \frac{\Gamma(K_L \to \pi^- \ell^+ \nu_\ell) - \Gamma(K_L \to \pi^+ \ell^- \bar{\nu_\ell})}{\Gamma(K_L \to \pi^- \ell^+ \nu_\ell) + \Gamma(K_L \to \pi^+ \ell^- \bar{\nu}_\ell)}. \tag{21.82}$$

The experimental numbers are (Workman *et al.* 2022)

$$|\eta_{00}| = (2.220 \pm 0.011) \times 10^{-3}, \quad |\eta_{+-}| = (2.232 \pm 0.011) \times 10^{-3} \tag{21.83}$$

$$\mathrm{Arg}\,\eta_{00} \approx 43.5°, \quad \mathrm{Arg}\,\eta_\pm \approx 43.5° \tag{21.84}$$

and

$$\delta_{\mathrm{L}} = (3.32 \pm 0.06) \times 10^{-3}. \tag{21.85}$$

It is useful to describe the final 2π states in terms of their isospin (although this requires the neglect of electromagnetic interactions), which then have a definite strong interaction phase. As noted in connection with the B decays, the allowed isospin states are only $I = 0$ and $I = 2$, and one has

$$A_{\pm} \equiv A_{\mathrm{K}^0 \to \pi^+\pi^-} = \sqrt{\frac{2}{3}} |A_0| e^{i(\delta_0 + \phi_0)} + \sqrt{\frac{1}{3}} |A_2| e^{i(\delta_2 + \phi_2)} \tag{21.86}$$

$$\bar{A}_{\pm} \equiv A_{\bar{\mathrm{K}}^0 \to \pi^+\pi^-} = -\sqrt{\frac{2}{3}} |A_0| e^{i(\delta_0 - \phi_0)} - \sqrt{\frac{1}{3}} |A_2| e^{i(\delta_2 - \phi_2)} \tag{21.87}$$

where the minus sign arises from our choice $\mathbf{CP}|\mathrm{K}^0\rangle = -|\mathrm{K}^0\rangle$, and where δ_I and ϕ_I are the strong and weak phases, respectively, for the state with isospin I. Also,

$$A_{00} \equiv A_{\mathrm{K}^0 \to \pi^0\pi^0} = \sqrt{\frac{1}{3}} |A_0| e^{i(\delta_0 + \phi_0)} - \sqrt{\frac{2}{3}} |A_2| e^{i(\delta_2 + \phi_2)} \tag{21.88}$$

$$\bar{A}_{00} \equiv \bar{A}_{\bar{\mathrm{K}}^0 \to \pi^0\pi^0} = -\sqrt{\frac{1}{3}} |A_0| e^{i(\delta_0 - \phi_0)} + \sqrt{\frac{2}{3}} |A_2| e^{i(\delta_2 - \phi_2)}. \tag{21.89}$$

The significant fact experimentally is that $|A_2|/|A_0| \sim 1/22$, a manifestation of the "$\Delta I = 1/2$" rule in this case (i.e. $\Delta I = 3/2$ is suppressed; see, for example, Donoghue *et al.* 1992, section VIII-4). Inserting (21.86), and (21.87), (21.88), and (21.89), into (21.81) and retaining only first order terms in $|A_2|/|A_0|$, and treating ϕ_0 and ϕ_2 as small, we find (problem 21.5)

$$\eta_{00} = \bar{\epsilon} + i\phi_0 - \sqrt{2} \frac{|A_2|}{|A_0|} i(\phi_2 - \phi_0) e^{i(\delta_2 - \delta_0)} \tag{21.90}$$

$$\eta_{+-} = \bar{\epsilon} + i\phi_0 + \frac{1}{\sqrt{2}} \frac{|A_2|}{|A_0|} i(\phi_2 - \phi_0) e^{i(\delta_2 - \delta_0)}. \tag{21.91}$$

These relations are usually written as

$$\eta_{00} = \epsilon - 2\epsilon', \quad \eta_{+-} = \epsilon + \epsilon', \tag{21.92}$$

where

$$\epsilon = \bar{\epsilon} + i\phi_0, \quad \epsilon' = i \frac{e^{i(\delta_2 - \delta_0)}}{\sqrt{2}} \frac{|A_2|}{|A_0|} (\phi_2 - \phi_0). \tag{21.93}$$

The merit of this manoeuvering is that the parameter ϵ' involves only the **CP** violation in the transition amplitude ('direct **CP** violation'), while ϵ involves both a transition phase and the mixing parameter $\bar{\epsilon}$.

What can experiment tell us about ϵ and ϵ'? Consider first δ_{L}. Assuming that $|A(\mathrm{K}^0 \to \ell^+ \nu_\ell \pi^-)| = |A(\bar{\mathrm{K}}^0 \to \ell^- \bar{\nu}_\ell \pi^+)|$ and that $A(\mathrm{K}^0 \to \ell^- \bar{\nu}_\ell \pi^+) = A(\bar{\mathrm{K}}^0 \to \ell^+ \nu_\ell \pi^-) = 0$, we find

$$\delta_{\mathrm{L}} = 2\mathrm{Re}\,\bar{\epsilon}/(1 + |\bar{\epsilon}|^2) \approx 2\mathrm{Re}\,\epsilon, \tag{21.94}$$

so that δ_{L} is sensitive to the same parameter as appears in the $\mathrm{K}_{\mathrm{L}} \to \pi\pi$ decays. An interesting observable is the ratio between the ratios of the decay rates to $\pi^+\pi^-$ and $\pi^0\pi^0$ of K_{S} and K_{L}. One finds

$$\frac{1}{6}\left(1 - \frac{|\eta_{00}|^2}{|\eta_{+-}|^2}\right) \approx \mathrm{Re}\,(\epsilon'/\epsilon), \tag{21.95}$$

which from equation (21.83) is another small number, approximately equal to 1.64×10^{-3}. In the years before the B factories opened, ϵ' was the only window into **CP** violation in the transition amplitude. But all the branching ratios in (21.95) are of order 10^{-3}, and establishing a non-zero value of ϵ' was very difficult. The first claim for non-zero ϵ' was by the NA 31 experiment at CERN (Barr *et al.* 1993), a 3.5 standard deviation effect. But a contemporary experiment at Fermilab (Gibbons *et al.* 1993) found a result compatible with zero. The next generation of experiments produced agreement:

$$\mathrm{Re}\,(\epsilon'/\epsilon) \;=\; (2.07) \pm 0.28) \times 10^{-3}\ \mathrm{Alavi-Harati}\ \textit{et al.}\ 2003\ (\mathrm{KTeV})$$

(21.96)

$$\mathrm{Re}\,(\epsilon'/\epsilon) \;=\; (1.47 \pm 0.22) \times 10^{-3}\ \mathrm{Batley}\ \textit{et al.}\ 2002\ (\mathrm{NA\,48}).$$

(21.97)

The current world average is $(1.66 \pm 0.23) \times 10^{-3}$. Fits to all the data also yield (Workman *et al.* 2022)

$$|\epsilon| = (2.228 \pm 0.011) \times 10^{-3}.$$

(21.98)

The experimental value of δ_{L} gives us $\mathrm{Re}\,\epsilon \simeq 1.66 \times 10^{-3}$, and we can deduce that $\arg\epsilon \simeq \pi/4$. The phase of ϵ' is $\pi/2 + \delta_2 - \delta_0$ which happens also to be approximately $\pi/4$. It follows that ϵ'/ϵ is very nearly real.

Comparison of these small numbers with theoretical predictions is complicated by hadronic uncertainties, and it is beyond our scope to pursue that issue.

In closing this discussion of mesonic mixing and **CP** violation, we briefly discuss the charm sector. First, we note that $\mathrm{D}^0 - \bar{\mathrm{D}}^0$ mixing has been observed. The first measurements (Aubert *et al.* 2007c, Staric *et al.* 2007, Aaltonen *et al.* 2008) were marginally compatible with $\Delta m_{\mathrm{D}} = 0$, and consistent with **CP** symmetry. Subsequent measurements by LHCb (Aaij *et al.* 2021b) reported the values

$$x_{\mathrm{D}} \;=\; (0.397 \pm 0.046\ (\mathrm{stat.}) \pm 0.029\ (\mathrm{syst.})) \times 10^{-2}$$

(21.99)

$$y_{\mathrm{D}} \;=\; (0.459 \pm 0.0120\ (\mathrm{stat.}) \pm 0.085\ (\mathrm{syst.})) \times 10^{-2}$$

(21.100)

where $x_{\mathrm{D}} = \Delta m_{\mathrm{D}}/\Gamma_{\mathrm{D}}$ and $y_{\mathrm{D}} = \Delta\Gamma_{\mathrm{D}}/2\Gamma_{\mathrm{D}}$, which established mixing with a significance of 7σ. The data were consistent with **CP** symmetry. The current values for these parameters are (Amhis *et al.* 2021) $x_{\mathrm{D}} \approx 0.4 \times 10^{-2}$ and $y_{\mathrm{D}} = 0.6 \times 10^{-2}$, with an accuracy of about 10%. These results show that the **CP** even state is shorter lived, as in the kaon system, but it is heavier than the **CP** odd state, in contrast to the kaon case. Non-perturbative strong-interaction effects make it difficult to calculate SM predictions for these parameters. $\mathrm{D}^0 - \bar{\mathrm{D}}^0$ mixing is reviewed by Asner and Schwartz (2022) in section 70 of Workman *et al.* (2022).

Within the SM, **CP**-violating effects in charm decays have been generally expected to be very small, because the mixing and decays are very well described by the physics of the first two generations. A rough estimate of the direct **CP**-violating asymmetries in D decays can be made following the method of section 21.1. Consider, for example, the decays $\mathrm{D}^0 \to \mathrm{K}^+\mathrm{K}^-$ and $\bar{\mathrm{D}}^0 \to \mathrm{K}^-\mathrm{K}^+$. As in (21.5) and (21.10), the amplitude for the first decay is

$$A(\mathrm{D}^0 \to \mathrm{K}^+\mathrm{K}^-) \;=\; V_{\mathrm{cs}}^* V_{\mathrm{us}} T_{\mathrm{KK}} + V_{\mathrm{cb}}^* V_{\mathrm{ub}} P_{\mathrm{KK}}$$

(21.101)

$$= T(1 + r_{\mathrm{K}}\exp^{\mathrm{i}(\delta_{\mathrm{K}} - \gamma)}),$$

(21.102)

where r_{K} is the relative magnitude of the penguin contribution, and δ_{K} is the relative strong phase. The amplitude for the **CP**-conjugate process is the same, with γ replaced by $-\gamma$. The penguin contribution is CKM-suppressed by a factor $V_{\mathrm{cb}}^* V_{\mathrm{ub}}/V_{\mathrm{cs}}^* V_{\mathrm{us}} \sim \lambda^4$, and there is also a loop factor, so that r_K would seem to be of order 10^{-4}. The asymmetry is then

$$
\mathcal{A}_{KK}^D = \frac{|A(D^0 \to K^+K^-)|^2 - |A(\bar{D}^0 \to K^-K^+)|^2}{|A(D^0 \to K^+K^-)|^2 + |A(\bar{D}^0 \to K^-K^+)|^2} \tag{21.103}
$$

$$
= 2r_K \sin\gamma \sin\delta_K, \tag{21.104}
$$

which is indeed very small, probably of order 10^{-3} or less. A similar argument predicts the asymmetry in the decays $D^0 \to \pi^+\pi^-$ and $\bar{D}^0 \to \pi^-\pi^+$ to be

$$
\mathcal{A}_{\pi\pi}^D = -2r_K \sin\gamma \sin\delta_K. \tag{21.105}
$$

The LHCb collaboration published a measurement of the difference between the time-integrated **CP** asymmetries in the KK and $\pi\pi$ decays, which to a very good approximation can be identified with the difference between the direct asymmetries (21.104) and (21.105). The LHCb result was (Aaij *et al.* 2012)

$$
\Delta\mathcal{A}^D \equiv \mathcal{A}_{KK}^D - \mathcal{A}_{\pi\pi}^D = (-0.82 \pm 0.21 \text{ (stat.)} \pm 0.11 \text{ (syst.)}) \times 10^{-2}, \tag{21.106}
$$

which is substantially larger than the estimates (21.104) and (21.105). More recently, however, LHCb reported the new result (Aaij *et al.* 2019)

$$
\Delta\mathcal{A}^D = (-0.182 \pm 0.032 \text{ (stat)} \pm 0.09 \text{ (syst)}) \times 10^{-2} \tag{21.107}
$$

which establishes **CP** violation at the 5σ level. The current average value is (Amhis *et al.* 2021)

$$
\Delta\mathcal{A}^D = (-0.164 \pm 0.028) \times 10^{-2} \tag{21.108}
$$

where the error includes both statistical and systematic uncertainties.

The asymmetry may be somewhat larger than the theoretical expectation. However, it must be noted that the mass scale of the charm quark, $m_c \sim 1.3$ GeV, is not large enough to be safely in the perturbative QCD regime (as indicated by the parameter $\Lambda_{\overline{MS}}/m_c$), so that non-perturbative enhancements are possible. **CP**-violation in the charm sector continues to be an interesting area for experimental and theoretical exploration.

21.4 Neutrino mixing and oscillations

21.4.1 Neutrino mass and mixing

Experiments with solar, atmospheric, reactor, and accelerator neutrinos have established the phenomenon of neutrino oscillations caused by non-zero neutrino masses, and mixing. We shall give an elementary introduction to this topic, which is a highly active field of research in particle physics; there are analogies with the meson oscillations we have been considering.

In the original SM, the neutrinos were taken to be massless, although there was no compelling theoretical reason for this. The basic (gauge symmetry) framework of the SM can be extended to accommodate massive neutrinos, but how that is done depends on whether they are Dirac or Majorana fermions. As explained in section 20.3, we do not yet know the answer to that question, and it may be some time before we do. The way the mass terms enter the Lagrangian is different in the two cases. We are familiar with the Dirac mass term

$$
m\bar{\hat{\psi}}\hat{\psi} = m(\bar{\hat{\psi}}_R\hat{\psi}_L + \bar{\hat{\psi}}_L\hat{\psi}_R), \tag{21.109}
$$

where $\hat{\psi}$ is a 4-component Dirac field, and R and L refer to the chirality components. Such a term will not be invariant under the $\mathrm{SU}(2)_\mathrm{L} \times \mathrm{U}(1)$ gauge symmetry of the GSW electroweak theory (see chapter 22), because the $\mathrm{SU}(2)_\mathrm{L}$ transformations only act on the $\hat{\psi}_\mathrm{L}$ component. A Dirac mass term therefore breaks the electroweak gauge symmetry explicitly, and this cannot be tolerated, in the sense that it will lead to violations of unitarity and then of renormalizability, as will be discussed in section 22.5. However, Dirac mass terms can be accommodated, as we shall see, by introducing a Yukawa coupling of fermions to the Higgs field, whose vacuum value thereby gives masses to the fermions. On the other hand, we learned in section 7.5.2 that a Majorana mass term can be written in the form

$$m\hat{\chi}_\mathrm{L}^\mathrm{T} i\sigma_2 \hat{\chi}_\mathrm{L} + \mathrm{h.c.} \tag{21.110}$$

where $\hat{\chi}_\mathrm{L}$ is a 2-component field of L chirality. This term is invariant under the gauge transformations of $\mathrm{SU}(2)_\mathrm{L}$ — that is, it is an $\mathrm{SU}(2)_\mathrm{L}$ singlet — and so it is allowed as a 'bare' mass term in the Lagrangian. For completeness, we mention here also a third possibility, that of 'sterile' neutrinos. By this is meant hypothetical particles which are singlets under the gauge group, and so can have bare mass terms in the Lagrangian. We shall not discuss sterile neutrinos any further, referring the reader to the review by Gonzalez-Garcia and Yokoyama, section 14 in Workman *et al.* (2022).

The difference in form between the Dirac and Majorana mass terms leads to a difference in the parametrization of neutrino mixing. Suppose, first, that the neutrinos are Dirac particles, with both L and R chiralities (or equivalently either helicity) for a given mass. We remind the reader that this is not ruled out experimentally, since the non-observation of the 'wrong' helicity component may be accounted for by the appearance of a suppression factors (m/E), where m is a neutrino mass and E is an average neutrino energy (see section 20.2.2). We also assume that their interactions have the V-A structure indicated by the phenomenology of the previous chapter. Then only the L (R) chirality component of a neutrino (antineutrino) field feels the weak force; the R (L) component of a neutrino (antineutrino) field has no interactions of SM type. But it will in general be necessary to allow for the possibility that the L-components of the neutrino fields which enter into the charged current V-A interaction and which have definite flavours, namely $\hat{\nu}_{\mathrm{eL}}$, $\hat{\nu}_{\mu\mathrm{L}}$, and $\hat{\nu}_{\tau\mathrm{L}}$, are linear combinations (governed by a unitary matrix) of the states which have definite masses and which we will call $\hat{\nu}_{1\mathrm{L}}$, $\hat{\nu}_{2\mathrm{L}}$, $\hat{\nu}_{3\mathrm{L}}$. For Dirac neutrinos, we therefore write

$$\begin{pmatrix} \hat{\nu}_e \\ \hat{\nu}_\mu \\ \hat{\nu}_\tau \end{pmatrix}_\mathrm{L} = \begin{pmatrix} U_{e1} & U_{e2} & U_{e3} \\ U_{\mu1} & U_{\mu2} & U_{\mu3} \\ U_{\tau1} & U_{\tau2} & U_{\tau3} \end{pmatrix} \begin{pmatrix} \hat{\nu}_1 \\ \hat{\nu}_2 \\ \hat{\nu}_3 \end{pmatrix}_\mathrm{L} \equiv \mathbf{U} \begin{pmatrix} \hat{\nu}_1 \\ \hat{\nu}_2 \\ \hat{\nu}_3 \end{pmatrix}_\mathrm{L}, \tag{21.111}$$

where the unitary matrix $\mathbf{U}$ is the PMNS matrix, named after Pontecorvo (1957, 1958, 1967), and Maki, Nakagawa and Sakata (1962).

Now we showed in section 20.7.3 that the general 3×3 unitary matrix has three real (rotation angle) parameters, and 6 phase parameters, five of which we could get rid of by rephasing the quark fields by global $\mathrm{U}(1)$ transformations of the form $\hat{q}' = \exp(\mathrm{i}\theta)\hat{q}$. Such rephasing transformations are equally allowed for the charged leptons, and also for Dirac neutrinos, since evidently the mass term (21.109) is invariant under a global $\mathrm{U}(1)$ transformation $\hat{\psi}' = \exp(\mathrm{i}\theta)\hat{\psi}$. Hence the matrix $\mathbf{U}$ will, in this Dirac case, have a parametrization of the CKM form, with one **CP**-violating phase. For later convenience we reproduce the matrix $\mathbf{U}$ in the now conventional parametrization of (20.166):

$$\mathbf{U} = \begin{pmatrix} c_{12}c_{13} & s_{12}c_{13} & s_{13}\mathrm{e}^{-\mathrm{i}\delta} \\ -s_{12}c_{23} - c_{12}s_{23}s_{13}\mathrm{e}^{\mathrm{i}\delta} & c_{12}c_{23} - s_{12}s_{23}s_{13}\mathrm{e}^{\mathrm{i}\delta} & s_{23}c_{13} \\ s_{12}s_{23} - c_{12}c_{23}s_{13}\mathrm{e}^{\mathrm{i}\delta} & -c_{12}s_{23} - s_{12}c_{23}s_{13}\mathrm{e}^{\mathrm{i}\delta} & c_{23}c_{13} \end{pmatrix} \tag{21.112}$$

where $c_{ij} = \cos\theta_{ij}$, $s_{ij} = \sin\theta_{ij}$ with $i, j = 1, 2, 3$.

The mixing described by (21.111) implies that the individual lepton flavour numbers $L_{\text{e}}, L_\mu, L_\tau$ are no longer conserved. However, since we are here taking the neutrinos to be Dirac particles, there will be a quantum number carried by ν_{e}, ν_μ, and ν_τ which is conserved by the interactions. This could, for example, be the total lepton number $L_{\text{e}} + L_\mu + L_\tau$, assigning $L(\nu_\alpha) = 1$ for $\alpha = \text{e}, \mu, \tau$, which would follow from invariance under the global U(1) transformation $\hat{\ell}'_\alpha = \exp(\text{i}\delta)\hat{\ell}_\alpha, \hat{\nu}'_\alpha = \exp(\text{i}\delta)\hat{\nu}_\alpha$, where δ is independent of the flavour α.

This 'Dirac' option, though simple, may be thought uneconomical, however. As noted, the R components of neutrino fields have no interactions of SM type. The charged leptons do have electromagnetic interactions, of course, as do the quarks, which also have strong interactions. But the neutral neutrinos only have weak interactions, which involve only their L-components. Why, then, enlarge the field content to include hypothetical $\hat{\nu}_{\text{R}}$ fields, which don't have any SM interactions? It seems more economical to make do with only the $\hat{\nu}_{\text{L}}$ fields. In this case, the Dirac mass term (21.109) is not possible, but a Majorana mass term (21.110) can still exist. Clearly, such a mass term is *not* invariant under U(1) global phase transformations, and it breaks lepton number conservation explicitly. As in the Dirac case, the chiral L component will include a 'wrong' (i.e. positive) helicity component with an amplitude proportional to m/E.

The fact that global phase changes on the neutrino fields are now no longer freely available, because that symmetry is lost if they are Majorana fields, has implications for the mixing matrix, call it $\mathbf{U}_{\text{M}}$, in this case. Since the three Majorana neutrino fields can no longer absorb phases, we have only the three phases from the charged leptons at our disposal, which leaves three phase parameters in $\mathbf{U}_{\text{M}}$, after rephasing. The PMNS matrix in the Majorana case therefore has two more irreducible phase parameters than the CKM matrix, and is conventionally parametrized as

$$\mathbf{U}_{\text{M}} = \mathbf{U}(\text{CKM} - \text{type}) \times \text{diag.}(1, \text{e}^{\text{i}\alpha_{21}/2}, \text{e}^{\text{i}\alpha_{31}/2}). \tag{21.113}$$

There are three **CP**-violating phases in the Majorana neutrino case.

The only information at present (2023) concerning the entries in $\mathbf{U}$ comes from neutrino oscillation experiments, which we shall discuss in the next sections. We shall see that the Majorana phases α_{21} and α_{31} cancel in the probabilities calculated for neutrino transitions. We shall discuss how the values of the parameters θ_{12}, θ_{13}, and θ_{23} can be inferred from the observed oscillations, and also the differences in the squared masses of the neutrinos. We shall also be interested in the evidence for **CP** violation. Anticipating these results, we state here that the two independent squared mass differences, $m_2^2 - m_1^2$ and $m_3^2 - m_2^2$, turn out to be very small indeed, and rather different from each other: namely approximately 7.4×10^{-5} eV^2 and $\pm 2.5 \times 10^{-3}$ eV^2, respectively, the sign depending on the (as yet unsettled) ordering of the masses.

Data on the actual mass values are limited. There is a bound on the $\bar{\nu}_{\text{e}}$ mass from measurements of the electron spectrum near the end-point in tritium β-decay, which gives (Aker *et al.* 2019)

$$m^{\text{eff}}_{\bar{\nu}_{\text{e}}} < 1.1 \text{ eV} \quad 90\% \text{ CL}. \tag{21.114}$$

Note that we write $m^{\text{eff}}_{\bar{\nu}_{\text{e}}}$ here, rather than simply $m_{\bar{\nu}_{\text{e}}}$ because, in the presence of mixing, the effective neutrino mass which enters into decay spectra is the averaged quantity $m^{\text{eff}}_\alpha = [\sum_i m_i^2 |U_{\alpha i}|^2]^{1/2}$ (see the review by Gonzalez-Garcia and Yokoyama (2022), section 14 in Workman *et al.* 2022). A weaker limit on m_{ν_μ} comes from measurements of the muon spectrum in charged pion decay (Workman *et al.* 2022):

$$m_{\nu_\mu} < 0.19 \text{ MeV} \quad 90\% \text{ CL}, \tag{21.115}$$

and a still weaker one for $m_{\nu_{tau}}$ from the spectrum in τ decay:

$$m_{\nu_\tau}^{\text{eff}} < 18.2 \text{ MeV } 95\% \; CL. \tag{21.116}$$

The strongest upper bound comes from cosmology, assuming three neutrinos. The bounds vary somewhat depending on what data are included in the analysis, as discussed in the review by Lesourgues and Verde (2022), section 26 of Workman *et al.* (2022). One of the strongest is (Abbott *et al.* 2022)

$$\sum_{i=1}^{3} m_{\nu_i} < 0.13 \text{ eV}. \tag{21.117}$$

Taking the squared mass differences as indicative of the actual mass scale, neutrino masses are evidently very much smaller than the masses of the other fermions in the SM. We shall return to what this might tell us about the origin of neutrino mass in section 22.5.1, where we discuss how gauge-invariant masses are generated in the SM.

Returning to the question of **CP** violation, we noted in section 4.2.3 that the **CP** violation present in the SM was insufficient to account for the matter-antimatter asymmetry in the universe. However, we now see that it is possible to have **CP** violation in the lepton sector, in an extended SM with massive neutrinos. Leptonic matter-antimatter asymmetries can be converted into baryon asymmetries in the very hot early universe by a non-perturbative process predicted by SM dynamics—a process called leptogenesis (Fukugita and Yanagida 1986, Kuzmin, Rubakov and Shaposhnikov 1985, Davidson, Nardi and Nir 2008). It has been argued that the Dirac and/or Majorana phases in the neutrino matrix **U** or $\mathbf{U}_{\mathrm{M}}$ can provide the **CP** violation necessary in leptogenesis models for the generation of the observed baryon asymmetry of the universe (Pascoli *et al.* 2007a, 2007b). If such a proposal should prove to be the case, the reach of Pauli's 'desperate remedy' will have been vast indeed.

21.4.2   Neutrino oscillations in vacuum: formulae

We will restrict our discussion to the three known neutrinos of the SM. The existence of neutrino oscillations means that if a neutrino of a given flavour $\nu_\alpha (\alpha = \mathrm{e}, \mu, \tau)$ with energy E is produced in a charged current weak interaction process, such as $\pi^+ \to \mu^+ \nu_\mu$, then at a sufficiently large distance L from the ν_α source the probability $P(\nu_\alpha \to \nu_\beta; E, L)$ of detecting a neutrino of a different flavour ν_β is non-zero.[4] Such a flavour change will of course imply that the ν_α survival probability, $P(\nu_\alpha \to \nu_\alpha; E, L)$, is less than 1. We shall give a simplified version of the derivation of such probabilities, following the approach of the review by Nakamura and Petcov (2010) in section 13 of Nakamura *et al.* (2010). This review includes a large list of references to the time-dependent formalism; we mention here the contributions of Kayser (1981), Nauenberg (1999), and Cohen *et al.* (2009). We shall treat all the neutrinos as stable particles.

We consider the evolution of the state $|\nu_\alpha\rangle$ in the frame in which the detector which measures its flavour is at rest (the lab frame). As in the meson case, the states with simple space-time evolution in a vacuum are the mass eigenstates $|\nu_i\rangle$ $(i = 1, 2, 3)$, a superposition of which is equal to the flavour state $|\nu_\alpha\rangle$:

$$|\nu_\alpha\rangle = \sum_j \mathrm{U}_{\alpha i}^* |\nu_i, p_i\rangle, \tag{21.118}$$

the complex conjugation arising from taking the dagger of the relation (21.111) for the

---

[4]We shall not indicate the chirality explicitly from now on, it being assumed that we are referring to the L (R) component for neutrinos (antineutrinos).

field operators. Here $\mathbf{U}$ stands for either the Dirac or the Majorana matrix, and p_i is the four-momentum of ν_i. Similarly,

$$|\bar{\nu}_\alpha\rangle = \sum_i \mathrm{U}_{\alpha i}|\bar{\nu}_i, p_i\rangle. \tag{21.119}$$

We will consider highly relativistic neutrinos, as is the case for the experiments under discussion. We will assume that there are no degeneracies among the masses m_j. The states in the superpositions (21.118) and (21.119) will all have, in general, different energies and momenta E_i, p_i. We shall also treat the evolution as occurring in one spatial dimension, taking all the momenta to lie in the direction from the source to the detector. Note that the fractional deviation of E_i and p_i from the massless case $E = p$ is of order m^2/E^2 which will be extremely small, of order one part in 10^{16}, say.

Suppose now that the neutrinos of flavour ν_α started in the state (21.118) at time $t = 0$ in the detector frame are detected at time T after production, having travelled a distance L. Then the amplitude for finding a neutrino of flavour ν_β at (L, T) is

$$\begin{aligned} A(\nu_\alpha \to \nu_\beta; L, E) &= \sum_i \mathrm{U}_{\alpha i}^* \, e^{\mathrm{i} E_i T + \mathrm{i} p_i L} \langle \nu_\beta | \nu_i, p_i \rangle \\ &= \sum_i \mathrm{U}_{\alpha i}^* \, \mathrm{U}_{\beta i} e^{-\mathrm{i} E_i T + \mathrm{i} p_i L}. \end{aligned} \tag{21.120}$$

We make two immediate comments on (21.120). First, the Majorana phases in (21.55) cancel in $A(\nu_\alpha \to \nu_\beta; L, E) = \delta_{\alpha\beta}$, since the same phase appears in $\mathrm{U}_{\alpha i}$ and $\mathrm{U}_{\beta i}$. We conclude that oscillation experiments cannot distinguish Majorana from Dirac neutrinos. Second, if the neutrinos were massless, the phase factors in (21.120) would all be unity, and then $A(\nu_\alpha \to \nu_\beta; L, E) = \delta_{\alpha\beta}$, from the unitarity of the matrix $\mathbf{U}$, so there would be no flavour change.

Flavour oscillations come about via the interference in $|A(\nu_\alpha \to \nu_\beta; L, T)|^2$ between phase factors that are slightly different from one another, because of the different masses. A typical interference phase is then $\phi_{ij} = (E_i - E_j)T - (p_i - p_j)L$. Following the review by Nakamura and Petcov (2010) in Nakamura *et al.* (2010), we note that

$$\frac{m_i^2 - m_j^2}{p_i + p_j} = \frac{(E_i^2 - p_i^2) - (E_j^2 - p_j^2)}{p_i + p_j} = (E_i - E_j)\frac{(E_i + E_j)}{(p_i + p_j)} - (p_i - p_j) \tag{21.121}$$

so that

$$\phi_{ij} = (E_i - E_j)\left[T - \frac{E_i + E_j}{p_i + p_j}L\right] + \frac{m_i^2 - m_j^2}{p_i + p_j}L. \tag{21.122}$$

Bearing in mind that the energies differ from the momenta by terms of order m^2/E^2, we see that the first term in (21.122) can be dropped, and the interference phase is, to a very good approximation,

$$\phi_{ij} = \frac{m_i^2 - m_j^2}{2E} \equiv \frac{\Delta m_{ij}^2}{2E} \tag{21.123}$$

where E is the average energy, or momentum, of the neutrinos. We therefore obtain the probability

$$P(\nu_\alpha \to \nu_\beta; L, E) = \sum_i |\mathrm{U}_{\alpha i}|^2 |\mathrm{U}_{\beta i}|^2$$

$$+2\sum_{i>j} |\mathrm{U}_{\beta i} U_{\alpha i}^* \mathrm{U}_{\alpha j} U_{\beta j}^*| \cos\left[\left(\frac{\Delta m_{ij}^2}{2E}\right)L - \phi_{\beta\alpha;ij}\right] \tag{21.124}$$

where

$$\phi_{\alpha\beta;ij} = \text{Arg}\,(U_{\beta i}U^*_{\alpha i}U_{\alpha j}U^*_{\beta j}). \tag{21.125}$$

A more useful expression can be obtained by using the unitarity of $\mathbf{U}$ (problem 21.6):

$$P(\nu_\alpha \to \nu_\beta; L, E) = \delta_{\alpha\beta} - 4\sum_{i>j} \text{Re}\,(U_{\beta i}U^*_{\alpha i}U_{\alpha j}U^*_{\beta j}) \sin^2 \frac{\Delta m^2_{ij}L}{4E}$$

$$+2\sum_{i>j} \text{Im}\,(U_{\beta i}U^*_{\alpha i}U_{\alpha j}U^*_{\beta j}) \sin \frac{\Delta m^2_{ij}L}{2E} \tag{21.126}$$

The expression for $P(\bar\nu_\alpha \to \bar\nu_\beta; L, T)$ is the same, except for a change in sign of the last term in (21.126):

$$P(\bar\nu_\alpha \to \bar\nu_\beta; L, E) = \delta_{\alpha\beta} - 4\sum_{i>j} \text{Re}\,(U_{\beta i}U^*_{\alpha i}U_{\alpha j}U^*_{\beta j}) \sin^2 \frac{\Delta m^2_{ij}L}{4E}$$

$$-2\sum_{i>j} \text{Im}\,(U_{\beta i}U^*_{\alpha i}U_{\alpha j}U^*_{\beta j}) \sin \frac{\Delta m^2_{ij}L}{2E}. \tag{21.127}$$

For neutrino flavour oscillations to occur, neutrinos must have different masses, and there must be mixing.

It follows from (21.126) and (21.127) that $P(\nu_\alpha \to \nu_\beta; L, E) = P(\bar\nu_\beta \to \bar\nu_\alpha; L, E)$, a consequence of $\mathbf{CPT}$ invariance. $\mathbf{CP}$ alone requires $P(\nu_\alpha \to \nu_\beta; L, E) = P(\bar\nu_\alpha \to \bar\nu_\beta; L, E)$. A measure of $\mathbf{CP}$ violation is provided by

$$\begin{aligned} \mathcal{A}^{(\beta\alpha)}_{\mathbf{CP}} &= P(\nu_\alpha \to \nu_\beta; L, E) - P(\bar\nu_\alpha \to \bar\nu_\beta; L, E) \\ &= 4\sum_{i>j} \text{Im}\,(U_{\beta i}U^*_{\alpha i}U_{\alpha j}U^*_{\beta j}) \sin \frac{\Delta m^2_{ij}}{2E}L. \end{aligned} \tag{21.128}$$

The reader will recognize the Jarlskog (1985) invariants in (21.128). In this 3×3 mixing situation, which is exactly analogous to quark mixing, all these invariants are equal up to a sign, and (21.128) becomes (Krastev and Petcov 1988)

$$\mathcal{A}^{(\mu e)}_{\mathbf{CP}} = -\mathcal{A}^{(\tau e)}_{\mathbf{CP}} = \mathcal{A}^{(\tau\mu)}_{\mathbf{CP}} =$$
$$4J_\nu \left[\sin(\frac{\Delta m^2_{32}}{2E}L) + \sin(\frac{\Delta m^2_{21}}{2E}L) + \sin(\frac{\Delta m^2_{13}}{2E}L) \right] \tag{21.129}$$

where

$$J_\nu = \text{Im}(U_{\mu 3}U^*_{e3}U_{e2}U^*_{\mu 2}). \tag{21.130}$$

which takes the form (20.172) in the parametrization (21.112). If any one mass-squared difference is zero, say Δm^2_{21}, then $\Delta m^2_{32} = -\Delta m^2_{13}$, and the right-hand side of (21.129) vanishes: we need all three mass-squared differences to be non-zero, in order to get $\mathbf{CP}$ violation. Also, if any one angle θ_{ij} is zero, no $\mathbf{CP}$ violation will be observed.

In proceeding to discuss the experimental situation, it will be useful to define an "oscillation length" $\lambda_{ij}(E)$ given by

$$\lambda_{ij}(E) = 2E/\Delta m^2_{ij} \approx 0.4\frac{(E/\text{GeV})}{(\Delta m^2_{ij}/\text{eV}^2)} \text{ km}. \tag{21.131}$$

In practice, the three-state mixing formalism can often be simplified, making use of what is now known about the neutrino mass spectrum. As noted earlier, one squared mass difference is considerably smaller than the other:

$$|\Delta m_{21}^2| \sim 7.4 \times 10^{-5} \text{ eV}^2, \quad |\Delta m_{32}^2| \sim 2.5 \times 10^{-3} \text{eV}^2. \tag{21.132}$$

Suppose now that $L/|\lambda_{31}(E)| \sim 1$, while $L/|\lambda_{21}(E)| \ll 1$. Then expression (21.126) reduces to (problem 21.7)

$$
\begin{aligned}
P(\nu_\alpha \to \nu_\beta; L, E) &\approx \delta_{\alpha\beta} - 4|U_{\alpha3}|^2[\delta_{\alpha\beta} - |U_{\beta3}|^2]\sin^2\frac{\Delta m_{31}^2}{4E}L \\
&= P(\bar\nu_\alpha \to \bar\nu_\beta; L, E).
\end{aligned}
\tag{21.133}
$$

In particular,

$$P(\bar\nu_e \to \bar\nu_e; L, E) = 1 - 4|U_{e3}|^2(1 - |U_{e3}|^2)\sin^2\frac{\Delta m_{31}^2}{4E}L, \tag{21.134}$$

which can describe the survival probability of reactor $\bar\nu_e$s, for example.

Adopting the parametrization (21.112), $|U_{e3}|^2$ is $\sin^2\theta_{13}$, which is found experimentally to be small (see the following section). It is often a good approximation to set $|U_{e3}|$ to zero, in which case $|U_{\mu3}|^2 = \sin^2\theta_{23}$. Then (21.133) gives the ν_μ survival probability

$$P(\nu_\mu \to \nu_\mu; L, E) = P(\bar\nu_\mu \to \bar\nu_\mu; L, E) \approx 1 - \sin^2 2\theta_{23}\sin^2(L/2\lambda_{31}(E)) \tag{21.135}$$

and the flavour-change probability

$$P(\nu_\mu \to \nu_\tau; L, E) = P(\bar\nu_\mu \to \bar\nu_\tau; L, E) \approx \sin^2 2\theta_{23}\sin^2(L/2\lambda_{31}(E)); \tag{21.136}$$

In this approximation, $P(\nu_\mu \to \nu_e) = P(\bar\nu_\mu \to \bar\nu_e) = 0$. Formulae (21.135) and (21.136) can be used to describe the dominant atmospheric ν_μ and $\bar\nu_\mu$ oscillations (see the following section), and the parameters θ_{23} and Δm_{31}^2 (or Δm_{32}^2) are referred to as the atmospheric mixing angle and mass squared difference. The smaller mass squared difference Δm_{21}^2, and the angle θ_{12}, are associated with solar ν_e oscillations.

The formulae (21.135) and (21.136) are, in fact, exactly what a simple 2-state mixing model would give. Suppose that the effective mixing matrix for the 2-state system has the form (see problem 1.6)

$$\begin{pmatrix} -a\cos 2\theta & a\sin 2\theta \\ a\sin 2\theta & a\cos 2\theta \end{pmatrix}, \tag{21.137}$$

where rows are labelled by e, μ and columns by 1 and 3; then the survival probability is just

$$1 - \sin^2 2\theta \sin^2(La), \tag{21.138}$$

where we have taken $L \approx T$ as before. We can therefore identify the mixing parameter as

$$a = [2\lambda_{31}(E)]^{-1} = \frac{\Delta m_{31}^2}{4E}. \tag{21.139}$$

Note that the energies are here measured relative to a common average energy; if $|\ell\rangle$ is the lighter eigenstate and $|h\rangle$ the heavier, then

$$
\begin{aligned}
|\nu_e\rangle &= \cos\theta|\ell\rangle + \sin\theta|h\rangle \\
|\nu_\mu\rangle &= -\sin\theta|\ell\rangle + \cos\theta|h\rangle.
\end{aligned}
\tag{21.140}
$$

21.4.3   Neutrino oscillations: experimental results

Historically, the search for neutrino oscillations began when experiments by Davis *et al.* (1968) detected *solar neutrinos* (from ^{8}B decays) at a rate approximately one-third of that predicted by the solar model calculations of Bahcall *et al.* (1968). Pontecorvo (1946) had proposed the experiment, in which the neutrinos are detected by the inverse β-decay process $\nu_e + {}^{37}\text{Cl} \rightarrow e^- + {}^{37}\text{Ar}$. The Davis experiment used 520 metric tons of liquid tetrachloroethylene (C_2Cl_4), buried 4850 feet underground in the Homestake gold mine, in South Dakota. Davis' findings provided the impetus to study solar neutrinos using Kamiokande, a 3000 ton imaging water Cerenkov detector situated about one kilometre underground in the Kamioka mine in Japan. Indeed, ^{8}B solar neutrinos were observed, and at a rate consistent with that of the Davis experiment (Hirata *et al.* 1989). Later results from the Homestake mine (Cleveland *et al.* 1998) reported a solar neutrino detection rate almost exactly one-third of the updated calculations of Bahcall *et al.* (2001).

In a separate development, Kamiokande also reported (Hirata *et al.* 1988) an anomaly in the *atmospheric neutrino* flux. Atmospheric neutrinos are produced as decay products in hadronic showers which result from collisions of cosmic rays with nuclei in the upper atmosphere of the Earth. Production of electron and muon neutrinos is dominated by the decay chain $\pi^+ \rightarrow \mu^+ + \nu_\mu$, $\mu^+ \rightarrow e^+ + \bar{\nu}_\mu + \nu_e$ (and its charge-conjugate), which gives an expected value of about 2 for the ratio of $(\nu_\mu + \bar{\nu}_\mu)$ flux to $(\nu_e + \bar{\nu}_e)$ flux.[5] While the number of electron-like events was in good agreement with the Monte Carlo calculations based on atmospheric neutrino interactions in the detector, the number of muon-like events was about one half of the expected number, at the 4σ level.

This muon-like defect (and the lack of an electron-like defect) was later confirmed at the 9σ level by Super-Kamiokande (Fukuda *et al.* 1998). In this experiment, a marked dependence was observed on the zenith angle of the muon neutrinos. This angle is simply related to the distance travelled by the neutrinos from their point of production, which varies from about 20 km (from above the detector) to over 10,000 km (from below the detector). The muon-like deficit was associated with the upward-going events, in which the neutrinos had travelled further to reach the detector. The Super-Kamiokande data was the first compelling evidence for neutrino oscillations. Interpreting their data in terms of a simple 2-state $\nu_\mu \leftrightarrow \nu_\tau$ model, as in (21.136), Fukuda *et al.* (1998) reported the values $\sin^2 2\theta_{23} > 0.82$, and $5 \times 10^{-4} < \Delta m_{31}^2 < 6 \times 10^{-3}$ eV^2 at 90% CL.

To demonstrate that $\nu_\mu \leftrightarrow \nu_\tau$ oscillations were actually the reason for the muon-like deficit, the appearance of ν_τ's has to be detected. Super-Kamiokande (Li *et al.* 2018) reported evidence for ν_τ appearance in atmospheric neutrino data at the 4.6 σ level of significance. OPERA (Agafonova *et al.* 2018) have improved this to a significance of 6.1σ.

We will postpone further discussion of the solar neutrino deficit for the moment, since it is complicated by interactions of the neutrinos with the Sun's matter (see the following subsection). We proceed to describe some of the main results which have come from the analysis of data from neutrinos produced in terrestrial accelerators and reactors. A fuller discussion will be found in the review by Gonzalez-Garcia and Yokoyama (2022), section 14 in Workman *et al.* (2022).

We begin with the CHOOZ *reactor antineutrino experiment*, which was the first experiment to limit the value of θ_{13} (Apollonio *et al.* 1999, 2003). CHOOZ is the name of a nuclear power station situated near the French village of the same name. The experiment was designed to detect reactor $\bar{\nu}_e$s via the inverse β-decay reaction $\bar{\nu}_e + p \rightarrow e^+ + n$. The signature was a delayed coincidence between the prompt e^+ signal, and the signal from the neutron capture. The detector was located in an underground laboratory about 1 km

---

[5]The detector could not measure the charge of the final state leptons, and therefore ν and $\bar{\nu}$ events could not be discriminated.

from the neutrino source. It consisted of a central 5-ton target filled with 0.09% Gd-doped liquid scintillator; Gd-doping was chosen to maximize the capture of the neutrons. The neutrino energy E was a few MeV, and L was 1 km. For these values $2\lambda_{21}(E)$ is greater than about 10 km, while $2\lambda_{31}(E)$ is about 0.3 km. The neglect of $\sin^2 L/2\lambda_{21}(E)$ is justified, and formula (21.134) can be used for the $\bar{\nu}_e$ survival probability. The experiment found no evidence for $\bar{\nu}_e$ disappearance, and reported the 90% CL upper limit of $\sin^2 2\theta_{13} < 0.19$, for $|\Delta m_{31}^2| = 2 \times 10^{-3}$ eV^2.

A non-zero value of every θ_{ij} is required for the existence of **CP** violation, and it is therefore important to establish that θ_{13} is non-zero, though small. The Double Chooz (de Kerret *et al.* 2020), Daya Bay (Adey *et al.* 2018) and RENO (Bak *et al.* 2018) collaborations have reported convincing evidence for non-zero θ_{13}:

$$\sin^2 2\theta_{13} = 0.0856 \pm 0.0029 \quad \text{(Daya Bay)} \tag{21.141}$$

$$\sin^2 2\theta_{13} = 0.0896 \pm 0.0048 \pm 0.0047 \quad \text{(RENO)}, \tag{21.142}$$

$$\sin^2 2\theta_{13} = 0.105 \pm 0.014 \quad \text{(DoubleChooz)} \tag{21.143}$$

Another reactor experiment, KamLAND at Kamioka, was designed to be sensitive to the smaller squared mass difference Δm_{21}^2, and thus to θ_{12}. The Kamioka Liquid scintillator Anti-Neutrino Detector is at the site of the former Kamioka experiment. The detector is essentially one kiloton of highly purified liquid scintillator surrounded by photomultiplier tubes. $\bar{\nu}_e$s are detected as usual via the inverse β-decay reaction $\bar{\nu}_e + p \rightarrow e^+ + n$. KamLAND is surrounded by 55 nuclear power units, each an isotropic $\bar{\nu}_e$ source. The flux-weighted average path length is $L \sim 180$ km, and the energy E ranges from about 2 MeV to about 8 MeV. For $E = 3$ MeV, $2\lambda_{21}(E) \sim 30$ km, which allows for more than one oscillation. In this case (21.126) reduces to

$$P(\bar{\nu}_e \rightarrow \bar{\nu}_e; L, E) = 1 - 4|U_{e1}|^2 |U_{e2}|^2 \sin^2(L/2\lambda_{21}(E)) \tag{21.144}$$

assuming $|U_{e3}| \approx 0$. In a parametrization of the form (21.112), this becomes

$$P(\bar{\nu}_e \rightarrow \bar{\nu}_e; L, E) = 1 - \sin^2 2\theta_{12} \sin^2(L/2\lambda_{21}(E)), \tag{21.145}$$

again a simple 2-state mixing result. Data shown in figure 21.11 (Abe *et al.* 2008) gives

$$|\Delta m_{21}^2| = 7.58^{+0.14+0.15}_{-0.13-0.15} \times 10^{-5} \text{ eV}^2 \tag{21.146}$$

$$\tan^2 \theta_{12} = 0.56^{+0.10+0.10}_{-0.70-0.06}. \tag{21.147}$$

The KamLAND data showed for the first time the periodic behaviour of the $\bar{\nu}_e$ survival probability. Later data (Gando *et al.* 2013) are very similar.

The mass squared range $\Delta m^2 > 2 \times 10^{-3}$ eV^2 can be explored by *accelerator-based long-baseline experiments*, with typically $E \sim 1$ GeV and $L \sim$ several hundred kilometres. The K2K (KEK-to-Kamioka) experiment was the first accelerator-based experiment with a neutrino path length extending hundreds of kilometres. A horn-focused wide-band ν_μ beam with mean energy 1.3 GeV and path length 250 km was produced by 12 GeV protons from the KEK-PS and directed to the Super-Kamiokande detector. In this case, $L/2\lambda_{21}(E) \sim 10^{-2}$ may be neglected. Then formulae (21.135) and (21.136) may be used, in the approximation $U_{13} \approx 0$. The K2K data showed (Ahn *et al.* 2006) that $\sin^2 2\theta_{23} \approx 1 (\theta_{23} \approx \pi/4)$, and that $|\Delta m_{31}^2|$ had a value consistent with (21.132).

Another long baseline accelerator experiment was MINOS at Fermilab. Neutrinos were produced by the Neutrinos at the Main Injector facility (NuMI), using 120 GeV protons from the Fermilab main injector. The detector is a 5.4 kton iron-scintillator tracking calorimeter

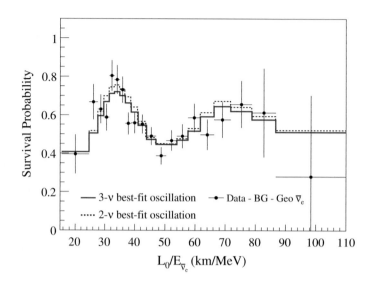

FIGURE 21.11

Ratio of the background and geo-neutrino subtracted $\bar{\nu}_e$ spectrum to the expectation for no-oscillation, as a function of L_0/E, where $L_0 = 180$ km. Figure reprinted with permission from S Abe *et al.* (KamLAND Collaboration) *Phys. Rev. Lett.* **100** 221803 (2008). Copyright 2008 by the American Physical Society.

with a toroidal magnetic field, situated underground in the Soudan mine, 735 km from Fermilab. The neutrino energy spectrum from a wide-band beam is horn-focused to be enhanced in the 1–5 GeV range. Early MINOS results yielded $|\Delta m_{31}^2| = (2.32_{-0.08}^{+0.12}) \times 10^{-3}$ eV^2, and $\sin^2 2\theta_{23} > 0.90$ at 90% CL (Adamson *et al.* 2011).

The major initial goal of second-generation long-baseline experiments was the observation of $\nu_\mu \leftrightarrow \nu_e$ oscillations. The first evidence for the appearance of ν_e in a ν_μ beam was obtained by the T2K collaboration (Abe *et al.* 2011). The ν_μ beam is produced using the high intensity proton accelerator at J-PARC, located in Tokai, Japan. The beam was directed 2.5° off-axis to the Super-Kamiokande detector at Kamioka, 295 km away. This configuration produces a narrow-band ν_μ beam, tuned at the first oscillation maximum $E_\nu = |\Delta m_{31}^2| L/2\pi \approx 0.6$ MeV, so as to reduce background from higher energy neutrino interactions. In the vacuum, the probability of the appearance of a ν_e in a ν_μ beam is given (in our customary effective 2-state mixing approximation) by (21.133) as

$$P(\nu_\mu \to \nu_e; L, E) = \sin^2 \theta_{23} \sin^2 2\theta_{13} \sin^2 \frac{\Delta m_{31}^2}{4E} L; \qquad (21.148)$$

$P(\bar{\nu}_\mu \to \bar{\nu}_e; L, E)$ is given by the same expression. Taking $|\Delta m_{31}^2| = 2.4 \times 10^{-3}$ eV^2 and $\sin^2 2\theta_{23} = 1$, the number of expected ν_e events was 1.5 ± 0.3(syst.) for $\sin^2 2\theta_{13} = 0$, and 5.5 ± 1.0 events if $\sin^2 \theta_{13} = 0.1$. Six events were observed which passed all the ν_e selection criteria, showing evidencefor ν_e appearance at the 2.5σ level. A later result from T2K (Abe *et al.* 2014) established $\nu_\mu \leftrightarrow \nu_e$ oscillations at more than 7σ significance.

NOνA is a long-baseline neutrino experiment, using the upgraded NuMI beamline with an off-axis configuration. The far detector is 810 km from the source, and the near detector is around 1 km from the source. The appearance of ν_e from a ν_μ beam was confirmed, and the appearance of $\bar{\nu}_e$ from a $\bar{\nu}_\mu$ beam was reported at the level of 4.4σ (Acero *et al.* 2019).

The ordering of the neutrino masses is not determined by the experiments described so far, nor has it yet (in 2023) been settled. There are two possibilities, according to the conventional classification. The first is the 'normal order' (NO), with $m_1 \ll m_2 < m_3$, which implies $m_2 \simeq (\Delta m_{21}^2)^{1/2} \sim 8.6 \times 10^{-3}$ eV, $m_3 \simeq (\Delta m_{32}^2 + \Delta m_{21}^2)^{1/2} \sim 0.5$ eV; the second is the "inverted order" (IO) with $m_3 \ll m_1 < m_2$, implying $m_1 \simeq (|\Delta m_{32}^2 + \Delta m_{21}^2|)^{1/2} \sim 0.05$ eV, $m_2 \simeq (|\Delta m_{32}^2|)^{1/2} \sim 0.05$ eV. The JUNO detector at Jiangmen in China has as its primary goal the determination of the mass ordering (An *et al.* 2016). A 20 kt liquid scintillator is located 53 km from two nuclear power plants. The energy resolution is designed to be extremely good (3% at 1 MeV), in order to resolve the small difference in the $\bar{\nu}_e$ energy spectrum in the NO and IO cases. JUNO should be able to determine the mass ordering at a significance of $3-4\sigma$ after 6 years of running. It is scheduled to begin taking data in 2023. JUNO will also provide precision measurements of neutrino mixing parameters, as well as a broad range of other observations in neutrino physics.

We now return to the *solar neutrino problem*, taking up the story after Davis' results. Some doubts remained as to whether the solar calculations could be absolutely relied upon, for example because of the extreme sensitivity to the core temperature ($\propto T^{18}$). One particular class of ν_e could, however, be reliably calculated, namely those associated with the initial reaction pp $\rightarrow^2$ H $+ e^+ + \nu_e$ of the pp cycle. Whereas the Davis experiments allowed detection of the higher energy ν_es (threshold 814 keV) from the B and Be stages of the cycle, the energy of the ν_es from the pp stage cuts off at around 400 keV. Detectors using the reaction $\nu_e + ^{71}$Ga $\rightarrow e^- + ^{71}$Ge, which has a 233 keV energy threshold, were built (GALLEX, GNO and SAGE); their results (Altman *et al.* 2005, Abdurashitov *et al.* 2009) are in agreement, and again much smaller than the (updated) Bahcall *et al.* (2005) prediction.

In 1999, the Sudbury Neutrino Observatory (SNO) in Canada began observation. This experiment used 1 kiloton of ultra-pure heavy water (D_2O). It measured ^{8}B solar ν_es via both the charged current (CC) reaction $\nu_e + d \rightarrow e^- + p + p$, and the neutral current (NC) reaction $\nu + d \rightarrow \nu + p + n$, as well as elastic νe^- scattering. The CC reaction is sensitive only to ν_e, while the NC reaction is sensitive to all active neutrinos, as is νe^- scattering. If the solar neutrino deficit were caused by neutrino oscillations, the solar neutrino fluxes measured by the CC and NC reactions would be significantly different. SNO found that, while the total neutrino flux was consistent with solar model expectations, the ratio of the ν_e flux to the total neutrino flux was about 1/3 (Ahmad *et al.* 2001, 2002). This number can be understood in terms of the effect of dense matter on the propagation of the ν_es, as we now discuss.

21.4.4 Matter effects in neutrino oscillations

We have assumed in the foregoing that neutrinos propagate in vacuum between the source and the detector. Since neutrinos interact only weakly, it might seem that this is always an excellent approximation. But in the same way that light travelling through a transparent medium can have its refractive index changed, so can a neutrino. In particular, the refractive index can be different for ν_e and ν_μ. The difference in refractive indices is determined by the difference in the real parts of the forward $\nu_e e^-$ and $\nu_\mu e^-$ elastic scattering amplitudes (Wolfenstein 1978). The essential point is that the scattering can be coherent, with the spins and momenta of the particles remaining unchanged. This means that the effect is going to be proportional to the density of electrons in the matter traversed, N_e. The scattering amplitude, in turn, is proportional to G_F, so that a figure of merit for the effect is given by the product $G_F N_e$. This has the dimensions of an energy, and can be interpreted as an addition to the effective 2-state mixing matrix of (21.137). Detailed analysis, which we omit, shows that the correct addition is actually $+\sqrt{2} G_F N_e$, so that (21.137) is modified to

$$
\begin{pmatrix} -\frac{\Delta m^2}{4E} \cos 2\theta + \sqrt{2} G_F N_e & \frac{\Delta m^2}{4E} \sin 2\theta \\ \frac{\Delta m^2}{4E} \sin 2\theta & \frac{\Delta m^2}{4E} \cos 2\theta \end{pmatrix}, \tag{21.149}
$$

where now $\Delta m^2 = m_2^2 - m_1^2$ and $\theta = \theta_{12}$. Two-state mixing now gives (problem 21.8) a new mixing angle θ_m such that

$$
\tan 2\theta_m = \frac{\tan 2\theta}{1 - N_e/N_{\text{res}}}, \quad N_{\text{res}} = \frac{\Delta m^2 \cos 2\theta}{2\sqrt{2} G_F E}, \tag{21.150}
$$

and the mass eigenstates $|1\rangle_m, |2\rangle_m$ correspond to the eigenvalue difference

$$
m_2 - m_1 = |\Delta m_{21}^2 / 2E| \left[\cos^2 2\theta (1 - N_e/N_{\text{res}})^2 + \sin^2 2\theta \right]^{1/2}. \tag{21.151}
$$

We see that although the new term is certainly very small, being proportional to G_F, nevertheless since Δm^2 is very small also, a significant effect can occur. In particular, if it should happen that $N_e \approx N_{\text{res}}$ for some (θ, E), then θ_m will be 'maximal' ($\theta_m = \pi/4$), irrespective of the value of the original θ. This is called 'resonant mixing' (Mikheev and Smirnov 1985, 1986). It implies that the probability for a $\nu_e \to \nu_\mu$ flavour change could be greatly enhanced over the vacuum value, which is proportional to $\sin^2 2\theta_{12}$. A point to note, also, is that the corresponding formulae for $\bar{\nu}_e$s are obtained by replacing N_e by $-N_e$; then, depending on the sign of $\Delta m^2 \cos 2\theta_{12}$, resonant mixing can occur for one or the other of ν_e or $\bar{\nu}_e$ as they pass through matter, but not both. Similar considerations apply to the propagation of neutrinos through the earth, but we shall not pursue this here (see Nakamura and Petcov 2010 in Nakamura *et al.* 2010).

In the case of solar neutrinos, the effect of the above modifications is quite simple. For the highest energy neutrinos, $N_e \gg N_{\text{res}}$ at the centre of the sun, so that $\theta_m \sim \pi/2$ at production in the core, and the ν_e is in the heavier mass state $|2\rangle_m$. On the way to the surface of the Sun, N_e will decrease, and a point will be reached when $N_e = N_{\text{res}}$. Here the mass difference (21.151) reaches its minimum, and two limiting cases may be distinguished depending on the scale of the variation in the electron density, which has been assumed constant in (21.149)–(21.151): (a) If the density variation is slow enough that at least one oscillation length fits into the resonant density region, then it can be shown that the state stays with state $|2\rangle_m$ ('adiabatic evolution') until it reaches the surface of the Sun, when $\theta_m \to \theta_{12}$. The probability that the neutrino will survive to the earth is then (using (21.140)) $|\langle \nu_e | 2 \rangle_m|^2 = \sin^2 \theta_{12}$, which has a value of about 1/3. In the alternative limit, (b), in which the oscillation length in matter is relatively large with respect to the scale of density variation, the state may 'jump' to the other mass state $|1\rangle_m$ ('extreme non-adiabatic evolution'), and then $|\langle \nu_e | 1 \rangle_m|^2 = \cos^2 \theta_{12}$. These are clearly extreme cases, and numerical work is required in the general case. However, the data from SNO and other water Cerenkov detectors are consistent with the first (adiabatic) alternative, and with the value $\sin^2 \theta_{12} \sim 1/3$. Note that the solar data imply that $(m_2^2 - m_1^2) \cos 2\theta_{12} > 0$.

By contrast, for the lowest energy neutrinos we can take $\theta_m \approx \theta$, so that the neutrinos are produced in the state $\cos \theta_{12} |1\rangle + \sin \theta_{12} |2\rangle$, and propagate as in a vacuum, oscillating with maximum excursion $\sin^2 2\theta_{12}$. The detectors average over many oscillations, giving a factor of 1/2, so that the survival probability for the low energy ν_es is $1 - \frac{1}{2} \sin^2 2\theta_{12} \sim 5/9$. The Gallium experiments are sensitive to the lower energy neutrinos, and indeed record some $60 - 70\%$ of the expected flux.

In summary, the solar neutrino data are consistent with the interpretation in terms of neutrino oscillations, as modified by the Wolfenstein-Mikheev-Smirnov (MSW) effect. A global solar + KamLAND analysis yielded best fit values (Aharmim *et al.* 2010)

$$\theta_{12} = 34.06^{+1.16}_{-0.84} \quad \Delta m^2_{21} = 7.59^{+0.20}_{-0.21} \times 10^{-5} \text{ eV}^2. \tag{21.152}$$

More recent data show rather little change (see the review by Gonzalez-Garcia and Yokoyama 2022 in Workman *et al.* 2022).

21.4.5 Summary

The determination of the leptonic parameters requires global analyses of the data from the different experiments. The most recent such analyses are those by Esteban *et al.* (2019), Esteban *et al.* (2020), Capozzi *et al.* (2018), and de Salas *et al.* (2018). The relatively better-known parameters are: $\theta_{12} \approx 34°$, $\theta_{13} \approx 8.5°$, $\Delta m^2_{21} \approx 7.4 \times 10^{-5} \text{ eV}^2$, and $|\Delta m^2_{32}| \approx 2.5 \times 10^{-3} \text{ eV}^2$. The parameter θ_{23} lies between $45°$ and $50°$ at the 3σ level. All analyses have a preference for the normal ordering of the neutrino masses, but the disfavouring is not yet worse than a 2–3 σ effect. The current best fit for the **CP**-violating weak phase is $\delta \approx 120°$, but **CP** conservation ($\delta \sim 180°$)) is still allowed at the 1–2 σ level.

Problems

21.1 Verify equation (21.35).

21.2 Verify equation (21.47) and (21.48).

21.3 Verify equations (21.49) and (21.50).

21.4 Verify equations (21.57).

21.5 Verify equations (21.90) and (21.91).

21.6 Verify equation (21.126).

21.7 Verify equation (21.133).

21.8 Verify equations (21.150)and (21.151).

22

The Glashow–Salam–Weinberg Gauge Theory of Electroweak Interactions

22.1 Difficulties with the current-current and 'naive' IVB models

In chapter 20 we developed the 'V-A current–current' phenomenology of weak interactions. We saw that this gives a remarkably accurate account of a wide range of data—so much so, in fact, that one might well wonder why it should not be regarded as a fully-fledged theory. One good reason for wanting to do this would be in order to carry out calculations beyond the lowest order, which is essentially all we have used it for so far (with the significant exceptions of the GIM argument, and box diagrams in $M - \bar{M}$ mixing). Such higher order calculations are indeed required by the precision attained in modern high energy experiments. But the electroweak theory of Glashow, Salam, and Weinberg, now recognized as one of the pillars of the standard model, was formulated long before such precision measurements existed, under the impetus of quite compelling theoretical arguments. These had to do, mainly, with certain in-principle difficulties associated with the current-current model, if viewed as a 'theory'. Since we now believe that the GSW theory is the correct description of electroweak interactions up to currently tested energies, further discussions of these old issues concerning the current-current model might seem irrelevant. However, these difficulties do raise several important points of principle. An understanding of them provides valuable motivation for the GSW theory—and some idea of what is 'at stake' in regard to experiments relating to those parts of it (notably the Higgs sector) which are still being intensively explored.

Before reviewing the difficulties, however, it is worth emphasizing once again a more positive motivation for a gauge theory of weak interactions (Glashow 1961). This is the remarkable 'universality' structure noted in chapter 20, not only as between different types of lepton, but also (within the context of CKM mixing) between the quarks and the leptons. This recalls very strongly the 'universality' property of QED, and the generalization of this property in the non-Abelian theories of chapter 13. A gauge theory would provide a natural framework for such universal couplings.

22.1.1 Violations of unitarity

We have seen several examples, in chapter 20, in which cross sections were predicted to rise indefinitely as a function of the invariant variable s, which is the square of the total energy in the CM frame. We begin by showing why this is ultimately an unacceptable behaviour.

Consider the process (figure 22.1)

$$\bar{\nu}_\mu + \mu^- \to \bar{\nu}_e + e^- \tag{22.1}$$

in the current-current model, regarding it as fundamental interaction, treated to lowest order in perturbation theory. A similar process was discussed in chapter 20. Since the troubles we

FIGURE 22.1
Current-current amplitude for $\bar{\nu}_\mu + \mu^- \to \bar{\nu}_e + e^-$.

shall find occur at high energies, we can simplify the expressions by neglecting the lepton masses without altering the conclusions. In this limit the invariant amplitude is (problem 22.1), up to a numerical factor,

$$\mathcal{M} = G_F E^2 (1 + \cos\theta) \tag{22.2}$$

where E is the CM energy, and θ is the CM scattering angle of the e^- with respect to the direction of the incident μ^-. This leads to the following behaviour of the cross section (cf (20.83), remembering that $s = 4E^2$):

$$\sigma \sim G_F^2 E^2. \tag{22.3}$$

The dependence on E^2 is a consequence of the fact that G_F is not dimensionless, having the dimensions of $[M]^{-2}$. Its value is (Workman *et al.* 2022)

$$G_F = 1.166388(6) \times 10^{-5} \text{ GeV}^{-2}. \tag{22.4}$$

The cross section has dimensions of $[L]^2 = [M]^{-2}$, but must involve G_F^2 which has dimension $[M]^{-4}$. It must also be relativistically invariant. At energies well above lepton masses, the only invariant quantity available to restore the correct dimensions to σ is s, the square of the CM energy E, so that $\sigma \sim G_F E^2$.

Consider now a partial wave analysis of this process. For spinless particles the total cross section may be written as a sum of partial wave cross sections

$$\sigma = \frac{4\pi}{k^2} \sum_J (2J + 1)|f_J|^2 \tag{22.5}$$

where f_J is the partial wave amplitude for angular momentum J and k is the CM momentum. It is a consequence of *unitarity*, or flux conservation (see, for example, Merzbacher (1998) chapter 13), that the partial wave amplitude may be written in terms of a phase shift δ_J:

$$f_J = e^{i\delta_J} \sin\delta_J \tag{22.6}$$

so that

$$|f_J| \leq 1. \tag{22.7}$$

Thus the cross section in each partial wave is bounded by

$$\sigma_J \leq 4\pi(2J + 1)/k^2 \tag{22.8}$$

which falls as the CM energy rises. By contrast, in (22.3) we have a cross section that rises with CM energy:

$$\sigma \sim E^2. \tag{22.9}$$

Moreover, since the amplitude (equation (22.2)) only involves $(\cos\theta)^0$ and $(\cos\theta)^1$ contributions, it is clear that this rise in σ is associated with only a few partial waves, and is

FIGURE 22.2
One-photon annihilation graph for $e^+e^- \to \mu^+\mu^-$.

not due to more and more partial waves contributing to the sum in σ. Therefore, at some energy E, the unitarity bound will be violated by this lowest-order (Born approximation) expression for σ.

This is the essence of the 'unitarity disease' of the current-current model. To fill in all the details, however, involves a careful treatment of the appropriate partial wave analysis for the case when all particles carry spin. We shall avoid those details. Instead we argue, again on dimensional grounds, that the dimensionless partial wave amplitude f_J (note the $1/k^2$ factor in (22.5)) must be proportional to $G_F E^2$, which violates the bound (22.7) for CM energies

$$E \geq G_F^{-1/2} \sim 300\text{GeV}. \tag{22.10}$$

At this point the reader may recall a very similar-sounding argument made in section 11.8, which led to the same estimate of the 'dangerous' energy scale (22.10). In that case, the discussion referred to a hypothetical '4-fermi' interaction without the V–A structure, and it was concerned with renormalization rather than unitarity. The gamma-matrix structure is irrelevant to these issues, which ultimately have to do with the dimensionality of the coupling constant, in both cases. In fact, as we shall see, unitarity and renormalizability are actually rather closely related.

Faced with this unitarity difficulty, we appeal to the most successful theory we have, and ask: what happens in QED? We consider an apparently quite similar process, namely $e^+e^- \to \mu^+\mu^-$ in lowest order (figure 22.2). In chapter 8 the total cross section for this process, neglecting lepton masses, was found to be (see problem 8.18 and equation (9.87))

$$\sigma = 4\pi\alpha^2/3E^2 \tag{22.11}$$

which obediently falls with energy as required by unitarity. In this case the coupling constant α, analogous to G_F, is dimensionless, so that a factor E^2 is required in the denominator to give $\sigma \sim [L]^2$.

If we accept this clue from QED, we are led to search for a theory of weak interactions that involves a dimensionless coupling constant. Pressing the analogy with QED further will help us to see how one might arise. Fermi's current-current model was, as we said, motivated by the vector currents of QED. But in Fermi's case the currents interact directly with each other, whereas in QED they interact only indirectly via the mediation of the electromagnetic field. More formally, the Fermi current-current interaction has the 'four point' structure

$$`G_F(\bar{\hat{\psi}}\hat{\psi}) \cdot (\bar{\hat{\psi}}\hat{\psi})' \tag{22.12}$$

while QED has the 'three-point' (Yukawa) structure

$$`e\bar{\hat{\psi}}\hat{\psi}\hat{A}.' \tag{22.13}$$

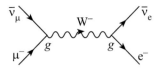

FIGURE 22.3
One-W^- annihilation graph for $\bar{\nu}_\mu + \mu^- \to \bar{\nu}_e + e^-$.

Dimensional analysis easily shows, once again, that $[G_F] = M^{-2}$ while $[e] = M^0$. This strongly suggests that we should take Fermi's analogy further, and look for a weak interaction analogue of (22.13), having the form

$$`g\bar{\hat{\psi}}\hat{\psi}\hat{W}` \qquad (22.14)$$

where $\hat{W}$ is a bosonic field. Dimensional analysis shows, of course, that $[g] = M^0$.

Since the weak currents are in fact vector-like, we must assume that the $\hat{W}$ fields are also vectors (spin-1) so as to make (22.14) Lorentz invariant. And because the weak interactions are plainly *not* long-range, like electromagnetic ones, the mass of the W quanta cannot be zero. So we are led to postulate the existence of a massive weak analogue of the photon, the 'intermediate vector boson' (IVB), and to suppose that weak interactions are mediated by the exchange of IVB's.

There is, of course, one further difference with electromagnetism, which is that the currents in β-decay, for example, carry charge (e.g. $\bar{\hat{\psi}}_e\gamma^\mu(1 - \gamma_5)\hat{\psi}_{\nu_e}$ creates negative charge or destroys positive charge). The 'companion' hadronic current carries the opposite charge (e.g. $\bar{\hat{\psi}}_p\gamma_\mu(1 - r\gamma_5)\hat{\psi}_n$ destroys negative charge or creates positive charge), so as to make the total effective interaction charge-conserving, as required. It follows that the $\hat{W}$ fields must then be charged, so that expressions of the form (22.14) are neutral. Because both charge-raising and charge-lowering currents exist, we need both W^+ and W^-. The reaction (22.1), for example, is then conceived as proceeding via the Feynman diagram shown in figure 22.3, quite analogous to figure 22.2.

Because we also have weak neutral currents, we need a neutral vector boson as well, Z^0. In addition to all these, there is the familiar massless neutral vector boson, the photon. Despite the fact that they are *not* massless, the $W^\pm$ and Z^0 can be understood as gauge quanta, thanks to the symmetry-breaking mechanism explained in section 19.6. For the moment, however, we are going to follow a more scenic route, and accept (as Glashow did in 1961) that we are dealing with ordinary 'unsophisticated' massive vector particles, charged and uncharged.

We now investigate whether the IVB model can do any better with unitarity than the current-current model. The analysis will bear a close similarity to the discussion of the renormalizability of the model in section 19.1, and we shall take up that issue again in section 22.1.2.

The unitarity-violating processes turn out to be those involving *external* W particles. Consider, for example, the process

$$\nu_\mu + \bar{\nu}_\mu \to W^+ + W^- \qquad (22.15)$$

proceeding via the graph shown in figure 22.4. The fact that this is experimentally a somewhat esoteric reaction is irrelevant for the subsequent argument: the proposed theory, represented by the IVB modification of the four-fermion model, will necessarily generate the

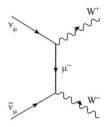

FIGURE 22.4

μ^--exchange graph for $\nu_\mu + \bar{\nu}_\mu \to W^+ + W^-$.

amplitude shown in figure 22.4, and since this amplitude violates unitarity, the theory is unacceptable. The amplitude for this process is proportional to

$$
\begin{aligned}
\mathcal{M}_{\lambda_1 \lambda_2} &= g^2 \epsilon_\mu^{-*}(k_2, \lambda_2) \epsilon_\nu^{+*}(k_1, \lambda_1) \bar{v}(p_2) \gamma^\mu (1 - \gamma_5) \\
&\quad \times \frac{(\not{p}_1 - \not{k}_1 + m_\mu)}{(p_1 - k_1)^2 - m_\mu^2} \gamma^\nu (1 - \gamma_5) u(p_1)
\end{aligned}
\tag{22.16}
$$

where the $\epsilon^\pm$ are the polarization vectors of the W's: $\epsilon_\mu^{-*}(k_2, \lambda_2)$ is that associated with the outgoing W^- with 4-momentum k_2 and polarization state λ_2, and similarly for ϵ_ν^{+*}.

To calculate the total cross section, we must form $|\mathcal{M}|^2$ and sum over the three states of polarization for each of the W's. To do this, we need the result

$$
\sum_{\lambda = 0, \pm 1} \epsilon_\mu(k, \lambda) \epsilon_\nu^*(k, \lambda) = -g_{\mu\nu} + k_\mu k_\nu / M_W^2
\tag{22.17}
$$

already given in (19.19). Our interest will as usual be in the high-energy behaviour of the cross section, in which regime it is clear that the $k_\mu k_\nu / M_W^2$ term in (22.17) will dominate the $g_{\mu\nu}$ term. It is therefore worth looking a little more closely at this term. From (19.17) and (19.18) we see that in a frame in which $k^\mu = (k^0, 0, 0, |\boldsymbol{k}|)$, the transverse polarization vectors $\epsilon^\mu(k, \lambda = \pm 1)$ involve no momentum dependence, which is in fact carried solely in the longitudinal polarization vector $\epsilon^\mu(k, \lambda = 0)$. We may write this as

$$
\epsilon(k, \lambda = 0) = \frac{k^\mu}{M_W} + \frac{M_W}{(k^0 + |\boldsymbol{k}|)} \cdot (-1, \hat{\boldsymbol{k}})
\tag{22.18}
$$

which at high energy tends to k^μ / M_W. Thus it is clear that it is the longitudinal polarization states which are responsible for the $k^\mu k^\nu$ parts of the polarization sum (12.21), and which will dominate real production of W's at high energy.

Concentrating therefore on the production of longitudinal W's, we are led to examine the quantity

$$
\frac{g^4}{M_W^4 (p_1 - k_1)^4} \mathrm{Tr}[\not{k}_2 (1 - \gamma_5)(\not{p}_1 - \not{k}_1) \not{k}_1 \not{p}_1 \not{k}_1 (\not{p}_1 - \not{k}_1) \not{k}_2 \not{p}_2]
\tag{22.19}
$$

where we have neglected m_μ, commuted the $(1 - \gamma_5)$ factors through, and neglected neutrino masses, in forming $\sum_{\text{spins}} |\mathcal{M}_{00}|^2$. Retaining only the leading powers of energy, we find (see problem 22.2)

$$
\sum_{\text{spins}} |\mathcal{M}_{00}|^2 \sim (g^4 / M_W^4)(p_1 \cdot k_2)(p_2 \cdot k_2) = (g^4 / M_W^4) E^4 (1 - \cos^2 \theta)
\tag{22.20}
$$

where E is the CM energy and θ the CM scattering angle. We see that the (unsquared) amplitude must behave essentially as $g^2 E^2 / M_W^2$, the quantity g^2 / M_W^2 effectively replacing G_F of the current-current model. The unitarity bound is violated for $E \geq M_W / g \sim 300\,\text{GeV}$, taking $g \sim e$.

Other unitarity-violating processes can easily be invented, and we have to conclude that the IVB model is, in this respect, no more fitted to be called a theory than was the four-fermion model. In the case of the latter, we argued that the root of the disease lay in the fact that G_F was not dimensionless, yet somehow this was not a good enough cure after all: perhaps(it is indeed so) 'dimensionlessness' is necessary but not sufficient (see the following section). Why is this? Returning to $\mathcal{M}_{\lambda_1, \lambda_2}$ for $\nu \bar{\nu} \to W^+ W^-$ (equation (22.16)) and setting $\epsilon = k_\mu / M$ for the *longitudinal* polarization vectors, we see that we are involved with an effective amplitude

$$\frac{g^2}{M_W^2} \bar{v}(p_2)\, \not{k}_2 (1 - \gamma_5) \frac{\not{p}_1 - \not{k}_1}{(p_1 - k_1)^2}\, \not{k}_1 (1 - \gamma_5) u(p_1). \qquad (22.21)$$

Using the Dirac equation $\not{p}_1 u(p_1) = 0$ and $p_1^2 = 0$, this can be reduced to

$$-\frac{g^2}{M_W^2} \bar{v}(p_2)\, \not{k}_2 (1 - \gamma_5) u(p_1). \qquad (22.22)$$

We see that the longitudinal ϵ's have brought in the factors M_W^{-2}, which are 'compensated' by the factor $\not{k}_2$, and it is this latter factor which causes the rise with energy. The longitudinal polarization states have effectively reintroduced a dimensional coupling constant g/M_W.

What happens in QED? We learnt in section 7.3 that, for real photons, the longitudinal state of polarization is absent altogether. We might well suspect, therefore, that since it was the longitudinal W's that caused the 'bad' high-energy behaviour of the IVB model, the 'good' high-energy behaviour of QED might have its origin in the absence of such states for photons. And this circumstance can, in its turn, be traced (cf section 7.3.1) to the *gauge invariance* property of QED.

Indeed, in section 8.6.3 we saw that in the analogue of (22.17) for photons (this time involving only the two transverse polarization states), the right hand side could be taken to be just $-g_{\mu\nu}$, *provided that* the Ward identity (8.166) held, a condition directly following from gauge invariance.

We have arrived here at an important theoretical indication that what we really need is a *gauge theory* of the weak interactions, in which the W's are gauge quanta. It must, however, be a peculiar kind of gauge theory, since normally gauge invariance requires the gauge field quanta to be massless. However, we have already seen how this 'peculiarity' can indeed arise, if the local symmetry is spontaneously broken (chapter 19). But before proceeding to implement that idea, in the GSW theory, we discuss one further disease (related to the unitarity one) possessed by both current-current and IVB models—that of non-renormalizability.

22.1.2 The problem of non-renormalizability in weak interactions

The preceding line of argument about unitarity violations is open to the following objection. It is an argument conducted entirely with in the framework of perturbation theory. What it shows, in fact, is simply that perturbation theory must fail, in theories of the type considered, at some sufficiently high energy. The essential reason is that the effective expansion parameter for perturbation theory is $E G_F^{1/2}$. Since $E G_F^{1/2}$ becomes large at high

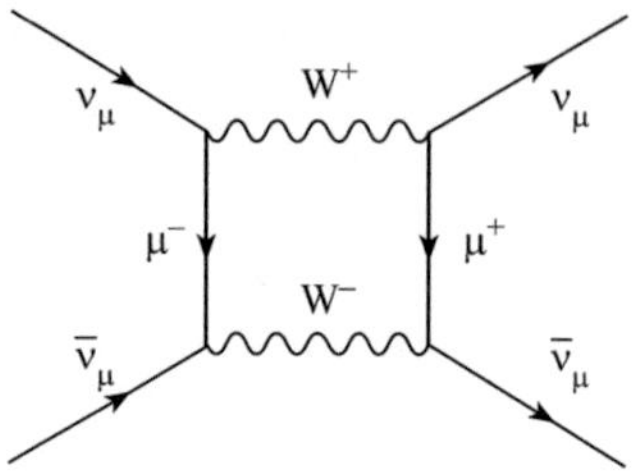

FIGURE 22.5

$O(g^4)$ contribution to $\nu_\mu \bar{\nu}_\mu \to \nu_\mu \bar{\nu}_\mu$.

energy, arguments based on lowest-order perturbation theory are irrelevant. The objection is perfectly valid, and we shall take account of it by linking high-energy behaviour to the problem of renormalizability, rather than unitarity. We might, however, just note in passing that yet another way of stating the results of the previous two sections is to say that, for both the current-current and IVB theories, 'weak interactions become strong at energies of order 1 TeV'.

We gave an elementary introduction to renormalization in chapters 10 and 11 of volume 1. In particular, we discussed in some detail, in section 11.8, the difficulties that arise when one tries to do higher order calculations in the case of a four-fermion interaction with the same form (apart from the V-A structure) as the current-current model. Its coupling constant, which we called G_F, also had dimension (mass)$^{-2}$. The 'non-renormalizable' problem was essentially that, as one approached the 'dangerous' energy scale (22.10), one needed to supply the values of an ever-increasing number of parameters from experiment, and the theory lost predictive power.

Does the IVB model fare any better? In this case, the coupling constant is dimensionless, just as in QED. 'Dimensionlessness' alone is not enough, it turns out: the IVB model is not renormalizable either. We gave an indication of why this is so in section 19.1, but we shall now be somewhat more specific, relating the discussion to the previous one about unitarity.

Consider, for example, the fourth-order processes shown in figure 22.5, for the IVB-mediated process $\nu_\mu \bar{\nu}_\mu \to \nu_\mu \bar{\nu}_\mu$. It seems plausible from the diagram that the amplitude must be formed by somehow 'sticking together' two copies of the tree graph shown in figure 22.4.[1] Now we saw that the high-energy behaviour of the amplitude $\nu\bar{\nu} \to W^+W^-$ (figure 22.4) grows as E^2, due to the k dependence of the longitudinal polarization vectors, and this turns out to produce, via figure 22.5, a non-renormalizable divergence, for the reason indicated in section 19.1—namely, the 'bad' behaviour of the $k^\mu k^\nu / M_W^2$ factors in the W-propagators, at large k.

So it is plain that, once again, the blame lies with the longitudinal polarization states for the W's. Let us see how QED—a renormalizable theory—manages to avoid this problem. In this case, there are two box graphs, shown in figures 22.6. There are also two corresponding tree graphs, shown in figures 22.7(a) and (b). Consider, therefore mimicking for figures 22.7(a) and (b) the calculation we did for figure 22.4. We would obtain the leading high-energy behaviour by replacing the photon polarization vectors by the corresponding momenta, and it can be checked (problem 21.3) that when this replacement is made for each photon the complete amplitude for the sum of figures 22.7(a) and (b) *vanishes*.

In physical terms, of course, this result was expected, since we knew in advance that it is always possible to choose polarization vectors for *real* photons such that they are purely

---

[1] The reader may here usefully recall the discussion of unitarity for one-loop graphs in section 13.3.3.

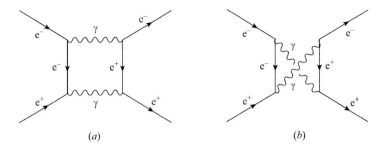

FIGURE 22.6
$O(e^4)$ contributions to $e^+e^- \to e^+e^-$.

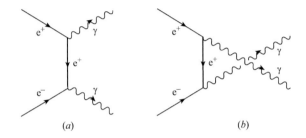

FIGURE 22.7
Lowest order amplitudes for $e^+e^- \to \gamma\gamma$: (a) direct graph, (b) crossed graph.

transverse, so that no physical process can depend on a part of ϵ_μ proportional to k_μ. Nevertheless, the calculation is highly relevant to the question of renormalizing the graphs in figure 22.6. The photons in this process are not real external particles, but are instead virtual, internal ones. This has the consequence that we should in general include their longitudinal ($\epsilon_\mu \propto k_\mu$) states as well as the transverse ones (see section 13.3.3 for something similar in the case of unitarity for 1-loop diagrams). The calculation of problem 22.3 then suggests that these longitudinal states are harmless, provided that both contributions in figure 22.7 are included.

Indeed, the sum of these two box graphs for $e^+e^- \to e^+e^-$ is *not divergent*. If it were, an infinite counter term proportional to a four-point vertex $e^+e^- \to e^+e^-$ (figure 22.8) would have to be introduced, and the original QED theory, which of course lacks such a fundamental interaction, would not be renormalizable. This is exactly what *does* happen in the case of figure 22.5. The bad high-energy behaviour of $\nu\bar{\nu} \to W^+W^-$ translates into a

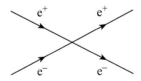

FIGURE 22.8
Four-point e^+e^- vertex.

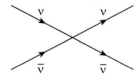

FIGURE 22.9
Four-point $\nu\bar{\nu}$ vertex.

divergence of figure 22.5—and this time there is no 'crossed' amplitude to cancel it. This divergence entails the introduction of a new vertex, figure 22.9, not present in the original IVB theory. Thus the theory without this vertex is non-renormalizable—and if we include it, we are landed with a four-field pointlike vertex which is non-renormalizable, as in the Fermi (current-current) case.

Our presentation hitherto has emphasized the fact that, in QED, the bad high-energy behaviour is rendered harmless by a cancellation between contributions from figures 22.7(a) and (b) (or figures 22.6(a) and (b)). Thus one way to 'fix up' the IVB theory might be to hypothesize a new physical process, to be added to figure 22.4, in such a way that a cancellation occurred at high energies. The search for such high-energy cancellation mechanisms can indeed be pushed to a successful conclusion (Llewellyn Smith 1973), given sufficient ingenuity and, arguably, a little hindsight. However, we are in possession of a more powerful principle. In QED, we have already seen (section 8.6.2) that the vanishing of amplitudes when an ϵ_μ is replaced by the corresponding k_μ is due to *gauge invariance*. In other words, the potentially harmful longitudinal polarization states are in fact harmless in a gauge-invariant theory.

We have therefore arrived once more, after a somewhat more leisurely discussion than that of section 19.1, at the idea that we need a *gauge* theory of massive vector bosons, so that the offending $k^\mu k^\nu$ part of the propagator can be 'gauged away' as in the photon case. This is precisely what is provided by the 'spontaneously broken' gauge theory concept, as developed in chapter 19. There we saw that, taking the U(1) case for simplicity, the general expression for the gauge boson propagator in such a theory (in a 't Hooft gauge) is

$$
\mathrm{i}\left[-g^{\mu\nu} + \frac{(1-\xi)k^\mu k^\nu}{k^2 - \xi M_W^2}\right] \bigg/ (k^2 - M_W^2 + \mathrm{i}\epsilon) \tag{22.23}
$$

where ξ is a gauge parameter. Our IVB propagator corresponds to the $\xi \to \infty$ limit, and with this choice of ξ all the troubles we have been discussing appear to be present. But for any finite ξ (for example $\xi = 1$) the high-energy behaviour of the propagator is actually $\sim 1/k^2$, the same as in the renormalizable QED case. This strongly suggests that such theories—in particular non-Abelian ones—are in fact renormalizable. 't Hooft's proof that they are ('t Hooft 1971b) triggered an explosion of theoretical work, as it became clear that, for the first time, it would be possible to make higher-order calculations for weak interaction processes using consistent renormalization procedures, of the kind that had worked so well for QED.

We now have all the pieces in place, and can proceed to introduce the GSW theory, based on the local gauge symmetry of SU(2) × U(1).

22.2 The SU(2) × U(1) electroweak gauge theory

22.2.1 Quantum number assignments; Higgs, W and Z boson masses

Given the preceding motivations for considering a gauge theory of weak interactions, the remaining question is this: what is the relevant symmetry group of local phase transformations, i.e. the relevant *weak gauge group*? Several possibilities were suggested, but it is now very well established that the one originally proposed by Glashow (1961), subsequently treated as a spontaneously broken gauge symmetry by Weinberg (1967) and by Salam (1968), and later extended by other authors, produces a theory which is in remarkable agreement with currently known data. We shall not give a critical review of all the experimental evidence, but instead proceed directly to an outline of the GSW theory, introducing elements of the data at illustrative points.

An important clue to the symmetry group involved in the weak interactions is provided by considering the transitions induced by these interactions. This is somewhat analogous to discovering the multiplet structure of atomic levels and hence the representations of the rotation group, a prominent symmetry of the Schrödinger equation, by studying electromagnetic transitions. However, there is one very important difference between the 'weak multiplets' we shall be considering, and those associated with symmetries which are not spontaneously broken. We saw in chapter 12 how an unbroken non-Abelian symmetry leads to multiplets of states which are degenerate in mass, but in section 17.1 we learned that that result only holds provided the vacuum is left invariant under the symmetry transformation. When the symmetry is spontaneously broken, the vacuum is *not* invariant, and we must expect that the degenerate multiplet structure will then, in general, disappear completely. This is precisely the situation in the electroweak theory.

Nevertheless, as we shall see, essential consequences of the weak symmetry group—specifically, the relations it requires between otherwise unrelated masses and couplings—are accessible to experiment. Moreover, despite the fact that members of a multiplet of a global symmetry which is spontaneously broken will, in general, no longer have even approximately the same mass, the concept of a multiplet is still useful. This is because when the symmetry is made a *local* one, we shall find (in sections 22.2.2 and 22.2.3) that the associated gauge quanta still mediate interactions between members of a given symmetry multiplet, just as in the manifest local non-Abelian symmetry example of QCD. Now, the leptonic transitions associated with the weak charged currents are, as we saw in chapter 20, $\nu_e \leftrightarrow e, \nu_\mu \leftrightarrow \mu$ etc. This suggests that these pairs should be regarded as *doublets* under some group. Further we saw in section 20.7 how weak transitions involving charged quarks suggested a similar doublet structure for them also. The simplest possibility is therefore to suppose that, in both cases, a 'weak SU(2) group' is involved, called 'weak isospin'. We emphasize once more that this weak isospin is distinct from the hadronic isospin of chapter 12, which is part of $SU(3)_f$. We use the symbols t, t_3 for the quantum numbers of weak isospin, and make the specific assignments for the leptonic fields

$$t = \frac{1}{2}, \quad \begin{cases} t_3 = +1/2 \\ t_3 = -1/2 \end{cases} \quad \begin{pmatrix} \hat{\nu}'_e \\ \hat{e}^- \end{pmatrix}_L, \quad \begin{pmatrix} \hat{\nu}'_\mu \\ \hat{\mu}^- \end{pmatrix}_L, \quad \begin{pmatrix} \hat{\nu}'_\tau \\ \hat{\tau}^- \end{pmatrix}_L \quad (22.24)$$

where $\hat{e}_L = \frac{1}{2}(1 - \gamma_5)\hat{e}$ etc, and for the quark fields

$$t = \frac{1}{2}, \quad \begin{cases} t_3 = +1/2 \\ t_3 = -1/2 \end{cases} \quad \begin{pmatrix} \hat{u} \\ \hat{d}' \end{pmatrix}_L, \quad \begin{pmatrix} \hat{c} \\ \hat{s}' \end{pmatrix}_L, \quad \begin{pmatrix} \hat{t} \\ \hat{b}' \end{pmatrix}_L. \quad (22.25)$$

As discussed in section 20.2.2, the subscript 'L' refers to the fact that only the left–handed chiral components of the fields enter, in consequence of the V–A structure. For this reason, the weak isospin group is referred to as $SU(2)_L$, to show that the weak isospin assignments and corresponding transformation properties apply only to these left–handed parts. Notice that, as anticipated for a spontaneously broken symmetry, these doublets all involve pairs of particles which are not mass degenerate. In (22.24) and (22.25), the primes indicate that these fields are related to the (unprimed) fields of definite mass by the unitary matrices $\mathbf{U}$ (for neutrinos) and $\mathbf{V}$ (for quarks), as discussed in sections 21.4.1 and 20.7.3 respectively.

Making this $SU(2)_L$ into a local phase invariance (following the logic of chapter 13) will entail the introduction of three gauge fields, transforming as a $t = 1$ multiplet (a triplet) under the group. Because (as with the ordinary $SU(2)_f$ of hadronic isospin) the members of a weak isodoublet differ by one unit of charge, the two gauge fields associated with transitions between doublet members will have charge ± 1. The quanta of these fields will, of course, be the now familiar $W^{\pm}$ bosons mediating the charged current transitions, and associated with the weak isospin raising and lowering operators $t_{\pm}$. What about the third gauge boson of the triplet? This will be electrically neutral, and a very economical and appealing idea would be to associate this neutral vector particle with the photon, thereby *unifying* the weak and electromagnetic interactions. A model of this kind was originally suggested by Schwinger (1957). Of course, the W's must somehow acquire mass, while the photon remains massless. Schwinger arranged this by introducing appropriate couplings of the vector bosons to additional scalar and pseudoscalar fields. These couplings were arbitrary and no prediction of the W masses could be made. We now believe, following the arguments of the preceding section, that the W mass must arise via the spontaneous breakdown of a non–Abelian gauge symmetry, and as we saw in section 19.6, this *does* constrain the W mass.

Apart from the question of the W mass in Schwinger's model, we now know (see chapter 20) that there exist *neutral current* weak interactions, in addition to those of the charged currents. We must also include these in our emerging gauge theory, and an obvious suggestion is to have these currents mediated by the neutral member W^0 of the $SU(2)_L$ gauge field triplet. Such a scheme was indeed proposed by Bludman (1958), again pre-Higgs, so that W masses were put in 'by hand'. In this model, however, the neutral currents will have the same pure left–handed V–A structure as the charged currents: but, as we saw in chapter 20, the neutral currents are *not* pure V-A. Furthermore, the attractive feature of including the photon, and thus unifying weak and electromagnetic interactions, has been lost.

A key contribution was made by Glashow (1961); similar ideas were also advanced by Salam and Ward (1964). Glashow suggested enlarging the Schwinger–Bludman $SU(2)$ schemes by inclusion of an additional $U(1)$ gauge group, resulting in an '$SU(2)_L \times U(1)$' group structure. The new Abelian $U(1)$ group is associated with a weak analogue of hypercharge— 'weak hypercharge'—just as $SU(2)_L$ was associated with 'weak isospin'. Indeed, Glashow proposed that the Gell–Mann-Nishijima relation for charges should also hold for these weak analogues, giving

$$eQ = e(t_3 + y/2) \tag{22.26}$$

for the electric charge Q (in units of e) of the t_3 member of a weak isomultiplet, assigned a weak hypercharge y. Clearly, therefore, the lepton doublets, (ν'_e, e^-), etc, then have $y = -1$, while the quark doublets (u, d'), etc, have $y = +\frac{1}{3}$. Now, when *this* group is gauged, everything falls marvellously into place: the charged vector bosons appear as before, but there are now *two* neutral vector bosons, which between them will be responsible for the weak neutral current processes, and for electromagnetism. This is exactly the piece of mathematics we went through in section 19.6, which we now appropriate as an important part of the SM.

For convenience, we reproduce here the main results of section 19.6. The Higgs field $\hat{\phi}$ is an SU(2) doublet

$$\hat{\phi} = \begin{pmatrix} \hat{\phi}^+ \\ \hat{\phi}^0 \end{pmatrix} \tag{22.27}$$

with an assumed vacuum expectation value (in unitary gauge) given by

$$\langle 0|\hat{\phi}|0\rangle = \begin{pmatrix} 0 \\ v/\sqrt{2} \end{pmatrix}. \tag{22.28}$$

Fluctuations about this value are parametrized in this gauge by

$$\hat{\phi} = \begin{pmatrix} 0 \\ \frac{1}{\sqrt{2}}(v + \hat{H}) \end{pmatrix} \tag{22.29}$$

where $\hat{H}$ is the (physical) Higgs field. The Lagrangian for the sector consisting of the gauge fields and the Higgs fields is

$$\mathcal{L}_{\mathrm{G\Phi}} = (\hat{D}_\mu \hat{\phi})^\dagger (\hat{D}^\mu \hat{\phi}) + \mu^2 \hat{\phi}^\dagger \hat{\phi} - \frac{\lambda}{4}(\hat{\phi}^\dagger \hat{\phi})^2 - \frac{1}{4}\hat{\boldsymbol{F}}_{\mu\nu} \cdot \hat{\boldsymbol{F}}^{\mu\nu} - \frac{1}{4}\hat{G}_{\mu\nu}\hat{G}^{\mu\nu}, \tag{22.30}$$

where $\hat{\boldsymbol{F}}_{\mu\nu}$ is the SU(2) field strength tensor (19.81) for the gauge fields $\hat{\boldsymbol{W}}^\mu$ and $\hat{G}_{\mu\nu}$ is the U(1) field strength tensor (19.82) for the gauge field B^μ, and $\hat{D}^\mu \hat{\phi}$ is given by (19.80). After symmetry breaking (i.e. the insertion of (22.29) in (22.30)) the quadratic parts of (22.30) can be written in unitary gauge as (see problem 19.9)

$$\hat{\mathcal{L}}_{\mathrm{G\Phi}}^{\mathrm{free}} = \frac{1}{2}\partial_\mu \hat{H} \partial^\mu \hat{H} - \mu^2 \hat{H}^2 \tag{22.31}$$

$$- \frac{1}{4}(\partial_\mu \hat{W}_{1\nu} - \partial_\nu \hat{W}_{1\mu})(\partial^\mu \hat{W}_1^\nu - \partial^\nu \hat{W}_1^\mu) + \frac{1}{8}g^2 v^2 \hat{W}_{1\mu}\hat{W}_1^\mu \tag{22.32}$$

$$- \frac{1}{4}(\partial_\mu \hat{W}_{2\nu} - \partial_\nu \hat{W}_{2\mu})(\partial^\mu \hat{W}_2^\nu - \partial^\nu \hat{W}_2^\mu) + \frac{1}{8}g^2 v^2 \hat{W}_{2\mu}\hat{W}_2^\mu \tag{22.33}$$

$$- \frac{1}{4}(\partial_\mu \hat{Z}_\nu - \partial_\nu \hat{Z}_\mu)(\partial^\mu \hat{Z}^\nu - \partial^\nu \hat{Z}^\mu) + \frac{v^2}{8}(g^2 + g'^2)\hat{Z}_\mu \hat{Z}^\mu \tag{22.34}$$

$$- \frac{1}{4}\hat{F}_{\mu\nu}\hat{F}^{\mu\nu} \tag{22.35}$$

where

$$\hat{Z}^\mu = \cos\theta_{\mathrm{W}} \hat{W}_3^\mu - \sin\theta_{\mathrm{W}} \hat{B}^\mu, \tag{22.36}$$

$$\hat{A}^\mu = \sin\theta_{\mathrm{W}} \hat{W}_3^\mu + \cos\theta_{\mathrm{W}} \hat{B}^\mu, \tag{22.37}$$

and

$$\hat{F}^{\mu\nu} = \partial^\mu A^\nu - \partial^\nu A^\mu, \tag{22.38}$$

with

$$\cos\theta_{\mathrm{W}} = g/(g^2 + g'^2)^{1/2}, \qquad \sin\theta_{\mathrm{W}} = g'/(g^2 + g'^2)^{1/2}. \tag{22.39}$$

Feynman rules for the vector boson propagators (in unitary gauge) and couplings, and for the Higgs couplings, can be read off from (22.30), and are given in Appendix Q.

Equations (22.31) – (22.35) give the tree-level masses of the Higgs boson and the gauge bosons: (22.31) tells us that the mass of the Higgs boson is

$$m_{\mathrm{H}} = \sqrt{2}\mu = \sqrt{\lambda}\, v/\sqrt{2}, \tag{22.40}$$

TABLE 22.1
Weak isospin and hypercharge assignments.

	t	t_3	y	Q
$\nu'_{eL},\ \nu'_{\mu L},\ \nu'_{\tau L}$	1/2	1/2	-1	0
$\nu'_{eR},\ \nu'_{\mu R},\ \nu'_{\tau R}$	0	0	0	0
$e_L,\ \mu_L,\ \tau_L$	1/2	-1/2	-1	-1
$e_R,\ \mu_R,\ \tau_R$	0	0	-2	-1
$u_L,\ c_L,\ t_L$	1/2	1/2	1/3	2/3
$u_R,\ c_R,\ t_R$	0	0	4/3	2/3
$d'_L,\ s'_L,\ b'_L$	1/2	-1/2	1/3	-1/3
$d'_R,\ s'_R,\ b'_R$	0	0	-2/3	-1/3
ϕ^+	1/2	1/2	1	1
ϕ^0	1/2	-1/2	1	0

where $v/\sqrt{2}$ is the (tree–level) Higgs vacuum value; (22.32) and (22.33) show that the charged W's have a mass

$$M_W = gv/2 \tag{22.41}$$

where g is the $SU(2)_L$ gauge coupling constant; (22.34) gives the mass of the Z^0 as

$$M_Z = M_W / \cos\theta_W \tag{22.42}$$

and (22.35) shows that the A^μ field describes a massless particle (to be identified with the photon).

Still unaccounted for are the *right–handed* chiral components of the fermion fields. There is at present no evidence for any weak interactions coupling to the right–handed field components, and it is therefore natural—and a basic assumption of the electroweak theory—that all 'R' components are singlets under the weak isospin group. Crucially, however, the 'R' components of the charged fermions do interact via the U(1) field $\hat{B}^\mu$; it is this that allows electromagnetism to emerge free of parity–violating γ_5 terms, as we shall see. With the help of the weak charge formula (equation (22.26)), we arrive at the assignments shown in table 22.1.

We have included 'R' components for the neutrinos in the table. It is, however, fair to say that in the original SM the neutrinos were taken to be massless, with only 'L' components and no neutrino mixing. We have seen in chapter 20 that it is for many purposes an excellent approximation to treat the neutrinos as massless, except when discussing neutrino oscillations. We shall mention their masses again in section 22.5.1, but for the moment we proceed in the 'massless neutrinos' approximation. In this case, there are *no* 'R' components for neutrinos, and no neutrino mixing.

We can now proceed to write down the *currents* of the electroweak theory. We will show that these dynamical symmetry currents are precisely the same as the phenomenological currents of the current-current model developed in chapter 20. The new feature here is that—as in the electromagnetic case—the currents interact with each other by the exchange of a gauge boson, rather than directly.

22.2.2 The leptonic currents (massless neutrinos): relation to current–current model

We write the $SU(2)_L \times U(1)$ covariant derivative, in terms of the fields $\hat{\boldsymbol{W}}^\mu$ and $\hat{B}^\mu$ of section 19.6, as

$$\hat{D}^\mu = \partial^\mu + ig\boldsymbol{\tau} \cdot \hat{\boldsymbol{W}}^\mu/2 + ig'y\hat{B}^\mu/2 \qquad \text{on 'L' } SU(2) \text{ doublets} \tag{22.43}$$

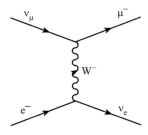

FIGURE 22.10
W-exchange process in $\nu_\mu + e^- \to \mu^- + \nu_e$.

and as

$$\hat{D}^\mu = \partial^\mu + ig'y\hat{B}^\mu/2 \qquad \text{on 'R' SU(2) singlets.} \tag{22.44}$$

The leptonic couplings to the gauge fields therefore arise from the 'gauge–covariantized' free leptonic Lagrangian:

$$\hat{\mathcal{L}}_{\text{lept}} = \sum_{f=e,\mu,\tau} \bar{\hat{l}}_{fL} i\hat{\slashed{D}}\hat{l}_{fL} + \sum_{f=e,\mu,\tau} \bar{\hat{l}}_{fR} i\hat{\slashed{D}}\hat{l}_{fR}, \tag{22.45}$$

where the $\hat{l}_{fL}$ are the left–handed doublets

$$\hat{l}_{fL} = \begin{pmatrix} \hat{\nu}_f \\ \hat{f}^- \end{pmatrix}_L \tag{22.46}$$

and $\hat{l}_{fR}$ are the singlets $\hat{l}_{eR} = \hat{e}_R$ etc.

Consider first the *charged leptonic currents*. The correct normalization for the charged fields is that $\hat{W}^\mu \equiv (\hat{W}_1^\mu - i\hat{W}_2^\mu)/\sqrt{2}$ destroys the W^+ or creates the W^- (cf (7.15)). The '$\boldsymbol{\tau} \cdot \hat{\boldsymbol{W}}/2$' terms can be written as

$$\boldsymbol{\tau} \cdot \hat{\boldsymbol{W}}^\mu/2 = \frac{1}{\sqrt{2}}\left\{ \tau_+ \frac{(\hat{W}_1^\mu - i\hat{W}_2^\mu)}{\sqrt{2}} + \tau_- \frac{(\hat{W}_1^\mu + i\hat{W}_2^\mu)}{\sqrt{2}} \right\} + \frac{\tau_3}{2}W_3^\mu, \tag{22.47}$$

where $\tau_\pm = (\tau_1 \pm i\tau_2)/2$ are the usual raising and lowering operators for the doublets. Thus the '$f=e$' contribution to the first term in (22.45) picks out the process $e^- \to \nu_e + W^-$ for example, with the result that the corresponding vertex is given by

$$-\frac{ig}{\sqrt{2}}\gamma^\mu\frac{(1-\gamma_5)}{2}. \tag{22.48}$$

The 'universality' of the single coupling constant 'g' ensures that (22.48) is also the amplitude for the $\mu-\nu_\mu-W$ and $\tau-\nu_\tau-W$ vertices. Thus the amplitude for the $\nu_\mu+e^- \to \mu^-+\nu_e$ process considered in section 20.8 is

$$\left\{ -\frac{ig}{\sqrt{2}}\bar{u}(\mu)\gamma_\mu\frac{(1-\gamma_5)}{2}u(\nu_\mu) \right\} \frac{i[-g^{\mu\nu} + k^\mu k^\nu/M_W^2]}{k^2 - M_W^2} \left\{ -i\frac{g}{\sqrt{2}}\bar{u}(\nu_e)\gamma_\nu\frac{(1-\gamma_5)}{2}u(e) \right\} \tag{22.49}$$

corresponding to the Feynman graph of figure 22.10.

For $k^2 \ll M_W^2$ we can replace the W–propagator by the constant value $g^{\mu\nu}/M_W^2$, leading to the amplitude

$$-\frac{ig^2}{8M_W^2}\bar{u}(\mu)\gamma_\mu(1-\gamma_5)u(\nu_\mu)\bar{u}(\nu_e)\gamma^\mu(1-\gamma_5)u(e), \tag{22.50}$$

which may be compared with the form we used in the current–current theory, equation (20.50). This comparison gives

$$\frac{G_{\mathrm{F}}}{\sqrt{2}} = \frac{g^2}{8M_{\mathrm{W}}^2}.$$

(22.51)

This is an important equation, giving the precise version, in the GSW theory, of the qualitative relation $g^2/M_{\mathrm{W}}^2 \sim G_{\mathrm{F}}$ introduced following equation (22.20), and in volume 1, at equation (1.32).

Putting together (22.41) and (22.51) we can deduce

$$G_{\mathrm{F}}/\sqrt{2} = 1/(2v^2)$$

(22.52)

so that from the known value (22.4) of G_{F} there follows the value of v:

$$v \simeq 246 \text{ GeV}.$$

(22.53)

Alternatively we may quote $v/\sqrt{2}$ (the vacuum value of the Higgs field) :

$$v/\sqrt{2} \simeq 174 \text{ GeV}.$$

(22.54)

This parameter sets the scale of electroweak symmetry breaking, but as yet no theory is able to predict its value. It is related to the parameters λ, μ of (22.30) by $v/\sqrt{2} = \sqrt{2}\mu/\lambda^{1/2}$ (cf (17.99)).

In general, the charge–changing part of (22.45) can be written as

$$-\frac{g}{\sqrt{2}}\left\{\bar{\hat{\nu}}_e\gamma^\mu\frac{(1-\gamma_5)}{2}\hat{e} + \bar{\hat{\nu}}_\mu\gamma^\mu\frac{(1-\gamma_5)}{2}\hat{\mu} + \bar{\hat{\nu}}_\tau\gamma^\mu\frac{(1-\gamma_5)}{2}\hat{\tau}\right\}\hat{W}_\mu$$

+hermitian conjugate,

(22.55)

where $\hat{W}^\mu = (\hat{W}_1^\mu - i\hat{W}_2^\mu)/\sqrt{2}$. (22.55) has the form

$$-\hat{j}_{\mathrm{CC}}^\mu(\mathrm{leptons})\hat{W}_\mu - j_{\mathrm{CC}}^{\mu\dagger}(\mathrm{leptons})\hat{W}_\mu^\dagger$$

(22.56)

where the *leptonic weak charged current* $\hat{j}_{\mathrm{CC}}^\mu(\mathrm{leptons})$ is precisely that used in the current–current model (equation (20.38)), up to the usual factors of g's and $\sqrt{2}$'s. Thus the dynamical symmetry currents of the $\mathrm{SU(2)_L}$ gauge theory are exactly the 'phenomenological' currents of the earlier current–current model. The Feynman rules for the lepton–W couplings (Appendix Q) can be read off from (22.55).

Turning now to the *leptonic weak neutral current*, this will appear via the couplings to the Z^0, written as

$$-\hat{j}_{\mathrm{NC}}^\mu(\mathrm{leptons})\hat{Z}_\mu.$$

(22.57)

Referring to (22.36) for the linear combination of $\hat{W}_3^\mu$ and $\hat{B}^\mu$ which represents $\hat{Z}^\mu$, we find (problem 22.4)

$$\hat{j}_{\mathrm{NC}}^\mu(\mathrm{leptons}) = \frac{g}{\cos\theta_{\mathrm{W}}}\sum_l \bar{\hat{\psi}}_l\gamma^\mu\left[t_3^l\left(\frac{1-\gamma_5}{2}\right) - \sin^2\theta_{\mathrm{W}}Q_l\right]\hat{\psi}_l,$$

(22.58)

where the sum is over the six lepton fields $\nu_e, e^-, \nu_\mu, \ldots \tau^-$. For the $Q = 0$ neutrinos with $t_3 = +\frac{1}{2}$,

$$\hat{j}_{\mathrm{NC}}^\mu(\mathrm{neutrinos}) = \frac{g}{2\cos\theta_{\mathrm{W}}}\sum_l \bar{\hat{\nu}}_l\gamma^\mu\frac{(1-\gamma_5)}{2}\hat{\nu}_l,$$

(22.59)

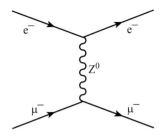

FIGURE 22.11
Z^0-exchange process in $e^-\mu^- \to e^-\mu^-$.

where now $l = e, \mu, \tau$. For the other (negatively charged) leptons, we shall have both L and R couplings from (22.58), and we can write

$$\hat{j}^\mu_{NC}(\text{charged leptons}) = \frac{g}{\cos\theta_W} \sum_{l=e,\mu,\tau} \bar{\hat{l}}\gamma^\mu \left[c^l_L \left(\frac{1-\gamma_5}{2}\right) + c^l_R \left(\frac{1+\gamma_5}{2}\right) \right] \hat{l}, \qquad (22.60)$$

where

$$c^l_L = t^l_3 - \sin^2\theta_W Q_l = -\frac{1}{2} + \sin^2\theta_W \qquad (22.61)$$

$$c^l_R = -\sin^2\theta_W Q_l = \sin^2\theta_W. \qquad (22.62)$$

As noted earlier, the Z^0 coupling is not pure 'V–A'. These relations (22.59) – (22.62) are exactly the ones given earlier, in (20.85) – (20.87); in particular, the couplings are independent of 'l' and hence exhibit lepton universality. The alternative notation

$$\hat{j}^\mu_{NC}(\text{charged leptons}) = \frac{g}{2\cos\theta_W} \sum_l \bar{\hat{l}}\gamma^\mu (g^l_V - g^l_A\gamma_5)\hat{l} \qquad (22.63)$$

is often used, where

$$g^l_V = -\frac{1}{2} + 2\sin^2\theta_W \qquad g^l_A = -\frac{1}{2}, \text{independent of } l. \qquad (22.64)$$

Note that g^l_V vanishes for $\sin^2\theta_W = 0.25$. Again, the Feynman rules for lepton-Z couplings (Appendix Q) are contained in (22.59) and (22.60).

As in the case of W-mediated charge–charging processes, Z^0-mediated processes reduce to the current–current form at low k^2. For example, the amplitude for $e^-\mu^- \to e^-\mu^-$ via Z^0 exchange (figure 22.11) reduces to

$$-\frac{ig^2}{4\cos^2\theta_W M_Z^2} \quad \bar{u}(e)\gamma_\mu[c^l_L(1-\gamma_5) + c^l_R(1+\gamma_5)]u(e)\bar{u}(\mu)\gamma^\mu$$

$$\times[c^l_L(1-\gamma_5) + c^l_R(1+\gamma_5)]u(\mu). \qquad (22.65)$$

It is customary to define the parameter

$$\rho = M_W^2/(M_Z^2\cos^2\theta_W), \qquad (22.66)$$

which is unity at tree–level, in the absence of loop corrections. The ratio of factors in front of the $\bar{u} \ldots u$ expressions in (22.65) and (22.50) (i.e. 'neutral current process'/'charged current process') is then 2ρ.

We may also check the electromagnetic current in the theory, by looking for the piece that couples to $\hat{A}^{\mu}$. We find

$$\hat{j}^{\mu}_{\text{emag}} = -g \sin\theta_{\text{W}} \sum_{l=\text{e},\mu,\tau} \bar{\hat{l}}\gamma^{\mu}\hat{l} \tag{22.67}$$

which allows us to identify the electromagnetic charge e as

$$e = g \sin\theta_{\text{W}} \tag{22.68}$$

as already suggested in (19.98) of chapter 19. Note that all the γ_5's cancel from (22.67), as is of course required.

22.2.3   The quark currents

The charge–changing quark currents, which are coupled to the $W^{\pm}$ fields, have a form very similar to that of the charged leptonic currents, except that the $t_3 = -\frac{1}{2}$ components of the L–doublets have to be understood as the flavour–mixed (weakly interacting) states

$$\begin{pmatrix} \hat{d}' \\ \hat{s}' \\ \hat{b}' \end{pmatrix}_{\text{L}} = \begin{pmatrix} V_{\text{ud}} & V_{\text{us}} & V_{\text{ub}} \\ V_{\text{cd}} & V_{\text{cs}} & V_{\text{cb}} \\ V_{\text{td}} & V_{\text{ts}} & V_{\text{tb}} \end{pmatrix} \begin{pmatrix} \hat{d} \\ \hat{s} \\ \hat{b} \end{pmatrix}_{\text{L}}, \tag{22.69}$$

where $\hat{d}, \hat{s}$, and $\hat{b}$ are the strongly interacting fields with masses $m_{\text{d}}, m_{\text{s}}$, and m_{b}, and the V–matrix is the CKM matrix used extensively in chapter 21. We shall discuss this matrix further in section 22.5.2. Thus the *charge–changing weak quark current* is

$$\hat{j}^{\mu}_{\text{CC}}(\text{quarks}) = \frac{g}{\sqrt{2}} \left\{ \bar{\hat{u}}\gamma^{\mu}\frac{(1-\gamma_5)}{2}\hat{d}' + \bar{\hat{c}}\gamma^{\mu}\frac{(1-\gamma_5)}{2}\hat{s}' + \bar{\hat{t}}\gamma^{\mu}\frac{(1-\gamma_5)}{2}\hat{b}' \right\}, \tag{22.70}$$

which generalizes (20.90) to three generations and supplies the factor $g/\sqrt{2}$, as for the leptons.

The neutral currents are diagonal in flavour if the matrix V is unitary (see also section 22.5.2). Thus $\hat{j}^{\mu}_{\text{NC}}(\text{quarks})$ will be given by the same expression as (20.103), except that now the sum will be over all six quark flavours. The *neutral weak quark current* is thus

$$\hat{j}^{\mu}_{\text{NC}}(\text{quarks}) = \frac{g}{\cos\theta_{\text{W}}} \sum_{q} \bar{\hat{q}}\gamma^{\mu} \left[c^q_{\text{L}}\frac{(1-\gamma_5)}{2} + c^q_{\text{R}}\frac{(1+\gamma_5)}{2} \right] \hat{q}, \tag{22.71}$$

where

$$c^q_{\text{L}} = t^q_3 - \sin^2\theta_{\text{W}}Q_q \tag{22.72}$$
$$c^q_{\text{R}} = -\sin^2\theta_{\text{W}}Q_q. \tag{22.73}$$

These expressions are exactly as given in (20.103) – (20.105). As for the charged leptons, we can alternatively write (22.71) as

$$\hat{j}^{\mu}_{\text{NC}}(\text{quarks}) = \frac{g}{2\cos\theta_{\text{W}}} \sum_{q} \bar{\hat{q}}\gamma^{\mu}(g^q_{\text{V}} - g^q_{\text{A}}\gamma_5)\hat{q}, \tag{22.74}$$

where

$$g_V^q = t_3^q - 2\sin^2\theta_W Q_q \tag{22.75}$$
$$g_A^q = t_3^q. \tag{22.76}$$

Before proceeding to discuss some simple phenomenological consequences, we remind the reader of one important feature of the SM currents in general. Reading (22.24) and (22.25) together 'vertically', the leptons and quarks are grouped in three *generations*, each with two leptons and two quarks. The theoretical motivation for such family grouping is that *anomalies* are cancelled within each complete generation, as discussed in section 18.4.

22.3   Simple (tree-level) predictions

The theory as so far developed has just 4 parameters: the gauge couplings g and g', and the parameters λ and μ of the Higgs potential. The previous two subsections show that all the couplings to fermions can be written in terms of the known quantities G_F and e (or α), and one free parameter which may be taken to be $\sin\theta_W$. We noted in section 20.9 that, before the discovery of the W and Z particles, the then known neutrino data were consistent with a single value of θ_W given by $\sin^2\theta_W \simeq 0.23$. Using (22.51) and (22.68), it was then possible to predict the value of M_W:

$$M_W = \left(\frac{\pi\alpha}{\sqrt{2}G_F}\right)^{1/2}\frac{1}{\sin\theta_W} \simeq \frac{37.28}{\sin\theta_W}\,\text{GeV} \simeq 77.73\,\text{GeV}. \tag{22.77}$$

Similarly, using (22.42) we predict

$$M_Z = M_W/\cos\theta_W \simeq 88.58\,\text{GeV}. \tag{22.78}$$

These predictions of the theory (at lowest order) indicate the power of the underlying symmetry to tie together many apparently unrelated quantities, which are all determined in terms of only a few basic parameters. We now present a number of other simple tree-level predictions.

The width for $W^- \to e^- + \bar{\nu}_e$ can be calculated using the vertex (22.48), with the result (problem 22.5)

$$\Gamma(W^- \to e^-\bar{\nu}_e) = \frac{1}{12}\frac{g^2}{4\pi}M_W = \frac{G_F}{2^{1/2}}\frac{M_W^3}{6\pi} \simeq 205\,\text{MeV}, \tag{22.79}$$

using (22.77). The widths to $\mu^-\bar{\nu}_\mu, \tau^-\bar{\nu}_\tau$ are the same. Neglecting CKM flavour mixing among the two energetically allowed quark channels $\bar{u}d$ and $\bar{c}s$, their widths would also be the same, apart from a factor of 3 for the different colour channels. The total W width for all these channels will therefore be about nine times the value in (22.79), i.e. 1.85 GeV, while the branching ratio for $W \to e\nu$ is

$$B(e\nu) = \Gamma(W \to e\nu)/\Gamma(\text{total}) \simeq 11\%. \tag{22.80}$$

In making these estimates we have neglected all fermion masses.

The width for $Z^0 \to \nu\bar{\nu}$ can be found from (22.79) by replacing $g/2^{1/2}$ by $g/2\cos\theta_W$, and M_W by M_Z, giving

$$\Gamma(Z^0 \to \nu\bar{\nu}) = \frac{1}{24}\frac{g^2}{4\pi}\frac{M_Z}{\cos^2\theta_W} = \frac{G_F}{2^{1/2}}\frac{M_Z^3}{12\pi} \simeq 152\,\text{MeV}, \tag{22.81}$$

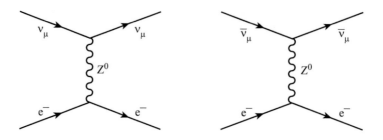

FIGURE 22.12
Neutrino-electron graphs involving Z^0 exchange.

using (22.78). Charged lepton pairs couple with both c_L^l and c_R^l terms, leading (with neglect of lepton masses) to

$$\Gamma(Z^0 \to l\bar{l}) = \left(\frac{|c_L^l|^2 + |c_R^l|^2}{6} \right) \frac{g^2}{4\pi} \frac{M_Z}{\cos^2 \theta_W}. \tag{22.82}$$

The values $c_L^\nu = \frac{1}{2}, c_R^\nu = 0$ in (22.82) reproduce (22.81). With $\sin^2 \theta_W \simeq 0.23$, we find

$$\Gamma(Z^0 \to l\bar{l}) \simeq 76.5 \text{ MeV}. \tag{22.83}$$

Quark pairs couple as in (22.71), the GIM mechanism ensuring that all flavour–changing terms cancel. The total width to $u\bar{u}, d\bar{d}, c\bar{c}, s\bar{s}$, and $b\bar{b}$ channels (allowing 3 for colour and neglecting masses) is then 1538 MeV, producing an estimated total width of approximately 2.22 GeV. (QCD corrections will increase these estimates by a factor of order 1.1). The branching ratio to charged leptons is approximately 3.4%, to the three (invisible) neutrino channels 20.5%, and to hadrons (via hadronization of the $q\bar{q}$ channels) about 69.3%. In section 22.4.3 we shall see how a precise measurement of the total Z^0 width at LEP determined the number of light neutrinos to be 3.

Cross sections for neutrino–lepton scattering proceeding via Z^0 exchange can be calculated (for $k^2 \ll M_Z^2$) using the currents (22.59) and (22.60), and the method of section 20.5. Examples are

$$\nu_\mu e^- \to \nu_\mu e^- \tag{22.84}$$

and

$$\bar{\nu}_\mu e^- \to \bar{\nu}_\mu e^- \tag{22.85}$$

as shown in figure 22.12. Since the neutral current for the electron is not pure V–A, as was the charged current, we expect to see terms involving both $|c_L^l|^2$ and $|c_R^l|^2$, and possibly an interference term. The cross section for (22.84) is found to be ('t Hooft 1971c)

$$d\sigma/dy = (2G_F^2 E m_e/\pi)[|c_L^l|^2 + |c_R^l|^2(1-y)^2 - \frac{1}{2}(c_R^{l\,*} c_L^l + c_L^{l\,*} c_R^l)y m_e/E], \tag{22.86}$$

where E is the energy of the incident neutrino in the 'laboratory' system, and $y = (E-E')/E$ as before, where E' is the energy of the outgoing neutrino in the 'laboratory' system[2]. Equation (22.86) may be compared with the $\nu_\mu e^- \to \mu^- \nu_e$ (charged current) cross section of (20.84) by noting that $t = -2m_e E y$: the $|c_L^l|^2$ term agrees with the pure V–A result (20.84), while the $|c_R^l|^2$ term involves the same $(1-y)^2$ factor discussed for $\nu\bar{q}$ scattering

---

[2]In the kinematics, lepton masses have been neglected wherever possible.

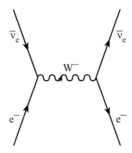

FIGURE 22.13
One-W annihilation graph in $\bar{\nu}_e e^- \to \bar{\nu}_e e^-$.

in section 20.7.2. The interference term is negligible for $E \gg m_e$. The cross section for the antineutrino process (22.85) is found from (22.86) by interchanging c_L^l and c_R^l.

A third neutrino–lepton process is experimentally available,

$$\bar{\nu}_e e^- \to \bar{\nu}_e e^-, \tag{22.87}$$

the cross section for which was measured by Reines, Gurr, and Sobel (1976), using electron anti-neutrinos from an 1800-MW fission reactor at Savannah River. In this case there is a single W intermediate state graph, shown in figure 22.13, to consider as well as the Z^0 one; the latter is similar to the right hand graph in figure 22.12, but with $\bar{\nu}_\mu$ replaced by $\bar{\nu}_e$. The cross section for (22.87) turns out to be given by an expression of the form (22.86), but with the replacements

$$c_L^l \to \frac{1}{2} + \sin^2\theta_W, c_R^l \to \sin^2\theta_W. \tag{22.88}$$

Reines, Gurr, and Sobel reported the result $\sin^2\theta_W = 0.29 \pm 0.05$.

We emphasize once more that all these cross sections are determined in terms of G_F, α and only one further parameter, $\sin^2\theta_W$. As mentioned in section 20.9, experimental fits to these predictions are reviewed by Commins and Bucksbaum (1983), Renton (1990), and Winter (2000).

Particularly precise initial determinations of the SM parameters were made at the e^+e^- colliders, LEP, and SLC. Consider the reaction $e^+e^- \to f\bar{f}$ where f is μ or τ, at energies where the lepton masses may be neglected in the final answers. In lowest order, the process is mediated by both γ–exchange and Z^0–exchange as shown in figure 22.14. Calculations of the cross section were made early on, by Budny (1975) for example. In modern notation, the differential cross section for the scattering of unpolarized e^- and e^+ is given by

$$\frac{d\sigma}{d\cos\theta} = \frac{\pi\alpha^2}{2s}[(1 + \cos^2\theta)A + \cos\theta B] \tag{22.89}$$

where θ is the CM scattering angle of the final state lepton, $s = (p_{e^-} + p_{e^+})^2$, and

$$A = 2g_V^e g_V^f \text{Re}\chi(s) + [(g_A^e)^2 + (g_V^e)^2][(g_A^f)^2 + (g_V^f)^2]|\chi(s)|^2 \tag{22.90}$$

$$B = 4g_A^e g_A^f \text{Re}\chi(s) + 8g_A^e g_V^e g_A^f g_V^f |\chi(s)|^2 \tag{22.91}$$

$$\chi(s) = s/[4\sin^2\theta_W \cos^2\theta_W(s - M_Z^2 + i\Gamma_Z M_Z)]. \tag{22.92}$$

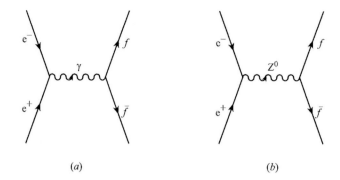

FIGURE 22.14

(a) One-γ and (b) one-W annihilation graphs in $e^+e^- \to f\bar{f}$.

Notice that the term surviving when all the g's are set to zero, which is therefore the pure single photon contribution, is exactly as calculated in problem 8.18. The presence of the $\cos\theta$ term leads to the forward–backward asymmetry noted in that problem.

The forward–backward asymmetry A_{FB} may be defined as

$$A_{\mathrm{FB}} \equiv (N_{\mathrm{F}} - N_{\mathrm{B}})/(N_{\mathrm{F}} + N_{\mathrm{B}}), \tag{22.93}$$

where N_{F} is the number scattered into the forward hemisphere $0 \leq \cos\theta \leq 1$, and N_{B} that into the backward hemisphere $-1 \leq \cos\theta \leq 0$. Integrating (22.89) one easily finds

$$A_{\mathrm{FB}} = 3B/8A. \tag{22.94}$$

For $\sin^2\theta_{\mathrm{W}} = 0.25$ we noted after (22.64) that the g_{V}^l's vanish, so they are very small for $\sin^2\theta_{\mathrm{W}} \simeq 0.23$. The effect is therefore controlled essentially by the first term in (22.91). At $\sqrt{s} = 29$ GeV, for example, the asymmetry is $A_{\mathrm{FB}} \simeq -0.063$.

This asymmetry was observed in experiments with PETRA at DESY and with PEP at SLAC (see figure 8.20(b)). These measurements, made at energies well below the Z^0 peak, were the first indication of the presence of Z^0 exchange in e^+e^- collisions.

However, QED alone produces a small positive A_{FB}, through interference between 1γ and 2γ annihilation processes (which have different charge conjugation parity), as well as between initial and final state bremsstrahlung corrections to figure 22.14(a). Indeed, *all* one-loop radiative effects must clearly be considered, in any comparison with modern high precision data.

At the CERN e^+e^- collider LEP, many such measurements were made 'on the Z peak', i.e. at $s = M_{\mathrm{Z}}^2$ in the parametrization (22.92). In that case, $\mathrm{Re}\chi(s) = 0$, and (22.94) becomes (neglecting the photon contribution)

$$A_{\mathrm{FB}}(Z^0 \text{ peak}) = \frac{3g_{\mathrm{A}}^{\mathrm{e}} g_{\mathrm{V}}^{\mathrm{e}} g_{\mathrm{A}}^f g_{\mathrm{V}}^f}{\{[(g_{\mathrm{A}}^{\mathrm{e}})^2 + (g_{\mathrm{V}}^{\mathrm{e}})^2][(g_{\mathrm{A}}^f)^2 + (g_{\mathrm{V}}^f)^2]\}}. \tag{22.95}$$

Another important asymmetry observable is that involving the difference of the cross sections for left–and right–handed incident electrons:

$$A_{\mathrm{LR}} \equiv (\sigma_{\mathrm{L}} - \sigma_{\mathrm{R}})/(\sigma_{\mathrm{L}} + \sigma_{\mathrm{R}}), \tag{22.96}$$

for which the tree-level prediction is

$$A_{\mathrm{LR}} = 2g_{\mathrm{V}}^{\mathrm{e}} g_{\mathrm{A}}^{\mathrm{e}}/[(g_{\mathrm{V}}^{\mathrm{e}})^2 + (g_{\mathrm{A}}^{\mathrm{e}})^2]. \tag{22.97}$$

This quantity is very sensitive to $\sin^2\theta_W$, and systematic uncertainties largely cancel. A similar combination of the g's for the final state leptons can be measured by forming the 'L–R F–B' asymmetry

$$A_{\mathrm{LR}}^{\mathrm{FB}} = [(\sigma_{\mathrm{LF}} - \sigma_{\mathrm{LB}}) - (\sigma_{\mathrm{RF}} - \sigma_{\mathrm{RB}})]/(\sigma_{\mathrm{R}} + \sigma_{\mathrm{L}}) \tag{22.98}$$

for which the tree-level prediction is

$$A_{\mathrm{LR}}^{\mathrm{FB}} = 2g_{\mathrm{V}}^f g_{\mathrm{A}}^f/[(g_{\mathrm{V}}^f)^2 + (g_{\mathrm{A}}^f)^2]. \tag{22.99}$$

The quantity on the right-hand side of (22.99) is usually denoted by A_f:

$$A_f = 2g_{\mathrm{V}}^f g_{\mathrm{A}}^f/[(g_{\mathrm{V}}^f)^2 + (g_{\mathrm{A}}^f)^2]. \tag{22.100}$$

More precise measurements of A_{LR} were made by the SLD Collaboration at the SLC (Abe *et al.* 2000a). SLD also extracted the quantities A_{b}, A_{c} (Schael *et al.* 2006), A_{s} (Abe *et al.* 2000b), A_τ and A_μ (Abe *et al.* 2001) from the L-R F-B asymmetry. These measurements provide important constraints on the SM, and their precision is such that comparison with theory requires the inclusion of higher-order corrections. As we shall indicate in section 22.6, these can be well approximated by replacing the couplings g_{V}^f and g_{A}^f by barred quantities, defined in (22.175) and (22.176).

The asymmetry A_{FB} is not, in fact, direct evidence for parity violation in $e^+e^- \to \mu^+\mu^-$, since we see from (22.90) and (22.91) that it is even under $g_{\mathrm{A}}^l \to -g_{\mathrm{A}}^l$, whereas a true parity–violating effect would involve terms odd (linear) in g_{A}^l. However, electroweak–induced parity violation effects in an apparently electromagnetic process were observed in a remarkable experiment by Prescott *et al.* (1978). Longitudinally polarized electrons were inelastically scattered from deuterium, and the flux of scattered electrons was measured for incident electrons of definite helicity. An asymmetry between the results, depending on the helicities, was observed—a clear signal for parity violation. This was the first demonstration of parity–violating effects in an 'electromagnetic' process; the corresponding value of $\sin^2\theta_W$ was in agreement with that determined from ν data.

We now turn to some of the main experimental evidence, beginning with the discoveries of the $W^\pm$ and Z^0 1983.

22.4 The discovery of the $W^\pm$ and Z^0 at the CERN $p\bar{p}$ collider

22.4.1 Production cross sections for W and Z in $p\bar{p}$ colliders

The possibility of producing the predicted $W^\pm$ and Z^0 particles was the principal motivation for transforming the CERN SPS into a $p\bar{p}$ collider using the stochastic cooling technique (Rubbia *et al.* 1977, Staff of the CERN $\bar{p}p$ project 1981). Estimates of W and Z^0 production in $\bar{p}p$ collisions may be obtained (see, for example, Quigg 1977) from the parton model, in a way analogous to that used for the Drell–Yan process in section 9.4 with γ replaced by W or Z^0, as shown in figure 22.15 (cf figure 9.11), and for two–jet cross sections in section 14.3.2. As in (14.51), we denote by $\hat{s}$ the subprocess invariant

$$\hat{s} = (x_1 p_1 + x_2 p_2)^2 = x_1 x_2 s \tag{22.101}$$

for massless partons. With $\hat{s}^{1/2} = M_W \sim 80$ GeV, and $s^{1/2} = 630$ GeV for the $\bar{p}p$ collider energy, we see that the x's are typically ~ 0.13, so that the valence q's in the proton and $\bar{q}$'s

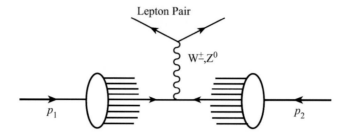

FIGURE 22.15
Parton model amplitude for $W^\pm$ or Z^0 production in $\bar{p}p$ collisions.

in the antiproton will dominate (at $\sqrt{s} = 1.8$ TeV, appropriate to the Fermilab Tevatron, $x \simeq 0.04$ and the sea quarks contribute). The parton model cross section $p\bar{p} \to W^\pm +$ anything is then (setting $V_{ud} = 1$ and all other $V_{ij} = 0$)

$$\sigma(p\bar{p} \to W^\pm + X) = \frac{1}{3} \int_0^1 dx_1 \int_0^1 dx_2 \hat{\sigma}(x_1, x_2) \left\{ \begin{array}{c} u(x_1)\bar{d}(x_2) + \bar{d}(x_1)u(x_2) \\ \bar{u}(x_1)d(x_2) + d(x_1)\bar{u}(x_2) \end{array} \right\} \quad (22.102)$$

where the $\frac{1}{3}$ is the same colour factor as in the Drell–Yan process, and the subprocess cross section $\hat{\sigma}$ for $q\bar{q} \to W^\pm + X$ is (neglecting the $W^\pm$ width)

$$\hat{\sigma} = 4\pi^2\alpha(1/4\sin^2\theta_W)\delta(\hat{s} - M_W^2) \quad (22.103)$$
$$= \pi 2^{1/2} G_F M_W^2 \delta(x_1 x_2 s - M_W^2). \quad (22.104)$$

QCD corrections to (22.102) must as usual be included. Leading logarithms will make the distributions Q^2–dependent, and they should be evaluated at $Q^2 = M_W^2$. There will be further $(O(\alpha_s^2))$ corrections, which are often accounted for by a multiplicative factor 'K', which is of order 1.5–2 at these energies. $O(\alpha_s^2)$ calculations are presented in Hamberg *et al.* (1991) and by van der Neerven and Zijlstra (1992); see also Ellis *et al.* (1996) section 9.4. The total cross section for production of W^+ and W^- at $\sqrt{s} = 630$ GeV is then of order 6.5 nb, while a similar calculation for the Z^0 gives about 2 nb. Multiplying these by the branching ratios gives

$$\sigma(p\bar{p} \to W + X \to e\nu X) \simeq 0.7 \text{ nb} \quad (22.105)$$
$$\sigma(p\bar{p} \to Z^0 + X \to e^+e^- X) \simeq 0.07 \text{ nb} \quad (22.106)$$

at $\sqrt{s} = 630$ GeV.

The total cross section for $p\bar{p}$ is about 70 mb at these energies. Hence (22.105) represents $\sim 10^{-8}$ of the total cross section, and (22.106) is 10 times smaller. The rates could, of course, be increased by using the $q\bar{q}$ modes of W and Z^0, which have bigger branching ratios. But the detection of these is very difficult, being very hard to distinguish from conventional two–jet events produced via the mechanism discussed in section 14.3.2, which has a cross section some 10^3 higher than (22.105). W and Z^0 would appear as slight shoulders on the edge of a very steeply falling invariant mass distribution, similar to that shown in figure 9.12, and the calorimetric jet energy resolution capable of resolving such an effect is hard to achieve. Thus despite the unfavourable branching ratios, the leptonic modes provide the better signatures, as discussed further in section 22.4.3.

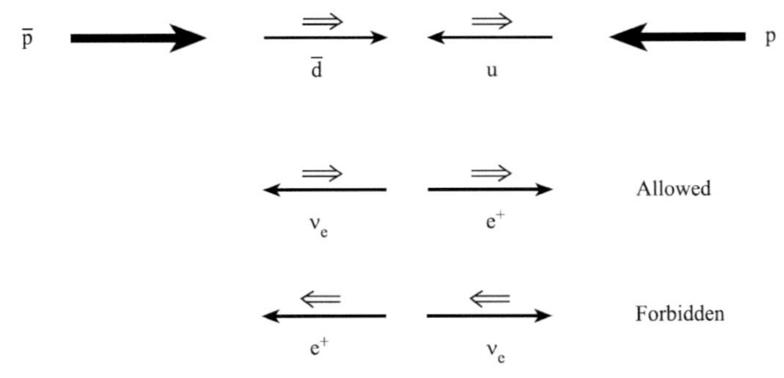

FIGURE 22.16
Preferred direction of leptons in W^+ decay.

22.4.2   Charge asymmetry in $W^{\pm}$ decay

At energies such that the simple valence quark picture of (22.102) is valid, the W^+ is created in the annihilation of a left–handed u quark from the proton and a right-handed $\bar{d}$ quark from the $\bar{p}$ (neglecting fermion masses). In the $W^+ \to e^+\nu_e$ decay, a right–handed e^+ and left–handed ν_e are emitted. Referring to figure 22.16, we see that angular momentum conservation allows e^+ production parallel to the direction of the antiproton, but forbids it parallel to the direction of the proton. Similarly, in $W^- \to e^-\bar{\nu}_e$, the e^- is emitted preferentially parallel to the proton (these considerations are exactly similar to those mentioned in section 20.7.2 with reference to νq and $\bar{\nu}$q scattering). The actual distribution has the form $\sim (1 + \cos\theta_e^*)^2$, where θ_e^* is the angle, in the rest frame of the W, between the e^- and the p (for $W^- \to e^-\bar{\nu}_e$) or the e^+ and the $\bar{p}$ (for $W^+ \to e^+\nu_e$).

22.4.3   Discovery of the $W^{\pm}$ and Z^0 at the $p\bar{p}$ collider and their properties

As already indicated in section 22.4.1, the best signatures for W and Z production in $p\bar{p}$ collisions are provided by the leptonic modes

$$p\bar{p} \to W^{\pm}X \to e^{\pm}\nu X \tag{22.107}$$

$$p\bar{p} \to Z^0X \to e^+e^-X. \tag{22.108}$$

Reaction (22.107) has the larger cross section, by a factor of 10 (cf (22.105) and (22.106)), and was observed first (UA1, Arnison *et al.* 1983a; UA2, Banner *et al.* 1983). However, the kinematics of (22.108) is simpler and so the Z^0 discovery (UA1, Arnison *et al.* (1983b); UA2, Bagnaia *et al.* 1983) will be discussed first.

The signature for (22.108) is an isolated, and approximately back–to–back, e^+e^- pair with invariant mass peaked around 90 GeV (cf (22.78)). Very clean events can be isolated by imposing a modest transverse energy cut—the e^+e^- pairs required are coming from the decay of a massive relatively slowly moving Z^0. Figure 22.17 shows the transverse energy distribution of a candidate Z^0 event from the first UA2 sample. Figure 22.18 shows (Geer 1986) the invariant mass distribution for a later sample of 14 UA1 events in which both electrons have well measured energies, together with the Breit–Wigner resonance curve appropriate to $M_Z = 93$ GeV/c^2, with experimental mass resolution folded in. The UA1 result for the Z^0 mass was

$$M_Z = 93.0 \pm 1.4(\text{stat}) \pm 3.2(\text{syst.})\ \text{GeV.} \tag{22.109}$$

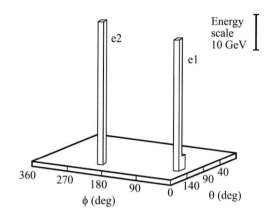

FIGURE 22.17

The cell transverse energy distribution for a $Z^0 \to e^+e^-$ event (UA2, Bagnaia *et al.* 1983) in the θ and ϕ plane, where θ and ϕ are the polar and azimuth angles relative to the beam axis.

The corresponding UA2 result (DiLella 1986), based on 13 well measured pairs, was

$$M_Z = 92.5 \pm 1.3(\text{stat.}) \pm 1.5(\text{syst.}) \quad \text{GeV.} \tag{22.110}$$

In both cases the systematic error reflects the uncertainty in the absolute calibration of the calorimeter energy scale. The agreement with (22.78) was acceptable, but there was a suggestion that the tree–level prediction was on the low side.

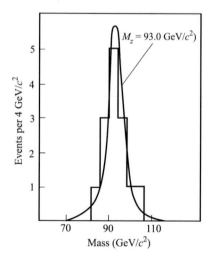

FIGURE 22.18

Invariant mass distribution for 14 well measured $Z^0 \to e^+e^-$ decays (UA1). Figure reprinted with permission from S Geer in *High Energy Physics 1985, Proc. Yale Theoretical Advanced Study Institute*, eds M J Bowick and F Gursey; copyright 1986 World Scientific Publishing Company.

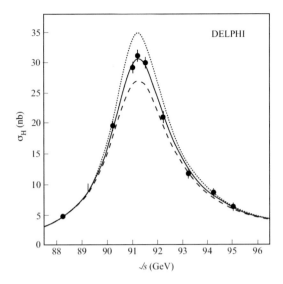

FIGURE 22.19
The cross–section for $e^+e^- \rightarrow$ hadrons around the Z^0 mass (Mönig, 1990). The dotted, continuous, and dashed lines are the predictions of the standard model assuming two, three, and four massless neutrino species respectively.

Forty years have passed since those historic experiments, and the current measured value for M_Z is

$$M_Z^{\mathrm{exp}} = 91.1876 \pm 0.0021 \ \mathrm{GeV}, \tag{22.111}$$

as determined from the Z lineshape scan at LEP1 (Schael *et al.* 2006). This is plainly in disagreement with the tree-level prediction (22.78), and indeed the current prediction including loop corrections (which we will briefly introduce in section 22.6) is (see the review by Erler and Freitas 2022, section 10 in Workman *et al.* 2022)

$$M_Z^{\mathrm{th}} = 91.1882 \pm 0.0020 \ \mathrm{GeV}, \tag{22.112}$$

in excellent agreement with (22.111).

The total Z^0 width Γ_Z is an interesting quantity. If we assume that, for any fermion family additional to the three known ones, only the neutrinos are significantly less massive than $M_Z/2$, we have

$$\Gamma_Z \simeq (2.5 + 0.16\Delta N_\nu) \ \mathrm{GeV} \tag{22.113}$$

from section 22.3, where ΔN_ν is the number of additional light neutrinos (i.e. beyond ν_e, ν_μ, and ν_τ) which contribute to the width through the process $Z^0 \rightarrow \nu\bar{\nu}$. Thus (22.113) can be used as an important measure of the number of such neutrinos (i.e. generations) if Γ_Z can be determined accurately enough. The mass resolution of the pp̄ experiments was of the same order as the total expected Z^0 width, so that (22.113) could not be used directly. The advent of LEP provided precision checks on (22.113); at the cost of departing from the historical development, we show data from DELPHI (Abreu *et al.* 1990, Abe 1991) in figure 22.19, which established $N_\nu = 3$.

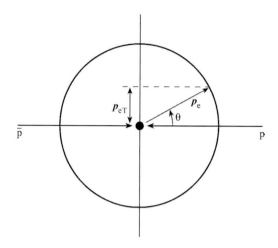

FIGURE 22.20
Kinematics of W $\to$ eν decay.

We turn now to the W$^\pm$. In this case an invariant mass plot is impossible, since we are looking for the eν ($\mu\nu$) mode, and cannot measure the ν's. However, it is clear that—as in the case of Z$^0 \to$ e$^+$e$^-$ decay—slow moving massive W's will emit isolated electrons with high transverse energy. Further, such electrons should be produced in association with large *missing* transverse energy (corresponding to the ν's), which can be measured by calorimetry, and which should balance the transverse energy of the electrons. Thus electrons of high E_T accompanied by balancing high missing E_T (i.e. similar in magnitude to that of the e$^-$ but opposite in azimuth) were the signatures used for the early event samples (UA1, Arnison *et al.* 1983a; UA2, Banner *et al.* 1983).

The determination of the mass of the W is not quite so straightforward as that of the Z, since we cannot construct directly an invariant mass plot for the eν pair: only the missing transverse momentum (or energy) can be attributed to the ν, since some unidentified longitudinal momentum will always be lost down the beam pipe. In fact, the distribution of events in p_{eT}, the magnitude of the transverse momentum of the e$^-$, should show a pronounced peaking towards the maximum kinematically allowed value, which is $p_{eT} \approx \frac{1}{2}M_W$, as may be seen from the following argument. Consider the decay of a W at rest (figure 22.20). We have $|\boldsymbol{p}_e| = \frac{1}{2}M_W$ and $|\boldsymbol{p}_{eT}| = \frac{1}{2}M_W \sin\theta \equiv p_{eT}$. Thus the transverse momentum distribution is given by

$$\frac{\mathrm{d}\sigma}{\mathrm{d}p_{eT}} = \frac{\mathrm{d}\sigma}{\mathrm{d}\cos\theta}\left|\frac{\mathrm{d}\cos\theta}{\mathrm{d}p_{eT}}\right| = \frac{\mathrm{d}\sigma}{\mathrm{d}\cos\theta}\left(\frac{2p_{eT}}{M_W}\right)\left(\frac{1}{4}M_W^2 - p_{eT}^2\right)^{-1/2}, \tag{22.114}$$

and the last (Jacobian) factor in (22.114) produces a strong peaking towards $p_{eT} = \frac{1}{2}M_W$. This peaking will be smeared by the width, and transverse motion, of the W. Early determinations of M_W used (22.114), but sensitivity to the transverse momentum of the W can be much reduced (Barger *et al.* 1983) by considering instead the distribution in 'transverse mass', defined by

$$M_T^2 = (E_{eT} + E_{\nu T})^2 - (\boldsymbol{p}_{eT} + \boldsymbol{p}_{\nu T})^2 \simeq 2p_{eT}p_{\nu T}(1 - \cos\phi), \tag{22.115}$$

where ϕ is the azimuthal separation between p_{eT} and $p_{\nu\text{T}}$. Here $E_{\nu\text{T}}$ and $\boldsymbol{p}_{\text{T}}$ are the neutrino transverse energy and momentum, measured from the missing transverse energy and momentum obtained from the global event reconstruction. This inclusion of additional measured quantities improves the precision as compared with the Jacobian peak method, using (22.114). A Monte Carlo simulation was used to generate M_{T} distributions for different values of M_{W}, and the most probable value was found by a maximum likelihood fit. The quoted results were

$$\text{UA1 (Geer 1986):} \quad M_{\text{W}} \;=\; 83.5 \pm^{1.1}_{1.0} \text{(stat.)} \pm 2.8 \text{(syst.) GeV} \qquad (22.116)$$

$$\text{UA2 (DiLella 1986):} \quad M_{\text{W}} \;=\; 81.2 \pm 1.1 \text{(stat.)} \pm 1.3 \text{(syst.) GeV} \qquad (22.117)$$

the systematic errors again reflecting uncertainty in the absolute energy scale of the calorimeters. The two experiments also quoted (Geer 1986, DiLella 1986)

$$\left.\begin{array}{ll} \text{UA1} & \Gamma_{\text{W}} < 6.5 \text{ GeV} \\ \text{UA2} & \Gamma_{\text{W}} < 7.0 \text{ GeV} \end{array}\right\} 90\% \text{ c.l.} \qquad (22.118)$$

We show in figure 22.21 a later determination of M_{W} by the CDF collaboration (Abe *et al.* (1995a)).

Once again, the agreement between the experiments, and of both with (22.77), was acceptable, the predictions again being on the low side. Loop corrections adjust (22.77) to the value (Workman *et al.* 2022)

$$M_{\text{W}}^{\text{th}} = 80.360 \pm 0.006 \text{ GeV}, \qquad (22.119)$$

which is in agreement with the current world average experimental value (Workman *et al.* 2022)

$$M_{\text{W}}^{\text{exp}} = 80.377 \pm 0.012 \text{ GeV}. \qquad (22.120)$$

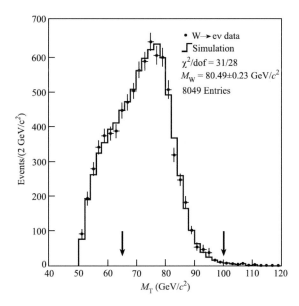

FIGURE 22.21

W $\rightarrow$ eν transverse mass distribution measured by the CDF collaboration. Figure reprinted with permission from F Abe *et al.* (CDF Collaboration) *Phys. Rev.* D **52** 4784 (1995). Copyright 1995 by the American Physical Society.

This world average, it should be noted, does not include the determination of the W mass by the CDF Collaboration (Aaltonen *et al.* 2022) based on their full Run-2 dataset, which was $M_W = 80.4335 \pm 0.0094$ MeV. This result is of higher precision than the world average quoted above, and is not in agreement with it, nor with the theoretical value (22.119). Subsequently, the ATLAS Collaboration (ATLAS 2023) reported a new (preliminary) result which is $M_W = 80.360 \pm 0.0016$ MeV. This is 10 MeV lower than the previous ATLAS result, and 16% more accurate. It is in agreement with (22.119). A significant discrepancy with (22.119) would imply new physics beyond the Standard Model.

The W and Z mass values may be used together with (22.42) to obtain $\sin^2 \theta_W$ via

$$\sin^2 \theta_W = 1 - M_W^2/M_Z^2. \tag{22.121}$$

The weighted average of UA(1) and UA(2) yielded

$$\sin^2 \theta_W = 0.212 \pm 0.022 \text{ (stat.)} \tag{22.122}$$

Radiative corrections have in general to be applied, but one renormalization scheme (see section 22.6) promotes (22.121) to a definition of the renormalized $\sin^2 \theta_W$ to all orders in perturbation theory. Using this scheme and the quoted values of M_W, M_Z one finds $\sin^2 \theta_W \simeq 0.223$.

Finally, figure 22.22 shows (Arnison *et al.* 1986) the angular distribution of the charged lepton in $W \to e\nu$ decay (see section); θ_e^* is the $e^+(e^-)$ angle in the W rest frame, measured

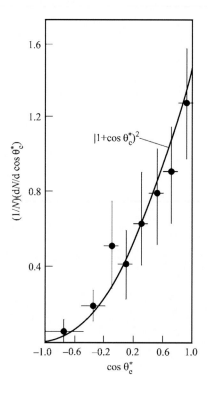

FIGURE 22.22
The W decay angular distribution of the emission angle θ_e^* of the positron (electron) with respect to the antiproton (proton) beam direction, in the rest frame of the W, for a total of 75 events; background subtracted and acceptance corrected (Arnison *et al.* 1986).

with respect to a direction parallel (antiparallel) to the $\bar{p}(p)$ beam. The expected form $(1 + \cos\theta_e^*)^2$ is followed very closely.

In summary, we may say that the early discovery experiments provided remarkably convincing confirmation of the principal expectations of the GSW theory, as outlined in section 22.3, while the later precision experiments have demonstrated the need for, and the remarkable success of, the renormalizable theory of the electroweak interactions.

We now consider some further aspects of the theory.

22.5 Fermion masses

22.5.1 One generation

The fact that the $SU(2)_L$ gauge group acts only on the L components of the fermion fields immediately appears to create a fundamental problem as far as the *masses* of these particles are concerned; we mentioned this briefly at the end of section 19.6. Let us recall first that the standard way to introduce the interactions of gauge fields with matter fields (e.g. fermions) is via the covariant derivative replacement

$$\partial^\mu \to D^\mu \equiv \partial^\mu + ig\boldsymbol{\tau}\cdot\boldsymbol{W}^\mu/2 \tag{22.123}$$

for $SU(2)$ fields $\boldsymbol{W}^\mu$ acting on $t = 1/2$ doublets. Now it is a simple exercise (compare problem 18.3) to check that the ordinary 'kinetic' part of a free Dirac fermion does not mix the L and R components of the field:

$$\bar{\hat{\psi}}\,\slashed{\partial}\hat{\psi} = \bar{\hat{\psi}}_R\,\slashed{\partial}\psi_R + \bar{\hat{\psi}}_L\,\slashed{\partial}\psi_L. \tag{22.124}$$

Thus we can in principle contemplate 'gauging' the L and the R components differently. Of course, in the case of QCD (c.f. (18.39)) the replacement $\slashed{\partial} \to \slashed{D}$ was made equally in each term on the right-hand side of (22.124). But this was because QCD conserves parity, and must therefore treat L and R components the same. Weak interactions are parity–violating, and the $SU(2)_L$ covariant derivative acts only in the *second* term of (22.124). On the other hand, a Dirac mass term has the form

$$-m(\bar{\hat{\psi}}_L\hat{\psi}_R + \bar{\hat{\psi}}_R\hat{\psi}_L) \tag{22.125}$$

(see equation (18.41) for example), and it precisely *couples* the L and R components. It is easy to see that if only $\hat{\psi}_L$ is subject to a transformation, then (22.125) is not invariant. Thus mass terms for Dirac fermions will *explicitly* break $SU(2)_L$. The same is not true for Majorana neutrinos, which are singlets under the weak gauge group, and can therefore have Lagrangian mass terms without violating the gauge symmetry (as noted in section 21.4.1).

Explicit breaking of the gauge symmetry by a Lagrangian mass term cannot be tolerated, in the sense that it will lead, once again, to violations of unitarity, and then of renormalizability. Consider, for example, a fermion-antifermion annihilation process of the form

$$f\bar{f} \to W_0^+ W_0^-, \tag{22.126}$$

where the subscript indicates the $\lambda = 0$ (longitudinal) polarization state of the $W^\pm$. We studied such a reaction in section 22.1.1 in the context of unitarity violations (in lowest order perturbation theory) for the IVB model. Appelquist and Chanowitz (1987) considered first the case in which 'f' is a lepton with $t = \frac{1}{2}, t_3 = -\frac{1}{2}$ coupling to W's, Z^0, and γ with the

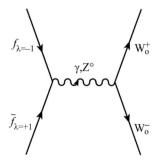

FIGURE 22.23

One-Z^0 and one-γ annihilation contribution to $f_{\lambda=-1}\bar{f}_{\lambda=1} \to W_0^+ W_0^-$.

usual $SU(2)_L \times U(1)$ couplings, but having an explicit (Dirac) mass m_f. They found that in the 'right' helicity channels for the leptons ($\lambda = +1$ for $\bar{f}$, $\lambda = -1$ for f) the bad high energy behaviour associated with a fermion–exchange diagram of the form of figure 22.4 was *cancelled* by that of the diagrams shown in figure 22.23. The sum of the amplitudes tends to a constant as s (or E^2) $\to \infty$. Such cancellations are a feature of gauge theories, as we indicated at the end of section 22.1.2, and represent one aspect of the renormalizability of the theory. But suppose, following Appelquist and Chanowitz (1987), we examine channels involving the 'wrong' helicity component, for example $\lambda = +1$ for the fermion f. Then it is found that the cancellation no longer occurs, and we shall ultimately have a 'non-renormalizable' problem on our hands, all over again.

An estimate of the energy at which this will happen can be made by recalling that the 'wrong' helicity state participates only by virtue of a factor (m_f/energy) (recall section 20.2.2), which here we can take to be $m_f/\sqrt{s}$. The typical bad high energy behaviour for an amplitude $\mathcal{M}$ was $\mathcal{M} \sim G_F s$, which we expect to be modified here to

$$\mathcal{M} \sim G_F s m_f/\sqrt{s} \sim G_F m_f \sqrt{s}. \tag{22.127}$$

The estimate obtained by Appelquist and Chanowitz differs only by a factor of $\sqrt{2}$. Attending to all the factors in the partial wave expansion gives the result that the unitarity bound will be saturated at $E = [E_f/(\text{TeV})] \sim \pi/[m_f/(\text{TeV})]$. Thus for $m_t \sim 175$ GeV, $E_t \sim 18$ TeV. This would constitute a serious flaw in the theory, even though the breakdown occurs at energies beyond those currently reachable.

However, in a theory with spontaneous symmetry breaking, there is a way of giving fermion masses without introducing an explicit mass term in the Lagrangian. Consider the electron, for example, and let us hypothesize a 'Yukawa'–type coupling between the electron–type SU(2) doublet

$$\hat{l}_{eL} = \begin{pmatrix} \hat{\nu}_e \\ \hat{e}^- \end{pmatrix}_L, \tag{22.128}$$

the Higgs doublet $\hat{\phi}$, and the R–component of the electron field:

$$\hat{\mathcal{L}}_{\text{Yuk}}^e = -g_e(\bar{\hat{l}}_{eL}\hat{\phi}\hat{e}_R + \bar{\hat{e}}_R\hat{\phi}^\dagger\hat{l}_{eL}). \tag{22.129}$$

In each term of (22.129), the two $SU(2)_L$ doublets are 'dotted together' so as to form an $SU(2)_L$ scalar, which multiplies the $SU(2)_L$ scalar R–component. Thus (22.129) is $SU(2)_L$–invariant, and the symmetry is preserved, at the Lagrangian level, by such a term. But now

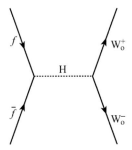

FIGURE 22.24
One-H annihilation graph.

insert just the vacuum value (22.28) of $\hat{\phi}$ into (22.129): we find the result

$$\hat{\mathcal{L}}^{\mathrm{e}}_{\mathrm{Yuk}}(\mathrm{vac}) = -g_{\mathrm{e}}\frac{v}{\sqrt{2}}(\bar{\hat{e}}_{\mathrm{L}}\hat{e}_{\mathrm{R}} + \bar{\hat{e}}_{\mathrm{R}}\hat{e}_{\mathrm{L}}) \tag{22.130}$$

which is exactly a (Dirac) mass of the form (22.125), allowing us to make the identification

$$m_{\mathrm{e}} = g_{\mathrm{e}}v/\sqrt{2}. \tag{22.131}$$

When oscillations about the vacuum value are considered via the replacement (22.29), the term (22.129) will generate a coupling between the electron and the Higgs fields of the form

$$-g_{\mathrm{e}}\bar{\hat{e}}\hat{e}\hat{H}/\sqrt{2} = -(m_{\mathrm{e}}/v)\bar{\hat{e}}\hat{e}\hat{H} \tag{22.132}$$

$$= -(gm_{\mathrm{e}}/2M_{\mathrm{W}})\bar{\hat{e}}\hat{e}\hat{H}. \tag{22.133}$$

The presence of such a coupling, if present for the process $f\bar{f} \rightarrow \mathrm{W}_0^+\mathrm{W}_0^-$ considered earlier, will mean that, in addition to the f-exchange graph analogous to figure 22.4 and the annihilation graphs of figure 22.23, a further graph shown in figure 22.24, must be included. The presence of the fermion mass in the coupling to H suggests that this graph might be just what is required to cancel the 'bad' high energy behaviour found in (22.127)—and by this time the reader will not be surprised to be told that this is indeed the case.

At first sight it might seem that this stratagem will only work for the $t_3 = -\frac{1}{2}$ components of doublets, because of the form of $\langle 0|\hat{\phi}|0\rangle$. But we learned in section 12.1.3 that if a pair of states $\begin{pmatrix} u \\ d \end{pmatrix}$ forming an SU(2) doublet transform by

$$\begin{pmatrix} u \\ d \end{pmatrix}' = e^{-\mathrm{i}\boldsymbol{\alpha}\cdot\boldsymbol{\tau}/2}\begin{pmatrix} u \\ d \end{pmatrix}, \tag{22.134}$$

then the charge conjugate states $\mathrm{i}\tau_2\begin{pmatrix} u^* \\ d^* \end{pmatrix}$ transform in exactly the same way. Thus if, in our case, $\hat{\phi}$ is the SU(2) doublet

$$\hat{\phi} = \begin{pmatrix} \frac{1}{\sqrt{2}}(\hat{\phi}_1 - \mathrm{i}\hat{\phi}_2) \equiv \hat{\phi}^+ \\ \frac{1}{\sqrt{2}}(\hat{\phi}_3 - \mathrm{i}\hat{\phi}_4) \equiv \hat{\phi}^0 \end{pmatrix}, \tag{22.135}$$

then the charge conjugate field

$$\hat{\phi}_{\mathbf{C}} \equiv \mathrm{i}\tau_2\hat{\phi}^* = \left(\begin{array}{c} \frac{1}{\sqrt{2}}(\hat{\phi}_3 + \mathrm{i}\hat{\phi}_4) \\ -\frac{1}{\sqrt{2}}(\hat{\phi}_1 + \mathrm{i}\hat{\phi}_2) \end{array} \right) \equiv \left(\begin{array}{c} \hat{\bar{\phi}}^0 \\ -\hat{\phi}^- \end{array} \right) \tag{22.136}$$

is also an SU(2) doublet, transforming in just the same way as $\hat{\phi}$. ((22.135) and (22.136) may be thought of as analogous to the $(\mathrm{K}^+, \mathrm{K}^0)$ and $(\bar{\mathrm{K}}^0, \mathrm{K}^-)$ isospin doublets in $\mathrm{SU}(3)_{\mathrm{f}}$). Note that the vacuum value (22.28) will now appear in the upper component of (22.136). With the help of $\hat{\phi}_{\mathbf{C}}$ we can write down another SU(2)-invariant coupling in the $\nu_{\mathrm{e}} - \mathrm{e}$ sector, namely

$$-g_{\nu_{\mathrm{e}}}(\hat{\bar{l}}_{\mathrm{eL}}\hat{\phi}_{\mathbf{C}}\hat{\nu}_{\mathrm{eR}} + \bar{\hat{\nu}}_{\mathrm{eR}}\hat{\phi}_{\mathbf{C}}^\dagger\hat{l}_{\mathrm{eL}}), \tag{22.137}$$

assuming now the existence of the field $\hat{\nu}_{\mathrm{eR}}$. In the Higgs vacuum (22.28), (22.137) then yields

$$-(g_{\nu_{\mathrm{e}}}v/\sqrt{2})(\bar{\hat{\nu}}_{\mathrm{eL}}\hat{\nu}_{\mathrm{eR}} + \bar{\hat{\nu}}_{\mathrm{eR}}\hat{\nu}_{\mathrm{eL}}) \tag{22.138}$$

which is precisely a (Dirac) mass for the neutrino, if we set $g_{\nu_{\mathrm{e}}}v/\sqrt{2} = m_{\nu_{\mathrm{e}}}$.

It is clearly possible to go on like this, and arrange for all the fermions, quarks as well as leptons, to acquire a mass by the same 'mechanism'. We will look more closely at the quarks in the next section. But one must admit to a certain uneasiness concerning the enormous difference in magnitudes represented by the couplings $g_{\nu_{\mathrm{e}}}, \ldots g_{\mathrm{e}}, \ldots g_{\mathrm{t}}$. If $m_{\nu_{\mathrm{e}}} < 1$ eV then $g_{\nu_{\mathrm{e}}} < 10^{-11}$, while $g_{\mathrm{t}} \sim 1$! Besides, whereas the use of the Higgs field 'mechanism' in the W–Z sector is quite economical, in the present case it seems rather unsatisfactory simply to postulate a different 'g' for each fermion–Higgs interaction. This does appear to indicate that we are dealing here with a 'phenomenological model', once more, rather than a 'theory'.

As far as the neutrinos are concerned, however, there is another possibility, already discussed in sections 7.5.2, 20.3, and 21.4.1, which is that they could be Majorana (not Dirac) fermions. In this case, rather than the four degrees of freedom ($\nu_{\mathrm{eL}}, \nu_{\mathrm{eR}}$, and their antiparticles) which exist for massive Dirac particles, only two possibilities exist for neutrinos, which we may take to be ν_{eL} and ν_{eR}. With these, it is certainly possible to construct a Dirac-type mass term of the form (22.138). But since, after all, the ν_{eR} component has been assigned zero quantum members both for $\mathrm{SU}(2)_{\mathrm{L}}$ W–interactions and for U(1) B–interactions (see table 22.1), we could consider economically dropping it altogether, making do with just the ν_{eL} component.

Suppose, then, that we keep only the field $\hat{\nu}_{\mathrm{eL}}$. We need to form a mass term for it. The charge–conjugate field is defined by (see (7.151))

$$(\hat{\nu}_{\mathrm{eL}})_{\mathbf{C}} = \mathrm{i}\gamma^2\gamma_0\bar{\hat{\nu}}_{\mathrm{eL}}^{\mathrm{T}} = \mathrm{i}\gamma^2\hat{\nu}_{\mathrm{L}}^{\dagger\mathrm{T}}, \tag{22.139}$$

and we know that the charge–conjugate field transforms under Lorentz transformations in the same way as the original field. So we can use $(\hat{\nu}_{\mathrm{eL}})_{\mathbf{C}}$ to form a Lorentz invariant

$$\overline{(\hat{\nu}_{\mathrm{eL}})_{\mathbf{C}}}\,\nu_{\mathrm{eL}} \tag{22.140}$$

which has mass dimension M^3. Hence we may write a mass term for $\hat{\nu}_{\mathrm{eL}}$ in the form

$$-\frac{1}{2}m_{\mathrm{M}}[\overline{(\hat{\nu}_{\mathrm{eL}})_{\mathbf{C}}}\,\hat{\nu}_{\mathrm{eL}} + \bar{\hat{\nu}}_{\mathrm{eL}}(\hat{\nu}_{\mathrm{eL}})_{\mathbf{C}}] \tag{22.141}$$

where the $\frac{1}{2}$ is conventional. Written out in more detail, we have

$$\overline{(\hat{\nu}_{\mathrm{eL}})_{\mathbf{C}}}\,\hat{\nu}_{\mathrm{eL}} = \hat{\nu}_{\mathrm{eL}}^{\mathrm{T}}(-\mathrm{i}\gamma^{2\dagger}\gamma_0)\hat{\nu}_{\mathrm{eL}} = \hat{\nu}_{\mathrm{eL}}^{\mathrm{T}}\mathrm{i}\gamma^2\gamma_0\hat{\nu}_{\mathrm{eL}}, \tag{22.142}$$

in the representation (20.14). Now

$$i\gamma^2\gamma_0 = \begin{pmatrix} -i\sigma_2 & 0 \\ 0 & i\sigma_2 \end{pmatrix}. \qquad (22.143)$$

But since $\hat{\nu}_{\mathrm{eL}}$ is an L–chiral field, only its 2 lower components are present (cf (20.26)) and (22.142) is effectively

$$\overline{(\hat{\nu}_{\mathrm{eL}})_{\mathbf{C}}}\,\hat{\nu}_{\mathrm{eL}} = \hat{\nu}_{\mathrm{eL}}^{\mathrm{T}}(i\sigma_2)\hat{\nu}_{\mathrm{eL}}. \qquad (22.144)$$

This is just the form of the mass term for a Majorana field, as we saw in equation (7.158) and equation (21.110). The two formalisms are equivalent.

As noted in section 21.4.1, the mass term (22.141) is not invariant under a global U(1) phase transformation

$$\hat{\nu}_{\mathrm{eL}} \to e^{-i\alpha}\hat{\nu}_{\mathrm{eL}} \qquad (22.145)$$

which would correspond to lepton number (if accompanied by a similar transformation for the electron fields): the Majorana mass term violates lepton number conservation.

There is a further interesting aspect to (22.144) which is that, since two $\hat{\nu}_{\mathrm{eL}}$ operators appear rather than a $\hat{\nu}_e$ and a $\hat{\nu}_e^\dagger$ (which would lead to L_e conservation), the (t, t_3) quantum numbers of the term are $(1,1)$. This means that we cannot form an SU(2)$_L$ invariant with it, using only the standard model Higgs $\hat{\phi}$, since the latter has $t = \frac{1}{2}$ and cannot combine with the $(1,1)$ operator to form a singlet. Thus we cannot make a 'tree–level' Majorana mass by the mechanism of Yukawa coupling to the Higgs field, followed by symmetry breaking.

However, we could generate suitable 'effective' operators via loop corrections, perhaps, much as we generated an effective operator representing an anomalous magnetic moment interaction in QED (cf section 11.7). But whatever it is, the operator would have to violate lepton number conservation, which is actually conserved by all the standard model interactions. Thus such an effective operator could not be generated in perturbation theory. It could arise, however, as a low energy limit of a theory defined at a higher mass scale, as the current–current model is the low energy limit of the GSW one. The typical form of such operator we need, in order to generate a term $\hat{\nu}_{\mathrm{eL}}^{\mathrm{T}}i\sigma_2\hat{\nu}_{\mathrm{eL}}$, is

$$-\frac{g_{\mathrm{eM}}}{\Lambda}(\bar{\hat{l}}_{\mathrm{eL}}\hat{\phi}_{\mathbf{C}})^{\mathrm{T}}i\sigma_2(\hat{\phi}_{\mathbf{C}}^\dagger\hat{l}_{\mathrm{eL}}). \qquad (22.146)$$

Note, most importantly, that the operator '$(l\phi)(\phi l)$' in (22.146) has mass dimension *five*, which is why we introduced the factor Λ^{-1} in the coupling; it is indeed a non–renormalizable effective interaction, just like the current–current one. We may interpret Λ as the mass scale at which 'new physics' enters, in the spirit of the discussion in section 11.7, and as we will discuss further in section 23.3 of the final chapter. After symmetry breaking (22.146) will generate the Majorana mass term, with

$$m_{\mathrm{M}} \sim g_{\mathrm{eM}}\frac{v^2}{2\Lambda}. \qquad (22.147)$$

From (21.132) we know that one neutrino mass must be at least of order 0.05 eV. This implies that $\Lambda/g_{\mathrm{eM}} \geq 10^{15}$ GeV (a scale not far from typical Grand Unified theories).

A more specific model can be constructed in which a relation of the form (22.147) can arise naturally. Suppose $\hat{\nu}_{\mathrm{R}}$ is an R-type neutrino field which is an SU(2)×U(1) singlet, and which has a gauge-invariant Yukawa coupling to the Higgs field, of the form (22.137). Then the Yukawa and the mass terms $\hat{\nu}_{\mathrm{R}}$ are

$$\mathcal{L}_{Y,R} = -g_{\mathrm{R}}(\bar{\hat{\ell}}_{\mathrm{eL}}\hat{\phi}_{\mathbf{C}}\hat{\nu}_{\mathrm{R}} + \bar{\hat{\nu}}_{\mathrm{R}}\hat{\phi}_{\mathbf{C}}^\dagger\hat{\ell}_{\mathrm{eL}}) - \frac{1}{2}m_{\mathrm{R}}[\overline{(\hat{\nu}_{\mathrm{R}})_{\mathbf{C}}}\,\hat{\nu}_{\mathrm{R}} + \mathrm{h.c.}]. \qquad (22.148)$$

Then, in the Higgs vacuum the first term in (22.148) becomes

$$-m_{\rm D}(\bar{\hat{\nu}}_{\rm eL}\hat{\nu}_{\rm R} + \bar{\hat{\nu}}_{\rm R}\hat{\nu}_{\rm eL}) \tag{22.149}$$

where $m_{\rm D} = g_{\rm R}v/\sqrt{2}$. The term (22.149) couples the fields $\hat{\nu}_{\rm R}$ and $\hat{\nu}_{\rm eL}$, so that we need to do a diagonalization to find the true mass eigenvalues and eigenstates. The combined mass terms from (22.148) and (22.149) can be written as

$$-\frac{1}{2}\overline{(\hat{N}_{\rm L})_{\bf C}}\,{\bf M}\,\hat{N}_{\rm L} + {\rm h.c.} \tag{22.150}$$

where

$$\hat{N}_{\rm L} \equiv \left(\begin{array}{c} \hat{\nu}_{\rm eL} \\ (\hat{\nu}_{\rm R})_{\bf C} \end{array} \right), \quad {\bf M} = \left(\begin{array}{cc} 0 & m_{\rm D} \\ m_{\rm D} & m_{\rm R} \end{array} \right). \tag{22.151}$$

CP invariance would imply that the parameters $m_{\rm R}$ and $m_{\rm D}$ are real, as we will assume, for simplicity.

Suppose now that $m_{\rm D} \ll m_{\rm R}$. Then the eigenvalues of **M** are approximately

$$m_1 \approx m_{\rm R}, \quad m_2 \approx -m_{\rm D}^2/m_{\rm R}. \tag{22.152}$$

The apparently troubling minus sign can be absorbed into the mixing parameters. Thus one eigenvalue is (by assumption) very large compared to $m_{\rm D}$, and one is very much smaller. The vanishing of the first element in **M** ensures that the lepton number violating term (22.141) is characterized by a large mass scale $m_{\rm R}$. It may be natural to assume that $m_{\rm D}$ is a 'typical' quark or lepton mass term, which would then imply that m_2 of (22.152) is very much lighter than that—as appears to be true for the neutrinos. This is the famous 'see-saw' mechanism of Minkowski (1977), Gell-Mann *et al.* (1979), Yanagida (1979) and Mohapatra and Senjanovic (1980, 1981). If in fact $m_{\rm R} \sim 10^{16}$Gev, we recover an estimate for m_2 which is similar to that in (22.147). It is worth emphasizing that the Majorana nature of the massive neutrinos is an essential part of the see-saw mechanism.

These considerations are tending to take us 'beyond the Standard Model', so we shall not pursue them at any greater length. Instead, we must now generalize the discussion of fermion masses to the three–generation case.

22.5.2 Three-generation mixing

We introduce three doublets of left handed quark fields

$$\hat{q}_{\rm L1} = \left(\begin{array}{c} \hat{u}_{\rm L1} \\ \hat{d}_{\rm L1} \end{array} \right), \qquad \hat{q}_{\rm L2} = \left(\begin{array}{c} \hat{u}_{\rm L2} \\ \hat{d}_{\rm L2} \end{array} \right), \qquad \hat{q}_{\rm L3} = \left(\begin{array}{c} \hat{u}_{\rm L3} \\ \hat{d}_{\rm L3} \end{array} \right) \tag{22.153}$$

and the corresponding six singlets

$$\hat{u}_{\rm R1}, \hat{d}_{\rm R1}, \hat{u}_{\rm R2}, \hat{d}_{\rm R2}, \hat{u}_{\rm R3}, \hat{d}_{\rm R3}, \tag{22.154}$$

which transform in the now familiar way under $SU(2)_L \times U(1)$. The $\hat{u}$-fields correspond to the $t_3 = +\frac{1}{2}$ components of $SU(2)_L$, the $\hat{d}$ ones to the $t_3 = -\frac{1}{2}$ components, and to their 'R' partners. The labels 1, 2, and 3 refer to the family number; for example, with no mixing at all, $\hat{u}_{\rm L1} = \hat{u}_{\rm L}, \hat{d}_{\rm L1} = \hat{d}_{\rm L}$, etc. We have to consider what is the most general $SU(2)_L \times U(1)$–invariant interaction between the Higgs field (assuming we can still get by with only one) and these various fields. Apart from the symmetry, the only other theoretical requirement is renormalizability—for, after all, if we drop this we might as well abandon the whole motivation for the 'gauge' concept. This implies (as in the discussion of the

Higgs potential $\hat{V}$) that we cannot have terms like $(\bar{\hat{\psi}}\hat{\psi}\hat{\phi})^2$ appearing—which would have a coupling with dimensions $(\text{mass})^{-4}$ and would be non–renormalizable. In fact the only renormalizable Yukawa coupling is of the form '$\bar{\hat{\psi}}\hat{\psi}\hat{\phi}$', which has a dimensionless coupling (as in the g_e and g_{ν_e} of (22.129) and (22.137)). However, there is no *a priori* requirement for it to be 'diagonal' in the weak interaction family index i. The allowed generalization of (22.129) and (22.137) is therefore an interaction of the form (summing on repeated indices)

$$\hat{\mathcal{L}}_{\psi\phi} = a_{ij}\bar{\hat{q}}_{Li}\hat{\phi}_{C}\hat{u}_{Rj} + b_{ij}\bar{\hat{q}}_{Li}\hat{\phi}\hat{d}_{Rj} + \text{h.c.} \tag{22.155}$$

where

$$\hat{q}_{Li} = \begin{pmatrix} \hat{u}_{Li} \\ \hat{d}_{Li} \end{pmatrix}, \tag{22.156}$$

and a sum on the family indices i and j (from 1 to 3) in (22.155) is assumed. After symmetry breaking, using the gauge (22.29), we find (problem 22.6)

$$\hat{\mathcal{L}}_{f\phi} = -\left(1 + \frac{\hat{H}}{v}\right)[\bar{\hat{u}}_{Li}m_{ij}^{u}\hat{u}_{Rj} + \bar{\hat{d}}_{Li}m_{ij}^{d}\hat{d}_{Rj} + \text{h.c.}], \tag{22.157}$$

where the 'mass matrices' are

$$m_{ij}^{u} = -\frac{v}{\sqrt{2}}a_{ij}, \qquad m_{ij}^{d} = -\frac{v}{\sqrt{2}}b_{ij}. \tag{22.158}$$

Although we have not indicated it, the m^u and m^d matrices could involve a 'γ_5' part as well as a '1' part in Dirac space. It can be shown (Weinberg (1973), Feinberg *et al.* (1959)) that m^u and m^d can both be made Hermitean, γ_5–free, and diagonal by making four separate unitary transformations on the 'generation triplets'

$$\hat{u}_L = \begin{pmatrix} \hat{u}_{L1} \\ \hat{u}_{L2} \\ \hat{u}_{L3} \end{pmatrix}, \qquad \hat{d}_L = \begin{pmatrix} \hat{d}_{L1} \\ \hat{d}_{L2} \\ \hat{d}_{L3} \end{pmatrix}, \text{etc} \tag{22.159}$$

via

$$\hat{u}_{L\alpha} = (U_L^{(u)})_{\alpha i}\hat{u}_{Li}, \qquad \hat{u}_{R\alpha} = (U_R^{(u)})_{\alpha i}\hat{u}_{Ri} \tag{22.160}$$

$$\hat{d}_{L\alpha} = (U_L^{(d)})_{\alpha i}\hat{d}_{Li}, \qquad \hat{d}_{R\alpha} = (U_R^{(d)})_{\alpha i}\hat{d}_{Ri}. \tag{22.161}$$

In this notation, 'α' is the index of the 'mass diagonal' basis, and 'i' is that of the 'weak interaction' basis.[3] Then (22.157) becomes

$$\hat{\mathcal{L}}_{qH} = -\left(1 + \frac{\hat{H}}{v}\right)[m_u\bar{\hat{u}}\hat{u} + \ldots + m_b\bar{\hat{b}}\hat{b}]. \tag{22.162}$$

Rather remarkably, we can still manage with only the one Higgs field. It couples to each fermion with a strength proportional to the mass of that fermion, divided by M_W.

Now consider the $SU(2)_L \times U(1)$ gauge invariant interaction part of the Lagrangian. Written out in terms of the 'weak interaction' fields $\hat{u}_{L,Ri}$ and $\hat{d}_{L,Ri}$ (cf (22.43) and (22.44)), it is

$$\begin{aligned} \hat{\mathcal{L}}_{f,W,B} &= i(\bar{\hat{u}}_{Lj}, \bar{\hat{d}}_{Lj})\gamma^\mu(\partial_\mu + ig\boldsymbol{\tau}\cdot\hat{\boldsymbol{W}}_\mu/2 + ig'y\hat{B}_\mu/2)\begin{pmatrix} \hat{u}_{Lj} \\ \hat{d}_{Lj} \end{pmatrix} \\ &+ i\bar{\hat{u}}_{Rj}\gamma^\mu(\partial_\mu + ig'y\hat{B}_\mu/2)\hat{u}_{Rj} + i\bar{\hat{d}}_{Rj}\gamma^\mu(\partial_\mu + ig'y\hat{B}_\mu/2)\hat{d}_{Rj} \end{aligned} \tag{22.163}$$

---

[3]So, for example, $\hat{u}_{L\alpha=t} \equiv \hat{t}_L, \hat{d}_{L\alpha=s} \equiv \hat{s}_L$, etc.

where a sum on j is understood. This now has to be rewritten in terms of the mass–eigenstate fields $\hat{u}_{\mathrm{L,R}\alpha}$ and $\hat{d}_{\mathrm{L,R}\alpha}$.

Problem 22.7 shows that the neutral current part of (22.163) is diagonal in the mass basis, provided the U matrices of (22.160) and (22.161) are unitary; that is, the neutral current interactions do not change the flavour of the physical (mass eigenstate) quarks. The charged current processes, however, involve the *non*–diagonal matrices τ_1 and τ_2 in (22.163), and this spoils the argument used in problem 22.7. Indeed, using (22.47) we find that the charged current piece is

$$
\begin{aligned}
\hat{\mathcal{L}}_{\mathrm{CC}} &= -\frac{g}{\sqrt{2}}(\bar{\hat{u}}_{\mathrm{L}j}, \bar{\hat{d}}_{\mathrm{L}j})\gamma_\mu\tau_+\hat{W}_\mu\begin{pmatrix}\hat{u}_{\mathrm{L}j}\\\hat{d}_{\mathrm{L}j}\end{pmatrix} + \mathrm{h.c.}\\
&= -\frac{g}{\sqrt{2}}\bar{\hat{u}}_{\mathrm{L}j}\gamma^\mu\hat{d}_{\mathrm{L}j}\hat{W}_\mu + \mathrm{h.c.}\\
&= -\frac{g}{\sqrt{2}}\bar{\hat{u}}_{\mathrm{L}\alpha}[(U_{\mathrm{L}}^{(\mathrm{u})})_{\alpha j}(U_{\mathrm{L}}^{(\mathrm{d})\dagger})_{j\beta}]\gamma^\mu\hat{d}_{\mathrm{L}\beta}\hat{W}_\mu + \mathrm{h.c.},
\end{aligned}
\tag{22.164}
$$

where the matrix

$$
V_{\alpha\beta} \equiv [U_{\mathrm{L}}^{(\mathrm{u})}U_{\mathrm{L}}^{(\mathrm{d})\dagger}]_{\alpha\beta}
\tag{22.165}
$$

is not diagonal, though it is unitary. This is the CKM matrix (Cabibbo (1963), Kobayashi and Maskawa (1973)), originally introduced by Kobayashi and Maskawa in the context of their three-generation extension of the then-developing Standard Model, in order to provide room for **CP** violation within the SU(2)$\times$ U(1) gauge theory framework. The interaction (22.164) then has the form

$$
-\frac{g}{\sqrt{2}}\hat{W}_\mu[\bar{\hat{u}}_{\mathrm{L}}\gamma^\mu\hat{d}'_{\mathrm{L}} + \bar{\hat{c}}_{\mathrm{L}}\gamma^\mu\hat{s}'_{\mathrm{L}} + \bar{\hat{t}}_{\mathrm{L}}\gamma^\mu\hat{b}'_{\mathrm{L}}] + \mathrm{h.c.},
\tag{22.166}
$$

where

$$
\begin{pmatrix}\hat{d}'_{\mathrm{L}}\\\hat{s}'_{\mathrm{L}}\\\hat{b}'_{\mathrm{L}}\end{pmatrix} = \begin{pmatrix}V_{\mathrm{ud}} & V_{\mathrm{us}} & V_{\mathrm{ub}}\\V_{\mathrm{cd}} & V_{\mathrm{cs}} & V_{\mathrm{cb}}\\V_{\mathrm{td}} & V_{\mathrm{ts}} & V_{\mathrm{tb}}\end{pmatrix}\begin{pmatrix}\hat{d}_{\mathrm{L}}\\\hat{s}_{\mathrm{L}}\\\hat{b}_{\mathrm{L}}\end{pmatrix},
\tag{22.167}
$$

with the phenomenology described in the previous chapter.

An analysis similar to the above can be carried out in the leptonic sector. We would then have leptonic flavour mixing in charged current processes, via the leptonic analogue of the CKM matrix, namely the PMNS matrix (Pontecorvo (1957, 1958, 1967), Maki, Nakagawa and Sakata (1962)); this is the matrix whose elements are probed in neutrino oscillations, as we saw in chapter 21.

22.6 Higher order corrections, custodial SU(2) symmetry and status of the electroweak theory

The remarkable level of precision now being reached by experiments requires that, in order to compare with the theoretical predictions, sophisticated radiative corrections have to be applied. Such calculations are well beyond the intended scope of this book. Suitable introductions to the topic include Altarelli *et al.* (1989), the pedagogical account by Consoli *et al.* (1989), and the equally approachable lecture notes by Hollik (1991). Current data require the inclusion of complete one-loop, dominant two-loop, and partial three- and four-loop radiative corrections. A recent discussion is provided by Erler and Schott (2019); see also the review by Erler and Freitas (2022), section 10 in Workman *et al.* (2022). We shall return to the topic of higher-order corrections in the following chapter.

It is worth recalling that, before the discovery of the top quark and the Higgs boson, global fits of the GSW predictions for precision electroweak data led to predictions for the masses of the top quark and the Higgs boson, from their presence as *virtual* particles in loops. We will touch on just a few of the simpler and more important aspects of one-loop corrections (without detailed calculations), especially as regards those predictions for m_t and m_H. With the discovery and precise measurements of those two particles, the focus of electroweak precision fits has moved to testing the full consistency of the electroweak renormalizable theory, and on constraining models which go beyond that framework.

Apart from the fermion masses and mixing angles, and the Higgs mass, the electroweak theory has three remaining parameters, namely g, g' and one combination of the λ and μ (for example, the tree-level vacuum value v). These three parameters are usually replaced by an equivalent set with smaller experimental errors: the Z^0 mass M_Z, the Fermi constant G_F, and the fine structure constant α. These are, of course, related to g, g' and v; for example, at tree level we have

$$\alpha = g^2 g'^2/(g^2 + g'^2)4\pi, \quad M_\mathrm{Z} = \frac{1}{2}v\sqrt{g^2 + g'^2}, \quad G_\mathrm{F} = \frac{1}{\sqrt{2}v^2}, \tag{22.168}$$

but these relations become modified in higher order. The renormalized parameters will 'run' in the way described in chapters 15 and 16; the running of α, for example, has been observed directly, as noted in section 11.5.

After renormalization one can derive radiatively corrected values for physical quantities. But a renormalization scheme has to be specified, at any finite order (though in practice the differences between schemes are usually small). One commonly used scheme is the modified minimal subtraction ($\overline{\mathrm{MS}}$) scheme (appendix O) which introduces the quantity $\sin^2 \hat{\theta}_\mathrm{W}(\mu) \equiv \hat{g}'^2(\mu)/[\hat{g}'^2(\mu) + \hat{g}^2(\mu)]$ where the couplings $\hat{g}$ and $\hat{g}'$ are defined in the $\overline{\mathrm{MS}}$ scheme and μ is chosen to be M_Z for most electroweak processes. Attention is then focused on $\hat{s}_\mathrm{Z}^2 \equiv \sin^2 \hat{\theta}_\mathrm{W}(M_\mathrm{Z})$. This is the scheme used by Erler and Langacker (2010) in Nakamura *et al* (2010).

Another, conceptually simple, scheme is the 'on-shell' one (Sirlin (1980, 1984); Kennedy *et al.* (1989); Kennedy and Lynn (1989); Bardin *et al.* (1989); Hollik (1990); for reviews see Langacker 1995). In this scheme, the tree-level formaula

$$\sin^2 \theta_\mathrm{W} = 1 - M_\mathrm{W}^2/M_\mathrm{Z}^2 \tag{22.169}$$

is promoted into a *definition* of the renormalized $\sin^2 \theta_\mathrm{W}$ to all orders in perturbation theory, it being then denoted by s_W^2:

$$s_\mathrm{W}^2 = 1 - M_\mathrm{W}^2/M_\mathrm{Z}^2 \approx 0.223. \tag{22.170}$$

The radiatively corrected value for M_W is then

$$M_\mathrm{W}^2 = \frac{(\pi\alpha/\sqrt{2}G_\mathrm{F})}{s_\mathrm{W}^2(1 - \Delta r)} \tag{22.171}$$

where Δr includes the radiative corrections relating $\alpha, \alpha(M_\mathrm{Z}), G_\mathrm{F}, M_\mathrm{W}$, and M_Z.

We shall continue here with the scheme defined by (22.170). We cannot go into detail about all the contributions to Δr, but we do want to highlight two features of the result—which are surprising, important phenomenologically, and related to an interesting symmetry. It turns out (Consoli *et al.* 1989, Hollik 1991) that the leading terms in Δr have the form

$$\Delta r = \Delta r_0 - \frac{(1 - s_\mathrm{W}^2)}{s_\mathrm{W}^2}\Delta\rho + (\Delta r)_\mathrm{rem}. \tag{22.172}$$

In (22.172), $\Delta r_0 = 1 - \alpha/\alpha(M_Z)$ is due to the running of α, and has the value $\Delta r_0 = 0.06630(7)$ (see section 11.5.3). $\Delta \rho$ is given by (Veltman (1977))

$$\Delta \rho = \frac{3G_F(m_t^2 - m_b^2)}{8\pi^2\sqrt{2}} \approx \frac{0.00936\, m_t^2}{(172.83\,\text{GeV})^2} \tag{22.173}$$

while the 'remainder' $(\Delta r)_{\text{rem}}$ contains a significant term proportional to $\ln(m_t/m_Z)$, and a contribution from the Higgs boson which is (for $m_H \gg M_W$)

$$(\Delta r)_{\text{rem,H}} \approx \frac{\sqrt{2}G_F M_W^2}{16\pi^2}\frac{11}{3}\left[\ln\left(\frac{m_H^2}{M_W^2}\right) - \frac{5}{6}\right]. \tag{22.174}$$

As the notation suggests, $\Delta \rho$ is a leading contribution to the parameter ρ introduced in (22.66). As explained there, ρ measures the strength of neutral current processes relative to charged current ones, and has the value 1 at tree level. $\Delta \rho$ is then a radiative correction, with $\rho = 1 + \Delta \rho$. .

Some of the leading electroweak radiative corrections in $e^+e^- \to Z^0 \to f\bar{f}$ can be captured by the running of α and the radiative corrections employed in extracting the Fermi constant G_F. The remaining corrections can be included by replacing the fermionic couplings g_V^f and g_A^f (see (22.64), (22.75) and (22.76)) by

$$\bar{g}_V^f = \sqrt{\rho_f}(t_3^{(f)} - 2Q_f\kappa_f s_W^2) \tag{22.175}$$

and

$$\bar{g}_A^f = \sqrt{\rho_f}t_3^{(f)}, \tag{22.176}$$

together with corrections to the Z^0–propagator. The corrections have the form (in the on-shell scheme) $\rho_f \approx 1 + \Delta\rho$ (of equation (22.173)) and $\kappa_f \approx 1 + \frac{s_W^2}{(1-s_W^2)}\Delta\rho$, for $f \neq$ b, t. For the b-quark there is an additional contribution coming from the presence of the virtual top quark in vertex corrections to $Z \to b\bar{b}$ (Akhundov *et al.* (1986), Beenakker and Hollik (1988)).

We now discuss the physics of (22.172), (22.173) and (22.174). The running of α in Δr_0 is expected, but (22.173) and (22.174) contain surprising features. As regards (22.173), it is associated with top-bottom quark loops in vacuum polarization amplitudes, of the kind discussed for $\bar{\Pi}_\gamma^{[2]}$ in section 11.5, but this time in weak boson propagators. In the QED case, referring to equation (11.39) for example, we saw that the contribution of heavy fermions '($|q^2| \ll m_f^2$)' was suppressed, appearing as $O(|q^2|/m_f^2)$. In such a situation (which is the usual one) the heavy particles are said to 'decouple'. But the correction (22.173) is quite different, the fermion masses being in the *numerator*. Clearly, with a large value m_t, this can make a relatively big difference. This is why some precision measurements are surprisingly sensitive to the value of m_t, in the range near (as we now know) the physical value. Secondly, as regards the dependence on m_H, we might well have expected it to involve m_H^2 in the numerator if we considered the typical divergence of a scalar particle in a loop (we shall return to this after discussing (22.173)). Δr would then have been very sensitive to m_H, but in fact the sensitivity is only logarithmic.

We can understand the appearance of the fermion masses (squared) in the numerator of (22.173) as follows. The shift $\Delta \rho$ is associated with vector boson vacuum polarization contributions, for example the one shown in Figure 22.25. Consider in particular the contribution from the longitudinal polarization components of the W's. As we have seen, these components are nothing but three of the four Higgs components which the $W^\pm$ and Z^0 'swallowed' to become massive. But the couplings of these 'swallowed' Higgs fields to fermions are determined by just the same Higgs-fermion Yukawa couplings as we introduced to generate the

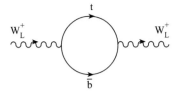

FIGURE 22.25

t - $\bar{\text{b}}$ vacuum polarization contribution.

fermion masses via spontaneous symmetry breaking. Hence we expect the fermion loops to contribute (to these longitudinal W states) something of order $g_f^2/4\pi$ where g_f is the Yukawa coupling. Since $g_f \sim m_f/v$ (see (22.131)) we arrive at an estimate $\sim m_f^2/4\pi v^2 \sim G_F m_f^2/4\pi$ as in (22.173). An important message is that *particles which acquire their mass spontaneously do not 'decouple' from all loops.*

But we now have to explain why $\Delta\rho$ in (22.173) would vanish if $m_t^2 = m_b^2$, and why only $\ln m_H^2$ appears in (22.174). Let us first consider the situation at tree level, where $\rho = 1$. It may be shown (Ross and Veltman 1975) that $\rho = 1$ is a natural consequence of having the symmetry broken by an SU(2)$_L$ doublet Higgs field (rather than a triplet, say)—or indeed by any number of doublets. The nearness of the measured ρ parameter to 1 is, in fact, good support for the hypothesis that there are only doublet Higgs fields. Problem 22.8 explores a simple model with a Higgs field in the triplet representation.

At tree level, it is simplest to think of ρ in connection with the mass ratio (22.66). To see the significance of this, let us go back to the Higgs-gauge field Lagrangian $\hat{\mathcal{L}}_{G\Phi}$ of (22.30) which produced the gauge boson masses. With the doublet Higgs of the form (22.135), it is a striking fact that the Higgs potential only involves the highly symmetrical combination of fields

$$\hat{\phi}_1^2 + \hat{\phi}_2^2 + \hat{\phi}_3^2 + \hat{\phi}_4^2. \tag{22.177}$$

This suggests that there may be some extra symmetry in (22.30) which is special to the doublet structure. But of course, to be of any interest, this symmetry has to be present in the $(\hat{D}_\mu\hat{\phi})^\dagger(\hat{D}^\mu\hat{\phi})$ term as well.

The nature of this symmetry is best brought out by introducing a change of notation for Higgs doublet $\hat{\phi}^+$ and $\hat{\phi}^0$: instead of (22.135), we now write (cf (18.70))

$$\hat{\phi} = \begin{pmatrix} (\hat{\pi}_2 + i\hat{\pi}_1)/\sqrt{2} \\ (\hat{\sigma} - i\hat{\pi}_3)/\sqrt{2} \end{pmatrix} \tag{22.178}$$

while the $\hat{\phi}_C$ field of (22.136) becomes

$$\hat{\phi}_C = \begin{pmatrix} (\hat{\sigma} + i\hat{\pi}_3)/\sqrt{2} \\ -(\hat{\pi}_2 - i\hat{\pi}_1)/\sqrt{2} \end{pmatrix}. \tag{22.179}$$

We then find that these can be written as

$$\hat{\phi} = \frac{1}{\sqrt{2}}(\hat{\sigma} + i\boldsymbol{\tau} \cdot \hat{\boldsymbol{\pi}}) \begin{pmatrix} 0 \\ 1 \end{pmatrix}, \qquad \hat{\phi}_C = \frac{1}{\sqrt{2}}(\hat{\sigma} + i\boldsymbol{\tau} \cdot \hat{\boldsymbol{\pi}}) \begin{pmatrix} 1 \\ 0 \end{pmatrix}. \tag{22.180}$$

Consider now the covariant SU(2)$_L\times$ U(1) derivative acting on $\hat{\phi}$, as in (22.30), and suppose to begin with that $g' = 0$. Then

$$\hat{D}_\mu\hat{\phi} = \frac{1}{\sqrt{2}}(\partial_\mu + ig\boldsymbol{\tau} \cdot \hat{\boldsymbol{W}}_\mu/2)(\hat{\sigma} + i\boldsymbol{\tau} \cdot \hat{\boldsymbol{\pi}}) \begin{pmatrix} 0 \\ 1 \end{pmatrix}$$

$$= \frac{1}{\sqrt{2}} \left\{ \partial_\mu \hat{\sigma} + \mathrm{i}\boldsymbol{\tau} \cdot \partial_\mu \hat{\boldsymbol{\pi}} + \mathrm{i}\frac{g}{2}\hat{\sigma}\boldsymbol{\tau} \cdot \hat{\boldsymbol{W}}_\mu \right.$$
$$\left. - \frac{g}{2}[\hat{\boldsymbol{\pi}} \cdot \hat{\boldsymbol{W}}_\mu + \mathrm{i}\boldsymbol{\tau} \cdot \hat{\boldsymbol{W}}_\mu \times \hat{\boldsymbol{\pi}}] \right\} \begin{pmatrix} 0 \\ 1 \end{pmatrix} \tag{22.181}$$

using $\tau_i\tau_j = \delta_{ij} + \mathrm{i}\epsilon_{ijk}\tau_k$. Now the vacuum choice (22.28) corresponds to $\hat{\sigma} = v, \hat{\boldsymbol{\pi}} = 0$, so that when we form $(\hat{D}_\mu\hat{\phi})^\dagger(\hat{D}^\mu\hat{\phi})$ from (22.181) we will get just

$$\frac{1}{2}(0,1) \left\{ \frac{g^2}{4}v^2(\boldsymbol{\tau} \cdot \hat{\boldsymbol{W}}_\mu)(\boldsymbol{\tau} \cdot \hat{\boldsymbol{W}}^\mu) \right\} \begin{pmatrix} 0 \\ 1 \end{pmatrix} = \frac{1}{2}M_W^2 \hat{\boldsymbol{W}}_\mu \cdot \hat{\boldsymbol{W}}^\mu \tag{22.182}$$

with $M_W = gv/2$ as usual. The condition $g' = 0$ corresponds (cf (22.39)) to $\theta_W = 0$, and thus to $\hat{W}_{3\mu} = \hat{Z}_\mu$, and so (22.182) says that in the limit of $g' \to 0$, $M_W = M_Z$, as expected if $\cos\theta_W = 1$. It is clear from (22.181) that the three components $\hat{\boldsymbol{W}}_\mu$ are treated on a precisely equal footing by the Higgs field (22.178), and indeed the notation suggests that $\hat{\boldsymbol{W}}_\mu$ and $\hat{\boldsymbol{\pi}}$ should perhaps be regarded as some kind of *new* triplets.

It is straightforward to calculate $(\hat{D}_\mu\hat{\phi})^\dagger(\hat{D}^\mu\hat{\phi})$ from (22.181); one finds (problem 22.9)

$$(D_\mu\hat{\phi})^\dagger D^\mu\hat{\phi} = \frac{1}{2}(\partial_\mu\hat{\sigma})^2 + \frac{1}{2}(\partial_\mu\hat{\boldsymbol{\pi}})^2 - \frac{g}{2}\partial_\mu\hat{\sigma}\hat{\boldsymbol{\pi}} \cdot \hat{\boldsymbol{W}}^\mu$$
$$+ \frac{g}{2}\hat{\sigma}\partial_\mu\hat{\boldsymbol{\pi}} \cdot \hat{\boldsymbol{W}}^\mu + \frac{g}{2}\partial_\mu\hat{\boldsymbol{\pi}} \cdot (\hat{\boldsymbol{\pi}} \times \hat{\boldsymbol{W}}^\mu)$$
$$+ \frac{g^2}{8}\hat{\boldsymbol{W}}_\mu^2(\hat{\sigma}^2 + \hat{\boldsymbol{\pi}}^2). \tag{22.183}$$

This expression now reveals what the symmetry is: (22.183) is invariant under global SU(2) transformations under which $\hat{\boldsymbol{W}}_\mu$ and $\hat{\boldsymbol{\pi}}$ are vectors—that is

$$\left. \begin{array}{c} \hat{\boldsymbol{W}}_\mu \to \hat{\boldsymbol{W}}_\mu + \boldsymbol{\epsilon} \times \hat{\boldsymbol{W}}_\mu \\ \hat{\boldsymbol{\pi}} \to \hat{\boldsymbol{\pi}} + \boldsymbol{\epsilon} \times \hat{\boldsymbol{\pi}} \\ \hat{\sigma} \to \hat{\sigma} \end{array} \right\}. \tag{22.184}$$

This is why, from the term $\hat{\boldsymbol{W}}_\mu^2\hat{\sigma}^2$, all three W fields have the same mass in this $g' \to 0$ limit.

If we now reinstate g', and use (22.36) and (22.37) to write $\hat{W}_{3\mu}$ and $\hat{B}_\mu$ in terms of the physical fields $\hat{Z}_\mu$ and $\hat{A}_\mu$ as in (19.97), (22.181) becomes

$$\frac{1}{\sqrt{2}} \left\{ \partial_\mu + \mathrm{i}g\frac{\tau_1}{2}\hat{W}_{1\mu} + \mathrm{i}g\frac{\tau_2}{2}\hat{W}_{2\mu} + \mathrm{i}g\frac{\tau_3}{2}\frac{\hat{Z}_\mu}{\cos\theta_W} + \mathrm{i}g\sin\theta_W\left(\frac{1+\tau_3}{2}\right)\hat{A}_\mu \right.$$
$$\left. - \frac{\mathrm{i}g}{\cos\theta_W}\sin^2\theta_W\left(\frac{1+\tau_3}{2}\right)\hat{Z}_\mu \right\}(\hat{\sigma} + \mathrm{i}\boldsymbol{\tau} \cdot \hat{\boldsymbol{\pi}}) \begin{pmatrix} 0 \\ 1 \end{pmatrix}. \tag{22.185}$$

We see from (22.185) that $g' \neq 0$ has two effects. First, there is a '$\boldsymbol{\tau} \cdot \hat{\boldsymbol{W}}$'–like term, as in (22.181), except that the '$\hat{W}_3$' part of it is now $\hat{Z}/\cos\theta_W$. In the vacuum $\hat{\sigma} = v, \hat{\boldsymbol{\pi}} = 0$ which simply means that the mass of the Z is $M_Z = M_W/\cos\theta_W$ i.e. $\rho = 1$; and this relation is preserved under 'rotations' of the form (22.184), since they do not mix $\hat{\boldsymbol{\pi}}$ and $\hat{\sigma}$. Hence this mass relation (and $\rho = 1$) is a consequence of the global SU(2) symmetry of the interactions and the vacuum under (22.184), and of the relations (22.36) and (22.37) which embody the requirement of a massless photon.

On the other hand, there are additional terms in (22.185) which single out the 'τ_3' component, and therefore break this global SU(2). These terms vanish as $g' \to 0$, and do not contribute at tree level, but we expect that they will cause $O(g'^2)$ corrections to $\rho = 1$ at the one loop level. Indeed, the Higgs contribution (22.174) to $\Delta\rho$ is $(\Delta\rho)_{\mathrm{H}} = -11G_{\mathrm{F}}m_Z^2\sin^2\theta_{\mathrm{W}}\ln(m_{\mathrm{H}}^2/m_{\mathrm{W}}^2)$ which vanishes as $g' \to 0$.

None of the above, however, yet involves the quark masses, and the question of why $m_{\mathrm{t}}^2 - b_{\mathrm{b}}^2$ appears in the numerator in (22.173). We can now answer this question. Consider a typical mass term, of the form discussed in section 22.5.2, for a quark doublet of the ith family

$$\hat{\mathcal{L}}_m = -g_+(\bar{\hat{u}}_{\mathrm{L}i}\bar{\hat{d}}_{\mathrm{L}i})\hat{\phi}_{\mathrm{C}}\hat{u}_{\mathrm{R}i} - g_-(\bar{\hat{u}}_{\mathrm{L}i}\bar{\hat{d}}_{\mathrm{L}i})\hat{\phi}\hat{d}_{\mathrm{R}i}. \tag{22.186}$$

Using (22.178) and (22.179), this can be written as

$$\begin{aligned}
\hat{\mathcal{L}}_m &= \frac{-g_+}{\sqrt{2}}(\bar{\hat{u}}_{\mathrm{L}i}\bar{\hat{d}}_{\mathrm{L}i})(\hat{\sigma} + \mathrm{i}\boldsymbol{\tau}\cdot\hat{\boldsymbol{\pi}})\begin{pmatrix} \hat{u}_{\mathrm{R}i} \\ 0 \end{pmatrix} - \frac{g_-}{\sqrt{2}}(\bar{\hat{u}}_{\mathrm{L}i}\bar{\hat{d}}_{\mathrm{L}i})(\hat{\sigma} + \mathrm{i}\boldsymbol{\tau}\cdot\hat{\boldsymbol{\pi}})\begin{pmatrix} 0 \\ \hat{d}_{\mathrm{R}i} \end{pmatrix} \\
&= -\frac{(g_+ + g_-)}{2\sqrt{2}}(\bar{\hat{u}}_{\mathrm{L}i}\bar{\hat{d}}_{\mathrm{L}i})(\hat{\sigma} + \mathrm{i}\boldsymbol{\tau}\cdot\hat{\boldsymbol{\pi}})\begin{pmatrix} \hat{u}_{\mathrm{R}i} \\ \hat{d}_{\mathrm{R}i} \end{pmatrix} \\
&\quad - \frac{(g_+ - g_-)}{2\sqrt{2}}(\bar{\hat{u}}_{\mathrm{L}i}\bar{\hat{d}}_{\mathrm{L}i})(\hat{\sigma} + \mathrm{i}\boldsymbol{\tau}\cdot\hat{\boldsymbol{\pi}})\tau_3\begin{pmatrix} \hat{u}_{\mathrm{R}i} \\ \hat{d}_{\mathrm{R}i} \end{pmatrix}.
\end{aligned} \tag{22.187}$$

Consider now a simultaneous (infinitesimal) global SU(2) transformation on the two doublets $(\hat{u}_{\mathrm{L}i}, \hat{d}_{\mathrm{L}i})^{\mathrm{T}}$ and $(\hat{u}_{\mathrm{R}i}, \hat{d}_{\mathrm{R}i})^{\mathrm{T}}$:

$$\begin{pmatrix} \hat{u}_{\mathrm{L}i} \\ \hat{d}_{\mathrm{L}i} \end{pmatrix} \to (1 - \mathrm{i}\boldsymbol{\epsilon}\cdot\boldsymbol{\tau}/2)\begin{pmatrix} \hat{u}_{\mathrm{L}i} \\ \hat{d}_{\mathrm{L}i} \end{pmatrix}, \qquad \begin{pmatrix} \hat{u}_{\mathrm{R}i} \\ \hat{d}_{\mathrm{R}i} \end{pmatrix} \to (1 - \mathrm{i}\boldsymbol{\epsilon}\cdot\boldsymbol{\tau}/2)\begin{pmatrix} \hat{u}_{\mathrm{R}i} \\ \hat{d}_{\mathrm{R}i} \end{pmatrix}. \tag{22.188}$$

Under (22.188), the first term of (22.187) becomes (to first order in $\boldsymbol{\epsilon}$)

$$-\frac{(g_+ + g_-)}{2\sqrt{2}}(\bar{\hat{u}}_{\mathrm{L}i}\bar{\hat{d}}_{\mathrm{L}i})[\hat{\sigma} + \mathrm{i}\boldsymbol{\tau}\cdot(\hat{\boldsymbol{\pi}} + \hat{\boldsymbol{\pi}}\times\boldsymbol{\epsilon})]\begin{pmatrix} \hat{u}_{\mathrm{R}i} \\ \hat{d}_{\mathrm{R}i} \end{pmatrix}. \tag{22.189}$$

From (22.189) we see that if, at the same time as (22.188), we *also* make the transformation of $\boldsymbol{\pi}$ given in (22.184), then this first term in $\hat{\mathcal{L}}_m$ will be invariant under these combined transformations. The second term in (22.187), however, will not be invariant under (22.188), but only under transformations with $\epsilon_1 = \epsilon_2 = 0, \epsilon_3 \neq 0$. We conclude that the global SU(2) symmetry of (22.184), which was responsible for $\rho = 1$ at the tree level, can be extended also to the quark sector; but—because the $g_{\pm}$ in (22.186) are proportional to the masses of the quark doublet—this symmetry is explicitly broken by the quark mass difference. This is why a t–$\bar{\mathrm{b}}$ loop in a W vacuum polarization correction can produce the 'non–decoupled' contribution (22.173) to ρ, which grows as $m_{\mathrm{t}}^2 - m_{\mathrm{b}}^2$ and produces quite detectable shifts from the tree–level predictions, given the accuracy of the data.

Returning to (22.188), the transformation on the L–components is just the same as a standard SU(2)$_{\mathrm{L}}$ transformation, except that it is global; so the gauge interactions of the quarks obey this symmetry also. As far as the R–components are concerned, they are totally decoupled in the gauge dynamics, and we are free to make the transformation (22.188) if we wish. The resulting complete transformation, which does the same to both the L and R components, is a non–chiral one—in fact it is precisely an ordinary 'isospin' transformation of the type

$$\begin{pmatrix} \hat{u}_i \\ \hat{d}_i \end{pmatrix} \to (1 - \mathrm{i}\boldsymbol{\epsilon}\cdot\boldsymbol{\tau}/2)\begin{pmatrix} \hat{u}_i \\ \hat{d}_i \end{pmatrix}. \tag{22.190}$$

The reader will recognize that the mathematics here is exactly the same as that in section 18.3, involving the SU(2) of isospin in the σ–model. This analysis of the symmetry of the

FIGURE 22.26

One-boson self-energy graph in $(\hat{\phi}^\dagger\hat{\phi})^2$.

Higgs (or a more general symmetry breaking sector) was first given by Sikivie *et al.* (1980). The isospin–SU(2) is frequently called 'custodial SU(2) symmetry' since it 'protects' $\rho = 1$.

What about the *absence* of m_H^2 corrections? Here the position is rather more subtle. Without the Higgs boson H the theory is non–renormalizable, and hence one might expect to see some radiative correction becoming very large ($O(m_H^2)$) as one tried to 'banish' H from theory by sending $m_H \to \infty$ (m_H would be acting like a cut–off). The reason is that in such a $(\hat{\phi}^\dagger\hat{\phi})^2$ theory, the simplest loop we meet is that shown in figure 22.26, and it is easy to see by counting powers, as usual, that it diverges as the square of the cut–off. This loop contributes to the Higgs self–energy, and will be renormalized by taking the value of the coefficient of $\hat{\phi}^\dagger\hat{\phi}$ in (22.30) from experiment. We will return to this important topic in section 23.3.

Even without a Higgs boson contribution however, it turns out that the electroweak theory is renormalizable at the one–loop level if the fermion masses are zero (Veltman 1968,1970). Thus one suspects that the large m_H^2 effects will not be so dramatic after all. In fact, calculation shows (Veltman 1977; Chanowitz *et al.* 1978, 1979) that one–loop radiative corrections to electroweak observables grow at most like $\ln m_H^2$ for large m_H. While there are finite corrections which are approximately $O(m_H^2)$ for $m_H^2 \ll M_{W,Z}^2$, for $m_H^2 \gg M_{W,Z}^2$ the $O(m_H^2)$ pieces cancel out from all observable quantities[4], leaving only $\ln m_H^2$ terms. This is just what we have in (22.174), and it means that the sensitivity of the data to this important parameter of the SM is only logarithmic. Fits to data, before the discovery of the Higgs boson, typically gave m_H in the region of 90 GeV at the minimum of the χ^2 curve, with an uncertainty of the order of ± 25 GeV. With the top quark, however, the situation was very different, as we shall see in the next section.

Before leaving the topic of radiative corrections, it is appropriate to summarize the current status of the electroweak theory, according to the global fit reported by Erler and Freitas (2022) in section 10 of Workman *et al.* (2022). These authors found that the fit describes the data generally very well, with $\chi^2/\text{d.o.f.} = 46.7/43$. The only conflict above the 3σ level is the muon $g - 2$ value, already discussed in section 11.7. The parameter A_b (see (22.100)) deviates by somewhere between 2σ and 3σ, but there are uncertainties in extracting the value from experiment. One other possible discrepancy (not included in the fit) concerns the W mass, discussed in section 22.4.3, which remains to be resolved.

22.7 The top quark

22.7.1 Introduction

The large mass of the top quark ($m_t \approx 173$ GeV) means that it plays a special role in the SM, for several reasons. First of all, unlike the mass of all the other quarks, m_t is greater

---

[4]Apart from the $\hat{\phi}^\dagger\hat{\phi}$ coefficient! See section 23.3.

than M_W, and this means that it can decay to b + W via *real* W emission:

$$t \rightarrow W^+ + b. \tag{22.191}$$

In contrast, the b quark itself decays by the usual *virtual* W processes. Now we have seen that the virtual process is suppressed by $\sim 1/M_W^2$ if the energy release (as in the case of b–decay) is well below M_W. But the real process (22.191) suffers no such suppression and proceeds very much faster. In fact (problem 22.10) the top quark lifetime τ_t for the decay (22.191) is estimated to be $\sim 5 \times 10^{-25}$ s. In this time, a top quark will travel a distance of order $c\tau_t \sim 10^{-16}$ m, which is substantially less than a typical hadronic size. This suggests that, when produced, t quarks will decay before they experience the strong confining QCD interactions. (In contrast, hadronization will be associated with the b quark, which has a more typical weak interaction lifetime of about 1.5×10^{-12} s.) One might therefore conclude that the mass of the top quark would be straightforward to measure experimentally, and interpret theoretically, free of difficulties associated with non-perturbative effects. However, there are subtle issues relating to the definition of the top quark mass, which have to be addressed when interpreting the implications of high precision data, as we shall discuss in section 22.7.3.

The large top mass has two further consequences to be noted. First, like all the other massive particles of the SM, the top quark gets its mass (at tree level) from its Yukawa coupling y_t to the Higgs boson, via the relation $m_t = y_t v/\sqrt{2}$ (c.f. (22.131)), which implies that $y_t \sim 1$. This effectively strong coupling will dominate in many dynamical calculations, one of which concerns the stability of the Higgs potential, as we will explain in section 22.8.3. The other consequence of the large top mass we have already noted in section 22.6, namely that the quantum (loop) corrections to the W boson mass M_W are very sensitive to the value of m_t.

Indeed, it is worth recalling the historical situation before the experimental discovery of the top quark in 1994. The W and Z particles were, as we have seen, discovered in 1983 and at that time, and for some years subsequently, the data were not precise enough to be sensitive to virtual t–effects. In the late 1980's and early 1990's, LEP at CERN and SLC at Stanford began to produce new and highly accurate data which did allow increasingly precise predictions to be made for the top quark mass, m_t. Thus a kind of race began, between experimentalists searching for the real top, and theorists fitting ever more precise data to get tighter and tighter limits on m_t, from its effect as a virtual particle in loops.

In fact, by the time of the actual experimental discovery of the top quark, the experimental error in m_t was just about the same as the theoretical one (and—of course—the central values were consistent). Thus, in their May 1994 review of the electroweak theory (contained in Montanet *et al.* 1994, p. 1304) Erler and Langacker (1994) gave the result of a fit to all electroweak data as

$$m_t = 169 \pm_{18}^{16} \pm_{20}^{17} \text{ GeV}, \tag{22.192}$$

the central figure and first error being based on $m_H = 300$ GeV, the second (+) error assuming $m_H = 1000$ GeV, and the second (−) error assuming $m_H = 60$ GeV.[5] At about the same time, Ellis *et al.* (1994) gave the extraordinarily precise value

$$m_t = 162 \pm 9 \text{ GeV} \tag{22.193}$$

without any assumption for m_H.

---

[5]The relatively small effect of large variations in m_H illustrates the lack of sensitivity to virtual Higgs effects, noted in the preceding section.

A month or so earlier, the CDF collaboration (Abe *et al.* (1994a,b)) announced 12 events consistent with the hypothesis of production of a $t\bar{t}$ pair, and on this hypothesis the mass was found to be

$$m_t = 174 \pm 10 \pm^{13}_{12} \text{ GeV}, \tag{22.194}$$

and this was followed by nine similar events from D0 (Abachi *et al.* 1995a). By February 1995, both groups had amassed more data and the discovery was announced (Abe *et al.* 1995b, Abachi *et al.* 1995b).

It is surely remarkable how the quantum fluctuations of a yet–to–be detected new particle could pin down its mass so precisely. The result demonstrates the power of a renormalizable quantum field theory, in this case one involving all the subtle intricacies of a spontaneously broken non-Abelian gauge theory. The 2022 experimental value for m_t is 172.69 ± 0.3 GeV (Workman *et al.* (2022)) as compared to the value predicted by fits to the electroweak data of 173.13 ± 0.56 GeV, given in the review of the electroweak theory by Erler and Freitas (2022), section 10 of Workman *et al.* (2022). Probing the consistency between the measured value of M_W and the value predicted by the SM after inclusion of full quantum corrections (which depend strongly on m_t) remains a major goal of high precision experiments.

Clearly the mass of the top quark is a very important parameter in the SM. But what exactly do we mean by m_t? Because of its significance in the interpretation of current and future high precision experiments, this question has received considerable attention from theorists. After first describing the basic features of top quark production at the Tevatron and the LHC, we will give a qualitative introduction to the top mass issue in section 22.7.3.

22.7.2 Top quark production at the Tevatron and the LHC

Top quarks are produced at hadron colliders primarily in pairs via the parton processes $q\bar{q} \to t\bar{t}$ and $gg \to t\bar{t}$. The leading order (LO) QCD diagrams are shown in figures 22.27 and 22.28. Figure 22.27 is an $\hat{s}$-channel process, while figures 22.28(a), (b), and (c) are, respectively, $\hat{s}$-, $\hat{t}$-, and $\hat{u}$-channel processes (recall from section 14.3.2 that hatted quantities denote invariant variables at the parton level). The total cross section for the process of figure 22.27 is (Glück, Owens and Reya 1978; Combridge 1979; Jones and Wyld 1978)

$$\sigma^{(\text{LO})}_{q\bar{q}}(\hat{s}) = \frac{8\pi\alpha_s^2}{27\hat{s}^2}(\hat{s} + 2m_t^2)\beta \tag{22.195}$$

where $\beta = (1 - 4m_t^2/\hat{s})$ and m_t is the top quark mass. The cross section calculated from the sum of the three diagrams of figure 22.28 is (Glück, Owens and Reya 1978; Combridge 1979; Jones and Wyld 1978)

$$\sigma^{(\text{LO})}_{gg}(\hat{s}) = \frac{\pi\alpha_s^2}{3\hat{s}}\left[(1 + \rho + \rho^2/16)\ln\left(\frac{1+\beta}{1-\beta}\right) - \beta\left(\frac{7}{4} + \frac{31}{16}\rho\right)\right] \tag{22.196}$$

where $\rho = 4m_t^2/\hat{s}$. These parton-level cross sections have to be combined with the distribution functions of the partons in the colliding hadrons, so as to generate the physical total

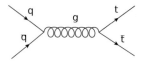

FIGURE 22.27
Leading order (LO) process for $q\bar{q} \to t\bar{t}$.

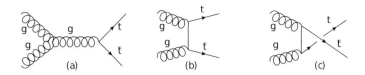

FIGURE 22.28
Leading order (LO) processes for gg $\to$ t$\bar{\text{t}}$.

t$\bar{\text{t}}$ cross section. This is similar to what was done in the case of deep inelastic scattering in sections 9.2 and 15.6, but here there are two initial state partons instead of just one.

At the Tevatron (p$\bar{\text{p}}$ at 1.96 TeV), where the total top pair production cross section is about 7 pb, the valence quarks in the antiproton contribute to the dominant (85%) q$\bar{\text{q}}$ production mechanism, the rest coming from the gg channel. At the LHC (pp at 13 Tev) about 90% of the cross section (roughly 830 pb) is from the gg channel, as the increased energy probes smaller x-values, where the gluon PDFs dominate.

Top quarks can also be produced singly, as was first observed at the Tevatron in 2009 by D0 (Abazov *et al.* 2009) and by CDF (Aaltonen *et al.* 2009). The cross section at the Tevatron is roughly half that of the pair production cross section. Single top quarks are produced via the electroweak LO processes shown in figure 22.29(a),(b), and (c). Process (a) is $\hat{s}$-channel virtual W exchange in q$\bar{\text{q}}' \to$ t$\bar{\text{b}}$; process (b) is $\hat{t}$-channel virtual W exchange in qb $\to$ q$'$t; and process (c) is W t associated production via bg $\to$ W$^-$t. Of these, the dominant process at LHC energies is the $\hat{t}$-channel process (b). At the Tevatron, the cross sections for processes (a) and (b) are roughly comparable, and they are the same for top quarks and antitop quarks. The latter is not the case at the LHC, because the initial state is not charge-symmetric. At the LHC, the total single top (t $+$ $\bar{\text{t}}$) production is roughly one quarter of the pair production cross section.

Before leaving the subject of top quark production, we should note that the cross sections for single top production are proportional to the modulus squared of the CKM matrix element V_{tb}, given that $|V_{\text{tb}}|$ is very much greater than $|V_{\text{ts}}|$ and $|V_{\text{td}}|$ (see section 20.7). This is one of the few direct ways to measure this CKM matrix element. The average of the determinations at the Tevatron (D0 and CMS) and at the LHC (ATLAS and CDF) is (Ceccucci, Ligeti and Sakai (2022), section 12 in Workman *et al.* 2022)

$$|V_{\text{tb}}| = 1.014 \pm 0.029, \tag{22.197}$$

to be compared to the current global fit value of $|V_{\text{tb}}| = 0.999118^{+0.000031}_{-0.000036}$.

In the years since the early discoveries, enormous progress has been made in improving the precision of the theoretical predictions, by including higher-order QCD corrections. This has gone hand-in-hand with a corresponding increase in the accuracy of the experiments. Both theory and experiment now quote results at the few % level, where there is hope of

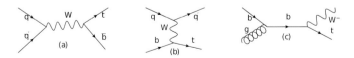

FIGURE 22.29
Three leading order mechanisms for single top production.

finding small discrepancies with the predictions, and hints of new physics (though so far none have appeared). We shall return to this topic in the following chapter.

22.7.3 The mass of the top quark

As we have seen, all fermion masses appear as fundamental parameters in the SM Lagrangian, after the Higgs field acquires a vacuum value. But unlike the leptons, which exist as unconfined free particles, the quarks are confined and not observed as free particles. In the case of all the quarks *except* the t quark, they are confined inside hadrons, and their masses must be determined by their influence on hadronic properties, as calculated in some theoretical framework. The dominant such framework now is lattice QCD, to which we gave an introduction in chapter 16. The definitions and determinations of the masses of these five quarks are discussed by Barnett, Lellouch, and Manohar (2022), in section 60 of Workman *et el.* (2022). But the top quark is different. Its lifetime is much shorter than a typical strong interaction lifetime (such as that of the ρ meson , $\tau_\rho \sim 10^{-23}$ s), which means that there will be no detectable hadronic states carrying t quarks, such as a 'toponium' $t\bar{t}$ state, or mesons or baryons with top constituents. Indeed, our qualitative discussion in section 22.7.1 suggested that its short lifetime might 'protect' it from non-perturbative QCD effects — in which case, the meaning of its mass should be straightforward. Unfortunately, that is not the case.

A simple approach would be to define m_t as the position of the peak in the invariant mass distribution of the top's decay products, W and b. This is what we do for the mass of hadronic resonance states, such as the ρ meson for example, via the peak in the invariant mass of the $\pi\pi$ state to which it decays. But the W+b state is crucially different from the $\pi\pi$ one. Like the t quark, the b quark carries colour charge, and it will not appear as an asymptotic state, travelling to detectors, because of colour confinement. The b quark must be linked to at least one other colour-carrying particle in the full amplitude, so that its colour can be neutralized. In fact, the b quark will be manifested experimentally in a jet of colourless hadrons, which cannot be accurately regarded as the true decay products of the b quark. This will lead to an uncertainty in the W+b invariant mass, due to uncalculated non-perturbative effects, which may be estimated to be of order Λ_{QCD} in the mass.

In practice, such a kinematical reconstruction of the top quark mass has been reported using various decay channels (see table 61.1 of the review of the top quark by Liss, Maltoni and Quadt (2022), section 61 of Workman *et al.* 2022). The mass measured in this direct way is commonly referred to as the 'pole' mass, and we shall continue to denote it by m_t. We learned in section 6.3.2 of volume 1 that the momentum space propagator of a free spin-zero particle has a pole at the on-shell point $p^2 = m^2$, where p is the 4-momentum carried by the particle. In the case of an unstable particle, the pole moves off the real axis to the complex point $\sqrt{p^2} = m - i\Gamma/2$, and the distribution of decay products follows a Breit-Wigner resonance shape (or that of a more sophisticated resonance paramatrization). The same is true for a free spin-1/2 particle, with propagator $(\not{p} - m_f - i\epsilon)^{-1}) = (\not{p} + m_f)/(p^2 - m_f^2 - i\epsilon)$. The top quark pole mass would then be defined as the real part of the complex pole position $\sqrt{p^2} = m_t - i\Gamma_t/2$.

Our previous qualitative discussion, however, suggested that m_t can only be defined up to an uncertainty of order Λ_{QCD}. Is there then no precise definition of the top mass? Within the framework of quantum field theory, there certainly is. The mass of an SM particle is a parameter in the SM Lagrangian, and like all such parameters is subject to renormalization. We recall that this will involve the choice of a renormalization scheme, and that the value of the renormalized mass will depend on the energy scale μ at which it is measured — in other words, the mass will be a 'running mass', as explained in section 15.5. It is standard now to use the $\overline{\text{MS}}$ (modified minimal subtraction) renormalization scheme for quark masses, so we denote the renormalized top quark mass in this scheme by $\overline{m}_{\text{top}}(\mu)$. If the scale μ is chosen sufficiently high, we may expect $\overline{m}_{\text{top}}(\mu)$ to be free of non-perturbative ambiguities, and therefore a good candidate for precision physics.[6] A suitable choice for μ would be the renormalized top mass itself, leading to the parameter $\overline{m}_{\text{top}}(\overline{m}_{\text{top}})$. This may be referred to as the 'short distance' mass. The currently quoted value for this mass is (Workman *et al.* 2022) $162.5^{+2.1}_{-1.5}$ GeV.

So comparison between theory and experiment could be done entirely in terms of the precisely definable short distance mass. But it is still rather more appealing to think in terms of the pole mass m_{t}. The connection between the two masses can be calculated in perturbation theory, and at one loop order it is

$$m_{\text{t}} = \overline{m}_{\text{top}}(\overline{m}_{\text{top}}) \left(1 + \frac{4}{3}\frac{\overline{\alpha}_{\text{s}}(\overline{m}_{\text{top}})}{\pi} + \ldots \right) \tag{22.198}$$

where $\overline{\alpha}_{\text{s}}$ is the strong coupling constant in the $\overline{\text{MS}}$ scheme, and the dots indicate higher order contributions, which are known to fourth order in $\overline{\alpha}_{\text{s}}$ (Marquard *et al.* 2015). But now we seem to have a contradiction: the pole mass was argued to have a non-perturbative uncertainty, while the right-hand side of (22.198) appears to be perfectly definite. The resolution is subtle. It is known that the perturbation series starts to diverge at some order, due to terms which grow factorially large, and are of infra-red origin (Beneke and Braun 1994, Bigi *et al.* 1994, Beneke 1995). Treated as an asymptotic expansion, there are mathematical techniques for estimating the ambiguity in the resummation, which go under the name of the 'renormalon ambiguity' (Beneke 1999). So the right-hand side of (22.198) does after all have an ambiguity, like the left-hand side. It has been estimated to be around 110 MeV (Beneke *et al.* 2017), and these authors further estimate the additional contribution to the mass relation from the five loop correction and beyond to be around 300 MeV.

We have discussed the mass of the top quark at some length. A review of the many other theoretical and experimental issues in top quark physics is provided by Liss and Maltoni (2022), section 61 of Workman *et al.* (2022). We move on to consider the other part of the SM which has been under intense scrutiny, the Higgs boson and its couplings in the Higgs sector.

22.8 The Higgs boson

It is worth noting that an essential feature of the type of theory which has been described in this note is the prediction of incomplete multiplets of scalar and vector bosons.

P W Higgs (1964)

22.8.1 Introduction

The Lagrangian for an *unbroken* $\text{SU}(2)_{\text{L}} \times \text{U}(1)$ gauge theory of vector bosons and fermions is rather simple and elegant, all the interactions being determined by just two Lagrangian parameters g and g' in a 'universal' way. All the particles in this hypothetical world are,

---

[6]The CMS Collaboration (Sirunyan *et al.* 2020) has made the first measurement of the running of the top quark mass in the $\overline{\text{MS}}$ scheme. The variation of the extracted mass is in good agreement with the one-loop prediction given in section 15.5.

however, massless. In the real world, while the electroweak interactions are undoubtedly well described by the $SU(2)_L \times U(1)$ theory, neither the mediating gauge quanta (apart from the photon) nor the fermions are massless. They must acquire mass in some way that does not break the gauge symmetry of the Lagrangian, or else the renormalizability of the theory is destroyed, and its remarkable empirical success (at a level which includes loop corrections) would be hard to explain. In chapter 19 we discussed how such a breaking of a gauge symmetry does happen, dynamically, in a superconductor. In that case 'electron pairing' was a crucial ingredient. In particle physics, while a lot of effort has gone into examining various analogous 'dynamical symmetry breaking' theories, none has yet emerged as both theoretically compelling and phenomenologically viable; we will briefly discuss such theories in section 23.4.2. However, a simple count of the number of degrees of freedom in a massive vector field, as opposed to a massless one, indicates that *some* additional fields must be present in order to give mass to the originally massless gauge bosons. As Higgs noted in the final paragraph of his 2-page Letter (Higgs 1964), an essential feature of the spontaneous symmetry breaking mechanism, in a gauge theory, is the appearance of incomplete multiplets of both scalar and vector bosons. Let us just rehearse this once more, in the $SU(2) \times U(1)$ case. We started with 4 massless gauge fields, belonging to an $SU(2)$ triplet and a $U(1)$ singlet; and, in addition, 4 scalar fields of equal mass, in an $SU(2)$ doublet. After symmetry breaking, three massive vector bosons emerged, leaving the photon massless. In the scalar sector, three of the scalars became the longitudinal components of the three massive vector bosons, and one lone massive scalar field survived, all that remained of the original scalar doublet. Its mass is a free parameter of the theory, being given at tree level in (22.40) by $m_H = \sqrt{2}\mu = \sqrt{\lambda}v/\sqrt{2}$.

In the SM, therefore, it is simply *assumed*, following the original ideas of Higgs and others (Higgs 1984, Englert and Brout 1964, Guralnik *et al.* 1964; Higgs 1966) that a suitable scalar ('Higgs') field exists, with a potential which causes the ground state (the vacuum) to break the symmetry spontaneously. Furthermore, rather than (as in BCS theory) obtaining the fermion mass gaps dynamically, they too are put in 'by hand' via Yukawa–like couplings to the Higgs field. The discovery of this *Higgs boson*, and the exploration of its couplings in the *Higgs sector*, has been a vital goal in particle physics for many years. Let us turn now to some simple aspects of Higgs boson production and decay processes at hadron colliders, as predicted by the Standard Model.

22.8.2 Higgs boson production at colliders and the 2012 discovery

Prior to the discovery of the Higgs boson in 2012, expectations regarding its mass m_H were less precise than they had been for the mass of the top quark. As we saw in section 22.6, the sensitivity of the electroweak radiative corrections to m_H is rather weak (logarithmic): fits to the data indicated a mass of about 90 GeV, with an error of roughly ± 25 GeV. Also, a lower bound on m_H had been established by LEP (LEP 2003):

$$m_H \geq 114.4 \text{ GeV (95\% Confidence Level)}. \tag{22.199}$$

This already excluded many possibilities in both production and decay. Subsequent searches were carried out at the hadron colliders. At both the Tevatron and the LHC, the dominant production mechanism was expected to be 'gluon fusion' via an intermediate top quark loop as shown in figure 22.30 (Georgi *et al.* (1978), Glashow *et al.* (1978), Stange *et al.* (1994a,b)). The intermediate t quark dominates, since the Higgs couplings to fermions are proportional to the fermion mass. Since the gluon probability distribution rises rapidly at small x values, which are probed at larger collider energy $\sqrt{s}$, the cross section for this process will rise with energy. At the Tevatron with $\sqrt{s} = 1.96$ TeV, the production cross section is about 1pb for $m_H = 125$ GeV, rising to about 50 pb at the LHC with $\sqrt{s} = 13$

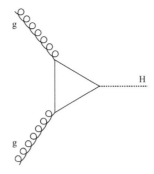

FIGURE 22.30
Higgs boson production process by 'gluon fusion'.

TeV. These numbers include QCD corrections (to be discussed in the following chapter), which increase the parton-level cross sections by a factor of about 2. The cross section is the same for pp and for p$\bar{\text{p}}$ colliders.

The next largest cross sections, some ten to twenty times smaller, are for 'vector boson fusion' (gq$'$ $\to$ qq$'$H) via the diagram of figure 22.31, and for associated production of a Higgs boson with a vector boson (q$\bar{\text{q}}$ $\to$ WH, ZH), shown in figure 22.32. The latter process can also proceed via a top quark box diagram, from an initial gg state. A fourth possibility is 'associated production with top quarks' as shown in figure 22.33, for example.

The Higgs boson has of course to be detected via its decays. For $m_H = 125$ GeV, decays to fermion–antifermion pairs dominate, of which b$\bar{\text{b}}$ has the largest branching ratio because of the larger value of m_b. The width of H $\to$ $f\bar{f}$ is easily calculated to lowest order, and is (problem 22.11)

$$\Gamma(\text{H} \to f\bar{f}) = \frac{CG_{\text{F}}m_f^2 m_{\text{H}}}{4\pi\sqrt{2}}\left(1 - \frac{4m_f^2}{m_{\text{H}}^2}\right)^{3/2}, \qquad (22.200)$$

where the colour factor C is 3 for quarks and 1 for leptons. QCD corrections are largely accounted for by replacing m_f^2 in the first factor on the right-hand side of (22.200) by the $\overline{\text{MS}}$ running mass value $\overline{m}_f^2(m_{\text{H}})$.

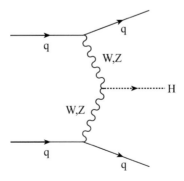

FIGURE 22.31
Higgs boson production process by 'vector boson fusion'.

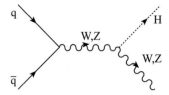

FIGURE 22.32

Higgs boson production in association with W or Z.

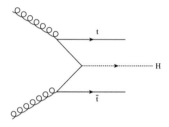

FIGURE 22.33

Higgs boson production in association with a t t̄ pair.

However, the large rate for the process gg → H → bb̄ has to compete against a very large background from the inclusive production of pp (or pp̄) → bb̄+X via the strong interaction. The Higgs signal can be separated from such a background by using a subleading decay mode such as H → γγ, where the final state particles can be very precisely measured, giving excellent resolution for m_H. There is no tree-level Hγγ coupling, and the lowest order coupling to photons is mediated by a top quark triangle loop (figure 22.34(a)), and by a W loop (figure 22.34(b), (c)), the W loop contributions dominating (Ellis, Gaillard, and Nanopoulos 1976, Shifman *et al.* 1979), and interfering destructively with the top fermion loop. In a similar way, the associated production modes W±H, Z H, allow use of the leptonic W and Z decays to reject QCD backgrounds.

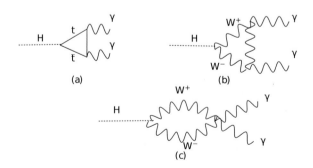

FIGURE 22.34

Higgs boson decay via (a) top quark loop; (b) and (c) W-boson loop.

Decays to a pair of vector bosons are also important. The tree-level width for H $\rightarrow$ W$^+$W$^-$ is (problem 22.11)

$$\Gamma(\text{H} \rightarrow \text{W}^+\text{W}^-) = \frac{G_{\text{F}} m_{\text{H}}^3}{8\pi\sqrt{2}} \left(1 - \frac{4M_{\text{W}}^2}{m_{\text{H}}^2}\right)^{1/2} \left(1 - \frac{4M_{\text{W}}^2}{m_{\text{H}}^2} + 12\frac{M_{\text{W}}^4}{m_{\text{H}}^4}\right), \tag{22.201}$$

and the width for H $\rightarrow$ ZZ is the same with $M_{\text{W}} \rightarrow M_{\text{Z}}$ and a factor of $\frac{1}{2}$ to allow for the two identical bosons in the final state. The branching ratio for H $\rightarrow$ W$^+$W$^-$ is about one-third of that for the b$\bar{\text{b}}$ channel. Again, the decay of the ZZ final state to four charged leptons gives excellent resolution for m_{H}.

By early 2012, the ATLAS and CMS experiments at the LHC had excluded an m_{H} value in the interval 129 GeV to 539 GeV at the 95% CL, and the mass region 120–130 GeV was under intensive experimental study, excesses of events having been reported by both ATLAS and CMS in the region 124–126 GeV with the hint of a signal in the $\gamma\gamma$ channel at 126.5 GeV (Aad *et al.* 2012a, Chatrchyan *et al.* 2012a). Then, on July 4, 2012, the ATLS and CMS collaborations simultaneously announced the observation (at a significance greater than 5σ) of a new boson with a mass in the range 125–126 GeV, and with properties compatible with those of a SM Higgs boson. These results (updated) were reported in Aad *et al.* (2012b) and Chatrchyan *et al.* (2012b). The crucial channels in the discovery were the decay modes H $\rightarrow$ $\gamma\gamma$ and H $\rightarrow$ ZZ* $\rightarrow$ four leptons, both of which provide a high-resolution invariant mass for fully reconstructed candidates. The channel H $\rightarrow$ WW* $\rightarrow$ $\ell\nu\ell\nu$ is equally sensitive, but has low resolution because the neutrinos are not reconstructed. The ATLAS result for the mass of the boson was (Aad *et al.* 2012b)

$$126.0 \pm 0.4(\text{stat.}) \pm 0.4(\text{syst.}) \, \text{GeV} \tag{22.202}$$

and the CMS result was (Chatrchyan *et al.* 2012b)

$$125.3 \pm 0.4(\text{stat.}) \pm 0.5(\text{syst.}) \, \text{GeV}. \tag{22.203}$$

At about the same time, the CDF and D0 collaborations at the Tevatron reported the combined results of their searches for a SM Higgs boson produced in association with a W or a Z boson, and subsequently decaying to a b$\bar{\text{b}}$ pair. The data corresponded to an integrated luminosity of 9.7 fb^{-1}. An excess of events was observed in the mass range 120–135 GeV, at a significance of 3σ, which was interpreted as evidence for a new particle, consistent with the SM Higgs boson (Aaltonen *et al.* 2012). This provided the first evidence for the decay of the new particle to a fermion-antifermion pair, at a rate consistent with the SM prediction.

A measure of the compatibility of the observed boson with the SM Higgs boson is provided by the best-fit value of the signal strength parameter μ defined by

$$\mu = (\sigma \cdot \text{BR})_{\text{observed}}/(\sigma \cdot \text{BR})_{\text{SM}} \tag{22.204}$$

where σ is the boson production cross section and BR is the branching ratio of the boson to the relevant final state. 'SM' denotes the SM predicted value, so that $\mu = 1$ is the SM hypothesis. In those first discoveries, ATLAS reported a best-fit μ-value of $\mu = 1.4 \pm 0.3$ for $m_{\text{H}} = 126$ GeV; the μ-values for the individual channels were all within one standard deviation of unity. CMS reported a best-fit μ-value of $\mu = 0.87 \pm 0.23$ at $m_{\text{H}} = 125.5$ GeV, and again the individual values in the observed channels were within one standard deviation of unity. The conclusion was, therefore, that the results were consistent, within the quoted uncertainties, with the predictions for the SM Higgs boson.

More recent data from the LHC has amply confirmed this conclusion. The mass is now known to an ccuracy of 0.1% : $m_{\text{H}} = 125.25 \pm 0.17$ GeV (Workman *et al.* 2022). As an

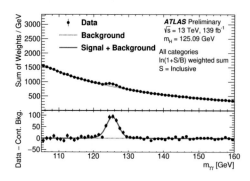
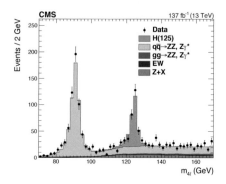

FIGURE 22.35

(Left) The invariant mass distribution of $\gamma\gamma$ candidates, with each event weighted by the ration of signal-to-background in each event category, observed by ATLAS (ATLAS 2020a); the residuals of the data with respect to the fitted background are shown in the lower plot. (Right) the invariant mass distribution in the four lepton channel observed by CMS (Sirunyan *et al.* 2021).

illustration of the relevant data, we show in Figure 22.35 (taken from the review of Higgs boson physics by Carena *et al.* (2022), section 11 of Workman *et al.* 2022) the invariant mass distributions from more recent data from Run 2 of the LHC, the left portion for the $\gamma\gamma$ channel (ATLAS 2020a), and the right portion for the four lepton channel (Sirunyan *et al.* 2021a) (the peak at the Z mass is due to production of an on-shell Z boson which decays to two charged leptons, one of which radiates a photon which then produces a second charged lepton pair.) The signal strength in Run 2 quoted by ATLAS (ATLAS 2019) was 1.02 ± 0.14, and by CMS was (Sirunyan *et al.* 2021b) 1.12 ± 0.09.

The width Γ_H the Higgs boson is determined once the mass is known, and is predicted to be about 4 MeV. Fermionic decays account for about 75% of the rate, the remainder coming from decays to vector bosons. However, the experimental mass resolution in the decay channels with the best resolution (those shown in figure 22.35) is about 1–2 GeV, much larger than the expected width. As a result, direct measurements of the width from the invariant mass distributions give only very weak upper limits, $\Gamma_H <$ a few GeV. With further assumptions, indirect constraints have been obtained by both CMS (Sirunyan *et al.* 2019b) and ATLAS (Aaboud *et al.* 2018b):

$$\Gamma_H < 14.4\,\text{MeV (ATLAS)}, \quad \Gamma_H = 3.2^{+2.8}_{-2.2}\,\text{MeV (CMS)}. \tag{22.205}$$

According to the SM, the Higgs boson should have J^{PC} quantum numbers $J^{PC} = 0^{++}$. The decay to two photons fixes C to be even, given that photons are C-odd eigenstates and that C is a multiplicative quantum number, assumed conserved in this decay. This also implies that the spin cannot be unity (Landau 1948, Yang 1950). Various alternative models for the spin and parity have been experimentally probed. In the H $\to \gamma\gamma$ and H $\to ZZ^* \to 4\ell$ channels, ATLAS reported (Aad *et al.* 2015) that all alternatives to $J^P = 0^+$ were excluded at 99.9% confidence level. A similar study by CMS (Khachatryan *et al.* 2015) found that all observations were consistent with the quantum numbers $J^{PC} = 0^{++}$.

Apart from these properties of the Higgs boson itself, the experimental confirmation (or otherwise) of the couplings of the boson to the other SM particles is of fundamental importance, because they are determined by the electroweak symmetry breaking mechanism. This was a major goal of the LHC, and will be discussed in the following chapter.

22.8.3 The stability of the electroweak vacuum

An essential feature of the GSW theory is that the potential $V(\hat{\phi}) = -\mu^2(\hat{\phi}^\dagger\hat{\phi}) + (\lambda/4)(\hat{\phi}^\dagger\hat{\phi})^2$ should have a minimum at a non-zero expectation value of the field: $\langle 0|\hat{\phi}|0\rangle = v/\sqrt{2}$. This configuration is interpreted as the ground state, or vacuum, of the theory. At tree level, v is given by $v = 2\mu/\sqrt{\lambda}$. But as soon as loop corrections are included, and renormalization undertaken, the Lagrangian parameters will become energy-dependent — i.e. they will 'run', as usual. A question then arises: will the potential continue to have the required minimum, or will the shape of the potential change, so as to fail to have a minimum? In particular, if the running coupling $\lambda(E)$ should evolve so as to pass through zero and become negative, the potential will clearly have no minimum. This possibility was raised by Cabibbo *et al.* (1979), whose very clear presentation we now follow.

At one loop order, and in our notation, the evolution of λ is given by the equation

$$8\pi^2\frac{d\lambda}{dt} = 3\lambda^2 + 6\lambda y_t^2 - \frac{3\lambda}{2}(3g^2 + g'^2) - \left[12y_t^4 - \frac{3}{4}(2g^4 + (g'^2 + g^2)^2)\right] \qquad (22.206)$$

where y_t is the Yukawa coupling of the top quark (only the largest fermion coupling is retained), g and g' are the usual gauge couplings, and $t = \ln(E/v)$. The other couplings y_t, g, and g' will also run, but we do not give their evolution equations here. The situation is easy to analyze qualitatively if we neglect the running of y_t, g, and g'. Then, if the term in square brackets in (22.206) is positive, the quadratic function of λ on the right-hand side of (22.206) will have one positive and one negative root, $\lambda = -a$ and $\lambda = b$, where a and b are positive. It is then easy to see (Problem 22.13) by integrating

$$8\pi^2\frac{d\lambda}{dt} = (\lambda + a)(\lambda - b) \qquad (22.207)$$

that the point $\lambda = -a$ is an ultraviolet ($E \to \infty$) fixed point. Under these conditions, for any value of λ in the range $0 < \lambda \le b$, λ will evolve through zero and become negative at some fixed value of E, and the potential will have no minimum. A lower bound for $\lambda(v)$, and hence for $m_H = \sqrt{\lambda}v/\sqrt{2}$, can then be obtained by requiring that this does not happen for E less than some energy E_c.

The stated condition is

$$12y_t^4 - \frac{3}{4}(2g^4 + (g^2 + g'^2)^2) > 0 \qquad (22.208)$$

or equivalently

$$12m_t^4 - 3(2m_W^4 + m_Z^4) > 0, \qquad (22.209)$$

which is satisfied for the physical masses. In the more complicated case, where all couplings evolve with E, the equations are solved numerically, and the result displayed as a plot showing the lower bound on m_H as a function of the top quark mass $y_t(v)v/\sqrt{2}$, for a given choice of E_c.

In 1979, neither m_H nor m_t was known, of course. But the qualitative picture was clear: the larger the top quark mass, the higher the lower bound on m_H. Subsequently, the calculation was improved by including higher order corrections. For example, Degrassi *et al.* (2012) presented the complete NNLO (two-loop) analysis of the SM Higgs potential, obtaining two-loop corrections to the relation between λ and m_H. The evolution of λ shows that it becomes negative at energies of the order of 10^{11} GeV, given the now accurately measured top quark mass (a crucial parameter in this context). The relative nearness of this energy to the Planck scale led to speculation that the near instability of the electroweak vacuum might play a role in the inflationary stage of the universe. The vacuum is most likely metastable, its lifetime being determined by the rate of quantum tunnelling from the unstable vacuum to the true vacuum, and has been estimated to be more than 10^{56} years (Andreassen, Frost, and Schwartz, 2018). Of course, it should be borne in mind that all the foregoing presupposes that no new physics intervenes between the electroweak scale v and the instability scale $\sim 10^{11}$ GeV — i.e. that the SM is valid up to a scale many times that of v.

Problems

22.1

(a) Using the representation for $\boldsymbol{\alpha}, \beta$, and γ_5 introduced in section 20.2.2 (equation (20.14)), massless particles are described by spinors of the form

$$u = E^{1/2} \begin{pmatrix} \phi_+ \\ \phi_- \end{pmatrix} \qquad \text{(normalized to } u^\dagger u = 2E)$$

where $\boldsymbol{\sigma} \cdot \hat{\boldsymbol{p}} \phi_\pm = \pm \phi_\pm, \hat{\boldsymbol{p}} = \boldsymbol{p}|\boldsymbol{p}|$. Find the explicit form of u for the case $\hat{\boldsymbol{p}} = (\sin\theta, 0, \cos\theta)$.

(b) Consider the process $\bar{\nu}_\mu + \mu^- \to \bar{\nu}_e + e^-$, discussed in section 22.1, in the limit in which all masses are neglected. The amplitude is proportional to

$$G_F \bar{v}(\bar{\nu}_\mu, R)\gamma_\mu(1 - \gamma_5)u(\mu^-, L)\bar{u}(e^-, L)\gamma^\mu(1 - \gamma_5)v(\bar{\nu}_e, R)$$

where we have explicitly indicated the appropriate helicities R or L (note that, as explained in section 20.2.2, $(1-\gamma_5)/2$ is the projection operator for a right-handed antineutrino). In the CM frame, let the initial μ^- momentum be $(0, 0, E)$ and the final e^- momentum be $E(\sin\theta, 0, \cos\theta)$. Verify that the amplitude is proportional to $G_F E^2(1 + \cos\theta)$. (Hint: evaluate the 'easy' part $\bar{v}(\bar{\nu}_\mu)\gamma_\mu(1 - \gamma_5)u(\mu^-)$ first; this will show that the components $\mu = 0, z$ vanish, so that only the $\mu = x, y$ components of the dot product need to be calculated.)

22.2 Verify equation (22.20).

22.3 Check that when the polarization vectors of each photon in figures 22.7(a) and (b) is replaced by the corresponding photon momentum, the sum of these two amplitudes vanishes.

22.4 By identifying the part of (22.45) which has the form (22.57), derive (22.58).

22.5 Using the vertex (22.48), verify (22.79).

22.6 Insert (22.29) into (22.155) to derive (22.157).

22.7 Verify that the neutral current part of (22.163) is diagonal in the 'mass' basis.

22.8 Suppose that the Higgs field is a triplet of $SU(2)_L$ rather than a doublet; and suppose that its vacuum value is

$$\langle 0|\hat{\phi}|0\rangle = \begin{pmatrix} 0 \\ 0 \\ f \end{pmatrix}$$

in the gauge in which it is real. The non-vanishing component has $t_3 = -1$, using

$$t_3 = \begin{pmatrix} 1 & 0 & 0 \\ 0 & 0 & 0 \\ 0 & 0 & -1 \end{pmatrix}$$

in the 'angular-momentum-like' basis. Since we want the charge of the vacuum to be zero, and we have $Q = t_3 + y/2$, we must assign $y(\hat{\phi}) = 2$. So the covariant derivative on $\hat{\phi}$ is

$$(\partial_\mu + \mathrm{i}g\boldsymbol{t} \cdot \hat{\boldsymbol{W}}_\mu + \mathrm{i}g'\hat{B}_\mu)\hat{\phi},$$

where

$$t_1 = \begin{pmatrix} 0 & \frac{1}{\sqrt{2}} & 0 \\ \frac{1}{\sqrt{2}} & 0 & \frac{1}{\sqrt{2}} \\ 0 & \frac{1}{\sqrt{2}} & 0 \end{pmatrix}, \quad t_2 = \begin{pmatrix} 0 & \frac{-\mathrm{i}}{\sqrt{2}} & 0 \\ \frac{\mathrm{i}}{\sqrt{2}} & 0 & \frac{-\mathrm{i}}{\sqrt{2}} \\ 0 & \frac{\mathrm{i}}{\sqrt{2}} & 0 \end{pmatrix}$$

and t_3 is as above (it is easy to check that these three matrices do satisfy the required SU(2) commutation relations $[t_1, t_2] = \mathrm{i}t_3$). Show that the photon and Z fields are still given by (22.36) and (22.37), with the same $\sin\theta_{\mathrm{W}}$ as in (22.39), but that now

$$M_Z = \sqrt{2}M_{\mathrm{W}}/\cos\theta_{\mathrm{W}}.$$

What is the value of the parameter ρ in this model?

22.9 Use (22.181) to verify (22.183).

22.10 Calculate the lifetime of the top quark to decay via $\mathrm{t} \to \mathrm{W}^+ + \mathrm{b}$.

22.12 Using the Higgs couplings given in appendix Q, verify (22.200) and (22.201).

22.13 By integrating (22.207), show that $\lambda = -a$ is an ultraviolet fixed point.

23

Further Developments

In the years following the discoveries of the top quark and the Higgs boson, enormous efforts, both experimental and theoretical, have been made to probe ever more precisely the predictions of the Standard Model (SM). For reasons which we shall discuss in section 23.3, the prevailing expectation has been that evidence would be found for discrepancies that indicated the presence of new physics, going beyond the SM, at a mass scale not much higher than a few TeV. It has turned out, however, that the SM has performed extremely well, as we shall discuss in sections 23.1 and 23.2, which appears to imply that the mass scale of any possible new physics is substantially higher. This in turn creates a problem for what is called the 'naturalness' of the Higgs sector of the SM, as we shall explain in section 23.3. In section 23.4 we provide elementary introductions to the two main approaches to resolving the naturalness problem. The first is supersymmetry, in which the SM Higgs boson is treated as an elementary spin-0 particle, and naturalness is maintained by virtue of the ultraviolet properties of a supersymmetric theory. The second approach treats the Higgs field, not as elementary, but as dynamically generated by new strong interactions at the TeV scale ('dynamical symmetry breaking'). Both approaches have been extensively pursued, both theoretically and experimentally, but without any clear experimental support as yet. This has led to a more general, less model-dependent, approach, in which the SM is extended to include additional (non-renormalizable) terms, so as to constitute a Standard Model Effective Field Theory (SMEFT) for systematically probing possible discrepancies between SM predictions and experiment. This will be the topic of the last section.

23.1 The couplings of the Higgs boson to SM particles

The couplings of the SM Higgs boson to the SM gauge bosons are determined by the $SU(2)_L \times U(1)$ covariant derivative in (22.30) and by (22.29) in unitary gauge:

$$\mathcal{L}_{WH,ZH} = \frac{2m_W^2}{v} \hat{W}_\mu^+ \hat{W}^{-\mu} \hat{H} + \frac{m_Z^2}{v} \hat{Z}_\mu \hat{Z}^\mu \hat{H} + \frac{m_W^2}{v^2} \hat{W}_\mu^+ \hat{W}^{-\mu} \hat{H}^2 + \frac{m_Z^2}{2v^2} \hat{Z}_\mu \hat{Z}^\mu \hat{H}^2. \quad (23.1)$$

The Higgs boson couplings to fermions are, of course, not determined by the gauge symmetry, but are simply parameters (masses and mixing matrices) taken from experiment. In particular, they are by no means universal, being indeed very different as between the three generations. These (Yukawa-like) couplings generate the masses via (22.132):

$$\mathcal{L}_{fH} = -\frac{m_f}{v} \bar{\hat{f}} \hat{f} \hat{H} \quad (23.2)$$

The Higgs boson self-couplings are determined from the Higgs potential in (22.30) after the insertion of (22.29), and are

$$\mathcal{L}_{3H,4H} = \frac{m_H^2}{2v} \hat{H}^3 + \frac{m_H^2}{8v^2} \hat{H}^4. \quad (23.3)$$

DOI: 10.1201/9781003411666-23

It is noteworthy that all the couplings are fixed once all the particle masses are known. The largest and experimentally most accessible couplings are those to the vector bosons and to the third generation fermions: the b and t quarks, and the τ lepton. One of the most prominent goals, and achievements, of the LHC and the ATLAS and CMS detectors has been the direct observation of the Yukawa couplings to b quarks, τ leptons, and t quarks. Here we shall give a brief summary of some characteristic results, referring the reader interested in more detail to the extensive discussion by Carena *et al.* (2022) in section 11 of Workman *et al.* (2022), from which our account is largely drawn.

The coupling to b quarks is accessed via the decay $H \rightarrow b\bar{b}$, which has a branching fraction of about 58%. The most sensitive production modes are the associated WH and ZH production processes, where the leptonic decays of the vector bosons can be used for triggering, to reject QCD backgrounds, and to reconstruct the vector bosons. The Higgs boson candidate mass is reconstructed from two b-tagged jets. Due to limited mass resolution, the SM Higgs boson signal is expected to appear as a broad enhancement in the reconstructed dijet mass distribution. Results from ATLAS and CMS are reported in Aad *et al.* (ATLAS) (2021a) and Sirunyan *et al.* (CMS) (2018a), which quote signal strengths of $1.02^{+0.12\ +0.14}_{-0.11\ -0.13}$ (ATLAS) and 1.01 ± 0.22 (CMS) respectively. Recalling that a signal strength of unity is equivalent to the SM value, these results constitute convincing evidence for this coupling. Both experiments have also searched for $H \rightarrow b\bar{b}$ in the vector boson fusion mode. Here the sensitivity is less good than in the VH mode,

The branching fraction for $H \rightarrow \tau^+\tau^-$ is about 6%. The vector boson fusion production process, with a pair of τs and two energetic jets separated by a large pseudo-rapidity, has the best sensitivity. The $\tau^+\tau^-$ invariant mass is reconstructed from a kinematic fit to the visible charged products from the τ decays, but the unseen neutrinos make the mass resolution poor (about 15%). Again, the signal is a broad excess over the expected background in the $\tau^+\tau^-$ invariant mass. Analyzing early data from Run 2 at the LHC, Aaboud *et al.* (ATLAS) (2019) reported an excess at a significance of $4.4\ \sigma$, which rose to $6.4\ \sigma$ when combined with the Run 1 results. Analogous results were reported by Sirunyan *et al.* (CMS) (2018b), the corresponding significances being $4.9\ \sigma$ and $5.9\ \sigma$. These results constitute unambiguous observation of this process.

Substantial indirect evidence for the coupling of the Higgs boson to top quarks is provided by the agreement with the SM prediction for production of H via gluon fusion, which is dominated by the top quark loop. Direct access to this coupling is provided by the process in which H is produced in association with a $t\bar{t}$ pair (figure 22.33), with H detected via the $H \rightarrow \gamma\gamma$ decay. ATLAS (Aaboud *et al.* 2018a) reached an observed significance of $5.8\ \sigma$ with the Run 2 partial data set, which increased to 6.3σ when combined with the Run 1 results. CMS (Sirunyan *et al.* 2018c) observed an excess of events over background at the 5.2σ level, combining Run 1 and Run 2 results. With the larger Run 2 dataset, ATLAS (Aad *et al.* 2020) reached an observed sensitivity of 5.2σ, and CMS (Sirunyan *et el.* 2020) reported a sensitivity of 6.6σ. These results provide very strong direct evidence for this coupling.

The next most massive fermion is the muon. The branching fraction in the $H \rightarrow \mu^+\mu^-$ channel for a SM Higgs boson of mass 125 GeV is 2.2×10^{-4}, about one tenth of that for $H \rightarrow \gamma\gamma$. The $(\mu^+\mu^-)$ invariant mass resolution is a few percent for both ATLAS and CMS. Both ATLAS (Aad *et al.* 2021b) and CMS (Sirunyan *et al.* 2021c) have observed an excess of events in the $m_{\mu^+\mu^-}$ distribution over a smooth background, with signal strengths of 1.2 ± 0.6 and $1.19 \pm 0.40(\text{stat.}) \pm 0.15(\text{syst.})$ respectively, and with significances of 2.0σ and 3.0σ respectively. These results are the first direct evidence for the Yukawa coupling of the Higgs boson to second generation fermions.

The results of these fundamentally important measurements of Higgs boson couplings to fermions are all consistent with the SM predictions. An equally important coupling is the trilinear self coupling (the $\hat{H}^3$ term in (23.3)). This can be accessed through the process

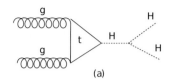 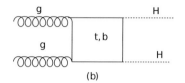

FIGURE 23.1
Feynman diagrams contributing at leading order to Higgs boson pair production through (a) trilinear self-coupling of the Higgs boson and (b) through a t and b quark loop.

shown in figure 23.1(a), which is Higgs boson pair production via the trilinear self-coupling to a Higgs boson produced in gluon fusion. This process competes with the Higgs boson pair production via the fermion (mainly top quark) loop diagram shown in figure 23.1(b). These processes interfere negatively, reducing the production rate from that of figure 23.1(a) alone. Only rather weak constraints have so far been obtained on the trilinear coupling. At the High Luminosity LHC, the significance of the HH process is expected to reach 4σ. Constraints on the quartic coupling, accessed from the HHH final state, are out of reach at the LHC, due to the very small production rates and the complicated final states.

Beyond establishing direct evidence for the various couplings of the Higgs boson to SM particles, one might like to have a more quantitative measure of the agreement (or otherwise) between the data and the SM predictions. A conceptually simple scheme is one in which the couplings are modified by parameters κ_i, defined as the ratios of the measured couplings to their SM values. Thus when all the κ_i are unity, the SM is reproduced. (The idea is reminiscent of the parameters ρ_f and κ_f introduced in the discussion of electroweak radiative corrections in section 22.6). In a simple example, one restricts the relative coupling of the Higgs boson to W and Z bosons to be fixed at its SM value, so that $\kappa_W = \kappa_Z \equiv \kappa_V$. In addition, one assigns the same modifier κ_F to all fermion couplings. Data can then be fitted in this two-dimensional parameter space. The ATLAS and CMS combined measurements with the Run 1 dataset led to (ATLAS and CMS 2015)

$$\kappa_V = 1.03 \pm 0.03, \quad \kappa_F = 0.97 \pm 0.07. \tag{23.4}$$

The ATLAS Run 2 dataset yielded (ATLAS 2020b)

$$\kappa_V = 1.05 \pm 0.04, \quad \kappa_F = 1.05 \pm 0.09. \tag{23.5}$$

A more elaborate scheme of the same type introduces effective couplings to gluons and photons with modifiers κ_g and κ_γ, as well as $\kappa_W, \kappa_Z, \kappa_t, \kappa_b$, and κ_τ. We quote here the results for the fermion modifiers as reported by ATLAS (ATLAS 2020b) and CMS (Sirunyan *et al.* 2019a) using a partial LHC Run 2 dataset:

modifier	ATLAS	CMS	HL − LHC (expected)
κ_t	1.00 ± 0.12	$1.01^{+0.06\;+0.09}_{-0.06\;-0.08}$	3.4%
κ_b	$0.98^{+0.14}_{-0.13}$	$1.18^{+0.14\;+0.13}_{-0.13\;-0.24}$	3.7%
κ_τ	$1.05^{+0.15}_{-0.14}$	$0.94^{+0.08\;+0.09}_{-0.11\;-0.06}$	1.9%

The last column gives the expected precision at the High Luminosity LHC (Cepeda *et al.* 2019).

Several other possible parametrizations are discussed by Carena *et al.* (2022), in Workman *et al.* (2022). In summary, one can conclude that all experimental results are consistent, so far, with the electroweak symmetry breaking mechanism of the SM.

The coupling modifier scheme was a practical and simple way to represent the extent to which the experimental data aligned with the SM predictions. Yet it has several flaws. One is that it does not probe any possible physics beyond the standard model (BSM). Another is that it cannot be systematically improved by perturbative corrections. It also cannot relate LHC to LEP data. One approach which overcomes these limitations is to characterize the possible deviations from SM physics by the use of Effective Field Theories, and in particular the SMEFT, which has the same field content as the SM, and the same linearly realized $SU(3)_c \times SU(2)_L \times U(1)_y$ symmetry. We shall give an introduction to this topic in section 23.5. The relation of the κ formalism to the SMEFT (Standard Model Effective Field Theory) is discussed in section 9.2 of Brivio and Trott (2019).

23.2 Precision physics in top quark production at the LHC

As an example of the precision tests now being applied to the SM, we consider the case of top quark production at the LHC. The leading order processes were discussed in section 22.7.2. Higher order perturbative QCD corrections to the inclusive total cross section for top quark pair production take the form (following Czakon, Fiedler, and Mitov 2013, and as we wrote in (15.63))

$$\sigma_{ij}(\hat{s}) = \frac{\alpha_s^2}{m_t^2} \left(\sigma_{ij}^{(0)} + \alpha_s \sigma_{ij}^{(1)} + \alpha_s^2 \sigma_{ij}^{(2)} + O(\alpha_s^3) \right) \tag{23.6}$$

where α_s is the $\overline{MS}$ QCD coupling constant with 5 active massless quarks at a renormalization scale $\mu_R^2 = m_t^2$. Here, for $ij = q\bar{q}$, the quantity $\frac{\alpha_s^2}{m_t^2}\sigma_{ij}^{(0)}$ is the LO cross section of (22.195), and similarly for $ij = gg$ and (22.196). The quantity $\sigma_{ij}^{(1)}$ is the one-loop next-to-leading order (NLO) contribution, $\sigma_{ij}^{(2)}$ is the two-loop next-to-next-to-leading order (NNLO) contribution, and so on. Apart from $\sigma_{ij}^{(0)}$, all the other parton cross sections are functions of $\hat{s}$ and of the factorization scale μ_F (see section 15.6.2), as well as of μ_R. Frequently both μ_R and μ_F are set equal to m_t.

The first step in calculating the higher-order contributions in (23.6) was taken by Nason, Dawson, and Ellis (1988), and by Beenakker *et al.* (1989), who calculated the NLO terms; the amplitudes were evaluated numerically. Some twenty years later, exact analytic results were obtained (Czakon and Mitov 2010). Remarkably, the very complicated NNLO corrections are now available (Baernreuther, Czakon and Mitov 2012, Czakon and Mitov 2012, Czakon and Mitov 2013, Czakon, Fiedler and Mitov 2013, Czakon, Heymes and Mitov 2016, Catani *et al.* 2019). Let us first consider results reported by Catani *et al.* (2019), for the total t$\bar{\text{t}}$ cross section in pp collisions at $\sqrt{s} = 13$ TeV. They fix $m_t = 173.3$ GeV, consider 5 massless quark flavours, and use the corresponding NNPDF31 (Ball *et al.* 2017) sets of parton distribution functions (PDFs), with $\alpha_s(m_Z) = 0.118$. The predictions for LO, NLO, and NNLO are made by using the PDFs at the corresponding perturbative order and the evolution of α_s at the next higher order. QCD scale uncertainties are estimated by varying the renormalization scale μ_R and the factorization scale μ_F by a factor of 2 around their common central value μ_0, which is taken to be m_t. These authors present the following results for the total cross section at the different orders, and the corresponding scale uncertainties:

$$\sigma_{\text{LO}}^{t\bar{t}} = 478.9(1)^{+29.6\%}_{-21.4\%}\text{pb}, \sigma_{\text{NLO}}^{t\bar{t}} = 726.9(1)^{+11.7\%}_{-11.9\%}\text{pb}, \sigma_{\text{NNLO}}^{t\bar{t}} = 794.0(8)^{+3,5\%}_{-5.7\$}\text{pb}. \tag{23.7}$$

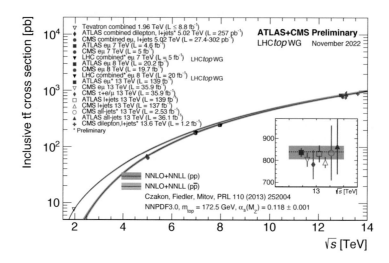

FIGURE 23.2
Measured and predicted $t\bar{t}$ cross section from the Tevatron energy in $p\bar{p}$ collisions to LHC energies in pp collisions. Retrieved on 15 May 2023 from https://twiki.cern.ch/twiki/bin/view/LHCPhysics/LHCTopWGSummaryPlots.

Several points are worth noting. First, the NLO result is a large correction to the LO (tree graph) result. But, secondly, the NNLO result is relatively smaller, about a 10 % effect. This is reassuring, as it indicates that the terms in the series, at least as carried thus far, are becoming substantially smaller. Thirdly, although the LO and the NLO numbers are not fully consistent with the corresponding quoted uncertainties, the NLO and the NNLO numbers are consistent, suggesting that the scale variations can be used to estimate the size of contributions beyond NNLO (recall that if corrections at all orders were included, there should be no scale dependence).

The cross section is substantially smaller at the lower Tevatron energy. Czakon, Fiedler and Mitov (2013) reported the value $\sigma^{t\bar{t}} = 7.16^{+0.11+0.17}_{-0.20-0.12}$pb for $p\bar{p}$ collisions at $\sqrt{s} = 1.96$ TeV, where the first uncertainty is from scale dependence and the second is from the PDFs. This calculation set $m_t = 173.3$ GeV, and used the MSTW2008nnlo68cl PDF set (Martin *et al.* 2009). It also included soft gluon resummation at next-to-text-to-leading-logarithm (NNLL) order (Cacciari *et al.* 2012, Beneke, Falgari and Schwinn 2010, Czakon, Mitov and Sterman 2009).

Figure 23.2 shows the measured and predicted $t\bar{t}$ cross section from the Tevatron energy in $p\bar{p}$ collisions to LHC energies in pp collisions. The agreement is excellent. One can see the importance of the valence antiquarks in the $p\bar{p}$ case. At LHC energies, the dominant production mechanism is via the gg channel, and the $pp - p\bar{p}$ difference nearly disappears. These precision measurements provide very stringent tests of the perturbative QCD calculations at the NNLO and NNLL level.

As for $t\bar{t}$ production, predictions are available beyond LO for single top production, as illustrated by the following examples. The NNLO cross section for $\hat{t}$-channel $t + \bar{t}$ production at $\sqrt{s} = 1.96$ TeV with $m_t = 173.2$ GeV, was calculated (Campbell, Neumann, and Sullivan 2021) to be $2.08^{+0.04+0.05}_{-0.03-0.10}$ pb, where the first uncertainty is from scale dependence and the second is from the PDFs. Also at the Tevatron energy, an NNLO calculation of the $\hat{s}$-channel process gave $1.03^{+0.05}_{-0.05}$ pb (Kidonakis 2010a). The W t associated production is

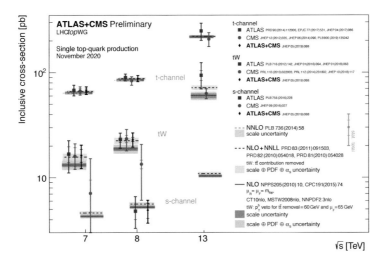

FIGURE 23.3
Comparison of measured and predicted single top production cross sections at LHC energies in p p collisions. Retrieved on 15 May 2023 from https://twiki.cern.ch/twiki/bin/view/ LHCPhysics/LHCTopWGSummaryPlot.

negligible at the Tevatron energy. For the LHC, the NNLO cross section for the $\hat{t}$-channel process at $\sqrt{s} = 14\,\text{TeV}$ was predicted (Campbell, Neumann, and Sullivan 2021, Berger, Gao, and Zhu 2017) to be $245^{+2.7}_{-1.3}$ pb, where the uncertainty is due to scale dependence. At $\sqrt{s} = 13\,\text{TeV}$, an approximate NNLO calculation predicted the $t + \bar{t}$ cross section to be $71^{+1.8\,+3.40}_{-1.8\,-3.00}$ pb (Kidonakis 2010b).

The experiments measuring single top production are reviewed by Lees, Maltoni, and Quadt (2022) in section 61 ("Top Quark") of Workman *et al.* 2022. Figure 23.3 summarizes the comparison between the theoretical predictions of perturbative QCD (up to NNLO) and the values measured at the LHC, as a function of the centre-of-mass energy. All cross section measurements are very well described by the theoretical calculations, within the quoted uncertainties.

The NNLO calculations for top quark pair and single production, to which we have referred, are just two examples of what has been called the 'two loop explosion' (see the Cern Courier, 17 March 2017). The level of precision now attained by the LHC experiments requires that these very complex calculations have to be undertaken in order to fully test the theory. This theoretical effort began in 1991, with the first NNLO calculation, of inclusive Drell-Yan production (recall section 9.4) by Hamberg, van Neerven and Matsuura (1991). This was followed in 2002 by the NNLO calculation of inclusive Higgs production in gluon-gluon fusion (Harlander and Kilgore 2002). These calculations revealed that, while the previous NLO corrections had been relatively large, the NNLO corrections were sensibly smaller, as we also noted concerning the results of the top quark pair production cross sections quoted in (23.7). They also indicated, similarly, that the scale dependence uncertainty was much reduced by going to NNLO, thereby substantially improving the accuracy of the theoretical calculations. Starting from about 2015, the results of the calculations of a large number of processes began to appear, and today all LHC cross sections involving $2 \rightarrow 2$ parton processes are known to NNLO accuracy. In the future, we may expect theory and experiment to push each other to new levels of precision.

23.3 Sensitivity of the Higgs potential to higher mass scales: the 'naturalness' problem

Somewhat paradoxically, the very success of the SM has brought into sharper focus an issue which has long troubled theorists: that of large quantum corrections (self-energy contributions) to the mass parameter of a scalar field theory — in this case, the parameter multiplying the $\hat{\phi}^\dagger \hat{\phi}$ term in the Higgs potential.

The problem arises when we regard the SM as a successful effective theory, valid for a limited energy range. Indeed, it seems reasonable to believe that the SM, despite its impressive consistency with the data, is not the 'Final Theory'; after all, there are too many obvious questions to which it provides no answers (to mention just four: what causes spontaneous electroweak symmetry breaking? why are there three generations? why do the fermion masses have the values they do? how is baryogenesis accounted for?) Granted, the theory is renormalizable, which means that in principle it is formally valid up to arbitrarily high energies. But in practice we believe that a more complete theory will intervene at some energy scale Λ beyond the electroweak scale set by v. At the very least, for example, one would expect that some kind of new physics will be necessary at the scale where quantum gravity becomes important, which is thought to be the Planck scale $M_{\mathrm{P}} = (G_{\mathrm{N}})^{-1/2} \simeq 1.2 \times 10^{16}$ TeV. From this perspective, it is plausible to regard the SM as a 'low energy' effective theory, as discussed in section 11.8 of volume 1. We saw there that terms in a Lagrangian density could be usefully characterized by their mass dimension. Non-renormalizable interactions had couplings carrying negative mass dimension $-p$, and would be suppressed at low energies E by powers $(E/\Lambda)^p$, for $E \ll \Lambda$. Couplings of renormalizable interactions would be dimensionless, and would tend to dominate at energies well below Λ. This idea was further discussed from the Wilsonian renormalization group point of view, in section 16.4.3. In that language, the free-field (kinetic) term was defined to be unchanged under the scale transformation $\mathbf{x}' = \mathbf{x}/f$, so that a scalar field ϕ scaled as $\phi' = f\phi$, allowing us to identify f with the mass dimension of the field. Scaling laws then followed for terms in the Lagrangian. The mass dimension of the coupling of a term such as ϕ^6 in the Lagrangian density is -2, which scales as $1/f^2$, and tends to zero at large values of f, and was therefore called an 'irrelevant' term. This exactly parallels the discussion in section 11.8, where such a non-renormalizable term would scale as $(E/\Lambda)^2$. Renormalizable terms were characterized as 'marginal', and scaled as the kinetic terms. But there was also a class of terms called 'relevant', to which a scalar mass term $m^2\phi^2$ belongs. Here the coupling constant m^2 scales as the positive power f^2, corresponding to an energy scaling of $(\Lambda/E)^2$. Such a term, which is of course present in the Higgs potential, creates a problem for regarding the SM as an effective theory, as already suggested at the end of section 16.4.

Consider the mass term for the Higgs boson in the SM, and assume that such a term survives as part of the fuller theory at scale Λ. At that scale, we can write it as $g_m \Lambda^2 \hat{\phi}^\dagger \hat{\phi}$, on dimensional grounds. Its contribution will then scale as $g_m \Lambda^2 / E^2$ relative to the kinetic term. At the weak scale, we will arrive at a term of order $g_m (\Lambda/v)^2 \hat{\phi}^\dagger \hat{\phi}$, meaning that m_{H} will be of order $(g_m)^{1/2}(\Lambda/v)$. But we know that the physical m_{H} is of order v. Hence if, as we are assuming, there is a substantial energy gap — perhaps several orders of magnitude — between the more complete theory and the SM, the coupling g_m will have to be delicately tuned to a small value, which appears contrived, or 'unnatural'.

We can reach the same conclusion by considering just the one-loop corrections to the $\hat{\phi}^\dagger \hat{\phi}$ term, according to the SM. In section 22.6 we noted that figure 22.26 gives a quadratically divergent ($O(\Lambda^2)$) and positive contribution to the $\hat{\phi}^\dagger \hat{\phi}$ term in the Lagrangian, at one loop order. This term would ordinarily, of course, be just a contribution to the mass term of the scalar field, and it would be added to the bare mass term to give the renormalized mass, taken from experiment in a standard renormalizable scalar field theory. But when the SM is regarded as a low energy effective theory, valid up to some energy scale Λ, the matter is more delicate. The whole phenomenology depends on the coefficient of the $\hat{\phi}^\dagger\hat{\phi}$ term having a negative value, of order $-v^2$, triggering the spontaneous breaking of the symmetry at the scale v. This means that the $O(\Lambda^2)$ one–loop correction must combine with the 'bare' term $\frac{1}{2}m_{\mathrm{H},0}^2\hat{\phi}^\dagger\hat{\phi}$ so as to achieve a negative coefficient of order $-v^2$. This cancellation between $m_{\mathrm{H},0}^2$ and Λ^2 will have to be very precise if Λ—the scale of 'new physics'—is very high.

This is not quite the whole story, however. In the SM, the Higgs boson also has trilinear couplings to the W and Z bosons, and to the fermions. So in addition to the Higgs boson loop of figure 22.26, there are contributions to the Higgs boson self-energy from W, Z, and fermion loops (of which much the most significant is that of the top quark, as usual). The total of all these contributions turns out to be (Veltman 1981, Decker and Pestieau 1980)

$$-\frac{3}{8\pi^2 v^2}(4M_t^2 - 2M_W^2 - M_Z^2 - M_H^2)\Lambda^2 \simeq -0.05\Lambda^2. \tag{23.8}$$

If this correction is not to overwhelm the desired scale $-v^2$, then the scale of the new physics should not be much greater than a few TeV. New physics on a scale much greater than this would not be 'natural'.

This is the problem of 'fine-tuning', or 'unnaturalness', once again. The serious problem posed for the SM by this unnatural situation, which is caused by quadratic sensitivity to high mass scales in quantum corrections in scalar theories, was pointed out by Wilson (1971c). Gildener (1976) and Weinberg (1979b) emphasized the difficulty of finding a natural theory in which some scalar fields are associated with symmetry breaking at a 'Grand Unified' scale (of order 10^{16} GeV), while others are associated with breaking at the much lower electroweak scale; this is usually referred to as the 'gauge hierarchy' problem. 't Hooft (1980) also drew attention to difficulties posed by theories with unnaturally light scalars.

The reader may wonder why attention should *now* be drawn to this particular piece of renormalization: aren't all divergences handled by cancellations? In a sense they are, and (to repeat) the SM is renormalizable. But the fact is that this is the first case we have had in which we have to deal with a *quadratic* divergence. The other mass–corrections have all been logarithmic, for which there is nothing like such a dramatic 'fine–tuning' problem. There is a good reason for this in the case of the electron mass, which we remarked on in section 11.2. Chiral symmetry forces self–energy corrections for fermions to be proportional to their mass, and hence to contain only logarithms of the high scale Λ. Similarly, gauge invariance for the vector bosons prohibits any $O(\Lambda^2)$ corrections in perturbation theory. But there is no symmetry, within the SM, which 'protects' the coefficient of $\hat{\phi}^\dagger\hat{\phi}$ in this way. It is hard to understand what can be stopping it from being of order Λ^2, if we take the apparently reasonable point of view that the standard model will ultimately fail at some scale Λ where new physics enters.

We may place the naturalness concept in a wider context by saying that it amounts to the idea that phenomena occurring at one characteristic length (or energy) scale should not be sensitively dependent on phenomena occurring at a very different scale. Such an effective *separation of scales* does not seem to be an *a priori* necessity for a physical theory, but it is hard to imagine how centuries of progress in understanding the physical world could have been made if — for example — one had had to know about atomic physics in order to understand the dynamics of the solar system. Quantum field theory, however, adds a new twist to the story: the effect of virtual particles. We already pointed out in section 22.7.1 how the sensitivity to the top quark mass of electroweak radiative (loop) corrections allowed a remarkably accurate prediction to be made of this mass, even before any real top quarks were produced. It was conceivable that there might have existed a complicated dynamics at a higher energy scale which conspired to produce the apparent effect of a particle with that mass, but the natural assumption (soon proved correct) was that the observations could be explained by the indicated particle. An earlier example relates to the $K_L^0 - K_S^0$ mass difference Δm. Even before Glashow, Iliopoulos and Maiani (GIM) proposed the existence of the charm quark in 1970, Mohapatra *et al.* (1968) had estimated that the most divergent contribution to the relevant amplitude would be of order $G_F^2 \sin\theta_C \cos\theta_C \Lambda^2$ where Λ is the ultraviolet cutoff. This estimate implied that $\Lambda \sim 2 - 3\,\mathrm{GeV}$, much lower than the expected cutoff $G_F^{-1/2} \sim 300\,\mathrm{GeV}$. In their 1970 paper, GIM suggested that their fourth quark would enter into loop calculations for Δm, and identified m_c with the cutoff Λ, implying that $m_c \sim 2 - 3\,\mathrm{GeV}$, providing a natural explanation for Δm. A complete calculation of Δm was then made by Gaillard and Lee (1974), using the two-generation SM, enabling a firm prediction to be made for m_c.

On the other hand, some physical quantities seem to have no natural explanation — for example, the cosmological constant (Giudice 2013). This has the value $3.35\,\mathrm{GeV}/m^3$, which can be written in energy units as $(2.3 \times 10^{-3}\,\mathrm{eV})^4$. But there is no evidence that current theories start to fail at this low energy scale. Perhaps, then, it is best to regard naturalness as a heuristic aid — a useful approach that might not produce a unique solution, but which has a good enough track record to be taken seriously.

23.4   Maintaining naturalness

Gregory (Scotland Yard Detective): Is there any other point to which you would wish to draw my attention?

(Sherlock) Holmes: To the curious incident of the dog in the nighttime.

Gregory: The dog did nothing in the nighttime.

Holmes: That was the curious incident.

From *The Adventure of Silver Blaze* by Arthur Conan Doyle.

One might consider that violating naturalness is not quite as wicked as violating unitarity. Be that as it may, the desire to maintain naturalness in the Higgs sector of the SM is the paradigm that has driven much of particle physics research for the last forty years. The basic assumption has been that new 'beyond the standard model (BSM) physics' is likely to exist at an energy scale Λ not too far above the weak scale v, and so alleviate the naturalness problem. Two main approaches have been pursued. The first aims to find models where the quadratic divergences are absent, due to cancellations between the contributions of boson and fermion loops. This requires relations among the coupling constants, and leads to the exploration of supersymmetric theories (SUSY), in which bosons and fermions are grouped together in symmetry multiplets. Such theories remain perturbative up to high mass scales, and in them the Higgs-like boson is elementary. An introduction to SUSY and the Minimal Supersymmetric Standard Model (MSSM) is provided by Aitchison (2007). We shall introduce SUSY in section 23.4.1.

As an aside, it should be noted that, historically, supersymmetry was not originally studied as a response to the naturalness problem. Supersymmetric field theories in two dimensions had been discussed since the early 1970s (Ramond 1971, Neveu and Schwarz 1971, Gervais and Sakita 1971) in the context of string theory. Four-dimensional SUSY theories were also the subject of research (Wess and Zumino 1974a, 1974b, 1974c, Salam and Strathdee 1974). Indeed, Fayet (1975, 1976, 1977, 1978) pioneered SUSY extensions of the SM, and Farrar and Fayet (1978a, 1978b, 1980) had begun to explore the phenomenology of SUSY extensions of the SM. In 1981 an influential paper by Witten (Witten 1981) pointed out how SUSY could solve the naturalness problem, and soon afterwards Dimopoulos and Georgi (1981) developed a simple and realistic grand unified supersymmetric model. This led to the modern intense interest in SUSY theories.

A second approach (section 23.4.2) to maintaining naturalness rejects the idea of a fundamental scalar (Higgs) field, and seeks a dynamical mechanism for the spontaneous symmetry breaking in the SM, viewing the Higgs sector as a phenomenological model, rather than as part of a fundamental theory. This point of view harks back to the BCS mechanism for superconductivity (see chapter 17), or to the way in which the strong interactions of QCD cause spontaneous breaking of chiral symmetry (see chapter 18). In this case the Higgs boson is a composite bound state rather than elementary, and its internal structure provides a natural ultraviolet cutoff.

We may characterize these as 'top-down' approaches, in which specific models are advanced for the new physics at the scale Λ. But these attempts have, as we shall see, not yet received any clear empirical support. A more modest, 'bottom-up', approach is to parametrize a broad range of possible BSM physcs, through the framework of effective field theories, and in particular the Standard Model Effective Field Theory (SMEFT). We shall describe this approach in section 23.5.

23.4.1 Supersymmetry and the MSSM

A supersymmetry transformation 'rotates' a spin-0 field $\hat{\phi}$ into a spin-1/2 field $\hat{\chi}$, and vice versa, with transformations of the form

$$\delta_\xi \hat{\phi}(x) \overset{?}{=} \xi \hat{\chi}(x) \tag{23.9}$$

$$\delta_\xi \hat{\chi}(x) \overset{?}{=} \xi \hat{\phi}(x) \tag{23.10}$$

where ξ is some kind of infinitesimal parameter, like the ones introduced in chapter 12. But the ξ in (23.9) and (23.10) is very different. First of all, it must be a spinor (spin-1/2) object, which combines with the spin-1/2 $\hat{\chi}$ to make a spin-0 $\hat{\phi}$ on the left-hand side of (23.9). Next, consider the degrees of freedom in (23.9), which must match on both sides. If $\hat{\phi}$ is a real scalar field $\hat{\phi} = \hat{\phi}^\dagger$, then it has only one degree of freedom. But no spin-1/2 field has only one degree of freedom: the minimum is two, as in a 2-component Weyl spinor (see section 7.5.2). So (23.9) will involve a complex scalar field $\hat{\phi}$ and a 2-component spinor $\hat{\chi}$. We take $\hat{\chi}$ to be an L-type spinor, because the weak isospin of the SM acts on the L-components of the fermions. ξ will also be an L-type spinor. The parameters in ξ will be taken to be x-independent, but the components ξ_1 and ξ_2 will be anticommuting, like those of $\hat{\chi}$. Finally, consider the dimensions of ξ. The mass dimension of $\hat{\phi}$ is M^1, and that of $\hat{\chi}$ is $M^{3/2}$, from which it follows that ξ must have dimension $M^{-1/2}$. We shall write (23.9) as

$$\delta_\xi \hat{\phi}(x) = \xi \cdot \hat{\chi}(x) \tag{23.11}$$

where the dot product is the spin-0 combination of two spin-1/2 objects: $\xi \cdot \hat{\chi} = \xi_1 \hat{\chi}_2 - \xi_2 \hat{\chi}_1$. This combination is Lorentz invariant.

We can now move on to the transformation (23.10). The left-hand side has mass dimension $M^{3/2}$, so we need something with mass dimension M^1 on the right-hand side. As in the SM, the fields are all initially massless, and the only available quantity is the derivative

∂_μ. The required transformation is

$$\delta_\xi \hat{\chi}(x) = \sigma^\mu \sigma_2 \xi^* \partial_\mu \hat{\phi}(x) \tag{23.12}$$

where $\sigma^\mu = (1, \boldsymbol{\sigma})$. The rather complicated-looking right-hand side of (23.12) is needed to make it transform as an L-type spinor.

Just as with all the symmetries we have met so far, we can introduce the generators $\hat{Q}_1, \hat{Q}_2$ of the transformations (23.11) and (23.12), which will in this case be fermionic operators obeying the anticommutator algebra

$$\{\hat{Q}_a, \hat{Q}_b\} = (\sigma^\mu)_{ab} \hat{P}_\mu \quad a, b = 1, 2. \tag{23.13}$$

The momentum operator $\hat{P}_\mu$ has its origin in the ∂_μ of (23.12). Equation (23.13) is quite remarkable. It tells us that the $\hat{Q}$'s have a status on a par with the space-time translation operator $\hat{P}_\mu$. Indeed, the $\hat{Q}$'s are a kind of 'square-root' of a translation.

As in the case of the Lie groups discussed in Appendix M, the SUSY algebra determines the structure of SUSY multiplets, which for massless fields have the form

$$\{|\text{helicity} = -j\rangle, \ |\text{helicity} = -j + 1/2\rangle\} \tag{23.14}$$

together with the TCP-conjugate states. Taking $j = 1/2$, we have the multiplet

$$\{|\text{helicity} = -1/2, \ |\text{helicity} = 0\rangle\} \tag{23.15}$$

which is the $\hat{\chi}, \hat{\phi}$ multiplet already introduced, and which is called the '(left) chiral supermultiplet'. Taking $j = 1$, we have the multiplet

$$\{|\text{helicity} = -1\rangle, \ |\text{helicity} = -1/2\rangle\} \tag{23.16}$$

which consists of a massless gauge field and a massless spin-1/2 partner. This is the 'gauge supermultiplet'. (We shall not give the transformation laws for the gauge supermultiplet here.) These two types of multiplet are all we need to describe the Minimal Supersmmetric Standard Model (MSSM), which has formed the basis of most SUSY extensions of the SM, and for which there is a well-developed phenomenology. It is conventional to denote the superpartners of SM fields by adding a tilde: thus the superpartner of the gluon g is the gluino g̃.

Although the $\hat{Q}$'s change the spin of states on which they act, they do not change the internal quantum numbers. This means that we can immediately extend the known multiplets of the SM into SUSY supermultiplets by just adding scalar partners to the SM fermions, and fermionic partners to the SM gauge bosons. We also add fermionic partners to scalar Higgs fields ('Higgsinos'). Thus we will have a colour octet of 'gluinos', an $SU(2)_L$ triplet of 'winos', and a $U(1)_y$ singlet 'bino', together with scalar fields with the SM quantum numbers of all the SM fermions. A technical detail concerns the R parts of the SM fermions. To accommodate them in this framework, which uses only L-component fields, the R parts are represented as the charge conjugates of the corresponding L-type antiparticle fields.

Remarkably, it is possible to construct a fully supersymmetric version of the SM, using these multiplets, with the same gauge couplings, and with the addition of only one extra parameter denoted by μ (a regrettably overused symbol). But there is one very important difference between this minimally extended SM, and the SM itself, which concerns the Higgs sector. In the SM, Yukawa interactions involving the Higgs doublet $(\hat{\phi}^+, \hat{\phi}^0)$ give mass to the $t_3 = -1/2$ components of the $SU(2)_L$ doublet fermions when $\hat{\phi}$ develops a vacuum expectation value $\langle 0|\hat{\phi}^0|0\rangle = v$. The $t_3 = +1/2$ components are given masses via corresponding Yukawa interactions with the charge conjugate field $\hat{\phi}_C = i\tau_2\hat{\phi}^{\dagger T}$ (see section 22.5). But supersymmetric theories are not allowed to contain the complex conjugates of fields. This is related to the fact that the complex variable $z^* = x - iy$ is not an analytic function of z (the Cauchy-Reimann relations are not satisfied — see equation (F.12) in Appendix F of volume 1). Hence *two separate* Higgs doublets are necessary to give masses to both the t_3 components of the $SU(2)_L$ doublets. The two Higgs fields are $\hat{H}_u = (\hat{H}_u^+, \hat{H}_u^0)$ and $\hat{H}_d = (\hat{H}_d^0, \hat{H}_d^-)$, and there will be two corresponding doublets of spin-1/2 Higgsinos. The two Higgs doublets contain altogether 8 degrees of freedom. After electroweak symmetry breaking, 3 will become the longitudinal spin components of the massive $W^\pm$ bosons and Z boson, as usual. The remaining 5 degrees of freedom will be one charged Higgs boson pair $H^\pm$, one CP-odd neutral scalar A, and two neutral CP-even states, H and $\mathcal{H}$, of which H is the lighter and will correspond to the SM Higgs boson.[1] Benchmark scenarios have been defined to highlight interesting conditions for the MSSM Higgs boson searches (Bagnaschi *et al.* 2019).

Elegant though such a theory might be, it cannot represent reality. For one thing, the vacuum energy in an exactly supersymmetric theory is zero. In the present case, this is evident from the form of the quadratic terms in the potential for the neutral Higgs fields, which are

$$V_\mu = |\mu|^2(|\hat{H}_u|^2 + |\hat{H}_d|^2) \tag{23.17}$$

and which can never be negative, ruling out the possibility of non-zero vacuum values for the scalar fields. Further, in the exact theory, all members of a supermultiplet would have the same mass, but no such superpartners of the SM particles have been observed. So supersymmetry must be broken, and in such a way as to generate a mass gap between the SM particles and their superpartners. Since there is as yet no satisfactory theory of spontaneous SUSY breaking, the elegant SUSY-invariant Lagrangian, which had only the one additional parameter μ, must be extended to include SUSY-breaking terms, with additional parameters. It is important that such terms should not spoil the ultraviolet behaviour of the theory (on which the naturalness depends), and this means that their coefficients should have mass dimension 1 or greater (i.e. they should be super-renormalizable). These are called 'soft' breaking terms, and the possible gauge-invariant soft SUSY-breaking terms include the following: gaugino masses, squark (mass)^2 terms, slepton (mass)^2 terms, Higgs (mass)^2 terms, and triple scalar couplings. It is interesting that gauge invariant mass terms are possible for all these superpartners, in marked contrast to their SM counterparts. Unfortunately, though, there are over 100 possible soft SUSY-breaking terms.

Let us now consider how electroweak symmetry breaking can be accommodated in the MSSM. The potential for the neutral scalar fields in the MSSM is

$$V = (|\mu|^2 + m_{H_u}^2)|\hat{H}_u^0|^2 + (|\mu|^2 + m_{H_d}^2)|\hat{H}_d^0|^2 - b(\hat{H}_u^0\hat{H}_d^0) + \frac{1}{8}(g^2 + g'^2)(|\hat{H}_u^0|^2 - |\hat{H}_d^0|^2)^2. \tag{23.18}$$

Note that, despite appearances, the SUSY-breaking mass parameters $m_{H_u}^2$ and $m_{H_d}^2$ can be negative. The parameter b is SUSY-breaking also. We draw attention to the fact that the coefficient of the quartic term in (23.18) is determined by the gauge coupling constants, and is not a free parameter as in the SM. This will lead to a lower bound on the MSSM Higgs boson, as we shall see. Both $|\hat{H}_u^0|$ and $|\hat{H}_d^0|$ are required to have non-zero vacuum values $v_u/\sqrt{2}$ and $v_d/\sqrt{2}$ respectively. This leads to the conditions

$$(|\mu|^2 + m_{H_u}^2)v_u = bv_d + \frac{1}{4}(g^2 + g'^2)(v_d^2 - v_u^2)v_u \tag{23.19}$$

---

[1]This notation is unconventional in the SUSY literature. In the usual notation, the lighter state is h, and the heavier is H. But we are accustomed to using H for the SM Higgs field, and we are going to stick with that.

and

$$(|\mu|^2 + m_{H_d}^2)v_d = bv_u + \frac{1}{4}(g^2 + g'^2)(v_d^2 - v_u^2)v_d. \tag{23.20}$$

The combination of $v_u^2 + v_d^2$ is fixed by experiment, since it determines the mass of the mass of the W and Z bosons. These masses emerge as in the SM, and one finds

$$m_Z^2 = \frac{1}{2}(g^2 + g'^2)(v_u^2 + v_d^2) \tag{23.21}$$

and

$$m_W^2 = \frac{1}{2}g^2(v_u^2 + v_d^2). \tag{23.22}$$

From this we deduce $(v_u^2 + v_d^2)^{1/2} = (2m_W^2/g^2)^{1/2} = 246\,\text{GeV}$. Equations (23.19) and (23.20) can now be written as

$$(|\mu|^2 + m_{H_u}^2) = b\cot\beta + (m_Z^2/2)\cos 2\beta \tag{23.23}$$

and

$$(|\mu|^2 + m_{H_d}^2) = b\tan\beta - (m_Z^2/2)\cos 2\beta, \tag{23.24}$$

where $\tan\beta = v_u/v_d$. We may use these last equations to eliminate one of the parameters $|\mu|$ and b in favour of $\tan\beta$, but the phase of μ is not determined. As both v_u and v_d are real and positive, β lies between 0 and $\pi/2$.

With all this in place, one can calculate the mass spectrum in this neutral part of the enlarged Higgs sector (note that a diagonalization has to be performed, because of the b term in (23.18)). The mass of the CP-odd state A is

$$m_A = (2b/sin2\beta)^{1/2}, \tag{23.25}$$

that of the lighter CP-even state is

$$m_H^2 = \frac{1}{2}\{m_A^2 + m_Z^2 - [(m_A^2 + m_Z^2)^2 - 4m_A^2 m_Z^2 \cos^2 2\beta]^{1/2}\} \tag{23.26}$$

and of the heavier is

$$m_{\mathcal{H}}^2 = \frac{1}{2}\{m_A^2 + m_Z^2 + [(m_A^2 + m_Z^2)^2 - 4m_A^2 m_Z^2 \cos^2 2\beta]^{1/2}\}. \tag{23.27}$$

A crucial point now is that, whereas m_A and $m_{\mathcal{H}}$ can in principle be arbitrarily large, the mass m_H is bounded below, reaching the maximum value of $m_Z|\cos 2\beta| \leq m_Z$ for very large m_A.

This bound was of course exceeded in 2012. But radiative corrections to these tree-level results can be large. One important contribution to m_H^2 arises from the incomplete cancellation of top quark and top squark loops, which would cancel in the exact SUSY limit. The magnitude of this contribution depends on the top squark mass, and results in the corrected bound (neglecting certain squark mixing effects)

$$m_H^2 \leq m_Z^2 + \frac{3g^2 m_t^4}{4\pi^2 m_W^2}\ln(m_S/m_t) \tag{23.28}$$

where m_t is the top quark mass, as usual, and m_S^2 is the average of the squared masses of the two top squarks, which correspond to the L and R components of the SM top quark. This implies a rather large value for m_S, of order 5 TeV, if this H is to be the state at 125 GeV. Such a value for the average stop mass seems uncomfortably larger than the electroweak symmetry breaking scale of v. However, the large size of this correction suggests that a more refined calculation is necessary, involving higher effects. A great deal of work has been done to obtain an accurate estimate of the radiative corrections to the H mass; a recent review is provided by Slavich *et al.* (2021).

The MSSM provides formulae for the masses and couplings of all the newly introduced sparticle fields, which depend on the SM parameters, the electroweak symmetry breaking parameters, and the soft SUSY breaking parameters. In the Higgs sector, equations (23.25) – (23.27) give the masses of the neutral members; the two charged Higgs bosons have mass $(m_W^2 + m_A^2)^{1/2}$. The Higgs sector is therefore determined, at tree level, by the SM parameters and the two new parameters $\tan\beta$ and m_A. In a study by the ATLAS Collaboration (ATLAS 2020c) properties of the measured 125 GeV Higgs boson are compared to six MSSM benchmark scenarios, assuming the observed 125 GeV state is the light CP-even state H of the MSSM. The compatibility of each scenario with the observed Higgs data is tested for a range of the parameters $\tan\beta$ and m_A, setting exclusion limits for these parameters. In the scenario labelled M_h^{125} (alignment), for $\tan\beta$ around 7 the properties of the lighter scalar are in agreement with those of the observed Higgs boson, for relatively low values of M_A.

The gluino is the only colour octet fermion, and since $SU(3)_c$ is unbroken, it cannot mix with any other MSSM particle. Its mass arises solely from the soft SUSY-breaking mass term. There are four 'neutralinos': the two Higgsinos and the wino and bino. In the absence of electroweak symmetry breaking, the wino and bino masses also would be given by just the soft SUSY-breaking mass terms. However, mixing involving all four states occurs after electroweak symmetry breaking. Also, the SUSY-invariant μ term induces mixing between the two neutral Higgsinos. A simple scenario is for the Higgsino masses to be largely determined by μ, which on naturalness grounds cannot be far from the scale of v (Dimopoulos and Giudice 1995, Papucci, Ruderman and Weiler 2012). Consequently, a SUSY spectrum with stop masses in the range of a few TeV, and light Higgsinos, is under close scrutiny at the LHC.

One further detail about the MSSM must be noted. The SM Lagrangian as it stands automatically conserves baryon number B and lepton number L, due to the fact that possible B- and L-violating terms in the Lagrangian, made from gauge-invariant combinations of SM fields, must have mass dimension 5 or higher (Weinberg 1979c, Wilczek and Zee 1979, Weldon and Zee 1980), and are therefore not renormalizable; they will be suppressed by powers of a high mass scale, and do not appear in the renormalizable SM Lagrangian. But with the new sparticle fields introduced in the MSSM, it is possible to construct renormalizable gauge-invariant terms which violate B and L. Such terms would allow proton decay at a rate far in excess of the experimental bound, unless the coupling was extraordinarily (and unnaturally) small. These terms can be excluded by imposing an additional symmetry called R-parity (Fayet 1977, 1978, Farrar and Fayet 1978), where $R = (-)^{3B-L-2s}$, and s is the spin of a particle. R is multiplicatively conserved. One quickly finds that $R = +1$ for all conventional matter particles, and (because of the $(-)^{2s}$ factor) $R = -1$ for all their superpartners.

The conservation of R in production or decay processes then has important phenomenological consequences. For example, any initial state in a production process will involve $R = +1$ particles only, from which it follows that sparticles must be produced in pairs. The sparticles will decay into lighter states, the decay chain leading in the end to the lightest supersymmetric particle (LSP), which would be absolutely stable, even if heavier than all SM particles. If electrically neutral and a colour singlet, a stable LSP has long been considered a promising candidate for dark matter (see for example section 3 of Steffen 2009). Such a particle — a 'neutralino', denoted by $\tilde{\chi}_1^0$ — is in general a linear combination of neutral wino, bino, and Higgsino states, which undergo mass-mixing (see section 11.2 of Aitchison 2007). The $\tilde{\chi}_1^0$ is a spin-1/2 fermion with no strong or electromagnetic interactions, and is therefore a WIMP (weakly interacting massive particle). Its mass is expected to be of the order of a few hundred GeV in the MSSM. Many direct experimental searches for neutralino candidates have been made, without success so far, leading to the exclusion of large areas of the parameter spaces of many MSSM scenarios. Nevertheless, viable possibilities remain (Barman *et al.* 2023, Delgado and Quirós 2021, Baer *et al.* 2018, Bramante *et al.* 2016, Kowalska and Sessolo 2018, Krall and Reece 2018).

The MSSM had two early qualitative successes. The first relates to the possible eventual unification of the three gauge coupling constants as they evolve ('run') up to a high energy scale (Georgi, Quinn and Weinberg 1974). This is not the case, in fact, for the SM itself, but it does work very convincingly in the MSSM without any changes (Dimopoulos, Raby and Wilczek 1981, Ibanez and Ross 1981, Einhorn and Jones 1982). The second positive indication for the MSSM concerns the evolution of the mass parameter $m_{H_u}^2$. Calculations show (Ibanez and Ross 1982, Ibanez 1982, Ellis, Nanopoulos and Tamvakis 1983, Alvarez-Gaume, Polchinski and Wise 1983) that in the MSSM this parameter will evolve from a positive value at a high mass scale to a negative value at the electroweak scale, thus tending to trigger electroweak symmetry breaking. Such breaking, however, would also depend on the values of other parameters in the Higgs potential.

It may therefore seem disappointing that no superpartners have yet been observed. For example, the LHC data have excluded coloured sparticles (primarily gluinos and first generation squarks) with masses below about 2 TeV. However, the precise limits are very model dependent. A general difficulty in interpreting the data is that in order to avoid an unduly large number of parameters, associated with all the possible SUSY-breaking terms, simplifying assumptions are usually made so as to reduce them to a manageable number. But such restrictions amount to particular models. Furthermore, there exist many extensions and variations of the MSSM which evade many of the current experimental constraints. The field is surveyed by Allanach and Haber (2022), section 88 in Workman *et al* (2022); the experimental situation is reviewed by D'Onofrio and Moortgat (2022) in section 89 of Workman *et al.* (2022). It is therefore premature to declare the demise of supersymmetry, though it may be fair to say that it, and the hope for naturalness, are being put under ever increasing stress.

23.4.2   Dynamical electroweak symmetry breaking

In contrast to supersymmetric versions of the SM, which remain perturbative, models of dynamical electroweak symmetry breaking generally postulate new strongly interacting gauge theories operating at a scale near, but above, the weak scale v. An immediate problem is that we lack reliable methods for solving strongly interacting quantum field theories, though lattice gauge theory is beginning to make interesting contributions (Pica 2016, Svetitsky 2018, Witzel 2019, Drach 2020, Appelquist *et al.* 2016). However, we do know of one strongly interacting gauge theory, namely QCD, where the non-Abelian global chiral symmetry of massless quarks is dynamically broken, as discussed in chapter 18. (This, in turn, had parallels with the BCS theory of superconductivity, covered in chapter 17.) The basic idea of the early dynamical approaches to electroweak symmetry breaking was to assume that the dynamics of the new strong interactions is essentially analogous to QCD. It is as well to bear in mind, however, that other strongly interacting theories may differ qualitatively from QCD, and indeed such differences could be essential in finding a phenomenologically viable model.

At the outset, it is convenient to distinguish the two roles which the SM Higgs boson plays: first, its couplings to the SM gauge bosons generate their masses; and secondly its Yukawa couplings to the SM fermions generate their masses and mixing matrices. The first role appears to be relatively easy to replace by a dynamical theory with no elementary Higgs field, but the second is far more challenging.

The first, and simplest, examples of such theories were proposed by Weinberg (1976), and by Susskind (1979). Consider, following Susskind, a flavour doublet of massless fermions (U,D) interacting via a non-Abelian gauge field, analogous to QCD, which Susskind called 'technicolour'. The Lagrangian would then have the form of equation (18.12), which is invariant under the chiral symmetry of the S(2)$_I$ of isospin, and the SU(2)$_{I5}$ of axial isospin. In the case of QCD, there are good reasons to believe (see chapter 18) that the axial isospin symmetry is spontaneously broken, and we make the assumption that the same will be true in the technicolour case. This breaking will occur at a certain energy scale v_T, analogous to the chiral symmetry breaking parameter, let us now call it v_{ch}, which was identified with the pion decay constant f_π (see (18.102)). There will be three massless Nambu-Goldstone bosons ('technipions'), analogous to the pions of QCD. The dynamics of these massless modes can be captured by a low energy effective theory of the σ-model type, as described in section 18.3.1, which involves a doublet boson field $\hat\phi_T$, as in section 17.6. As we saw in (18.81), $\hat\phi_T$ can be parametrized by

$$\hat\phi_T = \exp(-i\hat{\boldsymbol\pi}_T(x) \cdot \boldsymbol\tau/2v_T) \begin{pmatrix} 0 \\ \frac{1}{\sqrt{2}}(v_T + \hat h_T(x)) \end{pmatrix}, \qquad (23.29)$$

where $\hat{\boldsymbol\pi}_T(x)$ are the three massless technipion fields.

So far, all this just proceeds by analogy with QCD. The new feature here is that we now suppose that this $\hat\phi_T$, associated with the breaking of the global chiral symmetry of the techniquarks, is itself a doublet with $y = 1$ under the electroweak gauge group SU(2)$_L \times$ U(1)$_y$. The effective Lagrangian for the electroweak gauge fields is then given in (19.79), and the calculation of section 19.6 shows that the field $\hat\phi_T$ serves to also break the electroweak gauge group spontaneously. Indeed, by an electroweak gauge transformation we can transform away the $\hat{\boldsymbol\pi}_T$ fields (passing to the unitary gauge), which will form the longitudinal polarization states of the W and Z vector bosons, as usual; the charged W bosons will gain the mass $m_W = gv_T/2$, and the Z boson mass will be $m_Z = v_T(g^2 + g'^2)^{1/2}/2$. In particular, note that $m_W/m_Z = \cos\theta_W$, exactly as in the SM.

In this simple model, then, the masses of the SM vector bosons (the first role of the SM Higgs boson) are generated dynamically via the (assumed) breaking of the global chiral symmetry at a scale v_T in the technicolour theory, v_T being identified with the parameter v of the SM, so that $v_T \simeq 246$ GeV, as in (22.53). In a more complete version of this simplest model, there will also be the usual SM Lagrangian, the two (quark and techniquark) sectors being linked by the electroweak gauge fields. Note that this simple model incorporates the custodial SU(2) symmetry which protects the parameter ρ from large electroweak radiative corrections (see section 22.6).

It is important to stress that, although we have framed the discussion in terms of the scalar field $\hat\phi_T$, this is not an elementary field as in the SM, or as in a supersymmetric theory. It conveniently parametrizes the chiral symmetry breaking in the technicolour sector: the technipions are composites, like the pions in QCD, and so is the residual field $\hat h_T$, to which we shall return in a moment.

The simple model which we have outlined is effectively a scaled up (in energy) version of QCD, with scale factor $v_T/v_{ch} \simeq 2,700$. Taken literally, this would imply the existence of a large spectrum of technihadrons in the few TeV region, analogues of the hadron spectrum of QCD. One would expect technibaryons, and analogues of the vector mesons such as the ρ and ω. WW and ZZ scattering amplitudes involving the longitudinal polarization states would be expected to become strongly interacting in this energy region. Technicolour theories, and their phenomenology, are reviewed by Hill and Simmons (2003); see also the review by Black, Sekhar Chivukula, and Narain (2022), section 92 of Workman *et al.* (2022).

The technicolour idea, though attractive, and not suffering from the naturalness problem, is far from realistic. First of all, as it stands, it has nothing to say about the second role played by the SM Higgs boson, namely the generation of the fermionic masses and mixing matrices. We will discuss attempts to address this issue presently. A second more immediate point is that the QCD analogue of the h_T particle is the $f_0(500)$, a very broad and poorly defined s-wave resonance, which decays to the light pseudo Nambu-Goldstone pions. It is hard to see how the observed narrow Higgs boson, with a mass well below what should be the technihadron threshold, could arise through the dynamical scheme as envisaged so far.

However, the possibility of dynamically generating an SM-like Higgs boson may be emerging from lattice gauge theory calculations (Pica 2016, Svetitsky 2018, Witzel 2019, Drach 2020, Appelquist *et al.* 2016). In particular, novel strong dynamics may arise in theories with a large number N_f of light fermions transforming under the fundamental representation of the gauge group SU(3). It will be recalled that the asymptotic freedom property is traced to the behaviour of the beta function β_s, which depends on N_f (see (15.49) for example). Asymptotic freedom is lost if N_f is too large, but the possibility exists that at some intermediate value of N_f, say $N_{f,c}$, the theory might exhibit an infrared stable fixed point, with a beta function of the form shown in figure 15.6(b). The coupling constant would then flow in the infrared limit to a zero of the beta function, rendering the theory scale invariant, and therefore massless. This is not desirable, but perhaps when N_f is somewhat less than $N_{f,c}$ there would be a 'window' (Svetitsky 2018) where the running of the coupling slows down[2] , and the theory is nearly scale invariant over some range of scales. The scale invariance may be spontaneously broken, and the corresponding pseudo Nambu-Goldstone boson could be identified with the Higgs boson. Results supporting the existence of a light 0^{++} state in a strongly interacting SU(3) gauge theory with $N_f = 8$ light fermions have been reported by the LaKMI collaboration (Aoki *et al.* 2017) and by the LSI collaboration (Appelquist *et al.* 2014); see also Appelquist *et al.* (2016). Such a theory is, of course, hardly realistic: there will be 63 Nambu-Goldstone bosons to be disposed of, for example. Nevertheless, such explorations of strongly interacting theories which behave differently from QCD, and may provide alternative paradigms, are valuable. We may expect further progress in this area.

An alternative approach to understanding the existence of a Higgs boson which is light compared to the technicolour scale is to regard it, and the rest of the Higgs doublet, as approximate (or pseudo) Nambu-Goldstone bosons (NGBs) of a symmetry broken at an energy scale significantly higher than the electroweak scale v (Kaplan and Georgi 1984, Georgi and Kaplan 1984, Kaplan, Georgi and Dimopoulos 1984). Such theories generally treat the NGBs as associated with a broken symmetry which is realized *non-linearly*. Once again, QCD provides an analogy. We saw in section 18.3.1 how the low energy dynamics of the pions associated with chiral symmetry breaking could be most economically expressed in terms of the non-linear σ-model of (18.95), which exhibits the desired symmetry and correctly contains only the derivatives of the Nambu-Goldstone fields. In particular, there is *no* $\hat{\phi}^\dagger \hat{\phi}$ term, in contrast to the linear σ-model Lagrangian of (18.67), and consequently there is no corresponding naturalness problem. These theories (called composite Higgs models) typically require a large global symmetry of the new strongly interacting high energy theory, with a correspondingly large number of light pseudo Nambu-Goldstone bosons, and also additional electroweak vector bosons, as well as the expected spectrum of new bound states.

---

[2]In the context of Extended Technicolour Theories (to be discussed presently), this property was called 'walking', and was introduced (Holdom 1985) to resolve the problem of generating correct heavy quark masses while also suppressing flavour-changing neutral current processes.

A comprehensive review of composite Higgs models is provided by Bellazzini, Csáki, and Serra (2014).

Let us return now to the other role of a dynamically generated Higgs field: that of generating the SM fermion masses and mixing matrices. Any such mechanism would require the SM fermions to couple to the technifermion condensate that breaks the electroweak symmetry. One approach, referred to as 'extended technicolour' (ETC), introduced by Dimopoulos and Susskind (1979) and by Eichten and Lane (1980), is to extend the technicolour gauge interactions to include additional heavy gauge bosons mediating interactions between the technifermions $\hat{Q}$ and the SM fermions $\hat{\psi}$. These new interactions are part of a larger gauge group which breaks down at some scale Λ_{ETC} to the technicolour subgroup and the SM gauge groups, where the scale Λ_{ETC} is higher than the technicolour scale $\Lambda_{\mathrm{T}} \sim v_{\mathrm{T}}$. A characteristic difficulty of such theories is that the scale Λ_{ETC} has to be as high as 1000 TeV or higher in order to suppress the induced flavour changing neutral current transitions between SM fermions (Eichten and Lane 1980), but then the dynamically generated SM fermion masses are too small, at least for the second and third generations.

One may understand this qualitatively as follows. Just as in the case of the SM itself, at energies well below M_{W}, the interactions mediated by the heavy gauge bosons will become, at energies below Λ_{ETC}, effective current-current couplings of the form $C\hat{\bar{Q}}\Gamma^{\mu}\hat{Q}\,\hat{\bar{\psi}}\Gamma_{\mu}\hat{\psi}/\Lambda_{\mathrm{ETC}}^2$, where the Γ's include symmetry group matrices, and C is expected to be of order unity, on natural dimensional grounds. The spinor fields can be rearranged to produce equivalent operators which include terms of the form $C\hat{\bar{Q}}\hat{Q}\,\hat{\bar{\psi}}\hat{\psi}/\Lambda_{\mathrm{ETC}}^2$. Now, in a manner analogous to the role of the BCS condensate (see (17.121) and (17.122)) the technifermion condensate $\langle \hat{\bar{Q}}\hat{Q}\rangle$ which breaks the technicolour chiral symmetry will generate SM fermion masses which are of order

$$m_{\mathrm{q}} \sim C\langle \hat{\bar{Q}}\hat{Q}\rangle/\Lambda_{\mathrm{ETC}}^2. \tag{23.30}$$

On dimensional grounds we expect $\langle \hat{\bar{Q}}\hat{Q}\rangle \sim \Lambda_{\mathrm{T}}^3$ at the scale Λ_{T}, and so

$$m_{\mathrm{q}} \sim C\Lambda_{\mathrm{T}}^3/\Lambda_{\mathrm{ETC}}^2 \sim C(1\mathrm{TeV})^3/(1000\mathrm{TeV})^2 \sim 1\mathrm{MeV} \tag{23.31}$$

which is too small for the second, and especially the third, generation.of quarks.

The 'walking technicolour' idea alluded to earlier was introduced (Holdom 1985) to alleviate this problem. We have estimated $\langle \hat{\bar{Q}}\hat{Q}\rangle$ at the technicolour scale Λ_{T}. But the condensation in question occurs in the ETC dynamics at the scale Λ_{ETC}. Renormalization group considerations (see section 15.6) enable us to relate the condensate at different scales according to (see (15.82))

$$\langle \hat{\bar{Q}}\hat{Q}\rangle_{\mathrm{ETC}} = \langle \hat{\bar{Q}}\hat{Q}\rangle_{\mathrm{T}} \exp \int_{\Lambda_{\mathrm{T}}}^{\Lambda_{\mathrm{ETC}}} \mathrm{d}\ln\mu\,\gamma(\alpha_{\mathrm{T}}(\mu)) \tag{23.32}$$

where γ is the operator's anomalous dimension, and α_{T} is the relevant coupling constant. If the theory behaves qualitatively like QCD, the renormalization corrections from (23.32) will grow like powers of logarithms, making little difference. But if the theory is in fact close to a zero α_{T}^* of the beta function (as suggested earlier), then γ will be approximately constant, and (23.32) will lead to a power law enhancement $(\Lambda_{\mathrm{ETC}}/\Lambda_{\mathrm{T}})^{\gamma(\alpha^*)}$. For $\gamma(\alpha^*) \sim 1$ this would give $m_{\mathrm{q}} \sim 1\mathrm{GeV}$. But of course we must remember that none of this begins to address the range of masses seen in the quark spectrum.

We have sketched, in broad outline, some of the approaches that have been proposed to the problem of dynamical electroweak symmetry breaking (DSB) — or, as it is often stated, to finding an 'ultraviolet completion' of the SM. With the discovery of the Higgs boson, with properties and couplings now known to be remarkably close to the SM predictions, it remains a major challenge to find a model which correctly reproduces all the successful phenomenology of the SM, particularly in the fermionic sectors. The field is reviewed by Black, Sekhar Chivukula, and Narain (2022) in section 92 of Workman *et al.* (2022), who also discuss the many searches that have been made at the LHC for new particles suggested by the various DSB theories.

23.5   The Standard Model as an Effective Field Theory (SMEFT)

"Well! I've often seen a cat without a grin", thought Alice, "but a grin without a cat! That's the most curious thing I ever saw in my life!"

From *Alice's Adventures in Wonderland* by Lewis Carroll

Arguments from naturalness, and perhaps also a reasonable expectation that the SM is unlikely to be the final theory, suggest that some kind of *new physics* should exist at an energy scale Λ which is in the TeV region or higher. How can experiments at lower energies gain access to this new physics, if the particles involved have masses too high to be produced? In the case of the top quark, and to a lesser extent the Higgs boson, precision electroweak data was sensitive to their *virtual* effects, prior to their discovery through production processes. But that was dependent upon having a definite renormalizable theory (the SM) for calculating radiative corrections. As we have seen, however, the number of possibilities for such 'ultraviolet completions' of the SM itself is discouragingly large. The phenomenology of each of these models can of course be pursued, as can their effects in radiative corrections, but it would be desirable to have a more model-independent way of parametrizing such a wide range of candidates.

The way this can be done is illustrated by a familiar example. In section 22.2.2 we saw how, at energies or momentum transfers well below the W boson mass M_W, the W boson propagator reduces to the constant $g^{\mu\nu}/M_W^2$, and the tree-level SM amplitude (22.49) for W exchange in the process $\nu_\mu + e^- \rightarrow \mu^- + \nu_e$ reduces to the point-like four-fermion amplitude (22.50). This latter amplitude could itself have been generated by a four-fermion interaction term of the form (compare (20.43))

$$\frac{G_F}{\sqrt{2}} \hat{\bar{\mu}} \gamma_\mu (1 - \gamma_5) \hat{\nu}_\mu \bar{\hat{\nu}}_e \gamma^\mu (1 - \gamma_5) \hat{e} \tag{23.33}$$

where

$$\frac{G_F}{\sqrt{2}} = \frac{g^2}{8 M_W^2}. \tag{23.34}$$

At energies well below M_W, then, the W boson has disappeared from the theory, but its effect remains in the *effective interaction* of (23.33). It is very characteristic that (23.33) is a non-renormalizable interaction, the four-fermion operator product having mass dimension 6, which gives G_F the dimension -2, as discussed in chapter 11 of volume 1. In this case, since we are in possession of the theory at energies of order M_W, we can *match* the coefficient G_F of the low-energy theory (23.33) to parameters in the high-energy theory by condition (23.34), which reveals the origin of the dimensionality of G_F.

In a similar way, we can imagine new massive particles interacting with SM particles at an energy scale Λ large compared to the electroweak scale v. At energies well below Λ, the propagators of the massive particles will be replaced by constants, resulting in a variety of operator products constructed out of SM fields, which represent the effective low-energy version of the Λ-scale physics. These effective interactions will be added to the SM Lagrangian, creating thereby a 'Standard Model Effective Lagrangian' (SMEFT). The interactions of particles too massive to be produced in the laboratory can be (partially) accessed through their low-energy effective interactions.

What can be said about these effective interactions, in general? First of all, they should respect the linearly realized $SU(3)_c \times SU(2)_L \times U(1)_y$ gauge symmetries of the SM, which are unbroken above the scale v. Secondly, the effective interactions will be non-renormalizable, because the SM is the most general renormalizable theory incorporating those symmetries, given the SM field content. The effective interactions can then be organized according to their mass dimension D, beginning with D $= 5$. Those with D $= 5$ will have coefficients $c_i^{(5)}/\Lambda$, those with D $= 6$ will have coefficients $c_i^{(6)}/\Lambda^2$ and so on. Other things being equal, we should expect the D $= 5$ interactions to be suppressed at energies $E \ll \Lambda$ by a factor E/Λ, and those with D $= 6$ to be suppressed by a factor $(E/\Lambda)^2$ (the latter being the same as the Fermi theory). The case D $= 5$ is special, and one we have already met in section 22.5.1. As we saw there, a Majorana mass term can be generated for ν_e with the D $= 5$ operator in expression (22.146). The very small mass for ν_e implies that the associated energy scale for such an interaction would be very high, with $\Lambda/c_\nu^{(5)}$ of order 10^{15} GeV. This puts D $= 5$ interactions out of reach of collider experiments.

The next most important effective interactions are those with D $= 6$. Here, in addition to effective four-fermion interactions like (23.33) interactions which violate baryon number and lepton number would be possible (but are not allowed in the SM, as we noted in section 23.4.1). We can estimate a lower bound on the associated scale Λ_{BV} for interactions which violate baryon number from the proton lifetime, which is greater that 10^{33} years. The rate would be proportional to $(c_{BV}^{(6)}/\Lambda_{BV}^2)^2$, and the correct dimensions must be restored by a factor of order m_p^5, giving a rate of order $(c_{BV}^{(6)}/\Lambda_{BV}^2)^2 m_p^5$, from which we estimate that Λ_{BV} would be greater than 10^{15} GeV. The upshot is that the D $= 6$ interactions which conserve baryon number and lepton number will dominate at collider energies. In a similar way, interactions which allow flavour-changing neutral currents, which are known to be highly suppressed, will be associated with a relatively large scale Λ_{FCNC}. We have therefore arrived at an effective Lagrangian of the form

$$\hat{\mathcal{L}}_{\text{eff}} = \hat{\mathcal{L}}_{\text{SM}} + \frac{1}{\Lambda^2} \sum_i c_i^{(6)} \hat{\mathcal{O}}_i^{(6)} + \dots \qquad (23.35)$$

where interactions with D > 6 are not included at this stage, the interactions $\hat{\mathcal{O}}_i^{(6)}$ are gauge invariant products of SM fields which conserve baryon and lepton number, and the corresponding coefficients $c_i^{(6)}$ are called *Wilson coefficients*.

If we knew the precise form of the interactions at the scale Λ, we could match the Wilson coefficients to the parameters in those interactions, as we did in (23.34) for the Fermi theory. That would be a 'top down' approach. Lacking such knowledge, we have to work 'from the bottom up'. The best strategy is to include the effects of all possible such interactions by constructing a complete set of effective interactions $\hat{\mathcal{O}}_i^{(6)}$. This may sound straightforward, but it is not. The problem is *redundancy*: seemingly quite different D $= 6$ interactions can lead to the same S-matrix elements describing scattering and decay processes involving the SM particles. First of all, integration by parts in the expression for the action is allowed, and can transform one interaction into another. Next, the classical equations of motion can be used to relate different interactions to each other, since it has been shown (Arzt 1995) that shifting operators by a term proportional to the classical equations of motion does not change S-matrix elements, even when using the operators inside loops. Other field redefinitions are also allowed (Chisholm 1961, Coleman, Wess, and Zumino 1969)). Finally, Fierz transformations (Fierz 1937, Itzykson and Zuber 1980) can be performed to change the ordering of fermion operators in quadrilinear products of two fermion bilinears.

When all this is taken into account, for a single generation of fermions there are a minimum of 59 independent $D = 6$ interaction terms, which form a *basis*. These can be chosen in arbitrarily many different ways, any of which may be used to describe the data. The first complete construction of a basis was done by Grzadowski *et al.* (2010) for one generation, and was extended to three generations by Alonso *et al.* (2014). This widely used basis is called the *Warsaw basis*. Another choice is the *strongly interacting light Higgs (SILH) basis* (Giudice *et al.* 2007, Contino *et al.* 2013). When considering data in the Higgs sector, it is possible to reduce the number of operators substantially. The *Higgs basis*, introduced by the LHC Higgs Cross Section Working Group (LHCHXSWG 2015), uses just 9 operators; it is discussed by Falkowski (2016).

It is clear that the great merit of the SMEFT approach is that it is capable of relating different observables in different sectors, and at different energies, so as to constrain a finite set of effective interactions among the SM fields. It is also systematically improvable: as we noted in section 11.8, a non-renormalizable theory can be used in loops consistently, provided suitable counter terms are added. Here, using the $D = 6$ operators in second order (one loop) will generate amplitudes with a factor Λ^{-4}, which will require the addition of a basis for $D = 8$ operators as counter terms. It should always be noted, though, that measurements can only return values of the ratios $c_i^{(D)}/\Lambda^{D-4}$, and an assumption needs to be made about Λ before the Wilson coefficients themselves can be determined.

Unfortunately there is a price to pay for the generality of the SMEFT. A $D = 6$ basis with three generations requires 1250 CP-even parameters and 1149 CP-odd parameters, for a total of 2499 parameters (Alonso *et al.* 2014). This can be substantially reduced, to 76 parameters, if the restriction known as Minimal Flavour Violation (MFV) (D'Ambrosio *et al.* 2002, Cirigliano *et al.* 2005) is imposed. This essentially means that all flavour and CP violating interactions are linked to the known structure of the SM Yukawa couplings. If CP violating effects are also neglected, 53 parameters remain.

The $D = 6$ interactions lead to two kinds of deviations from the SM. They can modify the couplings and parameters (including those entering input parameters like $G_{\rm F}$ and $m_{\rm Z}$), and they can lead to new vertices which are not present in the SM, but which affect observables at collider energies. We shall only give a few simple illustrative examples. First, following Alonso *et al.* (2014), the $D = 6$ interaction $c_{\rm H}^{(6)}/\Lambda^2 \, (\hat{\phi}^\dagger \hat{\phi})^3$ will change the Higgs potential to

$$\hat{V}(\hat{\phi}) = \frac{\lambda}{4}(\hat{\phi}^\dagger\hat{\phi} - \frac{1}{2}v^2)^2 - \frac{c_{\rm H}^{(6)}}{\Lambda^2}(\hat{\phi}^\dagger\hat{\phi})^3, \tag{23.36}$$

leading to the new minimum at

$$\langle 0|\hat{\phi}^\dagger\hat{\phi}|0\rangle = \frac{v^2}{2}\left(1 + \frac{3c_{\rm H}^{(6)}v^2}{4\lambda\Lambda^2}\right) \equiv \frac{1}{2}v_{\rm eff}^2 \tag{23.37}$$

to first order in $c_{\rm H}^{(6)}/\Lambda^2$. Next, the interactions $c_{\rm H\square}^{(6)}/\Lambda^2 \, (\hat{\phi}^\dagger\hat{\phi})\square(\hat{\phi}^\dagger\hat{\phi})$ and $c_{{\rm H}D}^{(6)}/\Lambda^2 \, (\hat{\phi}^\dagger D_\mu\hat{\phi})^*$ $(\hat{\phi}^\dagger D^\mu\hat{\phi})$ will alter (22.28) to

$$\hat{\phi} = \frac{1}{\sqrt{2}}\left(\begin{array}{c} 0 \\ v_{\rm eff} + [1 + \frac{c_{\rm H,kin}^{(6)}}{\Lambda^2}]\hat{H} \end{array}\right) \tag{23.38}$$

where

$$\frac{c_{H,\text{kin}}^{(6)}}{\Lambda^2} = \left(\frac{c_{H\square}^{(6)}}{\Lambda^2} - \frac{1}{4} \frac{c_{HD}^{(6)}}{\Lambda^2} \right) v^2. \tag{23.39}$$

and $v_{\text{eff}} = (1 + 3c_H^{(6)} v^2/8\lambda\Lambda^2)v$. The coefficient of $\hat{H}$ in (23.38) is no longer unity, and this means that the Higgs field will have to be rescaled in order to restore its kinetic term to the correct normalization. The expression for the Higgs boson mass will also be altered. As a third example, the term

$$\frac{c_{He}^{(6)}}{\Lambda^2} i [\hat{\phi}^\dagger D_\mu \hat{\phi} - (D_\mu \hat{\phi}^\dagger)] \bar{\hat{e}}_R \gamma^\mu \hat{e}_R \tag{23.40}$$

leads, after inserting the SM vacuum value (22.28), to a coupling of the Z boson to right-handed electrons:

$$-\frac{c_{He}^{(6)}}{\Lambda^2} (g^2 + g'^2)^{1/2} Z_\mu \bar{\hat{e}}_R \gamma^\mu \hat{e}_R. \tag{23.41}$$

In the SM, this coupling is given by (22.60) and (22.62) as $(g^2 + g'^2)^{1/2} \sin^2\theta_W$, so the modification of this coupling by (23.41) will change the SM predictions for $e^+e^- \to Z \to e^+e^-$, for example. A detailed review of the SMEFT is provided by Brivio and Trott (2019), who supply an Appendix with formulae for cross sections and decay widths of selected processes as calculated in the SMEFT to leading order.

Possible deviations from the SM predictions which may be revealed by an analysis using the SMEFT framework will be measured at current collider energies, and to interpret these deviations precisely in terms of the dynamics presumed to exist at the scale Λ will require a renormalization group (RG) evolution of the parameters as controlled by their anomalous dimensions. The anomalous dimensions of all 2499 parameters have in fact been calculated (Alonso *et al.* 2014 — this paper contains further references). Because of operator mixing, the pattern of Wilson coefficients that is observed at the low scale is not identical to the pattern at the scale Λ. The results of Alonso *et al.* (2014) determine the RG running of all the logarithmically enhanced terms between the two scales. This general framework also permits a model-independent test of the MFV hypothesis.

Two promising areas to look for deviations from the SM are in top quark physics and the Higgs sector. The top quark plays a special role in the SM, being the heaviest SM particle with a strong Yukawa coupling, which has important implications for radiative corrections in the SM bearing on the naturalness problem. The Higgs sector, of course, figured prominently in the discussion of possible ultraviolet completions of the SM, in section 23.4. Both the top and the Higgs sectors are under intense study at the LHC, using the SMEFT framework, in addition to conventional searches for new particles.

There is now an extensive literature on SMEFT applications in the top quark sector; a review is provided in section 10 of Brivio and Trott (2019). Here we shall reference just one recent investigation at the LHC by the CMS Collaboration (CMS PAS TOP-22-006). Top quarks are produced with additional leptons, and the leptonic final state decays of the top quarks provide a clean environment for EFT treatments. A search for new physics in top quark production with additional leptons was performed with 138 fb^{-1} of p-p collisions at $\sqrt{s} = 13$ TeV. Potential new physics effects were parametrized in terms of 26 D=6 interactions. Kinematic variables corresponding to the transverse momentum (p_T) of the leading pair of leptons and jets, as well as the p_T of the on-shell Z bosons, were used to extract the 95% confidence intervals of the 26 D=6 interactions. No significant deviations from the SM predictions were found.

We end this section with an analysis by the ATLAS Collaboration (ATLAS 2020c) of Higgs boson production and decay rate measurements, made with up to 139 fb^{-1} of 13

TeV p-p collision data collected during Run 2 from 2015 to 2018, using the SMEFT to interpret the measurements. In this analysis the new physics scale was set at 1 TeV. The decay modes used were $H \to \gamma\gamma$ (all production modes), $H \to ZZ^* \to 4\ell$ (all production modes), and $H \to b\bar{b}$ (vector boson associated production). An initial set of 32 effective interaction operators was chosen from the Warsaw basis, which either directly or indirectly impacted the Higgs boson couplings to SM particles (Brivio and Trott 2019). However, the available data samples contained insufficient information to constrain the corresponding Wilson coefficients. Consequently, a reduced set of 10 operators was used, certain linear combinations of the original operators, which offered greater sensitivity in the analysis. The reported result was that all the measured operator coefficients were consistent with zero (i.e. consistent with the SM expectations), within their uncertainties.

The hunt for new physics will continue, with more precise measurements and corresponding theoretical treatments, and with higher luminosity. The discovery of significant evidence for $D = 6$ operators in the SMEFT would surely signal the opening of an exciting new era in particle physics.

M

Group Theory

DOI: 10.1201/9781003411666-M

M.1 Definition and simple examples

A group $\mathcal{G}$ is a set of elements $(a, b, c, \ldots)$ with a law for combining any two elements a, b so as to form their ordered 'product' ab, such that the following four conditions hold:

(i) For every $a, b \in \mathcal{G}$, the product $ab \in \mathcal{G}$ (the symbol '$\in$' means 'belongs to', or 'is a member of').

(ii) The law of combination is associative, ie

$$(ab)c = a(bc). \tag{M.1}$$

(iii) $\mathcal{G}$ contains a unique identity element, e, such that for all $a \in \mathcal{G}$,

$$ae = ea = a. \tag{M.2}$$

(iv) For all $a \in \mathcal{G}$, there is a unique inverse element, a^{-1}, such that

$$aa^{-1} = a^{-1}a = e. \tag{M.3}$$

Note that in general the law of combination is not commutative — i.e. $ab \neq ba$; if it is commutative, the group is *Abelian*; if not, it is *non-Abelian*. Any finite set of elements satisfying the conditions (i) – (iv) forms a finite group, the *order* of the group being equal to the number of elements in the set. If the set does not have a finite number of elements it is an infinite group.

As a simple example, the set of four numbers $(1, i, -1, -i)$ form a finite Abelian group of order 4, with the law of combination being ordinary multiplication. The reader may check that each of (i) – (iv) is satisfied, with e taken to be the number 1, and the inverse being the algebraic reciprocal. A second group of order 4 is provided by the matrices

$$\begin{pmatrix} 1 & 0 \\ 0 & 1 \end{pmatrix}, \begin{pmatrix} 0 & 1 \\ -1 & 0 \end{pmatrix}, \begin{pmatrix} -1 & 0 \\ 0 & -1 \end{pmatrix}, \begin{pmatrix} 0 & -1 \\ 1 & 0 \end{pmatrix}, \tag{M.4}$$

with the combination law being matrix multiplication, 'e' being the first (unit) matrix, and the inverse being the usual matrix inverse. Although matrix multiplication is not commutative in general, it happens to be so for these particular matrices. In fact, the way these four matrices multiply together is (as the reader can verify) exactly the same as the way the four numbers $(1, i, -1, -i)$ (in that order) do. Further, the correspondence between the elements of the two groups is 'one to one': that is, if we label the two sets of group elements by (e, a, b, c) and (e', a', b', c'), we have the correspondences $e \leftrightarrow e'$, $a \leftrightarrow a'$, $b \leftrightarrow b'$, $c \leftrightarrow c'$. Two groups with the same multiplication structure, and with a one-to-one correspondence between their elements, are said to be *isomorphic*. If they have the same multiplication structure but the correspondence is not one-to-one, they are *homomorphic*.

DOI: 10.1201/9781003411666-M

M.2   Lie Groups

We are interested in *continuous groups*—that is, groups whose elements are labelled by a number of continuously variable real parameters $\alpha_1, \alpha_2, \ldots, \alpha_r : g(\alpha_1, \alpha_2, \ldots, \alpha_r) \equiv g(\boldsymbol{\alpha})$. In particular, we are concerned with various kinds of 'coordinate transformations' (not necessarily space-time ones, but including also 'internal' transformations such as those of SU(3)). For example, rotations in three dimensions form a group, whose elements are specified by three real parameters (e.g. two for defining the axis of the rotation, and one for the angle of rotation about that axis). Lorentz transformations also form a group, this time with six real parameters (three for 3-D rotations, three for pure velocity transformations). The matrices of SU(3) are specified by the values of eight real parameters. By convention, parametrizations are arranged in such a way that $g(\boldsymbol{0})$ is the identity element of the group. For a continuous group, condition (i) takes the form

$$g(\boldsymbol{\alpha})g(\boldsymbol{\beta}) = g(\boldsymbol{\gamma}(\boldsymbol{\alpha}, \boldsymbol{\beta})), \tag{M.5}$$

where the parameters $\boldsymbol{\gamma}$ are continuous functions of the parameters $\boldsymbol{\alpha}$ and $\boldsymbol{\beta}$. A more restrictive condition is that $\boldsymbol{\gamma}$ should be an *analytic* function of $\boldsymbol{\alpha}$ and $\boldsymbol{\beta}$; if this is the case, the group is a *Lie group*.

The analyticity condition implies that if we are given the form of the group elements in the neighbourhood of any one element, we can 'move out' from that neighbourhood to other nearby elements, using the mathematical procedure known as 'analytic continuation' (essentially, using a power series expansion); by repeating the process, we should be able to reach all group elements which are 'continuously connected' to the original element. The simplest group element to consider is the identity, which we shall now denote by I. Lie proved that the properties of the elements of a Lie group which can be reached continuously from the identity I are determined from elements lying in the neighbourhood of I.

M.3   Generators of Lie groups

Consider (following Lichtenberg 1970, chapter 5) a group of transformations defined by

$$x_i' = f_i(x_1, x_2, \ldots, x_N; \alpha_1, \alpha_2, \ldots, \alpha_r), \tag{M.6}$$

where the x_i's $(i = 1, 2, \ldots, N)$ are the 'coordinates' on which the transformations act, and the α's are the (real) parameters of the transformations. By convention, $\boldsymbol{\alpha} = \boldsymbol{0}$ is the identity transformation, so

$$x_i = f_i(\boldsymbol{x}, \boldsymbol{0}). \tag{M.7}$$

A transformation in the neighbourhood of the identity is then given by

$$\mathrm{d}x_i = \sum_{\nu=1}^{r} \frac{\partial f_i}{\partial \alpha_\nu} \mathrm{d}\alpha_\nu, \tag{M.8}$$

where the $\{\mathrm{d}\alpha_\nu\}$ are infinitesimal parameters, and the partial derivative is understood to be evaluated at the point $(\boldsymbol{x}, \boldsymbol{0})$.

Consider now the change in a function $F(\boldsymbol{x})$ under the infinitesimal transformation (M.8). We have

$$
\begin{aligned}
F \to F + \mathrm{d}F &= F + \sum_{i=1}^{N} \frac{\partial F}{\partial x_i} \mathrm{d}x_i \\
&= F + \sum_{i=1}^{N} \left[\sum_{\nu=1}^{r} \frac{\partial f_i}{\partial \alpha_\nu} \mathrm{d}\alpha_\nu \right] \frac{\partial F}{\partial x_i} \\
&\equiv \{ 1 - \sum_{\nu=1}^{r} \mathrm{d}\alpha_\nu \mathrm{i} \hat{X}_\nu \} F,
\end{aligned}
\tag{M.9}
$$

where

$$
\hat{X}_\nu \equiv \mathrm{i} \sum_{i=1}^{N} \frac{\partial f_i}{\partial \alpha_\nu} \frac{\partial}{\partial x_i}
\tag{M.10}
$$

is a *generator of infinitesimal transformations*[1]. Note that in (M.10) ν runs from 1 to r, so there are as many generators as there are parameters labelling the group elements. Finite transformations are obtained by 'exponentiating' the quantity in braces in (M.9) (compare (12.30)):

$$
\hat{U}(\boldsymbol{\alpha}) = \exp\{ -\mathrm{i}\boldsymbol{\alpha} \cdot \hat{\boldsymbol{X}} \},
\tag{M.11}
$$

where we have written $\sum_{\nu=1}^{r} \alpha_\nu \hat{X}_\nu = \boldsymbol{\alpha} \cdot \hat{\boldsymbol{X}}$.

An important theorem states that the commutator of any two generators of a Lie group is a linear combination of the generators:

$$
[\hat{X}_\lambda, \hat{X}_\mu] = c_{\lambda\mu}^{\nu} \hat{X}_\nu,
\tag{M.12}
$$

where the constants $c_{\lambda\mu}^{\nu}$ are complex numbers called the *structure constants* of the group; a sum over ν from 1 to r is understood on the right-hand side. The commutation relations (M.12) are called the *algebra* of the group.

M.4 Examples

M.4.1 SO(3) and three-dimensional rotations

Rotations in three dimensions are defined by

$$
\boldsymbol{x}' = R\,\boldsymbol{x},
\tag{M.13}
$$

where R is a real 3×3 matrix such that the length of $\boldsymbol{x}$ is preserved, i.e. $\boldsymbol{x}'^{\mathrm{T}}\boldsymbol{x}' = \boldsymbol{x}^{\mathrm{T}}\boldsymbol{x}$. This implies that $R^{\mathrm{T}}R = I$, so that R is an orthogonal matrix. It follows that

$$
1 = \det(R^{\mathrm{T}}R) = \det R^{\mathrm{T}} \det R = (\det R)^2,
\tag{M.14}
$$

and so $\det R = \pm 1$. Those R's with $\det R = -1$ include a parity transformation ($\boldsymbol{x}' = -\boldsymbol{x}$), which is not continuously connected to the identity. Those with $\det R = 1$ are 'proper

---

[1] Clearly there is lot of 'convention' (the sign, the i) in the definition of $\hat{X}_\nu$. It is chosen for convenient consistency with familiar generators, for example those of SO(3) (see section M.4.1).

rotations', and they form the elements of the group SO(3): the *S*pecial *O*rthogonal group in *3* dimensions.

An R close to the identity matrix I can be written as $R = I + \delta R$ where

$$(I + \delta R)^{\mathrm{T}}(I + \delta R) = I. \tag{M.15}$$

Expanding this out to first order in δR gives

$$\delta R^{\mathrm{T}} = -\delta R, \tag{M.16}$$

so that δR is an antisymmetric 3×3 matrix (compare (12.19)). We may parametrize δR as

$$\delta R = \begin{pmatrix} 0 & \epsilon_3 & -\epsilon_2 \\ -\epsilon_3 & 0 & \epsilon_1 \\ \epsilon_2 & -\epsilon_1 & 0 \end{pmatrix}, \tag{M.17}$$

and an infinitesimal rotation is then given by

$$\boldsymbol{x}' = \boldsymbol{x} - \boldsymbol{\epsilon} \times \boldsymbol{x}, \tag{M.18}$$

(compare (12.64)), or

$$\mathrm{d}x_1 = -\epsilon_2 x_3 + \epsilon_3 x_2, \quad \mathrm{d}x_2 = -\epsilon_3 x_1 + \epsilon_1 x_3, \quad \mathrm{d}x_3 = -\epsilon_1 x_2 + \epsilon_2 x_1. \tag{M.19}$$

Thus in (M.8), identifying $\mathrm{d}\alpha_1 \equiv \epsilon_1$, $\mathrm{d}\alpha_2 \equiv \epsilon_2$, $\mathrm{d}\alpha_3 \equiv \epsilon_3$, we have

$$\frac{\partial f_1}{\partial \alpha_1} = 0, \quad \frac{\partial f_1}{\partial \alpha_2} = -x_3, \quad \frac{\partial f_1}{\partial \alpha_3} = x_2, \quad \text{etc.} \tag{M.20}$$

The generators (M.10) are then

$$\left. \begin{array}{rcl} \hat{X}_1 &=& ix_3 \frac{\partial}{\partial x_2} - ix_2 \frac{\partial}{\partial x_3} \\ \hat{X}_2 &=& ix_1 \frac{\partial}{\partial x_3} - ix_3 \frac{\partial}{\partial x_1} \\ \hat{X}_3 &=& ix_2 \frac{\partial}{\partial x_1} - ix_1 \frac{\partial}{\partial x_2} \end{array} \right\} \tag{M.21}$$

which are easily recognized as the quantum-mechanical angular momentum operators

$$\hat{\boldsymbol{X}} = \boldsymbol{x} \times -i\boldsymbol{\nabla}, \tag{M.22}$$

which satisfy the *SO(3) algebra*

$$[\hat{X}_i, \hat{X}_j] = i\epsilon_{ijk}\hat{X}_k. \tag{M.23}$$

The action of finite rotations, parametrized by $\boldsymbol{\alpha} = (\alpha_1, \alpha_2, \alpha_3)$, on functions F is given by

$$\hat{U}(\boldsymbol{\alpha}) = \exp\{-i\boldsymbol{\alpha} \cdot \hat{\boldsymbol{X}}\}. \tag{M.24}$$

The operators $\hat{U}(\boldsymbol{\alpha})$ form a group which is isomorphic to SO(3). The structure constants of SO(3) are $i\epsilon_{ijk}$, from (M.23).

M.4.2   SU(2)

We write the infinitesimal SU(2) transformation (acting on a general complex two-component column vector) as (cf (12.27))

$$\begin{pmatrix} q_1' \\ q_2' \end{pmatrix} = (1 + i\boldsymbol{\epsilon} \cdot \boldsymbol{\tau}/2) \begin{pmatrix} q_1 \\ q_2 \end{pmatrix}, \tag{M.25}$$

so that

$$dq_1 = \frac{i\epsilon_3}{2}q_1 + \left(\frac{i\epsilon_1}{2} + \frac{\epsilon_2}{2}\right)q_2$$

$$dq_2 = \frac{-i\epsilon_3}{2}q_2 + \left(\frac{i\epsilon_1}{2} - \frac{\epsilon_2}{2}\right)q_1. \tag{M.26}$$

Then (with $d\alpha_1 \equiv \epsilon_1$ etc)

$$\frac{\partial f_1}{\partial \alpha_1} = \frac{iq_2}{2}, \quad \frac{\partial f_1}{\partial \alpha_2} = \frac{q_2}{2}, \quad \frac{\partial f_1}{\partial \alpha_3} = \frac{iq_1}{2}, \tag{M.27}$$

$$\frac{\partial f_2}{\partial \alpha_1} = \frac{iq_1}{2}, \quad \frac{\partial f_2}{\partial \alpha_2} = -\frac{q_2}{2}, \quad \frac{\partial f_2}{\partial \alpha_3} = -\frac{iq_2}{2}, \tag{M.28}$$

and (from (M.10))

$$\hat{X}_1' = -\frac{1}{2}\left\{q_2\frac{\partial}{\partial q_1} + q_1\frac{\partial}{\partial q_2}\right\} \tag{M.29}$$

$$\hat{X}_2' = \frac{i}{2}\left\{q_2\frac{\partial}{\partial q_1} - q_1\frac{\partial}{\partial q_2}\right\} \tag{M.30}$$

$$\hat{X}_3' = \frac{1}{2}\left\{-q_1\frac{\partial}{\partial q_1} + q_2\frac{\partial}{\partial q_2}\right\}. \tag{M.31}$$

It is an interesting exercise to check that the commutation relations of the $\hat{X}_i'$'s are exactly the same as those of the $\hat{X}_i$'s in (M.23). The two groups are therefore said to have the same algebra, with the same structure constants, and they are in fact isomorphic in the vicinity of their respective identity elements. They are not the same for 'large' transformations, however, as we discuss in section M.7.

M.4.3 SO(4): The special orthogonal group in four dimensions

This is the group whose elements are 4×4 matrices S such that $S^\mathrm{T}S = I$, where I is the 4×4 unit matrix, with the condition $\det S = +1$. The Euclidean (length)^2 $x_1^2 + x_2^2 + x_3^2 + x_4^2$ is left invariant under SO(4) transformations. Infinitesimal SO(4) transformations are characterized by the 4-D analogue of those for SO(3), namely by 4×4 real antisymmetric matrices δS, which have 6 real parameters. We choose to parametrize δS in such a way that the Euclidean 4-vector $(\boldsymbol{x}, x_4)$ is transformed to (cf (18.76) and (18.77))

$$\boldsymbol{x}' = \boldsymbol{x} - \boldsymbol{\epsilon} \times \boldsymbol{x} - \boldsymbol{\eta}x_4,$$
$$x_4' = x_4 + \boldsymbol{\eta} \cdot \boldsymbol{x}, \tag{M.32}$$

where $\boldsymbol{x} = (x_1, x_2, x_3)$ and $\boldsymbol{\eta} = (\eta_1, \eta_2, \eta_3)$. Note that the first three components transform by (M.18) when $\boldsymbol{\eta} = 0$, so that SO(3) is a *subgroup* of SO(4). The six generators are (with $d\alpha_1 \equiv \epsilon_1$ etc)

$$\hat{X}_1 = ix_3\frac{\partial}{\partial x_2} - ix_2\frac{\partial}{\partial, x_3}, \tag{M.33}$$

and similarly for $\hat{X}_2$ and $\hat{X}_3$ as in (M.21), together with (defining $d\alpha_4 = \eta_1$ etc)

$$\hat{X}_4 = i\left(-x_4\frac{\partial}{\partial x_1} + x_1\frac{\partial}{\partial x_4}\right) \tag{M.34}$$

$$\hat{X}_5 = i\left(-x_4\frac{\partial}{\partial x_2} + x_2\frac{\partial}{\partial x_4}\right) \tag{M.35}$$

$$\hat{X}_6 = i\left(-x_4\frac{\partial}{\partial x_3} + x_3\frac{\partial}{\partial x_4}\right). \tag{M.36}$$

Relabelling these last three generators as $\hat{Y}_1 \equiv \hat{X}_4$, $\hat{Y}_2 \equiv \hat{X}_5$, $\hat{Y}_3 \equiv \hat{X}_6$, we find the following algebra:

$$[\hat{X}_i, \hat{X}_j] = i\epsilon_{ijk}\hat{X}_k \tag{M.37}$$

$$[\hat{X}_i, \hat{Y}_j] = i\epsilon_{ijk}\hat{Y}_k \tag{M.38}$$

$$[\hat{Y}_i, \hat{Y}_j] = i\epsilon_{ijk}\hat{X}_k, \tag{M.39}$$

together with

$$[\hat{X}_1, \hat{Y}_1] = [\hat{X}_2, \hat{Y}_2] = [\hat{X}_3, \hat{Y}_3] = 0. \tag{M.40}$$

(M.37) confirms that the three generators controlling infinitesimal transformations among the first three components $\boldsymbol{x}$ obey the angular momentum commutation relations. (M.37) – (M.40) constitute the algebra of SO(4).

This algebra may be simplified by introducing the linear combinations

$$\hat{M}_i = \frac{1}{2}(\hat{X}_i + \hat{Y}_i) \tag{M.41}$$

$$\hat{N}_i = \frac{1}{2}(\hat{X}_i - \hat{Y}_i), \tag{M.42}$$

which satisfy

$$[\hat{M}_i, \hat{M}_j] = i\epsilon_{ijk}\hat{M}_k \tag{M.43}$$

$$[\hat{N}_i, \hat{N}_j] = i\epsilon_{ijk}\hat{N}_k \tag{M.44}$$

$$[\hat{M}_i, \hat{N}_j] = 0. \tag{M.45}$$

From (M.43) – (M.45) we see that, in this form, the six generators have separated into two sets of three, each set obeying the algebra of SO(3) (or of SU(2)), and commuting with the other set. They therefore behave like two *independent* angular momentum operators. The algebra (M.43) – (M.45) is referred to as SU(2)×SU(2).

M.4.4   The Lorentz group

In this case the quadratic form left invariant by the transformation is the Minkowskian one $(x^0)^2 - \boldsymbol{x}^2$ (see Appendix D of volume 1). We may think of infinitesimal Lorentz transformations as corresponding physically to ordinary infinitesimal 3-D rotations, together with infinitesimal pure velocity transformations ('boosts'). The basic 4-vector then transforms by

$$\left.\begin{array}{rcl} x^{0\prime} &=& x^0 - \boldsymbol{\eta} \cdot \boldsymbol{x} \\ \boldsymbol{x}' &=& \boldsymbol{x} - \boldsymbol{\epsilon} \times \boldsymbol{x} - \boldsymbol{\eta}x^0 \end{array}\right\} \tag{M.46}$$

where $\boldsymbol{\eta}$ is now the infinitesimal velocity parameter (the reader may check that $(x^0)^2 - \boldsymbol{x}^2$ is indeed left invariant by (M.46), to first order in $\boldsymbol{\epsilon}$ and $\boldsymbol{\eta}$). The six generators are then $\hat{X}_1, \hat{X}_2, \hat{X}_3$ as in (M.21), together with

$$\hat{K}_1 = -i\left(x^1\frac{\partial}{\partial x^0} + x^0\frac{\partial}{\partial x^1}\right) \tag{M.47}$$

$$\hat{K}_2 = -i\left(x^2\frac{\partial}{\partial x^0} + x^0\frac{\partial}{\partial x^2}\right) \tag{M.48}$$

$$\hat{K}_3 = -i\left(x^3\frac{\partial}{\partial x^0} + x^0\frac{\partial}{\partial x^3}\right). \tag{M.49}$$

The corresponding algebra is

$$[\hat{X}_i, \hat{X}_j] = i\epsilon_{ijk}\hat{X}_k \tag{M.50}$$

$$[\hat{X}_i, \hat{K}_j] = i\epsilon_{ijk}\hat{K}_k \tag{M.51}$$

$$[\hat{K}_i, \hat{K}_j] = -i\epsilon_{ijk}\hat{X}_k. \tag{M.52}$$

Note the minus sign on the right-hand side of (M.52) as compared with (M.39).

M.4.5 SU(3)

A general infinitesimal SU(3) transformation may be written as (cf (12.71) and (12.72))

$$\begin{pmatrix} q_1 \\ q_2 \\ q_3 \end{pmatrix}' = \left(1 + i\frac{1}{2}\boldsymbol{\eta}\cdot\boldsymbol{\lambda}\right)\begin{pmatrix} q_1 \\ q_2 \\ q_3 \end{pmatrix}, \tag{M.53}$$

where there are now 8 of these $\boldsymbol{\eta}$'s, $\boldsymbol{\eta} = (\eta_1, \eta_2, \ldots, \eta_8)$, and the λ-matrices are the Gell-Mann matrices

$$\lambda_1 = \begin{pmatrix} 0 & 1 & 0 \\ 1 & 0 & 0 \\ 0 & 0 & 0 \end{pmatrix}, \; \lambda_2 = \begin{pmatrix} 0 & -i & 0 \\ i & 0 & 0 \\ 0 & 0 & 0 \end{pmatrix}, \; \lambda_3 = \begin{pmatrix} 1 & 0 & 0 \\ 0 & -1 & 0 \\ 0 & 0 & 0 \end{pmatrix} \tag{M.54}$$

$$\lambda_4 = \begin{pmatrix} 0 & 0 & 1 \\ 0 & 0 & 0 \\ 1 & 0 & 0 \end{pmatrix}, \; \lambda_5 = \begin{pmatrix} 0 & 0 & -i \\ 0 & 0 & 0 \\ i & 0 & 0 \end{pmatrix}, \; \lambda_6 = \begin{pmatrix} 0 & 0 & 0 \\ 0 & 0 & 1 \\ 0 & 1 & 0 \end{pmatrix} \tag{M.55}$$

$$\lambda_7 = \begin{pmatrix} 0 & 0 & 0 \\ 0 & 0 & -i \\ 0 & i & 0 \end{pmatrix}, \; \lambda_8 = \begin{pmatrix} \frac{1}{\sqrt{3}} & 0 & 0 \\ 0 & \frac{1}{\sqrt{3}} & 0 \\ 0 & 0 & -\frac{2}{\sqrt{3}} \end{pmatrix}. \tag{M.56}$$

In this parametrization the first three of the eight generators $\hat{G}_r$ ($r = 1, 2, \ldots, 8$) are the same as $\hat{X}_1', \hat{X}_2', \hat{X}_3'$ of (M.29) – (M.30). The others may be constructed as usual from (M.10); for example,

$$\hat{G}_5 = \frac{i}{2}\left(q_3\frac{\partial}{\partial q_1} - q_1\frac{\partial}{\partial q_3}\right), \; \hat{G}_7 = \frac{i}{2}\left(q_3\frac{\partial}{\partial q_2} - q_2\frac{\partial}{\partial q_3}\right). \tag{M.57}$$

The SU(3) algebra is found to be

$$[\hat{G}_a, \hat{G}_b] = if_{abc}\hat{G}_c, \tag{M.58}$$

where a, b, and c each run from 1 to 8. The structure constants are if_{abc}, and the non-vanishing f's are as follows:

$$f_{123} = 1, \; f_{147} = 1/2, \; f_{156} = -1/2, \; f_{246} = 1/2, \; f_{257} = 1/2 \tag{M.59}$$

$$f_{345} = 1/2, \; f_{367} = -1/2, \; f_{458} = \sqrt{3}/2, \; f_{678} = \sqrt{3}/2. \tag{M.60}$$

Note that the f's are antisymmetric in all pairs of indices (Carruthers (1966) chapter 2).

M.5 Matrix representations of generators and Lie groups

We have shown how the generators $\hat{X}_1, \hat{X}_2, \ldots, \hat{X}_r$ of a Lie group can be constructed as differential operators, understood to be acting on functions of the 'coordinates' to which the transformations of the group refer. These generators satisfy certain commutation relations, the Lie algebra of the group. For any given Lie algebra, it is also possible to find sets of *matrices* $X_1, X_2, \ldots, X_r$ (without hats) which satisfy the same commutation relations as the $\hat{X}_\nu$'s — that is, they have the same algebra. Such matrices are said to form a (matrix) representation of the Lie algebra, or equivalently of the generators. The idea is familiar from the study of angular momentum in quantum mechanics (Schiff (1968) section 27), where the entire theory may be developed from the commutation relations (with $\hbar = 1$)

$$[\hat{J}_i, \hat{J}_j] = \mathrm{i}\epsilon_{ijk}\hat{J}_k \qquad (M.61)$$

for the angular momentum operators $\hat{J}_i$, together with the physical requirement that the $\hat{J}_i$'s (and the matrices representing them) must be Hermitian. In this case the matrices are of the form (in quantum-mechanical notation)

$$\left(J_i^{(J)}\right)_{M_J' M_J} \equiv \langle JM_J'|\hat{J}_i|JM_J\rangle, \qquad (M.62)$$

where $|JM_J\rangle$ is an eigenstate of $\hat{\boldsymbol{J}}^2$ and of $\hat{J}_3$ with eigenvalues $J(J+1)$ and M_J, respectively. Since M_J and M_J' each run over the $2J + 1$ values defined by $-J \leq M_J$, $M_J' \leq J$, the matrices $J_i^{(J)}$ are of dimension $(2J + 1) \times (2J + 1)$. Clearly, since the generators of SU(2) have the same algebra as (M.61), an identical matrix representation may be obtained for them; these matrices were denoted by $T_i^{(T)}$ in section 12.1.2. It is important to note that J (or T) can take an infinite sequence of values $J = 0, 1/2, 1, 3/2, \ldots$, corresponding physically to various 'spin' magnitudes. Thus there are infinitely many sets of three matrices $(J_1^{(J)}, J_2^{(J)}, J_3^{(J)})$ all with the same commutation relations as (M.61).

A similar method for obtaining matrix representations of Lie algebras may be followed in other cases. In physical terms, the problem amounts to finding a correct labelling of the base states, analogous to $|JM\rangle$. In the latter case, the quantum number J specifies each different representation. The reason it does so is because (as should be familiar) the corresponding operator $\hat{\boldsymbol{J}}^2$ commutes with every generator:

$$[\hat{\boldsymbol{J}}^2, \hat{J}_i] = 0. \qquad (M.63)$$

Such an operator is called a *Casimir operator*, and by a lemma due to Schur (Hammermesh (1962) pages 100-101) it must be a multiple of the unit operator. The numerical value it has is different for each different representation, and may therefore be used to characterize a representation (namely as '$J = 0$', '$J = 1/2$', etc.).

In general, more than one such operator is needed to characterize a representation completely. For example, in SO(4), the two operators $\hat{\boldsymbol{M}}^2$ and $\hat{\boldsymbol{N}}^2$ commute with all the generators, and take values $M(M + 1)$ and $N(N + 1)$ respectively, where $M, N = 0, 1/2, 1, \ldots$. Thus the labelling of the matrix elements of the generators is the same as it would be for two independent particles, one of spin M and the other of spin N. For given M, N the matrices are of dimension $[(2M + 1) + (2N + 1)] \times [(2M + 1) + (2N + 1)]$. The number of Casimir operators required to characterize a representation is called the *rank* of the group (or the algebra). This is also equal to the number of independent mutually commuting generators (though this is by no means obvious). Thus SO(4) is a rank two group, with two commuting generators $\hat{M}_3$ and $\hat{N}_3$; so is SU(3), since $\hat{G}_3$ and $\hat{G}_8$ commute. Two Casimir operators are therefore required to characterize the representations of SU(3), which may be taken to be the 'quadratic' one

$$\hat{C}_2 \equiv \hat{G}_1^2 + \hat{G}_2^2 + \ldots + \hat{G}_8^2, \tag{M.64}$$

together with a 'cubic' one

$$\hat{C}_3 \equiv d_{abc}\hat{G}_a\hat{G}_b\hat{G}_c, \tag{M.65}$$

where the coefficients d_{abc} are defined by the relation

$$\{\lambda_a, \lambda_b\} = \frac{4}{3}\delta_{ab}I + 2d_{abc}\lambda_c, \tag{M.66}$$

and are symmetric in all pairs of indices (they are tabulated in Carruthers (1966) table 2.1). In practice, for the few SU(3) representations that are actually required, it is more common to denote them (as we have in the text) by their dimensionality, which for the cases **1** (singlet), **3** (triplet), **3*** (antitriplet), **8** (octet), and **10** (decuplet) is in fact a unique labelling. The values of $\hat{C}_2$ in these representations are

$$\hat{C}_2(\mathbf{1}) = 0, \ \hat{C}_2(\mathbf{3}, \mathbf{3}^*) = 4/3, \ \hat{C}_2(\mathbf{8}) = 3, \ \hat{C}_2(\mathbf{10}) = 6. \tag{M.67}$$

Having characterized a given representation by the eigenvalues of the Casimir operator(s), a further labelling is then required to characterize the states within a given representation (the analogue of the eigenvalue of $\hat{J}_3$ for angular momentum). For SO(4) these further labels may be taken to be the eigenvalues of $\hat{M}_3$ and $\hat{N}_3$; for SU(3) they are the eigenvalues of $\hat{G}_3$ and $\hat{G}_8$—i.e. those corresponding to the third component of isospin and hypercharge, in the flavour case (see figures 12.3 and 12.4).

In the case of groups whose elements are themselves matrices, such as SO(3), SO(4), SU(2), SU(3), and the Lorentz group, one particular representation of the generators may always be obtained by considering the general form of a matrix in the group which is infinitesimally close to the unit element. In a suitable parametrization, we may write such a matrix as

$$1 + i\sum_{\nu=1}^{r} \epsilon_\nu X_\nu^{(\mathcal{G})}, \tag{M.68}$$

where $(\epsilon_1, \epsilon_2, \ldots, \epsilon_r)$ are infinitesimal parameters, and $(X_1^{(\mathcal{G})}, X_2^{(\mathcal{G})}, \ldots, X_r^{(\mathcal{G})})$ are matrices representing the generators of the (matrix) group $\mathcal{G}$. This is exactly the same procedure we followed for SU(2) in section 12.1.1, where we found from (12.26) that the three $X_\nu^{(\mathrm{SU}(2))}$'s were just $\boldsymbol{\tau}/2$, satisfying the SU(2) algebra. Similarly, in section 12.2 we saw that the eight SU(3) $X_\nu^{(\mathrm{SU}(3))}$'s were just $\boldsymbol{\lambda}/2$, satisfying the SU(3) algebra. These particular two representations are called the *fundamental* representations of the SU(2) and SU(3) algebras, respectively; they are the representations of lowest dimensionality. For SO(3), the three $X_\nu^{(\mathrm{SO}(3))}$'s are (from (M.17))

$$X_1^{(\mathrm{SO}(3))} = \begin{pmatrix} 0 & 0 & 0 \\ 0 & 0 & -\mathrm{i} \\ 0 & \mathrm{i} & 0 \end{pmatrix}$$

$$\Xi_2^{(\mathrm{SO}(3))} = \begin{pmatrix} 0 & 0 & \mathrm{i} \\ 0 & 0 & 0 \\ -\mathrm{i} & 0 & 0 \end{pmatrix}$$

$$X_3^{(\mathrm{SO}(3))} = \begin{pmatrix} 0 & -\mathrm{i} & 0 \\ \mathrm{i} & 0 & 0 \\ 0 & 0 & 0 \end{pmatrix} \tag{M.69}$$

which are the same as the 3×3 matrices $T_i^{(1)}$ of (12.48):

$$\left(T_i^{(1)}\right)_{jk} = -i\epsilon_{ijk}. \tag{M.70}$$

The matrices $\tau_i/2$ and $T_i^{(1)}$ correspond to the values $J = 1/2$, $J = 1$, respectively, in angular momentum terms.

It is not a coincidence that the coefficients on the right-hand side of (M.70) are (minus) the SO(3) structure constants. One representation of a Lie algebra is always provided by a set of matrices $\{X_\nu^{(R)}\}$ whose elements are defined by

$$\left(X_\lambda^{(R)}\right)_{\mu\nu} = -c_{\lambda\mu}^\nu, \tag{M.71}$$

where the c's are the structure constants of (M.12), and each of μ, ν, λ runs from 1 to r. Thus these matrices are of dimensionality $r \times r$, where r is the number of generators. That this prescription works is due to the fact that the generators satisfy the *Jacobi identity*

$$[\hat{X}_\lambda, [\hat{X}_\mu, \hat{X}_\nu]] + [\hat{X}_\mu, [\hat{X}_\nu, \hat{X}_\lambda]] + [\hat{X}_\nu, [\hat{X}_\lambda, \hat{X}_\mu]] = 0. \tag{M.72}$$

Using (M.12) to evaluate the commutators, and the fact that the generators are independent, we obtain

$$c_{\mu\nu}^\alpha c_{\lambda\alpha}^\beta + c_{\nu\lambda}^\alpha c_{\mu\alpha}^\beta + c_{\lambda\mu}^\alpha c_{\nu\alpha}^\beta = 0. \tag{M.73}$$

The reader may fill in the steps leading from here to the desired result:

$$\left(X_\lambda^{(R)}\right)_{\nu\alpha} \left(X_\mu^{(R)}\right)_{\alpha\beta} - \left(X_\mu^{(R)}\right)_{\nu\alpha} \left(X_\lambda^{(R)}\right)_{\alpha\beta} = c_{\lambda\mu}^\alpha \left(X_\alpha^{(R)}\right)_{\nu\beta}. \tag{M.74}$$

(M.74) is precisely the $(\nu\beta)$ matrix element of

$$[X_\lambda^{(R)}, X_\mu^{(R)}] = c_{\lambda\mu}^\alpha X_\alpha^{(R)}, \tag{M.75}$$

showing that the $X_\mu^{(R)}$'s satisfy the group algebra (M.12), as required. The representation in which the generators are represented by (minus) the structure constants, in the sense of (M.71), is called the *regular* or *adjoint* representation.

Having obtained any particular matrix representation $\mathbf{X}^{(P)}$ of the generators of a group $\mathcal{G}$, a corresponding *matrix representation of the group elements* can be obtained by exponentiation, via

$$D^{(P)}(\boldsymbol{\alpha}) = \exp\{i\boldsymbol{\alpha} \cdot \mathbf{X}^{(P)}\}, \tag{M.76}$$

where $\boldsymbol{\alpha} = (\alpha_1, \alpha_2, \ldots, \alpha_r)$ (see (12.31) and (12.49) for SU(2), and (12.74) and (12.81) for SU(3)). In the case of the groups whose elements are matrices, exponentiating the generators $\mathbf{X}^{(\mathcal{G})}$ just recreates the general matrices of the group, so we may call this the 'self-representation': the one in which the group elements are represented by themselves. In the more general case (M.76), the crucial property of the matrices $D^{(P)}(\boldsymbol{\alpha})$ is that they obey the same group combination law as the elements of the group $\mathcal{G}$ they are representing: that is, if the group elements obey

$$g(\boldsymbol{\alpha})g(\boldsymbol{\beta}) = g(\boldsymbol{\gamma}(\boldsymbol{\alpha}, \boldsymbol{\beta})), \tag{M.77}$$

then

$$D^{(P)}(\boldsymbol{\alpha})D^{(P)}(\boldsymbol{\beta}) = D^{(P)}(\boldsymbol{\gamma}(\boldsymbol{\alpha}, \boldsymbol{\beta})). \tag{M.78}$$

It is a rather remarkable fact that there are certain, say, 10×10 matrices which multiply together in exactly the same way as the rotation matrices of SO(3).

M.6 The Lorentz group

Consideration of matrix representations of the Lorentz group provides insight into the equations of relativistic quantum mechanics, for example the Dirac equation. Consider the infinitesimal Lorentz transformation (M.46). The 4×4 matrix corresponding to this may be written in the form

$$1 + i\boldsymbol{\epsilon} \cdot \boldsymbol{X}^{(\mathrm{LG})} - i\boldsymbol{\eta} \cdot \boldsymbol{K}^{(\mathrm{LG})}, \tag{M.79}$$

where

$$X_1^{(\mathrm{LG})} = \begin{pmatrix} 0 & 0 & 0 & 0 \\ 0 & 0 & 0 & 0 \\ 0 & 0 & 0 & -i \\ 0 & 0 & i & 0 \end{pmatrix} \quad \text{etc,} \tag{M.80}$$

(as in (M.69) but with an extra border of 0's), and

$$K_1^{(\mathrm{LG})} = \begin{pmatrix} 0 & -i & 0 & 0 \\ -i & 0 & 0 & 0 \\ 0 & 0 & 0 & 0 \\ 0 & 0 & 0 & 0 \end{pmatrix}$$

$$K_2^{(\mathrm{LG})} = \begin{pmatrix} 0 & 0 & -i & 0 \\ 0 & 0 & 0 & 0 \\ -i & 0 & 0 & 0 \\ 0 & 0 & 0 & 0 \end{pmatrix}$$

$$K_3^{(\mathrm{LG})} = \begin{pmatrix} 0 & 0 & 0 & -i \\ 0 & 0 & 0 & 0 \\ 0 & 0 & 0 & 0 \\ -i & 0 & 0 & 0 \end{pmatrix}. \tag{M.81}$$

In (M.80) and (M.81) the matrices are understood to be acting on the 4-component vector

$$\begin{pmatrix} x^0 \\ x^1 \\ x^2 \\ x^3 \end{pmatrix}. \tag{M.82}$$

It is straightforward to check that the matrices $X_i^{(\mathrm{LG})}$ and $K_i^{(\mathrm{LG})}$ satisfy the algebra (M.50) – (M.52) as expected.

An important point to note is that the matrices $K_i^{(\mathrm{LG})}$, in contrast to $X_i^{(\mathrm{LG})}$ or $X_i^{(\mathrm{SO}(3))}$, and to the corresponding matrices of SU(2) and SU(3), are *not* Hermitian. A theorem states that only the generators of *compact* Lie groups can be represented by finite-dimensional Hermitian matrices. Here 'compact' means that the domain of variation of all the parameters is bounded (none exceeds a given positive number p in absolute magnitude) and closed (the limit of every convergent sequence of points in the set also lies in the set). For the Lorentz group, the limiting velocity c is not included (the γ-factor goes to infinity), and so the group is non-compact.

In a general representation of the Lorentz group, the generators X_i, K_i will obey the algebra (M.50) – (M.52). Let us introduce the combinations

$$\boldsymbol{P} \equiv \frac{1}{2}(\boldsymbol{X} + i\boldsymbol{K}) \tag{M.83}$$

$$\boldsymbol{Q} \equiv \frac{1}{2}(\boldsymbol{X} - i\boldsymbol{K}). \tag{M.84}$$

Then the algebra becomes

$$[P_i, P_j] = i\epsilon_{ijk}P_k \tag{M.85}$$

$$[Q_i, Q_j] = i\epsilon_{ijk}Q_k \tag{M.86}$$

$$[P_i, Q_j] = 0, \tag{M.87}$$

which are apparently the same as (M.43) – (M.45). We can see from (M.81) that the matrices $i\boldsymbol{K}^{(\mathrm{LG})}$ *are* Hermitian, and the same is in fact true in a general finite-dimensional representation. So we can appropriate standard angular momentum theory to set up the representations of the algebra of the $\boldsymbol{P}$'s and $\boldsymbol{Q}$'s—namely, they behave just like two independent (mutually commuting) angular momenta. The eigenvalues of $\boldsymbol{P}^2$ are of the form $P(P+1)$, for $P = 0, 1/2, \ldots$, and similarly for $\boldsymbol{Q}^2$; the eigenvalues of P_3 are M_P where $-P \leq M_P \leq P$, and similarly for Q_3.

Consider the particular case where the eigenvalue of $\boldsymbol{Q}^2$ is zero $(Q = 0)$, and the value of P is $1/2$. The first condition implies that the $\boldsymbol{Q}$'s are identically zero, so that

$$\boldsymbol{X} = i\boldsymbol{K} \tag{M.88}$$

in this representation, while the second condition tells us that

$$\boldsymbol{P} = \frac{1}{2}(\boldsymbol{X} + i\boldsymbol{K}) = \frac{1}{2}\boldsymbol{\sigma}, \tag{M.89}$$

the familiar matrices for spin-1/2. We label this representation by the values of P $(1/2)$ and Q (0) (these are the eigenvalues of the two Casimir operators). Then using (M.88) and (M.89) we find

$$\boldsymbol{X}^{(\frac{1}{2},0)} = \frac{1}{2}\boldsymbol{\sigma} \tag{M.90}$$

and

$$\boldsymbol{K}^{(\frac{1}{2},0)} = -\frac{i}{2}\boldsymbol{\sigma}. \tag{M.91}$$

Now recall that the general infinitesimal Lorentz transformation has the form

$$1 + i\boldsymbol{\epsilon} \cdot \boldsymbol{X} - i\boldsymbol{\eta} \cdot \boldsymbol{K}. \tag{M.92}$$

In the present case this becomes

$$1 + i\boldsymbol{\epsilon} \cdot \boldsymbol{\sigma}/2 - \boldsymbol{\eta} \cdot \boldsymbol{\sigma}/2. \tag{M.93}$$

These matrices are of dimension 2×2, and act on 2-component spinors, which therefore transform under an infinitesimal Lorentz transformation by (cf (4.19) and (4.42))

$$\phi' = (1 + i\boldsymbol{\epsilon} \cdot \boldsymbol{\sigma}/2 - \boldsymbol{\eta} \cdot \boldsymbol{\sigma}/2)\phi. \tag{M.94}$$

We say that ϕ 'transforms as the $(1/2, 0)$ representation of the Lorentz group'. The '$1 + i\boldsymbol{\epsilon} \cdot \boldsymbol{\sigma}/2$' part is the familiar (infinitesimal) rotation matrix for spinors, first met in section 4.4; it exponentiates to give $\exp(i\boldsymbol{\alpha} \cdot \boldsymbol{\sigma}/2)$ for finite rotations. The '$-\boldsymbol{\eta} \cdot \boldsymbol{\sigma}/2$' part shows how such a spinor transforms under a pure (infinitesimal) velocity transformation. The finite transformation law is

$$\phi' = \exp(-\boldsymbol{\vartheta} \cdot \boldsymbol{\sigma}/2)\phi \tag{M.95}$$

where the three real parameters $\boldsymbol{\vartheta} = (\vartheta_1, \vartheta_2, \vartheta_3)$ specify the direction and magnitude of the boost.

There is, however, a second two-dimensional representation, which is characterized by the labelling $P = 0, Q = 1/2$, which we denote by $(0, 1/2)$. In this case, the previous steps yield

$$\boldsymbol{X}^{(\frac{1}{2},0)} = \frac{1}{2}\boldsymbol{\sigma} \tag{M.96}$$

as before, but

$$\boldsymbol{K}^{(0,\frac{1}{2})} = \frac{i}{2}\boldsymbol{\sigma}. \tag{M.97}$$

So the corresponding 2-component spinor χ transforms by (cf (4.19) and (4.42))

$$\chi' = (1 + i\boldsymbol{\epsilon} \cdot \boldsymbol{\sigma}/2 + \boldsymbol{\eta} \cdot \boldsymbol{\sigma}/2)\chi. \tag{M.98}$$

We see that ϕ and χ behave the same under rotations, but 'oppositely' under boosts.

These transformation laws are exactly what we used in section 4.1.2 when discussing the behaviour of the Dirac wavefunction ψ under Lorentz transformations, where ψ is put together from one ϕ and one χ via

$$\psi = \begin{pmatrix} \phi \\ \chi \end{pmatrix}, \tag{M.99}$$

and describes a *massive* spin-1/2 particle according to the equations

$$E\phi = \boldsymbol{\sigma} \cdot \boldsymbol{p}\phi + m\chi$$
$$E\chi = -\boldsymbol{\sigma} \cdot \boldsymbol{p}\chi + m\phi, \tag{M.100}$$

consistent with the representation (3.40) of the Dirac matrices.

M.7 The relation between SU(2) and SO(3)

We have seen (sections M.4.1 and M.4.2) that the algebras of these two groups are identical. So the groups are isomorphic in the vicinity of their respective identity elements. Furthermore, matrix representations of one algebra automatically provide representations of the other. Since exponentiating these infinitesimal matrix transformations produces matrices representing group elements corresponding to finite transformations in both cases, it might appear that the groups are fully isomorphic. But actually they are not, as we shall now discuss.

We begin by re-considering the parameters used to characterize elements of SO(3) and SU(2). A general 3-D rotation is described by the SO(3) matrix $R(\hat{\boldsymbol{n}}, \theta)$, where $\hat{\boldsymbol{n}}$ is the axis of the rotation and θ is the angle of rotation. For example,

$$R(\hat{z}, \theta) = \begin{pmatrix} \cos\theta & \sin\theta & 0 \\ -\sin\theta & \cos\theta & 0 \\ 0 & 0 & 1 \end{pmatrix}. \tag{M.101}$$

On the other hand, we can write the general SU(2) matrix $\boldsymbol{V}$ in the form

$$\boldsymbol{V} = \begin{pmatrix} a & b \\ -b^* & a^* \end{pmatrix}, \tag{M.102}$$

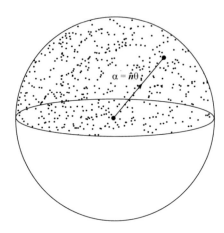

FIGURE M.1
The parameter spaces of SO(3) and SU(2): the whole sphere is the parameter space of SU(2), the upper (stippled) hemisphere that of SO(3).

where $|a|^2 + |b|^2 = 1$ from the unit determinant condition. It therefore depends on three real parameters, the choice of which we are now going to examine in more detail than previously. In (12.32) we wrote V as $\exp(i\boldsymbol{\alpha} \cdot \boldsymbol{\tau}/2)$, which certainly involves three real parameters $\alpha_1, \alpha_2, \alpha_3$; and below (12.35) we proposed, further, to write $\boldsymbol{\alpha} = \hat{\boldsymbol{n}}\theta$, where θ is an angle and $\hat{\boldsymbol{n}}$ is a unit vector. Then, since (as the reader may verify)

$$\exp(i\theta\boldsymbol{\tau} \cdot \hat{\boldsymbol{n}}/2) = \cos\theta/2 + i\boldsymbol{\tau} \cdot \hat{\boldsymbol{n}}\sin\theta/2, \tag{M.103}$$

it follows that this latter parametrization corresponds to writing, in (M.102),

$$a = \cos\theta/2 + in_z\sin\theta/2, \quad b = (n_y + in_x)\sin\theta/2, \tag{M.104}$$

with $n_x^2 + n_y^2 + n_z^2 = 1$. Clearly the condition $|a|^2 + |b|^2 = 1$ is satisfied, and one can convince oneself that the full range of a and b is covered if $\theta/2$ lies between 0 and π (in particular, it is not necessary to extend the range of $\theta/2$ so as to include the interval π to 2π, since the corresponding region of a, b can be covered by changing the orientation of $\hat{\boldsymbol{n}}$, which has not been constrained in any way). It follows that the parameters $\boldsymbol{\alpha}$ satisfy $\boldsymbol{\alpha}^2 \leq 4\pi^2$; that is, the space of the α's is the interior, and surface, of a sphere of radius 2π, as shown in figure M.1.

What about the parameter space of SO(3)? In this case, the same parameters $\hat{\boldsymbol{n}}$ and θ specify a rotation, but now θ (rather than $\theta/2$) runs from 0 to π. However, we may allow the range of θ to extend to 2π, by taking advantage of the fact that

$$R(\hat{\boldsymbol{n}}, \pi + \theta) = R(-\hat{\boldsymbol{n}}, \theta). \tag{M.105}$$

Thus if we agree to limit $\hat{\boldsymbol{n}}$ to directions in the upper hemisphere of figure M.1, for 3-D rotations, we can say that the whole sphere represents the parameter space of SU(2), but that of SO(3) is provided by the upper half only.

Now let us consider the correspondence—or *mapping*—between the matrices of SO(3) and SU(2): we want to see if it is one-to-one. The notation strongly suggests that the matrix $\boldsymbol{V}(\hat{\boldsymbol{n}}, \theta) \equiv \exp(i\theta\hat{\boldsymbol{n}} \cdot \boldsymbol{\tau}/2)$ of SU(2) corresponds to the matrix $R(\hat{\boldsymbol{n}}, \theta)$ of SO(3), but the way it actually works has a subtlety.

We form the quantity $\boldsymbol{x} \cdot \boldsymbol{\tau}$, and assert that

$$\boldsymbol{x}' \cdot \boldsymbol{\tau} = \boldsymbol{V}(\hat{\boldsymbol{n}}, \theta) \, \boldsymbol{x} \cdot \boldsymbol{\tau} \, \boldsymbol{V}^{\dagger}(\hat{\boldsymbol{n}}, \theta), \tag{M.106}$$

where $\boldsymbol{x}' = R(\hat{\boldsymbol{n}}, \theta)\boldsymbol{x}$. We can easily verify (M.106) for the special case $R(\hat{z}, \theta)$, using (M.101); the general case follows with more labour (but the general infinitesimal case should by now be a familiar manipulation). (M.106) establishes a precise mapping between the elements of SU(2) and those of SO(3), but it is not one-to-one (i.e. not an isomorphism), since plainly $\boldsymbol{V}$ can always be replaced by $-\boldsymbol{V}$ and $\boldsymbol{x}'$ will be unchanged, and hence so will the associated SO(3) matrix $R(\hat{\boldsymbol{n}}, \theta)$. It is therefore a homomorphism.

Next, we prove a little theorem to the effect that the identity element e of a group $\mathcal{G}$ must be represented by the unit matrix of the representation: $D(e) = I$. For, let $D(a), D(e)$ represent the elements a, e of $\mathcal{G}$. Then $D(ae) = D(a)D(e)$ by the fundamental property (M.78) of representation matrices. On the other hand, $ae = a$ by the property of e. So we have $D(a) = D(a)D(e)$, and hence $D(e) = I$.

Now let us return to the correspondence between SU(2) and SO(3). $\boldsymbol{V}(\hat{\boldsymbol{n}}, \theta)$ corresponds to $R(\hat{\boldsymbol{n}}, \theta)$, but can an SU(2) matrix be said to provide a valid representation of SO(3)? Consider the case $\boldsymbol{V}(\hat{n} = \hat{z}, \theta = 2\pi)$. From (M.103) this is equal to

$$\begin{pmatrix} -1 & 0 \\ 0 & -1 \end{pmatrix}, \tag{M.107}$$

but the corresponding rotation matrix, from (M.101), is the identity matrix. Hence our theorem is violated, since (M.107) is plainly not the identity matrix of SU(2). Thus the SU(2) matrices cannot be said to represent rotations, in the strict sense. Nevertheless, spin-1/2 particles certainly do exist, so Nature appears to make use of these 'not quite' representations! The SU(2) identity element is $\boldsymbol{V}(\hat{n} = \hat{z}, \theta = 4\pi)$, confirming that the rotational properties of a spinor are quite other than those of a classical object.

In fact, two and only two distinct elements of SU(2), namely

$$\begin{pmatrix} 1 & 0 \\ 0 & 1 \end{pmatrix} \quad \text{and} \quad \begin{pmatrix} -1 & 0 \\ 0 & -1 \end{pmatrix}, \tag{M.108}$$

correspond to the identity element of SO(3) in the correspondence (M.106)—just as, in general, $\boldsymbol{V}$ and $-\boldsymbol{V}$ correspond to the same SO(3) element $R(\hat{\boldsymbol{n}}, \theta)$, as we saw. The failure to be a true representation is localized simply to a sign: we may indeed say that, up to a sign, SU(2) matrices provide a representation of SO(3). If we 'factor out' this sign, the groups are isomorphic. A more mathematically precise way of saying this is given in Jones (1990, chapter 8).

N

Geometrical Aspects of Gauge Fields

N.1 Covariant derivatives and coordinate transformations

Let us go back to the U(1) case, equations (13.4) – (13.7). There, the introduction of the (gauge) covariant derivative D^μ produced an object, $D^\mu\psi(x)$, which transformed like $\psi(x)$ under local U(1) phase transformations, unlike the ordinary derivative $\partial^\mu\psi(x)$ which acquired an 'extra' piece when transformed. This followed from simple calculus, of course—but there is a slightly different way of thinking about it. The derivative involves not only $\psi(x)$ at the point x, but also ψ at the infinitesimally close, but different, point $x + \mathrm{d}x$; and the transformation law of $\psi(x)$ involves $\alpha(x)$, while that of $\psi(x + \mathrm{d}x)$ would involve the different function $\alpha(x + \mathrm{d}x)$. Thus we may perhaps expect something to 'go wrong' with the transformation law for the gradient.

To bring out the geometrical analogy we are seeking, let us write $\psi = \psi_\mathrm{R} + i\psi_\mathrm{I}$ and $\alpha(x) = q\chi(x)$ so that (13.3) becomes (cf (2.64))

$$\psi'_\mathrm{R}(x) = \cos\alpha(x)\psi_\mathrm{R}(x) - \sin\alpha(x)\psi_\mathrm{I}(x)$$

$$\psi'_\mathrm{I}(x) = \sin\alpha(x)\psi_\mathrm{R}(x) + \cos\alpha(x)\psi_\mathrm{I}(x). \tag{N.1}$$

If we think of $\psi_\mathrm{R}(x)$ and $\psi_\mathrm{I}(x)$ as being the components of a 'vector' $\vec{\psi}(x)$ along the $\vec{e}_\mathrm{R}$ and $\vec{e}_\mathrm{I}$ axes, respectively, then (N.1) would represent the components of $\vec{\psi}(x)$ as referred to new axes $\vec{e}'_\mathrm{R}$ and $\vec{e}'_\mathrm{I}$, which have been rotated by $-\alpha(x)$ about an axis in the direction $\vec{e}_\mathrm{R} \times \vec{e}_\mathrm{I}$ (i.e. normal to the $\vec{e}_\mathrm{R} - \vec{e}_\mathrm{I}$ plane), as shown in figure N.1. Other such 'vectors' $\vec{\phi}_1(x), \vec{\phi}_2(x), \ldots$ (i.e. other wavefunctions for particles of the same charge q) *when evaluated at the same point x* will have 'components' transforming the same as (N.1) under the axis rotation $\vec{e}_\mathrm{R}, \vec{e}_\mathrm{I} \to \vec{e}'_\mathrm{R}, \vec{e}'_\mathrm{I}$. But the components of the vector $\vec{\psi}(x+\mathrm{d}x)$ will behave differently. The transformation law (N.1) when written at $x + \mathrm{d}x$ will involve $\alpha(x+\mathrm{d}x)$, which (to first order in $\mathrm{d}x$) is $\alpha(x) + \partial_\mu\alpha(x)\mathrm{d}x^\mu$. Thus for $\psi'_\mathrm{R}(x+\mathrm{d}x)$ and $\psi'_\mathrm{I}(x+\mathrm{d}x)$ the rotation angle is $\alpha(x) + \partial_\mu\alpha(x)\mathrm{d}x^\mu$ rather than $\alpha(x)$. Now comes the key step in the analogy: we may think of the additional angle $\partial_\mu\alpha(x)\mathrm{d}x^\mu$ as coming about because, in going from x to $x + \mathrm{d}x$, the coordinate basis vectors $\vec{e}_\mathrm{R}$ and $\vec{e}_\mathrm{I}$ have been rotated through $+\partial_\mu\alpha(x)\mathrm{d}x^\mu$ (see figure N.2)! But that would mean that our 'naive' approach to rotations of the derivative of $\vec{\psi}(x)$ amounts to using one set of axes at x, and another at $x + \mathrm{d}x$, which is likely to lead to 'trouble'. Consider now an elementary example (from Schutz 1988, chapter 5) where just this kind of problem arises, namely the use of polar coordinate basis vectors $\vec{e}_r$ and $\vec{e}_\theta$, which point in the r and θ directions respectively. We have, as usual,

$$x = r\cos\theta, \quad y = r\sin\theta \tag{N.2}$$

and in a (real!) Cartesian basis $\mathrm{d}\vec{r}$ is given by

$$\mathrm{d}\vec{r} = \mathrm{d}x\,\vec{i} + \mathrm{d}y\,\vec{j}. \tag{N.3}$$

DOI: 10.1201/9781003411666-N

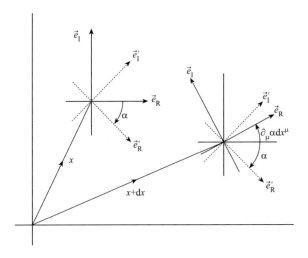

FIGURE N.1
Geometrical analogy for a U(1) gauge transformation.

Using (N.2) in (N.3) we find

$$
\begin{aligned}
\mathrm{d}\vec{r} &= (\mathrm{d}r\cos\theta - r\sin\theta \ \mathrm{d}\theta)\vec{i} + (\mathrm{d}r\sin\theta + r\cos\theta \ \mathrm{d}\theta)\vec{j} \\
&= \mathrm{d}r \ \vec{e}_r + \mathrm{d}\theta \ \vec{e}_\theta
\end{aligned}
\tag{N.4}
$$

where

$$
\vec{e}_r = \cos\theta \, \vec{i} + \sin\theta \, \vec{j}, \vec{e}_\theta = -r\sin\theta \, \vec{i} + r\cos\theta \, \vec{j}.
\tag{N.5}
$$

Plainly, $\vec{e}_r$ and $\vec{e}_\theta$ change direction (and even magnitude, for $\vec{e}_\theta$) as we move about in the $x - y$ plane, as shown in figure N.2. So at each point (r, θ) we have *different* axes $\vec{e}_r, \vec{e}_\theta$.

Now suppose that we wish to describe a vector field $\vec{V}$ in terms of $\vec{e}_r$ and $\vec{e}_\theta$ via

$$
\vec{V} = V^r \vec{e}_r + V^\theta \vec{e}_\theta \equiv V^\alpha \vec{e}_\alpha \ (\text{sum on } \alpha = r, \theta),
\tag{N.6}
$$

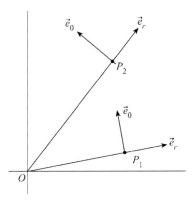

FIGURE N.2
Changes in the basis vectors $\vec{e}_r$ and $\vec{e}_\theta$ of polar coordinates.

and that we are also interested in the derivatives of $\vec{V}$, in this basis. Let us calculate $\frac{\partial \vec{V}}{\partial r}$, for example, by brute force:

$$\frac{\partial \vec{V}}{\partial r} = \frac{\partial V^r}{\partial r}\vec{e}_r + \frac{\partial V^\theta}{\partial r}\vec{e}_\theta + V^r \frac{\partial \vec{e}_r}{\partial r} + V^\theta \frac{\partial \vec{e}_\theta}{\partial r} \tag{N.7}$$

where we have included the derivatives of $\vec{e}_r$ and $\vec{e}_\theta$ to allow for the fact that *these vectors are not constant*. From (N.5) we easily find

$$\frac{\partial \vec{e}_r}{\partial r} = 0, \quad \frac{\partial \vec{e}_\theta}{\partial r} = -\sin\theta\,\vec{i} + \cos\theta\,\vec{j} = \frac{1}{r}\vec{e}_\theta, \tag{N.8}$$

which allows the last two terms in (N.7) to be evaluted. Similarly, we can calculate $\frac{\partial \vec{V}}{\partial \theta}$. In general, we may write these results as

$$\frac{\partial \vec{V}}{\partial q^\beta} = \frac{\partial V^\alpha}{\partial q^\beta}\vec{e}_\alpha + V^\alpha \frac{\partial \vec{e}_\alpha}{\partial q^\beta} \tag{N.9}$$

where $\beta = 1, 2$ with $q^1 = r, q^2 = \theta$, and $\alpha = r, \theta$.

In the present case, we were able to calculate $\partial \vec{e}_\alpha / \partial q^\beta$ explicitly from (N.5), as in (N.8). But whatever the nature of the coordinate system, $\partial \vec{e}_\alpha / \partial q^\beta$ is some vector and must be expressible as a linear combination of the basis vectors via an expression of the form

$$\frac{\partial \vec{e}_\alpha}{\partial q^\beta} = \Gamma^\gamma{}_{\alpha\beta}\vec{e}_\gamma \tag{N.10}$$

where the repeated index γ is summed over as usual ($\gamma = r, \theta$). Inserting (N.10) into (N.9) and interchanging the 'dummy' (i.e. summed over) indices α and γ gives finally

$$\frac{\partial \vec{V}}{\partial q^\beta} = \left(\frac{\partial V^\alpha}{\partial q^\beta} + \Gamma^\alpha{}_{\gamma\beta}V^\gamma\right)\vec{e}_\alpha. \tag{N.11}$$

This is a very important result. It shows that, whereas the components of $\vec{V}$ in the basis $\vec{e}_\alpha$ are just V^α, the components of the derivative of $\vec{V}$ are not simply $\partial V^\alpha / \partial q^\beta$, but *contain an additional term*: the 'components of the derivative of a vector' are not just the 'derivatives of the components of the vector'.

Let us abbreviate $\partial / \partial q^\beta$ to ∂_β; then (N.11) tells us that in the $\vec{e}_\alpha$ basis, as used in (N.11), the components of the ∂_β derivative of $\vec{V}$ are

$$\partial_\beta V^\alpha + \Gamma^\alpha{}_{\gamma\beta}V^\gamma \equiv D_\beta V^\alpha. \tag{N.12}$$

The expression (N.12) is called the 'covariant derivative' of V^α within the context of the mathematics of general coordinate systems. It is denoted (as in (N.12)) by $D_\beta V^\alpha$ or, often, by $V^\alpha{}_{;\beta}$ (in the latter notation, $\partial_\beta V^\alpha$ is $V^\alpha{}_{,\beta}$). The most important property of $D_\beta V^\alpha$ is its transformation character under general coordinate transformations. Crucially, it transforms as a *tensor* T^α_β (see Appendix D of volume 1) with the indicated 'one up, one down' indices; we shall not prove this here, referring instead to Schutz (1988), for example. This property is the reason for the name 'covariant derivative', meaning in this case essentially that it transforms the way its indices would have you believe it should. By contrast, and despite appearances, $\partial_\beta V^\alpha$ by itself does *not* transform as a 'T^α_β' tensor, and in a similar way $\Gamma^\alpha{}_{\gamma\beta}$ is *not* a '$T^\alpha_{\gamma\beta}$'-type tensor; only the combined object $D_\beta V^\alpha$ is a 'T^α_β'.

This circumstance is highly reminiscent of the situation we found in the case of gauge transformations. Consider the simplest case, that of U(1), for which $D_\mu\psi = \partial_\mu\psi + iqA_\mu\psi$.

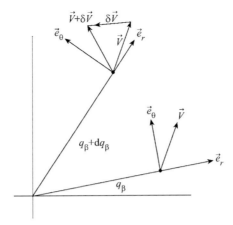

FIGURE N.3
Parallel transport of a vector $\vec{V}$ in a polar coordinate basis.

The quantity $D_\mu \psi$ transforms under a gauge transformation in the same way as ψ itself, but $\partial_\mu \psi$ does not. There is thus a close analogy between the 'good' transformation properties of $D_\beta V^\alpha$ and of $D_\mu \psi$. Further, the structure of $D_\mu \psi$ is very similar to that of $D_\beta V^\alpha$. There are two pieces, the first of which is the straightforward derivative, while the second involves a new field (Γ or A) and is also proportional to the original field. The 'i' of course is a big difference, showing that in the gauge symmetry case the transformations mix the real and imaginary parts of the wavefunction, rather than actual spatial components of a vector.

Indeed, the analogy is even closer in the non-Abelian—e.g. local SU(2)—case. As we have seen, $\partial^\mu \psi^{(\frac{1}{2})}$ does not transform as an SU(2) isospinor because of the extra piece involving $\partial^\mu \boldsymbol{\epsilon}$; nor do the gauge fields $\boldsymbol{W}^\mu$ transform as pure $T = 1$ states, also because of a $\partial^\mu \boldsymbol{\epsilon}$ term. But the gauge covariant combination $(\partial^\mu + \mathrm{i} g \boldsymbol{\tau} \cdot \boldsymbol{W}^\mu / 2)\psi^{(\frac{1}{2})}$ does transform as an isospinor under local SU(2) transformations, the two 'extra' $\partial^\mu \boldsymbol{\epsilon}$ pieces cancelling each other out.

There is a useful way of thinking about the two contributions to $D_\beta V^\alpha$ (or $D_\mu \psi$). Let us multiply (N.12) by $\mathrm{d}q^\beta$ and sum over β so as to obtain

$$DV^\alpha \equiv \partial_\beta V^\alpha \mathrm{d}q^\beta + \Gamma^\alpha_{\gamma\beta} V^\gamma \mathrm{d}q^\beta. \tag{N.13}$$

The first term on the right-hand side of (N.13) is $\frac{\partial V^\alpha}{\partial q^\beta}\mathrm{d}q^\beta$ which is just the conventional differential $\mathrm{d}V^\alpha$, representing the change in V^α in moving from q^β to $q^\beta + \mathrm{d}q^\beta$: $\mathrm{d}V^\alpha = [V^\alpha(q^\beta + \mathrm{d}q^\beta) - V^\alpha(q^\beta)]$. Again, despite appearances, the quantities $\mathrm{d}V^\alpha$ do not form the components of a vector, and the reason is that $V^\alpha(q^\beta + \mathrm{d}q^\beta)$ are components with respect to axes at $q^\beta + \mathrm{d}q^\beta$, while $V^\alpha(q^\beta)$ are components with respect to *different* axes at q^β. To form a 'good' differential DV^α, transforming as a vector, we must subtract quantities defined in the *same* coordinate system. This means that we need some way of 'carrying' $V^\alpha(q^\beta)$ to $q^\beta + \mathrm{d}q^\beta$, while keeping it somehow 'the same' as it was at q^β.

A reasonable definition of such a 'preserved' vector field is one that is unchanged in length, and has the same orientation relative to the axes at $q^\beta + \mathrm{d}q^\beta$ as it had relative to the axes at q^β (see figure N.3). In other words, $\vec{V}$ is 'dragged around' with the changing coordinate frame, a process called *parallel transport*. Such a definition of 'no change' of course implies that change *has* occurred, in general, with respect to the *original* axes at q^β. Let us denote by δV^α the difference between the components of $\vec{V}$ after parallel transport to $q^\beta + \mathrm{d}q^\beta$, and the components of $\vec{V}$ at q^β (see figure N.3). Then a reasonable definition of the 'good' differential of V^α would be $V^\alpha(q^\beta + \mathrm{d}q^\beta) - (V^\alpha(q^\beta) + \delta V^\alpha) = \mathrm{d}V^\alpha - \delta V^\alpha$. We interpret this as the covariant differential DV^α of (N.13), and accordingly, make the identification

$$\delta V^\alpha = -\Gamma^\alpha_{\ \gamma\beta}V^\gamma \mathrm{d}q^\beta. \tag{N.14}$$

On this interpretation, then, the coefficients $\Gamma^\alpha_{\ \gamma\beta}$ connect the components of a vector at one point with its components at a nearby point, after the vector has been carried by 'parallel transport' from one point to the other; they are often called 'connection coefficients', or just 'the connection'.

In an analogous way we can write, in the U(1) gauge case,

$$\begin{aligned} D\psi \equiv D^\mu\psi \mathrm{d}x_\mu &= \partial^\mu\psi \mathrm{d}x_\mu + \mathrm{i}eA^\mu\psi \mathrm{d}x_\mu \\ &\equiv \mathrm{d}\psi - \delta\psi \end{aligned} \tag{N.15}$$

with

$$\delta\psi = -\mathrm{i}eA^\mu\psi \mathrm{d}x_\mu. \tag{N.16}$$

Equation (N.16) has a very similar structure to (N.14), suggesting that the electromagnetic potential A^μ might well be referred to as a 'gauge connection', as indeed it is in some quarters. Equations (N.15) and (N.16) generalize straightforwardly for $D\psi^{(\frac{1}{2})}$ and $\delta\psi^{(\frac{1}{2})}$.

We can relate (N.16) in a very satisfactory way to our original discussion of electromagnetism as a gauge theory in chapter 2, and in particular to (2.83). For transport restricted to the three spatial directions, (N.16) reduces to

$$\delta\psi(\boldsymbol{x}) = \mathrm{i}e\boldsymbol{A} \cdot \mathrm{d}\boldsymbol{x}\psi(x). \tag{N.17}$$

However, the solution (2.83) gives

$$\psi(\boldsymbol{x}) = \exp\left(\mathrm{i}e \int_{-\infty}^{\boldsymbol{x}} \boldsymbol{A} \cdot \mathrm{d}\boldsymbol{\ell}\right) \psi(\boldsymbol{A} = \boldsymbol{0}, \boldsymbol{x}), \tag{N.18}$$

replacing q by e. So

$$\begin{aligned} \psi(\boldsymbol{x} + \mathrm{d}\boldsymbol{x}) &= \exp\left(\mathrm{i}e \int_{-\infty}^{\boldsymbol{x}+\mathrm{d}\boldsymbol{x}} \boldsymbol{A} \cdot \mathrm{d}\boldsymbol{\ell}\right) \psi(\boldsymbol{A} = \boldsymbol{0}, \boldsymbol{x} + \mathrm{d}\boldsymbol{x}) \\ &= \exp\left(\mathrm{i}e \int_{\boldsymbol{x}}^{\boldsymbol{x}+\mathrm{d}\boldsymbol{x}} \boldsymbol{A} \cdot \mathrm{d}\boldsymbol{\ell}\right) \exp\left(\mathrm{i}e \int_{-\infty}^{\boldsymbol{x}} \boldsymbol{A} \cdot \mathrm{d}\boldsymbol{\ell}\right) \psi(\boldsymbol{A} = \boldsymbol{0}, \boldsymbol{x} + \mathrm{d}\boldsymbol{x}) \\ &\approx (1 + \mathrm{i}e\boldsymbol{A} \cdot \mathrm{d}\boldsymbol{x})\exp\left(\mathrm{i}e \int_{-\infty}^{\boldsymbol{x}} \boldsymbol{A} \cdot \mathrm{d}\boldsymbol{\ell}\right) [\psi(\boldsymbol{A} = \boldsymbol{0}, \boldsymbol{x}) + \boldsymbol{\nabla}\psi(\boldsymbol{A} = \boldsymbol{0}, \boldsymbol{x}) \cdot \mathrm{d}\boldsymbol{x}] \\ &\approx \psi(\boldsymbol{x}) + \mathrm{i}e\boldsymbol{A} \cdot \mathrm{d}\boldsymbol{x}\psi(\boldsymbol{x}) + \exp\left(\mathrm{i}e \int_{-\infty}^{\boldsymbol{x}} \boldsymbol{A} \cdot \mathrm{d}\boldsymbol{\ell}\right) \boldsymbol{\nabla}\psi(\boldsymbol{A} = \boldsymbol{0}, \boldsymbol{x}) \cdot \mathrm{d}\boldsymbol{x}, \tag{N.19} \end{aligned}$$

to first order in $\mathrm{d}\boldsymbol{x}$. On the right-hand side of (N.19) we see (i) the change $\delta\psi$ of (N.17), due to 'parallel transport' as prescribed by the gauge connection $\boldsymbol{A}$ and (ii) the change in ψ viewed as a function of $\boldsymbol{x}$, in the absence of $\boldsymbol{A}$. The solution (N.18) gives, in fact, the 'integrated' form of the small displacement law (N.19).

At this point the reader might object, going back to the $\vec{e}_r, \vec{e}_\theta$ example, that we had made a lot of fuss about nothing. After all, no one forced us to use the $\vec{e}_r, \vec{e}_\theta$ basis, and if we had simply used the $\vec{i}, \vec{j}$ basis (which is constant throughout the plane) we would have had no such 'trouble'. This is a fair point, provided that we somehow knew that we are really doing physics in a 'flat' space, such as the Euclidean plane. But suppose instead that our

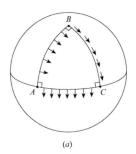

(a)

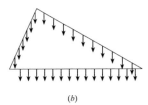

(b)

FIGURE N.4
Parallel transport (a) round a curved triangle on the surface of a sphere (b) round a triangle in a flat plane.

two-dimensional space was the surface of a sphere. Then, an intuitively plausible definition of parallel transport is shown in figure N.4(a), in which transport is carried out around a closed path consisting of three great circle arcs $A \to B, B \to C, C \to A$, with the rule that at each stage the vector is drawn 'as parallel as possible' to the previous one. It is clear from the figure that the vector we end up with at A, after this circuit, is no longer parallel to the vector we started with; in fact, it has rotated by $\pi/2$ in this example, in which $\frac{1}{8}$th of the surface area of the unit sphere is enclosed by the triangle ABC. By contrast, the parallel transport of a vector round a flat triangle in the Euclidean plane leads to no such net change in the vector (figure N.4(b)).

It seems reasonable to suppose that the information about whether the space we are dealing with is 'flat' or 'curved' is contained in the connection $\Gamma^\alpha{}_{\gamma\beta}$. In a similar way, in the gauge case the analogy we have built up so far would lead us to expect that there are potentials A^μ which are somehow 'flat' ($\boldsymbol{E} = \boldsymbol{B} = \boldsymbol{0}$) and others which represent 'curvature' (non-zero $\boldsymbol{E}, \boldsymbol{B}$). This is what we discuss next.

N.2 Geometrical curvature and the gauge field strength tensor

Consider a small closed loop in our (possibly curved) two-dimensional space—see figure N.5—whose four sides are the coordinate lines $q^1 = a, q^1 = a + \delta a, q^2 = b, q^2 = b + \delta b$. We want to calculate the net change (if any) in δV^α as we parallel transport $\vec{V}$ around the loop. The change along $A \to B$ is

$$(\delta V^\alpha)_{AB} = -\int_{q^2=b,q^1=a}^{q^2=b,q^1=a+\delta a} \Gamma^\alpha{}_{\gamma 1} V^\gamma dq^1$$

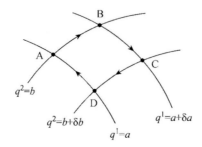

FIGURE N.5
Closed loop $ABCD$ in $q^1 - q^2$ space.

$$\approx \quad -\delta a \Gamma^\alpha{}_{\gamma 1}(a,b) V^\gamma(a,b) \tag{N.20}$$

to first order in δa, while that along $C \to D$ is

$$
\begin{aligned}
(\delta V^\alpha)_{CD} &= -\int_{q^2=b+\delta b, q^1=a+\delta a}^{q^2=b+\delta b, q^1=a} \Gamma^\alpha{}_{\gamma 1} V^\gamma dq^1 \\
&= +\int_{q^2=b+\delta b, q^1=a}^{q^2=b+\delta b, q^1=a+\delta a} \Gamma^\alpha{}_{\gamma 1} V^\gamma dq^1. \\
&\approx \delta a \Gamma^\alpha{}_{\gamma 1}(a, b+\delta b) V^\gamma(a, b+\delta b).
\end{aligned}
\tag{N.21}
$$

Now

$$\Gamma^\alpha{}_{\gamma 1}(a, b+\delta b) \approx \Gamma^\alpha{}_{\gamma 1}(a,b) + \delta b \frac{\partial \Gamma^\alpha{}_{\gamma 1}}{\partial q^2} \tag{N.22}$$

and, remembering that we are parallel-transporting $\vec{V}$,

$$V^\gamma(a, b+\delta b) \approx V^\gamma(a,b) - \Gamma^\gamma{}_{\delta 2} V^\delta \delta b. \tag{N.23}$$

Combining (N.20) and (N.21) to lowest order, we find

$$(\delta V^\alpha)_{AB} + (\delta V^\alpha)_{CD} \approx \delta a \delta b \left[\frac{\partial \Gamma^\alpha{}_{\gamma 1}}{\partial q^2} V^\gamma - \Gamma^\alpha{}_{\gamma 1} \Gamma^\gamma{}_{\delta 2} V^\delta \right] \tag{N.24}$$

or, interchanging dummy indices γ and δ in the last term,

$$(\delta V^\alpha)_{AB} + (\delta V^\alpha)_{CD} \approx \delta a \delta b \left[\frac{\partial \Gamma^\alpha{}_{\gamma 1}}{\partial q^2} - \Gamma^\alpha{}_{\delta 1} \Gamma^\delta{}_{\gamma 2} \right] V^\gamma. \tag{N.25}$$

Similarly,

$$(\delta V^\alpha)_{BC} + (\delta V^\alpha)_{DA} \approx \delta a \delta b \left[-\frac{\partial \Gamma^\alpha{}_{\gamma 2}}{\partial q^1} + \Gamma^\alpha{}_{\delta 2} \Gamma^\delta{}_{\gamma 1} \right] V^\gamma, \tag{N.26}$$

and so the net change around the whole small loop is

$$(\delta V^\alpha)_{ABCD} \approx \delta a \delta b \left[\frac{\partial \Gamma^\alpha{}_{\gamma 1}}{\partial q^2} - \frac{\partial \Gamma^\alpha{}_{\gamma 2}}{\partial q^1} + \Gamma^\alpha{}_{\delta 2} \Gamma^\delta{}_{\gamma 1} - \Gamma^\alpha{}_{\delta 1} \Gamma^\delta{}_{\gamma 2} \right] V^\gamma. \tag{N.27}$$

The indices '1' and '2' appear explicitly because the loop was chosen to go along these directions. In general, (N.27) would take the form

$$(\delta V^\alpha)_{loop} \approx \left[\frac{\partial \Gamma^\alpha{}_{\gamma \beta}}{\partial q^\sigma} - \frac{\partial \Gamma^\alpha{}_{\gamma \sigma}}{\partial q^\beta} + \Gamma^\alpha{}_{\delta \sigma} \Gamma^\delta{}_{\gamma \beta} - \Gamma^\alpha{}_{\delta \beta} \Gamma^\delta{}_{\gamma \sigma} \right] V^\gamma dA^{\beta \sigma} \tag{N.28}$$

where $\mathrm{d}A^{\beta\sigma}$ is the area element. The quantity in brackets in (N.28) is the *Reimann curvature tensor* $R^\alpha{}_{\gamma\beta\sigma}$ (up to a sign, depending on conventions), which can clearly be calculated once the connection coefficients are known. A flat space is one for which all components $R^\alpha{}_{\gamma\beta\sigma} = 0$; the reader may verify that this is the case for our polar basis $\vec{e}_r, \vec{e}_\theta$ in the Euclidean plane. A non-zero value for any component of $R^\alpha{}_{\gamma\beta\sigma}$ means the space is curved.

We now follow exactly similar steps to calculate the net change in $\delta\psi$ as given by (N.16), around the small two-dimensional rectangle defined by the coordinate lines $x_1 = a, x_1 = a + \delta a, x_2 = b, x_2 = b + \delta b$, labelled as in figure N.5 but with q^1 replaced by x_1 and q^2 by x_2. Then

$$(\delta\psi)_{AB} = -\mathrm{i}eA^1(a,b)\psi(a,b)\delta a \tag{N.29}$$

and

$$
\begin{aligned}
(\delta\psi)_{CD} &= +\mathrm{i}eA^1(a, b+\delta b)\psi(a, b+\delta b)\delta a \\
&\approx \mathrm{i}e\left(A^1(a,b) + \frac{\partial A^1}{\partial x_2}\delta b\right)[\psi(a,b) - \mathrm{i}eA^2(a,b)\psi(a,b)\delta b]\delta a \\
&\approx \mathrm{i}eA^1(a,b)\psi(a,b)\delta a \\
&\quad + \mathrm{i}e\left[\frac{\partial A^1}{\partial x_2}\psi(a,b) - \mathrm{i}eA^1(a,b)A^2(a,b)\psi(a,b)\right]\delta a\delta b.
\end{aligned}
\tag{N.30}
$$

Combining (N.29) and (N.30) we find

$$(\delta\psi)_{AB} + (\delta\psi)_{CD} \approx \left[\mathrm{i}e\frac{\partial A^1}{\partial x_2}\psi + e^2 A^1 A^2 \psi\right]\delta a\delta b. \tag{N.31}$$

Similarly,

$$(\delta\psi)_{BC} + (\delta\psi)_{DA} \approx \left[-\mathrm{i}e\frac{\partial A^2}{\partial x_1}\psi - e^2 A^1 A^2 \psi\right]\delta a\delta b, \tag{N.32}$$

with the result that the net change around the loop is

$$(\delta\psi)_{ABCD} \approx \mathrm{i}e\left(\frac{\partial A^1}{\partial x_2} - \frac{\partial A^2}{\partial x_1}\right)\psi\delta a\delta b. \tag{N.33}$$

For a general loop, (N.33) is replaced by

$$
\begin{aligned}
(\delta\psi)_{loop} &= \mathrm{i}e\left(\frac{\partial A^\mu}{\partial x_\nu} - \frac{\partial A^\nu}{\partial x_\mu}\right)\psi\mathrm{d}x_\mu\mathrm{d}x_\nu \\
&= -\mathrm{i}eF^{\mu\nu}\psi\mathrm{d}x_\mu\mathrm{d}x_\nu
\end{aligned}
\tag{N.34}
$$

where $F^{\mu\nu} = \partial^\mu A^\nu - \partial^\nu A^\mu$ is the familiar field strength tensor of QED.

The analogy we have been pursuing would therefore suggest that $F^{\mu\nu} = 0$ indicates 'no physical effect', while $F^{\mu\nu} \neq 0$ implies the presence of a physical effect. Indeed, when A^μ has the 'pure gauge' form $A^\mu = \partial^\mu\chi$ the associated $F^{\mu\nu}$ is zero; this is because such an A^μ can clearly be reduced to zero by a gauge transformation (and also, consistently, because $(\partial^\mu\partial^\nu - \partial^\nu\partial^\mu)\chi = 0$). If A^μ is not expressible as the 4-gradient of a scalar, then $F^{\mu\nu} \neq 0$ and an electromagnetic field is present, analogous to the spatial curvature revealed by $R^\alpha{}_{\gamma\beta\sigma} \neq 0$. Once again, there is a satisfying consistency between this 'geometrical' viewpoint and the discussion of the Aharonov-Bohm effect in section 2.6. As in our remarks at the end of the previous section, and equations (N.17) – (N.19), equation (2.83) can be regarded as the integrated form of (N.34), for spatial loops. Transport round such a loop results in a non-trivial net phase change if non-zero $\boldsymbol{B}$ flux is enclosed, and this can be observed.

From this point of view there is undoubtedly a strong conceptual link between Einstein's theory of gravity and quantum gauge theories. In the former, matter (or energy) is regarded as the source of curvature of space-time, causing the space-time axes themselves to vary from point to point, and determining the trajectories of massive particles; in the latter, charge is the source of curvature in an 'internal' space (the complex ψ-plane, in the U(1) case), a curvature which we call an electromagnetic field, and which has observable physical effects.

The reader may consider repeating, for the local SU(2) case, the closed-loop transport calculation of (N.29) – (N.33). For this calculation, the place of the Abelian vector potential is taken by the matrix-valued non-Abelian potential $A^\mu = \boldsymbol{\tau}/2 \cdot \boldsymbol{A}^\mu$. It will lead to the expression for the non-Abelian field strength tensor as calculated in section 13.1.2.

O

Dimensional Regularization

After combining propagator denominators of the form $(p^2 - m^2 + i\epsilon)^{-1}$ by Feynman parameters (cf (10.40)), and shifting the origin of the loop momentum to complete the square (cf (10.42) and (11.16)), all one-loop Feynman integrals may be reduced to evaluating an integral of the form

$$I_d(\Delta, n) \equiv \int \frac{d^d k}{(2\pi)^d} \frac{1}{[k^2 - \Delta + i\epsilon]^n}, \tag{O.1}$$

or to a similar integral with factors of k (such as $k_\mu k_\nu$) in the numerator. We consider (O.1) first.

For our purposes, the case of physical interest is $d = 4$, and n is commonly 2 (e.g. in one-loop self-energies). Power-counting shows that (O.1) diverges as $k \to \infty$ for $d \geq 2n$. The idea behind *dimensional regularization* ('t Hooft and Veltman (1972)) is to treat d as a variable parameter, taking values smaller than $2n$, so that (O.1) converges and can be evaluated explicitly as a function of d (and of course the other variables, including n).[1] Then the nature of the divergence as $d \to 4$ can be exposed (much as we did with the cut-off procedure in section 10.3), and dealt with by a suitable renormalization scheme. The crucial advantage of dimensional regularization is that it preserves gauge invariance, unlike the simple cut-off regularization we used in chapters 10 and 11.

We write

$$I_d = \frac{1}{(n-1)!} \left(\frac{\partial}{\partial \Delta} \right)^{n-1} \int \frac{d^d k}{(2\pi)^d} \frac{1}{[k^2 - \Delta + i\epsilon]}. \tag{O.2}$$

The d dimensions are understood as one time-like dimension k^0, and $d - 1$ spacelike dimensions. We begin by 'Euclideanizing' the integral, by setting $k^0 = i k^e$ with k^e real. Then the Minkowskian square k^2 becomes $-(k^e)^2 - \boldsymbol{k}^2 \equiv -k_{\mathrm{E}}^2$, and $d^d k$ becomes $i d^d k_{\mathrm{E}}$, so that now

$$I_d = \frac{-i}{(n-1)!} \left(\frac{\partial}{\partial \Delta} \right)^{n-1} \int \frac{d^d k_{\mathrm{E}}}{(2\pi)^d} \frac{1}{(k_{\mathrm{E}}^2 + \Delta)}; \tag{O.3}$$

the 'iϵ' may be understood as included in Δ. The integral is evaluated by introducing the following way of writing $(k_{\mathrm{E}}^2 + \Delta)^{-1}$:

$$(k_{\mathrm{E}}^2 + \Delta)^{-1} = \int_0^\infty d\beta e^{-\beta(k_{\mathrm{E}}^2 + \Delta)}, \tag{O.4}$$

which leads to

$$I_d = \frac{1}{(n-1)!} \left(\frac{\partial}{\partial \Delta} \right)^{n-1} \int_0^\infty d\beta \int \frac{d^d k_{\mathrm{E}}}{(2\pi)^d} e^{-\beta(k_{\mathrm{E}}^2 + \Delta)}. \tag{O.5}$$

---

[1] We concentrate here on ultraviolet divergences, but infrared ones (such as those met in section 14.4.2) can be dealt with too, by choosing d larger than $2n$

The interchange of the orders of the β and $k_{\rm E}$ integrations is permissible since I_d is convergent. The $k_{\rm E}$ integrals are, in fact, a series of Gaussians:

$$\int \frac{d^d k_{\rm E}}{(2\pi)^d} e^{-\beta(k_{\rm E}^2+\Delta)} = e^{-\beta\Delta}\left\{\prod_{j=1}^{d}\int \frac{dk_j}{(2\pi)}e^{-\beta k_j^2}\right\}$$

$$= \frac{e^{-\beta\Delta}}{(2\pi)^d}\left(\frac{\pi}{\beta}\right)^{d/2}. \qquad (\text{O.6})$$

Hence

$$I_d = \frac{-i}{(n-1)!}\frac{1}{(4\pi)^{d/2}}\left(\frac{\partial}{\partial\Delta}\right)^{n-1}\int d\beta e^{-\beta\Delta}\beta^{-d/2}$$

$$= \frac{-i}{(n-1)!}\frac{(-1)^{n-1}}{(4\pi)^{d/2}}\int d\beta e^{-\beta\Delta}\beta^{n-(d/2)-1}. \qquad (\text{O.7})$$

The last integral can be written in terms of Euler's integral for the *gamma function* $\Gamma(z)$ defined by (see, for example, Boas 1983, chapter 11)

$$\Gamma(z) = \int_0^\infty x^{z-1}e^{-x}dx. \qquad (\text{O.8})$$

Since $\Gamma(n) = (n-1)!$, it is convenient to write (O.8) entirely in terms of Γ functions as

$$I_d = i\frac{(-1)^n}{(4\pi)^{d/2}}\frac{\Gamma(n-d/2)}{\Gamma(n)}\Delta^{(d/2)-n}. \qquad (\text{O.9})$$

Equation (O.9) gives an explicit definition of I_d which can be used for any value of d, not necessarily an integer. As a function of z, $\Gamma(z)$ has isolated poles (see Apppendix F of volume 1) at $z = 0, -1, -2, \ldots$. The behaviour near $z = 0$ is given by

$$\Gamma(z) = \frac{1}{z} - \gamma + O(z), \qquad (\text{O.10})$$

where γ is the Euler-Mascheroni constant having the value $\gamma \approx 0.5772$. Using

$$z\Gamma(z) = \Gamma(z+1), \qquad (\text{O.11})$$

we find the behaviour near $z = -1$:

$$\Gamma(-1+z) = \frac{-1}{1-z}\Gamma(z)$$

$$= -[\frac{1}{z}+1-\gamma+O(z)]; \qquad (\text{O.12})$$

similarly

$$\Gamma(-2+z) = \frac{1}{2}[\frac{1}{z}+\frac{3}{2}-\gamma+O(z)]. \qquad (\text{O.13})$$

Consider now the case $n = 2$, for which $\Gamma(n-d/2)$ in (O.9) will have a pole at $d = 4$. Setting $d = 4 - \epsilon$, the divergent behaviour is given by

$$\Gamma(2-d/2) = \frac{2}{\epsilon} - \gamma + O(\epsilon) \qquad (\text{O.14})$$

from (O.10). $I_d(\Delta, 2)$ is then given by

$$I_d(\Delta, 2) = \frac{i}{(4\pi)^{2-\epsilon/2}} \Delta^{-\epsilon/2} \left[\frac{2}{\epsilon} - \gamma + O(\epsilon)\right]. \tag{O.15}$$

When $\Delta^{-\epsilon/2}$ and $(4\pi)^{-2+\epsilon/2}$ are expanded in powers of ϵ, for small ϵ, the terms linear in ϵ will produce terms independent of ϵ when multiplied by the ϵ^{-1} in the bracket of (O.15). Using $x^\epsilon \approx 1 + \epsilon \ln x + O(\epsilon^2)$ we find

$$I_d(\Delta, 2) = \frac{i}{(4\pi)^2} \left[\frac{2}{\epsilon} - \gamma + \ln 4\pi - \ln \Delta + O(\epsilon)\right]. \tag{O.16}$$

Another source of ϵ-dependence arises from the fact (see problem 15.7) that a gauge coupling which is dimensionless in $d = 4$ dimensions will acquire mass dimension $\mu^{\epsilon/2}$ in $d = 4 - \epsilon$ dimensions (check this!). A vacuum polarization loop with two powers of the coupling will then contain a factor μ^ϵ. When expanded in powers of ϵ, this will convert the $\ln \Delta$ in (O.16) to $\ln(\Delta/\mu^2)$.

Renormalization schemes will subtract the explicit pole pieces (which diverge as $\epsilon \to 0$), but may also include in the subtraction certain finite terms as well. For example, in the 'minimal subtraction' (MS) scheme, one subtracts just the pole pieces; in the 'modified minimal subtraction' or $\overline{\text{MS}}$ ('emm-ess-bar') scheme (Bardeen *et al.* 1978) one subtracts the pole and the '$-\gamma + \ln 4\pi$' piece.

The change from one scheme 'A' to another 'B' must involve a finite renormalization of the form (Ellis, Stirling, and Webber 1996, section 2.5)

$$\alpha_s^B = \alpha_s^A(1 + A_1 \alpha_s^A + \ldots). \tag{O.17}$$

Note that this implies that the first two coefficients of the β function are unchanged under this transformation, so that they are scheme-independent. Subsequent coefficients are scheme-dependent, as is the QCD parameter Λ introduced in section 15.3. From (15.54) the two corresponding values of Λ are related by

$$\ln\left(\frac{\Lambda_B}{\Lambda_A}\right) = \frac{1}{2} \int_{\alpha_s^A(|q^2|)}^{\alpha_s^B(|q^2|)} \frac{dx}{\beta_0 x^2(1 + \ldots)} \tag{O.18}$$

$$= \frac{A_1}{2\beta_0} \tag{O.19}$$

where we have taken $|q^2| \to \infty$ in (O.18) since the left-hand side is independent of $|q^2|$. Hence the relationship between the Λ's in different schemes is determined by the one-loop calculation which gives A_1 in (O.19). For example, changing from MS to $\overline{\text{MS}}$ gives (problem 15.8)

$$\Lambda_{\overline{\text{MS}}}^2 = \Lambda_{\text{MS}}^2 \exp(\ln 4\pi - \gamma), \tag{O.20}$$

as the reader may check.

Finally, consider the integral

$$I_d^{\mu\nu}(\Delta, n) \equiv \int \frac{d^d k}{(2\pi)^d} \frac{k^\mu k^\nu}{[k^2 - \Delta + i\epsilon]^n}. \tag{O.21}$$

From Lorentz covariance this must be proportional to the only second-rank tensor available, namely $g^{\mu\nu}$:

$$I_d^{\mu\nu} = A g^{\mu\nu}. \tag{O.22}$$

The constant 'A' can be determined by contracting both sides of (O.21) with $g_{\mu\nu}$, using $g^{\mu\nu}g_{\mu\nu} = d$ in d-dimensions. So

$$
\begin{aligned}
A &= \frac{1}{d}\int \frac{\mathrm{d}^d k}{(2\pi)^d} \frac{k^2}{(k^2 - \Delta + i\epsilon)^n} \\
&= \frac{1}{d}\left\{ \int \frac{\mathrm{d}^d k}{(2\pi)^d} \frac{1}{(k^2 - \Delta + i\epsilon)^{n-1}} + \Delta \int \frac{\mathrm{d}^d k}{(2\pi)^d} \frac{1}{(k^2 - \Delta + i\epsilon)^n} \right\} \\
&= \frac{i(-1)^n}{(4\pi)^{d/2}} \frac{\Delta^{(d/2)-n+1}}{d}\left\{ \frac{-\Gamma(n - 1 - d/2)}{\Gamma(n-1)} + \frac{\Gamma(n - d/2)}{\Gamma(n)} \right\} \\
&= \frac{i(-1)^n}{(4\pi)^{d/2}} \frac{\Delta^{(d/2)-n+1}}{d} \frac{\Gamma(n - 1 - d/2)}{\Gamma(n)}\{-n + (n - d/2)\} \\
&= \frac{i(-1)^{n-1}\Delta^{(d/2)-n+1}}{(4\pi)^{d/2}} \frac{1}{2}\frac{\Gamma(n - 1 - d/2)}{\Gamma(n)}.
\end{aligned}
\tag{O.23}
$$

Using these results, one can show straightforwardly that the gauge-non-invariant part of (11.18)—i.e. the piece in braces—vanishes. With the technique of dimensional regularization, starting from a gauge-invariant formulation of the theory the renormalization programme can be carried out while retaining manifest gauge invariance.

P

Grassmann Variables

In the path integral representation of quantum amplitudes (chapter 16) the fields are regarded as classical functions. Matrix elements of time-ordered products of bosonic operators could be satisfactorily represented (see the discussion following (16.79)). But something new is needed to represent, for example, the time-ordered product of two fermionic operators: there must be a sign difference between the two orderings, since the fermionic operators *anticommute*. Thus it seems that to represent amplitudes involving fermionic operators by path integrals we must think in terms of 'classical' anticommuting variables.

Fortunately, the necessary mathematics was developed by Grassmann in 1855, and applied to quantum amplitudes by Berezin (1966). Any two *Grassmann numbers* θ_1, θ_2 satisfy the fundamental relation

$$\theta_1\theta_2 + \theta_2\theta_1 = 0, \tag{P.1}$$

and of course

$$\theta_1^2 = \theta_2^2 = 0. \tag{P.2}$$

Grassmann numbers can be added and subtracted in the ordinary way, and multiplied by ordinary numbers. For our application, the essential thing we must be able to do with Grassmann numbers is to integrate over them. It is natural to think that, as with ordinary numbers and functions, integration would be some kind of inverse of differentiation. So let us begin with differentiation.

We define

$$\frac{\partial(a\theta)}{\partial\theta} = a, \tag{P.3}$$

where a is any ordinary number, and

$$\frac{\partial}{\partial\theta_1}(\theta_1\theta_2) = \theta_2 \, ; \tag{P.4}$$

then necessarily

$$\frac{\partial}{\partial\theta_2}(\theta_1\theta_2) = -\theta_1. \tag{P.5}$$

Consider now a function of one such variable, $f(\theta)$. An expansion of f in powers of θ terminates after only two terms because of the property (P.2):

$$f(\theta) = a + b\theta. \tag{P.6}$$

So

$$\frac{\partial f(\theta)}{\partial\theta} = b, \tag{P.7}$$

but also

$$\frac{\partial^2 f}{\partial\theta^2} = 0 \tag{P.8}$$

for any such f. Hence the operator $\partial/\partial\theta$ has no inverse (think of the matrix analogue $A^2 = 0$: if A^{-1} existed, we could deduce $0 = A^{-1}(A^2) = (A^{-1}A)A = A$ for all A). Thus we must approach Grassmann integration other than via an inverse of differentiation.

We only need to consider integrals over the complete range of θ, of the form

$$\int \mathrm{d}\theta\, f(\theta) = \int \mathrm{d}\theta(a + b\theta). \tag{P.9}$$

Such an integral should be linear in f; thus it must be a linear function of a and b. One further property fixes its value: we require the result to be *invariant under translations of θ by $\theta \to \theta + \eta$*, where η is a Grassmann number. This property is crucial to manipulations made in the path integral formalism, for instance in 'completing the square' manipulations similar to those in section 16.3, but with Grassmann numbers. So we require

$$\int \mathrm{d}\theta(a + b\theta) = \int \mathrm{d}\theta([a + b\eta] + b\theta). \tag{P.10}$$

This has changed the constant (independent of θ) term, but left the linear term unchanged. The only linear function of a and b which behaves like this is a multiple of b, which is conventionally taken to be simply b. Thus we define

$$\int \mathrm{d}\theta(a + b\theta) = b, \tag{P.11}$$

which means that integration is in some sense the same as differentiation!

When we integrate over products of different θ's, we need to specify a convention about the order in which the integrals are to be performed. We adopt the convention

$$\int \mathrm{d}\theta_1 \int \mathrm{d}\theta_2\, \theta_2\theta_1 = 1\,; \tag{P.12}$$

that is, the innermost integral is done first, then the next, and so on.

Since our application will be to Dirac fields, which are complex-valued, we need to introduce complex Grassmann numbers, which are built out of real and imaginary parts in the usual way (this would not be necessary for Majorana fermions). Thus we may define

$$\psi = \frac{1}{\sqrt{2}}(\theta_1 + i\theta_2), \quad \psi^* = \frac{1}{\sqrt{2}}(\theta_1 - i\theta_2), \tag{P.13}$$

and then

$$-i\mathrm{d}\psi\mathrm{d}\psi^* = \mathrm{d}\theta_1\mathrm{d}\theta_2. \tag{P.14}$$

It is convenient to define complex conjugation to include reversing the order of quantities:

$$(\psi\chi)^* = \chi^*\psi^*. \tag{P.15}$$

Then (P.14) is consistent under complex conjugation.

We are now ready to evaluate some Gaussian integrals over Grassmann variables, which is essentially all we need in the path integral formalism. We begin with

$$\int\int \mathrm{d}\psi^*\mathrm{d}\psi\, e^{-b\psi^*\psi} = \int\int \mathrm{d}\psi^*\mathrm{d}\psi(1 - b\psi^*\psi)$$

$$= \int\int \mathrm{d}\psi^*\mathrm{d}\psi(1 + b\psi\psi^*) = b. \tag{P.16}$$

Note that the analogous integral with ordinary variables is

$$\int\int \mathrm{d}x\mathrm{d}y\, e^{-b(x^2+y^2)/2} = 2\pi/b. \tag{P.17}$$

The important point here is that, in the Grassman case, b appears with a positive, rather than a negative, power. On the other hand, if we insert a factor $\psi\psi^*$ into the integrand in (P.16), we find that it becomes

$$\int\int \mathrm{d}\psi^*\mathrm{d}\psi\, \psi\psi^*(1+b\psi\psi^*) = \int\int \mathrm{d}\psi^*\mathrm{d}\psi\, \psi\psi^* = 1, \tag{P.18}$$

and the insertion has effectively produced a factor b^{-1}. This effect of an insertion is the same in the 'ordinary variables' case:

$$\int\int \mathrm{d}x\mathrm{d}y(x^2+y^2)/2\, e^{-b(x^2+y^2)/2} = 2\pi/b^2. \tag{P.19}$$

Now consider a Gaussian integral involving two different Grassmann variables:

$$\int \mathrm{d}\psi_1^*\mathrm{d}\psi_1\mathrm{d}\psi_2^*\mathrm{d}\psi_2\, e^{-\psi^{*\mathrm{T}}M\psi}, \tag{P.20}$$

where

$$\psi = \begin{pmatrix} \psi_1 \\ \psi_2 \end{pmatrix}, \tag{P.21}$$

and M is a 2×2 matrix, whose entries are ordinary numbers. The only terms which survive the integration are those which, in the expansion of the exponential, contain each of ψ_1^*, ψ_1, ψ_2^* and ψ_2 exactly once. These are the terms

$$\frac{1}{2}\left[M_{11}M_{22}(\psi_1^*\psi_1\psi_2^*\psi_2 + \psi_2^*\psi_2\psi_1^*\psi_1) + M_{12}M_{21}(\psi_1^*\psi_2\psi_2^*\psi_1 + \psi_2^*\psi_1\psi_1^*\psi_2) \right]. \tag{P.22}$$

To integrate (P.22) conveniently, according to the convention (P.12), we need to re-order the terms into the form $\psi_2\psi_2^*\psi_1\psi_1^*$; this produces

$$(M_{11}M_{22} - M_{12}M_{21})(\psi_2\psi_2^*\psi_1\psi_1^*), \tag{P.23}$$

and the integral (P.20) is therefore just

$$\int\int \mathrm{d}\psi_1^*\mathrm{d}\psi\mathrm{d}\psi_2^*\mathrm{d}\psi_2\, e^{-\psi^{*\mathrm{T}}M\psi} = \det M. \tag{P.24}$$

The reader may show, or take on trust, the obvious generalization to N independent complex Grassmann variables ψ_1, ψ_2, ψ_3, ..., ψ_N. This result is sufficient to establish the assertion made in section 16.4 concerning the integral (16.90), when written in 'discretized' form.

We may contrast (P.24) with an analogous result for two ordinary complex numbers z_1, z_2. In this case we consider the integral

$$\int\int \mathrm{d}z_1^*\mathrm{d}z_1\mathrm{d}z_2^*\mathrm{d}z_2\, e^{-z^{*}Hz}, \tag{P.25}$$

where z is a 2-component column matrix with elements z_1 and z_2. We take the matrix H to be Hermitian, with positive eigenvalues b_1 and b_2. Let H be diagonalized by the unitary transformation

$$\begin{pmatrix} z_1' \\ z_2' \end{pmatrix} = U \begin{pmatrix} z_1 \\ z_2 \end{pmatrix}, \tag{P.26}$$

with $UU^\dagger = I$. Then

$$\mathrm{d}z_1' \mathrm{d}z_2' = \det U \, \mathrm{d}z_1 \mathrm{d}z_2, \tag{P.27}$$

and so

$$\mathrm{d}z_1' \mathrm{d}z_1'^* \mathrm{d}z_2' \mathrm{d}z_2'^* = \mathrm{d}z_1 \mathrm{d}z_1^* \mathrm{d}z_2 \mathrm{d}z_2^*, \tag{P.28}$$

since $|\det U|^2 = 1$. The integral (P.25) then becomes

$$\int \mathrm{d}z_1' \mathrm{d}z_1'^* \mathrm{e}^{-b_1 z_1'^* z_1'} \int \mathrm{d}z_2' \mathrm{d}z_2'^* \mathrm{e}^{-b_2 z_2'^* z_2'}, \tag{P.29}$$

the integrals converging provided $b_1, b_2 > 0$. Next, setting $z_1 = (x_1 + \mathrm{i}y_1)/\sqrt{2}$, $z_2 = (x_2 + \mathrm{i}y_2)/\sqrt{2}$, (P.29) can be evaluated using (P.17), and the result is proportional to $(b_1 b_2)^{-1}$, which is the *inverse* of the determinant of the matrix H, when diagonalized. Thus—compare (P.16) and (P.17)—Gaussian integrals over complex Grassmann variables are proportional to the determinant of the matrix in the exponent, while those over ordinary complex variables are proportional to the inverse of the determinant.

Returning to integrals of the form (P.20), consider now a two-variable (both complex) analogue of (P.18):

$$\int \mathrm{d}\psi_1^* \mathrm{d}\psi_1 \mathrm{d}\psi_2^* \mathrm{d}\psi_2 \, \psi_1 \psi_2^* \, \mathrm{e}^{-\psi^{*\mathrm{T}} M \psi}. \tag{P.30}$$

This time, only the term $\psi_1^* \psi_2$ in the expansion of the exponential will survive the integration, and the result is just $-M_{12}$. By exploring a similar integral (still with the term $\psi_1 \psi_2^*$) in the case of three complex Grassmann variables, the reader should be convinced that the general result is

$$\prod_i \int \mathrm{d}\psi_i^* \mathrm{d}\psi_i \, \psi_k \psi_l^* \, \mathrm{e}^{-\psi^{*\mathrm{T}} M \psi} = (M^{-1})_{kl} \det M. \tag{P.31}$$

With this result we can make plausible the fermionic analogue of (16.87), namely

$$\langle \Omega | T \left\{ \psi(x_1) \bar{\psi}(x_2) \right\} | \Omega \rangle = \frac{\int \mathcal{D}\bar{\psi} \mathcal{D}\psi \, \psi(x_1) \bar{\psi}(x_2) \exp[-\int \mathrm{d}^4 x_{\mathrm{E}} \bar{\psi}(\mathrm{i}\,\partial\!\!\!/ - m)\psi]}{\int \mathcal{D}\bar{\psi} \mathcal{D}\psi \exp[-\int \mathrm{d}^4 x_{\mathrm{E}} \bar{\psi}(\mathrm{i}\,\partial\!\!\!/ - m)\psi]}; \tag{P.32}$$

note that $\bar{\psi}$ and ψ^* are unitarily equivalent. The denominator of this expression is [1] $\det(\mathrm{i}\,\partial\!\!\!/ - m)$, while the numerator is this same determinant multiplied by the inverse of the operator $(\mathrm{i}\,\partial\!\!\!/ - m)$; but this is just $(\not{p} - m)^{-1}$ in momentum space, the familiar Dirac propagator.

---

[1] The reader may interpret this as a finite-dimensional determinant, after discretization.

Q

Feynman Rules for Tree Graphs in QCD and the Electroweak Theory

Q.1 QCD

Q.1.1 External particles

Quarks

The SU(3) colour degree of freedom is not written explicitly; the spinors have 3 (colour) × 4 (Dirac) components. For each fermion or antifermion line entering the graph include the spinor

$$u(p, s) \quad \text{or} \quad v(p, s) \tag{Q.1}$$

and for spin-$\frac{1}{2}$ particles leaving the graph the spinor

$$\bar{u}(p', s') \quad \text{or} \quad \bar{v}(p', s'), \tag{Q.2}$$

as for QED

Gluons.

Besides the spin-1 polarization vector, external gluons also have a 'colour polarization' vector $a^c (c = 1, 2, \ldots, 8)$ specifying the particular colour state involved. For each gluon line entering the graph include the factor

$$\epsilon_\mu(k, \lambda) \, a^c \tag{Q.3}$$

and for gluons leaving the graph the factor

$$\epsilon_\mu^*(k', \lambda') \, a^{c*}. \tag{Q.4}$$

Q.1.2 Propagators

Quark

$$\xrightarrow{\hspace{2cm}} = \frac{\mathrm{i}}{\not{p} - m} = \mathrm{i} \frac{\not{p} + m}{p^2 - m^2}. \tag{Q.5}$$

Gluon

$$\text{OOOOOOOO} = \frac{\mathrm{i}}{k^2} \left(-g^{\mu\nu} + (1 - \xi) \frac{k^\mu k^\nu}{k^2} \right) \delta^{ab} \tag{Q.6}$$

for a general ξ gauge. Calculations are usually performed in Lorentz or Feynman gauge with $\xi = 1$ and gluon propagator equal to

$$\text{OOOOOOOO} = \mathrm{i} \frac{(-g^{\mu\nu}) \delta^{ab}}{k^2}. \tag{Q.7}$$

DOI: 10.1201/9781003411666-Q

Here a and b run over the 8 colour indices $1, 2, \ldots, 8$.

Q.1.3   Vertices

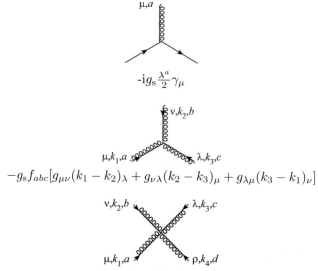

$$-g_s f_{abc}[g_{\mu\nu}(k_1 - k_2)_\lambda + g_{\nu\lambda}(k_2 - k_3)_\mu + g_{\lambda\mu}(k_3 - k_1)_\nu]$$

$$-ig_s^2[f_{abe}f_{cde}(g_{\mu\lambda}g_{\nu\rho} - g_{\mu\rho}g_{\nu\lambda}) + f_{ade}f_{bce}(g_{\mu\nu}g_{\lambda\rho} - g_{\mu\lambda}g_{\nu\rho}) + f_{ace}f_{dbe}(g_{\mu\rho}g_{\nu\lambda} - g_{\mu\nu}g_{\lambda\rho})]$$

It is important to remember that the rules given above are only adequate for tree diagram calculations in QCD (see section 13.3.3).

Q.2   The electroweak theory

For tree graph calculations, it is convenient to use the U gauge Feynman rules (sections 19.5 and 19.6) in which no unphysical particles appear. These U gauge rules are given below for the leptons $l = (e, \mu, \tau)$, $\nu_l = (\nu_e, \nu_\mu, \nu_\tau)$; for the $t_3 = +1/2$ quarks denoted by f, where f = u, c, t; and for the $t_3 = -1/2$ CKM-mixed quarks denoted by f' where f' = d', s', b'.

Note that for simplicity we do not include neutrino flavour mixing.

Q.2.1   External particles

Leptons and quarks

For each fermion or antifermion line entering the graph include the spinor

$$u(p, s) \quad \text{or} \quad v(p, s) \tag{Q.8}$$

and for spin-$\frac{1}{2}$ particles leaving the graph the spinor

$$\bar{u}(p', s') \quad \text{or} \quad \bar{v}(p', s'). \tag{Q.9}$$

Vector bosons

For each vector boson line entering the graph include the factor

$$\epsilon_\mu(k, \lambda) \tag{Q.10}$$

and for vector bosons leaving the graph the factor

$$\epsilon_\mu^*(k', \lambda').$$ (Q.11)

Q.2.2   Propagators

Leptons and quarks

$$\underrightarrow{\hspace{3cm}} = \frac{\mathrm{i}}{\not{p} - m} = \mathrm{i}\frac{\not{p} + m}{p^2 - m^2}.$$ (Q.12)

Vector bosons (U gauge)

$$\overset{W^\pm, Z^0}{\wwbar{\hspace{3cm}}} = \frac{\mathrm{i}}{k^2 - M_V^2}(-g_{\mu\nu} + k_\mu k_\nu / m_V^2)$$ (Q.13)

where 'V' stands for either 'W' (the W-boson) or 'Z' (the Z^0).

Higgs particle

$$\cdots\cdots\blacktriangleright\cdots\cdots = \frac{\mathrm{i}}{p^2 - m_H^2}$$ (Q.14)

Q.2.3   Vertices

Charged current weak interactions

Leptons

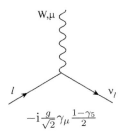

$$-\mathrm{i}\frac{g}{\sqrt{2}}\gamma_\mu\frac{1-\gamma_5}{2}$$

Quarks

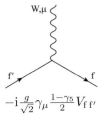

$$-\mathrm{i}\frac{g}{\sqrt{2}}\gamma_\mu\frac{1-\gamma_5}{2}V_{ff'}$$

Neutral current weak interactions (no neutrino mixing)

Fermions

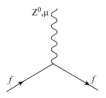

$$\frac{-ig}{\cos\theta_{\mathrm{W}}}\gamma_\mu\left(c_{\mathrm{L}}^f\frac{1-\gamma_5}{2}+c_{\mathrm{R}}^f\frac{1+\gamma_5}{2}\right),$$

where

$$c_{\mathrm{L}}^f = t_3^f - \sin^2\theta_{\mathrm{W}}Q_f \tag{Q.15}$$

$$c_{\mathrm{R}}^f = -\sin^2\theta_{\mathrm{W}}Q_f, \tag{Q.16}$$

and f stands for any fermion.

Vector boson couplings

(a) Trilinear couplings:

$\gamma\mathrm{W}^+\mathrm{W}^-$ vertex

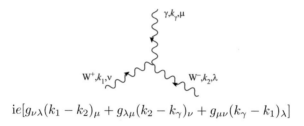

$$ie[g_{\nu\lambda}(k_1-k_2)_\mu + g_{\lambda\mu}(k_2-k_\gamma)_\nu + g_{\mu\nu}(k_\gamma-k_1)_\lambda]$$

$\mathrm{Z}^0\mathrm{W}^+\mathrm{W}^-$ vertex

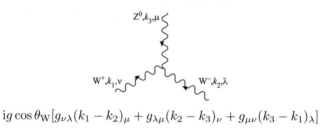

$$ig\cos\theta_{\mathrm{W}}[g_{\nu\lambda}(k_1-k_2)_\mu + g_{\lambda\mu}(k_2-k_3)_\nu + g_{\mu\nu}(k_3-k_1)_\lambda]$$

(b) Quadrilinear couplings:

$$-ie^2(2g_{\alpha\beta}g_{\mu\nu} - g_{\alpha\mu}g_{\beta\nu} - g_{\alpha\nu}g_{\beta\mu})$$

$$-ieg\cos\theta_{\mathrm{W}}(2g_{\alpha\beta}g_{\mu\nu} - g_{\alpha\mu}g_{\beta\nu} - g_{\alpha\nu}g_{\beta\mu})$$

$$-ig^2\cos^2\theta_{\mathrm{W}}(2g_{\alpha\beta}g_{\mu\nu} - g_{\alpha\mu}g_{\beta\nu} - g_{\alpha\nu}g_{\beta\mu})$$

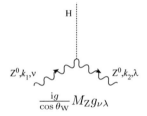

$$ig^2(2g_{\mu\alpha}g_{\nu\beta} - g_{\mu\beta}g_{\alpha\nu} - g_{\mu\nu}g_{\alpha\beta})$$

Higgs couplings

(a) Trilinear couplings
HW^+W^- vertex

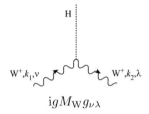

$$igM_{\rm W}g_{\nu\lambda}$$

HZ^0Z^0 vertex

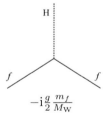

$$\frac{ig}{\cos\theta_{\rm W}}M_{\rm Z}g_{\nu\lambda}$$

Fermion Yukawa couplings (fermion mass m_f)

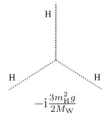

$$-i\frac{g}{2}\frac{m_f}{M_{\rm W}}$$

Trilinear self-coupling

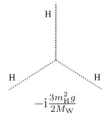

$$-i\frac{3m_{\rm H}^2 g}{2M_{\rm W}}$$

(b) Quadrilinear couplings:

HHW^+W^- vertex

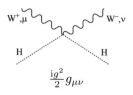

$$\frac{ig^2}{2} g_{\mu\nu}$$

HHZZ vertex

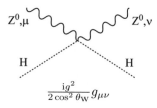

$$\frac{ig^2}{2\cos^2\theta_W} g_{\mu\nu}$$

Quadrilinear self-coupling

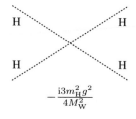

$$-\frac{i3m_H^2 g^2}{4M_W^2}$$

Bibliography

Aaboud M *et al.* 2018a (ATLAS Collaboration) *Phys. Lett.* B **784** 173

——2018b *Phys. Lett.* B **786** 223

——2019 *Phys. Rev.* D **99** 072001

Aad G *et al.* 2012a (ATLAS Collaboration) *Phys. Lett.* B **710** 49

——2012b *Phys. Lett.* B **716** 1

——2015 *Eur. Phys. J.* C **75** 476

——2020 *Phys. Rev. Lett.* **125** 061802

——2021a *Eur. Phys. J.* C **81** 178

——2021b *Phys. Lett.* B **812** 135980

Aaij R *et al.* 2012 (LHCb Collaboration) *Phys. Rev. Lett.* **108** 111602

——2015 *Phys. Rev. Lett.* **115** 031601

——2017 *Phys. Rev. Lett.* **119** 112001

——2019 *Phys. Rev. Lett.* **122** 211803

——2021a *JHEP* **02** 169

——2021b *Phys. Rev. Lett.* **127** 111801

Aaltonen T *et al.* 2008 (CDF Collaboration) *Phys. Rev. Lett.* **100** 121802

——2009 *Phys. Rev. Lett.* **103** 092002

——2012 (CDF and D0 Collaborations) *Phys. Rev. Lett.* **109** 121802

——2022 (CDF Collaboration) *Science* **376** 6589 170

Abachi S *et al.* 1995a (D0 Collaboration) *Phys. Rev. Lett.* **74** 2422

——1995b *Phys. Rev. Lett.* **74** 2632

Abashian A *et al.* 2001 (Belle Collaboration) *Phys. Rev. Lett.* **86** 2509

Abazov V M *et al* 2009 (D0 Collaboration) *Phys. Rev. Lett.* **103** 092001

Abbiendi G *et al.* 2001 *Eur. Phys. J.* C **20** 601

Abbott T M C *et al.* 2022 (DES Collaboration) *Phys. Rev.* D **105** 023520

Abdurashitov J N *et al.* 2009 *Phys. Rev.* C **80** 015807

Abe F *et al.* 1994a (CDF Collaboration) *Phys. Rev.* D **50** 2966

——1994b *Phys. Rev. Lett.* **73** 225

——1995a *Phys. Rev.* D **52** 4784

——1995b *Phys. Rev. Lett.* **74** 2626

Abe K 1991 *Proc. 25th Int. Conf. on High Energy Physics* eds K K Phua and Y Yamaguchi (Singapore: World Scientific) p 33

Abe K 2000a (SLD Collaboration) *Phys. Rev. Lett.* **84** 5945

——2000b *Phys. Rev. Lett.* **85** 5059

——2001 *Phys. Rev. Lett.* **86** 1162

Abe K *et al.* 2000 *Phys. Rev. Lett.* **84** 5945

——2011 (T2K Collaboration) *Phys. Rev. Lett.* **107** 041801

——2014 *Phys. Rev. Lett.* **112** 061802

Abe S *et al.* 2008 (KamLAND Collaboration) *Phys. Rev. Lett.* **100** 221803

Ablikin M *et al.* 2020a (BESIII Collaboration) *Phys. Rev.* D **101** 112002

——2020b *Phys. Rev.* **102** 052008

——2020c *Phys. Rev. Lett.* **124** 241802

Abramovicz H *et al.* 1982a *Z. Phys.* C **12** 289

——1982b *Z. Phys.* C **13** 199

——1983 *Z. Phys.* C **17** 283

Abreu P *et al.* 1990 *Phys. Lett.* B **242** 536

Acero M A *et al.* 2019 (NOvA Collaboration) *Phys. Rev. Lett.* 123 151803

Adachi I *et al.* 2012 (Belle Collaboration) *Phys. Rev. Lett.* **108** 171802

Adamson P *et al* 2011 (MINOS Collaboration) *Phys. Rev. Lett.* **106** 181801

Adey D *et al.* 2018 (Daya Bay Collaboration) *Phys. Rev. Lett.* **121** 241805

Adler S L 1963 *Phys. Rev.* **143** 1144

——1965 *Phys. Rev. Lett.* **14** 1051

——1969 *Phys. Rev.* **177** 2426

——1970 *Lectures on Elementary Particles and Quantum Field Theory (Proceedings of the Brandeis Summer Institute)* vol 1, ed S Deser *et al.* (Boston, MA: MIT)

Adler S L and Bardeen W A 1969 *Phys. Rev.* **182** 1517

Agafonova N *et al.* 2018 (OPERA Collaboration) *Phys. Rev. Lett.* **120** 211801

Agostini M *et al.* (GERDA Collaboration) 2020 *Phys. Rev. Lett.* **125** 252502

Aharmim B *et al.* 2010 (SNO Collaboration) *Phys. Rev.* C **81** 055504

Ahmad Q R *et al.* 2001 (SNO Collaboration) *Phys. Rev. Lett.* **87** 071301

——2002 *Phys. Rev. Lett.* **89** 011301

Ahn J K *et al.* 2012 (RENO Collaboration) *Phys. Rev. Lett.* **108** 191802

Ahn M H *et al.* 2006 (K2K Collaboration) *Phys. Rev.* D **74** 072003

Aitchison I J R 2007 *Supersymmetry in Particle Physics An Elementary Introduction* (Cambridge: Cambridge University Press)

Aitchison I J R *et al.* 1995 *Phys. Rev.* B **51** 6531

Aker M *et al.* 2019 (KATRIN Collaboration) *Phys. Rev. Lett.* **123** 221802

Akhundov A A *et al.* 1986 *Nucl. Phys.* B **276** 1

Akrawy M Z *et al.* 1990 (OPAL Collaboration) *Phys. Lett.* B **235** 389

Alavi-Harati A *et al.* 2003 (KTeV Collaboration) *Phys. Rev.* D **67** 012005; *ibid.* D **70** 079904 (erratum)

Alexandrou *et al.* 2014 *Phys. Rev.* D **90** 074501

Ali A and Kramer G 2011 *Eur. Phys. J.* H **36** 245

Allaby J *et al.* 1988 *J. Phys. C: Solid State Phys.* **38** 403

Allanach B C and Haber H E 2022 section 88 of Workman *et al.* 2022

Allasia D *et al.* 1984 *Phys. Lett.* B **135** 231

——1985 *Z. Phys.* C **28** 321

Allton C R *et al.* 1995 *Nucl. Phys.* B **437** 641

——2002 (UKQCD Collaboration) *Phys. Rev.* D **65** 054502

Alonso R *et al.* 2014 *JHEP* **04** 159

Alper B *et al.* 1973 *Phys. Lett.* B **44** 521

Altarelli G 1982 *Phys. Rep.* **81** 1

Altarelli G and Parisi G 1977 *Nucl. Phys.* B **126** 298

Altarelli G *et al.* 1978a *Nucl. Phys.* B **143** 521

——1978b *Nucl. Phys.* B **146** 544(E)

——1979 *Nucl. Phys.* B **157** 461

——1989 Z. Phys. *at LEP-1* CERN 89-08 (Geneva)

Altman M *et al.* 2005 *Phys. Lett.* B **616** 174

Alvarez-Gaume L, Polchinski J and Wise M B 1983 *Nucl. Phys.* B **221** 495

Amaudruz P *et al.* 1992 (NMC Collaboration) *Phys. Lett.* B **295** 159

Amhis Y S *et al.* 2021 (HFLAV Collaboration) *Eur. Phys. J.* C **81** 226

An F P *et al.* 2012 (Daya Bay Collaboration) *Phys. Rev. Lett.* **108** 171803

An F P *et al.* 2016 (JUNO Collaboration) *J. Phys.* **G 43** 030401

Andersen C W *et al.* 2018 *Phys. Rev.* D **97** 014506

Anderson P W 1963 *Phys. Rev.* **130** 439

Andreassen A, Frost D and Schwartz M D 2018 *Phys. Rev.* D **97** 056006

Andreotti E *et al.* (CUORCINO Collaboration) 2011 *Astropart. Phys.* **34** 822

Antoniadis I 2002 *2001 European School of High Energy Physics* ed N Ellis and J March-Russell CERN 2002-002 (Geneva) pp 301ff

Aoki Y *et al.* (LatKMI Collaboration) 2017 *Phys. Rev.* D **96** 014508

Aoki S *et al.* (Flavour Averaging Group) 2020 *Eur. Phys. J.* C **80** 113

Apollonio M *et al.* 1999 (CHOOZ Collaboration) *Phys. Lett.* **466** 415

——*Eur. Phys. J.* C **27** 331

Appel J A *et al.* 1986 *Z. Phys.* C **30** 341

Appelquist T and Chanowitz M S 1987 *Phys. Rev. Lett.* **59** 2405

Appelquist T *et al.* 2014 (LSD Collaboration) *Phys. Rev.* D **90** 114502

Appelquist T *et al.* 2016 *Phys. Rev.* D **93** 114514

Arnison G *et al.* 1983a *Phys. Lett.* B **122** 103

——1983b *Phys. Lett.* B **126** 398

——1984 *Phys. Lett.* B **136** 294

——1985 *Phys. Lett.* B **158** 494

——1986 *Phys. Lett.* B **166** 484

Arnold R *et al.* (NEMO Collaboration) 2006 *Nucl. Phys.* A **765** 483

Arzt C 1995 *Phys. Lett.* B **342** 189

Asner D M and Schwartz A J 2022 section 70 of Workman *et al.* 2022

ATLAS and CMS 2015 CMS-PAS-HIG-15-002

ATLAS 2019 (ATLAS Collaboration) ATLAS-CONF-2019-029

——2020a ATLAS-CONF-2020-026

——2020b ATLAS-CONF-2020-027

——2020c ATLAS-CONF-2020-053

——2023 ATLAS-CONF-2023-004

Attwood D *et al.* 2001 *Phys. Rev.* D **63** 036005

Aubert B *et al.* 2001 (BaBar Collaboration) *Phys. Rev. Lett.* **86** 2515

——2004 *Phys. Rev. Lett.* **93** 131801

——2007a *Phys. Rev. Lett.* **99** 171803

——2007b *Phys. Rev.* D **76** 102004

——2007c *Phys. Rev. Lett.* **98** 211802

——2008 *Phys. Rev.* D **78** 034023

——2009 *Phys. Rev.* D **79** 072009

——2010 *Phys. Rev. Lett.* **105** 121801

Aubin C *et al.* 2004 *Phys. Rev.* D **70** 09505

Ayres D S (NOνA Collaboration) 1995 arXiv:hep-ex/0503053

Baer H *et al.* 2018 *Eur. Phys. J.* C **78** 838

Baernreuther P, Czakon M and Mitov A 2012 *Phys. Rev. Lett.* **109** 132001

Bagnaia P *et al.* 1983 *Phys. Lett.* B **129** 130

——1984 *Phys. Lett.* B **144** 283

Bagnaschi *et al.* 2019 *Eur. Phys. J.* C **79** 617

Bahcall J N *et al.* 1968 *Phys. Rev. Lett.* **20** 1209

——2001 *Astrophys. J.* **555** 990

——2005 *ibid.* **621** L85

Baikov P A, Chetyrkin K G and Kühn J H 2017 *Phys. Rev. Lett.* **118** 082002

Baikov P A *et al.* 2008 *Phys. Rev. Lett.* **101** 012002

——2009 *Nucl. Phys. Proc. Suppl.* **189** 49

——2017 *Phys. Rev. Lett.* **118** 082002

Bailin D 1982 *Weak Interactions* (Bristol: Adam Hilger)

Bailey S *et al.* 2021 *Eur. Phys. J.* C **81** 341

Bak G *et al.* 2018 *Phys. Rev. Lett.* **121** 201801

Ball R D *et al.* 2017 (NNPDF Collaboration) *Eur. Phys. J.* C **77** 663

Bander M, Silverman D and Soni A 1979 *Phys. Rev. Lett.* **43** 242

Banks T *et al.* 1976 *Phys. Rev.* D **13** 1043

Banner M *et al.* 1982 *Phys. Lett.* B **118** 203

——1983 *Phys. Lett.* B **122** 476

Barber D P *et al.* 1979 *Phys. Rev. Lett.* **43** 830

Bardeen J 1957 *Nuovo Cimento* **5** 1766

Bardeen J, Cooper L N and Schrieffer J R 1957 *Phys. Rev.* **108** 1175

Bardeen W A, Fritzsch H and Gell-Mann M 1973 in *Scale and Conformal Symmetry in Hadron Physics* ed R Gatto (New York: Wiley) pp 139-151

Bardeen W A *et al.* 1978 *Phys. Rev.* D **18** 3998

Bardin D Yu *et al.* 1989 *Z. Phys.* C **44** 493

Barezyk A *et al.* 2008 (FAST Collaboration) *Phys. Lett.* B **663** 172

Barger V *et al.* 1983 *Z. Phys.* C **21** 99

Barman R K *et al.* 2023 *Phys. Rev. Lett.* **131** 011802

Barnett R M, Lellouch L P and Manohar A V 2022 section 60 in Workman *et al.* 2022

Barr G D *et al.* 1993 (NA31 Collaboration) *Phys. Lett.* B **317** 233

Bartel *et al.* 1986 (JADE Collaboration) *Z. Phys.* C **33** 23

Batley J R *et al.* 2002 (NA48 Collaboration) *Phys. Lett.* B **544** 97

Beenakker W and Hollik W 1988 *Z. Phys.* C **40** 569

Beenakker W *et al.* 1989 *Phys. Rev.* D **40** 54

Belavin A A *et al.* 1975 *Phys. Lett.* B **59** 85

Bell J S and Jackiw R 1969 *Nuovo Cimento* A **60** 47

Bellazzini B, Csáki C and Serra J 2014 *Eur. Phys. J.* C **74** 2766

Beneke M and Braun V M 1994 *Nucl. Phys.* B **426** 301

——1995 *Phys. Lett.* B **344** 341

——1999 *Phys. Rept.* **317** 1

Beneke M, Falgari P and Schwinn 2010 *Nucl. Phys.* B **828** 69

Beneke M *et al.* 2017 *Phys. Lett.* B **775** 63

Benvenuti A C *et al.* 1989 (BCDMS Collaboration) *Phys. Lett.* B **223** 485

Berends F A *et al.* 1981 *Phys. Lett.* B **103** 124

Berezin F A 1966 *The Method of Second Quantisation* (New York: Academic

Berger E L, Gao J and Zhu H X 2017 *JHEP* **11** 158

Bergsma F *et al.* 1983 *Phys. Lett.* B **122** 465

Berman S M, Bjorken J D and Kogut J B 1971 *Phys. Rev.* D **4** 3388

Bernard C W, Golterman M F and Shamir Y 2006 *Phys. Rev.* D **73** 114511

Bernreuther W and Wetzel W 1982 *Nucl. Phys.* B **197** 128

Bernstein J 1974 *Rev. Mod. Phys.* **46** 7

Bethke S 2009 *Eur. Phys. J.* C **64** 689

Bethke S *et al.* 1988 (JADE Collaboration) *Phys. Lett.* B **213** 235

Bettini A 2008 *Introduction to Elementary Particle Physics* (Cambridge: Cambridge University Press)

Bigi I I and Sanda A I 2000 *CP Violation* (Cambridge: Cambridge University Press)

Bigi I I *et al.* 1994 *Phys. Rev.* D **50** 2234

Bijnens J *et al.* 1996 *Phys. Lett.* B **374** 2010

Binney J J *et al.* 1992 *The Modern Theory of Critical Phenomena* (Oxford: Clarendon)

Bjorken J D 1973 *Phys. Rev.* D**8** 4098

Bjorken J D and Glashow S L 1964 *Phys. Lett.* **11** 255

Black K M, Sekhar Chivukula R and Narain M 2022 section 92 of Workman *et al.* 2022.

Blatt J M 1964 *Theory of Superconductivity* (New York: Academic)

Bloch F and Nordsieck A 1937 *Phys. Rev.* **52** 54

Bludman S A 1958 *Nuovo Cimento* **9** 443

Boas M L 1983 *Mathematical Methods in the Physical Sciences* (New York: Wiley)

Bogoliubov N N 1947 *J. Phys. USSR* **11** 23

——1958 *Nuovo Cimento* **7** 794

Bogoliubov N N *et al.* 1959 *A New Method in the Theory of Superconductivity* (New York: Consultants Bureau, Inc.)

Borici A 2000 *NATO Sci. Ser.* C **553** 41

Bouchiat C C *et al.* 1972 *Phys. Lett.* B **38** 519

Bouchiendra R *et al.* 2011 *Phys. Rev. Lett.* **106** 080801

Bramante J *et al.* 2016 *Phys. Rev.* D **93** 063525

Branco G C *et al.* 1999 *CP Violation* (Oxford: Oxford University Press)

Brandelik R *et al.* 1979 *Phys. Lett.* B **86** 243

Briceño R A *et al.* 2017 *Phys. Rev.* D **95** 074510

Briceño R A et al. 2018 *Rev. Mod. Phys.* **90** 025001

Brivio I and Trott M 2019 *Phys. Rept.* **703** 1

Brodsky S J, Lepage G P and Mackenzie P B 1983 *Phys. Rev.* D **28** 228

Budny R 1975 *Phys. Lett.* B **55** 227

Buleva J *et al.* 2016 *Nucl. Phys.* B **910** 842

Buras A J *et al.* 1994 *Phys. Rev.* D **50** 3433

Büsser F W *et al.* 1972 *Proc. XVI Int. Conf. on High Energy Physics (Chicago, IL)* vol 3 (Batavia: FNAL)

——1973 *Phys. Lett.* B **46** 471

Butterworth J M *et al.* 2008 *Phys. Rev. Lett.* **100** 242001

Cabibbo N 1963 *Phys. Rev. Lett.* **10** 531

Cabibbo N *et al.* 1979 *Nucl. Phys.* B **158** 295

Cacciari M 2015 *Int. J. Mod. Phys.* A **30** 1546001

Cacciari M and Salam G P 2006 *Phys. Lett.* B **641** 57

Cacciari M, Salam G P and Soyez 2012 *Eur. Phys. J.* C **72** 1896

Cacciari M *et al.* 2008 *JHEP* **0804** 063

——2012 *Phys. Lett.* B **710** 612

Callan C G 1970 *Phys. Rev.* D **2** 1541

Callan C, Coleman S, Wess J, and Zumino B 1969 *Phys. Rev.* **177** 2247

Campbell J, Neumann T and Sullivan Z 2021 *JHEP* **02** 040

Campbell J M *et al.* 2007 *Rept. on Prog. in Phys.* **70** 89

Capozzi F *et al.* 2018 *Prog. Part. Nucl. Phys.* **102** 48

Carena M *et al.* 2022 section 11 of Workman *et al.* 2022

Carruthers P A 1966 *Introduction to Unitary Symmetry* (New York: Wiley)

Carter A B and Sanda A I 1980 *Phys. Rev. Lett.* **45** 952

——1981 *Phys. Rev.* D **23** 1567

Caswell W E 1974 *Phys. Rev. Lett.* **33** 244

Catani S *et al.* 1991 *Phys. Lett.* B **269** 432

——1993 *Nucl. Phys.* B **406** 187

——2019 *JHEP* **07** 100

Ceccucci A, Ligeti Z and Sakai Y 2022: section 12 of Workman *et al.* 2022

Celmaster W and Gonsalves R J 1980 *Phys. Rev. Lett.* **44** 560

Cepeda M *et al.* 2019 HL/HE-LHCWG2group

Chadwick J 1932 *Proc. R. Soc.* A **136** 692

Chanowitz M *et al.* 1978 *Phys. Lett.* B **78** 285

——1979 *Nucl. Phys.* B **153** 402

Chao Y *et al.* 2005 (Belle Collaboration) *Phys. Rev.* D **71** 031502

Charles J *et al.* 2005 (CKMfitter Group) *Eur. Phys. J.* C **41** 1

Chatrchyan S *et al.* 2012a (CMS Collaboration) *Phys. Lett.* B **710** 26

——2012b *Phys. Lett.* B **716** 30

Chau L L and Keung W Y 1984 *Phys. Rev. Lett.* **53** 1802

Chen M-S and Zerwas P 1975 *Phys. Rev.* D **12** 187

Cheng T-P and Li L-F 1984 *Gauge Theory of Elementary Particle Physics* (Oxford: Clarendon)

Chetyrkin K G *et al.* 1979 *Phys. Lett.* B **85** 277

——1997 *Phys. Rev. Lett.* **79** 2184

——2017 *JHEP* **10** 170 [Addendum *JHEP* **10** 006 (2017)]

Chisholm J S R 1961 *Nucl. Phys.* **26** 469

Chitwood D B *et al.* 2007 (MuLan Collaboration) *Phys. Rev. Lett.* **99** 032001

Christenson J H *et al.* 1964 *Phys. Rev. Lett.* **13** 138

Cirigliano V *et al.* 2005 *Nucl. Phys.* B **728** 121

Cleveland B T *et al.* 1998 *Astrophys. J.* **496** 505

CMS PAS TOP-22-006

Cohen A G *et al.* 2009 *Phys. Lett.* B **678** 191

Colangelo G *et al.* 2001 *Nucl. Phys.* B **603** 125

Coleman S 1985 *Aspects of Symmetry* (Cambridge: Cambridge University Press)

——1966 *J. Math. Phys.* **7** 787

Coleman S and Gross D J 1973 *Phys. Rev. Lett.* 31 851

Coleman S, Wess J and Zumino B 1969 *Phys. Rev.* **177** 2239

Collins J C and Soper D E 1987 *Annu. Rev. Nucl. Part. Sci.* **37** 383

——1998 *Phys. Lett.* B **438** 184

Collins P D B and Martin A D 1984 *Hadron Interactions* (Bristol: Adam Hilger)

Combridge B L 1979 *Nucl. Phys.* B **151** 429

Combridge B L *et al.* 1977 *Phys. Lett.* B **70** 234

Combridge B L and Maxwell C J 1984 *Nucl. Phys.* B **239** 429

Commins E D and Bucksbaum P H 1983 *Weak Interactions of Quarks and Leptons* (Cambridge: Cambridge University Press)

Consoli M *et al.* 1989 Z. Phys. *at LEP-I* ed G Altarelli *et al.* CERN 89-08 (Geneva)

Contino R *et al.* 2013 *JHEP* 07 35

Cooper L N 1956 *Phys. Rev.* **104** 1189

Cornwall J M *et al.* 1974 *Phys. Rev.* D 10 1145

Cortina Gil E *et al.* 2021 (NA62) *JHEP* **06** 093

Cowan C L *et al.* 1956 Science **124** 103

Czakon M 2005 *Nucl. Phys.* B **710** 485

Czakon M and Mitov A 2010 *Nucl. Phys.* B **824** 111

——2012 *JHEP* **1212** 054

——2013 *JHEP* **1301** 080

Czakon M, Fiedler P and Mitov A 2013 *Phys. Rev. Lett.* **110** 252004

Czakon M, Heymes D and Mitov A 2016 *Phys. Rev. Lett.* **116** 082003

Czakon M, Mitov A and Sterman G F 2009 *Phys. Rev.* D **80** 074017

Dalitz R H 1953 *Phil. Mag.* **44** 1068

——1965 *High Energy Physics* ed C de Witt and M Jacob (New York: Gordon and Breach)

D'Ambrosio *et al.* 2002 *Nucl. Phys.* B **645** 155

Danby G *et al.* 1962 *Phys. Rev. Lett.* **9** 36

Davidson S, Nardi E and Nir Y 2008 *Phys. Rept.* **466** 105

Davies C T H *et al.* 2004 (HPQCD, UKQCD, MILC and Fermilab Collaborations) *Phys. Rev. Lett.* **92** 022001

Davies C T H *et al.* 2008 (HPQCD Collaboration) *Phys. Rev.* D **78** 11450

Davies C T H *et al.* 2010 *Phys. Rev.* D **82** 114504

Davies J *et al.* 2016 PoS **DIS2016** 059

Davis R 1955 *Phys. Rev.* **97** 766

——1964 *Phys. Rev. Lett.* **12** 303

Davis R *et al.* 1968 *Phys. Rev. Lett.* **20** 1205

Dawson S *et al.* 1990 *The Higgs Hunter's Guide* (Reading, MA: Addison-Wesley)

Decker R and Pestieau J 1980 *Lett. Nuovo Cim.* **29** 560

de Florian D *et al.* 2016 (LHC Higgs Cross Section Working Group), *CERN Report* **2017-002**

Degrassi G *et al.* 2012 *JHEP* **08** 098

de Groot J G H *et al.* 1979 *Z. Phys.* C **1** 143

de Kerret H *et al.* 2020 (Double Chooz Collaboration) *Nature Phys.* **16** 558

del Amo Sanchez P *et al.* 2010 (BaBar Collaboration) *Phys. Rev. Lett.* **105** 121801

Delgado A and Quirós M 2021 *Phys. Rev.* D **103** 105024

de Salas P F *et al.* 2018 *Phys. Lett.* B **782** 633

DeTar C *et al.* 2019 *Phys. Rev.* D **99** 034509

DiLella L 1985 *Annu. Rev. Nucl. Part. Sci.* **35** 107

——1986 *Proc. Int. Europhysics Conf. on High Energy Physics, Bari, Italy, July 1985* eds L Nitti and G Preparata (Bari: Laterza) pp 761ff

Dimopoulos S and Georgi H 1981 *Nucl. Phys.* B **193** 150

Dimopoulos S and Giudice G F 1995 *Phys. Lett.* B **357** 573

Dimopoulos S and Susskind L 1979 *Nucl. Phys.* B **155** 237

Dimopoulos S, Raby S and Wilczek F 1981 *Phys. Rev.* D **24** 1681

Dine M and Sapirstein J 1979 *Phys. Rev. Lett.* **43** 668

Dirac P A M 1931 *Proc. R. Soc.* A **133** 60

Dokshitzer Yu L 1977 *Sov. Phys.–JETP* **46** 641

Donald G C *et al.* 2012 *Phys. Rev.* D **86** 094501

D'Onofrio M and Moortgat F 2022 section 89 of Workman *et al.* 2022

Donoghue J F, Golowich E and Holstein B R 1992 *Dynamics of the Standard Model* (Cambridge: Cambridge University Press)

Dowdall R J *et al.* 2012 *Phys. Rev.* D **86** 094510

Drach V 2020 PoS **LATTICE2019** 242

Dudek J J, Edwards R G and Thomas C E 2013 *Phys. Rev.* D **87** 034505 [Erratum: *Phys. Rev.* D **90** 099902 (2014)]

Duke D W and Owens J F 1984 *Phys. Rev.* D **30** 49

Dürr S *et al.* 2008 (Budapest-Marseille-Wuppertal Collaboration) *Science* **322** 1224

Eden R J, Landshoff P V, Olive D I and Polkinghorne J C 1966 *The Analytic S-Matrix* (Cambridge: Cambridge University Press)

Eichten E and Lane K 1980 *Phys. Lett.* B **90** 125

Eichten E *et al.* 1980 *Phys. Rev.* D **21** 203

——1984 *Rev. Mod. Phys.* **56** 579 [Addendum: *Rev. Mod. Phys.* **58** 1065 (1986)]

Einhorn M B and Jones D R T 1982 *Nucl. Phys.* B **196** 475

Einhorn M B and Wudka J 1989 *Phys. Rev.* D **39** 2758

Eitel K *et al.* 2005 *Nucl. Phys. (Proc. Suppl.)* B **143** 197

Elias-Miro J *et al.* 2012 *Phys. Lett.* B **709** 222

Ellis J, Gaillard M K and Nanopoulos D 1976 *Nucl. Phys.* B **106** 292

Ellis J, Nanopoulos D V and Tamvakis K 1983 *Phys. Lett.* B **121** 123

Ellis J *et al.* 1976 *Nucl. Phys.* B **111** 253

——1977 Erratum *ibid.* B **130** 516

——1994 *Phys. Lett.* B **333** 118

Ellis R K, Stirling W J and Webber B R 1996 *QCD and Collider Physics* (Cambridge: Cambridge University Press)

Ellis S D and Soper D E 1993 *Phys. Rev.* D **48** 3160

Ellis S D *et al.* 2008 *Prog. Part. Nucl. Phys.* **60** 484

Englert F and Brout R 1964 *Phys. Rev. Lett.* **13** 321

Enz C P 1992 *A Course on Many-Body Theory Applied to Solid-State Physics (World Scientific Lecture Notes in Physics 11)* (Singapore: World Scientific)

——2002 *No Time to be Brief* (Oxford: Oxford University Press)

Erler J and Freitas A 2022 section 10 of Workman *et al.* 2022

Erler J and Schott M 2019 *Prog. Part. Nucl. Phys.* **106** 68

Erler J and Langacker P 2010 section 10 of Nakamura *et al.* 2010

Esteban I *et al.* 2019 *JHEP* **01** 106

——2020 *JHEP* **09** 178; NuFIT5.2 (2022) http://www.nu-fit.org

Fabri E and Picasso L E 1966 *Phys. Rev. Lett.* **16** 408

Faddeev L D and Popov V N 1967 *Phys. Lett.* B **25** 29

Falkowski A 2016 *Pramana Journal of Physics* **87** 39

Farrar G R and Fayet P 1978a *Phys. Lett.* B **76** 575

——1978b *Phys. Lett.* B **79** 442

——1980 *Phys. Lett.* b **89** 191

Fayet P 1975 *Nucl. Phys.* B **90** 104

——1976 *Phys. Lett.* B **64** 159

——1977 *Phys. Lett.* B **69** 489

——1978 in *New Frontiers in High-Energy Physics, Proc. Orbis Scientiae*, Coral Gables, FL, USA, eds. Perlmutter A and Scott L F (New York: Plenum) p. 413

Feinberg G *et al.* 1959 *Phys. Rev. Lett.* **3** 527, especially footnote 9

Fermi E 1934a *Nuovo Cimento* **11** 1

——1934b *Z. Phys.* **88** 161

Feynman R P 1963 *Acta Phys. Polon.* **26** 697

——1977 in *Weak and Electromagnetic Interactions at High Energies* ed R Balian and C H Llewellyn Smith (Amsterdam: North-Holland) p 121

Feynman R P and Gell-Mann M 1958 *Phys. Rev.* **109** 193

Feynman R P and Hibbs A R 1965 *Quantum Mechanics and Path Intergrals* (New York: McGraw-Hill)

Fierz M 1937 *Z. Physik* **104** 553

Follana E *et al.* 2007 (HPQCD, UKQCD) *Phys. Rev.* D **75** 054502

Fritzsch H and Gell-Mann M 1972 *Proc. XVI Int. Conf. on High Energy Physics, Batavia IL* eds J D Jackson and R G Roberts, pp 135-165

Fritzsch H, Gell-Mann M and Leutwyler H 1973 *Phys. Lett.* B **47** 365

Fukuda Y *et al.* 1998 (Super-Kamiokande Collaboration) *Phys. Rev. Lett.* **81** 1562

Fukugita M and Yanagida T 1986 *Phys. Lett.* B **174** 45

Gaillard M K and Lee B W 1974 *Phys. Rev.* D **10** 897

Gamow G and Teller E 1936 *Phys. Rev.* **49** 895

Gando A *et al.* 2013 (KamLAND Collaboration) *Phys. Rev.* D **88** 03301

Gasser J and Leutwyler H 1982 *Phys. Rep.* **87** 77

——1984 *Ann. Phys.* **158** 142

——1985 *Nucl. Phys.* B **250** 465

Geer S 1986 *High Energy Physics 1985, Proc. Yale Theoretical Advanced Study Institute* eds Bowick M J and Gursey F (Singapore: World Scientific)

Gell-Mann M 1961 *California Institute of Technology Report* CTSL-20 (reprinted in Gell-Mann and Ne'eman 1964)

——1972 *Acta Phys. Austriaca Suppl.* **9** 733

Gell-Mann M *et al.* 1979 *Supergravity* ed D Freedman and P van Nieuwenhuizen (Amsterdam: North-Holland) p 315

Gell-Mann M and Levy M 1960 *Nuovo Cimento* **16** 705

Gell-Mann M and Low F E 1954 *Phys. Rev.* **95** 1300

Gell-Mann M and Ne'eman 1964 *The Eightfold Way* (New York: Benjamin)

Gell-Mann M and Pais A 1955 *Phys. Rev.* **97** 1387

Georgi H *et al.* 1978 *Phys. Rev. Lett.* **40** 692

Georgi H and Kaplan D B 1984 *Phys. Lett.* B **145** 216

Georgi H and Politzer H D 1974 *Phys. Rev.* D **9** 416

Georgi H, Quinn H R and Weinberg S 1974 *Phys. Rev. Lett.* **33** 451

Gershon T and Gligorov V V 2017 *Rept. Prog. Phys.* **80** 046201

Gershon T, Kenzle M and Trabelsi K 2022 section 77 of Workman *et al.* 2022

Gervais J-L and Sakita B 1971 *Nucl. Phys.* B **34** 632

Gibbons L K 1993 (E731 Collaboration) *Phys. Rev. Lett.* **70** 1203

Gildener E 1976 *Phys. Rev.* D **14** 1667

Ginsparg P and Wilson K G 1982 *Phys. Rev.* D **25** 25

Ginzburg V I and Landau L D 1950 *Zh. Eksp. Teor. Fiz.* **20** 1064

Giri A *et al.* 2003 *Phys. Rev.* D **68** 054018

Giudice G F 2013 arXiv:hep-ph/1307.7879v2

Giudice G F *et al.* 2007 *JHEP* **06** 45

Glashow S L 1961 *Nucl. Phys.* **22** 579

Glashow S L *et al.* 1978 *Phys. Rev.* D **18** 1724

Glashow S L, Iliopoulos J and Maiani L 1970 *Phys. Rev.* D **2** 1285

Glück M, Owens J F and Reya E 1978 *Phys. Rev.* D **17** 2324

Goldberger M L and Treiman S B 1958 *Phys. Rev.* **95** 1300

Goldhaber M *et al.* 1958 *Phys. Rev.* **109** 1015

Goldstone J 1961 *Nuovo Cimento* **19** 154

Goldstone J, Salam A and Weinberg S 1962 *Phys. Rev.* **127** 965

Golteman M 2008 PoS CONFINEMENTS (2008) 014

Gonzalez-Garcia M C and Yokoyama M 2022: section 14 of Workman *et al.* 2022

Gorishnii S G and Larin S A 1986 *Phys. Lett.* **172** 109

Gorishnii S G *et al.* 1991 *Phys. Lett.* B **259** 144

Gorkov L P 1959 *Zh. Eksp. Teor. Fiz.* **36** 1918

Gottschalk T and Sivers D 1980 *Phys. Rev.* D **21** 102

Gray A *et al* 2005 (HPQCD and UKQCD Collaborations) *Phys. Rev.* D **72** 094507

Greenberg O W 1964 *Phys. Rev. Lett.* **13** 598

Gribov V N and Lipatov L N 1972 *Sov. J. Nucl. Phys.* **15** 438

Gronau M 1991 *Phys. Lett.* B **265** 389

Gronau M and London D 1990 *Phys. Rev. Lett.* **65** 3381

Gross D J and Llewellyn Smith C H 1969 *Nucl. Phys.* B **14** 337

Gross D J and Wilczek F 1973 *Phys. Rev. Lett.* **30** 1343

——1974 *Phys. Rev.* D **9** 980

Grossman Y *et al.* 2005 *Phys. Rev.* D **72** 031501

Grzadowski *et al.* 2010 *JHEP* **10** 85

Guest D, Cramer K and Whiteson D 2018 *Ann. Rev. Nucl. Part. Sci.* **68** 161

Guralnik G S *et al.* 1964 *Phys. Rev. Lett.* **13** 585

——1968 *Advances in Particle Physics* vol 2, ed R Cool and R E Marshak (New York: Interscience) pp 567ff

Haag R 1958 *Phys. Rev.* **112** 669

Hagiwara K *et al.* 2002 *Phys. Rev.* D **66** 010001

Halzen F and Martin A D 1984 *Quarks and Leptons* (New York: Wiley)

Hamberg R, van Neerven W L and Matsuura T 1991 *Nucl. Phys.* B **359** 343 [Erratum; 2002 *Nucl. Phys.* B **644** 403]

Hamberg R *et al.* 1991 *Nucl. Phys.* B **359** 343

Hambye T and Reisselmann K 1997 *Phys. Rev.* D **55** 7255

Hammermesh M 1962 *Group Theory and its Applications to Physical Problems* (Reading, MA: Addison-Wesley)

Han M Y and Nambu Y 1965 *Phys. Rev.* B **139** 1006

Harlander R V and Kilgore W B 2002 *Phys. Rev. Lett.* **88** 201801

Harrison P F and Quinn H R 1998 *The BaBar physics book: Physics at an asymmetric B factory* SLAC-R-0504

Hasenfratz P *et al.* 1998 *Phys. Lett.* B **427** 125

Hasenfratz P and Niedermayer F 1994 *Nucl. Phys.* B **414** 785

Hasert F J *et al.* 1973 *Phys. Lett.* B **46** 138

Hashimoto S and Sharpe S R 2022 *Lattice Quantum Chromodynamics*, section 17 in Workman *et al.* 2022

Heisenberg W 1932 *Z. Phys.* **77** 1

Herzog F *et al.* 2017 *JHEP* **02** 090

Higgs P W 1964 *Phys. Rev. Lett.* **13** 508

——1966 *Phys. Rev.* **145** 1156

Hill C T and Simmons E H 2003 *Phys. Rept.* **381** 235 [Erratum: *Phys. Rept.* **390** 553 (2004)]

Hirata K S *et al.* 1989 *Phys. Rev. Lett.* **63** 16

Höcker A *et al.* 2001 *Eur. Phys. J.* C **21** 225

Holdom B 1985 *Phys. Lett.* B **150** 301

Hollik W 1990 *Fortsch. Phys.* **38** 165

——1991 *1989 CERN-JINR School of Physics* CERN 91-07 (Geneva) p 50ff

Hornbostel K, Lepage G P and Morningstar C 2003 *Phys. Rev.* D **67** 034023

Hosaka J *et al.* 2006 *Phys. Rev.* D **73** 112001

Hughes R J 1980 *Phys. Lett.* B **97** 246

——1981 *Nucl. Phys.* B **186** 376

Huston J, Rabbertz K and Zanderighi G 2022 *Quantum Chromodynamics*, section 9 of Workman *et al.* 2022

Ibanez L E 1982 *Phys. Lett.* B **118** 73

Ibanez L E and Ross G G 1981 *Phys. Lett.* B **105** 439

——1982 *Phys. Lett.* B **110** 215

Isgur N and Wise M B 1989 *Phys. Lett.* B **232** 113

——1990 *Phys. Lett.* B **237** 527

Isidori G *et al.* 2001 *Nucl. Phys.* B **609** 387

Itzykson C and Zuber J.-B. 1980 *Quantum Field Theory* (New York, NY: McGraw Hill) p160

Jackiw R 1972 *Lectures in Current Algebra and its Applications* ed S B Treiman, R Jackiw and D J Gross (Princeton, NJ: Princeton University Press) pp 97–254

Jacob M and Landshoff P V 1978 *Phys. Rep.* C **48** 285

Jansen K *et al.* 1996 *Phys. Lett.* B **372** 275

Jarlskog C 1985 *Phys. Rev. Lett.* **55** 1039

Jones D R T 1974 *Nucl. Phys.* B **75** 531

Jones H F 1990 *Groups, Representations and Physics* (Bristol: IOP Publishing)

Jones L M and Wyld H 1978 *Phys. Rev.* D **17** 1782

Kabir P K 1968 *The CP Puzzle: Strange Decays of the Neutral Kaon* (London and New York: Academic Press)

Kadanoff L P 1977 *Rev. Mod. Phys.* **49** 267

Kaplan D B 1992 *Phys. Lett.* B **288** 342

Kaplan D B and Georgi H 1984 *Phys. Lett.* B **136** 183

Kaplan D B, Georgi H and Dimopoulos S 1984 *Phys. Lett.* B **136** 187

Kayser B 1981 *Phys. Rev.* D **24** 110

Kennedy D C *et al.* 1989 *Nucl. Phys.* B **321** 83

Kennedy D C and Lynn B W 1989 *Nucl. Phys.* B **322** 1

Khachatryan V *et al.* 2015 (CMS COllaboration) *Phys. Rev.* D **92** 012004

Kibble T W B 1967 *Phys. Rev.* **155** 1554

Kidonakis N 2010a *Phys. Rev.* D **81** 054028

——2010b *Phys. Rev.* D **82** 054018

Kim K J and Schilcher K 1978 *Phys. Rev.* D **17** 2800

Kinoshita T 1962 *J. Math. Phys.* **3** 650

Kittel C 1987 *Quantum Theory of Solids* second revised printing (New York: Wiley)

Klapdor-Kleingrothaus H V *et al.* (Heidelberg-Moscow Collaboration) 2001 *Eur. Phys. J.* A **12** 147

——2006 *Mod. Phys. Lett.* A **21** 1547

Klein O 1948 *Nature* **161** 897

Kluth S 2006 *Rept. on Prog. in Phys.* **69** 1771

Kobayashi M 2009 *Rev. Mod. Phys.* **81** 1019

Kobayashi M and Maskawa K 1973 *Prog. Theor. Phys.* **49** 652

Kogut J B and Susskind L 1975 *Phys. Rev.* D **11** 395

Kowalska K and Sessolo E M 2018 *Advances in High Energy Physics* **2018** 6828560

Krall R and Reece M 2018 *Chinese Physics* C **42** 043105

Kramer G and Lampe B 1987 *Z. Phys.* C **34** 497

Krastev P I and Petcov S T 1988 *Phys. Lett.* B **205** 84

Kugo T and Ojima I 1979 *Prog. Theor. Phys. Suppl.* **66** 1

Kunszt Z and Piétarinen E 1980 *Nucl. Phys.* B **164** 45

Kusaka A *et al.* 2007 (Belle Collaboration) *Phys. Rev. Lett.* **98** 221602

Kuzmin V A, Rubakov V A and Shaposhnikov M E 1985 *Phys. Lett.* B **155** 36

Landau L D 1948 *Dokl. Akad. Nauk. USSR* **60** 207

——1957 *Nucl. Phys.* **3** 127

Landau L D and Lifshitz E M 1980 *Statistical Mechanics* part 1, 3rd edn (Oxford: Pergamon)

Langacker P (ed) 1995 *Precision Tests of the Standard Electroweak Model* (Singapore: World Scientific)

Larin S A and Vermaseren J A M 1991 *Phys. Lett.* B **259** 345

——1993 *Phys. Lett.* B **303** 334

Larkoski A J, Moult I and Nachman B 2020 *Phys. Rept.* **841** 1

Lautrup B 1967 *Kon. Dan. Vid. Selsk. Mat.-Fys. Med.* **35** 1

Lee B W *et al.* 1977a *Phys. Rev. Lett.* **38** 883

——1977b *Phys. Rev.* D **16** 1519

Lee T D and Nauenberg M 1964 *Phys. Rev.* B **133** 1549

Lee T D, Rosenbluth R and Yang C N 1949 *Phys. Rev.* **75** 9905

Lee T D and Yang C N 1956 *Phys. Rev.* **104** 254

——1957 *Phys. Rev.* **105** 1671

——1962 *Phys. Rev.* **128** 885

Lees J P *et al.* 2013 (BaBar Collaboration) *Phys. Rev.* D **88** 012003

Lees T M, Maltoni F and Quadt A 2022, section 61 of Workman *et al.* 2022

LEP 2003 (The LEP working Group for Higgs Searches, ALEPH, DELPHI, L3 and OPAL Collaborations) *Phys. Lett.* B **565** 61

Lepage G P and Mackenzie P B 1993 *Phys. Rev.* **48** 2250

Lesourgues J and Verde L 2022 section 26 of Workman *et al.* 2022

Leutwyler H 1996 *Phys. Lett.* B **378** 313

LHCHXSWG 2015 LHCHXSWG-INT-2015-001

Li Z *et al.* 2018 (Super-Kamiokande Collaboration) *Phys. Rev.* D **98** 052006

Libby J *et al.* 2010 (CLEO Collaboration) *Phys. Rev.* D **82** 112006

Lichtenberg D B 1970 *Unitary Symmetry and Elementary Particles* (New York: Academic)

Lichtenberg L and Valencia G 2022 section 65 of Workman *et al.* 2022

Lipkin H J *et al.* 1991 *Phys. Rev.* D **44** 1454

Llewellyn Smith C H 1973 *Phys. Lett.* B **46** 233

Lobashev V *et al.* 2003 *Nucl. Phys.* A **719** 153c

London F 1950 *Superfluids Vol I, Macroscopic theory of Superconductivity* (New York: Wiley)

Louppe G *et al.* 2019 *JHEP* **01** 057

Lüscher M 1981 *Nucl. Phys.* B **180** 317

——1986 *Commun. Math. Phys.* **105** 153

——1991a *Nucl. Phys.* B **354** 531

——1991b *Nucl. Phys.* B **364** 237

——1998 *Phys. Lett.* B **428** 342

Lüscher M and Weisz P 1985a *Phys. Lett.* B **158** 250

——1985b *Commun. Math. Phys.* **97** 59 [Erratum: 1985 *Comm. Math. Phys.* **98** 433]

Lüscher M *et al.* 1980 *Nucl. Phys.* B **173** 365

Luthe T *et al.* 2016 *JHEP* **07** 127

——2017 *JHEP* **10** 166

Majorana E 1937 *Nuovo Cimento* **5** 171

Maki Z, Nakagawa M and Sakata S 1962 *Prog. Theor. Phys.* **28** 870

Mandelstam S 1976 *Phys. Rep.* C **23** 245

Mandl F 1992 *Quantum Mechanics* (New York: Wiley)

Marciano W J and Sirlin A 1988 *Phys. Rev. Lett.* **61** 1815

Marquard P *et al.* 2015 *Phys. Rev. Lett.* **114** 142002

Marshak R E *et al.* 1969 *Theory of Weak Interactions in Particle Physics* (New York: Wiley)

Martin A D *et al.* 1994 *Phys. Rev.* D **50** 6734

——2002 *Eur. Phys. J.* C **23** 73

——2009 *Eur. Phys. J.* C **63** 189

Maskawa T 2009 *Rev. Mod. Phys.* **81** 1027

McNeile C *et al.* 2010 *Phys. Rev.* D **82** 034512

Merzbacher E 1998 *Quantum Mechanics* 3rd edn (New York: Wiley)

Mikheev S P and Smirnov A Y 1985 *Sov. J. Nucl. Phys.* **42** 913

——1986 *Nuovo Cimento* **9** C 17

Minkowski P 1977 *Phys. Lett.* B **67** 421

Moch S, Vermaseren J A M and Vogt A 2009 *Nucl. Phys.* B **813** 220

Mohapatra R N *et al.* 1968 *Phys. Rev. Lett.* **20** 1081

Mohapatra R N and Senjanovic G 1980 *Phys. Rev. Lett.* **44** 912

——1981 *Phys. Rev.* D **23** 165

Montanet L *et al.* 1994 *Phys. Rev.* D **50** 1173

Mönig K 1990 in *Proc. 10th Moriond Workshop: New and Exotic Phenomena* pp 69–80 eds. Fackler O and Tran Thanh Van J (Gif-Sur-Yvette: Ed. Frontieres)

Montvay I and Münster G 1994 *Quantum Fields on a Lattice* (Cambridge: Cambridge University Press)

Morningstar C and Peardon M J 2004 *Phys. Rev.* D **69** 054501

Muta T 2010 *Foundations of Quantum Chromodynamics* 3rd edn (Singapore: World Scientific)

Nakamura K and Petcov S T 2010 section 13 of Nakamura *et al.* 2010

Nakamura K *et al.* 2010 *J.Phys.* G **37** 075021

Nambu Y 1960 *Phys. Rev. Lett.* **4** 380

——1974 *Phys. Rev.* D **10** 4262

Nambu Y and Jona-Lasinio G 1961a *Phys. Rev.* **122** 345

——1961b *Phys. Rev.* **124** 246

Nambu Y and Lurie D 1962 *Phys. Rev.* **125** 1429

Nambu Y and Schrauner E 1962 *Phys. Rev.* **128** 862

Narayanan R and Neuberger H 1993a *Phys. Lett.* B **302** 62

——1993b *Phys. Rev. Lett.* **71** 3251

——1994 *Nucl. Phys.* B **412** 574

——1995 *Nucl. Phys.* B **443** 305

Nason P, Dawson S and Ellis R K 1988 *Nucl. Phys.* B **303** 607

Nauenberg M 1999 *Phys. Lett.* B **447** 23

Ne'eman Y 1961 *Nucl. Phys.* **26** 222

Neuberger H 1998a *Phys. Lett.* B **417** 141

——1998b *Phys. Lett.* B **427** 353

Neveu A and Schwarz J H 1971 *Nucl. Phys.* B **31** 86

Nielsen N K 1981 *Am. J. Phys.* **49** 1171

Nielsen H B and Ninomaya M 1981a *Nucl. Phys.* B **185** 20

——1981b *Nucl. Phys.* B **193** 173

——1981c *Nucl. Phys.* B **195** 541

Nir Y 1989 *Phys. Lett.* B **221** 184

Noaki J *et al.* 2008 *Phys. Rev. Lett.* **101** 202004

Noether E 1918 *Nachr. Ges. Wiss. Göttingen* 171

Oddone P 1989 *Ann. N.Y. Acad. Sci.* **578** 237

Okubo S 1962 *Prog. Theor. Phys.* **27** 949

Pais A 2000 *The Genius of Science* (Oxford: Oxford University Press)

Pak A and Czarnecki A 2008 *Phys. Rev. Lett.* **100** 241807

Papucci M, Ruderman J T and Weiler A 2012 *JHEP* **09** 035

Parry W E 1973 *The Many Body Problem* (Oxford: Clarendon)

Pascoli S *et al.* 2007a *Phys. Rev.* D **75** 083511

——2007b *Nucl. Phys.* B **774** 1

Pauli W 1934 *Rapp. Septième Conseil Phys. Solvay, Brussels 1933* (Paris: Gautier-Villars), reprinted in Winter (2000) pp 7, 8

Peccei R D and Quinn H 1977a *Phys. Rev. Lett.* **38** 1440

——1977b *Phys. Rev.* D **16** 1791

Perkins D H 1975 in *Proc. Int. Symp. on Lepton and Photon Interactions at High Energies, Stanford, CA* p 571

Peskin M E 1997 in *1996 European School of High Energy Physics* ed N Ellis and M Neubert CERN 97-03 (Geneva) pp 49-142

Peskin M E and Schroeder D V 1995 *An Introduction to Quantum Field Theory* (Reading, MA: Addison-Wesley)

Pica C 2016 PoS **LATTICE2016** 015

Politzer H D 1973 *Phys. Rev. Lett.* **30** 1346

Poluektov *et al.* 2010 (Belle Collaboration) *Phys. Rev.* D **81** 112002

Pontecorvo B 1946 *Chalk River Laboratory Report* PD-205

——1947 *Phys. Rev.* **72** 246

——1957 *Zh. Eksp. Theor. Phys.* **33** 549

——1958 *ibid.* **34** 247

——1967 *ibid.* **53** 1717 (Engl. transl. *Sov. Phys.–JETP* **26** 984)

Prescott C Y *et al.* 1978 *Phys. Lett.* B **77** 347

Puppi G 1948 *Nuovo Cimento* **5** 505

Quigg C 1977 *Rev. Mod. Phys.* **49** 297

Rajaraman R 1982 *Solitons and Instantons* (Amsterdam: North-Holland)

Ramond P 1971 *Phys. Rev.* D **3** 2415

Reines F and Cowan C 1956 *Nature* **178** 446

Reines F, Gurr H and Sobel H 1976 *Phys. Rev. Lett.* **37** 315

Renton P 1990 *Electroweak Interactions* (Cambridge: Cambridge University Press)

Richardson J L 1979 *Phys. Lett.* B **82** 272

Ringwald A, Rosenberg L J and Rybka G 2022 section 90 of Workman *et al.* 2022

Rodrigo G and Santamaria A 1993 *Phys. Lett.* B **313** 441

Ross D A and Veltman M 1975 *Nucl. Phys.* B **95** 135

Rubbia C *et al.* 1977 *Proc. Int. Neutrino Conf., Aachen, 1976* (Braunschweig: Vieweg) p 683

Ryder L H 1996 *Quantum Field Theory* 2nd edn (Cambridge: Cambridge University Press)

Sakurai J J 1958 *Nuovo Cimento* **7** 649

——1960 *Ann. Phys., NY* **11** 1

Salam A 1957 *Nuovo Cimento* **5** 299

——1968 *Elementary Particle Physics* ed N Svartholm (Stockholm: Almqvist and Wiksells)

Salam A and Strathdee J 1974 *Nucl. Phys.* B **76** 477

Salam A and Ward J C 1964 *Phys. Lett.* **13** 168

Salam C P 2010 *Eur. Phys. J.* C **14** 47

Samuel M A and Surguladze L R 1991 *Phys. Rev. Lett.* **66** 560

Schael S *et al.* 2006 (ALEPH, DELPHI, L3, OPAL, SLD, LEP Electroweak Working Group, SLD Electroweak and Heavy Flavour Groups) *Phys. Reports* **427** 257

Schiff L I 1968 *Quantum Mechanics* 3rd edn (New York: McGraw-Hill)

Schrieffer J R 1964 *Theory of Superconductivity* (New York: Benjamin)

Schutz B F 1988 *A First Course in General Relativity* (Cambridge: Cambridge University Press)

Schwinger J 1957 *Ann. Phys., NY* **2** 407

——1962 *Phys. Rev.* **125** 397

Shaevitz M H *et al.* 1995 (CCFR Collaboration) *Nucl. Phys.* B *Proc. Suppl.* **38** 188

Shamir Y 1993 *Nucl. Phys.* B **406** 90

Sharpe S R 2006 *PoSLAT* 022 (hep-lat/0610094)

Shaw R 1955 The problem of particle types and other contributions to the theory of elementary particles *PhD Thesis* University of Cambridge

Sheikholeslami B and Wohlert R 1985 *Nucl. Phys.* B **259** 572

Shifman M A *et al.* 1979 *Sov. J. Nucl. Phys.* **30** 711

Sikivie P *et al.* 1980 *Nucl. Phys.* B **173** 189

Sirlin A 1980 *Phys. Rev.* D **22** 971

Sirunyan A M *et al.* 1984 (CMS Collaboration) *Phys. Rev.* D **29** 89

——2018a *Phys. Rev. Lett.* **121** 121801

——2018b *Phys. Lett.* B **779** 283

——2018c *Phys. Rev. Lett.* **120** 231802

——2019a *Eur. Phys. J.* C **79** 421

——2019b *Phys. Rev.* D **99** 112003

——2020 *Phys. Rev. Lett.* **125** 061801

——2021a *Eur. Phys. J.* C **81** 488

——2021b *JHEP* **07** 027

——2021c *JHEP* **01** 148

Slavich *et al.* 2021 *Eur. Phys. J.* C **81** 450

Slavnov A A 1972 *Teor. Mat. Fiz.* **10** 153 (Engl. transl. *Theor. and Math. Phys.* **10** 99)

Snyder A E and Quinn H R 1993 *Phys. Rev.* D **48** 2139

Sommer R 1994 *Nucl. Phys.* B **411** 839

Spergel D *et al.* 2007 *Astrophys. J. Supp.* **170** 377

Staffen F D 2009 *Eur. Phys. J.* C **59** 557

Staff of the CERN p̄p project 1981 *Phys. Lett.* B **107** 306

Stange A *et al.* 1994a *Phys. Rev.* D **49** 1354

——1994b *Phys. Rev.* D **50** 4491

Staric M *et al.* 2007 (Belle Collaboration) *Phys. Rev. Lett.* **98** 211803

Steinberger J 1949 *Phys. Rev.* **76** 1180

Steffen D 2009 *Eur. Phys. J.* C **59** 557

Sterman G and Weinberg S 1977 *Phys. Rev. Lett.* **39** 1436

Stueckelberg E C G and Peterman A 1953 *Helv. Phys. Acta* **26** 499

Sudarshan E C G and Marshak R E 1958 *Phys. Rev.* **109** 1860

Susskind L 1977 *Phys. Rev.* D **16** 3031

——1979 *Phys. Rev.* D **20** 2619

Sutherland D G 1967 *Nucl. Phys.* B **2** 433

Svetitsky B 2018 EPJ Web Conf. **175** 01017

Symanzik K 1970 *Commun. Math. Phys.* **18** 227

——1983 *Nucl. Phys.* B **226** 187, 205

Tarasov O V *et al.* 1980 *Phys. Lett.* B **93** 429

Tavkhelidze A 1965 *Seminar on High Energy Physics and Elementary Particles* (Vienna: IAEA) p 763

Taylor J C 1971 *Nucl. Phys.* B **33** 436

——1976 *Gauge Theories of Weak Interactions* (Cambridge: Cambridge University Press) 't Hooft G 1971a *Nucl. Phys.* B **33** 173

——1971b *Nucl. Phys.* B **35** 167

——1971c *Phys. Lett.* B **37** 195

——1976a *Phys. Rev.* D **14** 3432

——1976b *High Energy Physics, Proc. European Physical Society Int. Conf.* ed A Zichichi (Bologna: Editrice Composition) p 1225

——1980 *Recent Developments in Gauge Theories, Cargese Summer Institute 1979* ed G 't Hooft *et al.* (New York: Plenum)

——1986 *Phys. Rep.* **142** 357

't Hooft G and Veltman M 1972 *Nucl. Phys.* B **44** 189 *Rev. Mod. Phys.* **21** 153

Tiomno J and Wheeler J A 1949 *Rev. Mod. Phys.* **21** 153

Ur A C *et al.* (GERDA Collaboration) 2011 *Nucl. Phys. Proc. Suppl.* **217** 38

Valatin J G 1958 *Nuovo Cimento* **7** 843

van der Bij J J 1984 *Nucl. Phys.* B **248** 141

van der Bij J J and Veltman M 1984 *Nucl. Phys.* B **231** 205

van der Neerven W L and Zijlstra E B 1992 *Nucl. Phys.* B **382** 11

van Ritbergen T and Stuart R G 1999 *Phys. Rev. Lett.* **82** 488

van Ritbergen *et al.* 1997 *Phys. Lett.* B **400** 379

Veltman M 1967 *Proc. R. Soc.* A **301** 107

——1968 *Nucl. Phys.* B **7** 637

——1970 *Nucl. Phys.* B **21** 288

——1977 *Acta Phys. Polon.* B **8** 475

——1981 *Acta Phys. Polon.* B **12** 437

Vermaseren J A M, Vogt A and Moch S 2005 *Nucl. Phys.* B **724** 3

von Weiszäcker C F 1934 *Z. Phys.* **88** 612

Wegner F 1972 *Phys. Rev.* B **5** 4529

Weinberg S 1966 *Phys. Rev. Lett.* **17** 616

——1967 *Phys. Rev. Lett.* **19** 1264

——1973 *Phys. Rev.* D **8** 605, especially footnote 8

——1975 *Phys. Rev.* D **11** 3583

——1976 *Phys. Rev.* D **13** 974

——1978 *Phys. Rev. Lett.* **40** 223

——1979a *Physica* A **96** 327

——1979b *Phys. Rev.* D **19** 1277

——1979c *Phys. Rev. Lett.* **43** 1566

——1996 *The Quantum Theory of Fields Vol II Modern Applications* (Cambridge: Cambridge University Press)

Weisberger W 1965 *Phys. Rev. Lett.* **14** 1047

Weisskopf V F and Wigner E P 1930a *Z. Phys.* **63** 54

——1930b *Z. Phys.* **65** 18

Weldon H A and Zee A 1980 *Nucl. Phys.* B **173** 269

Wess J and Zumino B 1974a *Phys. Lett.* B **49** 52

——1974b *Nucl. Phys.* B **70** 39

——1974c *Nucl. Phys.* B **78** 1

Wilczek F 1978 *Phys. Rev. Lett.* **40** 279

Wilczek F and Zee A 1979 *Phys. Rev. Lett.* **43** 1571

Williams E J 1934 *Phys. Rev.* **45** 729

Wilson K G 1969 *Phys. Rev.* **179** 1499

——1971a *Phys. Rev.* B **4** 3174

——1971b *Phys. Rev.* B **4** 3184

——1971c *Phys. Rev.* D **3** 1818

——1974 *Phys. Rev.* D **10** 2445

——1975 *New Phenomena in Subnuclear Physics, Proc. 1975 Int. School on Subnuclear Physics 'Ettore Majorana'* ed A Zichichi (New York: Plenum)

Wilson K G and Kogut J 1974 *Phys. Rep.* **12C** 75

Winter K 2000 *Neutrino Physics* 2nd edn (Cambridge: Cambridge University Press)

Witten E 1981 *Nucl. Phys.* B **188** 1981

Witzel O 2019 PoS **LATTICE2018** 006

Wolfenstein L 1978 *Phys. Rev.* D **17** 2369

——1983 *Phys. Rev. Lett.* **51** 1945

Workman R L *et al.* 2022 (Particle Data Group) *Prog. Theor. Exp. Phys.* **2022** 083C01

Wu C S *et al.* 1957 *Phys. Rev.* **105** 1413

Yanagida T 1979 *Proc. Workshop on Unified Theory and Baryon Number in the Universe* ed O Sawada and A Sugamoto (Tsukuba: KEK)

Yang C N 1950 *Phys. Rev.* **77** 242

Yang C N and Mills R L 1954 *Phys. Rev.* **96** 191

Yosida K 1958 *Phys. Rev.* **111** 1255

Index